GROUND SOURCE HEAT PUMP
RESIDENTIAL AND
LIGHT COMMERCIAL

DESIGN AND INSTALLATION GUIDE

This manual is written for professionals or those who wish to become professionals in the Ground Source Heat Pump (GSHP) industry. The International Ground Source Heat Pump Association **(IGSHPA)** expects the reader to have the appropriate:

- Business and professional licenses/credentials,
- Knowledge and understanding of pertinent governing codes and/or regulations and adhere to same,
- Knowledge and understanding of pertinent federal, state, and local health and safety codes and/or regulations and adhere to same,
- Knowledge, understanding, and ability to evaluate job site conditions as they apply to your work,
- Equipment and tools necessary to perform your work under job site conditions, and perform same according to the equipment and tool manufacturer's operating manuals and/or recommendations.

Library of Congress Control Number: 2009928893
ISBN: 978-0-929974-07-1

2nd Printing August 2011

Prepared by:
Oklahoma State University
Division on Engineering Technology
Stillwater, Oklahoma 74078

Oklahoma State University was founded December 25, 1890 as the Oklahoma Agricultural and Mechanical College. It became Oklahoma State University on July 1, 1957. Oklahoma's land-grant institution has three goals: to instruct, to do research and to offer educational assistance to the public through extension.

The Division of Engineering Technology at Oklahoma State University is part of the College of Engineering, Architecture, and Technology.

The International Ground Source Heat Pump Association (IGSHPA) formed in 1987 was established as an integral part of Oklahoma State University. The purpose of IGSHPA is to promote the efficient use of ground source heat pumps, to develop and promote sound industry standards, to support the effective marketing of ground source heat pumps, to identify and to support industry related research, to develop and distribute internationally recognized training materials, to enable its members to have direct input into published materials, to provide a forum for information interchange, to enable association members to benefit from the advantage of large numbers, and to represent the association's members in matters of local, state, national and international interest.

Oklahoma State University, in compliance with Title VI of the Civil Rights Act of 1964 and Title IX of the Educational Amendments of 1972 (Higher Education Act), does not discriminate on the basis of race, color, national origin, or sex in any of its policies, practices or procedures. This provision includes, but is not limited to, admissions, employment, financial aid, and educational services.

Table of Contents

PAGE

CHAPTER 7 INSTALLING THE CLOSED-LOOP GHEX PIPING SYSTEM

CHAPTER 8 SYSTEM FLUSHING & PURGING

CHAPTER 9 DOMESTIC HOT WATER HEATING

CHAPTER 10 RESISTANCE HEAT

CHAPTER 11 GSHP STARTUP, PERFORMANCE CHECKING, AND TROUBLESHOOTING

APPENDIX A DESIGN EXAMPLES

APPENDIX B BLANK WORKSHEETS

ACKNOWLEDGMENTS

The development of this manual is the result of many years of effort by the members of the International Ground Source Heat Pump Association. The author is Charles Remund, Professor of Mechanical Engineering, South Dakota State University, with major contributions from Ryan Carda, Geo Connections, Inc. Other contributors include Phil Rawlings, Trison Construction, Inc. and James Bose, Oklahoma State University. The editor is Dara McCoy. Her assistance, comments and contributions are greatly appreciated.

Lastly, so many others contributed helpful suggestions, comments and criticisms that it is impossible to show a complete list. However, their assistance is acknowledged and appreciated.

Errors in the manuscript may have occurred in spite of our efforts. Any suggestions, corrections and/or comments should be made in writing to:

International Ground Source Heat Pump Association
Oklahoma State University
Stillwater, Oklahoma 74078
J.E. Bose, Project Director

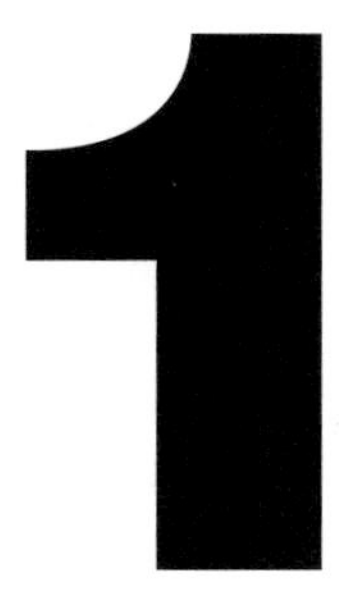

INTRODUCTION AND OVERVIEW

In This Section

This manual is a Design and Installation Guide for Residential and Light Commercial Ground Source Heat Pump Systems. The information contained herein is based on more than 30 years of practical experience, industry recognized and acceptable practice standards, and input from manufacturers, designers, and installers both in the United States and around the world. Ground heat exchange (GHEX) technologies addressed herein include vertically-bored, horizontally-trenched, and horizontally-bored closed-loop GHEX piping systems. This manual will provide design and installation recommendations for each of these three ground-coupling technologies.

1.1 The Case for Ground Source Heat Pumps

Consumers can choose from many space conditioning system options. Only one, the Ground Source Heat Pump System, was reported by *EPA (1993)* as "the most energy efficient, environmentally clean and cost effective space conditioning system available today." That quote is truer today because ground source heat pump equipment and ground coupling technologies continue to improve, while most conventional heating and cooling technologies reached their maximum efficiencies years ago. Many configurations of ground source heat pumps are now available (water-to-air, water-to-water, combined water-to-air and water-to-water, etc.) that offer flexibility in the design and installation of heating and cooling systems for residential and light commercial facilities, while providing efficiency levels that are unmatched by competing systems. Ground source heat pumps provide many other benefits including, but not limited to, the following:

- Economic Benefits
  - Reduced operating costs because of higher efficiencies in both heating and cooling modes when compared to competing systems
  - Reduced electrical peak demand in the cooling mode

  - Unlike air source heat pumps, there is no "defrost cycle" required to remove ice from the condenser coil in the heating mode
  - Reduced maintenance costs primarily due to all equipment being located in conditioned space or underground
  - Extended equipment life span and unbeatable long-term life cycle cost performance
  - Reduced replacement costs because ground coupling methods generally have a useful life in excess of 50 years
  - Optional auxiliary heat, if desired, for emergency purposes, but not required for supplemental heat like air source heat pumps
  - Optional desuperheater that can potentially meet a substantial portion of hot water needs year round
- Site Benefits
  - Improved aesthetics with no outdoor equipment, enclosures, concealments, etc.
  - No outdoor equipment subject to theft or vandalism
  - No outside power wiring, control wiring, or refrigerant line set piping
  - Reduced envelope penetrations because no flue or combustion air intakes are required
  - No accelerated deterioration of outdoor coils in adverse (salt laden, industrial, etc.) environments
  - Ground source piping and/or heat exchangers are not an impediment to landscaping, ground cover, etc., and can be placed under driveways and/or parking lots to further utilize otherwise unusable property
- Occupant Comfort and Safety Benefits
  - No combustion to generate carbon monoxide, thus eliminating associated occupant health and safety issues
  - Superior dehumidification as compared to air source heat pumps and air conditioners
  - Warmer supply-air temperatures in heating as compared to air source heat pumps and comparable to supply-air temperatures from "condensing" furnaces
  - Simple to sophisticated controls as specified by the designer or user
- Environmental Benefits
  - Package units require no field refrigerant charging and rarely require charge adjustments
  - Virtually no "point of use" pollutants
  - Significant reduction in generating facility fuel-related greenhouse gasses

1.2 Residential and Light Commercial Space Conditioning Systems

1.2.1 COMMON FUELS AND ENERGY EQUIVALENTS Residential and light commercial heating and cooling systems are generally fossil fuel based in heating and electricity based in cooling, except for air source heat pumps, which utilize electricity for both heating and cooling modes of operation. An understanding of the available fuels and their energy contents is important to determine the costs of heating and cooling with the various types of heating and cooling systems and their assorted efficiencies.

Electricity *(EIA, 2007a)* Electricity is delivered by a local utility through a network of existing transmission and distribution lines known as the electric grid. Most of that grid was initially constructed during a highly structured, highly regulated era designed to ensure that everyone in the United States had reasonable access to electricity service. Utilities built power plants, connected the plants to the grid, and transmitted electricity over the grid to the end-use customer. Some utilities generate all of the electricity they sell using their own power plants, and may have excess to sell, while others generate most, but purchase some of their supply from other utilities, power marketers, independent power producers, or from a "power pool." Some residential, industrial, commercial, and institutional users of electricity have the means to produce electricity for themselves and sell any excess back to their utility.

Power plants are grouped into the types of fuel or energy source they use to produce electricity, including fossil fuels (coal, natural gas, or a refined oil product), nuclear energy, and renewable energy sources (hydroelectric, biomass, waste-to-energy, geothermal, wind, solar, and alternative fuels). In the U.S. in 2003, approximately 71% of all electricity generated was from fossil fuels (coal-50.8%, natural gas-16.7%, petroleum-3.1% and other-0.2%), 19.7% was from nuclear, and 9.3% was from hydroelectric and other renewable sources. A major factor in the price of electricity is the cost and availability of the fuel used for electric generation and the transmission necessary to deliver that electricity, as well as the cost of constructing, operating, and maintaining the power generation plants. Accounting for inflation, the price of residential electricity in the United States in 2005 was more than four (4) cents per kWh less expensive than in 1960. Prices vary by state due to the cost and availability of fuels used to generate electricity, along with the existence of retail competition.

Supply and demand for fuel and transmission, international events, and changes in weather also affect the price of electricity. Fluctuations in prices have occurred due to high energy prices in the 1970s, a trend toward energy conservation that started in the 1980s, and energy distribution difficulties in the early 2000s. Daily and monthly demand has an effect on prices, resulting in peak demand rates during afternoon and early evening for some areas. Seasonal peaks occur due to regional weather and climatic conditions, with summer air conditioning being one of the driving factors.

Natural Gas *(EIA, 2007b)* Most of the natural gas used in the United States comes from domestic production, with the remainder being imported primarily from Canada. Generally, natural gas is produced or imported and put directly into pipelines to be transported to consumers all over the country. Excess production is placed into storage facilities for withdrawal in the winter, when the additional requirements for space heating cause total demand to exceed production and import capabilities. The gas is first transported via interstate pipelines owned by pipeline companies to local distribution companies that deliver to residential and other end-use customers through a complex network of piping systems.

The cost of natural gas to the end-user consists of the commodity cost, the transmission cost, and the distribution cost. The commodity cost (the cost of the gas itself) represented about 50% of the delivered cost in 2004. The distribution cost (the cost to bring the gas from the local gas company to the residence) represented about 40% of the cost. The transmission cost (the cost to pipe the gas from production point to the distributor) made up the remaining 10%. Recent trends have been for the commodity cost to remain high due to market conditions such as weak natural gas production, colder than normal weather during the heating season, declining net imports, and high crude oil prices. Factors that put upward price pressure on natural gas include weak domestic production compared to increasing demand, falling net imports, high oil prices that result in a shift toward use of natural gas by customers equipped to do so, and inadequate inventories to supply the winter demands. Natural gas prices fluctuate over the year, with prices being highest during the heating season. Forecasting of gas prices is uncertain and key factors that affect market prices include a prolonged cold spell or brief severe winter weather, disruptions in the pipeline delivery system, or problems in other energy supplies.

Propane *(EIA, 2007c)* Propane naturally occurs as a gas at atmospheric pressure, but turns to liquid when subjected to a moderate pressure increase. It is generally stored and transported in its compressed liquid form and, when released to atmospheric pressure, is vaporized into a gas for use. Propane is the byproduct of two other processes: natural gas processing and petroleum refining. Natural gas plant production of propane involves extracting the propane from the natural gas to prevent it from condensing and causing operational problems in natural gas pipelines. When oil refineries make major products, some propane is produced as a byproduct of those processes. The byproduct nature of propane means that the volume made available cannot be adjusted when prices and demand for it fluctuate. Imports make up about 10% of the United States propane supply and are very important when consumption exceeds available domestic supplies. Imports include pipeline and rail car delivery from Canada and ocean-going tankers from various countries.

Propane prices are influenced by many factors, including crude oil and natural gas prices, the balance of supply to demand, the proximity of the customer to the supply source, and the market served. Increases in natural gas and crude oil prices will result in a switch to propane by large industrial customers. The fairly constant domestic production of propane results in inventory buildup during low demand summer months and inventory drawdown during the winter months when residential heating demand is high. Low inventory levels and the start of the heating season coupled with colder-than-normal winter weather can result in upward price pressure on propane. Cold weather early in the heating season can lead to continued high prices throughout the entire heating season due to difficulty in making up inventory with imports that may be delayed several weeks. Customers farthest from the major propane supply sources will pay more due to increased distribution costs. Finally, other markets such as the petrochemical sector and the agricultural sector can cause supply demands, such as crop drying late into the fall.

Residential and commercial customers account for about 45% of all propane used in the United States. About 4.6 million households depend on propane as their primary source for space heating. Propane is most commonly used as a heating source in areas not served by natural gas distribution and competes mainly with heating oil for space heating purposes. Homeowners in the Midwest use propane primarily for space heating, while those in the Northeast use it more for cooking. Because of the intense use of propane for residential heating, it has a highly seasonal demand fluctuation. Propane prices will occasionally spike due to difficulties in obtaining resupply during the peak heating season.

Fuel Oil *(EIA, 2007d)* In 2001, approximately 8.1 million households in the United States used fuel oil as the primary heating source for space heating. Of those, about 82% or 6.3 million households were located in the Northeast United States. Because of this, the demand for fuel oil is highly seasonal, meaning that prices can fluctuate due to the severity of the heating season, the cost of crude oil, competition levels in a region, and the operating costs that exist within a given region. Within a given region, the cost is most sensitive to the severity of the heating season and the cost of crude oil.

Fuel oil is a petroleum product that results from the distillation of crude oil. The supply of fuel oil in the United States comes from both domestic refineries and imports from foreign countries. The fuel oil is distributed throughout the United States by pipelines, barges, tankers, trucks, and rail cars. During the refining process, the amount of fuel oil that can be produced is limited, meaning the refiners need to store fuel oil made during the summer months to help meet the winter months' demands. Inventory must be high enough to meet the winter heating demand or prices can be expected to rise. Refiners cannot increase fuel oil production in the winter unless they have demand for the other products that are produced. Also, if demand is high for a seasonal product such as gasoline, the producer may delay the production of fuel oil, which could lead to lower inventories at the start of the heating season.

In 2001, distribution and marketing accounted for 46% of the cost of fuel oil, followed by crude oil cost at 42%, and refinery processing at 12%. Therefore, fuel oil costs can be expected to follow closely the cost of crude oil.

Energy Equivalents of Common Fuels The energy equivalent of a fuel is the total energy contained in a specified volume or measurable quantity of that fuel. Liquid fuels (propane, fuel oil, kerosene, etc.) are generally measured on a volume basis such as a gallon, liter, or British Thermal Unit (Btu). Natural gas is also measured on a volume basis of ft^3 or m^3, with a Therm being the equivalent of 100 ft^3. Electricity is generally metered in kWhs. The energy equivalents of common fuels are provided in Table 1.1, including energy values for wood and shelled corn, which have gained popularity as an alternative heating fuel in the Midwestern U.S.

Table 1.1. Energy Equivalents of Common Fuels used with Heating and Cooling Systems

Fuel Type	Energy Equivalent	
	USCS Units	SI Units
Electricity	3,412 Btu/kWh	3,600 kJ/kWh
Natural Gas	100,000 Btu/Therm[1]	37,259 kJ/m^3
Propane	92,000 Btu/gal	25,642 kJ/L
Fuel Oil (No. 2)	140,000 Btu/gal	39,020 kJ/L
Kerosene (No. 1)	134,000 Btu/gal	37,348 kJ/L
Wood (Cord)	20,000,000 Btu/Cord[2]	
Wood (Pellets)	16,500,000 Btu/Ton	
Shelled Corn	392,000 Btu/Bushel	

1. 1 Therm = 100 ft^3 2. 1 Cord = 4 ft x 8 ft x 4 ft stack, tightly packed

1.2.2 TYPES OF HEATING AND COOLING SYSTEMS

1.2.2.1 HEATING SYSTEMS

Fossil Fuel Heating Systems burn fuel to create heat. Therefore, they can only deliver the amount of heat energy available in the fuel source less combustion related losses. There are two categories of fossil fuel systems (furnaces, which heat air and boilers, which heat water) based on the energy delivery method. Both of these are available as either standard combustion equipment or higher-efficiency "condensing" equipment (efficiencies approaching or exceeding 90%). An example of a common furnace/central air conditioning unit is shown in Figure 1.1. The three most widely utilized fuels by fossil fuel systems are natural gas, fuel oil, and propane. These fuels provide a varying amount of heat energy at the point of use as listed in Table 1.1.

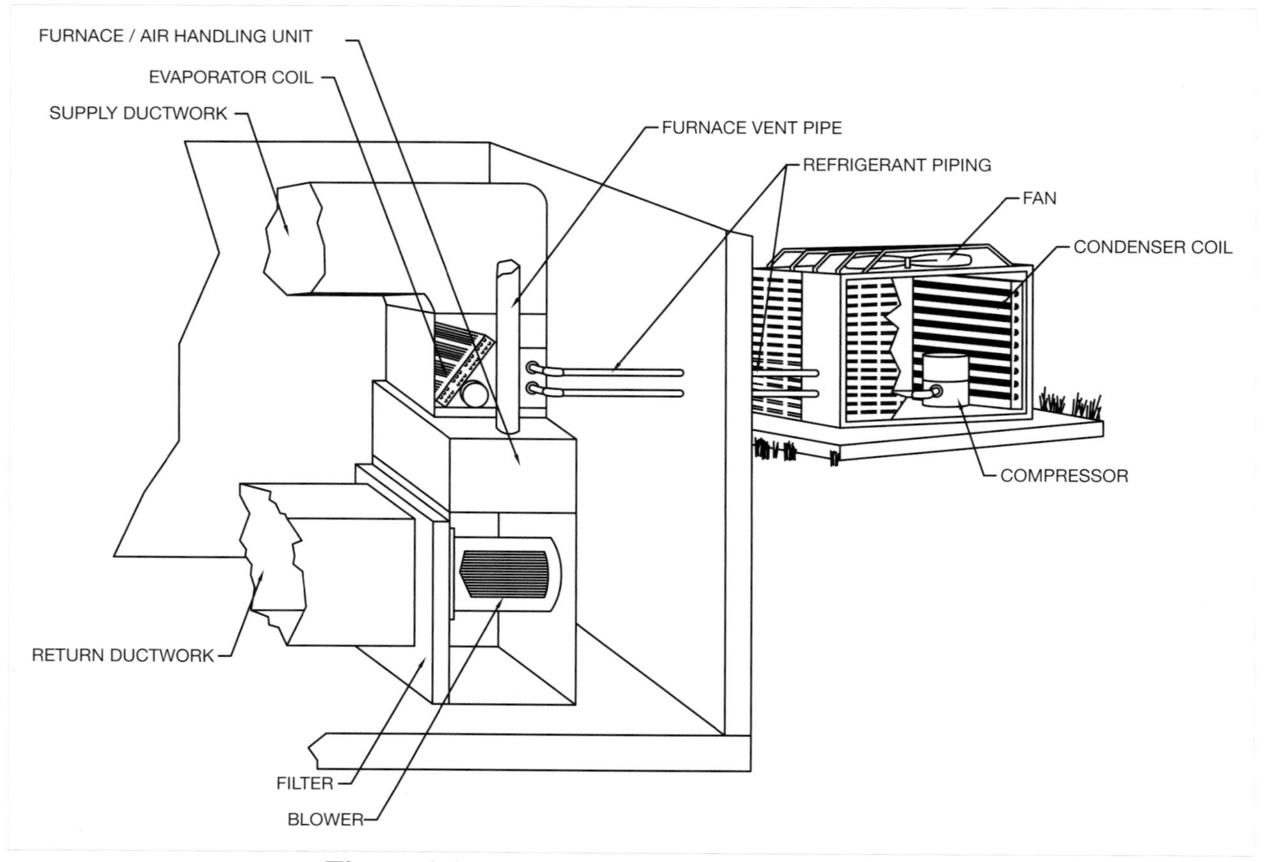

Figure 1.1. Furnace/Central Air Conditioning Unit

Because no fossil fuel equipment is 100% efficient, there is some percentage of heat lost to the environment during the combustion process. More importantly, all pollutants that are created during the combustion process are discharged to the environment at the point of use. For every 1,000,000 Btu of energy content burned at the point of use, the three major fuels provide "Greenhouse" pollutants to the atmosphere at the rate provided in Table 1.2 (The electric heat pump has no point of use emission of any kind.)

Table 1.2. Point of Use Greenhouse Gas Emissions Rates for Various Fuels.

Fuel Type	COx (Pounds)	SOx* (Grams)	NOx (Grams)
Natural Gas	11.71	0.0001S	0.015
Fuel Oil	150.33	1.014S	0.129
Propane	139.20	0.001S	0.015

*For SOx emission factors, S equals sulfur content expressed in gr/100 ft^3 gas vapor.

Electric Resistance Heating Systems consist of furnaces, boilers, or baseboard units that consume electricity to create heat. They deliver an amount of heat energy equal to the amount of electric energy consumed and are considered to be 100% efficient, providing 3,412 Btu of heat per kWh of electricity consumed (Table 1.1). Because there is no combustion process at the point of use, electric resistance heating systems have no point-of-use emissions.

Electric Heat Pump Heating Systems use electricity to power a vapor-compression refrigeration system to absorb energy from a low temperature source and deliver heat to a higher temperature indoor air or water load. Heat pump systems deliver more energy than the electricity they consume, generally by a factor of 2 to 5 times. There are two types of heat pump systems – air source and water source.

Air Source Heat Pumps use an outdoor evaporator to absorb heat energy from the surrounding air. The absorbed energy plus compressor energy is delivered to a condenser unit via a refrigerant line set. Two categories of air source heat pumps are air-to-air, for space heating via ducted air, and air-to-water, for heating water for domestic use or space heating through radiant panel or air-handling units. The air-to-air units utilize both an indoor condenser unit and outdoor evaporator unit fans to force air across each coil for adequate heat transfer (Figure 1.1). The air-to-water units utilize an indoor refrigerant-to-water coil along with a circulating pump to transfer heat to a circulating water load. Heating performance (heating capacity and efficiency) for both is directly affected by weather conditions and temperature, ambient air conditions conducive to corrosion of the outdoor coil, and other sources of air flow restriction and coil damage. When operating in the heating mode, the outdoor evaporator coil is subject to condensation of moisture as heat is absorbed from the air. The condensate will freeze on the coil at lower temperatures, requiring the heat pump unit to have a means of defrost to keep the coil free of ice. This always results in lower operating efficiency, the extent of which depends on the number of hours per year that temperatures are in the range that causes this problem.

Water Source Heat Pumps use a water-based source for heat addition to the evaporator coil. There are three general types of water source heat pump systems.

- Water loop heat pumps use boilers to maintain supply water to the heat pumps above a specified lower value. These systems are not addressed in this manual.
- Open-loop heat pumps use ground water, water from standing column wells, or surface water (lakes, ponds, etc.) directly as a heat source. These types of systems are not addressed in this manual. Examples of each of the prominent types of open-loop heat pump systems are shown in Figures 1.2a-1.2c.
- Closed-loop heat pumps use the ground or a body of water, indirectly, as a heat source by inserting a network of piping into the ground or body of water through which a circulating fluid flows. Examples of each of the prominent types of closed-loop heat pump systems are shown in Figures 1.3a-1.3c.

The configuration of a water source heat pump includes both water-to-air and water-to-water. Closed-loop heat pumps are the focus of this manual, and these systems will be discussed in more detail in later sections.

1.2.2.2 ELECTRIC COOLING SYSTEMS

Figure 1.2a. Ground Water Open-Loop GSHP System

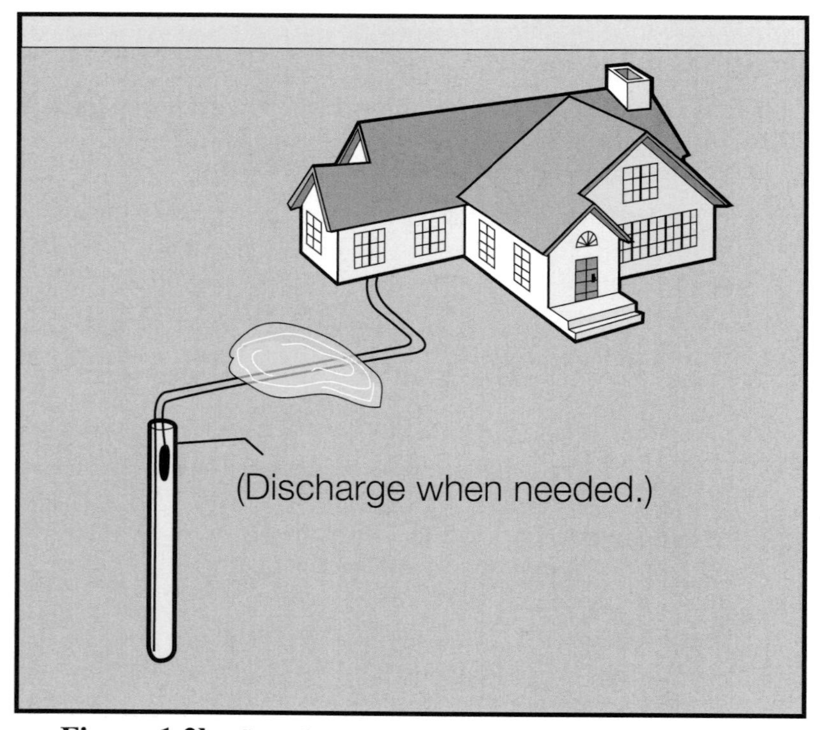

Figure 1.2b. Standing Column Open-Loop GSHP System

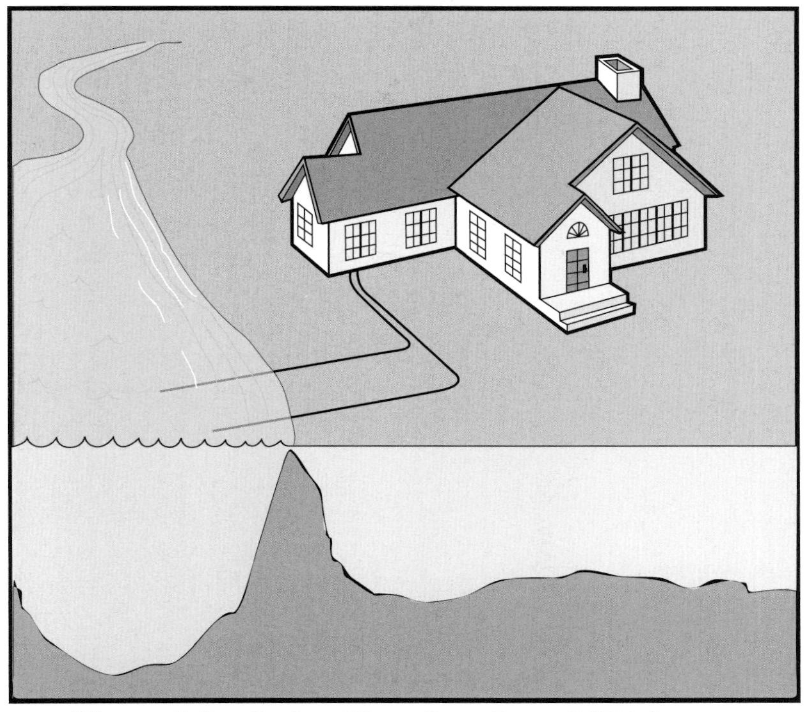

Figure 1.2c. Surface Water Open-Loop GSHP System

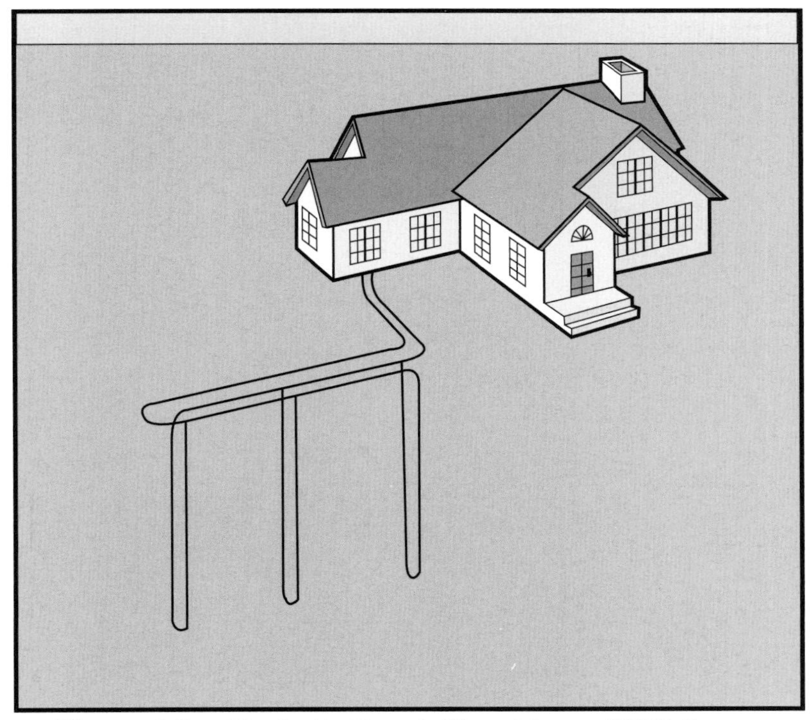

Figure 1.3a. Vertically-Bored, Closed-Loop GSHP System

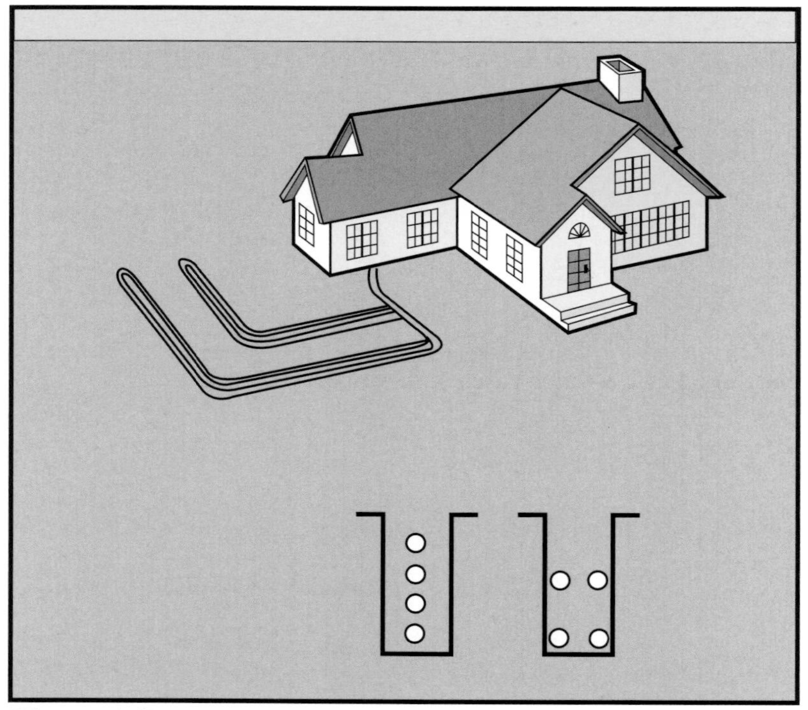

Figure 1.3b. Horizontally-Trenched, Closed-Loop GSHP System

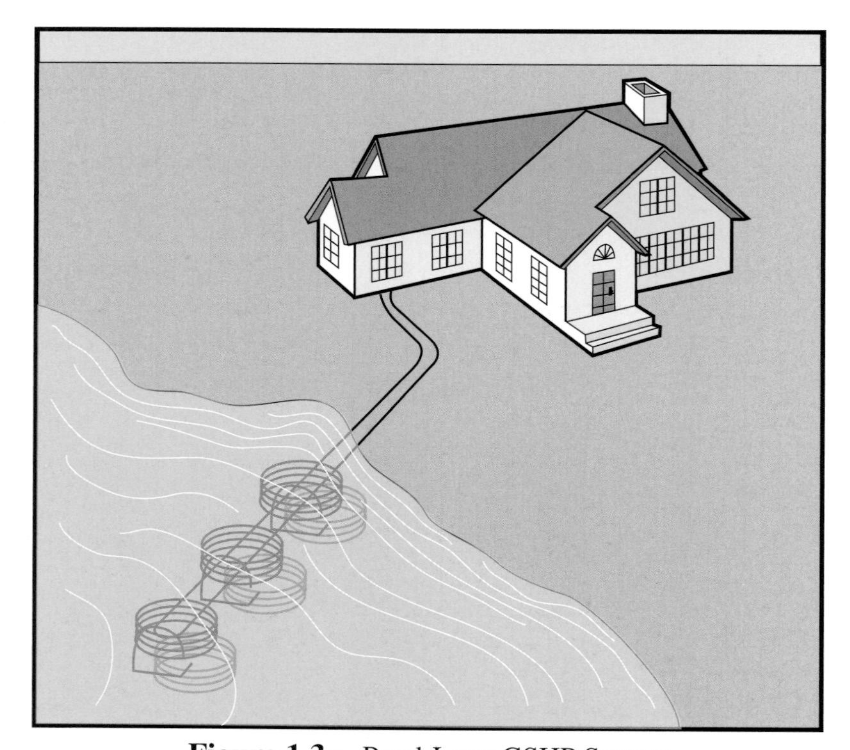

Figure 1.3c. Pond-Loop GSHP System

Air Conditioners and Air Source Heat Pumps use electricity to power a vapor-compression refrigeration system to absorb energy from a low-temperature indoor source and deliver that energy, along with the compressor and fan energy, to the higher outdoor air temperature sink. Two categories of air source heat pumps are air-to-air, for space cooling via ducted air, and air-to-water, for cooling water for domestic use or space conditioning through air-handling units. The air-to-air units utilize both an indoor evaporator unit and outdoor condenser unit fans to force air across each coil for adequate heat transfer. The air-to-water units utilize an indoor refrigerant-to-water coil along with a circulating pump to transfer heat from a circulating water load. The performance (cooling capacity and efficiency) of air-cooled air conditioning systems is directly affected by weather conditions and temperature, ambient air conditions conducive to corrosion of the outdoor coil, and other sources of air flow restriction and coil damage.

Water Source Heat Pumps use a water-based sink for heat rejection from the condenser coil. There are three general types of water source heat pump systems.

- Water-loop heat pumps use cooling towers to maintain supply water to the heat pumps below a specified upper value. These systems are not addressed in this manual.
- Open-loop heat pumps use ground water, water from standing column wells, or surface water (lakes, ponds, etc.) directly as a heat sink. These types of systems are not addressed in this manual.
- Closed-loop heat pumps use the ground or a body of water, indirectly, as a heat sink by inserting a network of piping into the ground or body of water through which a circulating fluid flows.

The configuration of a water source heat pump includes both water-to-air and water-to-water. Closed-loop heat pumps are the focus of this manual and these systems will be discussed in more detail in later sections.

1.2.3 EFFICIENCIES OF HEATING AND COOLING SYSTEMS

1.2.3.1 HEATING EFFICIENCY

Fuel Burning Heating Systems Efficiency is a measure of the output of a system divided by the input required to drive the system. Efficiency for any fuel burning equipment (furnaces or boilers) is expressed as the useful heating output of the system divided by the fuel input to the system, as expressed by Equation 1.1. Utilization efficiency describes the rate at which the equipment is converting chemical energy in the fuel to heat energy being delivered to the load and includes losses for latent and sensible exhaust heat, cyclic effects, infiltration, and pilot burner effect. Electric resistance furnaces, electric boilers, and electric strip heaters, which consume electricity directly to create heat, have utilization efficiencies of 100%.

Equation 1.1

$$\text{Eff} = \frac{\text{Heating Rate (Btu/hr)}}{\text{Fuel Consumption Rate (Btu/hr)}}$$

Heating efficiency for fuel burning equipment is commonly expressed by the Annual Fuel Utilization Efficiency (AFUE), which is an average efficiency for the entire heating season as expressed by Equation 1.2. AFUE consists of the utilization efficiency minus losses from a standing pilot during the non-heating season. ASHRAE Standard 103, U.S. Federal Trade Commission Code of Federal Regulations 16 Part 305, requires a minimum 78% AFUE for all furnaces manufactured after Jan. 1, 1992. AFUE values for several types of gas and fuel oil furnaces are provided in Table 1.3 (ASHRAE, 2004).

Equation 1.2

$$\text{AFUE} = \frac{\text{Seasonal Heating Energy Delivered (Btu)}}{\text{Seasonal Fuel Energy Consumed (Btu)}}$$

Table 1.3. AFUE for Indoor Combustion– ASHRAE (2004)

Type of Gas Furnace	AFUE (%)
1. Natural-draft with standing pilot	64.5
2. Natural-draft with intermittent ignition	69.0
3. Natural-draft with intermittent ignition and auto vent damper	78.0
4. Fan-assisted combustion with standing pilot or intermittent ignition	80.0
5. Same as 4, except with improved heat transfer	82.0
6. Direct vent, natural-draft with standing pilot, preheat	66.0
7. Direct vent, fan-assisted combustion, and intermittent ignition	80.0
8. Fan-assisted combustion (induced draft)	80.0
9. Condensing	90.0
Type of Oil Furnace	
1. Standard – pre-1992	71.0
2. Standard – post-1992	80.0
3. Same as 2, with improved heat transfer	81.0
4. Same as 3, with automatic vent damper	82.0
5. Condensing	91.0

Recent high fuel prices have led to an interest in alternative fuel heating systems that use either wood or corn as fuel. There is very little published data on average annual efficiencies for these types of heating systems, but efficiency ranges that can be expected for some of these heating systems are provided in Table 1.4 (eere.energy.gov). These efficiencies do not include the energy, work, and time required to procure and load the heating units with the fuel.

Table 1.4. Efficiency Ranges for Solid Fuel Stoves – www.eere.energy.gov

Type of Stove	Eff (%)
1. Franklin	30 - 40
2. Stoves w/ Circulating Fans	40 - 70
3. Catalytic Stoves	65 - 75
4. Pellet Stoves	85 - 90

Heat Pump Systems Heat pump systems, whether air source or water source, do not burn energy to produce heat, but use electricity to move heat. In fact, heat pumps move significantly more heat than the electric energy they consume. Therefore, heat pump systems are not rated for efficiency in the same manner as fuel burning equipment. In general, the heating efficiency of a heat pump is defined by the Coefficient of Performance (COP_H), which is defined by Equation 1.3.

$$COP_H = \frac{\text{Heating Rate (Btu/hr)}}{\text{Electric Energy Consumption Rate (kW)} \bullet 3{,}412\ \frac{\text{Btu/hr}}{\text{kW}}}$$ **Equation 1.3**

Electric energy consumption rate is generally measured in kilowatts (kW). Therefore, the electric conversion provided in Table 1.1 (1 kW=3,412 Btu/hr) is used in the bottom of Equation 1.3 to convert the units Btu/hr of heat into Btu/hr of electric consumption. Equation 1.3 can be applied to either air source or water source heat pumps because it is an instantaneous efficiency where the parameters that affect the efficiency remain constant.

Air source heat pumps have COP_H(s) that can range from below 2.0 to nearly 4.0, with the higher values corresponding to higher outdoor air temperatures (when heating loads in the structure are well below design values). When outdoor air temperature drops below 35 F, the COP_H of these units drops rapidly and, except for the very expensive units, the heating capacity is reduced to the point where auxiliary heat is commonly required. At low outdoor air temperatures, the lower heat pump efficiency is coupled with the need for auxiliary heat that operates with a COP_H of 1.0, dramatically increasing the cost of operating this type of heating system. In climatic zones where extremely low outdoor temperatures are common during the heating season, an air source heat pump is generally not a viable option because excessive auxiliary heat is required to meet the heating requirements of the structure.

Water source heat pumps have COP_H(s) that range from about 3.5 to 5.0, the value of which depends primarily on the temperature of the water entering the heat pump from the ground connection. A properly designed ground source heat pump system will be designed so that water temperatures entering the heat pump will be high enough to produce COP_H(s) of 3.5 or higher at design outdoor conditions, while requiring minimal auxiliary backup heat. Water source heat pump COP_H(s) with closed-loop ground connections will vary the most during the heating season as energy is drawn from the ground and ground temperatures around the buried piping are reduced. COP_H(s) with standing column well (SCW) ground connections vary less during the heating season because variations in the SCW water temperature are smaller. COP_H(s) for open-loop systems vary only slightly because there is little, if any, variation in the ground water temperature entering the heat pumps from the supply well during the heating season.

Both air source or water source heat pump efficiencies change, sometimes dramatically, with changes in ambient air or water supply temperatures, respectively, and efficiencies for these systems may be expressed for the entire heating season to account for those changes. Air source heat pumps operating in the heating mode use Heating Seasonal Performance Factor (HSPF), which is defined as seasonal Btus delivered divided by watt-hours consumed over the entire heating season (Equation 1.4). The HSPF reflects the efficiency level of the equipment, the integrated average heating season temperatures for the climatic zone for which the heat pump is rated, and the use of auxiliary backup resistance heat for extreme outdoor air temperatures. Values of 7.0 or better are not uncommon for very high quality air source equipment.

$$HSPF = \frac{\text{Seasonal Heating Energy Delivered (Btu)}}{\text{Seasonal Electrical Energy Consumption (Whr)}}$$ **Equation 1.4**

Water source heat pumps do not have a defined seasonal heating efficiency because the type of ground connection used will result in entering water temperatures to the heat pump that will vary significantly throughout the heating season based on location and ground connection design. For the location and ground connection selected, the average seasonal heating efficiency for the heat pump system can be calculated as the total energy supplied during the heating season divided by the total electrical energy consumed, and can be expressed as a seasonal average heating COP_a according to Equation 1.5.

$$COP_a = \frac{\text{Seasonal Heating Energy Delivered (Btu)}}{\text{Seasonal Electrical Energy Consumption (kWhr)} \bullet 3{,}412 \frac{\text{Btu/hr}}{\text{kW}}}$$ **Equation 1.5**

1.2.3.2 COOLING EFFICIENCY

Air Conditioners and Air Source Heat Pumps use electricity to power a vapor-compression refrigeration system to absorb energy from a low-temperature indoor source and deliver that energy, along with the compressor and fan energy, to the higher outdoor air temperature sink. The instantaneous efficiency of air conditioners and air source heat pumps operating in the cooling mode is the Energy Efficiency Ratio (EER) as expressed by Equation 1.6. The cooling season efficiency is defined by the Seasonal Energy Efficiency Ratio (SEER), as expressed by Equation 1.7.

$$EER = \frac{\text{Cooling Rate (Btu/hr)}}{\text{Electric Energy Consumption Rate (W)}}$$ **Equation 1.6**

$$SEER = \frac{\text{Seasonal Cooling Energy Removed (Btu)}}{\text{Seasonal Electrical Energy Consumption (Whr)}}$$ **Equation 1.7**

The instantaneous efficiency (EER) of air-cooled air conditioners and heat pumps is highly dependent on the outdoor air temperature being drawn across the outdoor condensing coil and decreases with increasing outdoor air temperature. Air conditioners and air source heat pumps are rated using the seasonal cooling efficiency (SEER), and a minimum value of 12 is required (Department of Energy).

Water Source Heat Pumps in cooling mode have efficiency expressed in terms of the Energy Efficiency Ratio (EER), an instantaneous efficiency that depends on the entering water temperature from the ground connection along with other performance parameters that will be discussed later. EER ratings ranging from a low of 12 to above 28 are possible and highly dependent on temperatures available from the ground coupling type. A properly designed ground source heat pump system will be designed so that water temperatures entering the heat pump will be limited on the high side to produce EERs of 12 or higher at design outdoor conditions, while meeting the entire cooling load of the structure. Water source heat pump EERs with closed-loop ground connections will vary the most during the cooling season as energy is rejected to the ground and ground temperatures around the buried piping are increased. EERs with standing column well (SCW) ground connections vary less during the cooling season because variation in the SCW water temperatures are smaller. EERs for open-loop systems vary only slightly because there is little, if any, variation in ground water temperature entering the heat pumps from the supply well during the cooling season.

Water source heat pumps do not have a defined seasonal cooling efficiency because the type of ground connection used will result in entering water temperatures to the heat pump that will vary significantly throughout the cooling season based on location and ground connection design. For the location and ground connection selected, the average seasonal cooling efficiency for the heat pump systems can be calculated as the total energy removed during the cooling season divided by the total electrical energy consumed, and can be expressed as a seasonal average cooling EER_a, according to Equation 1.8.

$$EER_a = \frac{\text{Seasonal Cooling Energy Removed (Btu)}}{\text{Seasonal Electrical Energy Consumption (Whr)}}$$ **Equation 1.8**

1.2.3.3 GENERAL OBSERVATIONS ON EFFICIENCY Efficiencies of all types of space conditioning systems are directly affected by factors other than the equipment or system itself. Fan motors (PSC versus ECM fan motors are discussed in detail in Chapter 3, Section 3.5.2), filter change time frames, ductwork quality, occupant operating characteristics, and maintenance related factors will all contribute to or detract from a system's efficiency. The effects of these influences are site specific and can not be fully quantified here. Aside from occupant comfort issues, key indicators include, but are not limited to, the following:

- **Supply Air Fan Motors:**
  - PSC motors have generally stable energy usage, but lose air flow (cubic feet per minute, CFM) capacity because of high ductwork static pressure caused by improper ductwork design, dirty filters, restrictions in ductwork, improper damper settings, etc.
  - ECM motors increase energy usage, but maintain CFM capacity (to a point) for the same reasons.
- **Filter Changes:** Space conditioning equipment manufacturers of ducted air distribution equipment typically recommend changing filters once a month. Actual occupant filter change intervals are typically longer, sometimes less than twice a year.
- **Ductwork Design/Quality** has a major impact on system efficiency. Typical problems are:
  - Poor air stream integrity allowing unconditioned air to enter the return air stream or losing conditioned supply air to unconditioned spaces.
  - Restrictions or improper design impedes supply and/or return air flow.
  - Improper damper settings (improper air balance and/or occupant changes) at branches and/or registers impede air flow.
- **Occupant System Operation Characteristics**
  - Failure to change filters as recommended
  - Closing off supply air diffusers
  - Locating furnishings over or in front of return air grilles
  - Open windows while operating the space conditioning system
  - Thermostat jockeying – frequent thermostat set point changes (excluding "set back" thermostatically controlled changes).
  - Concealments placed around outdoor condenser/evaporator that restricts air flow.
  - No preventative maintenance
- **Maintenance:**
  - No preventative maintenance program
  - Often the only time the filter is checked/replaced
  - Automatically putting gauges on the refrigerant circuit
  - Improper refrigerant charge
  - Improper refrigerant charge replacement after repairs
  - "Parts changer" technicians

These and similar impediments to optimum system performance are not considered in the previous efficiency discussions. However, the reader should also be aware that the performance data provided by manufacturers of all space conditioning equipment is laboratory data, not "real world" data. As a matter of necessity, all equipment is tested using defined application methods, under controlled conditions, to exacting requirements as specified by recognized testing agencies, such as the Air-Conditioning and Refrigeration Institute (ARI). Published data generated by this type of testing allows users to compare various manufacturers' equipment performance on an equal basis, or a "level playing field." After installation at an actual site, the operation of the equipment is rarely under the conditions in which ARI testing was performed. For example, for a ground source heat pump installation during continuous operation while commissioning or testing, CFM and gallons per minute (gpm) will typically be fixed, but not at ARI rating test flows. The same can be said for dry bulb/wet bulb air temperature conditions, entering water temperature, and electrical input. They will not be at ARI rating points and will change as air and water temperatures change over the testing time frame.

1.2.4 BASIC ECONOMICS OF ENERGY DELIVERY A simple way to compare the relative operating costs of alternative heating and cooling systems is to estimate the cost per 1,000,000 Btus of space conditioning ($C_{H\text{-}MBtu}$=cost/million heating Btu provided, $C_{C\text{-}MBtu}$= cost/million cooling Btu removed) based on average, or seasonal, efficiencies for each system. Methods to calculate $C_{H\text{-}MBtu}$ for major common fuel-type heating systems and a few alternate heating fuel systems are provided by Equation 1.9 through 1.17. Methods to calculate $C_{C\text{-}MBtu}$ for air conditioners/air source heat pumps and water source heat pumps are provided by Equation 1.18 and 1.19. Heating costs for natural gas, propane, fuel oil, electric resistance, and air and ground source heat pumps are summarized in Tables 1.5 through 1.8 for selected system efficiencies over a range of fuel and electricity costs. Cooling costs for air source and ground source heat pumps are summarized in Table 1.9 for selected system efficiencies over a range of electricity costs.

HEATING COSTS ($C_{H\text{-}MBTU}$):

- **Electric Resistance Heat** (1 kWhr=3,412 Btu, Eff=100%)

$$C_{H\text{-}MBtu} = \frac{1{,}000{,}000\ \text{Btu}}{3{,}412\ \dfrac{\text{Btu/hr}}{\text{kW}}} \bullet \text{Electric Energy Cost (\$/kWhr)} = \frac{\$}{\text{MBtu}} \qquad \textbf{Equation 1.9}$$

- **Natural Gas** (1 Therm=100,000 Btu, AFUE)

$$C_{H\text{-}MBtu} = \frac{1{,}000{,}000\ \text{Btu}}{100{,}000\ \dfrac{\text{Btu}}{\text{Therm}} \bullet \text{AFUE}} \bullet \text{Natural Gas Cost (\$/Therm)} = \frac{\$}{\text{MBtu}} \qquad \textbf{Equation 1.10}$$

- **Propane** (1 gal=92,000 Btu, AFUE)

$$C_{H\text{-}MBtu} = \frac{1{,}000{,}000\ \text{Btu}}{92{,}000\ \dfrac{\text{Btu}}{\text{gal}} \bullet \text{AFUE}} \bullet \text{Propane Cost (\$/gal)} = \frac{\$}{\text{MBtu}} \qquad \textbf{Equation 1.11}$$

- **Fuel Oil** (1 gal=140,000 Btu, AFUE)

$$C_{H\text{–}MBtu} = \frac{1{,}000{,}000 \text{ Btu}}{140{,}000 \frac{\text{Btu}}{\text{gal}} \bullet \text{AFUE}} \bullet \text{Fuel Oil Cost (\$/gal)} = \frac{\$}{\text{MBtu}}$$

Equation 1.12

- **Wood Stove** (1 Cord=20,000,000 Btu, Eff)

$$C_{H\text{–}MBtu} = \frac{1{,}000{,}000 \text{ Btu}}{20{,}000{,}000 \frac{\text{Btu}}{\text{Cord}} \bullet \text{Eff}} \bullet \text{Wood Cost (\$/Cord)} = \frac{\$}{\text{MBtu}}$$

Equation 1.13

- **Wood Pellet Stove** (1 Ton=16,500,000 Btu, Eff)

$$C_{H\text{–}MBtu} = \frac{1{,}000{,}000 \text{ Btu}}{16{,}500{,}000 \frac{\text{Btu}}{\text{Ton}} \bullet \text{Eff}} \bullet \text{Wood Pellet Cost (\$/Ton)} = \frac{\$}{\text{MBtu}}$$

Equation 1.14

- **Shelled Corn Stove** (1 Bushel=392,000 Btu, Eff)

$$C_{H\text{–}MBtu} = \frac{1{,}000{,}000 \text{ Btu}}{392{,}000 \frac{\text{Btu}}{\text{Bushel}} \bullet \text{Eff}} \bullet \text{Shelled Corn Cost (\$/Bushel)} = \frac{\$}{\text{MBtu}}$$

Equation 1.15

- **Air Source Heat Pump** (1 kWhr=3,412 Btu, HSPF)

$$C_{H\text{–}MBtu} = \frac{1{,}000{,}000 \text{ Btu}}{1{,}000 \frac{\text{W}}{\text{kW}} \bullet \text{HSPF} \left(\frac{\text{Btu}}{\text{Whr}}\right)} \bullet \text{Electric Energy Cost (\$/kWhr)} = \frac{\$}{\text{MBtu}}$$

Equation 1.16

- **Water Source Heat Pump** (1 kWhr=3,412 Btu, COP_a)

$$C_{H\text{–}MBtu} = \frac{1{,}000{,}000 \text{ Btu}}{3{,}412 \frac{\text{Btu}}{\text{kWhr}} \bullet COP_a} \bullet \text{Electric Energy Cost (\$/kWhr)} = \frac{\$}{\text{MBtu}}$$

Equation 1.17

COOLING COSTS ($C_{C\text{-}MBTU}$):

- **Air Conditioner and Air Source Heat Pump** (1 kWhr=3,412 Btu, SEER)

$$C_{C\text{–}MBtu} = \frac{1{,}000{,}000 \text{ Btu}}{1{,}000 \frac{\text{W}}{\text{kW}} \bullet \text{SEER} \left(\frac{\text{Btu}}{\text{Whr}}\right)} \bullet \text{Electric Energy Cost (\$/kWhr)} = \frac{\$}{\text{MBtu}}$$

Equation 1.18

- **Water Source Heat Pump** (1 kWhr=3,412 Btu, EER_a)

$$C_{C\text{–}MBtu} = \frac{1{,}000{,}000 \text{ Btu}}{1{,}000 \frac{\text{W}}{\text{kW}} \bullet EER_a \left(\frac{\text{Btu}}{\text{Whr}}\right)} \bullet \text{Electric Energy Cost (\$/kWhr)} = \frac{\$}{\text{MBtu}}$$

Equation 1.19

Table 1.5. Natural Gas Heating Costs - $/MBtu delivered for three AFUE values.

Gas Price ($/Therm)	Furnace or Boiler Efficiency		
	AFUE=65% Old Equipment	AFUE=78% Current Minimum	AFUE=95% High Efficiency
$1.00	$15.38	$12.82	$10.53
$1.05	$16.15	$13.46	$11.05
$1.10	$16.92	$14.10	$11.58
$1.15	$17.69	$14.74	$12.11
$1.20	$18.46	$15.38	$12.63
$1.25	$19.23	$16.03	$13.16
$1.30	$20.00	$16.67	$13.68
$1.35	$20.77	$17.31	$14.21
$1.40	$21.54	$17.95	$14.74
$1.45	$22.31	$18.59	$15.26
$1.50	$23.08	$19.23	$15.79
$1.55	$23.85	$19.87	$16.32
$1.60	$24.62	$20.51	$16.84
$1.65	$25.38	$21.15	$17.37
$1.70	$26.15	$21.79	$17.89
$1.75	$26.92	$22.44	$18.42
$1.80	$27.69	$23.08	$18.95
$1.85	$28.46	$23.72	$19.47
$1.90	$29.23	$24.36	$20.00
$1.95	$30.00	$25.00	$20.53
$2.00	$30.77	$25.64	$21.05
$2.05	$31.54	$26.28	$21.58
$2.10	$32.31	$26.92	$22.11
$2.15	$33.08	$27.56	$22.63
$2.20	$33.85	$28.21	$23.16
$2.25	$34.62	$28.85	$23.68

1 Therm = 100,000 Btu

Table 1.6. Propane Heating Costs - $/MBtu delivered for three AFUE values.

Propane Price ($/gal.)	Furnace or Boiler Efficiency		
	AFUE=65% Old Equipment	AFUE=78% Current Minimum	AFUE=95% High Efficiency
$1.50	$25.08	$20.90	$17.16
$1.60	$26.76	$22.30	$18.31
$1.70	$28.43	$23.69	$19.45
$1.80	$30.10	$25.08	$20.59
$1.90	$31.77	$26.48	$21.74
$2.00	$33.44	$27.87	$22.88
$2.10	$35.12	$29.26	$24.03
$2.20	$36.79	$30.66	$25.17
$2.30	$38.46	$32.05	$26.32
$2.40	$40.13	$33.44	$27.46
$2.50	$41.81	$34.84	$28.60
$2.60	$43.48	$36.23	$29.75
$2.70	$45.15	$37.63	$30.89
$2.80	$46.82	$39.02	$32.04
$2.90	$48.49	$40.41	$33.18
$3.00	$50.17	$41.81	$34.32
$3.10	$51.84	$43.20	$35.47
$3.20	$53.51	$44.59	$36.61
$3.30	$55.18	$45.99	$37.76
$3.40	$56.86	$47.38	$38.90
$3.50	$58.53	$48.77	$40.05
$3.60	$60.20	$50.17	$41.19
$3.70	$61.87	$51.56	$42.33
$3.80	$63.55	$52.95	$43.48
$3.90	$65.22	$54.35	$44.62
$4.00	$66.89	$55.74	$45.77

1 gal. = 92,000 Btu

Table 1.7. Fuel Oil Heating Costs - $/MBtu delivered for three AFUE values.

Fuel Oil Price ($/gal.)	Furnace or Boiler Efficiency		
	AFUE=65% Old Equipment	AFUE=78% Current Minimum	AFUE=95% High Efficiency
$3.00	$32.97	$27.47	$22.56
$3.10	$34.07	$28.39	$23.31
$3.20	$35.16	$29.30	$24.06
$3.30	$36.26	$30.22	$24.81
$3.40	$37.36	$31.14	$25.56
$3.50	$38.46	$32.05	$26.32
$3.60	$39.56	$32.97	$27.07
$3.70	$40.66	$33.88	$27.82
$3.80	$41.76	$34.80	$28.57
$3.90	$42.86	$35.71	$29.32
$4.00	$43.96	$36.63	$30.08
$4.10	$45.05	$37.55	$30.83
$4.20	$46.15	$38.46	$31.58
$4.30	$47.25	$39.38	$32.33
$4.40	$48.35	$40.29	$33.08
$4.50	$49.45	$41.21	$33.83
$4.60	$50.55	$42.12	$34.59
$4.70	$51.65	$43.04	$35.34
$4.80	$52.75	$43.96	$36.09
$4.90	$53.85	$44.87	$36.84
$5.00	$54.95	$45.79	$37.59
$5.10	$56.04	$46.70	$38.35
$5.20	$57.14	$47.62	$39.10
$5.30	$58.24	$48.53	$39.85
$5.40	$59.34	$49.45	$40.60
$5.50	$60.44	$50.37	$41.35

1 gal. = 140,000 Btu

Table 1.8. Electric Heating Costs - $/MBtu delivered for resistance, air source, and ground source systems at various performance levels.

Electricity Price ($/kWh)	Resistance	Air-Source Heat Pump Performance			Closed-Loop Heat Pump Performance			Water-Loop Heat Pump Performance		
	EFF=100%	HSPF=5.0 (COP_a=1.47) Old Equip	HSPF=6.8 (COP_a=2.00) Current Min	HSPF=9.4 (COP_a=2.75) High Eff	COP_a=3.1 Low	COP_a=3.5 Average	COP_a=4.2 High	COP_a=3.6 Low	COP_a=4.1 Average	COP_a=4.7 High
$0.020	$5.86	$4.00	$2.94	$2.13	$1.89	$1.67	$1.40	$1.63	$1.43	$1.25
$0.030	$8.79	$6.00	$4.41	$3.19	$2.84	$2.51	$2.09	$2.44	$2.14	$1.87
$0.040	$11.72	$8.00	$5.88	$4.26	$3.78	$3.35	$2.79	$3.26	$2.86	$2.49
$0.050	$14.65	$10.00	$7.35	$5.32	$4.73	$4.19	$3.49	$4.07	$3.57	$3.12
$0.060	$17.58	$12.00	$8.82	$6.38	$5.67	$5.02	$4.19	$4.88	$4.29	$3.74
$0.070	$20.51	$14.00	$10.29	$7.45	$6.62	$5.86	$4.88	$5.70	$5.00	$4.37
$0.080	$23.44	$16.00	$11.76	$8.51	$7.56	$6.70	$5.58	$6.51	$5.72	$4.99
$0.090	$26.37	$18.00	$13.24	$9.57	$8.51	$7.54	$6.28	$7.33	$6.43	$5.61
$0.100	$29.30	$20.00	$14.71	$10.64	$9.45	$8.37	$6.98	$8.14	$7.15	$6.24
$0.110	$32.23	$22.00	$16.18	$11.70	$10.40	$9.21	$7.68	$8.96	$7.86	$6.86
$0.120	$35.16	$24.00	$17.65	$12.77	$11.35	$10.05	$8.37	$9.77	$8.58	$7.48
$0.130	$38.09	$26.00	$19.12	$13.83	$12.29	$10.89	$9.07	$10.58	$9.29	$8.11
$0.140	$41.02	$28.00	$20.59	$14.89	$13.24	$11.72	$9.77	$11.40	$10.01	$8.73
$0.150	$43.95	$30.00	$22.06	$15.96	$14.18	$12.56	$10.47	$12.21	$10.72	$9.35
$0.160	$46.88	$32.00	$23.53	$17.02	$15.13	$13.40	$11.17	$13.03	$11.44	$9.98
$0.170	$49.81	$34.00	$25.00	$18.09	$16.07	$14.24	$11.86	$13.84	$12.15	$10.60
$0.180	$52.74	$36.00	$26.47	$19.15	$17.02	$15.07	$12.56	$14.65	$12.87	$11.22
$0.190	$55.67	$38.00	$27.94	$20.21	$17.96	$15.91	$13.26	$15.47	$13.58	$11.85
$0.200	$58.60	$40.00	$29.41	$21.28	$18.91	$16.75	$13.96	$16.28	$14.30	$12.47
$0.210	$61.53	$42.00	$30.88	$22.34	$19.85	$17.58	$14.65	$17.10	$15.01	$13.10
$0.220	$64.46	$44.00	$32.35	$23.40	$20.80	$18.42	$15.35	$17.91	$15.73	$13.72
$0.230	$67.39	$46.00	$33.82	$24.47	$21.74	$19.26	$16.05	$18.72	$16.44	$14.34
$0.240	$70.32	$48.00	$35.29	$25.53	$22.69	$20.10	$16.75	$19.54	$17.16	$14.97
$0.250	$73.25	$50.00	$36.76	$26.60	$23.64	$20.93	$17.45	$20.35	$17.87	$15.59
$0.260	$76.18	$52.00	$38.24	$27.66	$24.58	$21.77	$18.14	$21.17	$18.59	$16.21
$0.270	$79.11	$54.00	$39.71	$28.72	$25.53	$22.61	$18.84	$21.98	$19.30	$16.84

1 kWh = 3,412 Btu COP_A = HSPF/3.412

Table 1.9. Electric Cooling Costs - $/MBtu removed for air source and ground source systems at various performance levels.

Electricity Price ($/kWh)	Air-Source A/C or Heat Pump			Closed-Loop GS Heat Pump Performance			Water-Loop GS Heat Pump Performance		
	SEER=7 Old Equipment	SEER=12 Current Minimum	SEER=15 High Efficiency	EER_a=13 Low	EER_a=16 Average	EER_a=20 High	EER_a=16 Low	EER_a=19 Average	EER_a=24 High
$0.020	$2.86	$1.67	$1.33	$1.54	$1.25	$1.00	$1.25	$1.05	$0.83
$0.030	$4.29	$2.50	$2.00	$2.31	$1.88	$1.50	$1.88	$1.58	$1.25
$0.040	$5.71	$3.33	$2.67	$3.08	$2.50	$2.00	$2.50	$2.11	$1.67
$0.050	$7.14	$4.17	$3.33	$3.85	$3.13	$2.50	$3.13	$2.63	$2.08
$0.060	$8.57	$5.00	$4.00	$4.62	$3.75	$3.00	$3.75	$3.16	$2.50
$0.070	$10.00	$5.83	$4.67	$5.38	$4.38	$3.50	$4.38	$3.68	$2.92
$0.080	$11.43	$6.67	$5.33	$6.15	$5.00	$4.00	$5.00	$4.21	$3.33
$0.090	$12.86	$7.50	$6.00	$6.92	$5.63	$4.50	$5.63	$4.74	$3.75
$0.100	$14.29	$8.33	$6.67	$7.69	$6.25	$5.00	$6.25	$5.26	$4.17
$0.110	$15.71	$9.17	$7.33	$8.46	$6.88	$5.50	$6.88	$5.79	$4.58
$0.120	$17.14	$10.00	$8.00	$9.23	$7.50	$6.00	$7.50	$6.32	$5.00
$0.130	$18.57	$10.83	$8.67	$10.00	$8.13	$6.50	$8.13	$6.84	$5.42
$0.140	$20.00	$11.67	$9.33	$10.77	$8.75	$7.00	$8.75	$7.37	$5.83
$0.150	$21.43	$12.50	$10.00	$11.54	$9.38	$7.50	$9.38	$7.89	$6.25
$0.160	$22.86	$13.33	$10.67	$12.31	$10.00	$8.00	$10.00	$8.42	$6.67
$0.170	$24.29	$14.17	$11.33	$13.08	$10.63	$8.50	$10.63	$8.95	$7.08
$0.180	$25.71	$15.00	$12.00	$13.85	$11.25	$9.00	$11.25	$9.47	$7.50
$0.190	$27.14	$15.83	$12.67	$14.62	$11.88	$9.50	$11.88	$10.00	$7.92
$0.200	$28.57	$16.67	$13.33	$15.38	$12.50	$10.00	$12.50	$10.53	$8.33
$0.210	$30.00	$17.50	$14.00	$16.15	$13.13	$10.50	$13.13	$11.05	$8.75
$0.220	$31.43	$18.33	$14.67	$16.92	$13.75	$11.00	$13.75	$11.58	$9.17
$0.230	$32.86	$19.17	$15.33	$17.69	$14.38	$11.50	$14.38	$12.11	$9.58
$0.240	$34.29	$20.00	$16.00	$18.46	$15.00	$12.00	$15.00	$12.63	$10.00
$0.250	$35.71	$20.83	$16.67	$19.23	$15.63	$12.50	$15.63	$13.16	$10.42
$0.260	$37.14	$21.67	$17.33	$20.00	$16.25	$13.00	$16.25	$13.68	$10.83
$0.270	$38.57	$22.50	$18.00	$20.77	$16.88	$13.50	$16.88	$14.21	$11.25

1.3 The Ground Source Heat Pump Concept

Ground Source Heat Pumps (GSHPs) are self-contained, electrically powered systems that take advantage of the earth's relatively constant, moderate ground temperature to provide heating, cooling, and domestic hot water more efficiently and less expensively than would be possible through other conventional heating and air conditioning technologies. Two general categories of GSHPs are the closed-loop system and the open-loop system (not addressed in this manual). Closed-loop systems circulate a secondary fluid (water or a water-based antifreeze solution) through a sealed and buried piping arrangement to transfer heat to and from the earth. These systems can be utilized almost anywhere, and the configuration of the closed-loop ground heat exchanger depends primarily on space availability and cost of installation. Figure 1.3a shows one variation of vertically-bored ground loops. These ground loop configurations are utilized in situations with less available land area and are the most common, especially in urban environments. Figure 1.3b shows one variation of horizontally-trenched ground loops. These ground loop configurations are common when adequate land area is available with reasonable trenching conditions. Figure 1.3c shows one variation of pond-loop configurations, which are utilized in isolated ponds with adequate water volume and depth. The horizontally-bored ground loop configuration is one that is not shown in Figures 1.3a-1.3c, but is another configuration that is commonly used and is discussed later in this manual.

During the summer cooling season, the water-to-air GSHP system functions as an air conditioner, extracting heat from inside a building and delivering it (after being raised in temperature by the heat pump) to the fluid that is circulated through the cooler earth where it is rejected. (A refrigerator operates on the same principle, removing heat from the cooler refrigerator interior and rejecting it to the warmer kitchen space.) As an added benefit, surplus heat from the building that would otherwise be dumped to the ground in the summer can be used to heat water for domestic use at very minimal cost. In the winter, the GSHP utilizes the circulating fluid to absorb heat from the ground where it (after being raised in temperature by the heat pump) is transferred to the inside air or water-based heat delivery system. The GSHP can also heat water for domestic use during winter heating operation at a substantial savings over conventional water heating systems.

In general, heat pumps used for space and water heating efficiently move heat from a low temperature source (air, water, or the earth) to a high temperature load (air or water) using refrigeration principles as displayed conceptually in Figure 1.4a for a GSHP system.

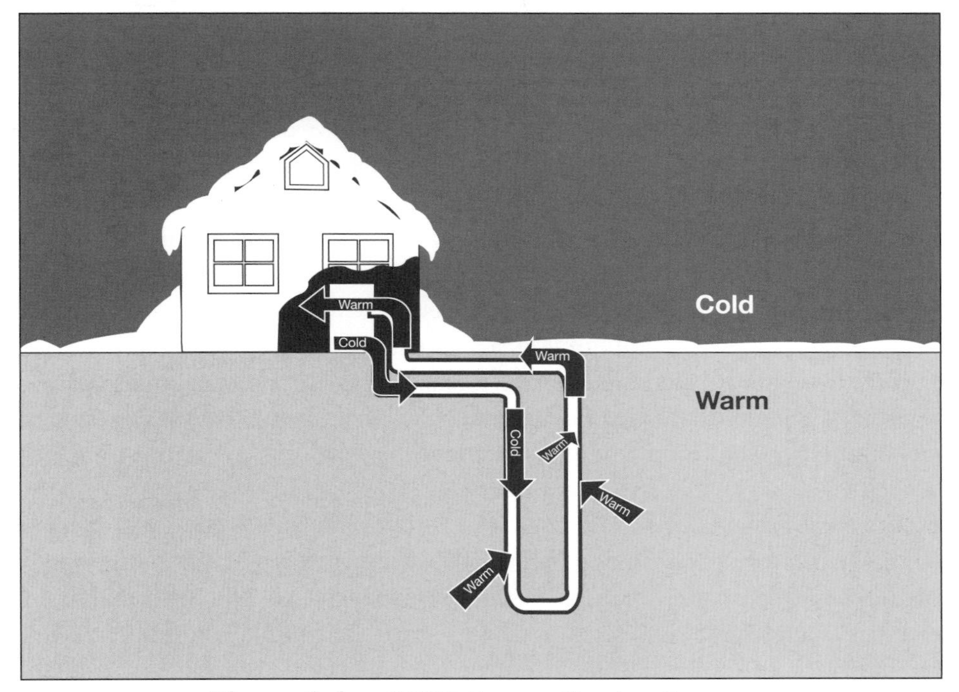

Figure 1.4a. GSHP System Heating Concept

This process is reversed for space cooling, removing heat from the low temperature load (air or water) and delivering heat to a high temperature sink (air, water, or the earth) as displayed conceptually in Figure 1.4b for a GSHP system.

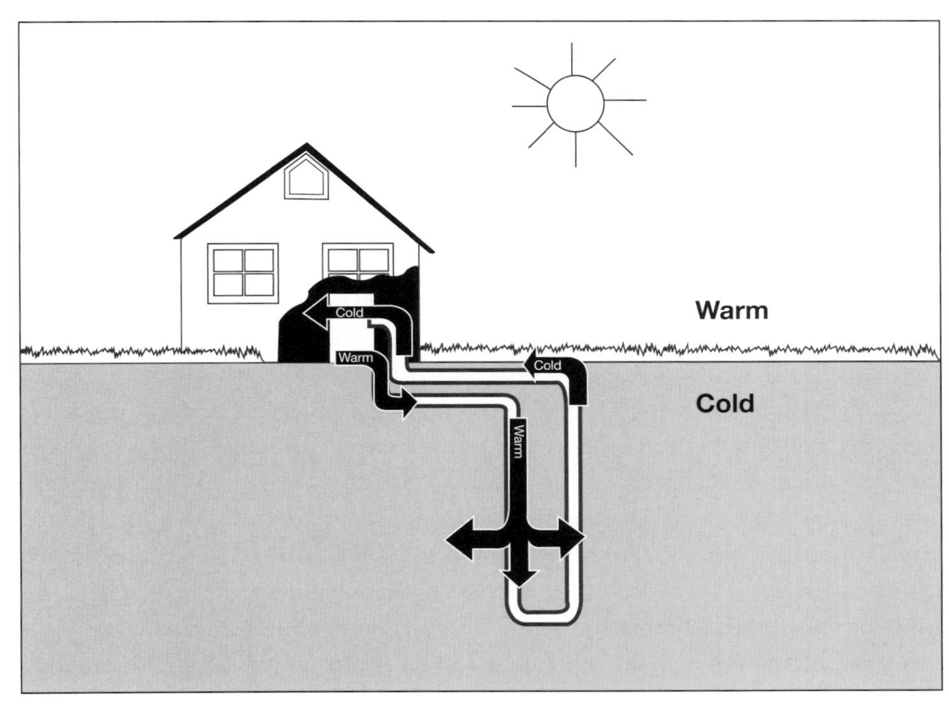

Figure 1.4b. GSHP System Cooling Concept

The efficiency of the heat pump system is related to the spread between the "source" and "sink" temperature, and the closer these two temperatures are to one another the more efficient the heat pump will be. For heating, it follows that higher source (air, water, or earth) temperatures and lower load (heated air or water) temperatures will lead to higher heating mode efficiencies. For cooling, it also follows that higher load (cooled air or water) temperatures and lower sink (air, water, or earth) temperatures will lead to higher cooling mode efficiencies. Space heating and cooling delivery temperatures vary little over the heating and cooling season, but the source and sink temperatures for a heat pump system will vary according to type. For air source heat pumps, the air temperatures can be very low during much of the heating season resulting in low heating mode efficiencies. For ground source heat pumps, the water temperatures will not be as low (depending on ground loop configuration), and the resulting operating efficiencies will be much higher. For air source heat pumps, the air temperatures can be very high during much of the cooling season resulting in low cooling mode efficiencies. For ground source heat pumps, the water temperatures will not be as high (depending on ground loop configuration) and the resulting cooling mode efficiencies will be much higher.

The efficiency of a heat pump depends on the temperature of the source energy for heating and the sink energy for cooling. A cost of operation comparison of various heating and cooling systems has already shown that ground source heat pumps can be much less expensive to heat with than conventional fuel-burning heating systems and offer significant advantage over air source heat pumps in both heating and cooling, or air conditioners in cooling. That advantage over air source heat pumps and air conditioners is derived primarily from the fact that air temperatures vary greatly during the year and throughout the day at any given location, while temperatures deep in the ground at the same location are constant throughout the year and vary to a much lesser extent than air temperatures near the surface. An air source heat pump must extract heat from outdoor air that can be very cold during the winter and must reject heat to outdoor air that can be very hot during the summer. A ground source heat pump can be designed to produce entering water temperatures from the ground connection that are well above the outside air temperatures during a majority of the heating season and well below outside air temperatures during a majority of the cooling season, resulting in much higher operating efficiencies than can be obtained with air source heat pumps. In addition, air source heat pump heating capacity drops off rapidly as outdoor air temperature drops below 35 F, resulting in air source heat pump systems to be sized such that a supplementary heat source (generally resistance heat with a COP_H=1.0) will be required for a significant portion of the heating season, lowering the overall efficiency of the air source heat pump system. Ground source heat pumps, especially with the introduction of two-capacity systems, can be sized to 100% of the design heating load, which minimizes the need for supplementary heat and keeps the heating season efficiency high. Finally, air source heat pumps utilize a condenser coil located outside which, during the heating mode, will experience significant icing when outdoor air temperatures and humidity levels promote this problem. Current air source heat pump technology implements a de-icing cycle, which both utilizes energy to melt the ice and requires a supplemental heat source to deliver heat during the de-icing cycle. The Iowa Energy Center (Klaassen, 2007) performed a study of current heat pump technologies including mature technology air source heat pumps, low temperature air source heat pumps, reverse cycle chillers, and ground source heat pumps for climates in Iowa. The results of that study (Figure 1.5) showed that newer technology air source heat pumps are making gains in heating mode efficiency for colder climates, but no current or advanced air source heat pump technology comes close to the efficiencies of ground source heat pump systems. It should be noted that improvements in air source heat pump technology ultimately end up being implemented in ground source heat pump equipment, meaning any advances made by the air source heat pump industry will result in comparable increases in the efficiency of ground source heat pump equipment.

For multiple heat pump systems, the use of a GSHP with a closed-loop configuration opens up many opportunities for energy management between zones in the facility along with potential for hot water generation from energy rejected in the cooling mode that would normally be rejected to the air for air-cooled air conditioning systems. These types of systems are becoming common in large homes and small commercial facilities. A common closed-loop system connecting all of the heat pumps together allows for maximum energy utilization efficiency, taking advantage of heating/cooling diversity in the facility and the opportunity to make hot water for domestic or other use with energy that would normally be rejected. The potential for highly efficient energy management systems is limited only by the imagination of the heating and cooling system designer.

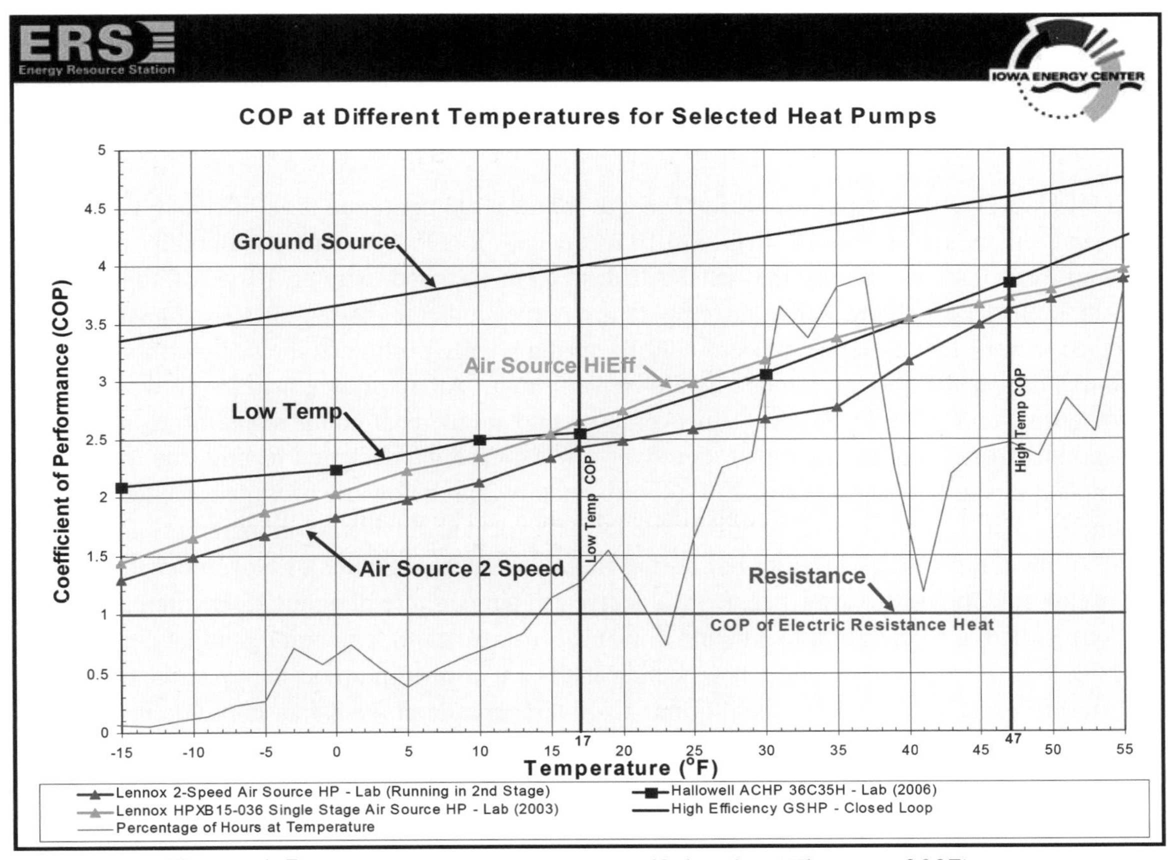

Figure 1.5. Comparison of heat pump efficiencies *(Klaassen. 2007)*

The magnitude of variation in the temperature of the energy "source" can be understood by considering four different locations in the U.S. ranging from the extremely cold Grand Forks, ND (Figure 1.6a), more moderate Salt Lake City, UT (Figure 1.6b), warmer and more humid Louisville, KY (Figure 1.6c) and extremely hot San Antonio, TX (Figure 1.6d). Each figure contains the average earth temperature from the surface to 10 feet deep, the earth temperatures as they vary throughout the year at depths of 6, 8, and 10 feet, the average annual air temperature, and the monthly average air temperature plotted on the accompanying monthly air temperature range. The ASHRAE 99% heating design and 0.4% cooling design temperatures, the annual heating and cooling extreme temperatures, and the 10-year heating and cooling extreme temperatures are also included. A summary of average annual air, deep earth, ASHRAE design, and extreme heating and cooling temperatures is provided in Table 1.10.

Table 1.10. Average Air, Average Earth[1], and Design Temperatures for Selected Locations in U.S.

			Heating Temperatures			Cooling Temperatures		
Location	T_{airavg}	T_{earth}	ASHRAE 99% Design	Annual Extreme	10-Yr Extreme	ASHRAE 0.4% Design	Annual Extreme	10-Yr Extreme
Grand Forks, ND	39.4	42	-16	-25	-31.4	91	98	104.1
Salt Lake City, UT	51.0	53	11	-3	-11.7	96	100	102.5
Louisville, KY	55.8	60	12	-1	-11.3	93	96	100.0
San Antonio, TX	68.6	72	32	20	10.0	100	104	107.4

1. Average temperature from the surface to 10-foot depth.

Grand Forks, ND (Figure 1.6a), where the average annual air temperature is about 39.4 F compared to an average earth temperature (from the surface to 10 feet deep) of 42 F, is a climate with both extreme heating and cooling loads, as indicated by the ASHRAE design, annual, and extreme 10-year temperatures summarized in Table 1.10. The monthly air temperature ranges indicate that an air source heat pump would be subjected to very low air temperatures (< 30 F) during a large portion of the heating months, while air temperatures during the cooling months would be very high. An air source heat pump would not be the most efficient system for this climate. However, a ground source heat pump would draw heat during the heating months from a ground loop that would, at a worst case, be designed to limit the lowest entering water temperatures to the heat pump at 25 F (approximately 20 F below the average earth temperature as a first estimate). This is 60 degrees higher than the lowest expected outside air temperatures and higher than nearly all of the temperatures experienced during the main heating months of November through March. A vertical ground loop would draw heat from a deep earth temperature of about 45 F, whereas a horizontal ground loop buried to a depth of 8 feet would draw heat from an earth temperature at that depth that would be about 37 F at the end of January, when heating loads are at maximum on the system. Because of the thermal storage capacity of the soil, the minimum soil temperature at the 8-foot burial depth would not be seen by the system until early April, when heating loads on the system have become much smaller than peak design loads. A ground loop sized for heating at this location would likely be oversized for the peak cooling load and would provide maximum entering water temperatures in cooling that are much less than most of the expected air temperatures during the main cooling months of June, July, and August, and may even be lower than the average monthly air temperatures for those months.

Salt Lake City, UT (Figure 1.6b) is a less extreme heating climate, with an average annual air temperature of 51 F and an average earth temperature (from the surface to 10 feet deep) of 53 F. This climate is not as extreme in heating as Grand Forks, ND, but has higher design cooling and annual extreme temperatures. Again, an air source heat pump would not be a good selection for this climate because of the significant time during the heating season with air temperatures below 30 F. In this case, a ground source heat pump would draw heat during the heating months from a ground loop that would, at a worst case, be designed to limit the lowest entering water temperatures to the heat pump at about 30 F (again, approximately 20 F below the deep earth temperature as a first estimate). This is 40 degrees higher than the lowest expected outside air temperatures and higher than nearly all of the temperatures experienced during the main heating months of November through March. A vertical ground loop would draw heat from a deep earth temperature of about 56 F, whereas a horizontal ground loop buried to a depth of 8 feet would be drawing heat from an earth temperature at that depth that would be about 50 F at the end of January, when heating loads are at maximum on the system. Because of the thermal storage capacity of the soil, the minimum soil temperature at the 8-foot burial depth would not be seen by the system until early April, when heating loads on the system have become much smaller than peak design loads. A ground loop that would be sized for heating at this location would still likely be oversized for the peak cooling load and would provide maximum entering water temperatures in cooling that are much less than most of the expected air temperatures during the main cooling months of June, July, and August, and likely lower than the average monthly air temperatures for those months.

Louisville, KY (Figure 1.6c) has a heating climate similar to Salt Lake City, but with slightly higher outdoor air temperatures during the summer cooling season, which are reflected in the somewhat higher average annual air temperature of 55.8 F and an average earth temperature (from the surface to 10 feet deep) of 60 F. In this climate, an air source heat pump may be an adequate selection because of the slightly higher outdoor air temperatures during the heating season, but the higher humidity of this climate may make air source heat pumps less efficient, as will be discussed later. In this case, a ground source heat pump would draw heat during the heating months from a ground loop that would, at a worst case, be designed to limit the lowest entering water temperatures to the heat pump at about 35-40 F (again, approximately 20 F below the deep earth temperature as a first estimate). This is 45 degrees higher than the lowest expected outside air temperatures and higher than nearly all of the temperatures experienced during the main heating months of November through March. A vertical ground loop would be drawing heat from a deep earth temperature of about 58 F, whereas a horizontal ground loop buried to a depth of 8 feet would be drawing heat from an earth temperature at that depth that would be about 57 F at the end of January, when heating loads are at maximum on the system Because of the thermal storage capacity of the soil, the minimum soil temperature at the 8-foot burial depth would not be seen by the system until early April, when heating loads on the system have become much smaller than peak design loads. A ground loop may be sized for either heating or cooling at this location, depending on the building heating and cooling loads. In any case, the ground temperatures and the efficiencies of ground source heat pumps would result in a very efficient and low cost of operation system in this climate.

San Antonio, TX (Figure 1.6d) is a very extreme cooling climate with an annual outside air temperature of 68.6 F and an average earth temperature (from the surface to 10 feet deep) of 72 F. A ground source heat pump for this climate would be designed based on the cooling load of the facility, and the ground loop would likely be designed to limit the entering water to the heat pump at a maximum of 95 F, or about 20-25 F above the average earth temperature. A loopfield designed for cooling in this climate would provide very warm entering water temperatures in the heating mode resulting in very efficient heating mode operation.

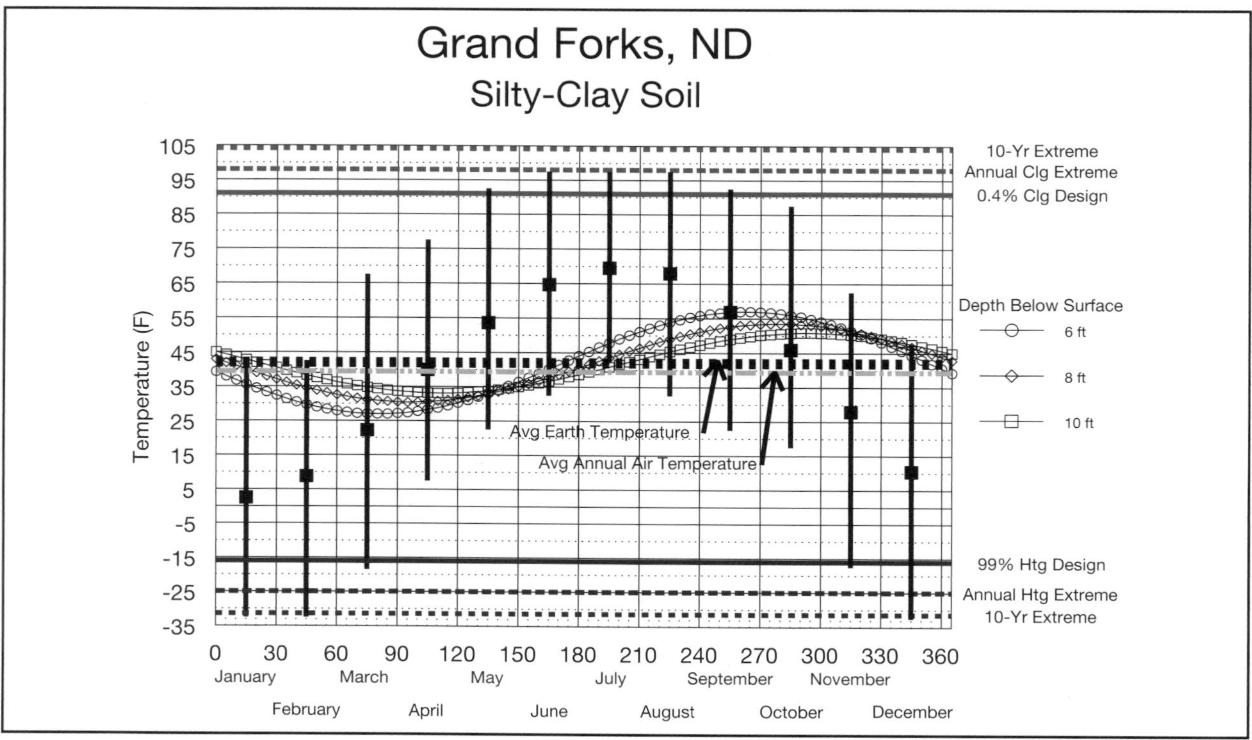

Figure 1.6a. Annual Air and Ground Temperature Variation-Grand Forks, ND

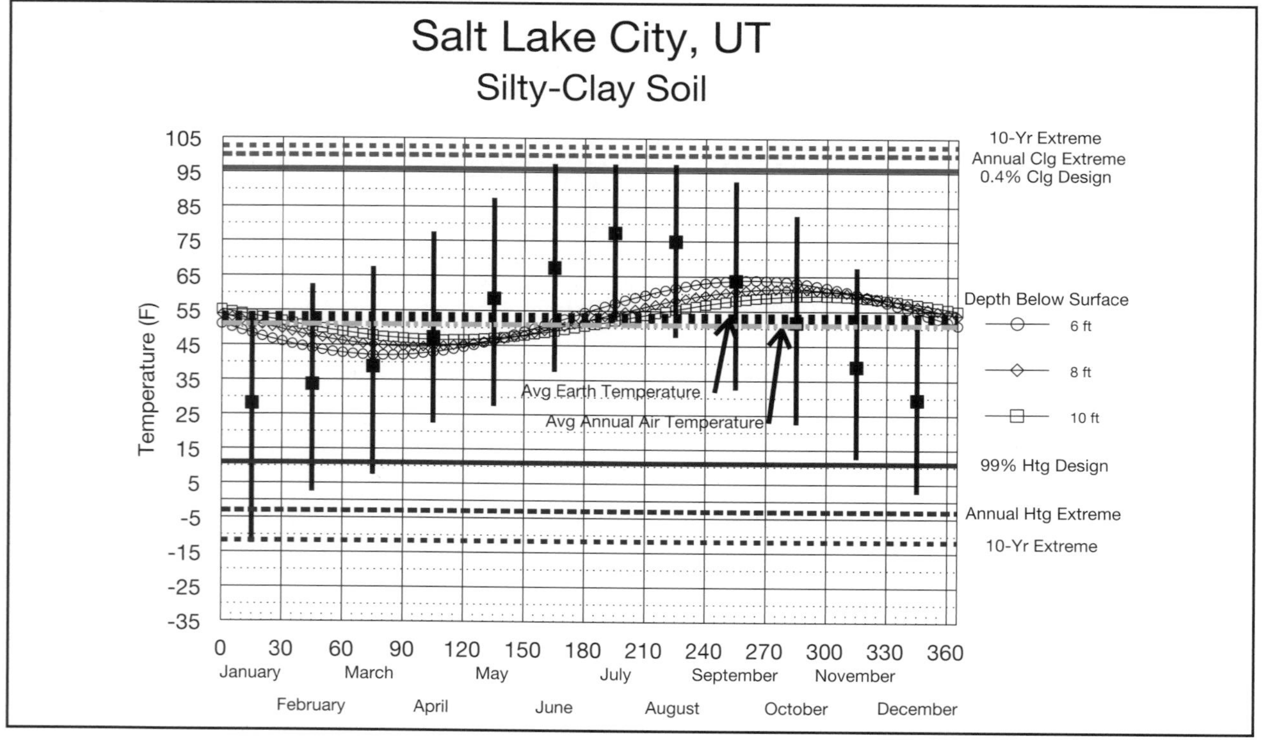

Figure 1.6b. Annual Air and Ground Temperature Variation-Salt Lake City, UT

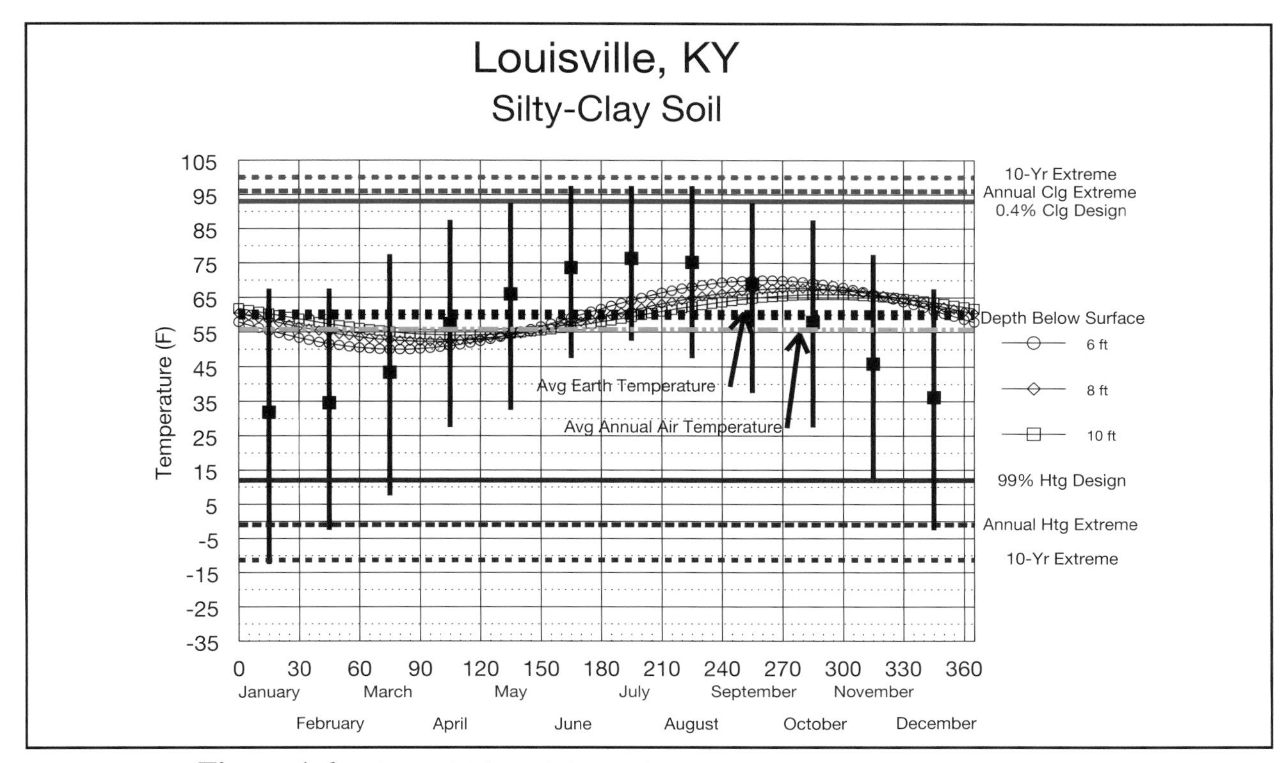

Figure 1.6c. Annual Air and Ground Temperature Variation-Louisville, KY

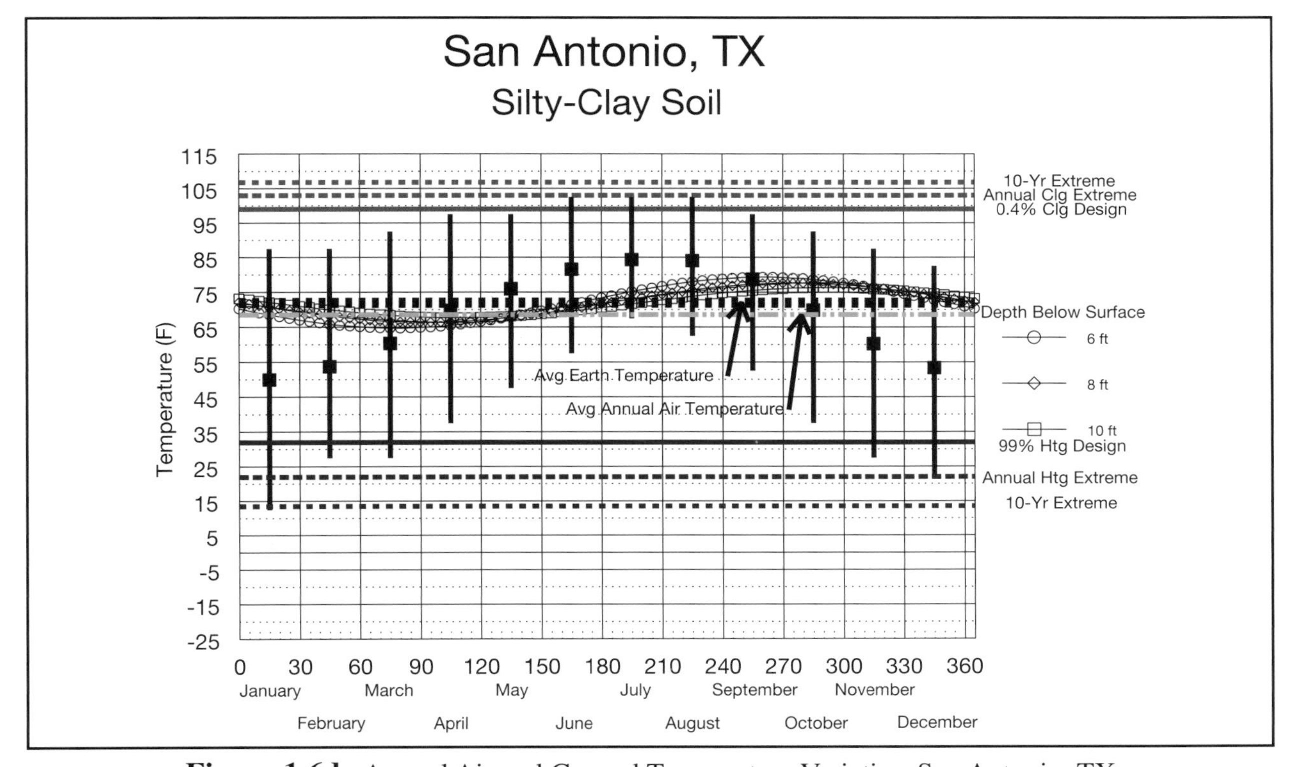

Figure 1.6d. Annual Air and Ground Temperature Variation-San Antonio, TX

1.4 Heat Pump Operation

1.4.1 BASIC HEAT PUMP SYSTEM In the GSHP system, a long-life, high-strength plastic pipe is buried beneath the earth's surface. Depending on available land area, the pipe will be placed in either a horizontal or vertical configuration (Figures 1.7a and 1.7b). Water or a water and antifreeze solution circulates through the closed-loop pipe, transporting heat energy to the heat pump in the winter and away from the heat pump during the summer. The heat energy exchange between the circulating fluid in the ground heat exchanger and the heat pump occurs in the heat pump's water-to-refrigerant heat exchanger. Hence the name water source heat pump since water or another fluid is used as the heat exchange medium. Other than this water heat exchanger, which replaces the outdoor air unit, the basic components in a water source heat pump are the same as air source heat pumps. Because there is no requirement for an outdoor condensing unit for heat exchange, all outdoor above ground components are eliminated. Also, there are no defrost components because defrost cycles are eliminated.

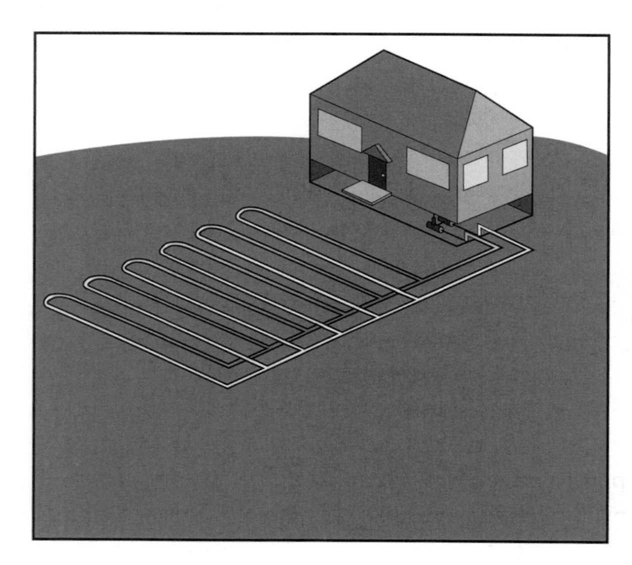

Figure 1.7a. GSHP System Horizontal Configuration Example

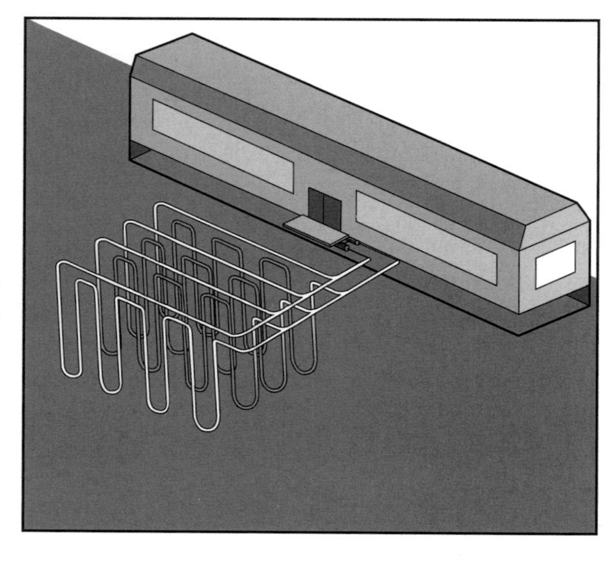

Figure 1.7b. GSHP System Vertical Configuration Example

1.4.2 HEAT PUMP CONFIGURATIONS Heat pumps are described in terms of their inputs and outputs. In this manual, the inputs or sources are air and water (liquid) and the outputs are air and/or water. Consequently, there are four types of heat pumps:

1. Air-to-air
2. Air-to-water
3. Water-to-air
4. Water-to-water

The most common type of heat pump to date is the air-to-air unit. Its early markets were in areas where outside air temperatures are moderate. Recent advances in air-to-air heat pumps have extended their use to more severe climates. The International Energy Agency (IEA) Web site (www.IEA.org) should be visited for recent advances.

Water-to-air is the most prevalent water source heat pump system in the United States where heating and cooling are required. The water source can be an open system (water well, direct use of pond or lake water, etc.) or a closed-loop ground heat exchanger (see Figures 1.2a through 1.3c).

Water-to-air heat pumps can be operated with a boiler and cooling tower. It is applied frequently in commercial buildings to "move" heat from one area to another. A water loop inside the building is connected to multiple heat pumps. When the loop temperature rises above a predetermined point, the cooling tower rejects the heat to the atmosphere. When the loop temperature drops below a predetermined point, the boiler rejects heat into the loop.

Water-to-water heat pump technology is used successfully in potable water heating and hydronic heating and cooling applications. Small and large commercial systems and larger homes have used water-to-water heat pumps in conjunction with a 4-pipe hydronic system. Heat delivery can be in-floor tubing or fan-coil units. Cooling where condensate must be removed from the air to meet comfort conditions is usually done with a fan-coil system. Some cooling with in-floor tubing has been used, but care must be taken not to form moisture on the floor surface. Water-to-water heat pumps have also been used in ice-making applications where the waste heat is used simultaneously for space heating (ice rinks). These systems have high efficiency since the required outlet condenser temperatures for hydronic systems is lower than forced air systems, which is an advantage to a heat pump system.

1.4.3 HEAT PUMP OPERATION The GSHP system consists of three loops that operate during all heat pump cycles and an optional fourth loop that preheats domestic hot water. (Figures 1.8-1.9)

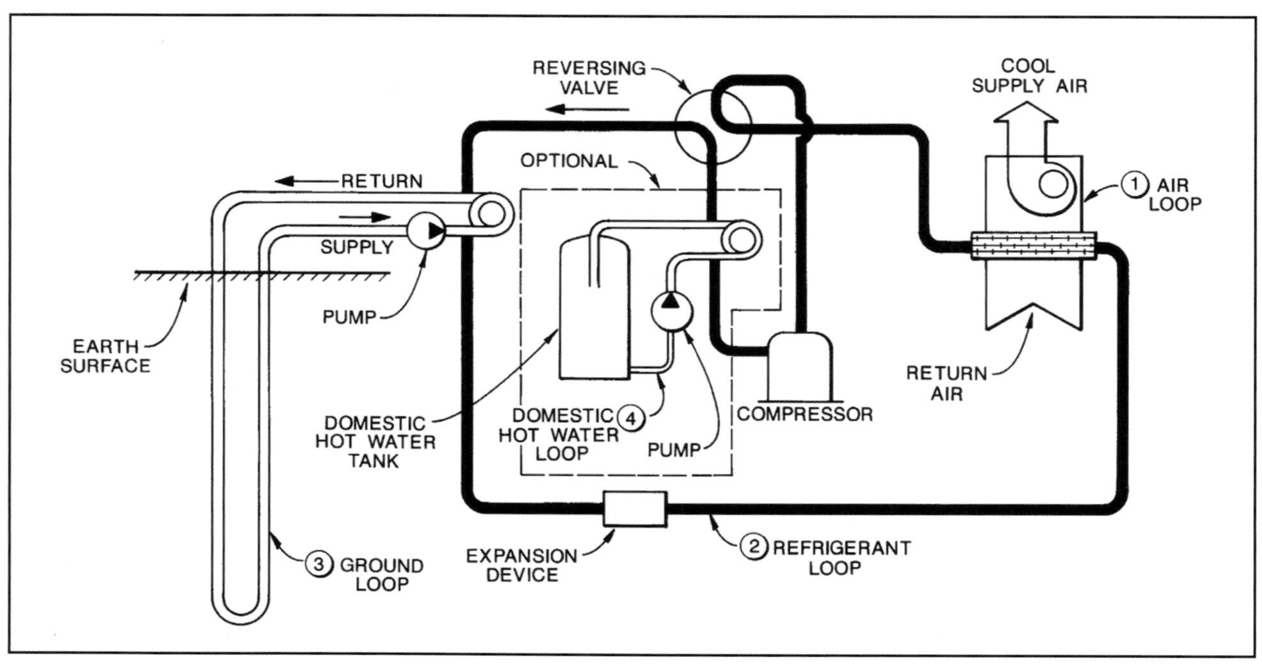

Figure 1.8. GSHP Operation in Cooling Mode

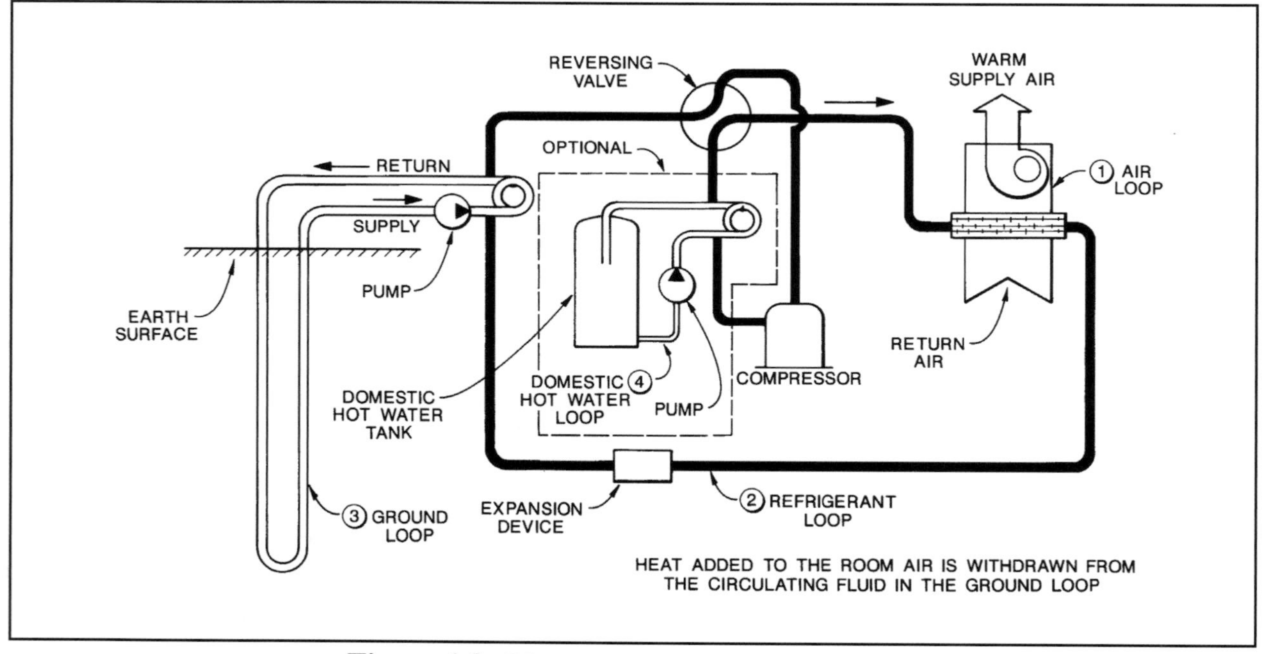

Figure 1.9. GSHP Operation in Warming Mode

Note: The loops apply to all technologies for ground-coupling, whether they are closed or open-loops.

Air loop or hydronic loop:

A loop used to distribute conditioned air or water to the building. A blower is used to move air through a duct distribution system. Duct and pipe sizes (diameter and length) are designed to distribute the conditioned air or water to specific locations depending on their heat loss or gain.

Refrigerant loop:

A sealed and pressurized loop that transports thermal energy from point to point in the circuit where refrigerant is forced through it by a compressor in the vapor portion of the loop.

Ground loop:

A sealed loop filled with water or antifreeze solution circulated below the earth's surface. It absorbs heat from the surrounding earth in the winter and rejects heat in the summer. The fluid is circulated by a low-power circulating pump.

Optional domestic hot water loop:

A sealed and pressurized loop that circulates water from the domestic hot water tank to a heat pump's desuperheater for heating domestic hot water. The water is circulated in this loop by a low-power circulating pump. Newer heat pumps may have full condensing loops that provide 100% of the domestic hot water.

During heat pump operation, thermal energy is transferred from one loop to another providing heating, cooling, dehumidification, and domestic hot water heating.

HEATING CYCLE

During heating cycles (see Figure 1.9) the heat inputs and outputs are as follows:

Inputs:
1. Thermal energy from the earth
2. Heat pump compressor energy
3. Circulating pump and blower energy

Outputs:
1. Space heating
2. Domestic hot water

The major heat inputs are the earth's thermal energy and compressor energy. For a 3-ton (36,000 Btu/hr) heat pump system operating with a COP of 3, two-thirds (24,000 Btu/hr) of the delivered energy will come from the earth and one-third (12,000 Btu/hr) from the electrical energy inputs. If the total energy required is greater than 36,000 Btu/hr, some form of supplemental heat is required. Normally the amount of heat input into the system from blower, pumps, etc., will be small compared with the earth and compressor energy. The blower motor energy will, however, be taken into consideration in the rating of the heat pump's capacity and performance.

The refrigerant vapor in the circuit is first compressed, which raises its temperature and pressure. This increased pressure forces the vapor through the refrigerant system.

The hot vapor continues to the refrigerant-to-air heat exchanger, which raises the temperature in the building because the cooler air absorbs heat from the hot refrigerant vapor. As heat is removed from the refrigerant in the refrigerant-to-air heat exchanger, the vapor condenses to a liquid. Thus, during the heating cycle, the indoor air coil serves as the system condenser.

The warm refrigerant returning from the condenser passes through a metering device. This reduces its pressure and causes a corresponding reduction in temperature. The low-pressure, low-temperature refrigerant then flows to the evaporator where thermal energy from the ground vaporizes the refrigerant and the cycle continues.

During the heating cycle, a circulating pump moves water or antifreeze solution through the ground heat exchanger. As this fluid circulates through the ground loop, it is warmed by the earth's higher temperature. The heat energy is transferred from the fluid to the refrigerant heat exchanger in the water source heat pump. During this part of the cycle, the water-to-refrigerant heat exchanger serves as an evaporator, changing the liquid refrigerant to vapor.

For heat pumps equipped with desuperheaters, the hot gas from the discharge of the compressor passes through a second water-to-refrigerant heat exchanger that heats domestic hot water. The domestic hot water, which is to be heated, is moved by a circulating pump. At this point, approximately 10% of the total heat energy available (the thermal energy of the vapor in a superheated condition) is removed.

COOLING CYCLE

During cooling cycles (see Figure 1.8), the heat inputs and outputs are as follows:

Inputs:
1. Building heat gain
2. Compressor energy
3. Pump and blower energy

Outputs:
1. Domestic hot water
2. Waste heat to the earth

The heat output during cooling generally will be much greater than that required for domestic hot water. For a nominal 3-ton (36,000 Btu/hr) cooling unit, the net amount of rejected thermal energy is approximately 48,000 Btu/hr (COP = 3). Of this, only about 12,000 to 15,000 Btu/hr is available or needed for domestic hot water if the unit has the water heating option. The remaining thermal energy must be rejected through the ground heat exchanger to the earth. Compared to the heating cycle, which absorbs about 24,000 Btu/hr, cooling cycles generally reject almost twice the thermal energy for each hour of operation.

During cooling operations (Figure 1.8), the hot gas leaving the compressor heats the domestic hot water (if the heat pump has the water heating option) and then is routed to the water-to-refrigerant heat exchanger where excess heat is rejected to the earth.

Space cooling is accomplished by the warm indoor air passing over the cold evaporator coil. The high pressure hot liquid in the condenser was forced through a metering device, which results in a low-pressure and low-temperature liquid. Evaporation of this liquid to a vapor is the cooling mechanism.

Figures 1.8 and 1.9 show the components and their arrangement in a commercial water source heat pump. Included in these drawings is the reversing valve that changes the direction of refrigerant flow to the indoor air coil and the water-to-refrigerant heat exchanger between the heating and cooling cycles. As noted in these figures, the domestic hot water heat exchanger is between the compressor and the reversing valve. This allows for heating domestic water in both the heating and cooling cycles.

1.4.4 DOMESTIC HOT WATER OPTIONS In addition to space heating and cooling, water source heat pump units are available with three specific domestic hot water options:

Option 1: None - No hot water heating capability
Option 2: Supplemental - Partial heating (desuperheater) during operation in the heating or cooling cycle
Option 3: Dedicated or on demand - Year-round total (full condensing water heating) water heating on a first priority basis

Option 1

This option uses a conventional water source heat pump without provision for domestic water heating. This configuration represents the lowest cost water source heat pump. It is used in buildings with more than one heat pump and where there is no requirement for domestic hot water.

Option 2

This option uses a refrigerant-to-water heat exchanger (desuperheater) installed in the discharge line of the heat pump compressor. The hot gas at this point is in a "superheated" condition giving rise to the common name desuperheater. In southern climates, "free hot water" is available during the cooling season and typically can supply 100% of the domestic hot water. In the winter, the desuperheater generating domestic hot water competes with the heating load. Because there is generally excess heat pump capacity for many winter hours, domestic hot water efficiency equals heat pump efficiency. In either mode, when the hot water demand has been met, the circulating pump will shut off.

Option 3

Total or 100% domestic hot water on a first priority basis simply means that, before any space heating or cooling is accomplished, all domestic hot water requirements are met by the heat pump system. These heat pumps have a completely different design. Heat pumps with this feature will have the longest-run cycles. This usually improves heat pump performance and generally reduces the total annual demand and energy costs.

Domestic hot water generation is discussed in detail in Chapter 9.

1.5 GSHP System Materials and Components

1.5.1 PLASTIC PIPE The most common material used in underground heat exchangers is high density polyethylene (HDPE). This material is flexible, allowing easy installation. The material can be joined by heat fusion or stab-in fittings, which give high reliability. The principal method is heat fusion. The grade of pipe used should be carefully chosen. See Chapter 4 for grade selection and the appropriate standards for pipe selection.

1.5.2 ANTIFREEZE SOLUTIONS

In designs where loop minimum entering fluid temperature will fall below 45 F, an antifreeze solution will be required to prevent freezing in the heat pump heat exchanger. The choices are:

1. Glycols-ethylene and propylene
2. Alcohols- methyl and ethyl

1.5.2.1 ANTIFREEZE CRITERIA

These solutions can and have been applied successfully. For any fluid to be used, it must be:

1. Safe
2. Non-toxic
3. Non-corrosive
4. Good heat-transfer medium
5. Low cost
6. Long lasting
7. Low viscosity

1.5.2.1.1 GLYCOLS: Ethylene and Propylene– Propylene glycols are relatively safe, generally non-corrosive, considered non-toxic, have fair heat transfer characteristics, medium cost, and a finite life. For low-temperature operation, 30 F and below, the glycols can become viscous, requiring greater pumping power and therefore, reduce the heat pump system efficiency. In loop designs where heat pump closed-loop operating temperatures remain above 35 F, glycols have been used successfully.

1.5.2.1.2 ALCOHOLS: Methyl and Ethyl– Improperly handled, alcohols are unsafe (will burn and explode when mixed with air), toxic, relatively non-corrosive, have fair heat transfer characteristics, medium cost, and a long life. Their major drawbacks are explosiveness and toxicity. Diluting (50% water and alcohol) the alcohol solution before taking it on site somewhat reduces the risk of explosion. Because alcohols are non-corrosive, their use is very popular. Alcohols are widely used in all climates.

CAUTION:

1. 100% alcohol solution burns invisible in bright sunlight.
2. Many anti-corrosion inhibitors are toxic and care must be taken to prevent a non-toxic antifreeze solution from becoming toxic after the solution is inhibited.

1.5.3 METAL COMPONENTS

The system's metal components must be compatible with the closed-loop GSHP circulating fluid. The components that must be carefully selected are:

1. Circulating pump and pump flanges
2. All metal piping
3. Sensing ports
4. Any metal component in contact with the circulating fluid

Depending on the antifreeze used, the metals chosen for the system must be carefully selected. The manufacturer's and dealer's recommendations must be followed carefully.

1.5.4 HEAT PUMPS A number of excellent water source heat pumps are on the market designed for closed-loop ground source operation. For closed-loop operation, the following characteristics should be considered:

1. Manufacturers' entering fluid temperature range
2. Features (desuperheater, insulated water lines TXV, etc.)
3. Safety listings (UL, ETL, or ARL)
4. Type (split, package, water-to-water, etc.)
5. Warranty
6. Performance rating (ARI/ISO 13256-1, 13256-2)

Water source heat pumps can be classified by their application and consequently their inlet fluid operating temperature range. For example, in commercial buildings the fluid temperature may range from 60 to 95 F. If the inlet fluid temperature drops below 60 F, a boiler is used to raise the temperature. If the temperature goes above 95 F, a cooling tower is used to reject heat. Heat pumps used for this type of application generally cannot be used for ground source systems where the inlet temperature may range from 25 F to 105 F.

A second type of heat pump is a groundwater heat pump that operates on well water temperatures ranging from 40 F to 80 F. The ground-coupled type of heat pump will operate from 25 F to 105 F. Care should be taken to pick a heat pump suitable for the expected or design inlet liquid temperature. The installing contractor should be aware of the differences and make recommendations accordingly.

1.6 Brief History of GSHP Systems

One of the fastest growing renewable energy applications worldwide today, ground source heat pumps (GSHP) have seen several revivals during the 150-year history of the technology. The first century from 1852 to 1950 saw what G. C. Groff has called a "heat pump awakening," largely beginning with Lord Kelvin's early proposal to use refrigeration systems for heat in 1852. One of the earliest known examples is that of Peter Ritter von Rittinger's engineered application for a salt works in an Austrian village, where he developed a thermo-compression plant.

Heat pump technology has since been refined and modified by an impressive list of engineers who have furthered the study of thermodynamics and designed even more efficient, readily accessible, and applicable systems.

Industrial, commercial, and residential growth in the 1950s fed an exploration for new energy sources. A number of U.S. manufacturers, both with and without experience in air conditioning and refrigeration products, began making heat pumps with names like Utopia and Miracula. Early enthusiasm came from both power utility companies looking to sell more electricity and product manufacturers who saw the marketing potential. Real growth of the technology was eventually thwarted by both a lack of trained installers and service technicians, and the production of products of questionable quality that led to failed installations.

The U.S. began to build a large number of homes and commercial buildings that were dependent on electricity without a backup fuel. Although installations continued for nearly 15 years, public interest in heat pumps waned. Until the mid-1970s, oil embargoes and the imperative to deal with scarce oil supplies, the technology was largely overlooked.

After establishment of the International Energy Agency (IEA) in 1974, heat pumps were identified as a key energy conservation strategy and renewed research became a priority for a number of agencies. It was at this time that the International Ground Source Heat Pump Association (IGSHPA) began looking into ways to improve and promote the technology at Oklahoma State University.

Efforts spearheaded by IGSHPA led to construction of a research and test facility on the OSU campus and by 1978, two residential installations were completed in Stillwater, OK. In 1978, the Department of Energy granted OSU a contract for the DOE Solar Assist project and heat pump research began in earnest.

Key industry developments that ensued from ongoing research include the use of in place or "in situ" testing, the development of manuals for design and installation of GSHP systems, improvement of thermal grouts, the use of polyethylene pipe and heat fusion joining, development of Slinky Heat Exchangers, and software design programs for both commercial and residential applications.

IGSHPA began holding annual conferences in 1987, with the first in Indianapolis, IN. Chapters have been established in Japan, Korea, and China, as well as other international locations. Six teleconferences have also been held to aid membership and promote the technology.

Global concerns over energy continue to support the market for GSHP as the foremost environmentally friendly and energy efficient technology available today. A renewed interest in applications for the technology, coupled with an increasing incentive and tax rebate structure in the U.S. at both state and federal levels have again brought growth to the industry. An estimated 10% annual increase of applications in nearly 30 countries including Austria, Canada, Germany, Sweden, Switzerland, and the United States accounts for more than 130 million heat pumps in use worldwide today.

2 PERFORMANCE OF GSHP EQUIPMENT

In This Section:

2.1 Types of GSHP Equipment

Ground Source Heat Pumps (GSHPs) utilize electrical power to move heat from a low temperature source to a high temperature sink. If the energy delivered to the high temperature sink is the desired output of the heat pump, the heat pump is in the heating mode. If the energy absorbed from the low temperature source is the desired effect, the heat pump is in the cooling mode. The medium from which the heat is removed from or delivered to also defines the type of heat pump. Figures 2.1(a-d) display multiple variations of GSHP applications that currently exist.

Figure 2.1(a) shows the most common GSHP application of ducted air heating or cooling (through a refrigerant-to-air heat exchanger using a ground connection or through a water-to-refrigerant heat exchanger, as a heat source for heating and a heat sink for cooling). The most common water-to-air heat pumps are single compressor-single capacity and single compressor-two capacity, although dual compressor-two capacity units are available in larger capacities. An optional heat exchanger, referred to as a desuperheater, is available that is connected to the domestic water heater to provide some portion of the domestic hot water requirements, depending on the time of year.

Figure 2.1(b) shows a water-to-water application with a water-to-refrigerant heat exchanger connecting to the ground source/sink and a refrigerant-to-water heat exchanger connecting to the load sink/source for heating/cooling, respectively. In heating mode, the heat pump extracts energy from the ground and provides energy to a circulating water distribution system. In the cooling mode, the heat pump extracts energy from the circulating water distribution system and rejects energy to the ground. Generally, small capacity units will be single compressor-single capacity or single compressor-two capacity, while larger capacity units may be dual compressor-two capacity. A desuperheater option is available on these units.

A third type of GSHP, which is becoming more popular, is the combination unit (Figure 2.1(c)). It has the ability to heat or cool air at full capacity or to heat or cool water at full capacity, but not both at the same time. These units have application in standard ducted heating or cooling situations where there is a water heating load that can be served whenever the heat pump is not serving the air-side load, such as domestic hot water generation or hot water for a radiant panel heating load. These units are available in single compressor single capacity units for water-to-air and water-to-water applications, and single compressor two-capacity water-to-air and single capacity water-to-water applications.

The fourth, and less common, type of GSHP is the dual-circuit unit (Figure 2.1(d)), which utilizes two compressors (generally of different capacities) connected to two refrigeration circuits to allow multiple modes of operation. This unit may use one compressor in heating or cooling of ducted air only, one compressor to heat water only, one compressor heating or cooling air while the other is heating water, or both compressors either heating or cooling air. Therefore, this heat pump is very versatile in its use, but is only available in limited heating and cooling capacities.

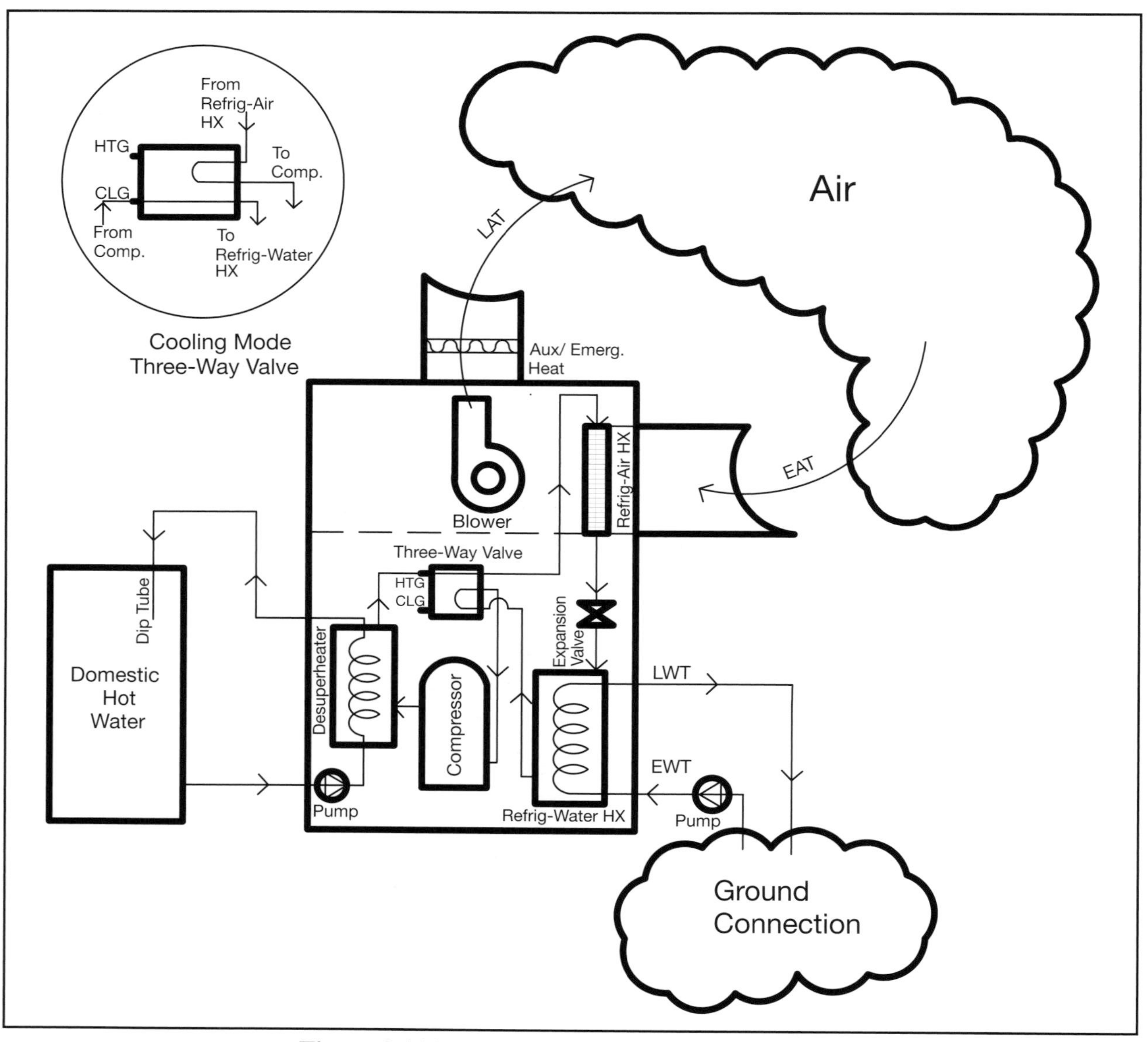

Figure 2.1(a) Water-to-Air GSHP – Heating Mode

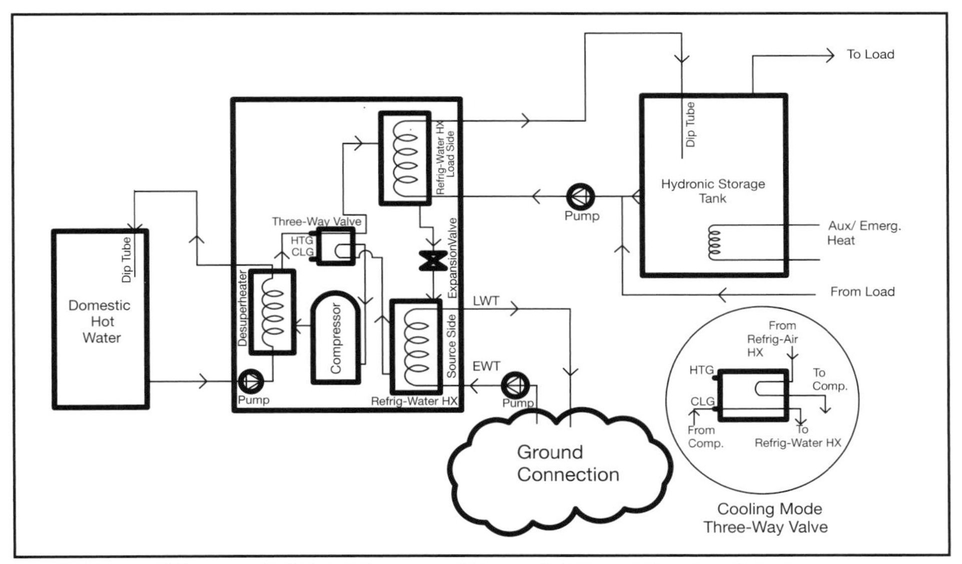

Figure 2.1(b) Water-to-Water GSHP – Heating Mode

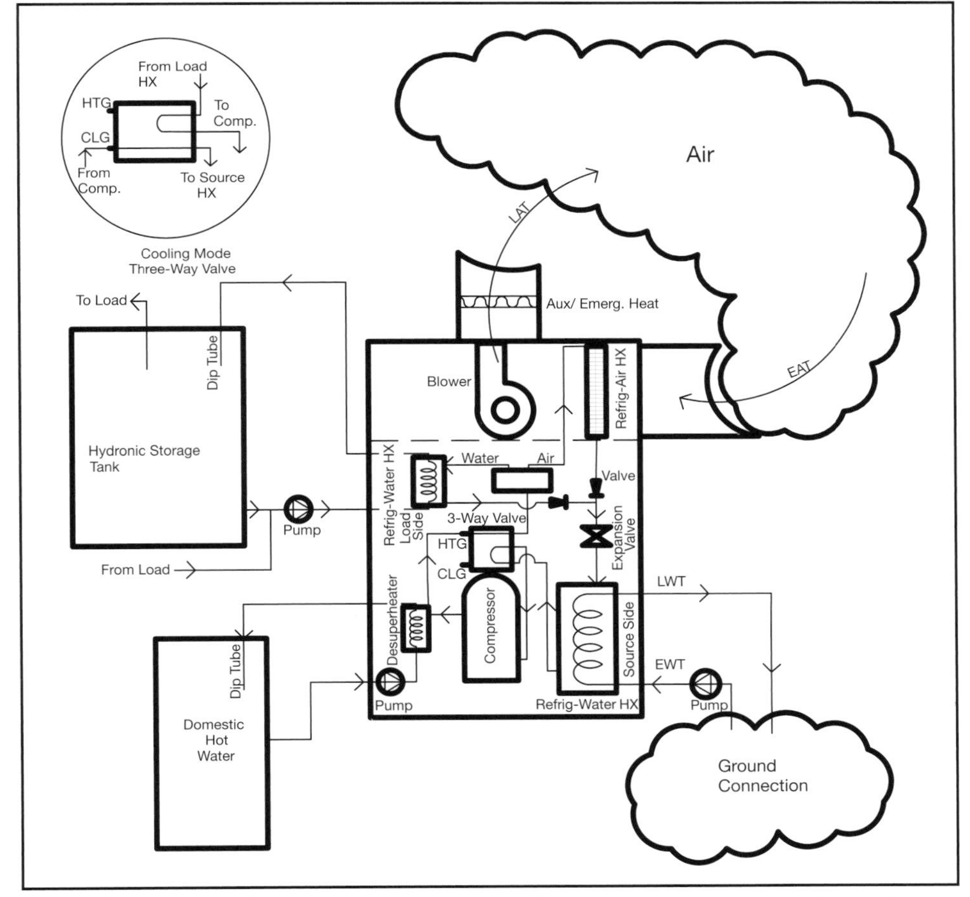

Figure 2.1(c) Combination GSHP (Water-to-Air or Water-to-Water)

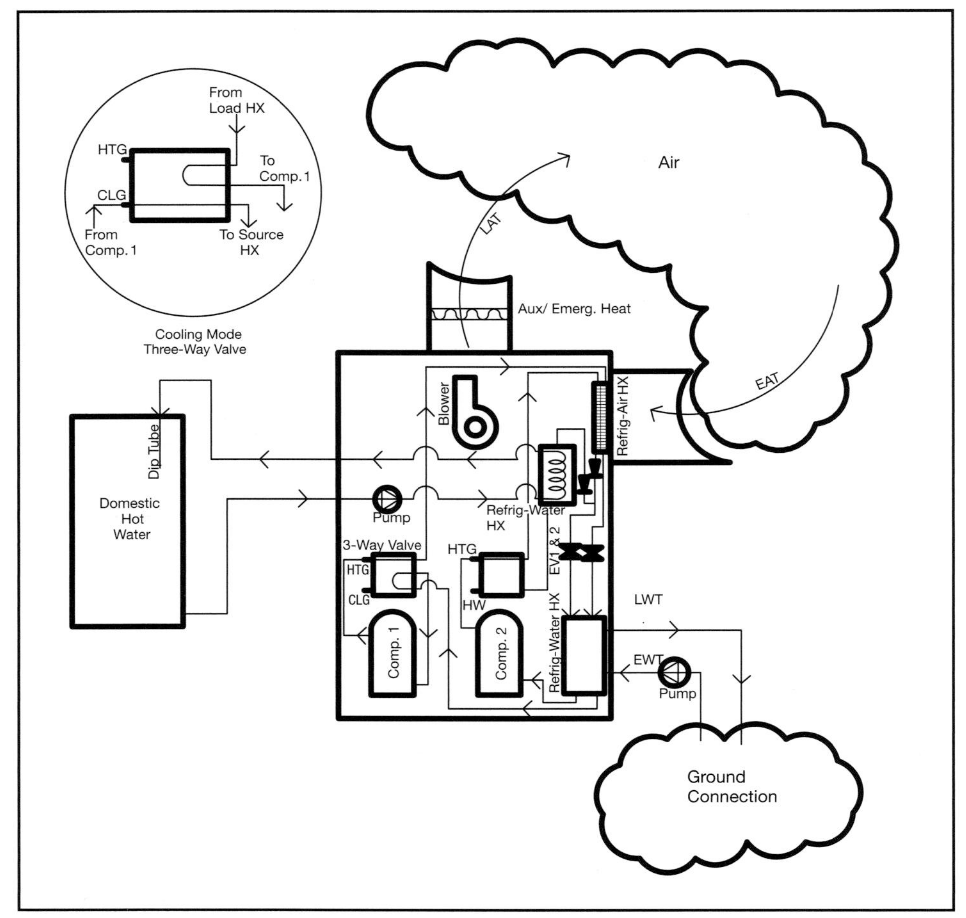

Figure 2.1(d) Dual-Circuit GSHP

2.2 Energy Balance on the GSHP and Ground Loads

2.2.1 HEATING MODE In general, the performance of a heat pump (in this case a ground source heat pump) is defined by its heating or cooling capacity (Btu/hr), its electrical demand (kW), and its efficiency. Figure 2.2(a) shows a GSHP in the heating mode providing heating output (HC) to a high temperature air or water sink. To provide that heating output, the heat pump utilizes an electric input (DMD) to run a vapor-compression refrigeration cycle to extract heat (HE) from a low temperature source (ground connection). Because heat pumps move heat instead of burning fuel to create heat, the heating efficiency is expressed in terms of a coefficient of performance (COP), as defined by Equation 2.1, which is the ratio of the heating output (HC) divided by the electrical demand (DMD) input converted to Btu/hr (1 kW=3,412 Btu/hr). The resulting units on COP are dimensionless, but could actually be interpreted as the ratio of Btu/hr of heating output to Btu/hr of equivalent electric input.

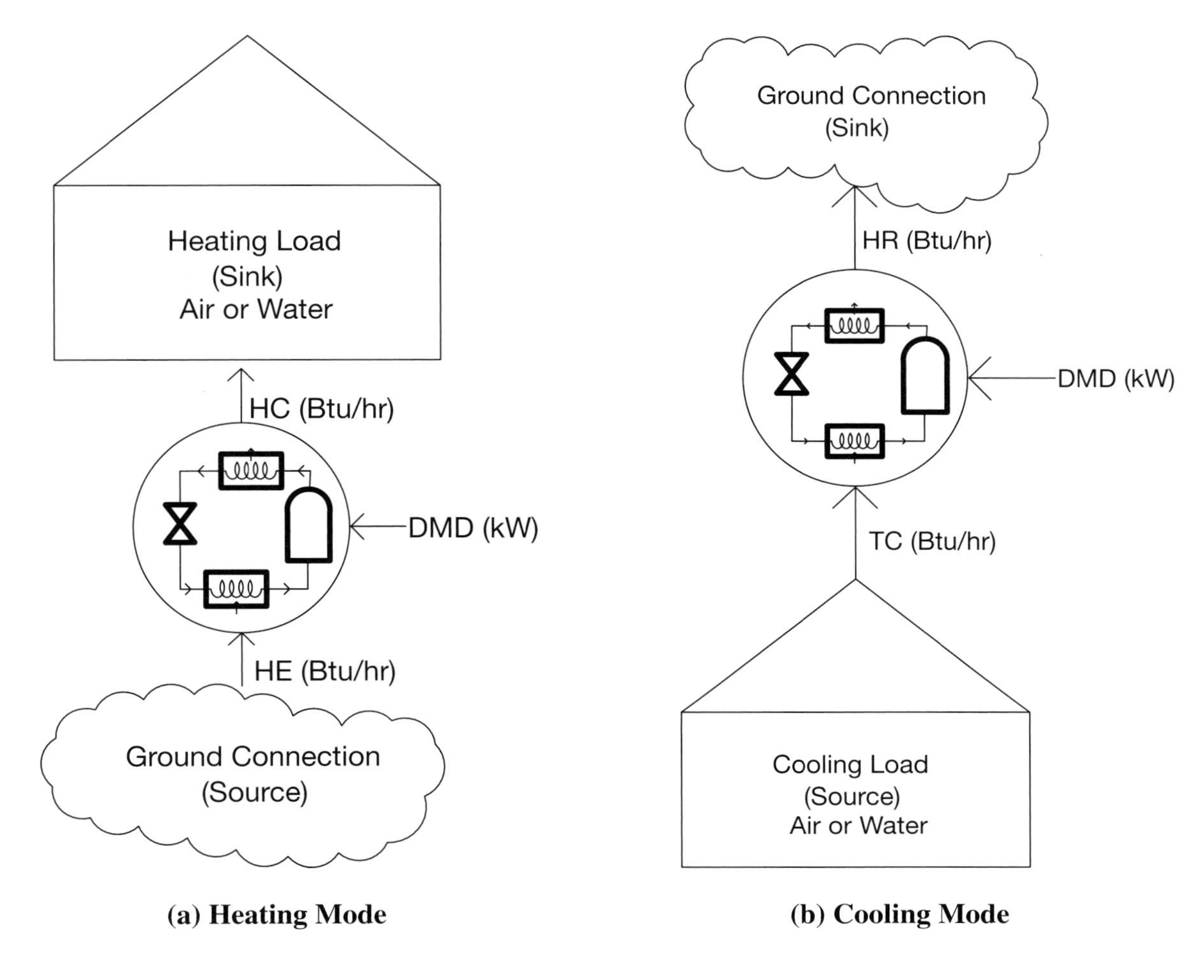

Figure 2.2 Ground source heat pump concept.

$$COP = \frac{HC}{DMD \cdot \frac{3{,}412\ \text{Btu/hr}}{\text{kW}}} \qquad \textbf{Equation 2.1}$$

Where:

COP = coefficient of performance (dimensionless)
HC = heating capacity (Btu/hr)
DMD = electric demand (kW)

An energy balance on the heat pump in Figure 2.2(a) yields Equation 2.2, which shows that the heating capacity of the heat pump is the sum of the heat of extraction from the ground plus the electrical demand of the heat pump. The heat of extraction is always smaller than the heating capacity of the heat pump because the electrical power consumption of the compressor, fan, and pumps add to the heating capacity of the heat pump.

$$HC = DMD \cdot \frac{3{,}412 \text{ Btu/hr}}{\text{kW}} + HE$$ **Equation 2.2**

Where:

HE = heat of extraction from the ground connection (Btu/hr)

Equations 2.1 and 2.2 can be combined (Equation 2.3) to express heat of extraction in terms of the heating capacity and the COP of the heat pump, which defines the portion of the heating capacity of the heat pump that is extracted from the ground based on heat pump efficiency.

$$HE = \left(\frac{COP - 1}{COP}\right) \cdot HC$$ **Equation 2.3**

Table 2.1 summarizes the heat of extraction for a nominal 3-ton (36,000 Btu/hr) heating capacity for heat pump COPs ranging from 2.0 to 5.0, which encompasses the heating efficiency of most available heat pump equipment over a wide range of operating conditions. As the COP increases, the heat of extraction also increases, indicating that a more efficient heat pump will require a higher heat of extraction from the ground loop and will generally result in a need for more ground loop capacity to maintain the design parameters for which the ground loop is designed. This will be discussed in detail in Chapter 5.

Table 2.1. Heat of Extraction[1] for Common COPs

COP	HE (Btu/hr)
2.0	18,000
2.5	21,600
3.0	24,000
3.5	25,714
4.0	27,000
4.5	28,000
5.0	28,800

1. Heating capacity (HC) =36,000 Btu/hr

2.2.2 COOLING MODE Figure 2.2(b) shows a GSHP in the cooling mode providing total cooling capacity (TC) to an air or water source. To provide that cooling, the heat pump utilizes an electrical input (DMD) to run a vapor-compression refrigeration cycle to reject heat (HR) to a high temperature sink (ground connection). In the case of cooling, efficiency is traditionally expressed in terms of an energy efficiency ratio (Equation 2.4), which is defined as the ratio of the rate of energy absorbed from the cooled load (TC) divided by the electrical input (DMD) expressed in watts (1 kW=1000 W).

$$EER = \frac{TC}{DMD \cdot \frac{1{,}000 \text{ W}}{\text{kW}}}$$ **Equation 2.4**

Where:

EER = energy efficiency ratio (Btu/hr W)
TC = total cooling capacity (Btu/hr)

An energy balance on the heat pump in Figure 2.2(b) yields Equation 2.5, which shows that the heat of rejection from the heat pump is the sum of the cooling capacity of the heat pump plus the electric demand of the heat pump. The heat of rejection is always larger than the cooling capacity of the heat pump because the electrical power consumption of the compressor, fan, and pumps must also be rejected to the heat sink (ground connection).

$$HR = DMD \cdot \frac{3{,}412\ \text{Btu/hr}}{\text{kW}} + TC$$ **Equation 2.5**

Where:

HR = heat of rejection to the ground connection (Btu/hr)

Equations 2.4 and 2.5 can be combined (Equation 2.6) to express heat of rejection in terms of the cooling capacity and the EER of the heat pump.

$$HR = \left(\frac{EER + 3.412\ \frac{\text{Btu/hr}}{\text{W}}}{EER} \right) \cdot TC$$ **Equation 2.6**

Table 2.2 summarizes the heat of rejection for a nominal 3-ton (36,000 Btu/hr) cooling capacity for heat pump EERs ranging from 10 to 22, which encompasses the cooling efficiency of most available heat pump equipment over a wide range of operating conditions. Table 2.2 is intended to show the effect of EER on heat of rejection, not the complete range of GSHP efficiencies, which can reach EER 30. As the EER increases, the heat of rejection decreases, indicating that the more efficient heat pump will lower the heat of rejection to the ground loop, which results in a need for less ground loop capacity to maintain the design parameters upon which the ground loop is designed. This will be discussed in detail in Chapter 5.

Table 2.2. Heat of Rejection[1] for Common EERs

EER	HR (Btu/hr)
10	48,283
12	46,236
14	44,774
16	43,677
18	42,824
20	42,142
22	41,583

1. Total Cooling Capacity (TC)=36,000 Btu/hr

2.3 Factors that Affect the Performance of a GSHP

2.3.1 BASIC VAPOR-COMPRESSION REFRIGERATION CYCLE-HEATING MODE GSHP operation is based on a vapor-compression refrigeration cycle. This cycle is shown schematically in Figure 2.3 for a water-to-air heat pump operating in the heating mode. The circled numbers represent "state" points of the refrigerant. The cycle utilizes two heat exchangers, a compressor, and an expansion valve to maintain high pressure (high temperature) and low pressure (low temperature) refrigerant sides in the cycle. The high pressure/temperature side rejects heat through a heat exchanger to a high temperature sink (air, in this case), and the low pressure/temperature side absorbs heat through a heat exchanger from a low temperature source (water, in this case). The cycle effectively moves heat from a low-temperature source (the earth) to a high-temperature sink (the home) by using electrical power input (DMD) to drive the compressor to create the high and low pressure/temperature sides to the system.

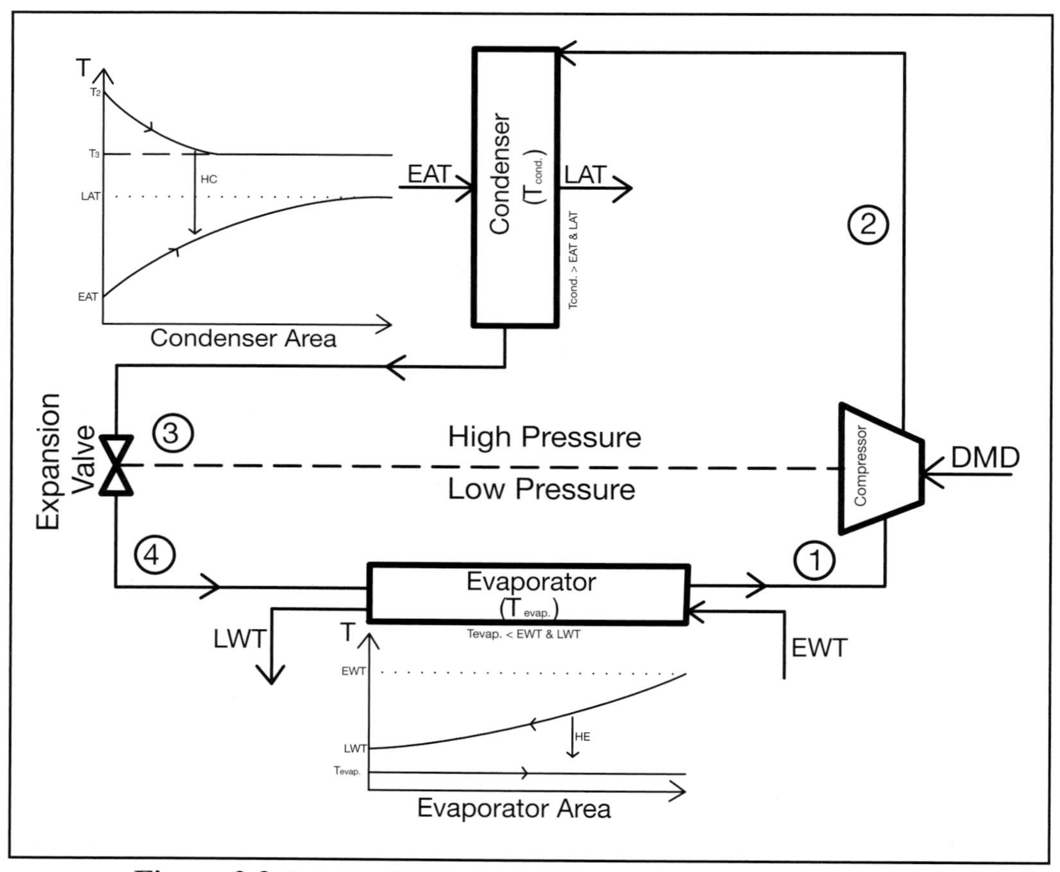

Figure 2.3. Schematic of a vapor-compression refrigeration cycle.

The operation of a vapor-compression refrigeration cycle is based on the pressure-temperature and phase characteristics of a refrigerant, most commonly R410A or R407C (R22 has been phased out). Figure 2.4 displays a pressure-enthalpy diagram, which shows the relationships between absolute pressure (psia), temperature (F), phase (liquid, liquid-vapor mixture, or vapor), and enthalpy (energy content per unit mass) for R410A, a refrigerant that has recently become common in GSHP equipment. The "dome-shaped" area between and underneath the saturated liquid (temperature and pressure where refrigerant is all liquid, but will immediately begin to evaporate with addition of heat) and saturated vapor (temperature and pressure where refrigerant is all vapor, but will immediately begin to condense with removal of heat) lines represents a mixture of liquid and vapor. Left-to-right movement along a constant pressure line (perpendicular to the Y-axis) in this region represents evaporation of the refrigerant while right-to-left movement represents condensation of the refrigerant. The temperature associated with any given pressure in this region is referred to as the saturation temperature, which stays constant during a constant-pressure evaporation or condensation process in the mixed region. The region to the left of the saturated liquid line contains properties for sub-cooled liquid (temperature below the boiling point at a given pressure), and the region to the right of the saturated vapor line contains properties for superheated vapor (temperature above the boiling point at a given pressure). The enthalpy at any given point on the chart is obtained by moving vertically downward to the X-axis.

The ideal vapor-compression refrigeration cycle represented by Figure 2.3 is plotted (solid thick lines) on Figure 2.4 for low-side and high-side refrigerant pressures of 85 psia and 360 psia, respectively. These pressures represent approximate mixed refrigerant temperatures in the evaporator and condenser of 15 F and 106 F, respectively, which may be representative of heating mode operation with a ground loop entering water temperature of 30 F to the evaporator and entering and leaving air temperatures to and from the condenser of about 70 F and 100 F, respectively. Note that the refrigerant temperature in the evaporator is about 15 F lower than the entering water temperature to promote heat transfer from the water to the refrigerant. Likewise, the refrigerant temperature in the condenser will be higher than the leaving air temperature and is significantly higher than the entering air temperature, which would be about 70 F in heating mode. The refrigerant at State 1 is ideally a saturated vapor at the low-side (suction) pressure, which resides directly on the saturated vapor line at the low-side pressure. The ideal compressor "squeezes" the refrigerant vapor into a much smaller volume, which is accompanied by a large increase in the temperature of the vapor. In this case, the refrigerant leaves the compressor at nearly 150 F (State 2), where it then enters the refrigerant-to-air heat exchanger (condenser). There, heat is transferred to the air, warming the air, and cooling the refrigerant until it reaches the saturated vapor point at the high side pressure, after which the refrigerant is condensed completely to a saturated liquid at the high side pressure (State 3). The refrigerant then flows through an expansion valve where the pressure is suddenly reduced to the low-side pressure, resulting in a mixture of vapor and liquid at a much lower temperature, but with an enthalpy that is the same as before the expansion valve because no energy was transferred in the process. The resulting low-temperature mixture (State 4) then flows into the evaporator where energy is picked up from the warmer water being circulated from the ground connection, cooling the circulating water, and vaporizing the refrigerant to the saturated vapor state at the low-side pressure where the process begins again.

The COP of the ideal vapor-compression refrigeration cycle operating in the heating mode can be determined using data from the R410A pressure-enthalpy diagram along with basic relations based on simple thermodynamics. For the low and high-side pressures identified in the preceding paragraph, the state point enthalpies are obtained from the cycle drawn on Figure 2.4 (solid thick lines). State 1 is located directly on the saturated vapor line at the low-side pressure. State 3 is located directly on the saturated liquid line at the high-side pressure. State 4 has the same enthalpy as State 3 because the expansion valve represents a constant enthalpy expansion. That is, the pressure is reduced while no energy is added or removed from the refrigerant and the enthalpy remains constant. State 2 represents the refrigerant output state from the

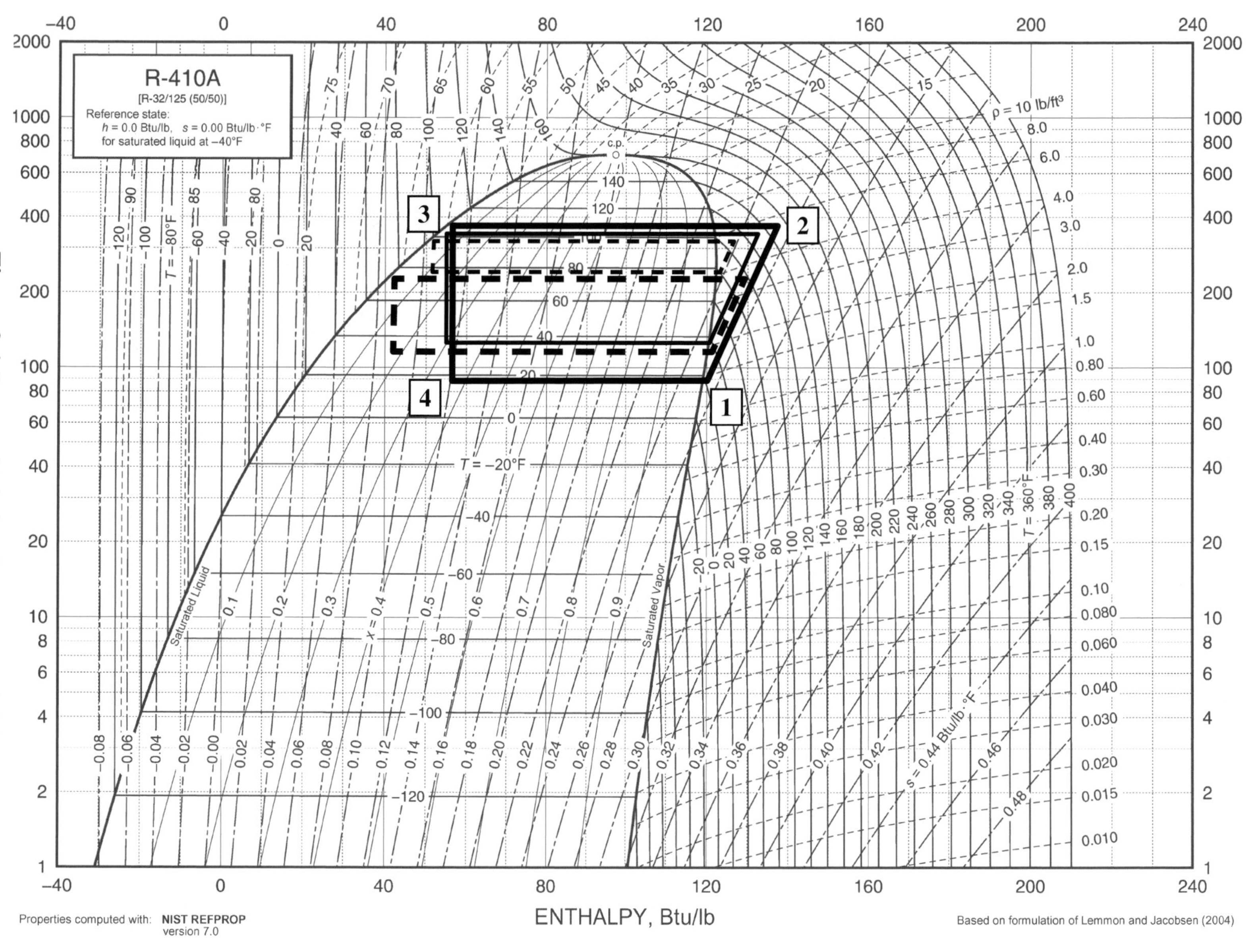

Figure 2.4. Pressure-Enthalpy diagram for R-410A.

compressor. For the ideal compression process, the compression is constant entropy, which means that the compression process line on Figure 2.4 follows a line of constant entropy(s) from State 1 to State 2 at the high-side pressure. The COP for the ideal cycle is calculated as the heating capacity of the condenser (HC - the desired effect) divided by the work of the compressor on the refrigerant ($Work_{Comp}$) or the enthalpy change across the condenser ($h_2 - h_3$) divided by the enthalpy change across the compressor ($h_2 - h_1$). The heat of extraction (HE) represents the amount of heat that is obtained from the ground connection through the evaporator ($h_1 - h_4$).

Example 2.1. Determine HC, $Work_{comp}$, COP, and HE for the ideal vapor-compression heating cycle discussed above (solid thick lines in Figure 2.4).

P_{Low} = 85 psia, T_{sat} = 15 F (approximately 15 F below EWT = 30 F)
P_{high} = 360 psia, T_{sat} = 106 F (approximately 5 F above LAT = 100 F)

State 1: P_1 = 85 psia & Saturated Vapor — $h_1 = 118.8$ Btu/lb_m
State 2: P_2 = 360 psia & Isentropic Compression — $h_2 = 136$ Btu/lb_m
State 3: P_3 = 360 psia & Saturated Liquid — $h_3 = 54.2$ Btu/lb_m
State 4: P_4 = 85 psia & Constant Enthalpy Expansion — $h_4 = 54.2$ Btu/lb_m

$$HC_{cond} = h_2 - h_3 = 136 - 54.2 = 81.8 \text{ Btu/lb}_m$$
$$Work_{Comp} = h_2 - h_1 = 136 - 118.8 = 17.2 \text{ Btu/lb}_m$$

$$COP_{ideal} = \frac{HC_{Cond}}{Work_{Comp}} = \frac{81.8}{17.2} = 4.8$$

$$HE_{evap} = h_1 - h_4 = 118.8 - 54.2 = 64.6 \text{ Btu/lb}_m$$

The COP for the ideal cycle is higher than would be expected for an actual ground source heat pump operating with the EWT and LAT identified in Example 2.1. Factors that contribute to non-ideal behavior and associated efficiency reduction of the actual vapor-compression cycle include:

- Compressor volumetric inefficiencies that will result in higher than ideal compressor power consumption
- Compressor motor inefficiency that will result in higher electric power consumption
- Pressure drops through condenser, evaporator, and refrigerant piping that require additional compressor power along with accompanying changes in state point enthalpies
- Superheat in the evaporator that will reduce the volumetric efficiency of the compressor and increase pressure drop losses in the suction line (minor for packaged systems)
- Heat transfer from environment that will add to superheat in the suction line and reduce heat delivered to the condenser through the high pressure line (both relatively minor for packaged systems, but can be significant for split systems)

The COP value for the ideal vapor compression cycle would even be higher if the refrigerant-to-air and the water-to-refrigerant heat exchangers were more effective, allowing the low-side pressure/temperature to be higher and the high-side pressure/temperature to be lower while still accomplishing the required heat transfer rates in each. In the limiting case, if the heat exchangers were perfect they could transfer from the entering water at 30 F to the refrigerant at 30 F in the evaporator and from the refrigerant at 70 F to the air at 70 F with no temperature changes in either the water or air during the heat transfer process. This would represent an ideal case of infinite heat capacity in the water and air that would be affected by infinite flow rates (which is not possible). Considering this change in the assumptions on the previous heating example (dashed thick lines on Figure 2.4), the result would be as shown in Example 2.2.

Example 2.2. Determine HC, $Work_{comp}$, COP, and HE for the ideal vapor-compression heating cycle discussed above, but with "perfect" heat exchangers on the water and air side of the cycle (dashed thick lines in Figure 2.4).

P_{Low} = 112.1 psia, T_{sat} = 30 F (T_{sat} = EWT = 30 F)
P_{high} = 208.3 psia, T_{sat} = 70 F (T_{sat} = EAT = 70 F)

State 1: P_1 = 112.1 psia & Saturated Vapor — $h_1 = 120.0$ Btu/lb_m
State 2: P_2 = 208.3 psia & Isentropic Compression — $h_2 = 126.5$ Btu/lb_m
State 3: P_3 = 208.3 psia & Saturated Liquid — $h_3 = 39.2$ Btu/lb_m
State 4: P_4 = 112.1 psia & Constant Enthalpy Expansion — $h_4 = 39.2$ Btu/lb_m

$HC_{Cond} = h_2 - h_3 = 126.5 - 39.2 = 87.3$ Btu/lb_m
$Work_{Comp} = h_2 - h_1 = 126.5 - 120.0 = 6.5$ Btu/lb_m

$$COP_{ideal} = \frac{HC_{Cond}}{Work_{Comp}} = \frac{87.3}{6.5} = 13.4$$

$HE_{Evap} = h_1 - h_4 = 120.0 - 39.2 = 80.8$ Btu/lb_m

The effect of eliminating the temperature differences across the evaporator and condenser is obvious and has consequences in the design of ground source heat pump equipment. In reality, there has to be a temperature difference across each to promote heat transfer. Using larger, more effective heat exchangers will make the cycle more efficient for given water source and air sink temperatures. Also, running the water and air at higher flow rates will increase the condenser and evaporator effectiveness by raising the average water-side refrigerant temperature and lowering the average air-side refrigerant temperature. However, increasing water flow rate comes with a penalty of increased pumping power while increasing air flow rate increases fan power and reduces delivery air temperature, which may lead to comfort issues.

2.3.2 BASIC VAPOR-COMPRESSION REFRIGERATION CYCLE-COOLING MODE A second vapor-compression refrigeration cycle is shown schematically in Figure 2.5 for a simple water-to-air heat pump operating in the cooling mode. Again, the circled numbers represent "state" points of the refrigerant. The cooling cycle is identical to the heating cycle from the refrigerant's perspective, except that the low-side evaporator is now extracting heat from the air-side load, and the high-side condenser is rejecting heat to the water circulating to the ground connection. The ideal vapor-compression refrigeration cycle represented by Figure 2.5 is plotted on the pressure-enthalpy diagram (solid thin lines in Figure 2.4) for low-side and high-side refrigerant pressures of 120 psia and 340 psia, respectively. These pressures represent approximate mixed refrigerant temperatures in the evaporator and condenser of 34 F and 102 F, respectively, which may be representative of cooling mode operation with a ground loop entering water temperature of 95 F to the condenser and entering and leaving air temperatures to and from the evaporator of about 75 F and 55 F, respectively. Note that the evaporator temperature is about 20 F lower than the leaving air temperature to promote heat transfer from the air to the refrigerant. Likewise, the mixed refrigerant temperature in the condenser is about 5 F higher than the entering water temperature. The refrigerant at State 1 is ideally a saturated vapor at the low-side (suction) pressure, which resides directly on the saturated vapor line at the low-side pressure. The ideal compressor "squeezes" the refrigerant vapor into a much smaller volume, which is accompanied by a large increase in the temperature of the vapor. In this case, the refrigerant leaves the compressor at about 145 F (State 2). Then, it enters the refrigerant-to-water heat exchanger (condenser). There, heat is transferred to the water, warming the water and cooling the refrigerant until it reaches the saturated vapor point at the high side pressure, after which the refrigerant is condensed completely to a saturated liquid at the high side pressure (State 3). The refrigerant then flows through an expansion valve where the pressure is suddenly reduced to the low-side pressure, resulting in a mixture of vapor and liquid at a much lower temperature, but with an enthalpy that is the same as before the expansion valve because no energy was transferred in the process. The resulting low-temperature mixture (State 4) then flows into the evaporator where energy is picked up from the warmer air being circulated from the building, cooling and dehumidifying the air, and vaporizing the refrigerant to the saturated vapor state at the low-side pressure where the process begins again.

The EER of the ideal vapor-compression refrigeration cycle operating in the cooling mode can be determined using data from the R410A pressure-enthalpy diagram along with basic relations based on simple thermodynamics. For the low and high-side pressures identified in the preceding paragraph, the state point enthalpies are obtained from Figure 2.4 (solid thin lines). State 1 is located directly on the saturated vapor line at the low-side pressure. State 3 is located directly on the saturated liquid line at the high-side pressure. State 4 has the same enthalpy as State 3 because the expansion valve represents a constant enthalpy expansion. That is, the pressure is reduced while no energy is added or removed from the refrigerant, and the enthalpy remains constant. State 2 represents the refrigerant output state from the compressor. For the ideal compression process, the compression is constant entropy, which means that the compression process line on Figure 2.4 follows a line of constant entropy(s) from State 1 to State 2 at the high-side pressure. The EER for the ideal cycle is calculated as the cooling capacity of the evaporator (TC - the desired effect) divided by the work of the compressor on the refrigerant ($Work_{Comp}$) or the enthalpy change across the evaporator (h_1 - h_4) divided by the enthalpy change across the compressor (h_2 - h_1) converted to Watts. The heat of rejection (HR) represents the amount of heat that is rejected to the ground connection by the condenser (h_2 - h_3).

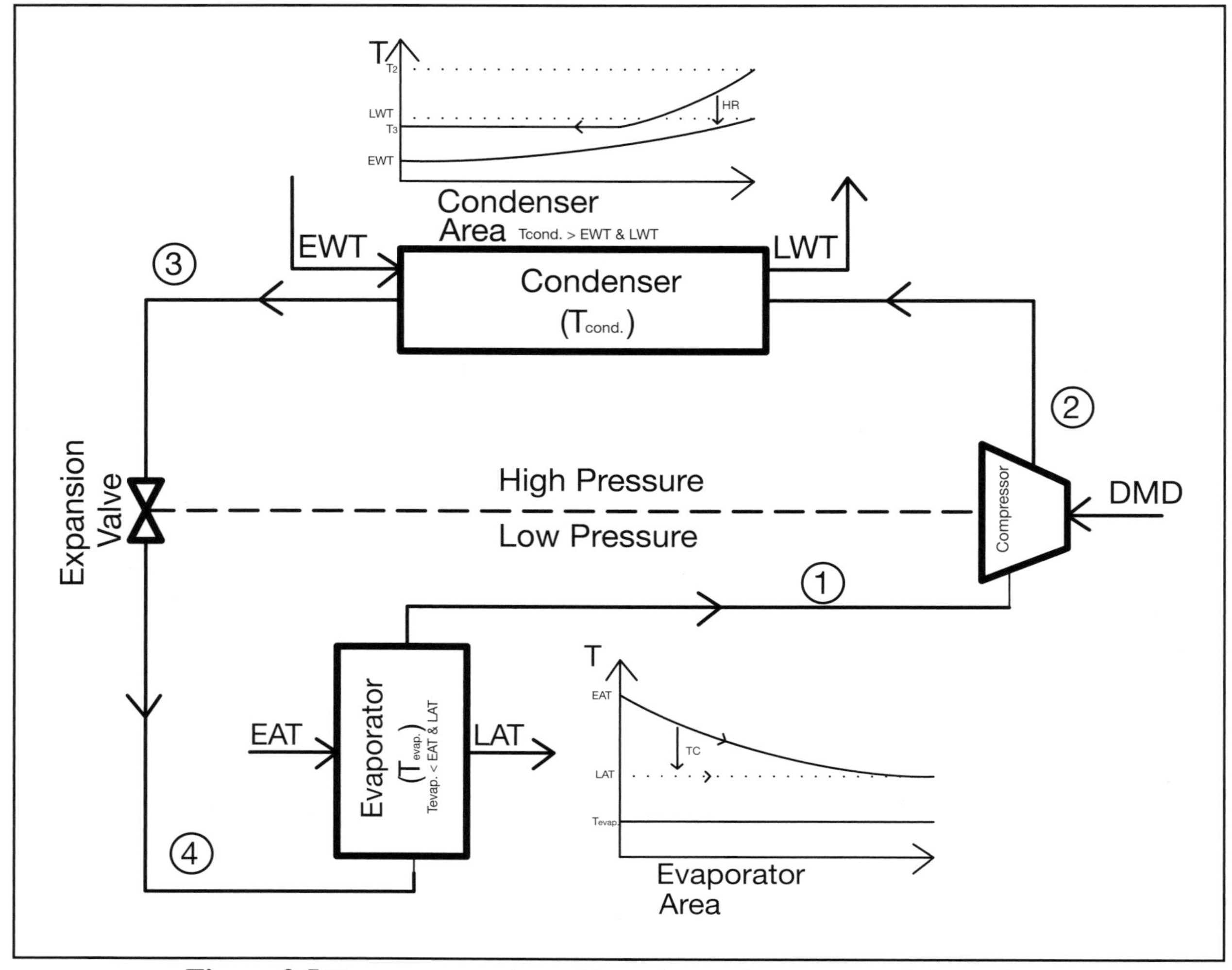

Figure 2.5 Vapor-compression refrigeration cycle – water-to-air in cooling.

Example 2.3. Determine TC, $Work_{comp}$, EER, and HR for the ideal vapor-compression cooling cycle discussed above (solid thin lines in Figure 2.4).

P_{Low} = 120 psia, T_{sat} = 34 F (approximately 30 F below EAT = 75 F)
P_{high} = 340 psia, T_{sat} = 102 F (approximately 5 F above EWT = 95 F)

State 1: P_1 = 120 psia & Saturated Vapor	h_1 = 120.2 Btu/lb_m
State 2: P_2 = 340 psia & Isentropic Compression	h_2 = 132 Btu/lb_m
State 3: P_3 = 340 psia & Saturated Liquid	h_3 = 52.3 Btu/lb_m
State 4: P_4 = 120 psia & Constant Enthalpy Expansion	h_4 = 52.3 Btu/lb_m

$$TC_{Evap} = h_1 - h_4 = 120.2 - 52.3 = 67.9 \text{ Btu/lb}_m$$

$$HR = \frac{h_2 - h_1}{3.412 \frac{\text{Btu/hr}}{\text{W}}} = \frac{132 - 120.2}{3.412} = 3.46 \text{ Whr/lb}_m$$

$$EER_{ideal} = \frac{TC_{Evap}}{Work_{Comp}} = \frac{67.9}{3.46} = 19.6 \frac{\text{Btu/hr}}{\text{W}}$$

$$HR_{Cond} = h_2 - h_3 = 132 - 52.3 = 79.7 \text{ Btu/lb}_m$$

The EER for the ideal cycle is higher than would be expected for an actual ground source heat pump operating with the EWT and LAT identified in the example for the same reasons identified for heating mode operation. The EER value for the ideal vapor-compression cycle would even be higher if the refrigerant-to-air and the water-to-refrigerant heat exchangers were more effective, allowing the low-side pressure/temperature to be higher and the high-side pressure/temperature to be lower while still accomplishing the required heat transfer rates in each. In the limiting case, if the heat exchangers were perfect they could transfer from the entering air at 75 F to the refrigerant at 75 F in the evaporator and from the refrigerant at 95 F to the water at 95 F with no temperature changes in either the air or water during the heat transfer process. This would represent an ideal case of infinite heat capacity in the air and water that would be affected by infinite flow rates (which is not possible). Considering this change in the assumptions on the previous cooling example (dashed thin lines on Figure 2.4), the result would be as shown in Example 2.4.

Example 2.4. Determine TC, $Work_{comp}$, EER, and HR for the ideal vapor-compression cooling cycle discussed above but with "perfect" heat exchangers on the water and air side of the cycle (dashed thin lines in Figure 2.4).

$P_{Low} = 233.5$ psia, $T_{sat} = 75$ F ($T_{sat} = EAT = 75$ F)
$P_{high} = 311.2$ psia, $T_{sat} = 95$ F ($T_{sat} = EWT = 95$ F)

State 1: $P_1 = 233.5$ psia & Saturated Vapor — $h_1 = 122.2 \text{ Btu/lb}_m$
State 2: $P_2 = 311.2$ psia & Isentropic Compression — $h_2 = 126 \text{ Btu/lb}_m$
State 3: $P_3 = 311.2$ psia & Saturated Liquid — $h_3 = 49.5 \text{ Btu/lb}_m$
State 4: $P_4 = 233.5$ psia & Constant Enthalpy Expansion — $h_4 = 49.5 \text{ Btu/lb}_m$

$$TC_{Evap} = h_1 - h_4 = 122.2 - 49.5 = 72.7 \text{ Btu/lb}_m$$

$$Work_{Comp} = \frac{h_2 - h_1}{3.412 \frac{\text{Btu/hr}}{\text{W}}} = \frac{126 - 122.2}{3.412} = 1.11 \text{ Whr/lb}_m$$

$$EER_{ideal} = \frac{TC_{Evap}}{Work_{Comp}} = \frac{72.7}{1.11} = 65.5 \frac{\text{Btu/hr}}{\text{W}}$$

$$HR_{Cond} = h_2 - h_3 = 126 - 49.5 = 76.5 \text{ Btu/lb}_m$$

The effect of eliminating the temperature differences across the evaporator and condenser is obvious and has consequences in the design of ground source heat pump equipment. In reality, there has to be a temperature difference across each to promote heat transfer. Using larger, more effective heat exchangers will make the cycle more efficient for given water source and air sink temperatures. Also, running the water and air at higher flow rates will increase the evaporator and condenser effectiveness by raising the average air-side refrigerant temperature and lowering the average water-side refrigerant temperature. However, increasing water flow rate will come with a penalty of increased pumping power. Increasing air flow rate will increase fan power, increase delivery air temperature, and reduce dehumidification, which may lead to comfort issues.

2.3.3 GENERAL OBSERVATIONS From the preceding discussion, the performance of the GSHP depends on factors that affect the performance of the refrigeration cycle and the heat exchanger on each side of the cycle. For a given type of refrigerant, the refrigeration cycle is affected by the pressure extremes that the compressor must maintain to achieve the high and low temperatures required to properly exchange heat with the heat sink and source. That is, a high temperature heat sink and a low temperature heat source will result in lower capacity and efficiency than when the temperatures of the sink and source are closer together. The heat exchangers (water-to-water or water-to-air) are sized to achieve proper heat transfer when acting as either a condenser or evaporator, and trade-offs are generally made in terms of sizing the heat exchangers and optimizing heat pump performance in either cooling or heating mode. Also, the performance of a refrigerant-to-air or a refrigerant-to-water heat exchanger is dependent on the flow of air or water, respectively, through each heat exchanger, and the design of the system must account for those flow rates for proper system performance.

2.3.3.1 PARAMETERS THAT AFFECT WATER-TO-AIR HEAT PUMP PERFORMANCE

The four parameters that have the largest effect on the performance of a water-to-air GSHP.

EWT-ENTERING WATER TEMPERATURE (SOURCE) The performance of a GSHP is highly dependent on the EWT because it directly affects the required low-side pressure in the refrigeration system in heating mode or the high-side pressure in cooling mode. In the heating mode (heat extraction from the ground), heat pump heating performance is reduced as EWT decreases and, in the cooling mode (heat rejection to the ground), heat pump cooling performance is reduced as EWT increases. Heating mode design EWT (EWT_{min}) and cooling mode design EWT (EWT_{max}) are designer-specified parameters that depend on location, the type of ground connection, and a trade-off between heat pump performance and ground heat exchanger size/cost. The temperature stability of the water or water and antifreeze solution entering the heat pump depends heavily on the type of ground connection. The EWT for closed-loop systems (whether ground heat exchangers or pond loops) will experience a large temperature variation from heating to cooling mode due to changing ground or pond temperatures as opposed to nearly constant entering water temperatures provided by open-loop systems, which are not addressed in this manual. For a buried heat exchanger, the soil/rock type, the ground loop configuration, and the extent of heat pump runtime that creates temperature buildup (cooling) or temperature drawdown (heating) around the buried pipes will result in a much larger swing in EWT during the year.

EAT-ENTERING AIR TEMPERATURE (LOAD) The performance of a GSHP is also dependent on the EAT because it directly affects the required high-side pressure in the refrigeration system in heating mode and the low-side pressure in cooling mode. In heating mode, the dry bulb temperature of the air entering the heat pump will depend on the heated space set point, which is generally between 65 and 75 F for most residential and light commercial applications. Ground source heat pumps are generally heating performance tested at 68 to 70 F EAT, with performance increasing as EAT decreases. A lower limit on EAT of approximately 55 F in heating mode exists due to resulting low-side refrigeration pressures and, if reduced to the setting of the low pressure switch on the unit for low pressure protection, the heat pump will lock out. In cooling mode, the dry and wet bulb temperatures of the air-vapor mixture directly affect both the sensible cooling and latent cooling capacity of the heat pump. Ground source heat pumps are generally cooling performance tested at 80 F dry bulb (db) and 67 F wet bulb (wb), or at a relative humidity of 50%, which is fairly warm and humid air. Reducing the entering air temperatures in cooling will significantly reduce both the sensible and latent cooling capacities of the heat pump, the extent of which depends on the relative change of each temperature.

GPM-WATER FLOW RATE (SOURCE) The water or water and antifreeze solution flow rate through the water-to-refrigerant heat exchanger affects heat pump performance by changing the effectiveness of the heat exchanger. The design flow rate is generally 2.5 to 3.0 gallons per minute (gpm) per nominal ton of heat pump capacity for closed-loop systems. Reducing the flow rate below that range will result in larger temperature changes in the circulating water in the ground loop, reducing the effectiveness of the heat exchanger and reducing the heat pump performance. For systems with EWTs where antifreeze is needed in the circulating fluid, low flow rates may result in icing of the evaporator coil and reduced heat pump performance or possible heat pump lock out on low head due to the low refrigerant pressures that will result. Increasing the water flow rate above the 2.5 to 3 gpm range results in minimal heat pump performance increases.

CFM-AIR FLOW RATE (LOAD) The air flow rate through the refrigerant-to-air heat exchanger affects heat pump performance by changing the effectiveness of the heat exchanger. Design air flow rates are generally 350 to 400 CFM per nominal ton of heat pump capacity. There are two types of fans that are utilized by GSHP manufacturers: the permanent split capacitor (PSC), which offers three fan speed settings and the variable speed electronically commutated motor (ECM), which offers several selectable fan speed settings that provide constant air flow rates over a wide range of external static pressures. For a PSC blower set on one of three speeds, reducing the air flow rate due to air flow restrictions, undersized ductwork, dirty air filter, etc., will significantly reduce the heating mode performance. The heating capacity of the heat pump will decrease while the electric demand will increase, thus decreasing the COP of the unit. For an ECM blower, the fan will adjust speed to maintain a set CFM up to a limiting static pressure. Therefore, performance reductions will be minor primarily due to increased fan electrical demand. In cooling mode, the reduction in air flow will reduce both sensible and latent capacity of the heat pump, but the electrical demand will also decrease (although by a smaller amount) resulting in an EER reduction that is smaller than expected.

2.3.3.2 PARAMETERS THAT AFFECT WATER-TO-WATER HEAT PUMP PERFORMANCE

The four parameters that have the largest effect on the performance of a water-to-water GSHP are discussed below.

EWT-ENTERING WATER TEMPERATURE (SOURCE) Same as Water-to-Air GSHP.

LWT-LEAVING WATER TEMPERATURE (LOAD) The performance of a water-to-water heat pump is

highly dependent on the temperature of the water leaving the unit on the load side. In the heating mode, high LWTs will require higher high-side refrigerant pressures, making the compressor work harder and lowering the heating capacity and COP of the unit. In the cooling mode, low LWTs will lower the low-side refrigerant pressures, again making the compressor work harder and lowering the cooling capacity and EER of the unit. In general, heating capacity is not very sensitive to LWT, but COP decreases rapidly with higher LWTs. In the cooling mode, both the cooling capacity and the EER are reduced significantly with lower LWTs.

GPM-WATER FLOW RATE (SOURCE) Same as Water-to-Air GSHP

GPM-WATER FLOW RATE (LOAD) Same type of effect as GPM of the source side of the heat pump.

An option on most water-to-air and water-to-water GSHPs is the desuperheater, which is a small refrigerant-to-water heat exchanger installed between the compressor and the condenser (Figure 2.4 or 2.5) and connected to the domestic water heater tank. During either heating or cooling mode operation, a pump circulates water from the tank through the heat exchanger to augment hot water production. The heat exchanger is sized to provide about 15% of the heat pump's capacity when the water is cold. As the water gets warmer, the heating capacity of the heat exchanger is reduced. The pump is controlled such that the water temperature is not raised to the point of being dangerous. In the heating mode, a desuperheater will always aid in the production of warm water, and its effectiveness is limited by the runtime of the heat pump and the temperature of the water in the tank. If the heat pump doesn't run when the water needs to be heated, it will not be useful. In the cooling mode, the same conditions exist, but the effectiveness of the desuperheater is also related to the ground loop temperature and the temperature of the refrigerant leaving the compressor. If the ground loop temperature is relatively low (early spring or northern climates, where loops are extremely oversized for cooling), the refrigerant will not be warm enough to add heat to the circulating domestic hot water and may even remove heat from the hot water if not controlled properly. In general, the more cooling dominant the climate the more effective the desuperheater becomes.

2.4 Performance of Single-Capacity, Water-to-Air GSHPs

Single-capacity, water-to-air GSHPs typically utilize a scroll compressor (smaller units may use reciprocating or rotary) to run at 100% capacity in terms of heating and cooling capacity. The single-capacity GSHP is controlled by a thermostat capable of one or two heat stages and one cooling stage causing full capacity heating or cooling mode operation when required. In the heating mode, a two stage heating thermostat can be used with the second stage controlling any supplemental electric heat used.

The performance (capacity, electric demand, and efficiency) of a single-capacity, water-to-air GSHP is measured by the heating capacity (Btu/hr), the electric demand (kW), and the COP in heating mode and by the cooling capacity (Btu/hr), the electric demand (kW), and the EER in cooling mode. Parameters that affect the heating and cooling performance of the water-to-air GSHP include entering water temperature (EWT) from the ground connection, entering air temperature (EAT) returning from the conditioned space, water flow rate (GPM) through the refrigerant-to-water coil, and air flow rate (CFM) through the refrigerant-to-air coil. Manufacturers of GSHP equipment generally express performance data for their equipment in a tabular format as a function of EWT, GPM, and CFM for a given EAT condition. Some manufacturers utilize only EWT and GPM in the performance table for given CFM and EAT conditions. Additional tables of performance correction factors are provided to account for the effect of EAT conditions and CFM on capacity, electric demand, and efficiency.

Performance data for a ClimateMaster Tranquility 20™ Model 048 PSC, a single-capacity, nominal 4-ton unit that utilizes R-410A refrigerant and a conventional 3-speed fan (PSC blower motor), are provided in the tabular format supplied by the manufacturer in Figure 2.6. Column numbers are placed above the table for easier identification during discussion of the data. Column 1 contains EWT from the ground connection over which performance data are included, and Column 2 contains three GPM levels for each EWT, except at EWT=20 F where only the highest water flow rate is recommended. For each GPM, the pressure drop in gauge pressure (psi) (Column 3) and head loss in feet of water equivalent pressure (Column 4) through the water-to-refrigerant coil are provided for water flowing through the coil for EWT above 40 F and for 15% antifreeze solution for EWT below 40 F. For each GPM in Column 2, the performance data are provided for two air flow rates (listed in Column 5 for cooling and Column 12 for heating). The higher air flow rate in each mode (1600 CFM in cooling, 1600 CFM in heating) are the air flow rates for which the GSHP is rated for ARI/ISO 13256-1 (Section 2.4.4). Manufacturer's data are not always presented in the same format as provided here and each should be considered individually.

2.4.1 HEATING MODE PERFORMANCE DATA Heating mode performance data are listed in columns 12 through 18 in Figure 2.6, with each line of data representing performance for a combination of EWT, GPM, and CFM as read proceeding from left to right across the table. Column 13 contains the heating capacity (HC - MBtu/hr) provided to the air returning from the conditioned space, Column 14 lists the electric demand (DMD - kW) of the unit, and Column 17 provides the COP. The heat of extraction (HE - MBtu/hr) from the ground connection is provided in Column 15, and the leaving air temperature back to the conditioned space (LAT - F) is shown in Column 16. The heating capacity of the optional desuperheater (HW - MBtu/hr) is available in Column 18. Most manufacturers express capacities in MBtu/hr (or MBH) to save space (1 MBtu/hr = 1000 Btu/hr).

Example 2.5 An example of using the performance data to determine heating mode performance is highlighted by solid outlines on Figure 2.6 for EWT=40 F, GPM=12, CFM=1600, and EAT=70 F. For those operating parameters:

HC = 46.1 MBtu/hr x 1000 Btu/MBtu = 46,100 Btu/hr	Column 13
DMD = 3.63 kW	Column 14
COP = 3.72	Column 17
HE = 33.8 MBtu/hr x 1000 = 33,800 Btu/hr	Column 15
LAT = 97 F	Column 16

Equation 2.3 provides HE as a function of COP and HC, and inserting those values obtained from the table in Figure 2.6 results in HE=33,707 Btu/hr, which is close to the tabled value of 33,800 Btu/hr and within round-off errors associated with converting HC from MBtu/hr (units in table) to Btu/hr.

Leaving air temperature (LAT) from the heat pump in heating mode can be calculated according to Equa-

$$HE = \left(\frac{COP - 1}{COP}\right) \cdot HC = \left(\frac{3.72 - 1}{3.72}\right) \cdot 46{,}100 \text{ Btu/hr} = 33{,}707 \text{ Btu/hr}$$

Column

1 2 3 4 5 6 7 8 9 10 11 12 13 14 15 16 17 18

1,600 CFM Nominal (Rated) Airflow Cooling, 1,600 CFM Nominal (Rated) Airflow Heating

Performance capacities shown in thousands of Btuh

EWT °F	GPM	WPD		Cooling - EAT 80/67°F							Heating - EAT 70°F						
		PSI	FT	Airflow CFM	TC	SC	kW	HR	EER	HW	Airflow CFM	HC	kW	HE	LAT	COP	HW
20	12.0	4.8	11.1	Operation not recommended							1200	34.4	3.80	22.2	97	2.65	4.2
	12.0	4.8	11.1								1600	35.3	3.47	23.6	90	2.98	3.6
30	6.0	1.3	3.0	1200	52.0	29.3	2.14	59.3	24.3	-	1200	37.1	3.84	24.7	99	2.83	4.9
	6.0	1.3	3.0	1600	54.2	35.1	2.22	61.7	24.4	-	1600	38.0	3.51	26.2	92	3.18	4.2
	9.0	2.6	6.0	1200	52.3	29.3	2.06	59.4	25.4	-	1200	38.5	3.86	26.0	100	2.92	4.9
	9.0	2.6	6.0	1600	54.5	35.1	2.13	61.7	25.6	-	1600	39.5	3.53	27.6	93	3.28	4.2
	12.0	4.5	10.4	1200	52.5	29.4	2.03	59.4	25.9	-	1200	39.3	3.87	26.8	100	2.98	4.9
	12.0	4.5	10.4	1600	54.7	35.2	2.09	61.8	26.1	-	1600	40.4	3.54	28.4	93	3.34	4.2
40	6.0	1.2	2.8	1200	53.2	30.3	2.32	61.1	22.9	-	1200	41.9	3.92	29.1	102	3.14	5.6
	6.0	1.2	2.8	1600	55.4	36.3	2.40	63.5	23.1	-	1600	43.0	3.58	30.9	95	3.52	4.8
	9.0	2.6	6.0	1200	53.4	30.4	2.24	61.1	23.9	-	1200	43.8	3.95	30.9	104	3.25	5.6
	9.0	2.6	6.0	1600	55.7	36.4	2.31	63.5	24.1	-	1600	45.0	3.61	32.8	96	3.65	4.8
	12.0	4.4	10.2	1200	53.7	30.4	2.18	61.1	24.6	-	1200	44.9	3.97	31.8	105	3.31	5.6
	12.0	4.4	10.2	1600	55.9	36.4	2.26	63.6	24.8	-	1600	46.1	3.63	33.8	97	3.72	4.8
50	6.0	1.1	2.5	1200	52.6	30.7	2.60	61.5	20.2	2.7	1200	47.2	4.01	33.9	106	3.45	6.3
	6.0	1.1	2.5	1600	54.8	36.8	2.69	63.9	20.4	2.8	1600	48.5	3.67	36.0	98	3.87	5.4
	9.0	2.5	5.8	1200	53.5	30.9	2.44	61.8	21.9	2.5	1200	49.6	4.06	36.1	108	3.58	6.3
	9.0	2.5	5.8	1600	55.7	37.0	2.52	64.3	22.1	2.6	1600	50.9	3.71	38.3	99	4.02	5.4
	12.0	4.2	9.7	1200	53.8	30.9	2.37	61.9	22.7	2.3	1200	50.9	4.08	37.3	109	3.65	6.2
	12.0	4.2	9.7	1600	56.0	37.0	2.45	64.3	22.9	2.4	1600	52.2	3.73	39.6	100	4.10	5.4
60	6.0	1.0	2.3	1200	50.7	30.2	2.85	60.4	17.8	3.2	1200	52.8	4.12	39.0	111	3.75	6.9
	6.0	1.0	2.3	1600	52.8	36.2	2.94	62.8	17.9	3.4	1600	54.2	3.77	41.4	101	4.22	5.9
	9.0	2.4	5.5	1200	52.1	30.6	2.67	61.2	19.5	3.0	1200	55.6	4.18	41.6	113	3.90	6.9
	9.0	2.4	5.5	1600	54.3	36.6	2.76	63.7	19.7	3.1	1600	57.1	3.82	44.1	103	4.38	5.9
	12.0	4.0	9.2	1200	52.7	30.7	2.59	61.5	20.4	2.8	1200	57.2	4.21	43.0	114	3.98	6.9
	12.0	4.0	9.2	1600	54.9	36.8	2.67	64.0	20.5	2.9	1600	58.8	3.85	45.6	104	4.47	5.9
70	6.0	1.0	2.3	1200	48.3	29.5	3.13	59.0	15.4	4.0	1200	58.6	4.24	44.2	115	4.05	7.5
	6.0	1.0	2.3	1600	50.3	35.3	3.24	61.3	15.5	4.1	1600	60.2	3.88	46.9	105	4.54	6.5
	9.0	2.3	5.3	1200	50.0	30.0	2.93	60.0	17.0	3.7	1200	61.9	4.32	47.2	118	4.20	7.5
	9.0	2.3	5.3	1600	52.1	35.9	3.03	62.4	17.2	3.8	1600	63.6	3.95	50.1	107	4.72	6.5
	12.0	3.8	8.8	1200	50.8	30.2	2.84	60.5	17.9	3.4	1200	63.8	4.36	48.9	119	4.29	7.5
	12.0	3.8	8.8	1600	52.9	36.2	2.93	62.9	18.0	3.5	1600	65.5	3.99	51.9	108	4.81	6.4
80	6.0	0.9	2.1	1200	45.7	28.7	3.45	57.5	13.2	4.9	1200	64.6	4.38	49.6	120	4.32	8.1
	6.0	0.9	2.1	1600	47.6	34.4	3.57	59.8	13.3	5.1	1600	66.3	4.01	52.6	108	4.85	7.0
	9.0	2.3	5.3	1200	47.5	29.2	3.23	58.5	14.7	4.5	1200	68.4	4.47	53.0	123	4.48	8.1
	9.0	2.3	5.3	1600	49.4	35.0	3.34	60.8	14.8	4.7	1600	70.2	4.09	56.2	111	5.03	7.0
	12.0	3.6	8.3	1200	48.3	29.5	3.13	59.0	15.5	4.2	1200	70.5	4.53	54.9	124	4.56	8.1
	12.0	3.6	8.3	1600	50.3	35.3	3.23	61.4	15.6	4.3	1600	72.4	4.14	58.3	112	5.12	6.9
85	6.0	0.9	2.1	1200	44.4	28.4	3.64	56.8	12.2	5.4	1200	67.6	4.46	52.3	122	4.45	8.4
	6.0	0.9	2.1	1600	46.2	34.0	3.76	59.1	12.3	5.6	1600	69.4	4.07	55.5	110	4.99	7.2
	9.0	2.3	5.2	1200	46.1	28.8	3.40	57.7	13.5	5.0	1200	71.7	4.56	55.9	125	4.60	8.4
	9.0	2.3	5.2	1600	48.0	34.5	3.52	60.0	13.6	5.2	1600	73.6	4.17	59.3	113	5.17	7.2
	12.0	3.6	8.2	1200	47.0	29.1	3.29	58.2	14.3	4.6	1200	73.9	4.62	57.9	127	4.69	8.4
	12.0	3.6	8.2	1600	48.9	34.8	3.40	60.6	14.4	4.8	1600	75.9	4.23	61.4	114	5.26	7.2
90	6.0	0.9	2.1	1200	43.1	28.1	3.82	56.1	11.3	5.9	1200	70.6	4.53	55.0	125	4.57	8.7
	6.0	0.9	2.1	1600	44.9	33.6	3.95	58.4	11.4	6.2	1600	72.5	4.14	58.4	112	5.13	7.5
	9.0	2.2	5.1	1200	44.8	28.4	3.57	57.0	12.5	5.5	1200	74.9	4.65	58.8	128	4.72	8.7
	9.0	2.2	5.1	1600	46.6	34.1	3.69	59.2	12.6	5.7	1600	76.9	4.25	62.4	115	5.30	7.5
	12.0	3.5	8.1	1200	45.6	28.7	3.46	57.4	13.2	5.1	1200	77.3	4.72	60.9	130	4.80	8.7
	12.0	3.5	8.1	1600	47.5	34.3	3.57	59.7	13.3	5.3	1600	79.3	4.31	64.6	116	5.39	7.4
100	6.0	0.8	1.8	1200	40.8	27.8	4.24	55.3	9.6	7.2	Operation not recommended						
	6.0	0.8	1.8	1600	42.5	33.3	4.39	57.5	9.7	7.4							
	9.0	2.1	4.9	1200	42.2	27.9	3.97	55.8	10.6	6.6							
	9.0	2.1	4.9	1600	44.0	33.4	4.10	58.0	10.7	6.9							
	12.0	3.3	7.6	1200	43.0	28.1	3.83	56.1	11.2	6.1							
	12.0	3.3	7.6	1600	44.8	33.6	3.96	58.3	11.3	6.3							
110	6.0	0.8	1.8	1200	39.2	27.8	4.68	55.2	8.4	8.6							
	6.0	0.8	1.8	1600	40.8	33.3	4.84	57.4	8.4	8.9							
	9.0	2.0	4.6	1200	40.1	27.9	4.42	55.2	9.1	7.9							
	9.0	2.0	4.6	1600	41.8	33.4	4.57	57.4	9.2	8.2							
	12.0	3.2	7.4	1200	40.7	27.9	4.27	55.3	9.5	7.3							
	12.0	3.2	7.4	1600	42.4	33.5	4.41	57.5	9.6	7.6							
120	6.0	0.7	1.6	1200	38.1	28.7	5.25	56.1	7.3	10.2							
	6.0	0.7	1.6	1600	39.7	34.4	5.42	58.3	7.3	10.6							
	9.0	1.9	4.4	1200	39.2	28.8	4.97	56.2	7.9	9.4							
	9.0	1.9	4.4	1600	40.8	34.5	5.14	58.4	7.9	9.8							
	12.0	3.0	6.9	1200	39.8	28.9	4.82	56.2	8.3	8.6							
	12.0	3.0	6.9	1600	41.4	34.6	4.98	58.5	8.3	9.0							

Figure 2.6. ClimateMaster Tranquility 20™ Model 048 PSC Performance Data

tion 2.7, which is based on thermal properties of air in the 50 to 100 F range.

$$LAT_h = EAT + \frac{HC}{1.08 \bullet CFM}$$ **Equation 2.7**

Where:

LAT_h = leaving air temperature from heat pump in heating mode (F)
EAT = entering dry bulb air temperature from conditioned space (F)
HC = heating capacity of heat pump (Btu/hr)
CFM = air flow rate (cfm)

Applying Equation 2.7 to the data in Example 2.5 yields 96.7 F, which agrees very closely to the tabled LAT=97 F.

$$LAT_h = EAT + \frac{HC}{1.08 \bullet CFM} = 70F + \frac{46{,}100 \text{ Btu/hr}}{1.08 \bullet 1600 \text{ cfm}} = 96.7F$$

Leaving water temperature (LWT) from the heat pump in heating mode can be calculated according to Equation 2.8, which is based on average thermal properties of water between 32 F and 105 F. If the circulating fluid is other than pure water, the constant of 500 in Equation 2.8 will change slightly (490 to 505 for most water-antifreeze mixtures) due to differences in mixture density and specific heat and will have only a minor influence on the predicted LWT.

$$LWT_h = EWT - \frac{HE}{500 \bullet GPM}$$ **Equation 2.8**

Where:

LWT_h = leaving water temperature from heat pump in heating mode (F)
EWT = entering water temperature from ground connection (F)
HE = heat of extraction from ground connection (Btu/hr)
GPM = water flow rate (gpm)

Using the data from the Example 2.5, the leaving water temperature from the heat pump can be calculated according to Equation 2.8.

$$LWT_h = EWT - \frac{HE}{500 \bullet GPM} = 40F - \frac{33{,}800 \text{ Btu/hr}}{500 \bullet 12.0 \text{ gpm}} = 34.4F$$

This represents a 5.6 F temperature drop (40 – 34.4 F) across the water-to-refrigerant coil, which is within the expected 4 to 6 F temperature drop range under design heating conditions.

Equation 2.8 for LWT_h has specific application (in the form of Equation 2.9) during system startup and performance verification, which is normally done in the heating mode. Typical installations include P/T (pressure/temperature) ports to allow for measurement of pressure and temperature change across the water-to-refrigerant coil for the purpose of checking for proper system performance. Pressure drop (psi) across the coil can be measured directly and compared to data in Column 3 of Figure 2.6 to estimate GPM. EWT and LWT can also be measured directly via the P/T ports, and Equation 2.9 can then be used to calculate the

measured HE from the ground loop, which is compared to the tabled HE (Column 15 in Figure 2.6). If the measured HE differs by a large amount from the tabled HE value, the system may not be working properly and further testing may be required. Performance checking will be discussed in detail in Chapter 11.

$$HE = 500 \cdot GPM \cdot (EWT - LWT_h)$$ **Equation 2.9**

The heating mode performance of a GSHP will vary during the heating season primarily based on the EWT from the ground connection. EAT, CFM, and GPM are design parameters that are set by the designer and around which ducting and piping/pumping systems are designed. EAT should remain in a small range depending on the type of heating load. CFM should remain relatively constant assuming that the air filter is kept clean and air flow remains unobstructed. GPM will also remain in a very narrow range as long as flow through the water-to-refrigerant coil does not become restricted due to fouling of the coil. The amount of flow required for optimum heat pump performance depends on the type of ground connection in use. Systems using open-loop/standing column ground connections require less system flow (as low as 2 gpm per ton of installed capacity) because EWT will be very stable in an operating water temperature range close to local ground water temperatures. Systems utilizing a closed-loop ground connection typically require 2.5-3.0 gpm per ton of installed heat pump capacity. For both heating and cooling modes of operation, lowering the system water flow rate will diminish heat pump capacity and efficiency. Increasing the system water flow rate above the recommended range of 3 gpm per ton of installed capacity will minimally improve performance, but not enough to justify doing so. Always refer to manufacturer performance data to determine water flow rates for selected equipment according to the type of earth connection used (3 gpm per nominal ton of installed heat pump capacity is the standard). As is the case in Figure 2.6, manufacturers will typically provide performance data for three different flow rates. The lowest flow rate given in the table is usually the minimum required flow rate for open-loop systems. The middle flow rate given is usually the minimum required flow rate for closed-loop systems. The highest flow rate is usually the flow rate required for optimum equipment performance in terms of capacity and efficiency.

EWT, on the other hand, will vary during the heating season based on the type of ground connection that is utilized and is particularly important in proper sizing of the GSHP in heating mode. HC and COP are graphed in Figure 2.7, as a function of EWT, for the performance data provided in Figure 2.6 at operating parameters of GPM=12, CFM=1600, and EAT=70. HC and COP are strongly influenced by EWT, with both decreasing as EWT decreases (a tendency of a GSHP due to the nature of the vapor-compression cycle). If the heat pump is to be sized for heating, the minimum EWT from the ground connection must be specified in order to determine the HC and COP of the heat pump at design conditions. HC and COP have been identified for an EWT of 32 F in Figure 2.7, which is the entering water temperature that ARI/ISO 13256-1 uses as the ground loop test condition (discussed in Section 2.4.4) for heating mode certification. Note that, although this is a nominal 4-ton GSHP, the HC is just under 42,000 Btu/hr at EWT=32 F and varies with EWT such that it must be specified to determine the actual HC of the GSHP. An EWT of about 43 F would be required to achieve a HC of 48,000 Btu/hr, which is considered 4 tons of heating capacity. More importantly, as EWT is further reduced (as would be the case in northern climates with low ground temperatures and high heating loads), it may not be possible to economically install a ground connection that provides EWTs above 30 F at design heating conditions, having direct consequence on the sizing of the GSHP equipment. The influence of design EWT on GSHP sizing and operating efficiency will be addressed in the selection of GSHP equipment to meet design heating loads (Chapter 3) and the design of the various ground coupling devices (Chapter 5).

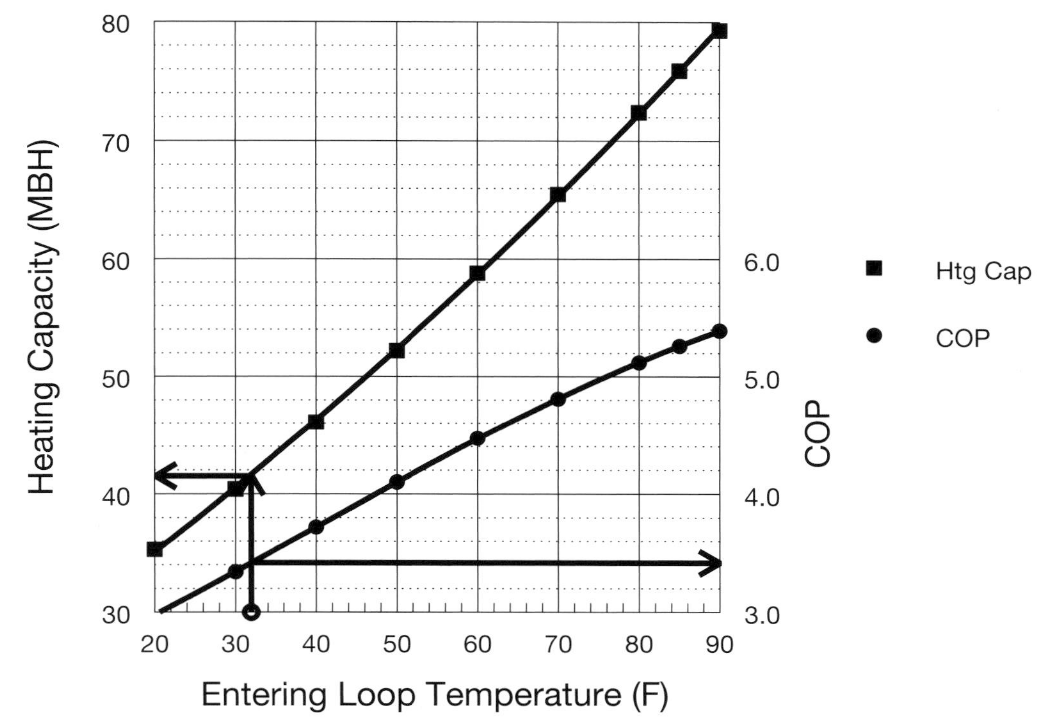

Figure 2.7. Heating Capacity and COP – ClimateMaster Tranquility 20 Model 048 PSC

2.4.2 COOLING MODE PERFORMANCE DATA Cooling mode performance data are listed in columns 5 through 11 in Figure 2.6, and each line of data represents performance for a combination of EWT, GPM, and CFM as read proceeding from left to right across the table. Column 6 contains the total cooling capacity (TC - MBtu/hr) provided to the air returning from the conditioned space, and Column 7 lists the associated sensible cooling capacity (SC - MBtu/hr). Column 8 lists the electric demand (DMD - kW) of the unit, and Column 10 provides the EER. The heat of rejection (HR - MBtu/hr) to the ground connection is provided in Column 9. The heating capacity of the optional desuperheater (HW - MBtu/hr) is available in Column 11. Most manufacturers express capacities in MBtu/hr (or MBH) to save space (1 MBtu/hr = 1,000 Btu/hr).

Example 2.6 An example of using the performance data to determine cooling mode performance is highlighted by dashed outlines on Figure 2.6 for EWT=90 F, GPM=12, CFM=1,600, and EAT=80 F db/67 F wb. For those operating parameters:

TC	= 47.5 MBtu/hr x 1,000 Btu/MBtu = 47,500 Btu/hr	Column 6
SC	= 34.3 MBtu/hr x 1,000 = 34,300 Btu/hr	Column 7
DMD	= 3.57 kW	Column 8
EER	= 13.3	Column 10
HR	= 59.7 MBtu/hr x 1,000 = 59,700 Btu/hr	Column 9

Equation 2.6 provides HR as a function of EER and TC, and inserting those values obtained from the table in Figure 2.6 results in HR=59,686 Btu/hr, which is close to the tabled value of 59,700 Btu/hr and within round-off errors associated with converting TC from MBtu/hr (units in table) to Btu/hr.

$$HR = \left(\frac{EER + 3.412 \frac{Btu/hr}{W}}{EER} \right) \cdot TC = \left(\frac{13.3 + 3.412 \frac{Btu/hr}{W}}{13.3} \right) \cdot 47{,}500\ Btu/hr = 59{,}686\ Btu/hr$$

Leaving air temperature (LAT) from the heat pump in the cooling mode can be calculated according to Equation 2.10, which is based on thermal properties of air in the 50 to 100 F range. In cooling mode, the leaving air temperature is a dry bulb temperature and will be based on sensible cooling capacity of the heat pump. The latent cooling capacity, which is the difference between TC and SC, is a measure of the energy associated with moisture removal from the air stream due to condensation and does not reduce the dry bulb temperature of the air stream.

$$LAT_c = EAT - \frac{SC}{1.08 \bullet CFM} \qquad \textbf{Equation 2.10}$$

Where:

LAT_c = leaving air dry bulb temperature from heat pump in cooling mode (F)
EAT = entering air dry bulb temperature from conditioned space (F)
SC = sensible cooling capacity of heat pump (Btu/hr)
CFM = air flow rate (cfm)

Applying Equation 2.10 to the example data above yields 60.2 F. LAT in cooling mode is generally not provided in performance data tables.

$$LAT_c = EAT - \frac{SC}{1.08 \bullet CFM} = 80F - \frac{34{,}300\ Btu/hr}{1.08 \bullet 1600\ cfm} = 60.2F$$

Leaving water temperature (LWT) from the heat pump in cooling mode can be calculated according to Equation 2.11, which is based on average thermal properties of water between 32 F and 105 F. If the circulating fluid is other than pure water, the constant of 500 in Equation 2.11 will change slightly (490 to 505 for most water-antifreeze mixtures) due to differences in mixture density and specific heat and will have only a minor influence on the predicted LWT.

$$LWT_c = EWT + \frac{HR}{500 \bullet GPM}$$ **Equation 2.11**

Where:
- LWT_c = leaving water temperature from heat pump in cooling mode (F)
- EWT = entering water temperature from ground connection (F)
- HE = heat of rejection to the ground connection (Btu/hr)
- GPM = water flow rate (gpm)

Using the data from the example above, the leaving water temperature from the heat pump can be calculated according to Equation 2.11.

$$LWT_c = EWT + \frac{HR}{500 \bullet GPM} = 90F + \frac{59{,}700 \text{ Btu/hr}}{500 \bullet 12.0 \text{ gpm}} = 99.95F$$

This represents a 9.95 F temperature rise (99.95 – 90) across the water-to-refrigerant coil, which is within the expected 8 to 10 F range under design cooling conditions.

The cooling mode performance of a GSHP will vary during the cooling season primarily based on the EWT from the ground connection. EAT, CFM, and GPM are design parameters that are set by the designer and around which ducting and piping/pumping systems are designed. EAT should remain in a small range depending on the type of cooling load. CFM should remain relatively constant assuming that the air filter is kept clean and air flow remains unobstructed. GPM will also remain in a very narrow range as long as flow through the water-to-refrigerant coil does not become restricted due to fouling of the coil. The amount of flow required for optimum heat pump performance depends on the type of ground connection in use. Systems using open-loop/standing column ground connections require less system flow (as low as 2 gpm per ton of installed capacity) because EWT will be very stable in an operating water temperature range close to local ground water temperatures. Systems utilizing a closed-loop ground connection typically require 2.5-3.0 gpm per ton of installed heat pump capacity. For both heating and cooling modes of operation, lowering the system water flow rate will diminish heat pump capacity and efficiency. Increasing the system water flow rate above the recommended range of 3 gpm per ton of installed capacity will minimally improve performance, but not enough to justify doing so. Always refer to manufacturer performance data to determine water flow rates for selected equipment according to the type of earth connection used (3 gpm per nominal ton of installed heat pump capacity is the standard). As is the case in Figure 2.6, manufacturers will typically provide performance data for three different flow rates. The lowest flow rate given in the table is usually the minimum required flow rate for open-loop systems. The middle flow rate given is usually the minimum required flow rate for closed-loop systems. The highest flow rate is usually the flow rate required for optimum equipment performance in terms of capacity and efficiency.

EWT, on the other hand, will vary during the cooling season based on the type of ground connection that is utilized and is particularly important in proper sizing of the GSHP in cooling mode. TC, SC, and EER are graphed as a function of EWT in Figure 2.8 for the performance data provided in Figure 2.6 for operating parameters of GPM=12, CFM=1600, and EAT=80/67. Also included is the latent cooling capacity (LC), which is the difference between TC and SC. TC, LC, and EER are strongly influenced by EWT, with all three decreasing as EWT increases (a tendency of a GSHP due to the nature of the vapor-compression cycle). However, SC varies only slightly over the entire EWT range, meaning that if TC decreases with increasing EWT and SC stays relatively constant, then LC will have to decrease with increasing EWT. (This has significance in warm, humid climates where the equipment and ground connection are designed based on cooling and attention must be paid to LC capacity of the installed GSHPs for high design EWTs.) If the heat pump is to be sized for cooling, the maximum EWT from the ground connection must be specified in order to determine the TC, SC, LC, and EER of the heat pump at design conditions. TC, SC, LC, and EER have been identified for an EWT of 77 F in Figure 2.8, which is the entering water temperature that ARI/ISO 13256-1 uses as the ground loop test condition (discussed in Section 2.4.4) for cooling mode certification. For this nominal, 4-ton GSHP, the TC is about 51,000 Btu/hr at EWT=77 F, which is slightly higher than the 48,000 Btu/hr considered to be 4 tons of nominal capacity. GSHPs are rated based on cooling mode performance at ARI/ISO conditions, thus the 4-ton rating for this unit. However, cooling performance varies considerably with EWT, which must be specified to determine the actual TC, SC, LC, and EER of the GSHP. As EWT is increased (as would be the case in southern climates with high ground temperatures and high cooling loads), it may not be possible to economically install a ground connection that provides EWTs lower than 95 F or more at design cooling conditions, having direct consequence on the sizing of the GSHP equipment relative to the total, sensible, and latent cooling design loads. The influence of design EWT on GSHP sizing and operating efficiency will be addressed in the selection of GSHP equipment to meet design cooling loads (Chapter 3) and the design of the various ground coupling devices (Chapter 5).

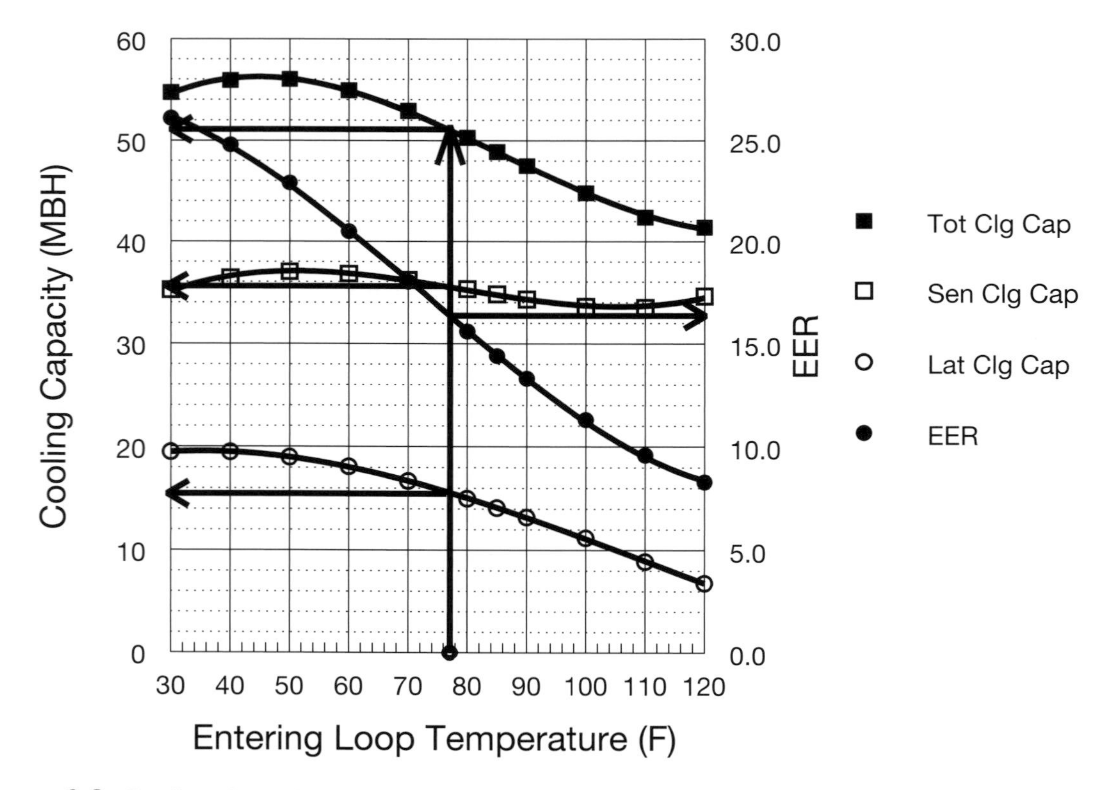

Figure 2.8. Cooling Capacities and EER – ClimateMaster Tranquility 20 Model 048 PSC

2.4.3 PERFORMANCE DATA CORRECTION FACTORS Heat pump performance is generally expressed as a function of EWT, GPM, EAT, and CFM, where the heating and cooling performance data supplied by the manufacturer covers a range of EWTs and GPMs, but may be tabled for a single EAT condition (70 F in heating and 80 F db/67 F wb in cooling) and CFM (350 to 400 CFM per nominal ton). To account for the effects of EAT and CFM (along with antifreeze in the ground loop circulating fluid) on the heating capacity (HC), electric demand (DMD), heat of extraction (HE), coefficient of performance (COP), leaving air temperature (LAT_h), and leaving water temperature (LWT_h), Equations 2.12 through 2.17 can be applied for the heating mode. To account for the effects of EAT and CFM (along with antifreeze in the ground loop circulating fluid) on the total cooling capacity (TC), electric demand (DMD), heat of rejection (HR), energy efficiency ratio (EER), leaving air temperature (LAT_c), and leaving water temperature (LWT_c), Equations 2.18 through 2.24 can be applied for the cooling mode. The various performance parameters (HC, DMD, and HE in heating mode and TC, SC, DMD, and HR in cooling mode) are corrected (parameters subscripted by cor) by multiplying each by correction factors that account for EAT, CFM, and antifreeze (AF) conditions that are not accounted for in the tabled data. The correction factors are obtained from the manufacturer in tabled format and are generally listed separately for heating mode or cooling mode of operation.

$$HC_{cor} = HC \cdot CF_{HC-EAT} \cdot CF_{HC-CFM} \cdot CF_{HC-AF} \qquad \textbf{Equation 2.12}$$

$$DMD_{cor} = DMD \cdot CF_{DMD-EAT} \cdot CF_{DMD-CFM} \cdot CF_{DMD-AF} \qquad \textbf{Equation 2.13}$$

$$HE_{cor} = HE \cdot CF_{HE-EAT} \cdot CF_{HE-CFM} \cdot CF_{HE-AF} \qquad \textbf{Equation 2.14}$$

$$COP_{cor} = \frac{HC_{cor}}{DMD_{cor} \cdot \dfrac{3{,}412 \text{Btu/hr}}{\text{kW}}} \qquad \textbf{Equation 2.15}$$

$$LAT_{h-cor} = EAT + \frac{HC_{cor}}{1.08 \cdot CFM} \qquad \textbf{Equation 2.16}$$

$$LWT_{h-cor} = EAT + \frac{HE_{cor}}{500 \cdot CFM} \qquad \textbf{Equation 2.17}$$

Where:

$CF_{HC\text{-}EAT}$ = heating capacity correction factor for EAT conditions
$CF_{HC\text{-}CFM}$ = heating capacity correction factor for CFM
$CF_{HC\text{-}AF}$ = heating capacity correction factor for antifreeze
$CF_{DMD\text{-}EAT}$ = electric demand correction factor for EAT conditions
$CF_{DMD\text{-}CFM}$ = electric demand correction factor for CFM
$CF_{DMD\text{-}AF}$ = electric demand correction factor for antifreeze
$CF_{HE\text{-}EAT}$ = heat of extraction correction factor for EAT conditions
$CF_{HE\text{-}CFM}$ = heat of extraction correction factor for CFM
$CF_{HE\text{-}AF}$ = heat of extraction correction factor for antifreeze

$$TC_{cor} = TC \bullet CF_{TC-EAT} \bullet CF_{TC-CFM} \bullet CF_{TC-AF}$$ **Equation 2.18**

$$SC_{cor} = SC \bullet CF_{SC-EAT} \bullet CF_{SC-CFM} \bullet CF_{SC-AF}$$ **Equation 2.19**

$$DMD_{cor} = DMD \bullet CF_{DMD-EAT} \bullet CF_{DMD-CFM} \bullet CF_{DMD-AF}$$ **Equation 2.20**

$$HR_{cor} = HR \bullet CF_{HR-EAT} \bullet CF_{HR-CFM} \bullet CF_{HR-AF}$$ **Equation 2.21**

$$EER_{cor} = \frac{TC_{cor}}{DMD_{cor} \bullet \frac{1{,}000W}{kW}}$$ **Equation 2.22**

$$LAT_{c-cor} = EAT - \frac{SC_{cor}}{1.08 \bullet CFM}$$ **Equation 2.23**

$$LWT_{c-cor} = EWT + \frac{HR_{cor}}{500 \bullet GPM}$$ **Equation 2.24**

Where:

CF_{TC-EAT} = total cooling capacity correction factor for EAT conditions
CF_{TC-CFM} = total cooling capacity correction factor for CFM
CF_{TC-AF} = total cooling capacity correction factor for antifreeze
CF_{SC-EAT} = sensible cooling capacity correction factor for EAT conditions
CF_{SC-CFM} = sensible cooling capacity correction factor for CFM
CF_{SC-AF} = sensible cooling capacity correction factor for antifreeze
$CF_{DMD-EAT}$ = electric demand correction factor for EAT conditions
$CF_{DMD-CFM}$ = electric demand correction factor for CFM
CF_{DMD-AF} = electric demand correction factor for antifreeze
CF_{HR-EAT} = heat of rejection correction factor for EAT conditions
CF_{HR-CFM} = heat of rejection correction factor for CFM
CF_{HR-AF} = heat of rejection correction factor for antifreeze

2.4.3.1 ENTERING AIR TEMPERATURE (EAT) CORRECTION FACTORS Correction factors for heating mode EAT conditions are provided in tabular format for the ClimateMaster Tranquility 20™ Series in the top table of Figure 2.9 (Heating Capacity (CF_{HC-EAT}), Power ($CF_{DMD-EAT}$), and Heat of Extraction (CF_{HE-EAT}). The EAT correction factors are applied to tabled performance data according to Equations 2.12 through 2.14 (which also include correction factors for CFM and antifreeze) to determine HC_{cor}, DMD_{cor}, and HE_{cor}. Then, COP_{cor}, LAT_{h-cor}, and LWT_{h-cor} can be determined utilizing Equations 2.15 through 2.17, which are based on the HC_{cor}, DMD_{cor}, and HE_{cor}. Looking at Figure 2.9, the effect of higher EAT in the heating mode is to decrease HC, increase DMD, and reduce HE, reducing the COP of the heat pump and resulting in a slightly lower LAT_h than would be expected for the given EAT. A lower EAT results in an increase in HC, a decrease in DMD, and an increase in HE, increasing the COP of the heat pump and resulting in a slightly higher LAT_h than would be expected for the given EAT. The Heating Correction Factors in Figure 2.9 are applied as shown in Example 2.7.

Heating			
Entering Air DB°F	Heating Capacity	Power	Heat of Extraction
45	1.0514	0.7749	1.1240
50	1.0426	0.8113	1.1032
55	1.0329	0.8525	1.0802
60	1.0224	0.8980	1.0551
65	1.0114	0.9473	1.0282
68	1.0046	0.9786	1.0115
70	1.0000	1.0000	1.0000
75	0.9883	1.0556	0.9706
80	0.9764	1.1135	0.9404

Cooling												
Entering Air WB°F	Total Capacity	Sensible Cooling Capacity Multiplier - Entering DB °F									Power	Heat of Rejection
		60	65	70	75	80	80.6	85	90	95		
50	0.7432	0.9111	*	*	*	*	*	*	*	*	0.9866	0.7901
55	0.8202	0.7709	0.8820	1.0192	*	*	*	*	*	*	0.9887	0.8527
60	0.8960		0.6702	0.8540	1.0473	*	*	*	*	*	0.9924	0.9146
65	0.9705			0.6491	0.8657	1.0809	1.1066	*	*	*	0.9975	0.9757
66.2	0.9882			0.5939	0.8152	1.0333	1.0592	1.2481	*	*	0.9990	0.9903
67	1.0000			0.5559	0.7801	1.0000	1.0261	1.2158	*	*	1.0000	1.0000
70	1.0438				0.6377	0.8645	0.8913	1.0847	1.2983	*	1.0042	1.0362
75	1.1159					0.6008	0.6289	0.8323	1.0578	1.2773	1.0123	1.0959

Figure 2.9. Entering Air Condition Correction Factors - ClimateMaster Tranquility 20™ Series

Example 2.7 Determine the corrected HC, DMD, HE, COP, and LAT for the data in Example 2.5 for an EAT of 55 F.

Based on Example 2.5, the performance data for EWT=40 F, GPM=12, CFM=1,600, and EAT=70 F were determined from Figure 2.6 to be:

HC = 46.1 MBtu/hr x 1,000 Btu/MBtu = 46,100 Btu/hr
DMD = 3.63 kW
COP = 3.72
HE = 33.8 MBtu/hr x 1,000 = 33,800 Btu/hr
LAT = 97 F

From Figure 2.9, correction factors (identified with solid outlines) for an EAT of 55 F are $CF_{HC\text{-}EAT}$=1.0329, $CF_{DMD\text{-}EAT}$=0.8525, and $CF_{HE\text{-}EAT}$=1.0802, which are then applied according to Equations 2.12 through 2.17 to provide corrected performance data for the heat pump in heating mode.

$$HC_{cor} = HC \cdot CF_{HC-EAT} \cdot CF_{HC-CFM} \cdot CF_{HC-AF} = 46{,}100 \text{ Btu/hr} \cdot 1.0329 \cdot 1 \cdot 1 = 47{,}617 \text{ Btu/hr} \quad (+\,3.29\%)$$

$$DMD_{cor} = DMD \cdot CF_{DMD-EAT} \cdot CF_{DMD-CFM} \cdot CF_{DMD-AF} = 3.63 \text{ kW} \cdot 0.8525 \cdot 1 \cdot 1 = 3.09 \text{ kW} \quad (-14.75\%)$$

$$HE_{cor} = HC \cdot CF_{HE-EAT} \cdot CF_{HE-CFM} \cdot CF_{HE-AF} = 33{,}800 \text{ Btu/hr} \cdot 1.0802 \cdot 1 \cdot 1 = 36{,}510 \text{ Btu/hr} \quad (+\,8.02\%)$$

$$COP_{cor} = \frac{HC_{cor}}{DMD_{cor} \cdot \dfrac{3{,}412 \text{ Btu/hr}}{\text{kW}}} = \frac{47{,}617 \text{ Btu/hr}}{3.09 \text{ kW} \cdot \dfrac{3{,}412 \text{ Btu/hr}}{\text{kW}}} = 4.52 \quad (+21.5\%)$$

$$LAT_{h-cor} = EAT + \frac{HC_{cor}}{1.08 \cdot CFM} = 55F + \frac{47{,}617 \text{ Btu/hr}}{1.08 \cdot 1600 \text{ cfm}} = 82.6 \text{ F}$$

$$LWT_{h-cor} = EWT - \frac{HE_{cor}}{500 \cdot GPM} = 40 - \frac{36{,}510 \text{ Btu/hr}}{500 \cdot 12.0 \text{ gpm}} = 33.9 \text{ F}$$

Correction factors for cooling mode EAT are provided for the ClimateMaster Tranquility 20™ Series in the bottom table of Figure 2.9 (Total Cooling Capacity ($CF_{TC\text{-}EAT}$), Sensible Cooling Capacity ($CF_{SC\text{-}EAT}$), Power ($CF_{DMD\text{-}EAT}$), and Heat of Rejection ($CF_{HR\text{-}EAT}$). The correction factors are applied to table performance data according to Equations 2.18 through 2.24 (which also include correction factors for CFM and anti-freeze) to determine TC_{cor}, SC_{cor}, DMD_{cor}, and HR_{cor}, EER_{cor}, LAT_{cor}, and LWT_{cor}, respectively. In the cooling mode, the EAT condition includes both the dry bulb and the wet bulb temperature. Looking at Figure 2.9, TC_{cor}, DMD_{cor}, and HR_{cor} are a function of entering wet bulb temperature only, while SC_{cor} depends both on the entering wet bulb and dry bulb temperatures. TC_{cor}, DMD_{cor}, and HR_{cor} increase with increasing entering wet bulb temperature and decrease with decreasing entering wet bulb temperature. However, because TC_{cor} changes faster than DMD_{cor}, the EER_{cor} of the heat pump will increase at entering wet bulb temperatures above 67 F and will decrease at entering wet bulb temperatures below that value. At a selected entering wet bulb temperature, SC_{cor} increases with increasing dry bulb temperature indicating that there is less moisture in the air and more of the cooling across the air coil is sensible cooling. The dry bulb/wet bulb temperature combinations that contain an asterisk have so little moisture available for condensation that the total cooling and sensible cooling capacities are equal and there is no latent cooling capacity. The Cooling Correction Factors in Figure 2.9 are applied as shown in Example 2.8.

Example 2.8 Determine the corrected TC, SC, DMD, HR, EER, and LAT for the data in Example 2.6 for EAT conditions of 70 F db and 60 F wb.

Based on Example 2.6, the performance data for EWT=90 F, GPM=12, CFM=1,600, and EAT=80 F db/67 F wb were determined from Figure 2.6 to be:

TC = 47.5 MBtu/hr x 1,000 Btu/MBtu = 47,500 Btu/hr
SC = 34.3 MBtu/hr x 1,000 = 34,300 Btu/hr
DMD = 3.57 kW
EER = 13.3
HR = 59.7 MBtu/hr x 1,000 = 59,700 Btu/hr

From Figure 2.9, correction factors (identified with dashed outlines) for EAT conditions of 70 F db/60 F wb are $CF_{TC\text{-}EAT}$=0.8960, $CF_{SC\text{-}EAT}$=0.8540, $CF_{DMD\text{-}EAT}$=0.9924, and $CF_{HR\text{-}EAT}$=0.9146, which are then applied according to Equations 2.18 through 2.24 to provide corrected performance data for the heat pump in cooling mode.

$$TC_{cor} = TC \bullet CF_{TC-EAT} \bullet CF_{TC-CFM} \bullet CF_{TC-AF} = 47{,}500 \text{ Btu/hr} \bullet 0.8960 \bullet 1 \bullet 1 = 42{,}560 \text{ Btu/hr} \quad (-10.4\%)$$

$$SC_{cor} = SC \bullet CF_{SC-EAT} \bullet CF_{SC-CFM} \bullet CF_{SC-AF} = 34{,}300 \text{ Btu/hr} \bullet 0.8540 \bullet 1 \bullet 1 = 29{,}292 \text{ Btu/hr} \quad (-14.6\%)$$

$$DMD_{cor} = DMD \bullet CF_{DMD-EAT} \bullet CF_{DMD-CFM} \bullet CF_{DMD-AF} = 3.57 \text{ kW} \bullet 0.9924 \bullet 1 \bullet 1 = 3.54 \text{ kW} \quad (-0.84\%)$$

$$HR_{cor} = HR \bullet CF_{HR-EAT} \bullet CF_{HR-CFM} \bullet CF_{HR-AF} = 59{,}500 \text{ Btu/hr} \bullet 0.9146 \bullet 1 \bullet 1 = 54{,}419 \text{ Btu/hr} \quad (-8.85\%)$$

$$EER_{cor} = \frac{TC_{cor}}{DMD_{cor} \bullet \frac{1{,}000 \text{ W}}{\text{kW}}} = \frac{42{,}560 \text{ Btu/hr}}{3.54 \text{ kW} \bullet \frac{1{,}000 \text{ W}}{\text{kW}}} = 12.02 \frac{\text{Btu/hr}}{\text{W}} \quad (-9.62\%)$$

$$LAT_{c-cor} = EAT - \frac{SC_{cor}}{1.08 \bullet CFM} = 70 - \frac{29{,}292 \text{ Btu/hr}}{1.08 \bullet 1600 \text{ cfm}} = 53.0 \text{ F}$$

$$LWT_{c-cor} = EWT + \frac{HR_{cor}}{500 \bullet GPM} = 90 + \frac{54{,}419 \text{ Btu/hr}}{500 \bullet 12.0 \text{ gpm}} = 99.1 \text{ F}$$

2.4.3.2 AIR FLOW RATE (CFM) CORRECTION FACTORS

The air flow rate across the refrigerant-to-air coil affects heat pump performance due to its impact on the effectiveness of the refrigerant-to-air heat exchanger. Typically, the fan in the heat pump is sized to supply an air flow rate of between 350 to 400 CFM per nominal ton of heat pump capacity against a static pressure in the ductwork (between 0.20 and 0.40 inches of water, but may vary depending on manufacturer). Lowering the system air flow rate below the manufacturer's recommended flow rate (due to high external static pressure) will reduce heat pump capacity and efficiency. Increasing air flow rate (by over sizing ductwork to reduce the external static pressure) will increase heat pump capacity and performance, but to a much lesser extent.

Blower performance data for the ClimateMaster Tranquility 20™ Series GSHP equipped with a PSC blower motor are provided in Figure 2.10, which is the table provided in the manufacturer's literature. For each heat pump model, there are three identified fans speeds labeled HI, MED, and LOW, one of which will be set through a hard-wire setting inside the heat pump. The air flow rate range (CFM) that each fan

Airflow in CFM with wet coil and clean air filter

Model	Fan Speed	Rated Airflow	Min CFM	Airflow (cfm) at External Static Pressure (in. wg)															
				0.00	0.05	0.10	0.15	0.20	0.25	0.30	0.35	0.40	0.45	0.50	0.60	0.70	0.80	0.90	1.00
018	HI	600	450	704	708	711	702	693	692	690	683	675	658	640	598	515			
	MED	600	450	602	601	599	590	581	583	585	579	573	560	547	492				
	LOW	600	450	531	529	527	522	517	512	506	501	495	479	462					
024	HI	850	600	965	960	954	943	931	923	914	898	882	862	842	794	725	635		
	MED	850	600	841	833	825	817	809	800	790	777	763	747	731	686	623			
	LOW	850	600	723	715	707	703	698	689	680	668	656	642	627					
030	HI	950	750	1271	1250	1229	1207	1185	1164	1143	1118	1093	1061	1029	953	875	753		
	MED	950	750	1048	1037	1025	1016	1007	994	981	962	943	915	886	822				
	LOW	950	750	890	887	884	879	874	865	855	842	829	809	789					
036	HI	1250	900	1411	1407	1402	1390	1378	1370	1361	1326	1290	1248	1205	1083	942			
	MED	1250	900	1171	1164	1156	1145	1133	1113	1092	1064	1035	997	958					
	LOW	1250	900	983	967	950	943	936	936										
042	HI	1400	1050	1634	1626	1618	1606	1594	1583	1571	1539	1507	1464	1420	1265	1078			
	MED	1400	1050	1332	1323	1314	1298	1282	1263	1243	1206	1169	1115	1060					
	LOW	1400	1050	1130	1109	1088	1086	1084	1066	1048	1052	1055							
048	HI	1600	1200	1798	1781	1764	1738	1711	1688	1665	1630	1595	1555	1514	1420	1239			
	MED	1600	1200	1384	1382	1379	1375	1371	1356	1341	1318	1294	1261	1227					
	LOW	1600	1200																
060	HI	1950	1500	2311	2306	2300	2290	2279	2268	2257	2233	2209	2175	2140	2088	1990	1901	1856	1752
	MED	1950	1500	2058	2049	2039	2028	2016	2000	1983	1966	1949	1935	1920	1874	1807	1750	1670	1582
	LOW	1950	1500	1868	1863	1858	1858	1858	1848	1838	1822	1806	1799	1792	1749	1699	1636	1570	
070	HI	2100	1800	2510	2498	2486	2471	2455	2440	2424	2401	2377	2348	2318	2247	2161	2078	1986	1855
	MED	2100	1800	2171	2167	2162	2162	2162	2158	2153	2135	2117	2101	2085	2024	1971	1891	1823	
	LOW	2100	1800	2010	2008	2006	2006	2006	2006	2006	1992	1977	1962	1947	1892	1851			

Figure 2.10. PSC Blower Performance Data - ClimateMaster Tranquility 20™ Series

setting provides is listed as a function of external static pressure (ESP) across the top of the table, which is expressed as inches wg (inches of water gauge pressure). For the Model 048 (solid outlines), the rated air flow rate of 1,600 CFM is achieved on HI fan speed setting with an ESP of between 0.35 and 0.40 inches wg. According to the HI fan speed data, the air flow rate can vary from 1,798 CFM at 0 inches wg to 1,239 CFM at 0.70 inches wg. ESP values higher than 0.70 are not recommended by the manufacturer (shaded black) and care must be taken in the duct sizing procedure to avoid static pressures above that value. The MED fan speed setting will produce air flow rates ranging from 1,384 CFM at 0 inches wg down to 1,227 CFM at 0.50 inches wg. The LOW fan speed setting is not recommended by the manufacturer for the Model 048.

Blower performance data for the ClimateMaster Tranquility 20™ Series GSHP equipped with an ECM blower motor are provided in Figure 2.11, which is the table provided in the manufacturer's literature for all ECM blower equipped heat pumps. The Max ESP column identifies the maximum external static pressure that the installed fan motor (third column) is capable of producing with the ECM controller, meaning that the fan speed will continue to be increased up to that ESP after which fan speed will remain constant and air flow rate will start to fall off of the rated values in the table. For each heat pump model there are four identified "Tap Settings" labeled 1, 2, 3, and 4, each of which is associated with a combination of air flow rates for fan only, stage 1, and stage 2 operation in cooling, dehumidification, and heating mode. The

Airflow in CFM with wet coil and clean air filter

Model	Max ESP (in. wg)	Fan Motor (hp)	Tap Setting	Cooling Mode			Dehumid Mode			Heating Mode			Residential Units Only	
				Stg 1	Stg 2	Fan	Stg 1	Stg 2	Fan	Stg 1	Stg 2	Fan	AUX CFM	Aux/ Emerg Mode
018	0.50	1/2	4	620	750	380	480	590	380	620	750	380	4	750
			3	570	700	350	450	550	350	570	700	350	3	700
			2	510	620	310	400	480	310	510	620	310	2	650
			1	430	530	270				430	530	270	1	650
024	0.50	1/2	4	780	950	470	610	740	470	870	1060	470	4	1060
			3	700	850	420	540	660	420	780	950	420	3	950
			2	630	770	360	490	600	360	670	820	390	2	820
			1	550	670	300				570	690	340	1	690
030	0.50	1/2	4	920	1130	560	720	880	560	1000	1230	560	4	1230
			3	820	1000	500	640	780	500	900	1100	500	3	1100
			2	740	900	450	580	700	450	800	980	450	2	980
			1	660	800	400				700	850	400	1	850
036	0.50	1/2	4	1150	1400	700	900	1090	700	1150	1400	700	4	1400
			3	1020	1250	630	800	980	630	1020	1250	630	3	1350
			2	890	1080	540	690	840	540	890	1080	540	2	1350
			1	740	900	450				750	920	450	1	1350
042	0.50	1/2	4	1290	1580	790	1010	1230	790	1290	1580	790	4	1580
			3	1150	1400	700	900	1090	700	1150	1400	700	3	1400
			2	1050	1280	640	820	1000	640	1020	1240	640	2	1350
			1	920	1120	560				900	1080	560	1	1350
048	0.75	1	4	1420	1730	870	1110	1350	870	1520	1850	865	4	1850
			3	1270	1550	780	990	1210	780	1350	1650	775	3	1650
			2	1180	1440	720	920	1120	720	1190	1450	720	2	1450
			1	1050	1280	640				1020	1250	640	1	1350
060	0.75	1	4	1680	2050	1030	1310	1600	1030	1870	2280	1030	4	2280
			3	1500	1830	910	1170	1420	910	1680	2050	910	3	2050
			2	1400	1700	850	1090	1330	850	1480	1800	850	2	1800
			1	1300	1580	790				1270	1550	790	1	1550
070	0.75	1	4	1830	2230	1100	1420	1740	1100	1830	2230	1100	4	2230
			3	1600	1950	980	1250	1520	980	1720	2100	980	3	2100
			2	1440	1750	880	1120	1360	880	1670	1950	880	2	1950
			1	1200	1580	790				1460	1780	790	1	1780

Figure 2.11. ECM Blower Performance Data - ClimateMaster Tranquility 20™ Series

Auxiliary/Emergency Mode column contains the air flow rate that the fan speed setting will provide at the specified Tap Setting. The Tranquility 20 Series is capable of single-capacity operation only, which would correspond to the tabled stage 2 values in heating or cooling mode and fan values, if the fan is "on," but the compressor is not running. Considering the Tranquility 20 Series Model 048 (solid outlines), for which the rated air flow rate is 1,550 CFM in cooling and 1,650 CFM in heating, the associated Tap Setting would be 3. The dehumidification mode of operation (preset or controlled by humidistat) is for cooling only where the fan speed is reduced 20 to 25% (in this case 1,550 CFM down to 1,210 CFM) to increase the latent cooling capacity of the heat pump. Auxiliary/Emergency Mode fan operation would provide 1,650 CFM.

Correction factors for CFM are provided in tabular format for the ClimateMaster Tranquility 20™ Series in Figure 2.12. PSC Fan Motor data are provided in the top table, and ECM Fan Motor data are provided in the bottom table. For each fan motor type, heating mode correction factors (Heating Capacity ($CF_{HC\text{-}CFM}$), Power ($CF_{DMD\text{-}CFM}$), and Heat of Extraction ($CF_{HE\text{-}CFM}$) are provided in the last three columns as a function of Percent of Rated Air flow (first column). The CFM correction factors are applied to tabled performance

PSC Fan Motor

Airflow	Cooling					Heating		
% of Rated	Total Capacity	Sensible Capacity	S/T	Power	Heat of Rejection	Heating Capacity	Power	Heat of Extraction
68.75%	0.9465	0.8019	0.8472	0.9614	0.9496			
75%	0.9602	0.8350	0.8696	0.9675	0.9617	0.9740	1.0936	0.9425
81.25%	0.9724	0.8733	0.8981	0.9744	0.9728	0.9810	1.0635	0.9592
87.50%	0.9831	0.9149	0.9306	0.9821	0.9829	0.9876	1.0379	0.9744
93.75%	0.9923	0.9578	0.9653	0.9906	0.9920	0.9940	1.0167	0.9880
100%	1.0000	1.0000	1.0000	1.0000	1.0000	1.0000	1.0000	1.0000
106.25%	1.0062	1.0392	1.0328	1.0102	1.0070	1.0057	0.9878	1.0105
112.50%	1.0109	1.0733	1.0617	1.0211	1.0130	1.0112	0.9800	1.0194
118.75%	1.0141	1.1001	1.0848	1.0329	1.0180	1.0163	0.9705	1.0284
125%	1.0159	1.1174	1.0999	1.0455	1.0220	1.0211	0.9614	1.0368
130%	1.0161	1.1229	1.1050	1.0562	1.0244	1.0247	0.9554	1.0430

Black area denotes where operation is not recommended.

ECM Fan Motor

Airflow	Cooling					Heating		
% of Rated	Total Capacity	Sensible Capacity	S/T	Power	Heat of Rejection	Heating Capacity	Power	Heat of Extraction
68.75%	0.9470	0.8265	0.8727	0.9363	0.9449			
75%	0.9619	0.8593	0.8933	0.9455	0.9587	0.9700	1.0822	0.9410
81.25%	0.9747	0.8943	0.9175	0.9564	0.9711	0.9775	1.0536	0.9579
87.50%	0.9853	0.9302	0.9441	0.9691	0.9821	0.9851	1.0304	0.9733
93.75%	0.9938	0.9659	0.9719	0.9837	0.9918	0.9925	1.0125	0.9874
100%	1.0000	1.0000	1.0000	1.0000	1.0000	1.0000	1.0000	1.0000
106.25%	1.0041	1.0313	1.0271	1.0181	1.0069	1.0074	0.9928	1.0112
112.50%	1.0060	1.0584	1.0522	1.0381	1.0123	1.0148	0.9909	1.0210
118.75%	1.0070	1.0815	1.0740	1.0598	1.0174	1.0222	0.9622	1.0377
125%	1.0076	1.0998	1.0916	1.0834	1.0225	1.0295	0.8681	1.0712
130%	1.0083	1.1110	1.1018	1.1035	1.0271	1.0354	0.8456	1.0844

Figure 2.12. Air flow Rate Correction Factors - ClimateMaster Tranquility 20™ Series

data according to Equations 2.12 through 2.14 to determine HC_{cor}, DMD_{cor}, and HE_{cor} from which COP_{cor}, LAT_{cor}, and LWT_{cor} can be determined (Equations 2.15 through 2.17). Looking at Figure 2.11, the effect of higher CFM in the heating mode is to increase HC, reduce DMD, and increase HE, increasing the COP of the heat pump and resulting in a slightly lower LAT due to the higher CFM. Lowering CFM results in a decrease in HC, an increase in DMD, and a decrease in HE, decreasing the COP of the heat pump and resulting in a slightly higher LAT due to the lower CFM. The Heating Correction Factors for a PSC Fan Motor in Figure 2.11 are applied as shown in Example 2.9.

Example 2.9 Determine the corrected HC, DMD, HE, COP, and LAT for the data in Example 2.5 for a CFM of 1200.

Based on Example 2.5, the performance data for EWT=40 F, GPM=12, CFM=1,600, and EAT=70 F were determined from Figure 2.6 to be:

HC = 46.1 MBtu/hr x 1,000 Btu/MBtu = 46,100 Btu/hr
DMD = 3.63 kW
COP = 3.72
HE = 33.8 MBtu/hr x 1000 = 33,800 Btu/hr
LAT = 97 F

From Figure 2.12, correction factors (identified with solid outlines) for a CFM of 1,200 (Percent of Rated Air flow = 1,200/1,600 = 0.75 = 75%) are $CF_{HC\text{-}CFM}$=0.9740, $CF_{DMD\text{-}CFM}$=1.0936, and $CF_{HE\text{-}CFM}$=0.9425, which are then applied according to Equations 2.12 through 2.17 to provide corrected performance data for the heat pump in heating mode.

$$HC_{cor} = HC \bullet CF_{HC-EAT} \bullet CF_{HC-CFM} \bullet CF_{HC-AF} = 46{,}100\ \text{Btu/hr} \bullet 1 \bullet 0.9740 \bullet 1 = 44{,}901\ \text{Btu/hr} \qquad (-2.60\%)$$

$$DMD_{cor} = DMD \bullet CF_{DMD-EAT} \bullet CF_{DMD-CFM} \bullet CF_{DMD-AF} = 3.63\ \text{kW} \bullet 1 \bullet 1.0936 \bullet 1 = 3.97\ \text{kW} \qquad (+9.36\%)$$

$$HE_{cor} = HE \bullet CF_{HE-EAT} \bullet CF_{HE-CFM} \bullet CF_{HE-AF} = 33{,}800\ \text{Btu/hr} \bullet 1 \bullet 0.9425 \bullet 1 = 31{,}857\ \text{Btu/hr} \qquad (-5.75\%)$$

$$COP_{cor} = \frac{HC_{cor}}{DMD_{cor} \bullet \dfrac{3{,}412\ \text{Btu/hr}}{\text{kW}}} = \frac{44{,}901\ \text{Btu/hr}}{3.97\ \text{kW} \bullet \dfrac{3{,}412\ \text{Btu/hr}}{\text{kW}}} = 3.31 \qquad (-11.0\%)$$

$$LAT_{h-cor} = EAT + \frac{HC_{cor}}{1.08 \bullet CFM} = 70F + \frac{44{,}901\ \text{Btu/hr}}{1.08 \bullet 1200\ \text{cfm}} = 104.6\ F$$

$$LWT_{h-cor} = EWT - \frac{HE_{cor}}{500 \bullet GPM} = 40 - \frac{31{,}857\ \text{Btu/hr}}{500 \bullet 12.0\ \text{gpm}} = 34.7\ F$$

Correction factors for cooling mode CFM are provided for the ClimateMaster Tranquility 20™ Series in the Cooling columns of Figure 2.12 (Total Cooling Capacity ($CF_{TC\text{-}CFM}$), Sensible Cooling Capacity ($CF_{SC\text{-}CFM}$), Power ($CF_{DMD\text{-}CFM}$), and Heat of Rejection ($CF_{HR\text{-}CFM}$). The correction factors are applied to tabled performance data according to Equations 2.18 through 2.24 to determine TC_{cor}, SC_{cor}, DMD_{cor}, and HR_{cor}, EER_{cor}, LAT_{cor}, and LWT_{cor}, respectively. The effect of reducing CFM is to decrease TC, decrease SC, decrease DMD, and decrease HR. For the PSC blower, the TC and DMD reductions are about the same, with TC reduction being only slightly larger, resulting in a very minimal reduction in EER. For the ECM blower, the DMD reduction is higher than the TC reduction, translating into a larger reduction in EER with reduced CFM. The Cooling Correction Factors for a PSC Fan Motor in Figure 2.12 are applied as shown in Example 2.10.

Example 2.10 Determine the corrected TC, SC, DMD, HR, EER, and LAT for the data in Example 2.6 for a CFM of 1,200.

Based on Example 2.6, the performance data for EWT=90 F, GPM=12, EAT=80 F db/67 F wb, and CFM=1,600 were determined from Figure 2.6 to be:

TC = 47.5 MBtu/hr x 1,000 Btu/MBtu = 47,500 Btu/hr
SC = 34.3 MBtu/hr x 1,000 = 34,300 Btu/hr
DMD = 3.57 kW
EER = 13.3
HR = 59.7 MBtu/hr x 1,000 = 59,700 Btu/hr

From Figure 2.12, correction factors (identified with dashed outlines) for a CFM of 1,200 (Percent of-Rated Air flow = 1,200/1,600 = 0.75 = 75%) are $CF_{TC\text{-}CFM}$=0.9602, $CF_{SC\text{-}CFM}$=0.8350, $CF_{DMD\text{-}CFM}$=0.9675, and $CF_{HR\text{-}CFM}$=0.9617, which are then applied according to Equations 2.18 through 2.24 to provide corrected performance data for the heat pump in cooling mode.

$$TC_{cor} = TC \bullet CF_{TC\text{–}EAT} \bullet CF_{TC\text{–}CFM} \bullet CF_{TC\text{–}AF} = 47{,}500\ \text{Btu/hr} \bullet 1 \bullet 0.9602 \bullet 1 = 45{,}610\ \text{Btu/hr} \qquad (-3.98\%)$$

$$SC_{cor} = SC \bullet CF_{SC\text{–}EAT} \bullet CF_{SC\text{–}CFM} \bullet CF_{SC\text{–}AF} = 34{,}300\ \text{Btu/hr} \bullet 1 \bullet 0.8350 \bullet 1 = 28{,}640\ \text{Btu/hr} \qquad (-16.5\%)$$

$$DMD_{cor} = DMD \bullet CF_{DMD\text{–}EAT} \bullet CF_{DMD\text{–}CFM} \bullet CF_{DMD\text{–}AF} = 3.57\ \text{kW} \bullet 1 \bullet 0.9675 \bullet 1 = 3.45\ \text{kW} \qquad (-3.33\%)$$

$$HR_{cor} = HR \bullet CF_{HR\text{–}EAT} \bullet CF_{HR\text{–}CFM} \bullet CF_{HR\text{–}AF} = 59{,}500\ \text{Btu/hr} \bullet 1 \bullet 0.9617 \bullet 1 = 57{,}221\ \text{Btu/hr} \qquad (-3.83\%)$$

$$EER_{cor} = \frac{TC_{cor}}{DMD_{cor} \bullet \frac{1{,}000\ \text{W}}{\text{kW}}} = \frac{45{,}610\ \text{Btu/hr}}{3.45\ \text{kW} \bullet \frac{1{,}000\ \text{W}}{\text{kW}}} = 13.2\ \frac{\text{Btu/hr}}{\text{W}} \qquad (-0.70\%)$$

$$LAT_{c\text{–}cor} = EAT - \frac{SC_{cor}}{1.08 \bullet CFM} = 80 - \frac{28{,}640\ \text{Btu/hr}}{1.08 \bullet 1200\ \text{cfm}} = 57.9\text{F}$$

$$LWT_{c\text{–}cor} = EWT + \frac{HR_{cor}}{500 \bullet GPM} = 90 + \frac{57{,}221\ \text{Btu/hr}}{500 \bullet 12.0\ \text{gpm}} = 99.5\text{F}$$

Further discussion of blower types, operation, and the effects of duct sizing on GSHP performance and selection can be found in the duct design section of Chapter 3.

2.4.3.3 ANTIFREEZE CORRECTION FACTORS The addition of antifreeze to the water in a closed-loop ground connection will act to increase the viscosity of the mixture. Higher viscosity will lead to less turbulence and less efficient heat transfer in the water-to-refrigerant coil in the GSHP. Most head loss and heat pump performance data provided by GSHP manufactures is for water flowing through the water-to-refrigerant coil. To account for the effect of antifreeze, correction factors for different antifreeze solutions at different percent concentrations are required, such as those provided in Figure 2.13 from ClimateMaster. It is obvious from the data in Figure 2.13 that for the percent antifreeze solutions commonly used in closed-loop GSHP systems (normally 20% or less) the effect of antifreeze is very minor and can generally be ignored. However, the addition of antifreeze will always result in a small reduction in TC and a small increase in

Antifreeze Type	Antifreeze %	Cooling EWT 90°F			Heating EWT 30°F		WPD Corr. Fct. EWT 30°F
		Total Cap	Sens Cap	Power	Htg Cap	Power	
Water	0	1.000	1.000	1.000	1.000	1.000	1.000
Propylene Glycol	5	0.995	0.995	1.003	0.989	0.997	1.070
	15	0.986	0.986	1.009	0.968	0.990	1.210
	25	0.978	0.978	1.014	0.947	0.983	1.360
Methanol	5	0.997	0.997	1.002	0.989	0.997	1.070
	15	0.990	0.990	1.007	0.968	0.990	1.160
	25	0.982	0.982	1.012	0.949	0.984	1.220
Ethanol	5	0.998	0.998	1.002	0.981	0.994	1.140
	15	0.994	0.994	1.005	0.944	0.983	1.300
	25	0.986	0.986	1.009	0.917	0.974	1.360
Ethylene Glycol	5	0.998	0.998	1.002	0.993	0.998	1.040
	15	0.994	0.994	1.004	0.980	0.994	1.120
	25	0.988	0.988	1.008	0.966	0.990	1.200

Figure 2.13. Antifreeze Correction Factors - ClimateMaster

DMD in cooling, thus reducing EER. In heating, the HC is reduced a little more than the DMD, resulting in a very small reduction in COP. If needed, apply correction factors in heating mode for HC ($CF_{HC\text{-}AF}$) and DMD ($CF_{DMD\text{-}AF}$) using Equations 2.12 through 2.17, and in cooling mode for TC ($CF_{TC\text{-}AF}$), SC ($CF_{SC\text{-}AF}$), and DMD ($CF_{DMD\text{-}AF}$) using Equations 2.18 through 2.24 along with correction factors for EAT and CFM.

2.4.4 ISO/ARI RATINGS The Air-Conditioning and Refrigeration Institute (ARI), American Society of Heating, Refrigeration, and Air-Conditioning Engineers (ASHRAE), and the International Standards Organization *(ISO, 1998a)* have developed certification standards *(ISO 13256-1)* to set requirements for testing and rating of water-to-air heat pump equipment to ensure that performance ratings published by manufacturers agree with the performance of their manufactured product. While compliance with the standards is completely voluntary, for any manufacturer to claim that they are "ARI certified" they must fully comply with the standards set forth by said organizations. The standards applying to water source heat pumps are as follows:

- WLHP (previously ARI 220) – Applies to water loop heat pump systems (boiler/tower)
- GWHP (previously ARI 225) – Applies to ground water/open loop heat pump systems
- GLHP (previously ARI 330) – Applies to closed-loop ground source heat pump systems

To be ISO/ARI certified, 30% of a manufacturer's models must be tested every year (randomly selected from stock by ARI). If a unit's tested performance (measured capacity and efficiency - TC and EER in cool-

ing and HC and COP in heating) is measured to be 90% or less of the published performance, the model from which the unit was selected must be either retested or made obsolete. A second failure of the same model requires all units within that model group to be made obsolete. The data that ARI certifies are heating and cooling capacities and associated COPs and EERs at the EWT and EAT conditions specified by the standard, as indicated by the table in Figure 2.14 for the water-to-air heat pump system type. As indicated, air flow and water flow rates are specified by the manufacturer for each heat pump model. ARI certifica-

	Water-loop heat pumps		Ground-water heat pumps		Ground-loop heat pumps	
	Cooling	Heating	Cooling	Heating	Cooling	Heating
Air entering indoor side - dry bulb, C (F) - wet bulb, C (F)	 27 (80.6) 19 (66.2)	 20 (68) 15 (59)	 27 (80.6) 19 (66.2)	 20 (68) 15 (59)	 27 (80.6) 19 (66.2)	 20 (68) 15 (59)
Air surrounding unit - dry bulb, C (F)	27 (80.6)	20 (68)	27 (80.6)	20 (68)	27 (80.6)	20 (68)
Standard Rating Test Liquid entering heat exchanger, C (F)	30 (86)	20 (69)	15 (59)	10 (50)	25 (77)	0 (32)
Part-load Rating Test Liquid entering heat exchanger, C (F)	30 (86)	20 (69)	15 (59)	10 (50)	20 (68)	5 (41)
Frequency*	Rated	Rated	Rated	Rated	Rated	Rated
Voltage**	Rated	Rated	Rated	Rated	Rated	Rated
* Equipment with dual-rated frequencies shall be tested at each frequency ** Equipment with dual-rated voltages shall be tested at both voltages or at the lower of the two voltages if only a single rating is published.						

Water and air flow rates specified by manufacturer.

Figure 2.14. ARI/ISO 13256-1 Test Condition Table

tion ensures that all heat pump manufacturers' performance data is submitted on a level playing field. It essentially prevents a manufacturer from manipulating operating points (via changing EWT or EAT to temperatures not commonly seen during operation, inflating levels of air or water flow, etc.) for the purpose of giving the impression of superior performance over a competitor.

In addition to the entering air and water temperature specifications, ARI certification ratings include fan power, corrected to zero external static pressure, and pumping power, to induce water flow through the water coil only. Heating and cooling capacity are directly affected by the amount of fan power consumed to generate the required amount of air flow for a specific heat pump unit. If a piece of equipment were tested in conjunction with undersized ductwork, the heating capacity would be inflated because higher levels of fan power would be necessary to push the required amount of air flow. Alternatively, if a piece of equipment were tested in conjunction with oversized ductwork, cooling capacity would be inflated because of lower levels of fan power being rejected to the ground loop through the heat pump to generate the required amount of air flow.

ARI/ISO performance ratings for the ClimateMaster Tranquility 20™ Series are provided in Figure 2.15. Performance ratings (HC and COP for heating and TC and EER for cooling) are provided for Water-Loop Heat Pumps (WLHP), Ground-Water Heat Pumps (GWHP), and Ground-Loop Heat Pumps (GLHP) entering water conditions for both the PSC and ECM blower options (for further discussion on the differences between the ARI/ISO ratings, refer to http://www.iso.org). For the Model 048 with PSC blower, the GLHP ARI performance ratings are circled with solid outlines for heating mode and with dashed outlines for cooling mode. These capacities and efficiencies include fan power and pumping power penalties minus the effects of the ductwork and the ground loop.

ASHRAE/ARI/ISO 13256-1. English (IP) Units

Model	Fan Motor	Water Loop Heat Pump				Ground Water Heat Pump				Ground Loop Heat Pump			
		Cooling 86°F		Heating 68°F		Cooling 59°F		Heating 50°F		Cooling 77°F		Heating 32°F	
		Capacity Btuh	EER Btuh/W	Capacity Btuh	COP	Capacity Btuh	EER Btuh/W	Capacity Btuh	COP	Capacity Btuh	EER Btuh/W	Capacity Btuh	COP
018	PSC	17,300	16.2	21,400	5.4	20,200	26.7	17,400	4.6	18,300	19.0	13,400	3.7
	ECM	17,700	16.8	21,700	5.9	20,500	28.1	17,500	4.9	18,600	19.8	13,500	4.0
024	PSC	25,100	16.2	29,600	4.9	28,600	25.7	25,000	4.3	26,300	19.1	19,000	3.7
	ECM	25,000	17.0	30,000	5.3	28,100	27.4	25,100	4.6	26,000	20.0	19,400	3.8
030	PSC	28,200	15.3	34,900	5.0	31,700	22.9	29,400	4.4	29,400	17.6	23,600	3.8
	ECM	28,600	15.6	35,200	5.3	32,200	23.9	29,400	4.6	29,800	18.0	23,700	3.9
036	PSC	33,000	16.6	39,800	5.5	37,300	25.1	32,900	4.8	34,500	19.2	25,700	3.9
	ECM	33,100	17.6	39,500	5.8	37,300	26.5	32,900	5.1	34,600	20.2	25,800	4.2
042	PSC	37,400	16.0	49,400	5.4	42,900	24.3	40,100	4.6	39,300	19.4	31,600	3.8
	ECM	37,800	17.1	48,600	5.7	44,200	27.1	39,300	4.9	40,000	20.0	30,400	4.0
048	PSC	47,000	15.3	60,000	5.0	53,900	23.3	49,000	4.4	49,900	17.6	39,000	3.7
	ECM	47,600	15.9	59,700	5.2	54,100	24.6	48,700	4.5	50,100	18.5	38,400	3.8
060	PSC	61,000	15.9	70,400	5.0	67,000	23.2	58,700	4.5	63,300	18.2	46,500	3.7
	ECM	61,000	16.4	70,800	5.2	67,200	24.3	59,100	4.6	64,000	19.0	46,700	3.8
070	PSC	67,500	14.4	85,800	5.0	77,100	21.6	69,400	4.3	70,800	16.6	54,000	3.6
	ECM	67,000	15.2	84,900	5.0	77,000	23.5	69,000	4.4	70,000	17.8	53,900	3.6

Cooling capacities based on 80.6°F DB, 66.2°F WB entering air temperature
Heating capacities based on 68°F DB, 59°F WB entering air temperature
All ratings based upon operation at lower voltage of dual-voltage rated models

Figure 2.15. ARI/ISO Ratings - ClimateMaster Tranquility 20™ Series

2.5 Performance of Two-Capacity, Water-to-Air GSHPs

Two-capacity, water-to-air GSHPs utilize a scroll compressor with two internal bypass ports that enable the system to run at 67% or 100% capacity, allowing for "part-load" or "full-load" operation in terms of heating and cooling capacities. The two-capacity GSHP is controlled by a thermostat capable of two or three heat stages and two cooling stages, allowing the heat pump to operate at part-load capacity until it can't meet the heating or cooling requirements of the zone where it is then stepped up to full-load heating or cooling capacity. In the heating mode, using a thermostat with a third stage in heating allows for control of any supplemental electric heat, if used.

The performance (capacity, electric demand, and efficiency) of a two-capacity, water-to-air GSHP is measured by the full-load and part-load heating capacities (Btu/hr), electric demands (kW), and COPs in heating mode, and by the full-load and part-load cooling capacities (Btu/hr), electric demands (kW), and EERs in cooling mode. Like single-capacity GSHPs, parameters that affect the heating and cooling performance of the two-capacity, water-to-air GSHP include entering water temperature (EWT) from the ground connection, entering air temperature (EAT) returning from the conditioned space, water flow rate (GPM) through the refrigerant-to-water coil, and air flow rate (CFM) through the refrigerant-to-air coil. Manufacturers of GSHP equipment generally express performance data for their equipment in a tabular format as a

function of EWT, GPM, and CFM for a given EAT condition. Two-capacity GSHPs are generally rated at different water flow rates (GPM) and air flow rates (CFM) at full-load and part-load heating and cooling. Some manufacturers utilize only EWT and GPM in the full-load and utilize part-load performance tables for given CFM and EAT conditions. Additional tables of performance correction factors are provided to account for the effect of EAT conditions and CFM on capacity, electric demand, and efficiency at both full-load and part-load operating mode.

Performance data for a ClimateMaster Tranquility 27™ Model 049 (a two-capacity, nominal 4-ton unit that utilizes R-410A refrigerant and an ECM blower motor) are provided in Figure 2.16 for full-load operation and Figure 2.17 for part-load operation in the tabular formats supplied by the manufacturer. The format of these two performance data tables is identical to the single-capacity ClimateMaster Tranquility 20™ Model 048 provided in Figure 2.6. Column numbers are placed above the table for easier identification during discussion of the data. Column 1 contains EWT from the ground connection over which performance data are included, and Column 2 contains three GPM levels for each EWT, except at EWT=20 F where only the highest water flow rate is recommended. For each GPM, the pressure drop in psi (Column 3) and head loss in feet of water equivalent pressure (Column 4) through the water-to-refrigerant coil are provided for water flowing through the coil for EWT above 40 F and for 15% antifreeze solution for EWT below 40 F. For each GPM in Column 2, the performance data are provided for two air flow rates (listed in Column 5 for cooling and Column 12 for heating). The higher air flow rate in each mode (1550 CFM in cooling, 1,650 CFM in heating) are the air flow rates for which the GSHP is rated for ARI/ISO 13256-1 (Section 2.4.4). Manufacturer's data are not always presented in the same format as provided here and each should be considered individually.

2.5.1 HEATING MODE PERFORMANCE DATA Full-load heating mode performance data are listed in Columns 12 through 18 in Figure 2.16, with each line of data representing performance for a combination of EWT, GPM, and CFM as read proceeding from left to right across the table. Column 13 contains the heating capacity (HC - MBtu/hr) provided to the air returning from the conditioned space, Column 14 lists the electric demand (DMD - kW) of the unit, and Column 17 provides the COP. The heat of extraction (HE - MBtu/hr) from the ground connection is provided in Column 15 and the leaving air temperature back to the conditioned space (LAT - F) is shown in Column 16. The heating capacity of the optional desuperheater (HWC - MBtu/hr) is available in Column 18. Most manufacturers express capacities in MBtu/hr (or MBH) to save space (1 MBtu/hr = 1,000 Btu/hr).

Column

1550 CFM Nominal (Rated) Airflow Cooling, 1650 CFM Nominal (Rated) Airflow Heating — Performance capacities shown in thousands of Btuh

1	2	3	4	5	6	7	8	9	10	11	12	13	14	15	16	17	18
EWT °F	GPM	WPD		Cooling - EAT 80/67°F							Heating - EAT 70°F						
		PSI	FT	Airflow CFM	TC	SC	kW	HR	EER	HWC	Airflow CFM	HC	kW	HE	LAT	COP	HWC
20	12.0	4.8	11.0	Operation not recommended							1430	31.6	2.90	22.1	90.5	3.20	3.8
	12.0	4.8	11.0								1650	32.3	2.78	22.9	88.1	3.40	3.3
30	6.0	1.3	2.9	1330	56.1	32.1	2.13	63.2	26.4	-	1430	34.7	2.98	24.9	92.5	3.41	4.0
	6.0	1.3	2.9	1550	57.4	35.1	2.22	65.0	25.9	-	1650	35.4	2.86	25.7	89.9	3.62	3.5
	9.0	2.7	6.1	1330	56.5	32.1	2.01	63.2	28.2	-	1430	36.3	3.03	26.3	93.5	3.51	3.9
	9.0	2.7	6.1	1550	57.9	35.2	2.09	65.0	27.6	-	1650	37.0	2.90	27.2	90.8	3.73	3.4
	12.0	4.6	10.5	1330	56.8	32.1	1.94	63.3	29.3	-	1430	37.2	3.05	27.0	94.1	3.57	3.9
	12.0	4.6	10.5	1550	58.2	35.2	2.03	65.1	28.7	-	1650	37.9	2.93	28.0	91.3	3.79	3.3
40	6.0	1.1	2.7	1330	55.8	33.0	2.32	63.6	24.1	-	1430	40.0	3.13	29.6	95.9	3.75	4.1
	6.0	1.1	2.7	1550	57.1	36.2	2.42	65.4	23.6	-	1650	40.8	3.00	30.6	92.9	3.99	3.6
	9.0	2.6	5.9	1330	56.4	33.0	2.19	63.7	25.8	-	1430	42.0	3.18	31.4	97.2	3.87	4.0
	9.0	2.6	5.9	1550	57.7	36.2	2.28	65.5	25.3	-	1650	42.8	3.05	32.4	94.0	4.11	3.5
	12.0	4.4	10.1	1330	56.6	33.0	2.12	63.8	26.7	-	1430	43.1	3.21	32.4	97.9	3.94	4.0
	12.0	4.4	10.1	1550	58.0	36.2	2.22	65.5	26.2	-	1650	43.9	3.08	33.5	94.6	4.19	3.4
50	6.0	1.1	2.5	1330	54.5	33.3	2.51	63.0	21.8	1.9	1430	45.5	3.27	34.6	99.5	4.08	4.3
	6.0	1.1	2.5	1550	55.8	36.5	2.62	64.7	21.3	2.0	1650	46.4	3.14	35.8	96.0	4.34	3.7
	9.0	2.5	5.7	1330	55.6	33.4	2.37	63.6	23.5	1.7	1430	48.0	3.33	36.8	101.1	4.22	4.2
	9.0	2.5	5.7	1550	56.9	36.6	2.47	65.4	23.0	1.7	1650	48.9	3.20	38.0	97.4	4.48	3.7
	12.0	4.2	9.6	1330	56.0	33.4	2.30	63.7	24.3	1.4	1430	49.3	3.37	38.0	101.9	4.29	4.2
	12.0	4.2	9.6	1550	57.3	36.7	2.40	65.5	23.8	1.4	1650	50.3	3.23	39.3	98.2	4.56	3.6
60	6.0	1.0	2.3	1330	52.5	32.6	2.71	61.7	19.4	2.8	1430	51.3	3.42	39.8	103.2	4.39	4.6
	6.0	1.0	2.3	1550	53.7	35.8	2.82	63.4	19.0	2.8	1650	52.3	3.28	41.1	99.3	4.67	4.0
	9.0	2.4	5.5	1330	54.0	33.1	2.56	62.7	21.1	2.4	1430	54.2	3.50	42.4	105.1	4.54	4.5
	9.0	2.4	5.5	1550	55.3	36.3	2.67	64.4	20.7	2.4	1650	55.2	3.36	43.8	101.0	4.83	3.9
	12.0	4.0	9.2	1330	54.7	33.3	2.49	63.1	21.9	2.0	1430	55.8	3.54	43.8	106.1	4.62	4.5
	12.0	4.0	9.2	1550	55.9	36.5	2.60	64.8	21.5	2.0	1650	56.9	3.40	45.3	101.9	4.91	3.9
70	6.0	0.9	2.2	1330	49.9	31.7	2.93	59.9	17.1	3.8	1430	57.2	3.58	45.1	107.1	4.68	5.0
	6.0	0.9	2.2	1550	51.1	34.8	3.06	61.6	16.7	3.8	1650	58.3	3.44	46.6	102.7	4.98	4.4
	9.0	2.3	5.4	1330	51.7	32.4	2.77	61.1	18.7	3.2	1430	60.6	3.68	48.1	109.2	4.83	5.0
	9.0	2.3	5.4	1550	53.0	35.5	2.89	62.8	18.3	3.3	1650	61.7	3.53	49.7	104.7	5.13	4.3
	12.0	3.8	8.8	1330	52.6	32.7	2.70	61.7	19.5	2.7	1430	62.4	3.73	49.7	110.4	4.90	4.9
	12.0	3.8	8.8	1550	53.8	35.8	2.81	63.4	19.1	2.7	1650	63.6	3.58	51.4	105.7	5.21	4.2
80	6.0	0.9	2.1	1330	47.1	30.6	3.18	57.9	14.8	4.9	1430	63.2	3.76	50.5	111.0	4.94	5.6
	6.0	0.9	2.1	1550	48.2	33.6	3.32	59.5	14.5	5.0	1650	64.5	3.60	52.2	106.2	5.25	4.8
	9.0	2.3	5.2	1330	49.0	31.4	3.01	59.3	16.3	4.2	1430	67.0	3.87	53.8	113.4	5.07	5.5
	9.0	2.3	5.2	1550	50.2	34.4	3.14	60.9	16.0	4.3	1650	68.3	3.71	55.7	108.3	5.39	4.8
	12.0	3.6	8.3	1330	50.0	31.7	2.92	59.9	17.1	3.5	1430	69.1	3.94	55.7	114.8	5.14	5.4
	12.0	3.6	8.3	1550	51.2	34.8	3.05	61.6	16.8	3.5	1650	70.5	3.78	57.6	109.5	5.47	4.7
85	6.0	0.9	2.0	1330	45.5	30.1	3.32	56.9	13.7	5.6	1430	66.3	3.85	53.2	112.9	5.04	5.9
	6.0	0.9	2.0	1550	46.6	33.0	3.47	58.5	13.4	5.7	1650	67.6	3.69	55.0	107.9	5.36	5.1
	9.0	2.2	5.1	1330	47.5	30.8	3.14	58.2	15.1	4.8	1430	70.3	3.98	56.7	115.5	5.18	5.8
	9.0	2.2	5.1	1550	48.6	33.8	3.28	59.8	14.8	4.9	1650	71.6	3.82	58.6	110.2	5.50	5.1
	12.0	3.6	8.2	1330	48.5	31.2	3.05	58.9	15.9	3.9	1430	72.5	4.06	58.6	116.9	5.24	5.7
	12.0	3.6	8.2	1550	49.6	34.2	3.19	60.5	15.6	4.0	1650	73.9	3.89	60.6	111.5	5.57	5.0
90	6.0	0.9	2.0	1330	44.0	29.5	3.47	55.9	12.7	6.2	1430	69.3	3.95	55.8	114.9	5.15	6.2
	6.0	0.9	2.0	1550	45.0	32.4	3.62	57.4	12.4	6.3	1650	70.7	3.78	57.7	109.7	5.47	5.4
	9.0	2.2	5.0	1330	46.0	30.2	3.28	57.2	14.0	5.3	1430	73.5	4.09	59.5	117.6	5.27	6.1
	9.0	2.2	5.0	1550	47.1	33.2	3.42	58.8	13.8	5.4	1650	74.9	3.92	61.6	112.1	5.60	5.3
	12.0	3.5	8.1	1330	47.0	30.6	3.18	57.9	14.8	4.3	1430	75.8	4.17	61.5	119.1	5.33	6.0
	12.0	3.5	8.1	1550	48.1	33.6	3.32	59.5	14.5	4.4	1650	77.3	4.00	63.6	113.4	5.66	5.2
100	6.0	0.8	1.9	1330	40.9	28.5	3.80	53.9	10.7	7.6	Operation not recommended						
	6.0	0.8	1.9	1550	41.8	31.2	3.97	55.4	10.5	7.6							
	9.0	2.1	4.8	1330	42.8	29.1	3.59	55.1	11.9	6.5							
	9.0	2.1	4.8	1550	43.8	31.9	3.74	56.6	11.7	6.6							
	12.0	3.3	7.7	1330	43.8	29.4	3.48	55.8	12.6	5.3							
	12.0	3.3	7.7	1550	44.9	32.3	3.64	57.3	12.3	5.4							
110	6.0	0.8	1.8	1330	37.8	27.6	4.19	52.2	9.0	9.2							
	6.0	0.8	1.8	1550	38.7	30.2	4.38	53.7	8.8	9.4							
	9.0	2.0	4.7	1330	39.6	28.1	3.95	53.2	10.0	7.8							
	9.0	2.0	4.7	1550	40.6	30.8	4.12	54.7	9.8	7.9							
	12.0	3.2	7.3	1330	40.6	28.4	3.83	53.7	10.6	6.4							
	12.0	3.2	7.3	1550	41.5	31.1	4.00	55.2	10.4	6.5							
120	6.0	0.8	1.7	1330	34.9	27.0	4.65	51.0	7.5	10.9							
	6.0	0.8	1.7	1550	35.8	29.6	4.86	52.4	7.4	11.1							
	9.0	2.0	4.5	1330	36.6	27.3	4.37	51.6	8.4	9.3							
	9.0	2.0	4.5	1550	37.4	29.9	4.57	53.1	8.2	9.4							
	12.0	3.0	7.0	1330	37.4	27.5	4.24	52.0	8.8	7.6							
	12.0	3.0	7.0	1550	38.3	30.1	4.43	53.5	8.7	7.7							

Figure 2.16. ClimateMaster Tranquility 27™ Model 049 – Full Load

Example 2.11 An example of using the performance data to determine full-load heating mode performance is highlighted with solid outlines on Figure 2.16 for EWT=40 F, GPM=12, CFM=1,650, and EAT=70 F. For those operating parameters:

HC = 43.9 MBtu/hr x 1,000 Btu/MBtu = 43,900 Btu/hr	Column 13
DMD = 3.08 kW	Column 14
COP = 4.19	Column 17
HE = 33.5 MBtu/hr x 1,000 = 33,500 Btu/hr	Column 15
LAT = 94.6 F	Column 16

The same calculations that were performed on single-capacity heating performance data to determine HE (Equation 2.3), LAT_h (Equation 2.7), and LWT_h (Equation 2.8) apply here for full-load operation. The difference in heating mode performance between the single-capacity and the two-capacity nominal 4-ton GSHPs in the heating mode is of interest. For EWT=40 F, GPM=12 and EAT=70 F, with the single-capacity unit at CFM=1,600 and the two-capacity unit at CFM=1,650, the full-load heating capacity of the two-capacity unit is lower than the single-capacity unit. However, the demand is lower by a larger percentage, resulting in a COP that is slightly higher for the two-capacity unit. This is the case for the entire range of EWTs for this particular model.

Part-load heating mode performance data are listed in Columns 12 through 18 in Figure 2.17, with each line of data representing performance for a combination of EWT, GPM, and CFM as read proceeding from left to right across the table. Column 13 contains the heating capacity (HC - MBtu/hr) provided to the air returning from the conditioned space, Column 14 lists the electric demand (DMD - kW) of the unit, and Column 17 provides the COP. The heat of extraction (HE - MBtu/hr) from the ground connection is provided in Column 15 and the leaving air temperature back to the conditioned space (LAT - F) is shown in Column 16. The heating capacity of the optional desuperheater (HWC - MBtu/hr) is available in Column 18. Most manufacturers express capacities in MBtu/hr (or MBH) to save space (1 MBtu/hr = 1,000 Btu/hr).

Example 2.12 An example of using the performance data to determine part-load heating mode performance is highlighted with solid outlines on Figure 2.17 for EWT=40 F, GPM=11, CFM=1,400, and EAT=70 F. For those operating parameters:

HC = 31.9 MBtu/hr x 1,000 Btu/MBtu = 31,900 Btu/hr	Column 13
DMD = 2.19 kW	Column 14
COP = 4.27	Column 17
HE = 24.8 MBtu/hr x 1,000 = 24,800 Btu/hr	Column 15
LAT = 91.1F	Column 16

Again, the same calculations that were performed on single-capacity performance data to determine HE (Equation 2.3), LAT_h (Equation 2.7), and LWT_h (Equation 2.8) apply here for part-load operation. The difference in heating mode performance between full-load and part-load operation in the heating mode is of more interest. For EWT=40 F, GPM=12, EAT=70 F, and CFM=1,650 the full-load HC was 43,900 Btu/hr. At the same EWT and EAT and for GPM=11 and CFM=1,400 the part-load HC was 31,900 Btu/hr, which is 72.7% of the full-load capacity. However, the part-load DMD is 71.1% of the full-load DMD, resulting in a COP that is slightly higher under part-load operation.

The differences between full-load (GPM=12, CFM=1,650, and EAT=70 F) and part-load (GPM=11, CFM=1,400, and EAT=70 F) operation can be seen clearly in Figure 2.18, which compares both the heating capacities and the COPs for each capacity as a function of EWT. Part-load HCs are always between 70 to 74% of the full-load HCs, but full-load COPs switch from being lower than part-load COPs at higher EWTs to being slightly higher at low EWTs. This is generally not an issue because closed-loop system EWTs to the heat pump will reach their lowest values only when the heat pump is placing the highest demand (HE) on it or at full-load heating mode of operation. HC and COP have been identified for an EWT of 32 F at full-load conditions and an EWT of 41 F at part-load conditions in Figure 2.18, which are the entering water temperatures that ARI/ISO 13256-1 uses as the ground loop test condition (discussed in Section 2.4.4) for heating mode certification for two-capacity GSHPs. Although this is a nominal 4-ton GSHP, the full-load HC is just under 39,000 Btu/hr at EWT=32 F and varies with EWT such that an EWT of about 47 F would be required to achieve a HC of 48,000 Btu/hr, which is considered 4 tons of heating capacity. Full-load heating capacity and its relation to EWT relates directly to GSHP selection to meet the design heating load and to the design of the ground connection. The influence of design EWT on two-capacity GSHP sizing and operating efficiency will be addressed in the selection of GSHP equipment to meet design heating loads (Chapter 3) and the design of the various ground coupling devices (Chapter 5).

Column

1300 CFM Nominal (Rated) Airflow Cooling, 1400 CFM Nominal (Rated) Airflow Heating

Performance capacities shown in thousands of Btuh

1	2	3	4	5	6	7	8	9	10	11	12	13	14	15	16	17	18
EWT °F	GPM	WPD		Cooling - EAT 80/67°F							Heating - EAT 70°F						
		PSI	FT	Airflow CFM	TC	SC	kW	HR	EER	HWC	Airflow CFM	HC	kW	HE	LAT	COP	HWC
20	11.0	4.0	9.3	Operation not recommended							1200	23.2	2.16	16.2	87.9	3.14	3.3
	11.0	4.0	9.3								1400	23.5	2.10	16.6	85.5	3.27	2.8
30	5.5	1.1	2.5	1120	38.6	24.0	1.20	42.6	32.0	-	1200	25.6	2.20	18.6	89.8	3.42	3.4
	5.5	1.1	2.5	1300	39.1	25.2	1.22	43.3	32.0	-	1400	25.9	2.14	18.9	87.2	3.56	2.9
	8.3	2.3	5.2	1120	38.8	24.0	1.14	42.6	34.2	-	1200	26.6	2.21	19.4	90.5	3.52	3.3
	8.3	2.3	5.2	1300	39.4	25.3	1.15	43.2	34.2	-	1400	26.9	2.15	19.8	87.8	3.67	2.9
	11.0	3.9	8.9	1120	39.0	24.0	1.10	42.7	35.4	-	1200	27.1	2.22	19.9	90.9	3.58	3.3
	11.0	3.9	8.9	1300	39.6	25.2	1.12	43.3	35.4	-	1400	27.4	2.15	20.3	88.1	3.73	2.9
40	5.5	1.0	2.3	1120	40.8	26.2	1.37	45.5	29.9	-	1200	29.6	2.24	22.4	92.9	3.88	3.5
	5.5	1.0	2.3	1300	41.4	27.6	1.38	46.1	29.9	-	1400	30.0	2.18	22.9	89.8	4.04	3.0
	8.3	2.2	5.0	1120	41.2	26.3	1.28	45.5	32.1	-	1200	30.9	2.25	23.6	93.8	4.02	3.4
	8.3	2.2	5.0	1300	41.8	27.6	1.30	46.2	32.1	-	1400	31.2	2.19	24.1	90.7	4.19	3.0
	11.0	3.7	8.6	1120	41.3	26.3	1.25	45.5	32.9	-	1200	31.6	2.26	24.3	94.4	4.10	3.4
	11.0	3.7	8.6	1300	41.9	27.6	1.27	46.1	32.9	-	1400	31.9	2.19	24.8	91.1	4.27	2.9
50	5.5	0.9	2.1	1120	40.8	26.8	1.49	45.9	27.3	0.9	1200	34.0	2.28	26.7	96.3	4.38	3.6
	5.5	0.9	2.1	1300	41.4	28.2	1.51	46.5	27.3	0.9	1400	34.4	2.21	27.2	92.8	4.56	3.1
	8.3	2.1	4.9	1120	41.2	26.9	1.41	45.9	29.1	0.8	1200	35.6	2.29	28.2	97.5	4.56	3.5
	8.3	2.1	4.9	1300	41.7	28.2	1.43	46.6	29.1	0.9	1400	36.0	2.23	28.8	93.8	4.74	3.1
	11.0	3.6	8.3	1120	41.3	26.9	1.40	46.0	29.4	0.8	1200	36.5	2.30	29.1	98.2	4.65	3.5
	11.0	3.6	8.3	1300	41.9	28.3	1.42	46.7	29.4	0.8	1400	36.9	2.23	29.6	94.4	4.85	3.0
60	5.5	0.8	2.0	1120	40.0	27.1	1.73	45.9	23.2	1.6	1200	38.7	2.32	31.2	99.8	4.89	3.7
	5.5	0.8	2.0	1300	40.6	28.5	1.75	46.5	23.2	1.6	1400	39.1	2.25	31.8	95.9	5.09	3.3
	8.3	2.0	4.7	1120	40.7	27.2	1.62	46.2	25.1	1.4	1200	40.6	2.33	33.0	101.3	5.10	3.7
	8.3	2.0	4.7	1300	41.3	28.6	1.64	46.9	25.1	1.5	1400	41.1	2.27	33.7	97.2	5.31	3.2
	11.0	3.5	8.1	1120	41.0	27.2	1.57	46.3	26.1	1.3	1200	41.7	2.34	34.1	102.1	5.21	3.7
	11.0	3.5	8.1	1300	41.5	28.7	1.59	46.9	26.1	1.3	1400	42.1	2.27	34.7	97.9	5.43	3.2
70	5.5	0.8	1.8	1120	38.2	26.5	1.94	44.8	19.7	2.2	1200	43.4	2.36	35.7	103.5	5.39	3.9
	5.5	0.8	1.8	1300	38.8	27.9	1.97	45.4	19.7	2.2	1400	43.9	2.29	36.4	99.0	5.62	3.4
	8.3	2.0	4.6	1120	39.3	26.9	1.82	45.5	21.6	2.0	1200	45.6	2.38	37.9	105.2	5.62	3.9
	8.3	2.0	4.6	1300	39.9	28.3	1.85	46.1	21.6	2.0	1400	46.1	2.31	38.6	100.5	5.85	3.4
	11.0	3.3	7.5	1120	39.8	27.1	1.76	45.7	22.6	1.8	1200	46.8	2.39	39.0	106.1	5.73	3.9
	11.0	3.3	7.5	1300	40.3	28.4	1.79	46.4	22.6	1.8	1400	47.3	2.32	39.7	101.3	5.97	3.4
80	5.5	0.7	1.7	1120	35.9	25.7	2.17	43.3	16.5	2.8	1200	48.0	2.40	40.2	107.1	5.86	4.2
	5.5	0.7	1.7	1300	36.5	27.0	2.20	43.9	16.5	2.8	1400	48.6	2.33	41.0	102.1	6.10	3.6
	8.3	1.9	4.5	1120	37.2	26.2	2.04	44.2	18.2	2.5	1200	50.4	2.43	42.5	108.9	6.07	4.1
	8.3	1.9	4.5	1300	37.7	27.5	2.07	44.8	18.2	2.6	1400	50.9	2.36	43.3	103.7	6.33	3.6
	11.0	3.2	7.3	1120	37.8	26.4	1.98	44.6	19.1	2.3	1200	51.6	2.45	43.7	109.8	6.18	4.1
	11.0	3.2	7.3	1300	38.4	27.8	2.01	45.2	19.1	2.3	1400	52.2	2.38	44.5	104.5	6.44	3.6
85	5.5	0.7	1.7	1120	34.7	25.2	2.30	42.5	15.0	3.1	1200	50.2	2.43	42.3	108.8	6.06	4.4
	5.5	0.7	1.7	1300	35.2	26.5	2.34	43.1	15.0	3.1	1400	50.8	2.36	43.1	103.6	6.31	3.8
	8.3	1.9	4.3	1120	36.0	25.7	2.17	43.4	16.6	2.8	1200	52.6	2.46	44.6	110.6	6.26	4.3
	8.3	1.9	4.3	1300	36.5	27.0	2.20	44.0	16.6	2.9	1400	53.2	2.39	45.4	105.2	6.52	3.8
	11.0	3.1	7.3	1120	36.6	25.9	2.10	43.8	17.4	2.5	1200	53.8	2.48	45.7	111.5	6.35	4.3
	11.0	3.1	7.3	1300	37.1	27.3	2.13	44.4	17.4	2.6	1400	54.4	2.41	46.6	106.0	6.62	3.7
90	5.5	0.7	1.6	1120	33.4	24.7	2.44	41.7	13.7	3.3	1200	52.4	2.46	44.4	110.5	6.25	4.5
	5.5	0.7	1.6	1300	33.9	25.9	2.47	42.3	13.7	3.4	1400	53.0	2.39	45.3	105.1	6.51	3.9
	8.3	1.8	4.2	1120	34.7	25.2	2.30	42.6	15.1	3.0	1200	54.8	2.50	46.7	112.3	6.44	4.4
	8.3	1.8	4.2	1300	35.2	26.5	2.33	43.2	15.1	3.1	1400	55.4	2.42	47.6	106.7	6.70	3.9
	11.0	3.1	7.2	1120	35.4	25.5	2.23	43.0	15.9	2.7	1200	56.0	2.52	47.8	113.2	6.52	4.4
	11.0	3.1	7.2	1300	35.9	26.8	2.26	43.6	15.9	2.8	1400	56.7	2.45	48.8	107.5	6.79	3.8
100	5.5	0.7	1.5	1120	30.9	23.7	2.73	40.2	11.3	3.8	Operation not recommended						
	5.5	0.7	1.5	1300	31.3	24.9	2.77	40.7	11.3	3.9							
	8.3	1.8	4.1	1120	32.1	24.2	2.58	40.9	12.5	3.5							
	8.3	1.8	4.1	1300	32.6	25.4	2.62	41.5	12.5	3.5							
	11.0	3.0	6.8	1120	32.8	24.4	2.50	41.3	13.1	3.1							
	11.0	3.0	6.8	1300	33.3	25.7	2.54	41.9	13.1	3.2							
110	5.5	0.6	1.5	1120	28.5	22.9	3.07	39.0	9.3	4.3							
	5.5	0.6	1.5	1300	28.9	24.1	3.11	39.6	9.3	4.3							
	8.3	1.7	4.0	1120	29.6	23.3	2.90	39.5	10.2	3.9							
	8.3	1.7	4.0	1300	30.0	24.5	2.94	40.1	10.2	3.9							
	11.0	2.8	6.6	1120	30.2	23.5	2.82	39.8	10.7	3.5							
	11.0	2.8	6.6	1300	30.6	24.7	2.86	40.4	10.7	3.5							
120	5.5	0.6	1.4	1120	26.7	22.7	3.45	38.4	7.7	4.7							
	5.5	0.6	1.4	1300	27.0	23.8	3.50	39.0	7.7	4.8							
	8.3	1.7	3.8	1120	27.5	22.7	3.26	38.6	8.4	4.2							
	8.3	1.7	3.8	1300	27.9	23.9	3.31	39.2	8.4	4.3							
	11.0	2.7	6.3	1120	27.9	22.8	3.17	38.8	8.8	3.8							
	11.0	2.7	6.3	1300	28.3	24.0	3.22	39.3	8.8	3.9							

Figure 2.17. ClimateMaster Tranquility 27™ Model 049 – Part Load

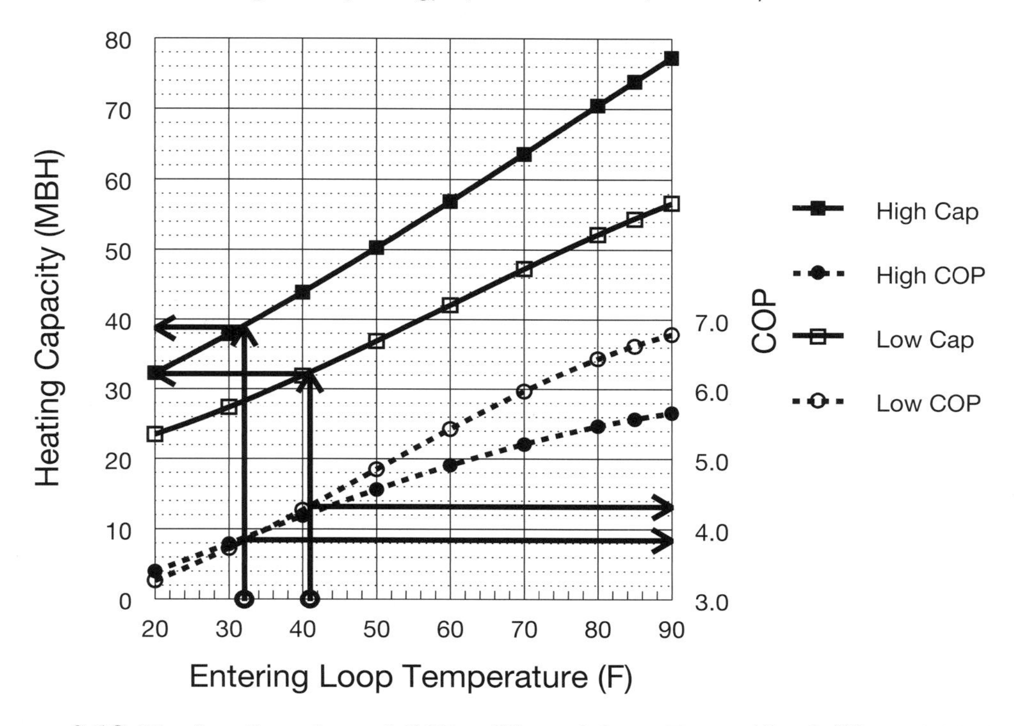

Figure 2.18. Heating Capacity and COP – ClimateMaster Tranquility 27™ Model 049

2.5.2 COOLING MODE PERFORMANCE DATA Full-load cooling mode performance data are listed in Columns 5 through 11 in Figure 2.16, and each line of data represents performance for a combination of EWT, GPM, and CFM as read proceeding from left to right across the table. Column 6 contains the total cooling capacity (TC - MBtu/hr) provided to the air returning from the conditioned space, and Column 7 lists the associated sensible cooling capacity (SC - Mbtu/hr). Column 8 lists the electric demand (DMD - kW) of the unit, and Column 10 provides the EER. The heat of rejection (HR - MBtu/hr) to the ground connection is provided in Column 9. The heating capacity of the optional desuperheater (HWC - MBtu/hr) is available in Column 11. Most manufacturers express capacities in MBtu/hr (or MBH) to save space (1 MBtu/hr = 1,000 Btu/hr).

Example 2.13 An example of using the performance data to determine full-load cooling mode performance is highlighted with dashed outlines on Figure 2.16 for EWT=90 F, GPM=12, CFM=1,550, and EAT=80 F db/67 F wb. For those operating parameters:

TC = 48.1 MBtu/hr x 1,000 Btu/MBtu = 48,100 Btu/hr	Column 6
SC = 33.6 MBtu/hr x 1,000 = 33,600 Btu/hr	Column 7
DMD = 3.32 kW	Column 8
EER = 14.5	Column 10
HR = 59.5 MBtu/hr x 1,000 = 59,500 Btu/hr	Column 9

The same calculations that were performed on single-capacity performance data to determine HR (Equation 2.6), LAT_c (Equation 2.10), and LWT_c (Equation 2.11) apply here for full-load operation. The difference in cooling mode performance between the single-capacity and the two-capacity nominal 4-ton GSHPs in the cooling mode is of interest. For EWT=90 F, GPM=12, and EAT=80 F db/67 F wb, with the single-capacity unit at CFM=1,600 and the two-capacity unit at CFM=1,550, the full-load cooling capacity of the two-capacity unit is higher than the single-capacity unit for all but the highest EWTs. However, the demand is always lower, resulting in EERs that are always slightly larger for the two-capacity unit.

Part-load cooling mode performance data are listed in columns 5 through 11 in Figure 2.17, and each line of data represents performance for a combination of EWT, GPM, and CFM as read proceeding from left to right across the table. Column 6 contains the total cooling capacity (TC - MBtu/hr) provided to the air returning from the conditioned space, and Column 7 lists the associated sensible cooling capacity (SC - Mbtu/hr). Column 8 lists the electric demand (DMD - kW) of the unit, and Column 10 provides the EER. The heat of rejection (HR - MBtu/hr) to the ground connection is provided in Column 9. For an optional desuperheater, the heating capacity of the desuperheater (HWC - MBtu/hr) is available in Column 11. Most manufacturers express capacities in MBtu/hr (or MBH) to save space (1 MBtu/hr = 1,000 Btu/hr).

Example 2.14 An example of using the performance data to determine part-load cooling mode performance is highlighted with dashed outlines on Figure 2.17 for EWT=90 F, GPM=11, CFM=1,300, and EAT=80 F db/67 F wb. For those operating parameters:

TC = 35.9 MBtu/hr x 1,000 Btu/MBtu = 35,900 Btu/hr	Column 6
SC = 26.8 MBtu/hr x 1,000 = 26,800 Btu/hr	Column 7
DMD = 2.26 kW	Column 8
EER = 15.9	Column 10
HR = 43.6 MBtu/hr x 1,000 = 43,600 Btu/hr	Column 9

The same calculations that were performed on single-capacity performance data to determine HR (Equation 2.6), LAT_c (Equation 2.10), and LWT_c (Equation 2.11) apply here for part-load operation. The difference in cooling mode performance between the single-capacity and the two-capacity nominal 4-ton GSHPs in the cooling mode is of more interest. For EWT=90 F, GPM=12, and EAT=80 F db/67 F wb, with the single-capacity unit at CFM=1,600 and the two-capacity unit at CFM=1,550, the full-load cooling capacity of the two-capacity unit is higher than the single-capacity unit. However, the demand is lower, resulting in an EER that is slightly higher for the two-capacity unit.

The differences between full-load (GPM=12, CFM=1,550, and EAT=80 F db/67 F wb) and part-load (GPM=11, CFM=1,300, and EAT=80 F db/67 F wb) operation can be seen clearly in Figure 2.19, which compares both the cooling capacities and the EERs for each capacity as a function of EWT. (Sensible and latent cooling capacities are not presented in Figure 2.19 to avoid cluttering the graph with too much information, but SC and LC cooling capacities will vary in a similar manner to single-capacity equipment that was discussed in Section 2.4.2.) Part-load TCs are always between 70 to 75% of the full-load TCs, but full-load EERs trend from being lower than part-load EERs at lower EWTs to being about equal at high EWTs. This is generally not an issue, though, because closed-loop system EWTs to the heat pump will reach their highest values only when the heat pump is placing the highest demand (HR) on it or at full-load cooling mode of operation. TC and EER have been identified for an EWT of 77 F at full-load conditions and an

EWT of 68 F at part-load conditions in Figure 2.19, which are the entering water temperatures that ARI/ ISO 13256-1 uses as the ground loop test condition (discussed in Section 2.4.4) for cooling mode certification for two-capacity GSHPs. This is a nominal 4-ton GSHP, and the full-load TC is about 52,000 Btu/hr at EWT=77 F, which is larger than the required 48,000 Btu/hr that is considered 4 tons of heating capacity. GSHPs are rated based on cooling mode performance at ARI/ISO conditions, thus the 4-ton rating for this unit. However, cooling performance varies considerably with EWT, which must be specified to determine the actual TC, SC, LC, and EER of the GSHP. As EWT is increased (as would be the case in southern climates with high ground temperatures and high cooling loads), it may not be possible to economically install a ground connection that provides EWTs lower than 95 F or more at design cooling conditions, having direct consequence on the sizing of the GSHP equipment relative to the total, sensible, and latent cooling design loads. The influence of design EWT on GSHP sizing and operating efficiency will be addressed in the selection of GSHP equipment to meet design cooling loads (Chapter 3) and the design of the various ground coupling devices (Chapter 5).

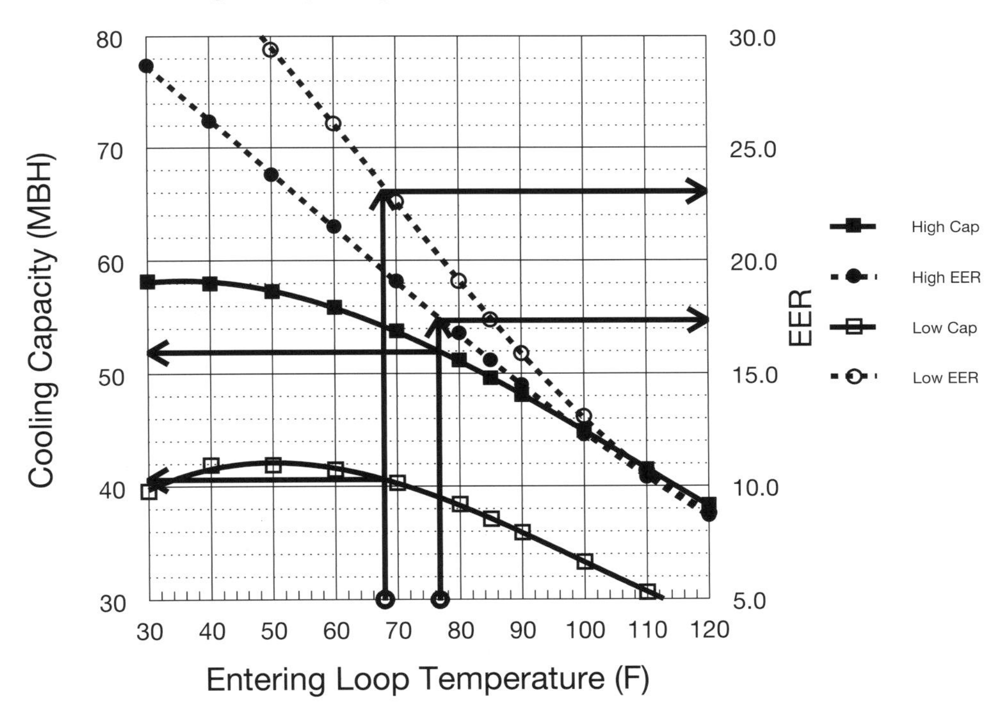

Figure 2.19. Cooling Capacity and EER – ClimateMaster Tranquility 27™ Model 049

2.5.3 PERFORMANCE DATA CORRECTION FACTORS Heat pump performance is generally expressed as a function of EWT, GPM, EAT, and CFM, where the heating and cooling performance data supplied by the manufacturer covers a range of EWTs and GPMs, but may be tabled for a single EAT condition (70 F in heating and 80 F db/67 F wb in cooling) and CFM (350 to 400 CFM per nominal ton). To account for the effects of EAT and CFM (along with antifreeze in the ground loop circulating fluid) on the heating capacity (HC), electric demand (DMD), heat of extraction (HE), coefficient of performance (COP), leaving air temperature (LAT_h), and leaving water temperature (LWT_h), Equations 2.12 through 2.17 can be applied for the heating mode. To account for the effects of EAT and CFM (along with antifreeze in the ground loop circulating fluid) on the total cooling capacity (TC), electric demand (DMD), heat of rejection (HR), energy efficiency ratio (EER), leaving air temperature (LAT_c), and leaving water temperature (LWT_c), Equations 2.18 through 2.24 can be applied for the cooling mode. The various performance parameters (HC, DMD, and HE in heating mode and TC, SC, DMD, and HR in cooling mode) are corrected (parameters subscripted by cor) by multiplying each by correction factors that account for EAT, CFM, and antifreeze (AF) conditions that are not accounted for in the tabled data. The correction factors are obtained from the manufacturer in tabled format and are generally listed separately for heating mode or cooling mode of operation. Separate tables may be provided for part-load and full-load heating and cooling modes of operation.

2.5.3.1 ENTERING AIR TEMPERATURE (EAT) CORRECTION FACTORS Heat pump performance is affected by EAT conditions returning from the conditioned space, but performance data supplied by manufacturers is generally for a single EAT condition (70 F in heating and 80 F db/67 F wb in cooling). Correction factors for heating mode EAT conditions are provided in tabular format for the ClimateMaster Tranquility 27™ Series at full-load operation in the top table of Figure 2.20 and at part-load operation in the top table of Figure 2.21 (Heating Capacity ($CF_{HC\text{-}EAT}$), Power ($CF_{DMD\text{-}EAT}$), and Heat of Extraction ($CF_{HE\text{-}EAT}$). The EAT correction factors are applied to tabled performance data according to Equations 2.12 through 2.14 (which also include correction factors for CFM and antifreeze) to determine HC_{cor}, DMD_{cor}, and HE_{cor}. Then, COP_{cor}, $LAT_{h\text{-}cor}$, and $LWT_{h\text{-}cor}$ can be determined utilizing Equations 2.15 through 2.17, which are based on the HC_{cor}, DMD_{cor}, and HE_{cor}. Looking at Figures 2.20 and 2.21, the effect of higher EAT in the heating mode is to decrease HC, increase DMD, and reduce HE, reducing the COP of the heat pump and resulting in a slightly lower LAT_h than would be expected for the given EAT. A lower EAT results in an increase in HC, a decrease in DMD, and an increase in HE, increasing the COP of the heat pump and resulting in a slightly higher LAT_h than would be expected for the given EAT. The Heating Correction Factors in Figures 2.20 and 2.21 are applied as shown in Examples 2.15 and 2.16 for full-load and part-load heating mode, respectively. Some manufacturers provide only one set of correction factors that are applied to both full-load and part-load performance data.

Heating			
Entering Air DB°F	Heating Capacity	Power	Heat of Extraction
40	1.052	0.779	1.120
45	1.043	0.808	1.102
50	1.035	0.841	1.084
55	1.027	0.877	1.065
60	1.019	0.915	1.045
65	1.010	0.957	1.023
68	1.004	0.982	1.010
70	1.000	1.000	1.000
75	0.989	1.045	0.974
80	0.976	1.093	0.946

* = Sensible capacity equals total capacity
ARI/ISO/ASHRAE 13256-1 uses entering air conditions of Cooling - 80.6°F DB/66.2°F WB, and Heating - 68°F DB/59°F WB entering air temperature

Cooling													
Entering Air WB°F	Total Capacity	Sensible Cooling Capacity Multiplier - Entering DB °F										Power	Heat of Rejection
		60	65	70	75	80	80.6	85	90	95	100		
45	0.832	1.346	1.461	1.603	*	*	*	*	*	*	*	0.946	0.853
50	0.850	1.004	1.174	1.357	*	*	*	*	*	*	*	0.953	0.870
55	0.880	0.694	0.902	1.115	1.331	*	*	*	*	*	*	0.964	0.896
60	0.922		0.646	0.875	1.103	1.329	1.356	*	*	*	*	0.977	0.932
65	0.975			0.639	0.869	1.096	1.123	1.320	*	*	*	0.993	0.979
66.2	0.990			0.582	0.812	1.039	1.066	1.262	1.482	*	*	0.997	0.991
67	1.000			0.545	0.774	1.000	1.027	1.223	1.444	*	*	1.000	1.000
70	1.040				0.630	0.853	0.880	1.075	1.297	1.517	*	1.011	1.035
75	1.117					0.601	0.627	0.821	1.046	1.275	1.510	1.033	1.101

Figure 2.20. Entering Air Condition Correction Factors - ClimateMaster Tranquility 27™ Series - Full Load

Example 2.15 Determine the corrected HC, DMD, HE, COP, and LAT for the data in Example 2.11 for an EAT of 55 F.

Based on Example 2.11, the full-load heating mode performance data for EWT=40 F, GPM=12, CFM=1,650, and EAT=70 F were determined from Figure 2.16 to be:

HC = 43.9 MBtu/hr x 1,000 Btu/MBtu = 43,900 Btu/hr
DMD = 3.08 kW
COP = 4.19
HE = 33.5 MBtu/hr x 1,000 = 33,500 Btu/hr
LAT = 94.6 F

From Figure 2.20, full-load heating mode correction factors (identified with solid outlines) for an EAT of 55 F are $CF_{HC\text{-}EAT}$=1.027, $CF_{DMD\text{-}EAT}$=0.877, and $CF_{HE\text{-}EAT}$=1.065, which are then applied according to Equations 2.12 through 2.17 to provide corrected performance data for the heat pump in heating mode.

$$HC_{cor} = HC \bullet CF_{HC\text{-}EAT} \bullet CF_{HC\text{-}CFM} \bullet CF_{HC\text{-}AF} = 43{,}900 \text{ Btu/hr} \bullet 1.027 \bullet 1 \bullet 1 = 45{,}085 \text{ Btu/hr} \quad (+\,2.70\%)$$

$$DMD_{cor} = DMD \bullet CF_{DMD\text{-}EAT} \bullet CF_{DMD\text{-}CFM} \bullet CF_{DMD\text{-}AF} = 3.08 \text{ kW} \bullet 0.877 \bullet 1 \bullet 1 = 2.70 \text{ kW} \quad (-12.3\%)$$

$$HE_{cor} = HE \bullet CF_{HE\text{-}EAT} \bullet CF_{HE\text{-}CFM} \bullet CF_{HE\text{-}AF} = 33{,}500 \text{ Btu/hr} \bullet 1.065 \bullet 1 \bullet 1 = 35{,}678 \text{ Btu/hr} \quad (+\,6.50\%)$$

$$COP_{cor} = \frac{HC_{cor}}{DMD_{cor} \bullet \dfrac{3{,}412 \text{ Btu/hr}}{\text{kW}}} = \frac{45{,}085 \text{ Btu/hr}}{2.70 \text{ kW} \bullet \dfrac{3{,}412 \text{ Btu/hr}}{\text{kW}}} = 4.89 \quad (+16.7\%)$$

$$LAT_{h\text{-}cor} = EAT + \frac{HC_{cor}}{1.08 \bullet CFM} = 55 \text{ F} + \frac{45{,}085 \text{ Btu/hr}}{1.08 \bullet 1650 \text{ cfm}} = 80.3 \text{ F}$$

$$LWT_{h\text{-}cor} = EWT - \frac{HE_{cor}}{500 \bullet GPM} = 40 - \frac{35{,}678 \text{ Btu/hr}}{500 \bullet 12.0 \text{ gpm}} = 34.1 \text{ F}$$

Heating			
Entering Air DB°F	Heating Capacity	Power	Heat of Extraction
40	1.084	0.732	1.161
45	1.073	0.764	1.140
50	1.060	0.802	1.117
55	1.046	0.846	1.090
60	1.031	0.893	1.061
65	1.016	0.945	1.031
68	1.006	0.978	1.013
70	1.000	1.000	1.000
75	0.984	1.058	0.968
80	0.968	1.117	0.936

* = Sensible capacity equals total capacity
ARI/ISO/ASHRAE 13256-1 uses entering air conditions of Cooling - 80.6°F DB/66.2°F WB, and Heating - 68°F DB/59°F WB entering air temperature

Cooling													
Entering Air WB°F	Total Capacity	Sensible Cooling Capacity Multiplier - Entering DB °F										Power	Heat of Rejection
		60	65	70	75	80	80.6	85	90	95	100		
45	0.876	1.286	1.302	1.389	*	*	*	*	*	*	*	0.981	0.895
50	0.883	1.002	1.099	1.241	*	*	*	*	*	*	*	0.985	0.901
55	0.903	0.706	0.871	1.060	1.271	*	*	*	*	*	*	0.989	0.918
60	0.935		0.617	0.844	1.079	1.319	1.349	*	*	*	*	0.993	0.945
65	0.979			0.595	0.849	1.096	1.128	1.342	*	*	*	0.998	0.982
66.2	0.991			0.531	0.789	1.040	1.070	1.284	1.522	*	*	0.999	0.993
67	1.000			0.486	0.747	1.000	1.030	1.245	1.481	*	*	1.000	1.000
70	1.035				0.583	0.842	0.873	1.090	1.327	1.552	*	1.003	1.030
75	1.105					0.552	0.584	0.811	1.057	1.290	1.510	1.008	1.088

Figure 2.21. Entering Air Condition Correction Factors - ClimateMaster Tranquility 27™ Series - Part Load

Example 2.16 Determine the corrected HC, DMD, HE, COP, and LAT for the data in Example 2.12 for an EAT of 55 F.

Based on Example 2.12, the part-load heating mode performance data for EWT=40 F, GPM=11, CFM=1,400, and EAT=70 F were determined from Figure 2.17 to be:

HC = 31.9 MBtu/hr x 1,000 Btu/MBtu = 31,900 Btu/hr
DMD = 2.19 kW
COP = 4.27
HE = 24.8 MBtu/hr x 1,000 = 24,800 Btu/hr
LAT = 91.1F

From Figure 2.21, part-load heating mode correction factors (identified with solid outlines) for an EAT of 55 F are $CF_{HC\text{-}EAT}$=1.046, $CF_{DMD\text{-}EAT}$=0.846, and $CF_{HE\text{-}EAT}$=1.090, which are then applied according to Equations 2.12 through 2.17 to provide corrected performance data for the heat pump in heating mode.

$$HC_{cor} = HC \bullet CF_{HC-EAT} \bullet CF_{HC-CFM} \bullet CF_{HC-AF} = 31{,}900 \text{ Btu/hr} \bullet 1.046 \bullet 1 \bullet 1 = 33{,}367 \text{ Btu/hr} \quad (+\,4.60\%)$$

$$DMD_{cor} = DMD \bullet CF_{DMD-EAT} \bullet CF_{DMD-CFM} \bullet CF_{DMD-AF} = 2.19 \text{ kW} \bullet 0.846 \bullet 1 \bullet 1 = 1.85 \text{ kW} \quad (-15.5\%)$$

$$HE_{cor} = HE \bullet CF_{HE-EAT} \bullet CF_{HE-CFM} \bullet CF_{HE-AF} = 24{,}800 \text{ Btu/hr} \bullet 1.090 \bullet 1 \bullet 1 = 27{,}032 \text{ Btu/hr} \quad (+\,9.00\%)$$

$$COP_{cor} = \frac{HC_{cor}}{DMD_{cor} \bullet \dfrac{3{,}412 \text{ Btu/hr}}{\text{kW}}} = \frac{33{,}367 \text{ Btu/hr}}{1.85 \text{ kW} \bullet \dfrac{3{,}412 \text{ Btu/hr}}{\text{kW}}} = 5.29 \quad (+23.9\%)$$

$$LAT_{h-cor} = EAT + \frac{HC_{cor}}{1.08 \bullet CFM} = 55 \text{ F} + \frac{33{,}367 \text{ Btu/hr}}{1.08 \bullet 1400 \text{ cfm}} = 77.1 \text{ F}$$

$$LWT_{h-cor} = EWT - \frac{HE_{cor}}{500 \bullet GPM} = 40 - \frac{27{,}032 \text{ Btu/hr}}{500 \bullet 11.0 \text{ gpm}} = 35.1 \text{ F}$$

Correction factors for cooling mode EAT are provided for the ClimateMaster Tranquility 27™ Series at full-load operation in the bottom table of Figure 2.20 (Total Cooling Capacity ($CF_{TC\text{-}EAT}$), Sensible Cooling Capacity ($CF_{SC\text{-}EAT}$), Power ($CF_{DMD\text{-}EAT}$), and Heat of Rejection ($CF_{HR\text{-}EAT}$). The correction factors are applied to table performance data according to Equations 2.18 through 2.24 (which also include correction factors for CFM and antifreeze) to determine TC_{cor}, SC_{cor}, DMD_{cor}, and HR_{cor}, EER_{cor}, LAT_{cor}, and LWT_{cor}, respectively. In the cooling mode, the EAT condition includes both the dry bulb and the wet bulb temperature. Looking at Figure 2.20, TC_{cor}, DMD_{cor}, and HR_{cor} are a function of entering wet bulb temperature only, while SC_{cor} depends both on the entering wet bulb and dry bulb temperatures. TC_{cor}, DMD_{cor}, and HR_{cor} increase with increasing entering wet bulb temperature and decrease with decreasing entering wet bulb temperature. However, because TC_{cor} changes faster than DMD_{cor}, the EER_{cor} of the heat pump will increase at entering wet bulb temperatures above 67 F and will decrease at entering wet bulb temperatures below that value. At a selected entering wet bulb temperature, SC_{cor} increases with increasing dry bulb temperature indicating that there is less moisture in the air and more of the cooling across the air coil is sensible cooling. The dry bulb/

wet bulb temperature combinations that contain an asterisk have so little moisture available for condensation that the total cooling and sensible cooling capacities are equal and there is no latent cooling capacity. The Cooling Correction Factors in Figures 2.20 and 2.21 are applied as shown in Examples 2.17 and 2.18 for full-load and part-load cooling mode, respectively. Some manufacturers provide only one set of correction factors that are applied to both full-load and part-load performance data.

Example 2.17 Determine the corrected TC, SC, DMD, HR, EER, and LAT for the data in Example 2.13 for EAT conditions of 70 F db and 60 F wb.

Based on Example 2.13, the full-load cooling mode performance data for EWT=90 F, GPM=12, CFM=1,550, and EAT=80 F db/67 F wb were determined from Figure 2.16 to be:

TC = 48.1 MBtu/hr x 1,000 Btu/MBtu = 48,100 Btu/hr
SC = 33.6 MBtu/hr x 1,000 = 33,600 Btu/hr
DMD = 3.32 kW
EER = 14.5
HR = 59.5 MBtu/hr x 1000 = 59,500 Btu/hr

From Figure 2.20, full-load cooling mode correction factors (identified with dashed outlines) for EAT conditions of 70 F db/60 F wb are $CF_{TC\text{-}EAT}$=0.922, $CF_{SC\text{-}EAT}$=0.875, $CF_{DMD\text{-}EAT}$=0.977, and $CF_{HR\text{-}EAT}$=0.932, which are then applied according to Equations 2.18 through 2.24 to provide corrected performance data for the heat pump in cooling mode.

$$TC_{cor} = TC \cdot CF_{TC-EAT} \cdot CF_{TC-CFM} \cdot CF_{TC-AF} = 48{,}100 \text{ Btu/hr} \cdot 0.922 \cdot 1 \cdot 1 = 44{,}348 \text{ Btu/hr} \qquad (-7.80\%)$$

$$SC_{cor} = SC \cdot CF_{SC-EAT} \cdot CF_{SC-CFM} \cdot CF_{SC-AF} = 33{,}600 \text{ Btu/hr} \cdot 0.875 \cdot 1 \cdot 1 = 29{,}400 \text{ Btu/hr} \qquad (-12.5\%)$$

$$DMD_{cor} = DMD \cdot CF_{DMD-EAT} \cdot CF_{DMD-CFM} \cdot CF_{DMD-AF} = 3.32 \text{ kW} \cdot 0.977 \cdot 1 \cdot 1 = 3.24 \text{ kW} \qquad (-2.41\%)$$

$$HR_{cor} = HR \cdot CF_{HR-EAT} \cdot CF_{HR-CFM} \cdot CF_{HR-AF} = 59{,}500 \text{ Btu/hr} \cdot 0.932 \cdot 1 \cdot 1 = 55{,}454 \text{ Btu/hr} \qquad (-6.80\%)$$

$$EER_{cor} = \frac{TC_{cor}}{DMD_{cor} \cdot \frac{1{,}000 \text{ W}}{\text{kW}}} = \frac{44{,}348 \text{ Btu/hr}}{3.24 \text{ kW} \cdot \frac{1{,}000 \text{ W}}{\text{kW}}} = 13.69 \frac{\text{Btu/hr}}{\text{W}} \qquad (-5.59\%)$$

$$LAT_{c-cor} = EAT - \frac{SC_{cor}}{1.08 \cdot CFM} = 70 - \frac{29{,}400 \text{ Btu/hr}}{1.08 \cdot 1550 \text{ cfm}} = 52.4 \text{ F}$$

$$LWT_{c-cor} = EWT + \frac{HR_{cor}}{500 \cdot GPM} = 90 + \frac{55{,}454 \text{ Btu/hr}}{500 \cdot 12.0 \text{ gpm}} = 99.2 \text{ F}$$

Example 2.18 Determine the corrected TC, SC, DMD, HR, EER, and LAT for the data in Example 2.14 for EAT conditions of 70 F db and 60 F wb.

Based on Example 2.14, the part-load cooling mode performance data for EWT=90 F, GPM=11, CFM=1,300, and EAT=80 F db/67 F wb were determined from Figure 2.17 to be:

TC = 35.9 MBtu/hr x 1,000 Btu/MBtu = 35,900 Btu/hr
SC = 26.8 MBtu/hr x 1,000 = 26,800 Btu/hr
DMD = 2.26 kW
EER = 15.9
HR = 43.6 MBtu/hr x 1,000 = 43,600 Btu/hr

From Figure 2.21, part-load cooling mode correction factors (identified with dashed outlines) for EAT conditions of 70 F db/60 F wb are $CF_{TC\text{-}EAT}=0.935$, $CF_{SC\text{-}EAT}=0.844$, $CF_{DMD\text{-}EAT}=0.993$, and $CF_{HR\text{-}EAT}=0.945$, which are then applied according to Equations 2.18 through 2.24 to provide corrected performance data for the heat pump in cooling mode.

$$TC_{cor} = TC \bullet CF_{TC-EAT} \bullet CF_{TC-CFM} \bullet CF_{TC-AF} = 35{,}900\ \text{Btu/hr} \bullet 0.935 \bullet 1 \bullet 1 = 33{,}567\ \text{Btu/hr} \qquad (-6.50\%)$$

$$SC_{cor} = SC \bullet CF_{SC-EAT} \bullet CF_{SC-CFM} \bullet CF_{SC-AF} = 26{,}800\ \text{Btu/hr} \bullet 0.844 \bullet 1 \bullet 1 = 22{,}619\ \text{Btu/hr} \qquad (-15.6\%)$$

$$DMD_{cor} = DMD \bullet CF_{DMD-EAT} \bullet CF_{DMD-CFM} \bullet CF_{DMD-AF} = 2.26\ \text{kW} \bullet 0.993 \bullet 1 \bullet 1 = 2.24\ \text{kW} \qquad (-0.70\%)$$

$$HR_{cor} = HR \bullet CF_{HR-EAT} \bullet CF_{HR-CFM} \bullet CF_{HR-AF} = 43{,}600\ \text{Btu/hr} \bullet 0.945 \bullet 1 \bullet 1 = 41{,}202\ \text{Btu/hr} \qquad (-5.50\%)$$

$$EER_{cor} = \frac{TC_{cor}}{DMD_{cor} \bullet \dfrac{1{,}000\ \text{W}}{\text{kW}}} = \frac{33{,}567\ \text{Btu/hr}}{2.24\ \text{kW} \bullet \dfrac{1{,}000\ \text{W}}{\text{kW}}} = 14.99\ \frac{\text{Btu/hr}}{\text{W}} \qquad (-5.72\%)$$

$$LAT_{c-cor} = EAT - \frac{SC_{cor}}{1.08 \bullet CFM} = 70 - \frac{22{,}619\ \text{Btu/hr}}{1.08 \bullet 1300\ \text{cfm}} = 53.9\ \text{F}$$

$$LWT_{c-cor} = EWT + \frac{HR_{cor}}{500 \bullet GPM} = 90 + \frac{41{,}202\ \text{Btu/hr}}{500 \bullet 11.0\ \text{gpm}} = 97.5\ \text{F}$$

2.5.3.2 AIR FLOW RATE (CFM) CORRECTION FACTORS

The air flow rate across the refrigerant-to-air coil affects heat pump performance due to its impact on the effectiveness of the refrigerant-to-air heat exchanger. Typically, the fan in the heat pump is sized to supply an air flow rate of between 350 to 400 CFM per nominal ton of heat pump capacity against a static pressure in the ductwork (between 0.2 and 0.4 inches of water, but may vary depending on manufacturer). Lowering the system air flow rate below the manufacturer's recommended flow rate (due to high external static pressure) will reduce heat pump capacity and efficiency. Increasing air flow rate (by over sizing ductwork to reduce the external static pressure) will increase heat pump capacity and performance, but to a much lesser extent. There are two types of fans that are utilized by GSHP manufacturers: the permanent split capacitor (PSC), which offers three speed fan speed settings, and the variable speed electronically commutated motor (ECM), which offers several selectable fan speed settings that provide constant air flow rates over a wide range of external static pressures. Two-capacity, water-to-air GSHPs utilize ECM blowers to provide different levels of air flow rate depending on mode of operation.

Blower performance data for the ClimateMaster Tranquility 27™ Series GSHP equipped with an ECM blower motor are provided in Figure 2.22, which is the table provided in the manufacturer's literature. The Max ESP column identifies the maximum external static pressure that the installed fan motor (third column) is capable of producing with the ECM controller, meaning that the fan speed will continue to be increased up to that ESP after which fan speed will remain constant and air flow rate will start to fall off of the rated values in the table. For each heat pump model, there are four identified "Tap Settings" labeled 1, 2, 3, and 4, each of which is associated with a combination of air flow rates for fan only, stage 1, and stage 2 operation in cooling, dehumidification, and heating mode. The Auxiliary/Emergency Mode column contains the air flow rate that the fan speed setting will provide at the specified Tap Setting. The Tranquility 27 Series is capable of two-capacity operation, and the fan performance data for the Model 049 for tap setting 3 has been highlighted in Figure 2.22. In heating mode, the fan will produce 780 CFM during "fan only" mode, 1,400 CFM for first stage (part-load) heating and 1,650 CFM for high capacity (full-load) heating. If auxiliary heat is required, the fan is programmed to provide 1660 CFM. In cooling mode, the fan will produce 780 CFM during "fan only" mode, 1,300 CFM for first stage (part-load) cooling and 1,550 CFM for high capacity (full-load) cooling. The dehumidification mode of operation (preset or controlled by humidistat) is for cooling only where the fan speed is reduced 20 to 25% (in this case 1,300 CFM down to 1,020 CFM for low capacity/part-load cooling mode) to increase the latent cooling capacity of the heat pump. Changing the tap settings internally in the heat pump will result in a different combination of air flow rates for each mode of operation.

Airflow in CFM with wet coil and clean air filter

Model	Max ESP (in. wg)	Fan Motor (hp)	Tap Setting	Cooling Mode			Dehumid Mode			Heating Mode			Residential Units Only	
				Stg 1	Stg 2	Fan	Stg 1	Stg 2	Fan	Stg 1	Stg 2	Fan	AUX CFM	Aux/ Emerg Mode
026	0.50	1/2	4	810	950	475	630	740	475	920	1060	475	4	1060
	0.50	1/2	3	725	850	425	560	660	425	825	950	425	3	950
	0.50	1/2	2	620	730	370	490	570	370	710	820	370	2	820
	0.50	1/2	1	520	610	300				600	690	300	1	690
038	0.50	1/2	4	1120	1400	700	870	1090	700	1120	1400	700	4	1400
	0.50	1/2	3	1000	1250	630	780	980	630	1000	1250	630	3	1350
	0.50	1/2	2	860	1080	540	670	840	540	860	1080	540	2	1350
	0.50	1/2	1	730	900	450				730	900	450	1	1350
049	0.75	1	4	1460	1730	870	1140	1350	870	1560	1850	870	4	1850
	0.75	1	3	1300	1550	780	1020	1210	780	1400	1650	780	3	1660
	0.75	1	2	1120	1330	670	870	1040	670	1200	1430	670	2	1430
	0.75	1	1	940	1120	560				1010	1200	560	1	1350
064	0.75	1	4	1670	2050	1020	1300	1600	1020	1860	2280	1020	4	2280
	0.75	1	3	1500	1825	920	1160	1430	920	1650	2050	920	3	2040
	0.75	1	2	1280	1580	790	1000	1230	790	1430	1750	790	2	1750
	0.75	1	1	1080	1320	660				1200	1470	660	1	1470
072	0.75	1	4	1620	2190	1050	1270	1650	1050	1690	2230	1050	4	2230
	0.75	1	3	1500	1950	980	1170	1520	980	1600	2100	980	3	2100
	0.75	1	2	1400	1830	910	1100	1420	910	1400	1850	910	2	1870
	0.75	1	1	1320	1700	850				1240	1620	850	1	1670

Factory shipped on Tap Setting 2
During Auxilliary operation (residential units only) the CFM will run at the higher of the heating (delay jumper) or AUX settings
Airflow is controlled within +/– 5% up to Max ESP shown with wet coil and standard 1" fiberglass filter
Do not select Dehumidification mode if HP CFM is on setting 1
All units ARI/ISO/ASHRAE 13256-1 rated HP (Cooling) Delay (Heating) CFM Setting 3

Figure 2.22. ECM Blower Performance Data - ClimateMaster Tranquility 27™ Series

Correction factors for CFM are provided in tabular format for the ClimateMaster Tranquility 27™ Series (which utilizes only the ECM blower motor) for full-load operation in Figure 2.23 and part-load operation in Figure 2.24. Heating mode correction factors (Heating Capacity ($CF_{HC\text{-}CFM}$), Power ($CF_{DMD\text{-}CFM}$), and Heat of Extraction ($CF_{HE\text{-}CFM}$) are provided in the last three columns as a function of Percent of Rated Air flow (first column). The CFM correction factors are applied to tabled performance data according to Equations 2.12 through 2.14 to determine HC_{cor}, DMD_{cor}, and HE_{cor} from which COP_{cor}, LAT_{cor}, and LWT_{cor} can be determined (Equations 2.15 through 2.17). Looking at Figures 2.23 and 2.24, the effect of higher CFM in the heating mode is to increase HC, reduce DMD, and increase HE, increasing the COP of the heat pump and resulting in a slightly lower LAT due to the higher CFM. Lowering CFM results in a decrease in HC, an increase in DMD, and a decrease in HE, decreasing the COP of the heat pump and resulting in a slightly higher LAT due to the lower CFM. The full-load and part-load Heating Correction Factors in Figures 2.23 and 2.24 are applied as shown in Examples 2.19 and 2.20.

Airflow	Cooling				Heating		
% of Rated	Total Capacity	Sensible Capacity	Power	Heat of Rejection	Heating Capacity	Power	Heat of Extraction
60%	0.925	0.788	0.913	0.922	0.946	1.153	0.896
69%	0.946	0.829	0.926	0.942	0.959	1.107	0.924
75%	0.960	0.861	0.937	0.955	0.969	1.078	0.942
81%	0.972	0.895	0.950	0.968	0.977	1.053	0.959
88%	0.983	0.930	0.965	0.979	0.985	1.032	0.974
94%	0.992	0.965	0.982	0.990	0.993	1.014	0.988
100%	1.000	1.000	1.000	1.000	1.000	1.000	1.000
106%	1.007	1.033	1.020	1.009	1.006	0.989	1.011
113%	1.012	1.064	1.042	1.018	1.012	0.982	1.019
119%	1.016	1.092	1.066	1.025	1.018	0.979	1.027
125%	1.018	1.116	1.091	1.032	1.022	0.977	1.033
130%	1.019	1.132	1.112	1.037	1.026	0.975	1.038

Figure 2.23. Air flow Rate Correction Factors - ClimateMaster Tranquility 27™ Series - Full Load

Airflow	Cooling				Heating		
% of Rated	Total Capacity	Sensible Capacity	Power	Heat of Rejection	Heating Capacity	Power	Heat of Extraction
60%	0.920	0.781	0.959	0.927	0.946	1.241	0.881
69%	0.942	0.832	0.964	0.946	0.960	1.163	0.915
75%	0.956	0.867	0.696	0.959	0.969	1.115	0.937
81%	0.969	0.901	0.975	0.970	0.978	1.076	0.956
88%	0.981	0.934	0.982	0.981	0.986	1.043	0.973
94%	0.991	0.967	0.990	0.991	0.993	1.018	0.988
100%	1.000	1.000	1.000	1.000	1.000	1.000	1.000
106%	1.007	1.033	1.011	1.008	1.006	0.990	1.010
113%	1.013	1.065	1.023	1.015	1.012	0.986	1.017
119%	1.018	1.098	1.036	1.021	1.017	0.983	1.024
125%	1.021	1.131	1.051	1.026	1.021	0.981	1.030
130%	1.023	1.159	1.063	1.030	1.024	0.979	1.034

Figure 2.24. Air flow Rate Correction Factors - ClimateMaster Tranquility 27™ Series - Part Load

Example 2.19 Determine the corrected HC, DMD, HE, COP, and LAT for the data in Example 2.11 for a CFM of 1,250.

Based on Example 2.11, the full-load heating mode performance data for EWT=40 F, GPM=12, CFM=1,650, and EAT=70 F were determined from Figure 2.16 to be:

HC = 43.9 MBtu/hr x 1,000 Btu/MBtu = 43,900 Btu/hr
DMD = 3.08 kW
COP = 4.19
HE = 33.5 MBtu/hr x 1,000 = 33,500 Btu/hr
LAT = 94.6 F

From Figure 2.23, full-load heating mode correction factors (identified with solid outlines) for a CFM of 1,250 (Percent Rated Air flow = 1,250/1,650 = 0.758 ≈ 75%) are $CF_{HC\text{-}CFM}$=0.969, $CF_{DMD\text{-}CFM}$=1.078, and $CF_{HE\text{-}CFM}$=0.942, which are then applied according to Equations 2.12 through 2.17 to provide corrected performance data for the heat pump in heating mode.

$$HC_{cor} = HC \cdot CF_{HC-EAT} \cdot CF_{HC-CFM} \cdot CF_{HC-AF} = 43{,}900\text{ Btu/hr} \cdot 1 \cdot 0.969 \cdot 1 = 42{,}539\text{ Btu/hr} \qquad (-3.10\%)$$

$$DMD_{cor} = DMD \cdot CF_{DMD-EAT} \cdot CF_{DMD-CFM} \cdot CF_{DMD-AF} = 3.08\text{ kW} \cdot 1 \cdot 1.078 \cdot 1 = 3.32\text{ kW} \qquad (+7.80\%)$$

$$HE_{cor} = HE \cdot CF_{HE-EAT} \cdot CF_{HE-CFM} \cdot CF_{HE-AF} = 33{,}500\text{ Btu/hr} \cdot 1 \cdot 0.942 \cdot 1 = 31{,}557\text{ Btu/hr} \qquad (-5.80\%)$$

$$COP_{cor} = \frac{HC_{cor}}{DMD_{cor} \cdot \dfrac{3{,}412\text{ Btu/hr}}{\text{kW}}} = \frac{42{,}539\text{ Btu/hr}}{3.32\text{ kW} \cdot \dfrac{3{,}412\text{ Btu/hr}}{\text{kW}}} = 3.76 \qquad (-10.3\%)$$

$$LAT_{h-cor} = EAT + \frac{HC_{cor}}{1.08 \cdot CFM} = 70F + \frac{42{,}539\text{ Btu/hr}}{1.08 \cdot 1250\text{ cfm}} = 101.5\text{ F}$$

$$LWT_{h-cor} = EWT - \frac{HE_{cor}}{500 \cdot GPM} = 40 - \frac{31{,}557\text{ Btu/hr}}{500 \cdot 12.0\text{ gpm}} = 34.7\text{ F}$$

Example 2.20 Determine the corrected HC, DMD, HE, COP, and LAT for the data in Example 2.12 for a CFM of 1,050.

Based on Example 2.12, the part-load heating mode performance data for EWT=40 F, GPM=11, CFM=1,400, and EAT=70 F were determined from Figure 2.17 to be:

HC = 31.9 MBtu/hr x 1,000 Btu/MBtu = 31,900 Btu/hr
DMD = 2.19 kW
COP = 4.27
HE = 24.8 MBtu/hr x 1,000 = 24,800 Btu/hr
LAT = 91.1F

From Figure 2.24, part-load heating mode correction factors (identified with solid outlines) for a CFM of 1,050 (Percent Rated Air flow = 1,050/1,400 = 0.75 = 75%) are $CF_{HC\text{-}CFM}$=0.969, $CF_{DMD\text{-}CFM}$=1.115, and $CF_{HE\text{-}CFM}$=0.937, which are then applied according to Equations 2.12 through 2.17 to provide corrected performance data for the heat pump in heating mode.

$$HC_{cor} = HC \bullet CF_{HC-EAT} \bullet CF_{HC-CFM} \bullet CF_{HC-AF} = 31{,}900\ \text{Btu/hr} \bullet 1 \bullet 0.969 \bullet 1 = 30{,}911\ \text{Btu/hr} \qquad (-3.10\%)$$

$$DMD_{cor} = DMD \bullet CF_{DMD-EAT} \bullet CF_{DMD-CFM} \bullet CF_{DMD-AF} = 2.19\ \text{kW} \bullet 1 \bullet 1.115 \bullet 1 = 2.44\ \text{kW} \qquad (+11.5\%)$$

$$HE_{cor} = HE \bullet CF_{HE-EAT} \bullet CF_{HE-CFM} \bullet CF_{HE-AF} = 24{,}800\ \text{Btu/hr} \bullet 1 \bullet 0.937 \bullet 1 = 23{,}238\ \text{Btu/hr} \qquad (-6.30\%)$$

$$COP_{cor} = \frac{HC_{cor}}{DMD_{cor} \bullet \dfrac{3{,}412\ \text{Btu/hr}}{\text{kW}}} = \frac{30{,}911\ \text{Btu/hr}}{2.44\ \text{kW} \bullet \dfrac{3{,}412\ \text{Btu/hr}}{\text{kW}}} = 3.71 \qquad (-13.0\%)$$

$$LAT_{h-cor} = EAT + \frac{HC_{cor}}{1.08 \bullet CFM} = 70\ \text{F} + \frac{30{,}911\ \text{Btu/hr}}{1.08 \bullet 1050\ \text{cfm}} = 97.3\ \text{F}$$

$$LWT_{h-cor} = EWT - \frac{HE_{cor}}{500 \bullet GPM} = 40 - \frac{23{,}238\ \text{Btu/hr}}{500 \bullet 11.0\ \text{gpm}} = 35.8\ \text{F}$$

Cooling mode correction factors are provided in the Cooling columns of Figures 2.23 and 2.24 (Total Cooling Capacity ($CF_{TC\text{-}CFM}$), Sensible Cooling Capacity ($CF_{SC\text{-}CFM}$), Power ($CF_{DMD\text{-}CFM}$), and Heat of Rejection ($CF_{HR\text{-}CFM}$). The correction factors are applied to tabled performance data according to Equations 2.18 through 2.24 to determine TC_{cor}, SC_{cor}, DMD_{cor}, and HR_{cor}, EER_{cor}, LAT_{cor}, and LWT_{cor}, respectively. The effect of reducing CFM is to decrease TC, decrease SC, decrease DMD, and decrease HR. The full-load and part-load Cooling Correction Factors in Figures 2.23 and 2.24 are applied as shown in Examples 2.21 and 2.22.

Example 2.21 Determine the corrected TC, SC, DMD, HR, EER, and LAT for the data in Example 2.13 for a CFM of 1,150.

Based on Example 2.13, the full-load cooling mode performance data for EWT=90 F, GPM=12, EAT=80 F db/67 F wb, and CFM=1,550 were determined from Figure 2.16 to be:

TC = 48.1 MBtu/hr x 1,000 Btu/MBtu = 48,100 Btu/hr
SC = 33.6 MBtu/hr x 1,000 = 33,600 Btu/hr
DMD = 3.32 kW
EER = 14.5
HR = 59.5 MBtu/hr x 1,000 = 59,500 Btu/hr

From Figure 2.23, full-load cooling mode correction factors (identified with dashed outlines) for a CFM of 1,150 (Percent Rated Air flow = 1,150/1,550 = 0.742 ≈ 75%) are $CF_{TC\text{-}CFM}$=0.960, $CF_{SC\text{-}CFM}$=0.861, $CF_{DMD\text{-}CFM}$=0.937, and $CF_{HR\text{-}CFM}$=0.955, which are then applied according to Equations 2.18 through 2.24 to provide corrected performance data for the heat pump in cooling mode.

$TC_{cor} = TC \cdot CF_{TC-EAT} \cdot CF_{TC-CFM} \cdot CF_{TC-AF} = 48{,}100\ Btu/hr \cdot 1 \cdot 0.960 \cdot 1 = 46{,}176\ Btu/hr$ (-4.00%)

$SC_{cor} = SC \cdot CF_{SC-EAT} \cdot CF_{SC-CFM} \cdot CF_{SC-AF} = 33{,}600\ Btu/hr \cdot 1 \cdot 0.861 \cdot 1 = 28{,}930\ Btu/hr$ (-13.9%)

$DMD_{cor} = DMD \cdot CF_{DMD-EAT} \cdot CF_{DMD-CFM} \cdot CF_{DMD-AF} = 3.32\ kW \cdot 1 \cdot 0.937 \cdot 1 = 3.11\ kW$ (-6.30%)

$HR_{cor} = HR \cdot CF_{HR-EAT} \cdot CF_{HR-CFM} \cdot CF_{HR-AF} = 59{,}500\ Btu/hr \cdot 1 \cdot 0.955 \cdot 1 = 56{,}823\ Btu/hr$ (-4.50%)

$$EER_{cor} = \frac{TC_{cor}}{DMD_{cor} \cdot \frac{1{,}000\ W}{kW}} = \frac{46{,}176\ Btu/hr}{3.11\ kW \cdot \frac{1{,}000\ W}{kW}} = 14.8\ \frac{Btu/hr}{W} \qquad (+2.40\%)$$

$$LAT_{c-cor} = EAT - \frac{SC_{cor}}{1.08 \cdot CFM} = 80 - \frac{28{,}930\ Btu/hr}{1.08 \cdot 1150\ cfm} = 56.7\ F$$

$$LWT_{c-cor} = EWT + \frac{HR_{cor}}{500 \cdot GPM} = 90 + \frac{56{,}823\ Btu/hr}{500 \cdot 12.0\ gpm} = 99.5\ F$$

Example 2.22 Determine the corrected TC, SC, DMD, HR, EER, and LAT for the data in Example 2.14 for a CFM of 975.

Based on Example 2.14, the part-load cooling mode performance data for EWT=90 F, GPM=11, EAT=80 F db/67 F wb, and CFM=1,300 were determined from Figure 2.17 to be:

TC = 35.9 MBtu/hr x 1,000 Btu/MBtu = 35,900 Btu/hr
SC = 26.8 MBtu/hr x 1,000 = 26,800 Btu/hr
DMD = 2.26 kW
EER = 15.9
HR = 43.6 MBtu/hr x 1,000 = 43,600 Btu/hr

From Figure 2.24, part-load cooling mode correction factors (identified with dashed outlines) for a CFM of 975 (Percent Rated Air flow = 975/1,300 = 0.75 = 75%) are CF_{TC-CFM}=0.956, CF_{SC-CFM}=0.867, $CF_{DMD-CFM}$=0.969 (typo in table), and CF_{HR-CFM}=0.959, which are then applied according to Equations 2.18 through 2.24 to provide corrected performance data for the heat pump in cooling mode.

$TC_{cor} = TC \cdot CF_{TC-EAT} \cdot CF_{TC-CFM} \cdot CF_{TC-AF} = 35{,}900\ Btu/hr \cdot 1 \cdot 0.956 \cdot 1 = 34{,}320\ Btu/hr$ (-4.40%)

$SC_{cor} = SC \cdot CF_{SC-EAT} \cdot CF_{SC-CFM} \cdot CF_{SC-AF} = 26{,}800\ Btu/hr \cdot 1 \cdot 0.867 \cdot 1 = 23{,}236\ Btu/hr$ (-13.3%)

$DMD_{cor} = DMD \cdot CF_{DMD-EAT} \cdot CF_{DMD-CFM} \cdot CF_{DMD-AF} = 2.26\ kW \cdot 1 \cdot 0.969 \cdot 1 = 2.19\ kW$ (-3.40%)

$HR_{cor} = HR \cdot CF_{HR-EAT} \cdot CF_{HR-CFM} \cdot CF_{HR-AF} = 43{,}600\ Btu/hr \cdot 1 \cdot 0.959 \cdot 1 = 41{,}812\ Btu/hr$ (-4.10%)

$$EER_{cor} = \frac{TC_{cor}}{DMD_{cor} \bullet \frac{1{,}000\ W}{kW}} = \frac{34{,}320\ Btu/hr}{2.19\ kW \bullet \frac{1{,}000\ W}{kW}} = 15.7\ \frac{Btu/hr}{W} \qquad (-1.00\%)$$

$$LAT_{c-cor} = EAT - \frac{SC_{cor}}{1.08 \bullet CFM} = 80 - \frac{23{,}236\ Btu/hr}{1.08 \bullet 975\ cfm} = 57.9\ F$$

$$LWT_{c-cor} = EWT + \frac{HR_{cor}}{500 \bullet GPM} = 90 + \frac{41{,}812\ Btu/hr}{500 \bullet 11.0\ gpm} = 97.6\ F$$

Further discussion of blower types, operation, and the effects of duct sizing on GSHP performance and selection can be found in the duct design section of Chapter 3.

2.5.3.3 ANTIFREEZE CORRECTION FACTORS SEE SECTION 2.4.3.3.

2.5.4 ISO/ARI RATINGS The Air-Conditioning and Refrigeration Institute (ARI), American Society of Heating, Refrigeration, and Air-Conditioning Engineers (ASHRAE), and the International Standards Organization *(ISO, 1998a)* have developed certification standards *(ISO 13256-1)* to set requirements for testing and rating of water-to-air heat pump equipment to ensure that performance ratings published by manufacturers agree with the performance of their manufactured product. A discussion of ISO/ARI ratings is provided in Section 2.4.4.

Figure 2.14 provided test conditions for both single-capacity and two-capacity water-to-air heat pumps. For two-capacity heat pumps, the full-load performance is tested at the same standard rating test conditions as single-capacity heat pumps. Part-load rating test conditions allow for the manufacturer to specify a lower GPM and CFM, and the entering water temperatures are set at 41 F for heating mode and 68 F for cooling mode for ground-loop heat pumps. Entering air conditions (EAT=68 F for heating and 80.6 F db/66.2 F wb for cooling) are the same for both full-load and part-load operation.

ARI/ISO performance ratings for the ClimateMaster Tranquility 27™ Series are provided in Figure 2.25. Performance ratings (HC and COP for heating and TC and EER for cooling) are provided for WLHP, GWHP, and GLHP entering water conditions. For the Model 049, the GLHP ARI full-load performance ratings are identified using solid thick outlines for heating mode and dashed thick outlines for cooling mode. The GLHP ARI part-load performance ratings are identified using solid thin outlines for heating mode and dashed thin outlines for cooling mode. All capacities and efficiencies include fan power and pumping power penalties minus the effects of the ductwork and the ground loop.

ASHRAE/ARI/ISO 13256-1. English (IP) Units

Model	Capacity Modulation	Water Loop Heat Pump				Ground Water Heat Pump				Ground Loop Heat Pump			
		Cooling 86°F		Heating 68°F		Cooling 59°F		Heating 50°F		Cooling Full Load 77°F Part Load 68°F		Heating Full Load 32°F Part Load 41°F	
		Capacity Btuh	EER Btuh/W	Capacity Btuh	COP	Capacity Btuh	EER Btuh/W	Capacity Btuh	COP	Capacity Btuh	EER Btuh/W	Capacity Btuh	COP
026	Full	25,300	15.9	30,800	5.3	28,900	24.5	25,700	4.8	26,600	18.5	19,800	4.0
	Part	19,400	18.3	22,400	6.1	22,200	30.8	18,600	5.1	21,300	26.0	16,500	4.6
038	Full	36,200	15.6	44,800	5.3	41,200	23.0	36,700	4.7	38,200	18.2	29,000	4.0
	Part	26,200	18.5	30,800	6.3	30,200	31.5	24,800	5.1	28,900	27.0	2,100	4.5
049	Full	48,400	15.7	59,900	5.2	54,600	22.5	48,300	4.7	50,600	17.9	37,500	4.0
	Part	36,100	18.0	44,300	6.2	40,700	28.7	35,400	5.1	39,600	24.9	31,200	4.6
064	Full	61,500	15.0	72,300	5.0	68,600	22.0	59,600	4.4	64,800	17.5	48,000	3.9
	Part	44,900	17.6	51,100	5.7	51,900	29.7	41,800	4.7	49,800	25.3	37,500	4.3
072	Full	68,700	14.2	88,600	4.9	77,100	19.9	70,200	4.3	71,600	16.2	54,100	3.6
	Part	52,800	16.0	65,200	5.1	59,800	24.5	51,700	4.3	57,700	21.4	45,400	3.9

Cooling capacities based upon 80.6F DB, 66.2F WB entering air temperature
Heating capacities based upon 68F DB, 59F WB entering air temperature
Ground Loop Heat Pump ratings based on 15% antifreeze solution
All ratings based upon operation at lower voltage of dual voltage rated models

Figure 2.25. ARI/ISO Ratings - ClimateMaster Tranquility 27™ Series

2.6 Performance of Water-to-Water GSHPs

Single-capacity, water-to-water GSHPs typically utilize a scroll compressor to run at 100% capacity in terms of heating and cooling capacity. The single-capacity, water-to-water GSHP is generally controlled by an aquastat capable of two heat stages and one cooling stage, causing full capacity heating or cooling mode operation when required. In the heating mode, the second stage of heating controls an auxiliary source of heat, if needed. If the water-to-water heat pump is serving one zone (load flow rate is sufficient for proper heat pump operation), then a thermostat with two heat stages and one cooling stage may be used to provide full capacity heating or cooling when required, and second stage heat can be utilized when needed.

The performance (capacity, electric demand, and efficiency) of a single-capacity, water-to-water GSHP is measured by the heating capacity (Btu/hr), the electric demand (kW), and the COP in heating mode, and by the cooling capacity (Btu/hr), the electric demand (kW), and the EER in cooling mode. Parameters that affect the heating and cooling performance of the water-to-water GSHP include entering water temperature (EWT) from the ground connection, entering water temperature (ELT) returning from the conditioned space or load, water flow rate (GPM) through the refrigerant-to-water coil connected to the ground source, and water flow rate (GPM_{load}) through the refrigerant-to-water coil connected to the load. Manufacturers of GSHP equipment generally express performance data for their equipment in a tabular format as a function of EWT, GPM, ELT, and GPM_{load}. Some manufacturers utilize the leaving water temperature to the load (LLT) to express GSHP performance.

Heating and cooling performance data for a WaterFurnace EW Model 060 (a single-capacity, nominal 5-ton water-to-water GSHP that utilizes R-410A refrigerant) are provided in the tabular format supplied by the manufacturer in Figures 2.26 and 2.27, respectively. Column numbers are placed above the table for easier identification during discussion of the data. Column 1 contains ELT from the conditioned space, and

Column 2 contains several EWTs (in this case EST - entering source temperature from the ground connection) for each ELT. Column 3 contains three GPM_{load} levels for each EWT, with Columns 4 and 5 providing the pressure drop in psi and head loss in feet of water equivalent pressure, respectively, through the water-to-refrigerant coil on the load side of the heat pump. For each GPM_{load} in Column 3, a set of performance data are provided for three source-side flow rates (GPM) as listed across the top of the table. For each GPM, there are eight columns of data (6-13, 14-21, and 22-29) that contain eight heating or cooling performance parameters. Manufacturer's data are not always presented in the same format as provided here and each should be considered individually.

2.6.1 HEATING MODE PERFORMANCE DATA Heating mode performance data for three EWTs are listed in Columns 6-13, 14-21, and 22-29 of Figure 2.26. Each line on the table represents three sets of performance data for a combination of ELT, EST, GPM_{load}, and GPM, as read proceeding from left to right across the table. Columns 7, 15, and 23 contain the heating capacities (HC - kBtuH) provided to the water returning from the conditioned space. Columns 8, 16, and 24 list the electric demands (DMD - kW) of the unit, and Columns 10, 18, and 26 provide the COPs for each of the three ground source GPMs. The heats of extraction (HE - kBtuH) from the ground connection are provided in Columns 9, 17, and 25. The leaving water temperatures back to the conditioned space (LLT - F) are shown in Columns 6, 14, and 22 for the three ground source GPMs. Other performance data that are available for each GPM include LSTs (leaving source temperatures) in Columns 11, 19, and 27, and the pressure drops in psi and head losses in feet of water equivalent pressure, respectively, through the water-to-refrigerant coil on the ground source side of the heat pump (Columns 12 and 13, 20 and 21, and 28 and 29). This manufacturer expresses energy rates in kBtuH to save space. (1 kBtuH = 1 MBtu/hr = 1,000 Btu/hr).

Column

1	2	3	4	5	6	7	8	9	10	11	12	13	14	15	16	17	18	19	20	21	22	23	24	25	26	27	28	29
ELT	EST	Load Flow			Source Flow - 7 GPM								Source Flow - 14 GPM								Source Flow - 21 GPM							
		Flow	PD		LLT	HC	Power	HE	COP	LST	Source PD		LLT	HC	Power	HE	COP	LST	Source PD		LLT	HC	Power	HE	COP	LST	Source PD	
°F	°F	GPM	PSI	FT HD	°F	kBTUH	kW	kBTUH		°F	PSI	FT HD	°F	kBTUH	kW	kBTUH		°F	PSI	FT HD	°F	kBTUH	kW	kBTUH		°F	PSI	FT HD
60	30	7	1.1	2.5	77	59.3	3.43	47.6	5.07	16	1.2	2.7	77	60.1	3.47	48.3	5.08	23	3.1	7.2	78	62.2	3.50	50.3	5.21	25	5.5	12.7
		14	2.8	6.6	69	62.1	3.47	50.2	5.24	16	1.2	2.7	69	62.9	3.51	50.9	5.26	23	3.1	7.2	69	65.1	3.54	53.0	5.39	25	5.5	12.7
		21	5.0	11.6	66	63.1	3.51	51.1	5.27	15	1.2	2.7	66	64.0	3.54	51.9	5.29	23	3.1	7.2	66	66.2	3.58	54.0	5.42	25	5.5	12.7
	50	7	1.1	2.5	82	76.0	3.49	64.1	6.38	32	1.1	2.6	82	77.1	3.53	65.0	6.40	41	2.9	6.8	83	79.7	3.56	67.6	6.56	44	5.2	12.0
		14	2.8	6.6	71	79.5	3.53	67.5	6.60	31	1.1	2.6	72	80.6	3.6	68.5	6.62	40	2.9	6.8	72	83.4	3.60	71.1	6.78	43	5.2	12.0
		21	5.0	11.6	68	80.9	3.57	68.7	6.64	30	1.1	2.6	68	82.0	3.61	69.7	6.66	40	2.9	6.8	68	84.8	3.64	72.4	6.82	43	5.2	12.0
	70	7	1.1	2.5	87	92.8	3.55	80.6	7.65	47	1.0	2.4	87	94.0	3.59	81.8	7.68	58	2.7	6.4	88	97.3	3.63	84.9	7.86	62	4.9	11.3
		14	2.8	6.6	74	97.0	3.59	84.8	7.91	46	1.0	2.4	74	98.3	3.63	86.0	7.94	58	2.7	6.4	75	101.7	3.67	89.2	8.13	62	4.9	11.3
		21	5.0	11.6	69	98.7	3.63	86.3	7.96	45	1.0	2.4	70	100.0	3.67	87.5	7.99	58	2.7	6.4	70	103.4	3.71	90.8	8.18	61	4.9	11.3
	90	7	1.1	2.5	88	98.6	3.60	86.4	8.03	65	1.0	2.2	89	100.0	3.64	87.6	8.06	77	2.6	6.0	90	103.4	3.67	90.9	8.25	81	4.5	10.5
		14	2.8	6.6	75	103.2	3.64	90.8	8.30	64	1.0	2.2	75	104.6	3.7	92.0	8.33	77	2.6	6.0	75	108.2	3.71	95.5	8.54	81	4.5	10.5
		21	5.0	11.6	70	104.9	3.68	92.4	8.35	64	1.0	2.2	70	106.4	3.72	93.7	8.38	77	2.6	6.0	70	110.0	3.76	97.2	8.59	81	4.5	10.5
	110	7	1.1	2.5	90	104.5	3.65	92.1	8.40	84	0.9	2.1	90	106.0	3.69	93.4	8.43	97	2.4	5.5	91	109.6	3.72	96.9	8.63	101	4.2	9.8
		14	2.8	6.6	76	109.3	3.69	96.7	8.68	82	0.9	2.1	76	110.8	3.73	98.1	8.72	96	2.4	5.5	76	114.6	3.76	101.8	8.93	100	4.2	9.8
		21	5.0	11.6	71	111.2	3.73	98.5	8.74	82	0.9	2.1	71	112.7	3.77	99.9	8.77	96	2.4	5.5	71	116.6	3.81	103.6	8.98	100	4.2	9.8
80	30	7	1.0	2.3	96	57.1	4.35	42.3	3.85	18	1.2	2.7	97	57.9	4.40	42.9	3.86	24	3.1	7.2	97	59.9	4.44	44.7	3.95	26	5.5	12.7
		14	2.7	6.2	89	59.8	4.40	44.7	3.98	17	1.2	2.7	89	60.6	4.45	45.4	3.99	24	3.1	7.2	89	62.7	4.49	47.3	4.09	25	5.5	12.7
		21	4.7	10.9	86	60.8	4.45	45.6	4.00	17	1.2	2.7	86	61.6	4.50	46.3	4.02	23	3.1	7.2	86	63.7	4.54	48.2	4.11	25	5.5	12.7
	50	7	1.0	2.3	101	74.1	4.42	59.0	4.91	33	1.1	2.6	101	75.1	4.47	59.9	4.93	41	2.9	6.8	102	77.7	4.51	62.3	5.05	44	5.2	12.0
		14	2.7	6.2	91	77.5	4.47	62.3	5.08	32	1.1	2.6	91	78.6	4.5	63.2	5.10	41	2.9	6.8	92	81.3	4.56	65.7	5.22	44	5.2	12.0
		21	4.7	10.9	88	78.8	4.52	63.4	5.11	32	1.1	2.6	88	79.9	4.57	64.3	5.13	41	2.9	6.8	88	82.7	4.61	66.9	5.25	44	5.2	12.0
	70	7	1.0	2.3	106	91.1	4.49	75.8	5.94	48	1.0	2.4	106	92.4	4.54	76.9	5.97	59	2.7	6.4	107	95.6	4.58	79.9	6.11	62	4.9	11.3
		14	2.7	6.2	94	95.3	4.54	79.8	6.15	47	1.0	2.4	94	96.6	4.59	81.0	6.17	58	2.7	6.4	94	100.0	4.63	84.1	6.32	62	4.9	11.3
		21	4.7	10.9	89	96.9	4.59	81.3	6.19	47	1.0	2.4	89	98.3	4.64	82.4	6.21	58	2.7	6.4	90	101.6	4.69	85.7	6.36	62	4.9	11.3
	90	7	1.0	2.3	109	102.3	4.57	86.7	6.56	65	1.0	2.2	110	103.7	4.62	87.9	6.58	77	2.6	6.0	111	107.2	4.66	91.3	6.74	81	4.5	10.5
		14	2.7	6.2	95	107.0	4.62	91.2	6.79	64	1.0	2.2	95	108.4	4.7	92.5	6.81	77	2.6	6.0	96	112.2	4.71	96.1	6.97	81	4.5	10.5
		21	4.7	10.9	90	108.8	4.67	92.8	6.83	63	1.0	2.2	91	110.3	4.72	94.2	6.85	77	2.6	6.0	91	114.1	4.77	97.8	7.02	81	4.5	10.5
	110	7	1.0	2.3	112	113.4	4.65	97.6	7.15	82	0.9	2.1	113	115.0	4.69	99.0	7.18	96	2.4	5.5	114	118.9	4.74	102.8	7.35	100	4.2	9.8
		14	2.7	6.2	97	118.6	4.70	102.6	7.40	81	0.9	2.1	97	120.3	4.75	104.1	7.43	95	2.4	5.5	98	124.4	4.79	108.0	7.61	100	4.2	9.8
		21	4.7	10.9	91	120.6	4.75	104.4	7.44	80	0.9	2.1	92	122.3	4.80	105.9	7.47	95	2.4	5.5	92	126.5	4.85	110.0	7.65	100	4.2	9.8
100	30	7	0.9	2.2	116	55.0	5.57	36.0	2.89	20	1.2	2.7	116	55.7	5.62	36.5	2.90	25	3.1	7.2	116	57.6	5.68	38.2	2.97	26	5.5	12.7
		14	2.5	5.8	108	57.5	5.63	38.3	2.99	19	1.2	2.7	108	58.3	5.69	38.9	3.00	24	3.1	7.2	109	60.3	5.74	40.7	3.08	26	5.5	12.7
		21	4.4	10.2	106	58.5	5.60	39.0	3.01	19	1.2	2.7	106	59.3	5.75	39.6	3.02	24	3.1	7.2	106	61.3	5.80	41.5	3.09	26	5.5	12.7
	50	7	0.9	2.2	120	71.2	5.44	52.7	3.84	35	1.1	2.6	121	72.2	5.49	53.5	3.85	42	2.9	6.8	121	74.7	5.55	55.8	3.95	45	5.2	12.0
		14	2.5	5.8	111	74.5	5.50	55.7	3.97	34	1.1	2.6	111	75.5	5.6	56.6	3.99	42	2.9	6.8	111	78.1	5.61	59.0	4.08	44	5.2	12.0
		21	4.4	10.2	107	75.8	5.56	56.8	4.00	34	1.1	2.6	107	76.8	5.61	57.6	4.01	42	2.9	6.8	108	79.4	5.67	60.1	4.11	44	5.2	12.0
	70	7	0.9	2.2	125	87.5	5.31	69.4	4.83	50	1.0	2.4	125	88.7	5.36	70.4	4.85	60	2.7	6.4	126	91.7	5.41	73.3	4.97	63	4.9	11.3
		14	2.5	5.8	113	91.5	5.37	73.2	5.00	49	1.0	2.4	113	92.8	5.42	74.3	5.02	59	2.7	6.4	114	96.0	5.47	77.3	5.14	63	4.9	11.3
		21	4.4	10.2	109	93.1	5.42	74.6	5.03	49	1.0	2.4	109	94.3	5.48	75.6	5.05	59	2.7	6.4	109	97.6	5.53	78.7	5.17	63	4.9	11.3
	90	7	0.9	2.2	129	100.6	5.60	81.5	5.27	67	1.0	2.2	129	101.9	5.65	82.7	5.29	78	2.6	6.0	130	105.5	5.71	86.0	5.41	82	4.5	10.5
		14	2.5	5.8	115	105.2	5.66	85.9	5.45	65	1.0	2.2	115	106.6	5.7	87.1	5.47	78	2.6	6.0	116	110.3	5.77	90.6	5.60	81	4.5	10.5
		21	4.4	10.2	110	107.0	5.72	87.5	5.48	65	1.0	2.2	110	108.4	5.78	88.7	5.50	77	2.6	6.0	111	112.2	5.83	92.3	5.63	81	4.5	10.5
	110	7	0.9	2.2	132	113.6	5.88	93.6	5.66	83	0.9	2.1	133	115.2	5.94	94.9	5.68	96	2.4	5.5	134	119.2	6.00	98.7	5.82	101	4.2	9.8
		14	2.5	5.8	117	118.9	5.95	98.6	5.86	82	0.9	2.1	117	120.5	6.01	100.0	5.88	96	2.4	5.5	118	124.6	6.07	103.9	6.02	100	4.2	9.8
		21	4.4	10.2	112	120.9	6.01	100.4	5.89	81	0.9	2.1	112	122.5	6.07	101.8	5.91	95	2.4	5.5	112	126.7	6.13	105.8	6.06	100	4.2	9.8
120	30	7	0.9	2.0	135	53.3	7.20	28.7	2.17	22	1.2	2.7	135	54.0	7.28	29.2	2.18	26	3.1	7.2	136	55.9	7.35	30.8	2.23	27	5.5	12.7
		14	2.3	5.4	128	55.7	7.28	30.9	2.24	21	1.2	2.7	128	56.5	7.36	31.4	2.25	26	3.1	7.2	128	58.4	7.43	33.1	2.31	27	5.5	12.7
		21	4.1	9.5	125	56.7	7.36	31.6	2.26	21	1.2	2.7	125	57.5	7.44	32.1	2.26	25	3.1	7.2	126	59.4	7.51	33.8	2.32	27	5.5	12.7
	50	7	0.9	2.0																								
		14	2.3	5.4	130	71.1	7.32	46.1	2.85	37	1.1	2.6	130	72.1	7.4	46.9	2.86	43	2.9	6.8	131	74.6	7.47	49.1	2.93	45	5.2	12.0
		21	4.1	9.5	127	72.3	7.40	47.1	2.86	37	1.1	2.6	127	73.3	7.48	47.8	2.87	43	2.9	6.8	127	75.8	7.55	50.1	2.94	45	5.2	12.0
	70	7	0.9	2.0																								
		14	2.3	5.4	132	86.5	7.36	61.4	3.44	52	1.0	2.4	133	87.7	7.44	62.3	3.45	61	2.7	6.4	133	90.7	7.51	65.1	3.54	64	4.9	11.3
		21	4.1	9.5	128	88.0	7.45	62.6	3.46	52	1.0	2.4	128	89.2	7.52	63.5	3.48	61	2.7	6.4	129	92.2	7.60	66.3	3.56	64	4.9	11.3
	90	7	0.9	2.0																								
		14	2.3	5.4																								
		21	4.1	9.5	130	100.1	7.35	75.0	3.99	69	1.0	2.2	130	101.5	7.43	76.1	4.00	79	2.6	6.0	130	105.0	7.50	79.4	4.10	82	4.5	10.5
	110	7	0.9	2.0																								
		14	2.3	5.4																								
		21	4.1	9.5	131	112.2	7.26	87.5	4.53	85	0.9	2.1	131	113.8	7.33	88.7	4.55	97	2.4	5.5	131	117.7	7.41	92.4	4.66	101	4.2	9.8

Notes: Multiple flow rates for source side and flow side are shown. The lowest source flow rate shown is used for geothermal open loop/well water systems with a minimum 50 F. The second source flow rate shown is the minimum closed loop flow rate. The third source flow rate shown is optimum for geothermal closed loop and the suggested flow rate for boiler/tower applications. When selecting units and designing the system, actual operating parameters must fall within the temperature and flow rate ranges shown on the table. Using temperature/flow rate combinations outside the range of the table will result in performance problems. *For 3 phase capacity, multiply above data by .948. For 3 phase power, multiply the above data by .943.

Rev, 8/05

Figure 2.26. WaterFurnace EW Model 060 Heating Mode Performance Data

Example 2.23 An example of using the performance data to determine heating mode performance is highlighted with solid outlines on Figure 2.26 for ELT=100 F, EWT=30 F, GPM_{load}=21, and GPM = 21. For those operating parameters:

HC = 61.3 kBtuH x 1,000 Btu/hr/kBtuH = 61,300 Btu/hr	Column 23
DMD = 5.80 kW	Column 24
COP = 3.09	Column 26
HE = 41.5 kBtuH x 1,000 = 41,500 Btu/hr	Column 25
LLT = 106 F	Column 22
LST = 26 F	Column 27

Equation 2.3 can be used to verify the tabled HE based on the HC and COP. LST (LWT_h)is calculated using Equation 2.8 along with the EWT, HE, and GPM from the table. Leaving load water temperature (LLT_h) from the heat pump in heating mode can be calculated according to Equation 2.25, which is based on thermal properties of water in the 32 to 105 F range.

$$LLT_h = ELT + \frac{HC}{500 \bullet GPM_{load}}$$ **Equation 2.25**

Where:
- LLT_h = leaving load water temperature from heat pump in heating mode (F)
- ELT = entering load water temperature from conditioned space (F)
- HC = heating capacity of heat pump (Btu/hr)
- GPM_{load} = water flow rate on load side of heat pump (gpm)

Applying Equation 2.25 to the data in Example 2.23 yields 105.8 F, which agrees very closely to the tabled LLT of 106 F.

$$LLT_h = ELT + \frac{HC}{500 \bullet GPM_{load}} = 100\text{ F} + \frac{61{,}300\text{ Btu/hr}}{500 \bullet 21\text{ gpm}} = 105.8\text{ F}$$

The heating mode performance of a water-to-water GSHP will vary during the heating season primarily based on the EWT from the ground connection. ELT, GPM_{load}, and GPM are design parameters that are set by the designer and around which piping/pumping systems are designed. ELT should remain in a small range depending on the type of heating load. GPM_{load} should remain relatively constant assuming that all piping and pumps in the system remain clear of debris. GPM will also remain in a very narrow range as long as flow through the water-to-refrigerant coil on the ground connection side does not become restricted due to fouling of the coil. EWT, on the other hand, will vary during the heating season based on the type of ground connection that is utilized and is particularly important in proper sizing of the GSHP in heating mode. HC and COP are graphed in Figure 2.27, as a function of EWT, for the performance data provided in Figure 2.26 at operating parameters of GPM_{load}=21 and GPM=21 and for ELTs of 60, 80, 100, and 120 F. HC and COP are both strongly influenced by EWT, with both decreasing as EWT decreases (a tendency of a GSHP due to the nature of the vapor-compression cycle). HC is not strongly influenced by ELT, but COP shows large reductions with increasing ELT. If the heat pump is to be sized for heating, the minimum

EWT from the ground connection and the maximum ELT from the heated load must both be specified in order to determine the HC and COP of the heat pump at design conditions. The influence of design EWT and ELT on GSHP sizing and operating efficiency will be addressed in the selection of GSHP equipment to meet design heating loads (Chapter 3) and the design of the various ground coupling devices (Chapter 5).

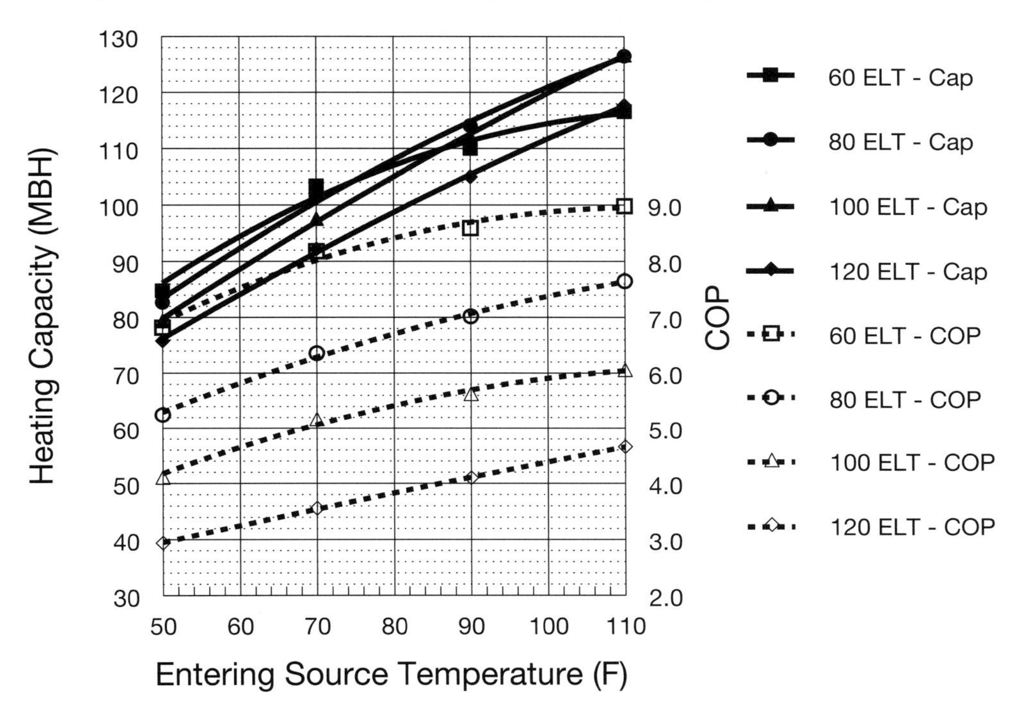

Figure 2.27. WaterFurnace EW Model 060 Heating Mode Performance Data

2.6.2 COOLING MODE PERFORMANCE DATA Cooling mode performance data for three EWTs are listed in Columns 6-13, 14-21, and 22-29 of Figure 2.28. Each line on the table represents three sets of performance data for a combination of ELT, EST, GPM_{load}, and GPM, as read proceeding from left to right across the table. Columns 7, 15, and 23 contain the cooling capacities (TC - kBtuH) provided to the water returning from the conditioned space. Columns 8, 16, and 24 list the electric demands (DMD - kW) of the unit, and Columns 10, 18, and 26 provide the EERs for each of the three ground source GPMs. The heats of rejection (HR - kBtuH) from the ground connection are provided in Columns 9, 17, and 25, and the leaving water temperatures back to the conditioned space (LLT - F) are shown in Columns 6, 14, and 22 for the three ground source GPMs. Other performance data that are available for each GPM include LSTs (leaving source temperatures) in Columns 11, 19, and 27 and the pressure drops in psi and head losses in feet of water equivalent pressure, respectively, through the water-to-refrigerant coil on the ground source side of the heat pump (Columns 12 and 13, 20 and 21, and 28 and 29). This manufacturer expresses energy rates in kBtuH to save space. (1 kBtuH = 1 MBtu/hr = 1,000 Btu/hr).

Column

1	2	3	4	5	6	7	8	9	10	11	12	13	14	15	16	17	18	19	20	21	22	23	24	25	26	27	28	29
ELT	EST	Load Flow			Source Flow - 7 GPM								Source Flow - 14 GPM								Source Flow - 21 GPM							
°F	°F	Flow	PD		LLT	TC	Power	HR	EER	LST	Source PD		LLT	TC	Power	HR	EER	LST	Source PD		LLT	TC	Power	HR	EER	LST	Source PD	
		GPM	PSI	FT HD	°F	kBTUH	kW	kBTUH		°F	PSI	FT HD	°F	kBTUH	kW	kBTUH		°F	PSI	FT HD	°F	kBTUH	kW	kBTUH		°F	PSI	FT HD
	50	7	1.2	2.7																								
		14	3.1	7.2																								
		21	5.5	12.7																								
	60	7	1.2	2.7																								
		14	3.1	7.2																								
30		21	5.5	12.7	26	43.2	3.03	53.5	14.2	75	1.1	2.5	26	46.7	3.20	57.6	14.6	68	2.8	6.6	25	49.1	3.31	60.4	14.9	66	5.0	11.6
	80	7	1.2	2.7																								
		14	3.1	7.2	24	39.4	4.28	54.0	9.2	95	1.0	2.3	24	42.5	4.52	58.0	9.4	88	2.7	6.2	24	44.7	4.66	60.6	9.6	86	4.7	10.9
		21	5.5	12.7	26	39.7	4.13	53.8	9.6	95	1.0	2.3	26	42.9	4.35	57.8	9.9	88	2.7	6.2	26	45.1	4.50	60.5	10.0	86	4.7	10.9
	100	7	1.2	2.7																								
		14	3.1	7.2	25	36.0	5.39	54.4	6.7	116	0.9	2.2	24	38.9	5.69	58.3	6.8	108	2.5	5.8	24	40.9	5.87	60.9	7.0	106	4.4	10.2
		21	5.5	12.7	27	36.3	5.20	54.0	7.0	115	0.9	2.2	26	39.2	5.48	57.9	7.2	108	2.5	5.8	26	41.2	5.66	60.6	7.3	106	4.4	10.2
	50	7	1.1	2.6	33	59.2	3.07	69.7	19.3	70	1.1	2.6	32	64.0	3.24	75.1	19.7	61	2.9	6.8	31	67.3	3.35	78.7	20.1	57	5.2	12.0
		14	2.9	6.8	41	59.9	2.93	69.9	20.4	70	1.1	2.6	41	64.8	3.09	75.4	20.9	61	2.9	6.8	40	68.1	3.20	79.0	21.3	58	5.2	12.0
		21	5.2	12.0	44	60.5	2.83	70.1	21.4	70	1.1	2.6	44	65.4	2.98	75.6	21.9	61	2.9	6.8	43	68.7	3.08	79.2	22.3	58	5.2	12.0
	60	7	1.1	2.6	34	55.1	3.33	66.5	16.5	79	1.1	2.5	33	59.6	3.52	71.6	16.9	70	2.8	6.6	32	62.6	3.63	75.0	17.2	67	5.0	11.6
		14	2.9	6.8	42	55.8	3.18	66.7	17.5	79	1.1	2.5	41	60.3	3.36	71.8	18.0	70	2.8	6.6	41	63.4	3.47	75.2	18.3	67	5.0	11.6
50		21	5.2	12.0	45	56.3	3.07	66.8	18.4	79	1.1	2.5	44	60.9	3.24	71.9	18.8	70	2.8	6.6	44	64.0	3.34	75.4	19.1	67	5.0	11.6
	80	7	1.1	2.6	35	51.6	4.61	67.3	11.2	99	1.0	2.3	34	55.8	4.86	72.4	11.5	90	2.7	6.2	33	58.6	5.02	75.8	11.7	87	4.7	10.9
		14	2.9	6.8	43	52.2	4.40	67.2	11.9	99	1.0	2.3	42	56.5	4.64	72.3	12.2	90	2.7	6.2	42	59.4	4.79	75.7	12.4	87	4.7	10.9
		21	5.2	12.0	45	52.7	4.24	67.2	12.4	99	1.0	2.3	45	57.0	4.47	72.2	12.7	90	2.7	6.2	44	59.9	4.62	75.6	13.0	87	4.7	10.9
	100	7	1.1	2.6	36	48.6	5.74	68.1	8.5	119	0.9	2.2	35	52.5	6.05	73.2	8.7	110	2.5	5.8	34	55.2	6.25	76.5	8.8	107	4.4	10.2
		14	2.9	6.8	43	49.2	5.48	67.9	9.0	119	0.9	2.2	42	53.2	5.78	72.9	9.2	110	2.5	5.8	42	55.9	5.97	76.2	9.4	107	4.4	10.2
		21	5.2	12.0	45	49.6	5.28	67.6	9.4	119	0.9	2.2	45	53.6	5.57	72.6	9.6	110	2.5	5.8	45	56.4	5.75	76.0	9.8	107	4.4	10.2
	50	7	1.0	2.4	49	73.7	3.12	84.4	23.6	74	1.1	2.6	47	79.7	3.30	90.9	24.2	63	2.9	6.8	46	83.8	3.40	95.4	24.6	59	5.2	12.0
		14	2.7	6.4	59	74.6	2.98	84.8	25.0	74	1.1	2.6	58	80.7	3.15	91.4	25.6	63	2.9	6.8	58	84.8	3.25	95.9	26.1	59	5.2	12.0
		21	4.9	11.3	63	75.3	2.87	85.1	26.2	74	1.1	2.6	62	81.4	3.03	91.7	26.8	63	2.9	6.8	62	85.6	3.13	96.2	27.3	59	5.2	12.0
	60	7	1.0	2.4	50	68.6	3.39	80.2	20.2	83	1.1	2.5	49	74.2	3.58	86.4	20.7	72	2.8	6.6	48	78.0	3.69	90.6	21.1	69	5.0	11.6
		14	2.7	6.4	60	69.5	3.23	80.5	21.5	83	1.1	2.5	59	75.1	3.41	86.8	22.0	72	2.8	6.6	59	79.0	3.52	91.0	22.4	69	5.0	11.6
70		21	4.9	11.3	63	70.1	3.12	80.7	22.5	83	1.1	2.5	63	75.8	3.29	87.0	23.0	72	2.8	6.6	62	79.7	3.40	91.3	23.4	69	5.0	11.6
	80	7	1.0	2.4	51	65.0	4.74	81.2	13.7	103	1.0	2.3	50	70.3	5.00	87.4	14.0	92	2.7	6.2	49	73.9	5.17	91.5	14.3	89	4.7	10.9
		14	2.7	6.4	61	65.8	4.53	81.3	14.5	103	1.0	2.3	60	71.2	4.78	87.5	14.9	92	2.7	6.2	59	74.8	4.93	91.6	15.2	89	4.7	10.9
		21	4.9	11.3	64	66.4	4.36	81.3	15.2	103	1.0	2.3	63	71.8	4.60	87.5	15.6	92	2.7	6.2	63	75.5	4.75	91.7	15.9	89	4.7	10.9
	100	7	1.0	2.4	52	62.5	5.81	82.3	10.7	124	0.9	2.2	51	67.5	6.13	88.4	11.0	113	2.5	5.8	50	71.0	6.33	92.6	11.2	109	4.4	10.2
		14	2.7	6.4	61	63.2	5.55	82.2	11.4	123	0.9	2.2	60	68.4	5.85	88.3	11.7	113	2.5	5.8	60	71.9	6.04	92.5	11.9	109	4.4	10.2
		21	4.9	11.3	64	63.8	5.35	82.0	11.9	123	0.9	2.2	63	69.0	5.64	88.2	12.2	113	2.5	5.8	63	72.5	5.83	92.4	12.4	109	4.4	10.2
	50	7	1.0	2.2	66	82.7	3.14	93.4	26.3	77	1.1	2.6	64	89.4	3.31	100.7	27.0	64	2.9	6.8	63	93.9	3.42	105.6	27.4	60	5.2	12.0
		14	2.6	6.0	78	83.7	3.00	93.9	27.9	77	1.1	2.6	77	90.5	3.16	101.3	28.6	64	2.9	6.8	76	95.1	3.27	106.2	29.1	60	5.2	12.0
		21	4.5	10.5	82	84.4	2.89	94.3	29.2	77	1.1	2.6	81	91.3	3.05	101.7	29.9	65	2.9	6.8	81	95.9	3.15	106.7	30.5	60	5.2	12.0
	60	7	1.0	2.2	68	77.0	3.41	88.6	22.6	85	1.1	2.5	66	83.2	3.60	95.5	23.1	74	2.8	6.6	65	87.5	3.71	100.1	23.6	70	5.0	11.6
		14	2.6	6.0	79	77.9	3.25	89.0	24.0	85	1.1	2.5	78	84.2	3.43	95.9	24.5	74	2.8	6.6	77	88.5	3.54	100.6	25.0	70	5.0	11.6
90		21	4.5	10.5	83	78.6	3.14	89.3	25.1	86	1.1	2.5	82	85.0	3.31	96.3	25.7	74	2.8	6.6	81	89.3	3.42	101.0	26.1	70	5.0	11.6
	80	7	1.0	2.2	69	73.2	4.84	89.7	15.1	106	1.0	2.3	67	79.2	5.10	96.6	15.5	94	2.7	6.2	66	83.2	5.27	101.2	15.8	90	4.7	10.9
		14	2.6	6.0	79	74.1	4.61	89.9	16.1	106	1.0	2.3	79	80.2	4.87	96.8	16.5	94	2.7	6.2	78	84.3	5.03	101.4	16.8	90	4.7	10.9
		21	4.5	10.5	83	74.8	4.45	90.0	16.8	106	1.0	2.3	82	80.9	4.69	96.9	17.2	94	2.7	6.2	82	85.0	4.85	101.5	17.5	90	4.7	10.9
	100	7	1.0	2.2	70	70.8	5.91	91.0	12.0	126	0.9	2.2	68	76.5	6.24	97.8	12.3	114	2.5	5.8	67	80.4	6.44	102.4	12.5	110	4.4	10.2
		14	2.6	6.0	80	71.7	5.64	90.9	12.7	126	0.9	2.2	79	77.5	5.95	97.8	13.0	114	2.5	5.8	78	81.4	6.15	102.4	13.2	110	4.4	10.2
		21	4.5	10.5	83	72.3	5.44	90.9	13.3	126	0.9	2.2	83	78.2	5.74	97.7	13.6	114	2.5	5.8	82	82.2	5.93	102.4	13.9	110	4.4	10.2
	50	7	0.9	2.1	82	98.5	3.31	109.8	29.8	81	1.1	2.6	80	106.5	3.49	118.4	30.5	67	2.9	6.8	78	112.0	3.61	124.3	31.0	62	5.2	12.0
		14	2.4	5.5	96	99.8	3.16	110.5	31.6	82	1.1	2.6	95	107.8	3.33	119.2	32.3	67	2.9	6.8	94	113.4	3.44	125.1	32.9	62	5.2	12.0
		21	4.2	9.8	100	100.6	3.05	111.0	33.0	82	1.1	2.6	100	108.8	3.21	119.8	33.8	67	2.9	6.8	99	114.4	3.32	125.7	34.5	62	5.2	12.0
	60	7	0.9	2.1	84	91.7	3.59	104.0	25.5	90	1.1	2.5	82	99.2	3.79	112.1	26.2	76	2.8	6.6	80	104.3	3.91	117.6	26.6	71	5.0	11.6
		14	2.4	5.5	97	92.9	3.43	104.6	27.1	90	1.1	2.5	96	100.4	3.62	112.8	27.8	76	2.8	6.6	95	105.6	3.73	118.3	28.3	71	5.0	11.6
110		21	4.2	9.8	101	93.7	3.31	105.0	28.3	90	1.1	2.5	100	101.3	3.49	113.2	29.0	76	2.8	6.6	100	106.5	3.60	118.8	29.6	71	5.0	11.6
	80	7	0.9	2.1	85	88.4	4.97	105.4	17.8	110	1.0	2.3	83	95.6	5.25	113.5	18.2	96	2.7	6.2	81	100.5	5.42	119.0	18.5	91	4.7	10.9
		14	2.4	5.5	97	89.5	4.75	105.7	18.9	110	1.0	2.3	96	96.8	5.01	113.9	19.3	96	2.7	6.2	95	101.7	5.17	119.4	19.7	91	4.7	10.9
		21	4.2	9.8	101	90.3	4.58	105.9	19.7	110	1.0	2.3	101	97.6	4.83	114.1	20.2	96	2.7	6.2	100	102.6	4.99	119.6	20.6	91	4.7	10.9
	100	7	0.9	2.1	85	86.5	6.00	106.9	14.4	131	0.9	2.2	83	93.5	6.33	115.0	14.8	116	2.5	5.8	82	98.2	6.53	120.5	15.0	111	4.4	10.2
		14	2.4	5.5	97	87.5	5.72	107.1	15.3	131	0.9	2.2	96	94.6	6.04	115.2	15.7	116	2.5	5.8	96	99.5	6.23	120.7	16.0	111	4.4	10.2
		21	4.2	9.8	102	88.3	5.52	107.1	16.0	131	0.9	2.2	101	95.5	5.82	115.3	16.4	116	2.5	5.8	100	100.3	6.01	120.8	16.7	112	4.4	10.2

Notes: Multiple flow rates for source side and flow side are shown. The lowest source flow rate shown is used for geothermal open loop/ well water systems with a minimum 50 F. The second source flow rate shown is the minimum closed loop flow rate. The third source flow rate shown is optimum for geothermal closed loop and the suggested flow rate for boiler/tower applications. When selecting units and designing the system, actual operating parameters must fall within the temperature and flow rate ranges shown on the table. Using temperature/flow rate combinations outside the range of the table will result in performance problems. *For 3 phase capacity, multiply above data by .948. For 3 phase power, multiply the above data by .943.

Rev, 8/05

Figure 2.28. WaterFurnace EW Model 060 Cooling Mode Performance Data

Example 2.24 An example of using the performance data to determine cooling mode performance is highlighted with dashed outlines on Figure 2.28 for ELT=50 F, EWT=80 F, GPM_{load}=21, and GPM = 21. For those operating parameters:

TC = 59.9 kBtuH x 1,000 Btu/hr/kBtuH = 59,900 Btu/hr	Column 23
DMD = 4.62 kW	Column 24
EER = 13.0	Column 26
HR = 75.6 kBtuH x 1,000 = 75,600 Btu/hr	Column 25
LLT = 44 F	Column 22
LST = 87 F	Column 27

Equation 2.4 can be used to verify the tabled HR based on the TC and EER. LST (LWT_c) is calculated using Equation 2.11 along with the EWT, HR, and GPM from the table. Leaving load water temperature (LLT_c) from the heat pump in cooling mode can be calculated according to Equation 2.26, which is based on thermal properties of water in the 32 to 105 F range.

$$LLT_c = ELT - \frac{TC}{500 \bullet GPM_{load}} \quad \textbf{Equation 2.26}$$

Where:

LLT_c = leaving load water temperature from heat pump in cooling mode (F)
ELT = entering load water temperature from conditioned space (F)
TC = cooling capacity of heat pump (Btu/hr)
GPM_{load} = water flow rate on load side of heat pump (gpm)

Applying Equation 2.26 to the data in Example 2.24 yields 44.3 F, which agrees very closely to the tabled LLT of 44 F.

$$LLT_c = ELT - \frac{TC}{500 \bullet GPM_{load}} = 50\ F - \frac{59{,}900\ Btu/hr}{500 \bullet 21\ gpm} = 44.3\ F$$

The cooling mode performance of a water-to-water GSHP will vary during the cooling season primarily based on the EWT from the ground connection. ELT, GPM_{load}, and GPM are design parameters that are set by the designer and around which piping/pumping systems are designed. ELT should remain in a small range depending on the type of cooling load. GPM_{load} should remain relatively constant assuming that piping and pumps in the system remain clear of debris. GPM will also remain in a very narrow range as long as flow through the water-to-refrigerant coil on the ground connection side does not become restricted due to fouling of the coil. EWT, on the other hand, will vary during the cooling season based on the type of ground connection that is utilized and is particularly important in proper sizing of the GSHP in cooling mode. TC and EER are graphed, as a function of EWT, in Figure 2.29 for the performance data provided in Figure 2.28 at operating parameters of GPM_{load}=21 and GPM=21 and for ELTs of 30, 50, 70, and 90 F. Unlike HC in the heating mode of operation, TC is not very sensitive to EWT. However, EER is very dependent on EWT and decreases rapidly with increasing EWT. Unlike heating mode of operation, both TC and EER are very sensitive to ELT, with both decreasing rapidly with smaller values of ELT. If the heat pump is to be sized for cooling, the maximum EWT from the ground connection and the minimum ELT from the cooling load must both be specified in order to determine the HC and COP of the heat pump at design conditions.

The influence of design EWT on GSHP sizing and operating efficiency will be addressed in the selection of GSHP equipment to meet design cooling loads (Chapter 3) and the design of the various ground coupling devices (Chapter 5).

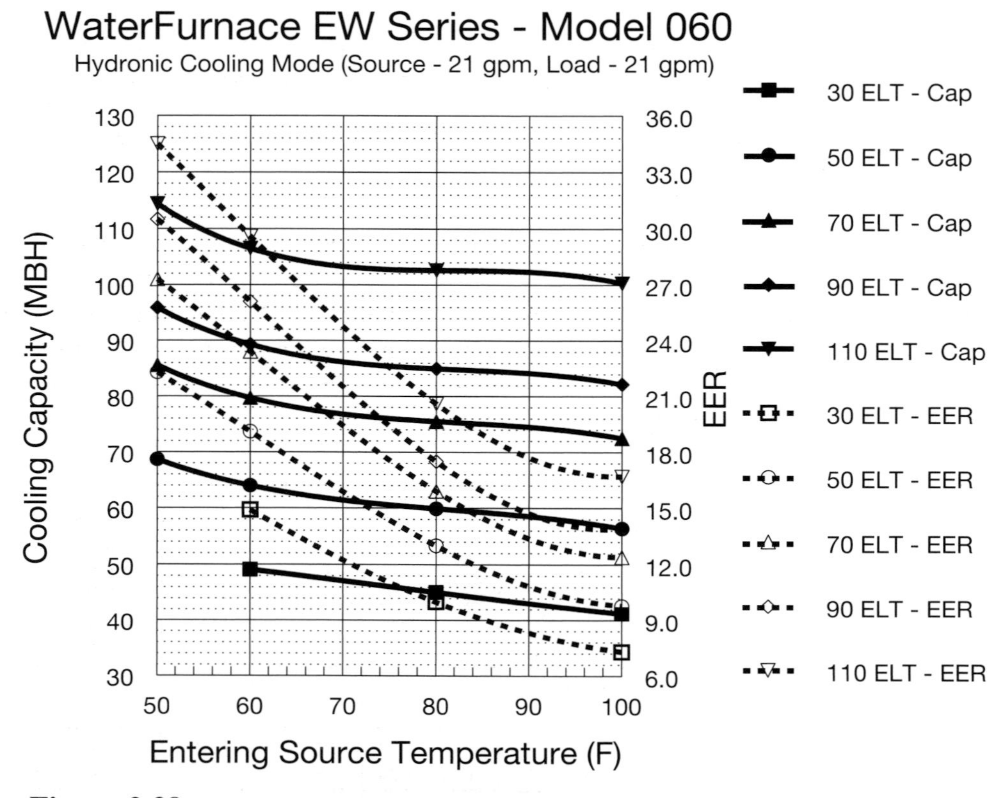

Figure 2.29. WaterFurnace EW Model 060 Cooling Mode Performance Data

2.6.3 ISO/ARI RATINGS The Air-Conditioning and Refrigeration Institute (ARI), American Society of Heating, Refrigeration, and Air-Conditioning Engineers (ASHRAE), and the International Standards Organization *(ISO, 1998b)* have developed certification standards *(ISO 13256-2)* to set requirements for testing and rating of water-to-water heat pump equipment to ensure that performance ratings published by manufacturers agree with the performance of their manufactured product.

Figure 2.30 provides test conditions for both single-capacity and two-capacity, water-to-water heat pumps. For two-capacity heat pumps, the full-load performance is tested at the same standard rating test conditions as single-capacity heat pumps. Part-load rating test conditions allow for the manufacturer to specify a lower GPM and CFM, and the entering water temperatures on the ground connection side are set at 41 F for heating mode and 68 F for cooling mode for ground-loop heat pumps. Entering water conditions on the load side for ground-loop heat pumps (ELT=104 F for heating and 53.6 F for cooling) are the same for both full-load and part-load operation.

	Water-loop heat pumps		Ground-water heat pumps		Ground-loop heat pumps	
	Cooling	Heating	Cooling	Heating	Cooling	Heating
Liquid entering indoor side	12 (53.6)	40 (104)	12 (53.6)	40 (104)	12 (53.6)	40 (104)
Air surrounding unit - dry bulb, C (F)	15-30 (59-86)	15-30 (59-86)	15-30 (59-86)	15-30 (59-86)	15-30 (59-86)	15-30 (59-86)
Standard Rating Test Liquid entering heat exchanger, C (F)	30 (86)	20 (69)	15 (59)	10 (50)	25 (77)	0 (32)
Part-load Rating Test Liquid entering heat exchanger, C (F)	30 (86)	20 (69)	15 (59)	10 (50)	20 (68)	5 (41)
Frequency*	Rated	Rated	Rated	Rated	Rated	Rated
Voltage**	Rated	Rated	Rated	Rated	Rated	Rated

* Equipment with dual-rated frequencies shall be tested at each frequency
** Equipment with dual-rated voltages shall be tested at both voltages or at the lower of the two voltages if only a single rating is published.

Water and load-side water flow rates specified by manufacturer.

Figure 2.30. ARI/ISO 13256-2 Test Condition Table

ARI/ISO performance ratings for the WaterFurnace EW Series water-to-water heat pumps are provided in Figure 2.31. Performance ratings (HC, DMD, and COP for heating and TC, DMD, and EER for cooling) are provided at two different source and load water flow rates for GLHP entering water conditions. For the Model 060, the GLHP ARI full-load performance ratings are identified using solid outlines for heating mode and dashed outlines for cooling mode. All capacities and efficiencies include pumping power penalties minus the effects of the inside piping and the ground loop.

Heating Capacity Data

MODEL	SOURCE						LOAD						CAPACITY		COP
	FLOW			SWPD			FLOW			LWPD					
	EWT	GPM	L/s	PSI	FT HD	BAR	EWT	GPM	L/s	PSI	FT HD	BAR	HEAT OUTPUT	kW	
EW020	32	6	0.38	4.3	9.9	0.3	104	3	0.19	1.2	2.8	0.1	22.2	6.5	2.8
	32	9	0.57	9.0	20.8	0.6	104	9	0.57	9	20.8	0.6	22.5	6.6	3.0
EW030	32	8	0.50	2.8	6.5	0.2	104	5	0.32	1.4	3.2	0.1	29.0	8.5	3.0
	32	12	0.76	5.6	12.9	0.4	104	12	0.76	5.6	12.9	0.4	29.5	8.6	3.1
EW042	32	11	0.69	2.5	5.8	0.2	104	7	0.44	1.2	2.8	0.1	44.9	13.2	2.9
	32	16.5	1.04	5.6	12.9	0.4	104	16.5	1.04	5.6	12.9	0.4	45.5	13.3	3.0
EW060*	32	14	0.88	2.7	6.2	0.2	104	11	0.69	1.8	4.2	0.1	59.2	17.4	2.8
	32	21	1.32	5.7	13.2	0.4	104	21	1.32	5.7	13.2	0.4	60.7	17.8	3.0

Notes: Based on 60 Hz., single phase. For more detailed data, refer to pages 14-20.
* For 3 phase EW060 capacity, multiply above data by .948. For 3 phase EW060 power, multiply above data by .943.

Cooling Capacity Data

MODEL	SOURCE						LOAD						CAPACITY		EER
	FLOW			SWPD			FLOW			LWPD					
	EWT	GPM	L/s	PSI	FT HD	BAR	EWT	GPM	L/s	PSI	FT HD	BAR	MBTU	kW	
EW030R	77	8	0.50	2.8	6.5	0.2	53.6	8.0	0.50	2.8	6.5	0.2	21.9	6.42	10.67
	77	12	0.76	5.6	12.9	0.4	53.6	12.0	0.76	5.6	12.9	0.4	22.1	6.48	10.93
EW042R	77	11	0.69	2.5	5.8	0.2	53.6	11.0	0.69	2.5	5.8	0.2	42.7	12.51	12.42
	77	16.5	1.04	5.6	12.9	0.4	53.6	16.5	1.04	5.6	12.9	0.4	43.2	12.66	12.86
EW060R*	77	14	0.88	2.7	6.2	0.2	53.6	14.0	0.88	2.7	6.2	0.2	58.0	17.00	12.54
	77	21	1.32	5.7	13.2	0.4	53.6	21.0	1.32	5.7	13.2	0.4	61.1	17.91	13.47

Notes: Based on 60 Hz., single phase. For more detailed data, refer to pages 14-20.
* For 3 phase EW060R capacity, multiply above data by .948. For 3 phase EW060R power, multiply above data by .943.

Figure 2.31. WaterFurnace EW Series Performance Data – Ground-loop application

2.7 Design Criteria for GSHP Equipment Selection

Sections 2.3.3.1 and 2.3.3.2 discussed the parameters that affect the performance of water-to-air and water-to-water GSHPs, respectively. For water-to-air GSHPs, those parameters included entering water temperature (EWT), entering air temperature (EAT), water flow rate (GPM), and air flow rate (CFM). For water-to-water GSHPs, those parameters include entering water temperature (EWT), entering water temperature – load side (ELT) (or leaving water temperature - load side (LLT)) water flow rate (GPM), and water flow rate - load side (GPM_{load}).

2.7.1 WATER-TO-AIR GSHP EQUIPMENT

2.7.1.1 HEATING MODE GSHP SELECTION Entering air temperature and air flow rate are generally dictated by the space temperature setpoint (usually about 70 F, but may vary considerably depending on the application) and the required air flow rate for the GSHP model (approximately 350 to 400 CFM per nominal ton), and the ductwork must be sized to obtain the required air flow rate for proper GSHP performance. Correction factors were introduced to account for entering air temperature and air flow rate other than what the GSHP heating mode performance data were reported for along with antifreeze. Water-to-air GSHP selection worksheets will be provided in Chapter 3 for single and dual-capacity GSHP equipment that allow for complete documentation of GSHP heating mode performance.

Entering water temperature from the closed-loop GHEX and water flow rate through the GSHP are two other parameters that affect GSHP performance, but also relate directly to the design of the closed-loop GHEX. The water flow rate is generally selected as the highest flow rate for which performance data is available (commonly 3 gpm per nominal ton), and the GHEX must be laid out to accommodate the required water flow rate (including antifreeze for most design heating mode applications) to ensure turbulence at design heating conditions while controlling the head loss through the GHEX to keep the required circulating pump power as small as possible.

The minimum entering water temperature used to size the GSHP and design the GHEX at design heating conditions is the one design parameter that may be varied by the designer. However, the location of the installation, the GHEX configuration, the soil or rock type into which the GHEX will be installed, and the economics of the GHEX installation will place practical limits on the highest minimum entering water temperature that can be designed for, while GSHP heating performance will limit the lowest minimum entering water temperature that can be designed for. As a starting point for design, the minimum entering water temperature should be 15 to 20 F below the deep-earth temperature at the installation location or 25 F, whichever is greater. In locations with higher deep-earth temperatures, the minimum entering temperatures may be set higher with resulting GHEX lengths being longer, but still of reasonable length (refer to Chapter 5 on loop design). Typically, for locations with higher deep-earth temperatures, GHEX design is dictated by the cooling load requirements, not heating load requirements. As deep-earth temperature gets lower, the minimum entering water temperature must be reduced to keep GHEX lengths reasonable, and the final design minimum entering water temperature value will be a trade-off between GHEX lengths and resulting cost of installation (higher values of EWT_{min}) and GSHP heating performance (lower values of EWT_{min}). It is not uncommon for the designer to start with a 30 F minimum entering water temperature and adjust downward or upward depending on location.

2.7.1.2 COOLING MODE GSHP SELECTION Entering dry bulb and wet bulb air temperature and air flow rate are generally dictated by the space temperature setpoint (usually about 75 F and 50% relative humidity, but may vary considerably depending on the application) and the required air flow rate for the GSHP model (approximately 350 to 400 CFM per nominal ton), and the ductwork must be sized to obtain the required air flow rate for proper GSHP performance. Correction factors were introduced to account for entering air temperatures and air flow rate other than what the GSHP cooling mode performance data were reported for along with antifreeze. Water-to-air GSHP selection worksheets will be provided in Chapter 3 for single and dual-capacity GSHP equipment that allow for the complete documentation of GSHP cooling mode performance.

Entering water temperature from the closed-loop GHEX and water flow rate through the GSHP are two other parameters that affect GSHP performance, but also relate directly to the design of the closed-loop GHEX. The water flow rate is generally selected as the highest flow rate for which performance data is available (commonly 3 gpm per nominal ton), and the GHEX must be laid out to accommodate the required water flow rate (no antifreeze for most design cooling mode applications) to ensure turbulence at worst-case heating conditions while controlling the head loss through the GHEX to keep the required circulating pump power as small as possible.

The maximum entering water temperature used to size the GSHP and design the GHEX at design cooling conditions is the one design parameter that may be varied by the designer. However, the location of the installation, the GHEX configuration, the soil or rock type into which the GHEX will be installed, and the economics of the GHEX installation will place practical limits on the lowest maximum entering water temperature that can be designed for, while GSHP cooling performance will limit the highest maximum entering water temperature that can be designed for. As a starting point for design, the maximum entering water temperature should be 30 to 40 F above the deep-earth temperature at the installation location or 95 F, whichever is less. In locations with lower deep-earth temperatures, the maximum entering temperatures may be set lower with resulting GHEX lengths being longer, but still of reasonable length. Typically, for locations with higher deep-earth temperatures, GHEX design is dictated by the cooling load requirements, not heating load requirements. As deep-earth temperature gets higher, the maximum entering water temperature must be increased to keep GHEX lengths reasonable, and the final design maximum entering water temperature value will be a trade-off between GHEX lengths and resulting cost of installation (lower values of EWT_{max}) and GSHP cooling performance (higher values of EWT_{max}). It is not uncommon for the designer to start with a 90 F maximum entering water temperature and adjust upward or downward depending on location.

2.7.2 WATER-TO-WATER GSHP EQUIPMENT

2.7.2.1 HEATING MODE GSHP SELECTION Leaving load water temperature and load water flow rate are generally dictated by the design of the heating distribution system (radiant floor distribution, fan coil units, etc.). The required load water flow rate for the GSHP model (approximately 2 to 3 gpm per nominal ton) and the piping distribution and pumping system must be sized to obtain the required water flow rate for proper GSHP performance. Correction factors were introduced to account for antifreeze. A water-to-water GSHP selection worksheet will be provided in Chapter 3 that allows for complete documentation of GSHP heating mode performance.

Entering water temperature from the closed-loop GHEX and water flow rate through the GSHP are two other parameters that affect GSHP performance, but also relate directly to the design of the closed-loop GHEX. The water flow rate is generally selected as the highest flow rate for which performance data is

available (commonly 3 gpm per nominal ton), and the GHEX must be laid out to accommodate the required water flow rate (including antifreeze for most design heating mode applications) to ensure turbulence at design heating conditions while controlling the head loss through the GHEX to keep the required circulating pump power as small as possible.

The minimum entering water temperature used to size the GSHP and design the GHEX at design heating conditions is the one design parameter that may be varied by the designer. However, the location of the installation, the GHEX configuration, the soil or rock type into which the GHEX will be installed, and the economics of the GHEX installation will place practical limits on the highest minimum entering water temperature that can be designed for, while GSHP heating performance will limit the lowest minimum entering water temperature that can be designed for. As a starting point for design, the minimum entering water temperature should be 15 to 20 F below the deep-earth temperature at the installation location or 25 F, whichever is greater. In locations with higher deep-earth temperatures, the minimum entering temperatures may be set higher with resulting GHEX lengths being longer, but still of reasonable length. Typically, for locations with higher deep-earth temperatures, GHEX design is dictated by the cooling load requirements, not heating load requirements. As deep-earth temperature gets lower, the minimum entering water temperature must be reduced to keep GHEX lengths reasonable, and the final design minimum entering water temperature value will be a trade-off between GHEX lengths and resulting cost of installation (higher values of EWT_{min}) and GSHP heating performance (lower values of EWT_{min}). It is not uncommon for the designer to start with a 30 F minimum entering water temperature and adjust downward or upward depending on location.

2.7.2.2 COOLING MODE GSHP SELECTION Leaving load water temperature and load water flow rate are generally dictated by the design of the cooling distribution system (fan coil units, etc.) and the required load water flow rate for the GSHP model (approximately 2 to 3 GPM per nominal ton), and the piping distribution and pumping system must be sized to obtain the required water flow rate for proper GSHP performance. Correction factors were introduced to account for antifreeze. A water-to-water GSHP selection worksheet will be provided in Chapter 3 that allows for complete documentation of GSHP cooling mode performance.

Entering water temperature from the closed-loop GHEX and water flow rate through the GSHP are two other parameters that affect GSHP performance, but also relate directly to the design of the closed-loop GHEX. The water flow rate is generally selected as the highest flow rate for which performance data is available (commonly 3 gpm per nominal ton), and the GHEX must be laid out to accommodate the required water flow rate (no antifreeze for most design cooling mode applications) to ensure turbulence at worst-case heating conditions while controlling the head loss through the GHEX to keep the required circulating pump power as small as possible.

The maximum entering water temperature used to size the GSHP and design the GHEX at design cooling conditions is the one design parameter that may be varied by the designer. However, the location of the installation, the GHEX configuration, the soil or rock type into which the GHEX will be installed, and the economics of the GHEX installation will place practical limits on the lowest maximum entering water temperature that can be designed for, while GSHP cooling performance will limit the highest maximum entering water temperature that can be designed for. As a starting point for design, the maximum entering water temperature should be 30 to 40 F above the deep-earth temperature at the installation location or 95 F, whichever is less. In locations with lower deep-earth temperatures, the maximum entering temperatures may be set lower with resulting GHEX lengths being longer, but still of reasonable length. Typically, for locations with higher deep-earth temperatures, GHEX design is dictated by the cooling load requirements, not heating load requirements. As deep-earth temperature gets higher, the maximum entering water temperature must be increased to keep GHEX lengths reasonable, and the final design maximum entering water temperature value will be a trade-off between GHEX lengths and resulting cost of installation (lower values of EWT_{max}) and GSHP cooling performance (higher values of EWT_{max}). It is not uncommon for the designer to start with a 90 F maximum entering water temperature and adjust upward or downward depending on location.

SELECTING A GSHP FOR HEATING & COOLING LOADS

In This Section

3.1 Overall System Design Procedure for the GSHP

Following a step-by-step procedure during the design phase of a GSHP installation not only minimizes overall process complication, but it also allows for exploration of alternative designs. The recommended procedure for designing a GSHP system is as follows:

- Determine the building's heating and cooling design loads
- Select a properly sized heat pump system and earth connection type (determination of proper connection type in addition to ground heat exchanger (GHEX) sizing are discussed in detail in Chapter 5)
- Select the type and location of the indoor distribution system
- Select the proper type and location of the air supply diffusers, registers, and return grilles for the air distribution system
- Size the indoor air distribution system
- Estimate the heat pump system's annual energy usage given:
  - The building's heating and cooling loads
  - The type and size of equipment selected and its performance as it relates to the building loads, GHEX design, distribution system design, weather conditions, and the type of earth connection used
  - The climatic conditions (as given by bin weather data for the area)
- Estimate the ground heat exchanger loads, both annually and for the design month
- Design the GHEX based on loads, earth connection type, soil properties, etc.

As just noted, the complete design of a building's GSHP system involves several steps. Each is important for satisfactory system performance. However, the first step in designing such a system is determining the building's heating and cooling design loads. This is because the calculation of these loads is the most crucial

step; everything else in the system is dependent on their values. Basic load calculations are referred to as design loads. Two other types of loads are involved in load calculations: energy loads and ground loads. The purposes of the different load types are as follows:

Design loads (Section 3.2) are used to select the equipment for a system (such as a heat pump) and to design the air distribution system (supply air diffusers, return air grilles, and the duct system). Design loads are based on standard or accepted conditions for a given locality (a design day).

Energy loads (Section 3.6) are used in predicting the energy necessary to operate the system for some prescribed time such as a month, year, or season. The calculation methodology may be the same as for the design load; however, the actual operating and weather data are used instead of design conditions.

Ground loads (Section 3.6) are associated with ground source systems and relate to the design of the GHEX. In principle, these calculations are similar to the energy loads except the ground load is heat rejected to the ground (cooling mode) or removed from it (heating mode).

3.2 Heating and Cooling Design Load Calculation Procedures

For any type of building, it is bad practice to use "rules of thumb" when calculating design loads. These loads depend on conditions that are individual to the building such as location, construction type and quality, orientation (with respect to the sun and wind direction), and use. Floor area to tonnage ratios for homes in the United States can vary from 500 ft^2 to 1,500 ft^2 per ton *(ACCA, 2006, Manual J)*. Simply rotating a home on the site can change this ratio due to differences in exposure to solar radiation and infiltration levels. The performance of a heating/cooling system and its ability to comfortably condition a building greatly depends on the accuracy of the load calculations used to size that equipment. Significant over sizing of a system can lead to poor temperature and humidity control, mold and mildew problems (in extreme cases and very humid climates), and lower overall system efficiency due to short-cycling. The problems associated with undersized equipment include poor temperature control and increased run hours (which dictates the need for more installed ground loop capacity in the case of a ground source heat pump) in addition to the need for a supplemental or backup heating system to handle the portion of the load that the primary system cannot in Northern climates. In some instances, problems associated with poorly sized equipment can result in legal action against responsible parties. The intention of this section is to explain all terms related to heat loss/gain calculations and to identify the parameters necessary to perform accurate design load calculations to be used in equipment sizing and selection.

3.2.1 ENVELOPE LOADS Envelope loads are the total heating or cooling loads through any building component surrounding a conditioned space, including fenestration and infiltration loads. Building loads due to heat transfer through the panels used to construct the home are due to the temperature difference between the indoor set point and the outdoor weather conditions. The greater the temperature difference between the conditioned space and outside weather conditions, the greater the building envelope load will be. In addition to outside weather conditions, the envelope load is affected by the construction quality of the home itself (insulation levels, tightness/leakiness of windows and doors, placement of windows or skylights relative to direct sunlight, etc). Fenestration loads are defined as heat transfer through glass panels (windows, glass doors, and skylights) and are obviously greatly affected by the location of a glass panel on a home with respect to the position of the sun. Infiltration loads are caused by outside air leaking into the home through cracks in windows, doors, wall panels, etc. Infiltration is clearly affected by wind speed and direction typical to the geographic location of a building. Many ways exist to measure/estimate infiltration levels for a specific home. Infiltration can be measured via a blower door test, or it can be estimated using several different methods (simplified ACH estimates, etc.). Care should be taken when estimating infiltration levels in a home because they can have significant impact on a building's peak loads. Simply changing the orientation of a home on site can affect infiltration loads because of typical wind direction and velocity for the area. A natural consequence of infiltration of outside air into the home is exfiltration of conditioned air out of the home.

A building's envelope load includes both the sensible load as well as the latent load. Typically, a building's heating load only includes the sensible load and a building's cooling load includes both the sensible and latent loads. Rare cases exist where latent load is considered in heating when air humidification equipment is utilized. Air humidification during the heating season will more than likely reduce peak heating requirements if humidification is to be consistently used throughout the season. The sensible load pertains to the dry bulb temperature of the conditioned space, which is the temperature displayed by the thermostat. The latent load in a building pertains to the amount of moisture in the air in the conditioned space, which is typically only a factor in the cooling season. The amount of moisture in the air is measured by both dry bulb and wet bulb temperature and may be expressed in terms of dry bulb temperature and relative humidity. Then, the total cooling load in a building is the total of the sensible and latent loads. Although the total cooling load is not used to size equipment, it is necessary to know because the ground heat exchanger for a ground source heat pump system will need to be sized to handle both the sensible and latent cooling loads (i.e. – the total cooling load). Sensible heat factor (SHF) is another term used in relation to a building's cooling load. The SHF for a building is the ratio of the building's sensible cooling load to its total cooling load.

Fenestration panels (glass windows, doors, etc.) will create both sensible and latent heating and cooling loads in a home. Infiltration into a building will also create both sensible and latent heating and cooling loads. Building occupants produce only cooling loads, both sensible and latent. The magnitude of the loads generated by the occupants depends on their activity levels (running versus sitting down, etc.). Appliances, computers, etc., produce only sensible cooling loads because they generate heat, but do not add moisture to the air *(ACCA, 2006, Manual J).*

When sizing equipment to serve a home, it should be sized to handle the greater of the peak heating or cooling load that it will need to satisfy. Generally, equipment is sized to handle either 100% of the home's peak cooling load or anywhere from 75-100% of its peak heating load. The determining factor depends on the type of equipment selected and which load (heating or cooling) is greater in magnitude and by how much. Equipment sizing is discussed later in this chapter.

3.2.1.1 HEATING LOADS When calculating design heating loads for a residence, the building envelope load is the only factor to be considered. The calculation of peak heating loads should not include heat gains from appliances, occupants, lighting, solar radiation, etc. The envelope heating load is the sum of the component loads for the building including heat loss through the floors, walls, ceilings, and doors in addition to fenestration and infiltration loads. Calculation of heating design loads is the same regardless of building type (houses, multi-family complexes, or commercial buildings).

3.2.1.2 COOLING LOADS When calculating design cooling loads for a building, not only do the envelope loads need to be considered, but also internal heat gains from appliances, occupants, lighting, solar radiation, etc. Similar to envelope heating loads, the envelope cooling load is the sum of the component loads

for the building including heat gain through the floors, walls, ceilings, and doors in addition to fenestration and infiltration loads. Gains due to solar radiation, especially fenestration gains, are sensitive to the orientation of the home with respect to the position of the sun. Calculation of cooling design envelope loads is the same regardless of building type. However, much more detail and attention must be paid to internal gain calculations for commercial structures because the contribution of such gains to a final design load estimate is usually very significant. Because of this, commercial buildings usually require hour-by-hour calculations that, when totaled, give all of the information necessary for each room, zone, and building. For residences and multi-family complexes, all reasonable internal gains need to be accounted for, such as gains from computers and printers in a home office, a reasonable number of occupants in a home (the family residing in the home plus a few guests), etc. However, equipment should not be sized to handle extreme situations such as large home gatherings.

3.2.2 ZONE LOADS A zone is generally a building area served by a single heating and cooling delivery system (air handler, heat pump, radiant piping from a flow center, etc). A zone load is the amount of heating or cooling that the delivery system must provide to satisfy the peak loads for that specific area, and a single thermostat is used to control the delivery system for that zone. An individual zone can be defined as a single room, a group of rooms, an entire building level, etc.

3.2.3 BLOCK LOADS The block load is defined to be the total of the zone loads. If a home is defined to be a single zone, then the peak load for the entire home will be both the peak zone and peak block loads. If a home is split up into several zones (main level and basement, for example), but is served by a single piece of equipment (two delivery systems, two thermostats, one heat pump), then two zone loads will exist and their total will be the block load for the heat pump system. The block load can vary throughout the year because zones in different locations of a building can be requiring different modes of equipment operation. The peak block load is the largest heating or cooling load that the piece of equipment will need to satisfy during the course of a year and is the load that should be used for equipment sizing.

3.2.4 EQUIPMENT LOADS Equipment loads are additional loads served by the heating/cooling system that are not included in peak heating/cooling block load calculations. These loads include duct and hydronic piping losses/gains as well as ventilation loads. In other words, the load that a piece of equipment will need to satisfy will be its peak block heating or cooling load in addition to heat losses or gains through ductwork or hydronic piping as well as loads imposed by ventilation.

3.2.4.1 DUCT/HYDRONIC PIPING LOADS Sometimes duct systems will be installed in unconditioned spaces such as a crawlspace, an unconditioned basement, or an unconditioned attic. In all such cases, a significant portion of the system's capacity can be lost through the ductwork before the air reaches the area that it serves, especially when uninsulated ductwork is utilized. Similar to a forced air ducted system, a hydronic system can experience significant losses from the piping to its surroundings. Careful consideration must be taken when such situations arise to ensure that the equipment is not undersized because of significant duct or piping losses/gains.

3.2.4.2 OUTSIDE AIR (VENTILATION) LOADS Modern construction practices have drastically reduced the infiltration rate for most buildings. Such practices aggravate indoor air quality (Section 3.3) issues because the best way to remove pollutants and odors from a home is to simply vent them to the outside. In some instances, the rate of infiltration for a building is not considered enough to satisfy fresh air requirements. It is in such instances that fresh air must be introduced into the building, imposing additional loads on the heating/cooling system. Many different types of ventilation systems exist, all of which affect peak loads differently. If outdoor air is simply brought into the return of a ducted system without any form of exhaust, the conditioned space will become pressurized. Such pressurization will serve to reduce the amount of infiltration allowed into the building. If a ventilation system is installed which draws outdoor air and exhausts it at the same rate, the building's infiltration level will be unchanged. Also, outside air can simply be brought into the home via a direct intake system (typically preconditioned by an auxiliary system), or it can be routed through an energy recovery heat exchanger system. If a recovery system is installed to reclaim some of the energy lost via exhausting conditioned air out of the building, the ventilation load will be reduced. Automatic dampers can also be utilized to stop outside air when the unit is not in operation. Innumerable options exist in regard to ventilation systems. Every situation needs to be closely scrutinized and analyzed to generate an accurate assessment of the ventilation load.

3.2.4.3 EQUIPMENT SIZING The peak load that a single piece of equipment will need to satisfy is the sum of the peak block heating/cooling loads and the equipment loads (duct, hydronic piping, and ventilation loads). Total system capacity must be sufficient to handle both the peak block heating and cooling loads and its additional equipment loads. Equipment selection is discussed in detail in Section 3.4.

3.2.5 LOAD CALCULATION PROCEDURES

Many acceptable methods exist to generate accurate estimates of peak heating and cooling loads. Load calculation procedures and software that follow recommendations set forth by *ASHRAE (2009) Handbook of Fundamentals* or *ACCA (2006) Manual J* or *Manual* J_{AE} are recommended. One must be careful that proper procedures are followed according to the type of building being analyzed. Different procedures are used for residential buildings than for commercial buildings (internal gain estimation procedures). Because most of the residential load calculation software available in the marketplace is written to comply with the standards and procedures set forth by *ACCA (2006) Manual J*, the suggested do's and don'ts for this methodology have been listed.

3.2.5.1 ACCA *MANUAL J* DO'S AND DON'TS Following are *Manual J* do's and don'ts per *ACCA (2006) Manual J 8th Edition Version 2*:

MANUAL J DO'S

- Use the outdoor design conditions recommended by Table 1A of *Manual J* (Software that utilizes *Manual J* to perform the calculations will comply with these recommendations).
- Use indoor conditions that are compatible with the comfort chart.
- Consider orientation of the structure on the site.
  - Use the actual orientation whenever possible.
  - Use "best case-worst case" load estimates for cookie cutter designs that may have varying site orientations when built.
- Verify all construction details prior to calculating loads.
- Take full credit for documented window, glass door and skylight U-values and SHGC values.
  - For clear glass, use the generic glass data provided in $MJ8_{AE}$, Tables 3A and 3C.

    - Use advanced *Manual J* procedures and tables for other types of generic glass and for rated (NFRC) glass.
- Take credit for overhangs (overhang adjustments shall be applied to all windows and glass doors, including purpose-built day-lighting windows).
- Take credit for internal shade (the default is a medium color blind with slats at 45 degrees, or use the actual device – this applies to all vertical glass and does not apply to purpose-built day-light windows).
- Take credit for insect screens when such devices are installed or specified.
- In general, take full credit for the rated (or tested) performance of construction materials, insulation materials, and construction features.
    - As specified for new construction
    - As installed (verify the installation conforms to methods and materials protocols)
    - As tested (see quality control programs for new construction, investigate existing construction)
- Take full credit for the tightness of the envelope construction.
    - As specified by builder or code
    - As installed (verify the installation conforms to methods and materials protocols)
    - As tested (see quality control programs for new construction, investigate existing construction)
- Follow *Manual J* procedures for infiltration and ventilation.
    - Use Table 8A procedure to evaluate the fresh air requirement
    - Use Table 5A to estimate infiltration rates for heating and cooling (ignore intermittent exhaust fans)
    - Decide on the installation of an engineered ventilation system (mandatory if the fresh air requirement is larger than an honest estimate of the *Manual J* infiltration rate)
    - Intermittent bathroom and kitchen exhaust fans are not "ventilation devices" or "ventilation systems"
- Take full credit for duct system sealing and duct insulation when such efforts are confidently anticipated or certifiable.
    - Use the default (0.12/0.24) scenario for untested ducts that are reasonably sealed
    - Take full credit for sealing efforts that are certifiably tighter than the default scenario for sealed ducts
    - If the duct sealing work is deficient – seal the ducts and take credit for sealed ducts (use unsealed options to show why the sealing work is required)
- Match location as close as possible when selecting a duct load table (use Table 7 unabridged if the $MJ8_{AE}$ tables do not provide a satisfactory match).
    - For attic locations, match roof material, roof color, use of radiant barrier, and attic ventilation
    - For closed crawl-space locations, match crawl-space tightness, crawl-space wall insulation, and crawl-space ceiling insulation
- Match duct system geometry (radial and spider systems tend to have less surface area than extended plenum and trunk and branch systems).
- Match return system geometry (use advanced *Manual J* procedures when the system has more than one or two large returns or when the returns are not located close to the air handler).
    - Be sure to use the duct wall insulation correction if the R-value of the insulation is not R-6

  - Be sure to use the surface adjustment factor for the exposed duct surface area when the surface area of the actual duct system is significantly different than the defaults listed in Table 7
  - Make sure the occupancy and internal loads are compatible with the $MJ8_{AE}$ defaults (use *Manual J*, Worksheet F, if $MJ8_{AE}$ does not provide a satisfactory match)
  - Add blower heat to the sensible load if equipment performance data is not adjusted for blower heat (if equipment manufacturer or blower power is unknown, assume 1,707 Btu/hr for indoor blower motor heat)
  - Educate consumers: Sit down with your customers or clients and educate them on these issues

MANUAL J DON'TS

- Do not use *Manual J* (any version) for:
  - Any type of commercial application (even if located in a residential structure)
  - Large multi-family buildings or residential high-rise structures
  - A room or space containing an indoor swimming pool or hot tub
- Do not use $MJ8_{AE}$ to estimate heating and cooling loads for applications that are not compatible with the "Abridged Edition Check List" (see the page that precedes Section 1).
- Do not design for record breaking (or news making) weather conditions.
- Do not add a "safety factor" to the Table 1A design conditions.
- Do not design for abnormally low or high indoor temperatures or humidity conditions (unless there is a certified medical reason for doing so).
- Do not assume that there will be no internal shade on ordinary windows and glass doors (bare glass is an acceptable assumption for glass specifically installed for "day-lighting").
- Do not fail to take credit for overhangs.
- Do not assume that the load for the worst case site orientation can be used for other orientations. Rotating the dwelling on a site can change the cooling load by a half a ton or more. Room air flow requirements change as the orientation changes. If the same design is used for any orientation, some rooms may have too much supply air and other rooms will not have enough supply air for temperature control and comfort.
- Do not reduce known ceiling, wall, or floor R-values "just to be safe."
- Do not fail to give credit for the builder's effort to produce a tight envelope.
- If a local code specifies a fresh air requirement (typically an air exchange per hour value), do not assume the infiltration rate will satisfy the requirement, and do not use the code ventilation requirement as the input value for the infiltration rate.
- Do not assume that windows and doors will be open when making the infiltration estimate.
- Do not make worst case "everything is going full blast" assumptions about internal loads (all assumptions must be defensible).
- Do not add extra occupancy loads for "entertaining groups of people."
- Do not add internal loads for special events.
- Do not arbitrarily assume that ducts are unsealed (i.e., do not assume that they are leaky).
- Do not fail to give full credit for efforts to provide tight, properly insulated ducts.
- Do not apply "safety factors" during any stage of the load calculation process.
- Do not apply a safety factor to the final answer or to the equipment selection procedure.

3.2.5.2 ACCA *MANUAL J* WORKSHEET EXPLANATIONS All of the worksheets described in this section are tools used in conjunction with *Manual J* for hand calculation purposes. These worksheets are proprietary to ACCA and can be purchased through their Web site. Following are worksheet descriptions per *ACCA Manual J 8th Edition Version 2*:

- Form J1 and Form $J1_{AE}$ produce heating and cooling load estimates for the collection of rooms served by central equipment (block load estimate) and a set of heating and cooling loads for each room
- Worksheet A holds geographical and climatic data from Table 1 and the indoor design conditions. It is also used to generate values for the heating temperature difference (HTD), cooling temperature difference (CTD). And to record the altitude correction factor (ACF) from Table 10A.
- Worksheet B produces heat transfer multiplier (HTM) values for generic windows and glass doors. (Advanced applications use U-values and SHGCs for rated glass and generic glass.)
- Worksheet C produces HTM values for generic skylights. (Advanced applications use U-values and SHGCs for rated glass and for generic glass, with custom curb options and light shaft options.)
- Worksheet D produces HTM values for opaque panels.
- Worksheet E uses Table 5A to evaluate infiltration loads. (Advanced applications support Tables 5A and 5B, the component leakage method and the blower door method.)
- Worksheet F evaluates internal loads. For $MJ8_{AE}$ simple defaults of 1,200 or 2,400 Btu/hr are used.
- Worksheet G produces default duct load factors for heating and cooling and a value for the latent load produced by the duct system.
- Worksheet H generates values for the ventilation loads. (Advanced *Manual J* applications support heat recovery equipment and ventilation dehumidifiers.)
- Worksheet I evaluates hot water piping loss, winter humidification load, and blower motor heat.

For more details on how to perform peak load calculations by hand, refer to either *ACCA (2006) Manual J* or *ASHRAE (2009) Handbook of Fundamentals*. For everyday use, load calculation software is recommended for increased accuracy, reliability, and repeatability of all calculations.

3.3 Indoor Air Quality (IAQ)

Supplying clean, appropriately conditioned air is extremely important in any HVAC installation, both commercial and residential. Indoor air quality concerns the amount of pollutants and odors contained in interior air that could affect occupant health and comfort levels. IAQ depends on many factors that include, but are not limited to, supply of suitable outdoor air, exhaust of unsuitable interior air, occupancy levels, occupant activity levels, and proper maintenance and control of the HVAC system as a whole. The primary methods of controlling indoor air quality include air filtration, purification, and ventilation.

Adequate filtration is one method of improving IAQ. The typical filter supplied with a unit will remove less than 20% of the small atmospheric dust particles that enter the filter. Filters which will remove 80 to 90% of pollutants in the size range of smoke particles are available at a reasonable cost. These filters are usually pleated to have large surface area and must be built into the system.

Another method of improving IAQ is the act of bringing in outdoor air to dilute/remove interior air odors and contaminants. As discussed in Section 3.2, in some instances, the rate of infiltration for a building is not considered enough to satisfy fresh air requirements. As a guideline, whenever the natural infiltration rate for a home is estimated to be less than 0.40-0.50 air changes per hour (ACH), natural infiltration rates are most likely inadequate and positive ventilation methods should be explored. Depending on the ventilation requirements and energy recovery provisions that are made, the energy required to condition outdoor air to proper levels can be a considerable portion of the total peak design loads. Air-to-air heat exchangers are available to recover energy from exhaust air in the winter and to moderate the temperature of incoming air in the cooling season as well. Whenever fresh air is introduced into a building, its effects on equipment heating and cooling design loads need to be considered, with or without energy recovery.

Figure 3.1 is a schematic displaying how a heat recovery ventilator may be connected to a GSHP system to provide fresh air to the space. This arrangement assures a clean, healthy environment in the space, provided the outdoor air is clean. When the weather is mild, this system also permits circulation of large quantities of outdoor air under controlled conditions. Care must be taken to properly size the outdoor air duct and the filter.

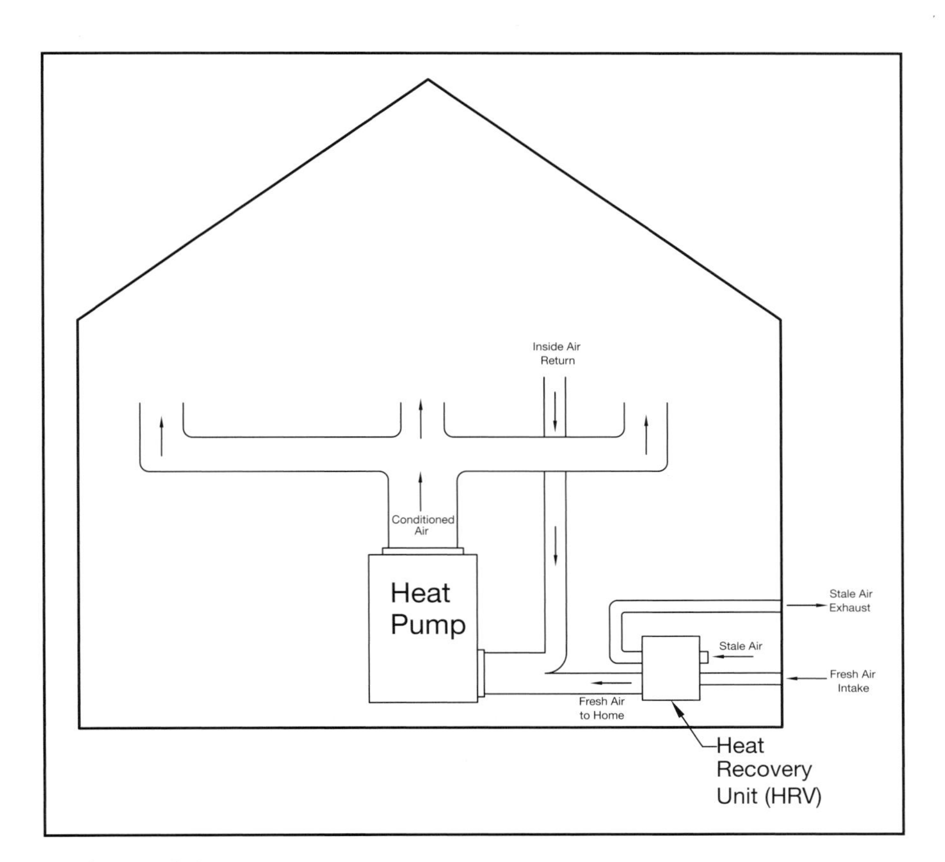

Figure 3.1. GSHP-Heat Recovery Ventilator Connection Schematic

3.4 Selection of GSHP Equipment

Once the peak block and equipment loads have been calculated, the equipment selection process can begin. The constraints placed on the equipment selection process are dependent on the building's location and climate. Sizing the heating/cooling equipment for a comfort conditioning system is fundamentally the most important step in the design process. This has already been stressed in Section 3.2 as equipment selection is based on the design loads.

In heating dominant climates, the total heating capacity of a system must be large enough to keep the area served at the thermostat set point under peak conditions. In some instances, the argument is made that a heat pump must be sized to handle 75% of the peak heating load in colder climates where the peak cooling load can be as little as one-fourth of the peak heating load. In such instances, a backup system, such as resistance heat, is needed to supplement the heat pump unit in times when it does not have sufficient capacity to satisfy the load. The rationale behind this argument is that if a single-capacity heat pump is sized to handle a heating load significantly larger than the cooling load, the system will not run enough in cooling to adequately remove humidity from the air. Utilizing two-stage or two-capacity units eliminates the need to size equipment to handle 75% of the load because they have the ability to operate in low speed for better moisture removal and humidity control in cooling. When using two-stage equipment, sizing for 100% of the peak heating load is recommended. In cooling dominated climates, heating capacity over sizing is not an issue because humidity is not a problem in the winter time. In all actuality, a system will not operate at peak conditions very many hours in a given year. For the case of a building in a heating dominant climate, the system will be oversized in heating and significantly oversized in cooling for the majority of the year, making proper equipment selection crucial for overall system performance, adequate comfort levels, and overall customer satisfaction. Dual-capacity equipment is superior in most climates and locations. The ability of such equipment to operate at part-load is not only well suited in situations where the cooling load is significantly less than the heating load, but also because it reduces short-cycling in any geographical location during mild-weather time periods, offers superior dehumidification due to longer run times in low speed (cooling), maximizes efficiency, and lowers annual operating cost. When dual-capacity equipment operates at low speed, efficiencies (measured by COP in heating and EER in cooling) are typically better compared to high-speed operation.

When sizing a heat pump for cooling, two things that must be considered are the peak sensible cooling load and the SHF for the building. The sensible cooling capacity of the unit must be large enough to keep the area served at the thermostat set point under peak conditions. Additionally, the SHF of the unit must be less than the SHF of the space to ensure that the unit has adequate latent capacity to remove moisture from the air. For single-capacity heat pumps installed in heating dominant climates, over sizing of no more than 25% of the peak block sensible cooling load is recommended. Two-capacity units can be oversized in total capacity by greater amounts because they have the ability to operate at low capacity for better moisture removal and humidity control.

Undersized equipment in either heating or cooling will not only generate the need for a backup system (heating only) and cause significant increases in total system operating cost and equipment full-load run hours (possibly shortening equipment life), but also will generate the need for more installed loopfield capacity to handle the substantial run hour increase. The total amount of energy rejected to or extracted from the loopfield (depending on mode of operation) is dependent on the capacity and efficiency of the equipment coupled to the ground loop as they relate to building load in addition to equipment full-load run hours.

The installed loopfield must have sufficient capacity to accommodate such rejection/extraction of energy, which will be magnified to some extent in the case of an undersized system. For example, if a system is severely undersized in a heating dominant climate where the peak cooling load is much smaller, it will run continuously during the periods for which it is undersized. The extended run periods in heating will draw large amounts of energy from the soil, potentially enough energy to draw down the average soil temperature in the loopfield over the long term (especially in large systems). Lower soil temperatures will cause lower entering water temperatures from the loopfield, diminishing heat pump capacity and efficiency in heating mode. This problem will be compounded because the cooling load (which is significantly smaller than the heating load) will not cause the GSHP system to reject a sufficient amount of heat to the soil to offset the large thermal drawdown in heating. This situation would be referred to as an "unbalanced ground load." Alternatively, undersized equipment in hot climates could raise the temperature of the loopfield, diminishing heat pump capacity and efficiency in cooling mode. In the case where a system is significantly undersized, special attention must be paid to the loopfield design to ensure that entering water temperatures from a closed-loop system will stay within an acceptable range. Utilizing the heat pump selection worksheets given in this manual will help ensure equipment sizing within reasonable limits.

3.4.1 SINGLE-CAPACITY, WATER-TO-AIR GSHP SELECTION The Single-Capacity, Water-to-Air Heat Pump Selection Worksheet is shown in Figure 3.2. Each input blank in Figure 3.2 has been labeled with a number identifying a short input description provided below. A blank copy of Figure 3.2 is provided in Appendix B.

1) Insert the peak heating load to be served by the heat pump
2) Insert the outdoor design air temperature for which the peak heating load was calculated, determined by geographical location
3) Insert the interior design temperature in heating (thermostat set point) for which the load was calculated, usually assumed to be 70 F (dry bulb) unless otherwise specified
4) Insert the interior design temperature in cooling (thermostat set point) for which the load was calculated, usually assumed to be 75 F (dry bulb) unless otherwise specified
5) Insert the interior level of relative humidity in cooling for which the sensible heat factor (SHF) was calculated, usually assumed to be 50% unless otherwise specified
6) Insert the peak total cooling load to be served by the heat pump
7) Insert the outdoor design air temperature for which the peak cooling load was calculated, determined by geographical location
8) Insert the peak sensible cooling load to be served by the heat pump
9) Insert the peak latent cooling load to be served by the heat pump
10) Calculate and enter the space SHF according to the equation given
11) Insert the proposed heat pump brand
12) Insert the proposed heat pump model and size
13) Insert the minimum design entering water temperature from the loopfield to ensure that equipment capacity complies with ground-loop design parameters in heating
14) Insert the fluid flow rate through the heat pump in heating, typically 2.5-3.0 gpm per nominal ton of installed heat pump capacity
15) Insert the entering air temperature from the space in heating, typically assumed to be the same as the thermostat set point (3)

Single Capacity Water-Air GSHP Selection Worksheet

Block Load Description =	Example Residence	Zone #	Single Zone

Total Heat Loss =	1	Btu/hr @ 2 F OAT	Total Heat Gain = 6 Btu/hr @ 7 F OAT
			Sen Heat Gain = 8 Btu/hr
			Lat Heat Gain = 9 Btu/hr
T-Stat Set Point =	3 / 4 / 5	(Htg dB / Clg dB / %-RH)	SHF_{space} = Sen Heat Gain / Total Heat Gain
			SHF_{space} = 10

Brand	11
Model	12

Heating Mode Performance Data

Performance Parameters			Correction Factors[A] HC	Correction Factors[A] DMD
EWT_{min}=	13	F		
GPM=	14	gpm		
EAT=	15	F	CF_{EAT}= 19	CF_{EAT}= 23
CFM=	16	cfm	CF_{CFM}= 20	CF_{CFM}= 24
Antifreeze / %[B]	17 / 18		CF_{AF}= 21	CF_{AF}= 25
CF_{EAT} x CF_{CFM} x CF_{AF}==>			CF_{HC}= 22	CF_{DMD}= 26

Cooling Mode Performance Data

Performance Parameters			Correction Factors[A] TC	Correction Factors[A] SC	Correction Factors[A] DMD
EWT_{max}=	27	F			
GPM=	28	gpm			
EAT=	29 / 30	F/F (dB/wB)	CF_{EAT}= 34	CF_{EAT}= 38	CF_{EAT}= 42
CFM=	31	cfm	CF_{CFM}= 35	CF_{CFM}= 39	CF_{CFM}= 43
Antifreeze / %[B]	32 / 33		CF_{AF}= 36	CF_{AF}= 40	CF_{AF}= 44
CF_{EAT} x CF_{CFM} x CF_{AF}==>			CF_{TC}= 37	CF_{SC}= 41	CF_{DMD}= 45

Tabled Heating Mode Performance Data

HC =	46	Btu/hr
DMD =	47	kW
COP =	48	
HE =	49	Btu/hr (= HC x (COP-1)/COP)[C]

Tabled Cooling Mode Performance Data

TC =	50	Btu/hr
SC =	51	Btu/hr (= TC x 0.75 approximately)[C]
DMD =	52	kW
EER =	53	
HR =	54	Btu/hr (= TC x (EER+3.412)/EER)[C]

Corrected Heating Mode Performance Data

HC_c=	HC x CF_{HC}=	55	Btu/hr
DMD_c=	DMD x CF_{DMD}=	56	kW
COP_c=	HC_c/(DMD_c x 3412) =	57	
HE_c =	HC_c x (COP_c-1)/COP_c =	58	Btu/hr
LWT_{min}=	EWT_{min} - HE_c/(500xGPM) =	59	F

Corrected Cooling Mode Performance Data

TC_c=	TC x CF_{TC}=	60	Btu/hr
SC_c=	SC x CF_{SC}=	61	Btu/hr
DMD_c=	DMD x CF_{DMD}=	62	kW
EER_c=	TC_c/(DMD_c x 1000) =	63	
HR_c =	TC_c x (EER_c+3.412)/EER_c =	64	Btu/hr
LWT_{max}=	EWT_{max} + HR_c/(500xGPM) =	65	F

% Sizing in Heating Mode

% Sizing = HC_c / Total Heat Loss x 100%

= 66 %

% Sizing in Cooling Mode

% Oversizing = (SC_c - Sen Heat Gain) / Sen Heat Gain x 100%

= 67 % (> 0 & ≤ 25%) 70 OK

SHF_{unit}= SC_c / TC_c

= 68 (≤ SHF_{space}) 71 OK

Latent Clg Cap = TC_c - SC_c

= 69 Btu/hr (> Lat Ht Gain) 72 OK

Blower Performance Summary

Type	Speed	CFM	Static (in)
PSC	L M H	73	74
ECM	On	75	76
	w/ Comp	77	78
	w/ Emer	79	80

Tabled Water Coil Performance Summary

EWT_{min}	GPM	WPD (psi)	WPD (ft)	Test Circulating Fluid
81	82	83	84	85

A. Apply correction factors for CFM and EAT as needed depending on format of Manufacturer's Literature. (No correction - CF=1)
B. CF_{AF} = CF_{Used}/CF_{Tested}. CF_{Used} for antifreeze being used. CF_{Tested} for circulating fluid performance was determined with.
C. Use this equation if performance parameter not provided by Manufacturer in the Engineering Specifications.

NOTES:

Figure 3.2. Single-Capacity, Water-to-Air Heat Pump Selection Worksheet

16) Insert the system total air flow rate in heating, typically 350-450 CFM per nominal ton of installed heat pump capacity
17) Insert the type of antifreeze to be used as the circulating fluid in the closed-loop ground heat exchanger, if necessary (EWT_{min}<45 F)
18) Insert the volumetric concentration of the antifreeze to be used, if necessary (EWT_{min}<45 F)
19) Insert the capacity correction factor if the system entering air temperature in heating differs from the entering air temperature given in performance data tables, provided by manufacturer
20) Insert the capacity correction factor if the system air flow rate in heating differs from the air flow rate given in performance data table, provided by manufacturer
21) Insert the capacity correction factor if the circulating fluid differs from what is utilized in the performance tables, provided by manufacturer
22) Calculate the total heating capacity correction factor according to provided equation
23) Insert the demand correction factor if the system entering air temperature in heating differs from the entering air temperature given in performance data tables, provided by manufacturer
24) Insert the demand correction factor if the system air flow rate in heating differs from the air flow rate given in performance data table, provided by manufacturer
25) Insert the demand correction factor if the circulating fluid differs from what is utilized in the performance tables, provided by manufacturer
26) Calculate the total heating demand correction factor according to provided equation
27) Insert the maximum design entering water temperature from the loopfield to ensure that equipment capacity complies with ground-loop design parameters in cooling
28) Insert the fluid flow rate through the heat pump in cooling, typically 2.5-3.0 gpm per nominal ton of installed heat pump capacity
29) Insert the entering air dry bulb temperature in cooling from the space, typically assumed to be the same as the thermostat set point (4)
30) Insert the entering air wet bulb temperature in cooling from the space, typically defined by the relative humidity for the space (5)
31) Insert the system total air flow rate in cooling, typically 350-450 CFM per nominal ton of installed heat pump capacity
32) Insert the type of antifreeze to be used as the circulating fluid in the closed-loop ground heat exchanger, if necessary (17) (EWT_{min}<45 F)
33) Insert the volumetric concentration of the antifreeze to be used, if necessary (18) (EWT_{min}<45 F)
34) Insert the total cooling capacity correction factor if the system entering air temperature differs from the entering air temperature given in performance data tables, provided by manufacturer
35) Insert the total cooling capacity correction factor if the system air flow rate differs from the air flow rate given in performance data table, provided by manufacturer
36) Insert the total cooling capacity correction factor if the circulating fluid differs from what is utilized in the performance tables, provided by manufacturer
37) Calculate the total cooling capacity correction factor according to provided equation
38) Insert the sensible cooling capacity correction factor if the system entering air temperature differs from the entering air temperature given in performance data tables, provided by manufacturer
39) Insert the sensible cooling capacity correction factor if the system air flow rate differs from the air flow rate given in performance data table, provided by manufacturer

40) Insert the sensible cooling capacity correction factor if the circulating fluid differs from what is utilized in the performance tables, provided by manufacturer
41) Calculate the sensible cooling capacity correction factor according to provided equation
42) Insert the demand correction factor if the system entering air temperature in cooling differs from the entering air temperature given in performance data tables, provided by manufacturer
43) Insert the demand correction factor if the system air flow rate in cooling differs from the air flow rate given in performance data table, provided by manufacturer
44) Insert the demand correction factor in cooling if the circulating fluid differs from what is utilized in the performance tables, provided by manufacturer
45) Calculate the cooling demand correction factor according to provided equation
46) Insert the unit heating capacity at the minimum design entering water temperature (13), provided in manufacturer performance data tables
47) Insert the heat pump electric demand in heating at the minimum design entering water temperature, provided in manufacturer performance data tables
48) Insert the heat pump coefficient of performance (COP) in heating at the minimum design entering water temperature, provided in manufacturer performance data tables
49) Insert the heat of extraction at the minimum entering water temperature, usually given in manufacturer performance data tables (if not given, calculate according to provided equation)
50) Insert the unit total cooling capacity at the maximum design entering water temperature (26), provided in manufacturer performance data tables
51) Insert the unit sensible cooling capacity at the maximum design entering water temperature, typically given in manufacturer performance data tables (if not provided, can be assumed to be approximately 75% of the total cooling capacity at the given EWT)
52) Insert the heat pump electric demand in cooling at the maximum design entering water temperature, provided in manufacturer performance data tables
53) Insert the heat pump energy efficiency rating (EER) in cooling at the maximum design entering water temperature, provided in manufacturer performance data tables
54) Insert the heat of rejection at the maximum entering water temperature, usually given in manufacturer performance data tables (if not given, calculate according to provided equation)
55) Calculate the equipment installed heating capacity by multiplying the tabled heating capacity (46) by the total heating capacity correction factor (22)
56) Calculate the installed heat pump electric demand in heating by multiplying the tabled heating demand (47) by the total heating demand correction factor (26)
57) Calculate the installed heat pump COP in heating according to the provided equation
58) Calculate the installed heat of extraction according to the provided equation
59) Calculate the minimum LWT in heating according to the provided equation
60) Calculate the equipment installed total cooling capacity by multiplying the tabled total cooling capacity (50) by the total cooling capacity correction factor (37)
61) Calculate the equipment installed sensible cooling capacity by multiplying the tabled sensible cooling capacity (51) by the sensible cooling capacity correction factor (41)
62) Calculate the installed heat pump electric demand in cooling by multiplying the tabled cooling demand (52) by the total cooling demand correction factor (45)
63) Calculate the installed heat pump EER in cooling according to the provided equation

64) Calculate the installed heat of rejection according to the provided equation
65) Calculate the maximum LWT in cooling according to the provided equation
66) Calculate the percent-sizing in heating according to provided equation (if less than 100%, supplemental heating will need to be utilized during peak conditions)
67) Calculate the percent-sizing in sensible cooling according to provided equation (if greater than 25%, insufficient levels of equipment run time may lead to air moisture removal issues)
68) Calculate the heat pump installed sensible heat factor (if greater than space SHF, installed latent capacity will be insufficient)
69) Calculate the heat pump installed latent cooling capacity to double check sufficient moisture removal capacity
70) Place a check mark on this line if equipment oversizing in sensible cooling capacity (as it relates to peak sensible cooling load) is between 0% and 25%
71) Place a check mark on this line if the equipment SHF is less than (or equal to) the space SHF under design conditions
72) Place a check mark on this line if equipment sizing in latent cooling capacity (as defined by the space SHF and the peak sensible cooling load) is greater than the peak latent heat gain for the space
73) For a heat pump equipped with a PSC blower, insert the greater of the total system air flow rates (heating air flow rate versus cooling air flow rate)
74) Insert the available external static pressure that the PSC blower can generate at the total system air flow rate (69), provided by manufacturer
75) For a heat pump equipped with an ECM blower, enter the system air flow rate for when the fan is on, but no heating or cooling is required, provided by manufacturer
76) Insert the maximum external static pressure that the ECM blower can generate, provided by manufacturer
77) For a heat pump equipped with an ECM blower, insert the greater of the total system air flow rates (heating air flow rate versus cooling air flow rate) for when the unit is in full-load (high-capacity) operation
78) Insert the maximum external static pressure that the ECM blower can generate, provided by manufacturer
79) For a heat pump equipped with an ECM blower, insert the system air flow rate for when the emergency resistance heat is in operation, provided by manufacturer
80) Insert the maximum external static pressure that the ECM blower can generate, provided by manufacturer
81) Insert the minimum design entering water temperature from the loopfield (13)
82) Insert the fluid flow rate through the heat pump in heating (14)
83) Insert the water pressure drop (psi) through the water-to-refrigerant coil, provided by manufacturer
84) Insert the water pressure drop (ft) through the water-to-refrigerant coil, provided by manufacturer
85) Insert the circulating fluid type used to figure pressure drop through the water-to-refrigerant coil and the GHEX, provided by manufacturer

Items 73-80 on the selection worksheet are necessary for the ductwork design process. Items 81-85 on the selection worksheet are necessary to perform system head loss calculations for pump sizing. For convenience, these items are found and recorded at this point of the design process.

A basic understanding of supplemental heating systems is also helpful when selecting equipment. For forced-air, ducted-delivery systems, supplemental backup heat can be provided in the form of strip-resistance heaters installed in the ductwork plenum or directly in the GSHP cabinet itself. Supplemental heating is typically staged and automatically controlled via a programmable thermostat. For example, residential, dual-capacity units are usually advertised to have three stages of heating (w/emergency) and two stages of cooling capacity. Such units are sold with strip-resistance plenum heaters included in the price. The heat pump unit itself can provide low-capacity and high-capacity heating and cooling via normal operation (two stages of heating and cooling). The third stage of heating is then provided in the form of the resistance heating elements for periods when the heat pump alone cannot satisfy the building load. Similar to heat pump compressor operation, supplemental resistance heat can also be staged. Resistance heaters are normally staged such that half of the elements turn on when required by insufficient heat pump capacity. Strip-resistance heating is typically programmed to switch on in full only when emergency heat is needed (compressor failure, for example). Designers should try to avoid the use of backup heat for large amounts of hours in a given year (severe under sizing) because it is much more expensive in terms of operating cost when compared to normal heat pump operation. Resistance heat has a COP of 1.0. This means that if a heat pump normally operates with a COP of 3.5-4.5, then operating the resistance heaters will be 2.5-3.5 times more expensive.

3.4.2 TWO-CAPACITY, WATER-TO-AIR GSHP SELECTION The Two-Capacity, Water-to-Air Heat Pump Selection Worksheet is shown in Figure 3.3. Each input blank in Figure 3.3 has been labeled with a number identifying a short input description provided below. A blank copy of Figure 3.3 is provided in Appendix B.

1) Insert the peak heating load to be served by the heat pump
2) Insert the outdoor design air temperature for which the peak heating load was calculated, determined by geographical location
3) Insert the interior design temperature in heating (thermostat set point) for which the load was calculated, usually assumed to be 70 F (dry bulb) unless otherwise specified
4) Insert the interior design temperature in cooling (thermostat set point) for which the load was calculated, usually assumed to be 75 F (dry bulb) unless otherwise specified
5) Insert the interior level of relative humidity in cooling for which the sensible heat factor (SHF) was calculated, usually assumed to be 50% unless otherwise specified
6) Insert the peak total cooling load to be served by the heat pump
7) Insert the outdoor design air temperature for which the peak cooling load was calculated, determined by geographical location
8) Insert the peak sensible cooling load to be served by the heat pump
9) Insert the peak latent cooling load to be served by the heat pump
10) Calculate and enter the space SHF according to the equation given
11) Insert the proposed heat pump brand
12) Insert the proposed heat pump model and size

At this point, specify on the worksheet whether the heat pump selection process will be dependent on heating or cooling mode of operation, determined by the peak loads given in (1) and (6). Next, specify whether the performance data listed on the selection worksheet is based on high or low-capacity operation for both heating and cooling. Record these distinctions by circling where appropriate. Typically, a heating dependent design will require the heat pump to operate in high capacity under peak conditions in heating, but will only require low-capacity operation under peak conditions in cooling. Alternatively, a cooling dependent design could require the heat pump to operate in high capacity under peak conditions in cooling, but will only require low-capacity operation under peak conditions in heating.

Two Capacity Water-Air GSHP Selection Worksheet

Block Load Description = Example Residence   Zone # Single Zone

Total Heat Loss = 1 Btu/hr @ 2 F OAT	Total Heat Gain = 6 Btu/hr @ 7 F OAT
	Sen Heat Gain = 8 Btu/hr
	Lat Heat Gain = 9 Btu/hr
T-Stat Set Point = 3 / 4 / 5 (Htg dB / Clg dB / %-RH)	SHF_{space} = Sen Heat Gain / Total Heat Gain
	SHF_{space} = 10

Brand 11

Model 12   Sized for High Capacity in Heating / Cooling

(High / Low) Capacity Heating Mode Performance Data

Performance Parameters			Correction Factors[A] HC	Correction Factors[A] DMD
EWT_{min}=	13	F		
GPM=	14	gpm		
EAT=	15	F	CF_{EAT}= 19	CF_{EAT}= 23
CFM=	16	cfm	CF_{CFM}= 20	CF_{CFM}= 24
Antifreeze / %[B]	17 / 18		CF_{AF}= 21	CF_{AF}= 25
CF_{EAT} x CF_{CFM} x CF_{AF}==>			CF_{HC}= 22	CF_{DMD}= 26

(High / Low) Capacity Cooling Mode Performance Data

Performance Parameters			Correction Factors[A] TC	Correction Factors[A] SC	Correction Factors[A] DMD
EWT_{max}=	27	F			
GPM=	28	gpm			
EAT=	29 / 30	F/F (dB/wB)	CF_{EAT}= 34	CF_{EAT}= 38	CF_{EAT}= 42
CFM=	31	cfm	CF_{CFM}= 35	CF_{CFM}= 39	CF_{CFM}= 43
Antifreeze / %[B]	32 / 33		CF_{AF}= 36	CF_{AF}= 40	CF_{AF}= 44
CF_{EAT} x CF_{CFM} x CF_{AF}==>			CF_{TC}= 37	CF_{SC}= 41	CF_{DMD}= 45

Tabled Heating Mode Performance Data

HC =	46	Btu/hr
DMD =	47	kW
COP =	48	
HE =	49	Btu/hr (= HC x (COP-1)/COP)[C]

Tabled Cooling Mode Performance Data

TC =	50	Btu/hr
SC =	51	Btu/hr (= TC x 0.75 approximately)[C]
DMD =	52	kW
EER =	53	
HR =	54	Btu/hr (= TC x (EER+3.412)/EER)[C]

Corrected Heating Mode Performance Data

HC_c=	HC x CF_{HC}=	55	Btu/hr
DMD_c=	DMD x CF_{DMD}=	56	kW
COP_c=	$HC_c/(DMD_c$ x 3412) =	57	
HE_c =	HC_c x $(COP_c-1)/COP_c$ =	58	Btu/hr
LWT_{min}=	EWT_{min} - HE_c/(500xGPM) =	59	F

Corrected Cooling Mode Performance Data

TC_c=	TC x CF_{TC}=	60	Btu/hr
SC_c=	SC x CF_{SC}=	61	Btu/hr
DMD_c=	DMD x CF_{DMD}=	62	kW
EER_c=	$TC_c/(DMD_c$ x 1000) =	63	
HR_c =	TC_c x $(EER_c+3.412)/EER_c$ =	64	Btu/hr
LWT_{max}=	EWT_{max} + HR_c/(500xGPM) =	65	F

% Sizing in Heating Mode

% Sizing = HC_c / Total Heat Loss x 100%

= 66 %

% Sizing in Cooling Mode

% Oversizing = (SC_c - Sen Heat Gain) / Sen Heat Gain x 100%

= 67 %   (> 0 & ≤ 25%) 70 OK

SHF_{unit}= SC_c / TC_c

= 68   (≤ SHF_{space}) 71 OK

Latent Clg Cap = TC_c - SC_c

= 69 Btu/hr   (> Lat Ht Gain) 72 OK

Blower Performance Summary

Type	Speed	CFM	Static (in)
ECM	On	73	74
	w/ Low	75	76
	w/ High	77	78
	w/ Emer	79	80

Tabled Water Coil Performance Summary

EWT_{min}	GPM	WPD (psi)	WPD (ft)	Test Circulating Fluid
81	82	83	84	85

A. Apply correction factors for CFM and EAT as needed depending on format of Manufacturer's Literature. (No correction - CF=1)

B. CF_{AF} = CF_{Used}/CF_{Tested}. CF_{Used} for antifreeze being used. CF_{Tested} for circulating fluid performance was determined with.

C. Use this equation if performance parameter not provided by Manufacturer in the Engineering Specifications.

NOTES:

Figure 3.3. Two-Capacity, Water-to-Air Heat Pump Selection Worksheet

13) Insert the minimum design entering water temperature from the loopfield to ensure that equipment capacity complies with ground-loop design parameters in heating
14) Insert the fluid flow rate through the heat pump in heating, typically 2.5-3.0 gpm per nominal ton of installed heat pump capacity (high capacity)
15) Insert the entering air temperature from the space in heating, typically assumed to be the same as the thermostat set point (3)
16) Insert the system total air flow rate in heating, typically 350-450 CFM per nominal ton of installed heat pump capacity
17) Insert the type of antifreeze to be used as the circulating fluid in the closed-loop ground heat exchanger, if necessary (EWT_{min}<45 F)
18) Insert the volumetric concentration of the antifreeze to be used, if necessary (EWT_{min}<45 F)
19) Insert the capacity correction factor if the system entering air temperature in heating differs from the entering air temperature given in performance data tables, provided by manufacturer
20) Insert the capacity correction factor if the system air flow rate in heating differs from the air flow rate given in performance data table, provided by manufacturer
21) Insert the capacity correction factor if the circulating fluid differs from what is utilized in the performance tables, provided by manufacturer
22) Calculate the total heating capacity correction factor according to provided equation
23) Insert the demand correction factor if the system entering air temperature in heating differs from the entering air temperature given in performance data tables, provided by manufacturer
24) Insert the demand correction factor if the system air flow rate in heating differs from the air flow rate given in performance data table, provided by manufacturer
25) Insert the demand correction factor if the circulating fluid differs from what is utilized in the performance tables, provided by manufacturer
26) Calculate the total heating demand correction factor according to provided equation
27) Insert the maximum design entering water temperature from the loopfield to ensure that equipment capacity complies with ground-loop design parameters in cooling
28) Insert the fluid flow rate through the heat pump in cooling, typically 2.5-3.0 gpm per nominal ton of installed heat pump capacity
29) Insert the entering air dry bulb temperature in cooling from the space, typically assumed to be the same as the thermostat set point (4)
30) Insert the entering air wet bulb temperature in cooling from the space, typically defined by the relative humidity for the space (5)
31) Insert the system total air flow rate in cooling, typically 350-450 CFM per nominal ton of installed heat pump capacity
32) Insert the type of antifreeze to be used as the circulating fluid in the closed-loop ground heat exchanger, if necessary (17) (EWT_{min}<45 F)
33) Insert the volumetric concentration of the antifreeze to be used, if necessary (18) (EWT_{min}<45 F)
34) Insert the total cooling capacity correction factor if the system entering air temperature differs from the entering air temperature given in performance data tables, provided by manufacturer
35) Insert the total cooling capacity correction factor if the system air flow rate differs from the air flow rate given in performance data table, provided by manufacturer
36) Insert the total cooling capacity correction factor if the circulating fluid differs from what is utilized in the performance tables, provided by manufacturer

37) Calculate the total cooling capacity correction factor according to provided equation
38) Insert the sensible cooling capacity correction factor if the system entering air temperature differs from the entering air temperature given in performance data tables, provided by manufacturer
39) Insert the sensible cooling capacity correction factor if the system air flow rate differs from the air flow rate given in performance data table, provided by manufacturer
40) Insert the sensible cooling capacity correction factor if the circulating fluid differs from what is utilized in the performance tables, provided by manufacturer
41) Calculate the sensible cooling capacity correction factor according to provided equation
42) Insert the demand correction factor if the system entering air temperature in cooling differs from the entering air temperature given in performance data tables, provided by manufacturer
43) Insert the demand correction factor if the system air flow rate in cooling differs from the air flow rate given in performance data table, provided by manufacturer
44) Insert the demand correction factor in cooling if the circulating fluid differs from what is utilized in the performance tables, provided by manufacturer
45) Calculate the cooling demand correction factor according to provided equation
46) Insert the unit heating capacity at the minimum design entering water temperature (13), provided in manufacturer performance data tables
47) Insert the heat pump electric demand in heating at the minimum design entering water temperature, provided in manufacturer performance data tables
48) Insert the heat pump coefficient of performance (COP) in heating at the minimum design entering water temperature, provided in manufacturer performance data tables
49) Insert the heat of extraction at the minimum entering water temperature, usually given in manufacturer performance data tables (if not given, calculate according to provided equation)
50) Insert the unit total cooling capacity at the maximum design entering water temperature (26), provided in manufacturer performance data tables
51) Insert the unit sensible cooling capacity at the maximum design entering water temperature, typically given in manufacturer performance data tables (if not provided, can be assumed to be approximately 75% of the total cooling capacity at the given EWT)
52) Insert the heat pump electric demand in cooling at the maximum design entering water temperature, provided in manufacturer performance data tables
53) Insert the heat pump energy efficiency rating (EER) in cooling at the maximum design entering water temperature, provided in manufacturer performance data tables
54) Insert the heat of rejection at the maximum entering water temperature, usually given in manufacturer performance data tables (if not given, calculate according to provided equation)
55) Calculate the equipment installed heating capacity by multiplying the tabled heating capacity (46) by the total heating capacity correction factor (22)
56) Calculate the installed heat pump electric demand in heating by multiplying the tabled heating demand (47) by the total heating demand correction factor (26)
57) Calculate the installed heat pump COP in heating according to the provided equation
58) Calculate the installed heat of extraction according to the provided equation
59) Calculate the minimum LWT in heating according to the provided equation
60) Calculate the equipment installed total cooling capacity by multiplying the tabled total cooling capacity (50) by the total cooling capacity correction factor (37)

61) Calculate the equipment installed sensible cooling capacity by multiplying the tabled sensible cooling capacity (51) by the sensible cooling capacity correction factor (41)
62) Calculate the installed heat pump electric demand in cooling by multiplying the tabled cooling demand (52) by the total cooling demand correction factor (45)
63) Calculate the installed heat pump EER in cooling according to the provided equation
64) Calculate the installed heat of rejection according to the provided equation
65) Calculate the maximum LWT in cooling according to the provided equation
66) Calculate the percent-sizing in heating according to provided equation (if less than 100%, supplemental heating will need to be utilized during peak conditions)
67) Calculate the percent-sizing in sensible cooling according to provided equation (if greater than 25%, insufficient levels of equipment run time may lead to air moisture removal issues)
68) Calculate the heat pump installed sensible heat factor (if greater than space SHF, installed latent capacity will be insufficient)
69) Calculate the heat pump installed latent cooling capacity to double check sufficient moisture removal capacity
70) Place a check mark on this line if equipment oversizing in sensible cooling capacity (as it relates to peak sensible cooling load) is between 0% and 25%
71) Place a check mark on this line if the equipment SHF is less than (or equal to) the space SHF under design conditions
72) Place a check mark on this line if equipment sizing in latent cooling capacity (as defined by the space SHF and the peak sensible cooling load) is greater than the peak latent heat gain for the space
73) For a heat pump equipped with a PSC blower, insert the greater of the total system air flow rates (heating air flow rate versus cooling air flow rate)
74) Insert the available external static pressure that the PSC blower can generate at the total system air flow rate (69), provided by manufacturer
75) For a heat pump equipped with an ECM blower, enter the system air flow rate for when the fan is on, but no heating or cooling is required, provided by manufacturer
76) Insert the maximum external static pressure that the ECM blower can generate, provided by manufacturer
77) For a heat pump equipped with an ECM blower, insert the greater of the total system air flow rates (heating air flow rate versus cooling air flow rate) for when the unit is in full-load (high- capacity) operation
78) Insert the maximum external static pressure that the ECM blower can generate, provided by manufacturer
79) For a heat pump equipped with an ECM blower, insert the system air flow rate for when the emergency resistance heat is in operation, provided by manufacturer
80) Insert the maximum external static pressure that the ECM blower can generate, provided by manufacturer
81) Insert the minimum design entering water temperature from the loopfield (13)
82) Insert the fluid flow rate through the heat pump in heating (14)
83) Insert the water pressure drop (psi) through the water-to-refrigerant coil, provided by manufacturer
84) Insert the water pressure drop (ft) through the water-to-refrigerant coil, provided by manufacturer
85) Insert the circulating fluid type used to figure pressure drop through the water-to-refrigerant coil and the GHEX, provided by manufacturer

As previously stated, items 73-80 on the selection worksheet are necessary for the ductwork design process. Items 81-85 on the selection worksheet are necessary to perform system head loss calculations for pump sizing. For convenience, these items are found and recorded at this point of the design process.

This worksheet will be filled out in the same manner as the single-capacity worksheet. Primary differences between the single-capacity and dual-capacity selection worksheets are the ability to specify and utilize data for low capacity in one mode of operation and high capacity for the other mode for dual-capacity equipment. This ability is especially useful for cases where a building is located in a heating dominant climate and a heat pump selected to serve the peak heating load is significantly oversized in cooling capacity. Special care should be taken to ensure that low-capacity/high-capacity data is properly selected and recorded according to building loads (distinguish whether low or high-capacity data is utilized for equipment selection for each mode of operation). Another difference between the two worksheets is because dual-capacity units are equipped with ECM blowers to accommodate the different levels of air flow necessary for different modes of operation. System air flow will be predetermined by manufacturer settings. Typically, ECM blowers will be set to provide a specific amount of air flow for when the unit is operating at low capacity, a higher amount of air flow for high-capacity operation, and an even higher amount of air flow during supplemental resistance heating operation. Actual air flow settings and blower external static pressure capabilities will vary from manufacturer to manufacturer. Ensure to note the maximum ESP that the selected heat pump with ECM blower can provide for duct design.

Typically, heat pump manufacturers provide performance data for low-capacity and high-capacity operation at different flow rates. However, unless the installed control system makes allowances for staged pumping, the flow rate will not change when equipment operation switches from high to low or vice versa. Typical residential installations are such that the flow center is full-on when the heat pump is operating independent of low or high speed operation. Head loss calculations for pump sizing should be performed at the lowest expected entering water temperature (EWT_{min} - #13 on Two-Capacity, Water-to-Air Heat Pump Selection Worksheet) and the highest expected system flow rate.

3.4.3 WATER-TO-WATER GSHP SELECTION

The Single-Capacity, Water-to-Water Heat Pump Selection Worksheet is shown in Figure 3.4. Each input blank in Figure 3.4 has been labeled with a number identifying a short input description provided below. A blank copy of Figure 3.4 is provided in Appendix B.

1) Insert the peak heating load to be served by the heat pump
2) Insert the outdoor design air temperature for which the peak heating load was calculated, determined by geographical location
3) Insert the design entering water temperature from the load in heating (the load side of the heat pump refers to water entering the GSHP unit from a storage tank, fan coil, radiant panel, etc., to be heated)
4) Insert the design load side water flow rate in heating
5) Insert the interior design temperature in heating (thermostat set point) for which the load was calculated, usually assumed to be 70 F (dry bulb) unless otherwise specified
6) Insert the interior design temperature in cooling (thermostat set point) for which the load was calculated, usually assumed to be 75 F (dry bulb) unless otherwise specified
7) Insert the interior level of relative humidity in cooling for which the sensible heat factor (SHF) was calculated, usually assumed to be 50% unless otherwise specified
8) Insert the peak total cooling load to be served by the heat pump
9) Insert the outdoor design air temperature for which the peak cooling load was calculated, determined by geographical location
10) Insert the design entering water temperature from the load in cooling (the load side of the heat pump refers to water entering the GSHP unit from a storage tank, fan coil, etc., to be chilled)

11) Insert the design load side water flow rate in cooling
12) Insert the proposed heat pump brand
13) Insert the proposed heat pump model and size
14) Insert the minimum design entering water temperature from the loopfield to ensure that equipment capacity complies with ground-loop design parameters in heating
15) Insert the fluid flow rate through the source side of the heat pump in heating, typically 2.5-3.0 gpm per nominal ton of installed heat pump capacity (the source side of the heat pump refers to water entering the unit from the ground connection)
16) Insert the design load side water flow rate in heating (4)
17) Insert the design entering load side water temperature from the space in heating (3)
18) Insert the type of antifreeze to be used as the circulating fluid in the closed-loop ground heat exchanger, if necessary (EWT_{min}<45 F)
19) Insert the volumetric concentration of the antifreeze to be used, if necessary (EWT_{min}<45 F)
20) Insert the capacity correction factor if the circulating fluid differs from what is utilized in the performance tables, provided by manufacturer
21) Insert the demand correction factor if the circulating fluid differs from what is utilized in the performance tables, provided by manufacturer
22) Insert the maximum design entering water temperature from the loopfield to ensure that equipment capacity complies with ground-loop design parameters in cooling
23) Insert the fluid flow rate through the source side of the heat pump in cooling, typically 2.5-3.0 gpm per nominal ton of installed heat pump capacity (the source side of the heat pump refers to water entering the unit from the ground connection)
24) Insert the design load side water flow rate in cooling (11)
25) Insert the design entering load side water temperature from the space in cooling (10)
26) Insert the type of antifreeze to be used as the circulating fluid in the closed-loop ground heat exchanger, if necessary (18) (EWT_{min}<45 F)
27) Insert the volumetric concentration of the antifreeze to be used, if necessary (19) (EWT_{min}<45 F)
28) Insert the antifreeze total cooling capacity correction factor if the circulating fluid differs from what is utilized in the performance tables, provided by manufacturer
29) Insert the antifreeze demand correction factor in cooling if the circulating fluid differs from what is utilized in the performance tables, provided by manufacturer
30) Insert the unit heating capacity at the minimum design source side entering water temperature (14) and the maximum design load side entering water temperature (17), at the selected flow rates (15) & (16) provided in manufacturer performance data tables
31) Insert the heat pump electric demand in heating at the selected design conditions, provided in manufacturer performance data tables
32) Insert the heat pump coefficient of performance (COP) in heating at the selected design conditions, provided in manufacturer performance data tables
33) Insert the heat of extraction at the selected design conditions, usually given in manufacturer performance data tables (if not given, calculate according to provided equation)
34) Insert the leaving load side water temperature in heating from the GSHP to the space (if not given, calculate according to provided equation)
35) Insert the unit total cooling capacity at the maximum design source side entering water temperature (22) and the minimum design load side entering water temperature (25), at the selected flow rates (23) & (24) provided

Single Capacity Water-Water GSHP Selection Worksheet

Block Load Description = Example Residence | Zone # Single Zone

Total Htg Load = 1 Btu/hr @ 2 F OAT		Total Clg Load = 8 Btu/hr @ 9 F OAT	
Htg Load Design Water Temp = 3 F (HLDWT)		Clg Load Design Water Temp = 10 F (CLDWT)	
Htg Load Water Flow Rate = 4 gpm		Clg Load Water Flow Rate = 11 gpm	
T-Stat Set Point = 5 / 6 / 7 (Htg dB / Clg dB / %-RH)			

Brand 12

Model 13

Heating Mode Performance Data			Cooling Mode Performance Data		
Performance Parameters	Correction Factors[A]		Performance Parameters	Correction Factors[A]	
	HC	DMD		TC	DMD
EWT_{min}= 14 F			EWT_{max}= 22 F		
GPM= 15 gpm			GPM= 23 gpm		
GPM_{load}= 16 gpm			GPM_{load}= 24 gpm		
ELT_{max}= 17 F			ELT_{min}= 25 F		
Antifreeze / %[B] 18 / 19	CF_{AF}= 20	CF_{AF}= 21	Antifreeze / %[B] 26 / 27	CF_{AF}= 28	CF_{AF}= 29

Tabled Heating Mode Performance Data		Tabled Cooling Mode Performance Data	
HC =	30 Btu/hr	TC =	35 Btu/hr
DMD =	31 kW	DMD =	36 kW
COP =	32	EER =	37
HE =	33 Btu/hr (= HC x (COP-1)/COP)[C]	HR =	38 Btu/hr (= TC x (EER+3.412)/EER)[C]
LLT_H =	34 F (= ELT_{max} + HC / (GPM_{load} x 500)) (≥ HLDWT)	LLT_C =	39 F (= ELT_{min} - TC / (GPM_{load} x 500)) (≤ CLDWT)

Corrected Heating Mode Performance Data			Corrected Cooling Mode Performance Data		
HC_c=	HC x CF_{AF} =	40 Btu/hr	TC_c=	TC x CF_{AF} =	45 Btu/hr
DMD_c=	DMD x C_{AF} =	41 kW	DMD_c=	DMD x CF_{AF} =	46 kW
COP_c=	HC_c/(DMD_c x 3412) =	42	EER_c=	HC_c/(DMD_c x 1000) =	47
HE_c =	HC_c x (COP_c-1)/COP_c =	43 Btu/hr	HR_c =	HC_c x (EER_c+3.412)/EER_c =	48 Btu/hr
LWT_{min}=	EWT_{min} - HE_c/(500xGPM) =	44 F	LWT_{max}=	EWT_{max} + HR_c/(500xGPM) =	49 F

% Sizing in Heating Mode

$$\% \text{ Sizing} = \frac{HC_c}{\text{Total Heat Loss}} \times 100\%$$

% Sizing = 50 %

% Sizing in Cooling Mode

$$\% \text{ Oversizing} = \frac{TC_c - \text{Total Clg Ld}}{\text{Total Clg Ld}} \times 100\%$$

= 51 %

Tabled Water Coil Performance Summary

Side	Mode	EWT_{min}	GPM	WPD (psi)	WPD (ft)	Test Circulating Fluid
Loop	52	53	54	55	56	57
Load	58	59	60	61	62	63

A. Apply correction factors for Antifreeze (AF) as needed depending on Manufacturer's Literature. (No correction - CF=1)
B. CF_{AF} = CF_{Used}/CF_{Tested}. CF_{Used} for antifreeze being used. CF_{Tested} for circulating fluid performance was determined with.
C. Use this equation if performance parameter not provided by Manufacturer in the Engineering Specifications.

NOTES:

Figure 3.4. Single-Capacity, Water-to-Water Heat Pump Selection Worksheet

in manufacturer performance data tables

36) Insert the heat pump electric demand in cooling at the selected design conditions, provided in manufacturer performance data tables
37) Insert the heat pump energy efficiency rating (EER) in cooling at the selected design conditions, provided in manufacturer performance data tables
38) Insert the heat of rejection at the selected design conditions, usually given in manufacturer performance data tables (if not given, calculate according to provided equation)
39) Insert the leaving load side water temperature in cooling from the GSHP to the space (if not given, calculate according to provided equation)
40) Calculate the equipment installed heating capacity by multiplying the tabled heating capacity (30) by the heating capacity antifreeze correction factor (20)
41) Calculate the installed heat pump electric demand in heating by multiplying the tabled heating demand (31) by the heating demand antifreeze correction factor (21)
42) Calculate the installed heat pump COP in heating according to the provided equation
43) Calculate the installed heat of extraction according to the provided equation
44) Calculate the minimum LWT in heating according to the provided equation
45) Calculate the equipment installed total cooling capacity by multiplying the tabled total cooling capacity (35) by the total cooling capacity antifreeze correction factor (28)
46) Calculate the installed heat pump electric demand in cooling by multiplying the tabled cooling demand (36) by the total cooling demand antifreeze correction factor (29)
47) Calculate the installed heat pump EER in cooling according to the provided equation
48) Calculate the installed heat of rejection according to the provided equation
49) Calculate the maximum LWT in cooling according to the provided equation
50) Calculate the percent-sizing in heating according to provided equation (if less than 100%, supplemental heating will need to be utilized during peak conditions)
51) Calculate the percent-sizing in total cooling capacity according to provided equation
52) Insert the mode of operation (heating/cooling) for which the highest source side flow rate will occur (if the flow rate is same for heating and cooling, head loss calculations will be based on heating mode because of lower entering water temperatures experienced)
53) Insert the minimum design source side entering water temperature from the loopfield (14)
54) Insert the maximum source side fluid flow rate through the heat pump (15) or (23)
55) Insert the water pressure drop (psi) through the source side water-to-refrigerant coil, provided by manufacturer
56) Insert the water pressure drop (ft) through the source side water-to-refrigerant coil, provided by manufacturer
57) Insert the circulating fluid type used to figure pressure drop through the source side water-to- refrigerant coil and the GHEX, provided by manufacturer
58) Insert the mode of operation (heating/cooling) for which the highest load side flow rate will occur (if the flow rate is same for heating and cooling, head loss calculations will be based on cooling mode because of lower entering water temperatures experienced)
59) Insert the minimum design load side entering water temperature from the space (25)
60) Insert the maximum load side fluid flow rate through the heat pump (16) or (24)
61) Insert the water pressure drop (psi) through the load side water-to-refrigerant coil, provided by manufacturer
62) Insert the water pressure drop (ft) through the load side water-to-refrigerant coil, provided by manufacturer
63) Insert the circulating fluid type through the load side water-to-refrigerant coil and the hydronic piping, provided by manufacturer

Items 52-63 on the selection worksheet are necessary to perform system head loss calculations for pump sizing. For convenience, these items are found and recorded at this point of the design process.

This worksheet will be filled out in the same manner as the Single-Capacity, Water-to-Air GSHP Selection Worksheet. Primary differences between the water-to-air and water-to-water selection worksheets are the fact that the load side of the refrigeration cycle is affected by water flow through the coil in the heat pump rather than forced air. Because of this, equipment performance must be specified according to load side entering water temperatures rather than entering air temperatures from the space. Sensible cooling capacity for water-to-water equipment is not considered until fan coil sizing is performed. Moisture condensation from the air (latent cooling) will not occur until the chilled water is routed through a fan coil where it comes into contact with the moist air from the space. Therefore, adequate sensible cooling capacity and SHF are factors considered only during the fan coil selection process and not during the water-to-water GSHP selection process.

For hydronic (radiant) heating systems, supplemental heat in the hydronic side is not usually included. It is for these systems that equipment sizing is especially critical. Without any form of supplemental heat, the heat pump unit itself will be relied upon to satisfy peak loads. If strip-resistance plenum heat is included with the heat pump, it cannot be used in conjunction with the radiant system because heat pump units only have the ability to heat air or heat water at any given time. Both tasks cannot be accomplished concurrently (dual-circuit equipment being the exception). If the system is undersized, it will not have the ability to maintain the space at the thermostat set point during peak conditions. In fact, equipment over sizing of 25% is recommended for water-to-water GSHP systems (heating or cooling-based designs) to allow for quicker storage tank temperature recovery and "ramp up" operation subsequent to thermostat setback periods.

Any other type of GSHP system (dual-capacity, water-to-water, combination, or dual-circuit equipment) is selected in the same manner as listed in the three types of systems discussed in detail in this section. The selection principles will simply be a combination of the principles for proper selection of single-capacity, water-to-air; dual-capacity, water-to-air; and single-capacity, water-to-water equipment as described.

3.5 Air Distribution System Layout and Design

After accurate load calculations have been performed, indoor air quality provisions have been made (ventilation system type, if necessary), and equipment has been properly selected, the air distribution system can be designed. Because heat pumps do not deliver air at temperatures as high as gas-fired furnaces and proper amounts of air flow are critical to heat pump performance, accurate duct design is essential. There are three basic steps in designing the air distribution system:

1) Select the best location for the duct system: crawl space, basement, or attic.
2) Select and locate the supply air diffusers and return air grilles based on individual room heat gain/loss calculations.
3) Lay out and size the ducts to match the heat pump air flow rate and static pressure capabilities

The air distribution system provides both the heating and cooling for the system. As a result, attention must be given to both. A system that does not provide a comfortable environment cannot be operated efficiently. Buoyancy of the air in the space and the manner in which air is introduced to the space play important roles with respect to comfort. The object is to introduce and mix the air in the space so the temperature will be uniform throughout, and occupants will be unaware of air movement.

3.5.1 AIR DISTRIBUTION SYSTEM LAYOUT Before a design can be attempted, the locations of the heat pump system, the air supply diffusers, registers, and return grilles must be specified. Once the location of each is acknowledged, the air distribution system location must be recognized as to whether it will be located in a crawl space, basement, attic, etc. In all actuality, the equipment and duct system locations will be known before heat loss/gain calculations are made because, if they are located in an unconditioned space, the additional equipment loads must be accounted for. The important factor in verifying duct location at this stage of the design process is to identify duct size limitations as well as air supply diffuser, register, and return grille location limitations, if any.

The number and location of the diffusers and grilles in a single room is determined by the amount of air flow necessitated by the heat loss/gain calculations for that room. Air flow should be divided in direct proportion to each room's load. Location and selection of the air diffusers and air supply registers can greatly enhance overall system performance. The importance of using high-quality diffusers and air supply registers rather than one-piece, stamped-out registers cannot be over emphasized. At a minimum, both the throw and pressure required should be supplied by the manufacturer. Through-the-floor, perimeter-type diffusers should be placed near the outer walls. Most rooms can be adequately served by one diffuser, with the exception of corner rooms with windows on each side and larger rooms. Diffuser throw should be about equal to ceiling height for conventional spreading-vane types. Regarding return grille locations, it is good practice to place a return grille in each bedroom in addition to having a centrally located return grille. In order to provide superior comfort levels throughout a home, the air distribution system should be designed such that there will be no more than a 2 F temperature difference between any of the rooms (a 2 F temperature difference is ideal, but the maximum allowable temperature difference is 4 F) *(ACCA (2009) Manual D)*. However, this task is a difficult one in the case of a single-zone system typical to most home installations. In theory, uniform room temperature in a residence could be attained by air flow adjustments made via balance dampers in conjunction with continuously mixing and moving air around the home via continuous blower operation.

3.5.2 DUCT SIZING Once the location of the supply air diffusers, registers, and return air grilles is known, the duct run placement and sizing must be determined. The next step in the design process is to lay out and size the ductwork to match the flow rate for each run and the static pressure capabilities of the blower on the selected equipment. As the air flow through a supply trunk is directed through its branches, the duct size should be gradually stepped down to maintain proper air velocity. The blower type, permanent split capacitor (PSC) versus electrically commutated motor (ECM) for the selected equipment must first be determined. PSC blowers generate a specific amount of external static pressure (ESP) at a given speed setting. A blower's available ESP is the pressure drop that a fan can overcome at a given air flow after pressure losses over the coil are accounted for. The amount of air flow that a PSC blower will produce is dependent on the amount of air flow resistance (static pressure drop) encountered in the distribution system.

ECM blowers use electronically controlled motors to produce a specific air flow rate in a given mode of operation and will adjust available ESP for the blower (by adjusting rotation speed) to do so. ECM blowers can typically generate external static pressures in the range of 0.10-0.80 inches of water column (IWC) to produce the air flow required for proper unit operation. The guidelines given for duct sizing in this section pertain to PSC blowers, but should be followed for equipment with ECM blowers to ensure that system performance is satisfactory with respect to overall operating efficiency in addition to fan noise and equipment life cycle.

The basic objective of designing an air distribution system is to design the ductwork so that the air flow resistance in that system matches the ESP produced by the blower when it generates the desired amount of air

Table 3.1. PSC Blower Data Example

Blower Data	
Air Flow (CFM)	Resistance (IWC)
1210	0.10
1150	0.15
1110	0.20
1070	0.25
1030	0.30
985	0.35
960	0.40
895	0.50

Source – ECONAR Engineering Specifications Manual – 3-ton unit with PSC Fan, High Setting

flow to the area served. It is good practice to base duct design calculations on a medium fan speed. Doing so allows for adjustments to be made to the fan performance after the system has been installed. Table 3.1 gives an example of manufacturer's blower data. Most commonly, ductwork is sized to cause static pressure drop of 0.20-0.30 IWC. Values below 0.10 IWC should not be commonly used unless the effective length of the longest circulation path is relatively short. Alternatively, values above 0.35 IWC should not be commonly used unless the effective length of the longest circulation path is relatively long *(ACCA (2009) Manual D).*

If this specific heat pump system has been selected according to the manufacturer's performance data at an air flow rate of 1,110 CFM, according to the blower performance data (Table 3.1), the air distribution system would have to be designed such that the longest circulation path (longest supply run plus the longest return run) would experience a pressure drop of no more than 0.20 IWC as the design air flow rate. The design duct pressure drop has to include pressure drop through the filters, air-side devices (dampers, junction boxes, etc.), secondary equipment (plenum heaters, humidifiers, ventilation devices, etc.), supply registers, and return grilles.

In addition to external static pressure drop, air velocity (feet per minute – FPM) as it travels through all parts of the ducted system must be taken into consideration. If air velocity is too high in the distribution system, the turbulence of the air will generate large amounts of noise. In some cases, a duct run will be sized such that it satisfies the friction rate requirement, but causes air velocity higher than the recommended maximum velocity. In such instances, the duct size will need to be increased to ensure than the air flow does not exceed maximum recommended velocity. In general, it is good practice to always insulate ductwork for noise reduction both on the supply and the return trunks. Table 3.2 gives air velocity recommendations for duct sizing purposes:

Table 3.2. Air Velocity Recommendations for Duct Sizing

Air Velocity Recommendations (fpm)								
	Supply Side				Return Side			
	Recommended		Maximum		Recommended		Maximum	
	Rigid	Flex	Rigid	Flex	Rigid	Flex	Rigid	Flex
Trunk Ducts	700	600	900	700	600	600	700	700
Branch Ducts	600	600	900	700	400	400	700	700
Supply Outlet Face Velocity	Size for Throw		700		--		--	
Return Grille Face Velocity	--		--		--		500	
Filter Grille Face Velocity	--		--		--		300	

Source – ACCA Manual D (page 3-7)

Figure 3.5 displays typical heat pump ductwork connections that would produce the best level of system noise reduction.

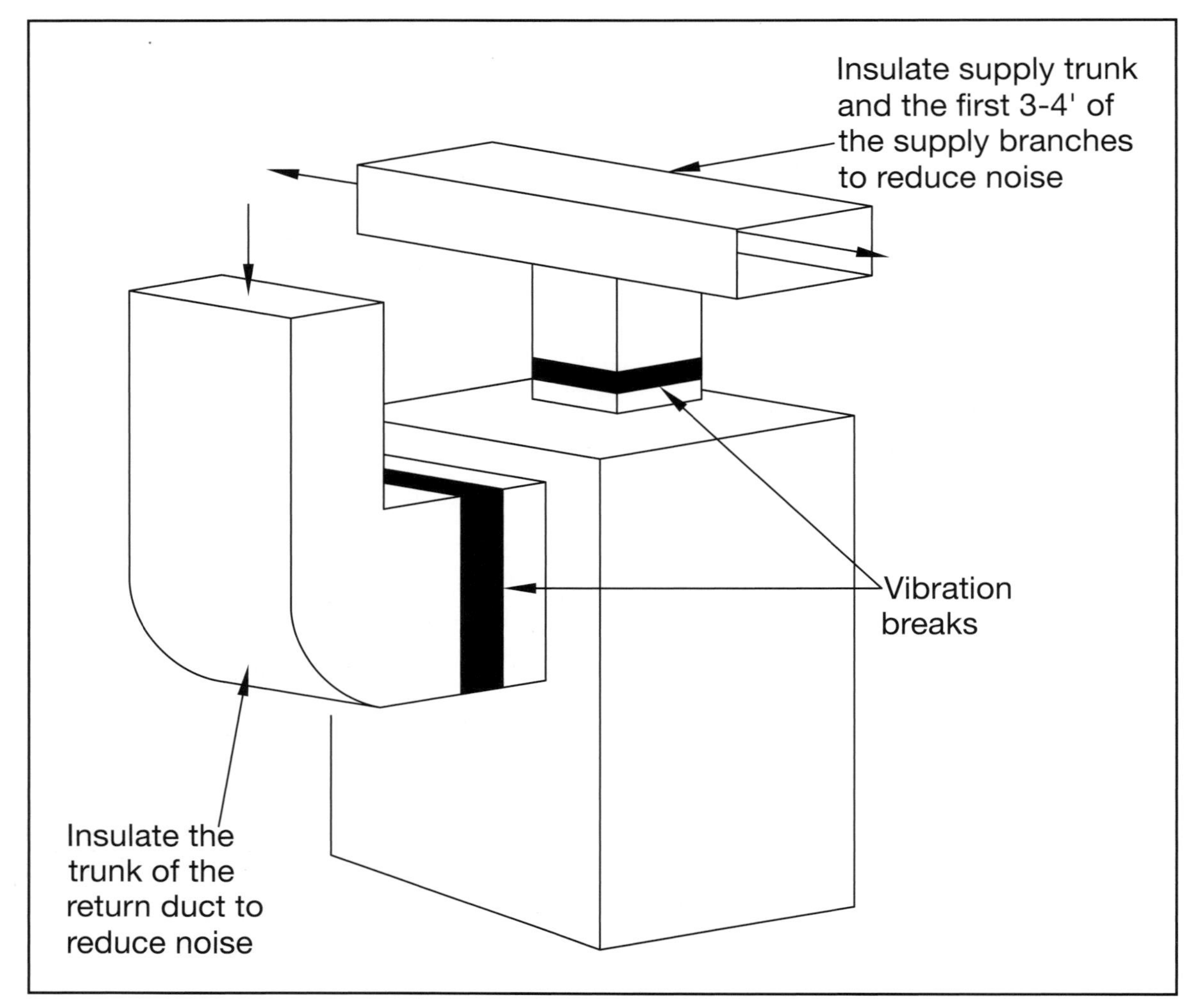

Figure 3.5. Typical Heat Pump Ductwork Installation

Many considerations exist which much be taken into account when performing duct design. Proper air flow delivery is critical for room comfort. A superior design will provide adequate amounts of air circulation and mixing at acceptable pressure drop levels and account for drafts, noise, and air stream integrity. Acceptable methods for duct design are given in *ACCA (2009) Manual D* in addition to the equal friction and balanced capacity methods given in *ASHRAE (2009) Handbook of Fundamentals.* Refer to either manual for detailed procedures, equations, and guidelines necessary for successful duct design.

3.6 Building/Zone Energy Load Analysis Using the Bin Method

Three basic ways exist to estimate the amount of energy a building's heating/cooling system will use in a given year: the degree day method, the bin method, and the hour-by-hour method. The importance of estimating the amount of energy a particular system will use lies in the ability to compare the operating cost over the life of the equipment versus its initial installation cost. Performing an accurate energy analysis allows contractors to calculate annual savings (in terms of operating cost) that a more expensive GSHP system offers over a cheaper, less efficient system. The annual savings can then be used to calculate payback period, in years, for a specific piece of equipment.

- The **degree day method** is the simplest, least reliable energy estimation method. This method is only accurate if the efficiency of the heating/cooling system is constant and not dependent on outdoor conditions, a very unrealistic constraint.
- The **bin method** is best suited to estimate the amount of energy a system will use in a residential setting. It takes into consideration outside temperature and part-load conditions.
- The **hour-by-hour** method is an extremely detailed energy estimation method best suited for large commercial systems where a considerable amount of precision is required.

The bin method, because of its high applicability in the residential sector, is discussed at length in this section. If further detail is needed for the degree day method, refer to Chapter 19 of the *ASHRAE (2009) Handbook of Fundamentals.* A free computer program capable of performing highly detailed hour-by-hour energy analysis for commercial buildings is provided by the DOE (eQUEST). Attempts to use this software should not be done without proper training.

The basic concept behind the bin method is the fact that all hours throughout the course of the year when a particular temperature band occurs can be grouped together. An energy calculation is then made for those hours based on the building load and equipment operating efficiency at the temperature band in question. For calculation purposes, temperature bands (called bins) are collected in 5 F increments and are typically divided into three daily 8-hour shifts for every month of the year. A weather bin data example is shown in Figure 3.6. The figure displays the hours spent in each temperature bin for every month, but is not broken down by time of day. Operating cost and energy usage analysis via the bin method can be performed for a building for a specific period during the day for the design month, or it can be performed for all hours of the day for an entire year.

The modified bin method is recommended for commercial buildings where internal heat gains tend to be large and can vary greatly. For most light commercial and residential applications, the simplified bin method is sufficient and is described herein. Up-to-date engineering weather data is available for purchase in CD-ROM format from NOAA. Necessary weather data are also available in *ASHRAE's Bin Weather Data (1997)* and *USAF (1978)*. Further explanation of the both the modified bin method and the simplified bin method are given in Chapter 19 of the *ASHRAE (2009) Handbook of Fundamentals.*

The first step in using the bin method is to gather information about the building in question. In particular, the amount of heating or cooling required to maintain the space at the thermostat set point (i.e. – the building's load profile) as a function of outdoor temperature is needed. Figure 3.7 illustrates the parameters necessary to understand and develop a building's load profile:

Temperature Bin	Grand Forks, ND												
	Month												
	Jan	Feb	Mar	Apr	May	Jun	Jul	Aug	Sept	Oct	Nov	Dec	Total
115/119													0
110/114													0
105/109							0						0
100/104						0	0	1	0				1
95/99				0	0	2	3	3	1				9
90/94				0	3	8	14	15	3	0			43
85/89				1	10	20	34	33	7	0			105
80/84				3	20	41	81	60	17	1			223
75/79				5	34	65	107	85	28	3			327
70/74				10	54	96	128	111	49	9	0		457
65/69			0	17	74	122	137	128	71	18	1		568
60/64			1	29	97	133	122	131	96	35	2		646
55/59		0	5	45	107	114	74	95	125	58	4		627
50/54	0	0	10	67	110	69	32	54	124	92	12	0	570
45/49	0	3	18	91	90	33	9	22	99	119	27	2	513
40/44	4	9	42	112	70	13	1	7	59	138	42	4	501
35/39	14	30	107	126	46	3		1	30	123	77	18	575
30/34	30	51	142	111	21			0	8	82	135	52	632
25/29	52	70	105	59	6				3	38	134	77	544
20/24	66	79	75	25	2				0	18	96	96	457
15/19	75	67	71	9						6	72	90	390
10/14	68	61	57	5						2	49	88	330
5/9	78	69	47	3						1	31	78	307
0/4	82	70	30	2						0	21	69	274
-5/-1	63	48	16	0							7	53	187
-10/-6	68	46	11								7	53	185
-15/-11	62	34	4								2	35	137
-20/-16	43	20	1								1	22	87
-25/-21	25	9									0	8	42
-30/-26	10	5										2	17
	---------	---------	---------	---------	---------	---------	---------	---------	---------	---------	---------	---------	---------
	740	671	742	720	744	719	742	746	720	743	720	747	8754

Figure 3.6. Weather Bin Data for Grand Forks, ND

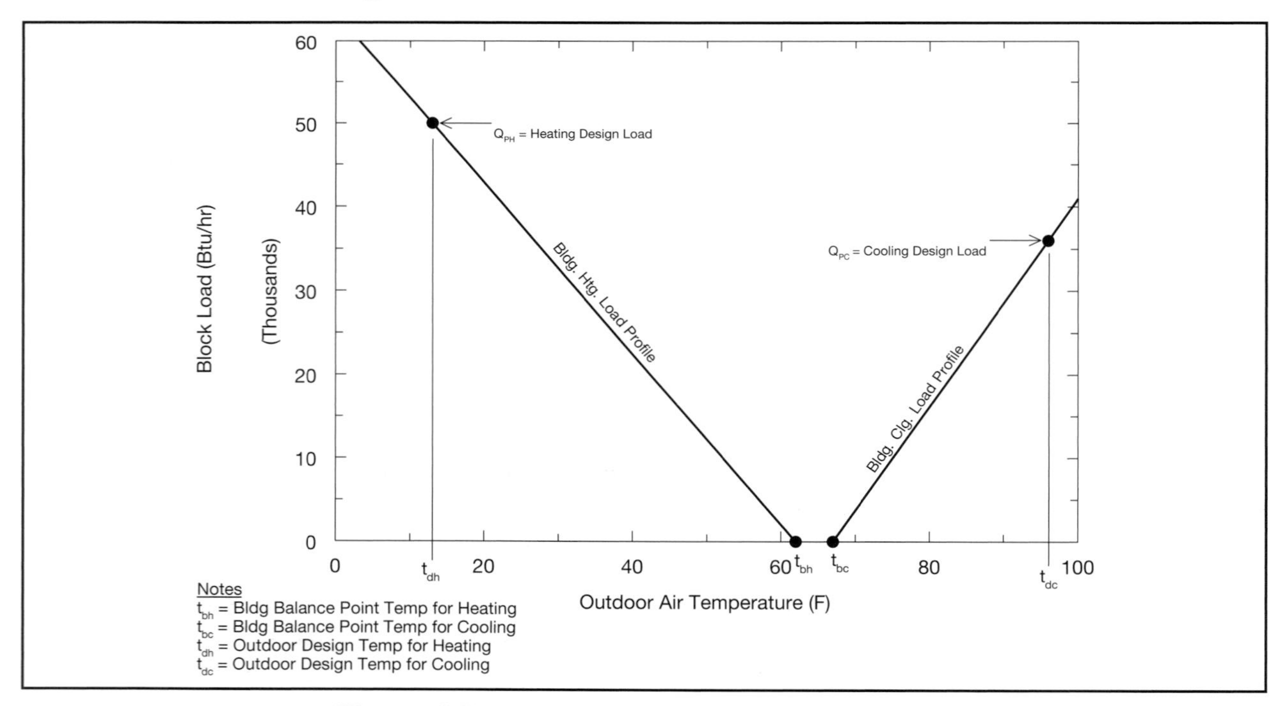

Figure 3.7. Building Load Profile for Bin Analysis

To develop the load profile shown by Figure 3.7, two points for each mode of operation must be located. First, the peak heating and cooling loads for a building at the outdoor design temperatures (determined by geographical location) must be calculated. Next, the building's balance point temperatures must be defined. The balance point temperatures are the outdoor air temperatures where internal heat gains from people, appliances, etc., offset the envelope heat loss to the atmosphere. It is at those temperatures where no indoor heating or cooling will be required to maintain the temperature of the home at the thermostat set point. It is well documented that neither heating nor cooling is generally required when the outdoor temperature is around 60-67 F. In some cases, more than one load profile may be necessary to accommodate a building's uses, such as separate profiles for occupied and unoccupied hours. As previously stated, bin weather data is divided in 8-hour shifts to make such an analysis possible.

Once the peak loads and balance temperatures are defined, the equations describing the building load profile (Figure 3.7) must be calculated. If the outdoor air temperature is lower than the balance point temperature for heating, then heating will be required to keep the space at the thermostat set point. If the outdoor air temperature is higher than the balance point temperature for cooling, then cooling will be required to keep the space at the thermostat set point. If the outdoor air temperature is between the balance point temperatures, no heating or cooling is required for the building. Equation 3.1 shows how to calculate the building heating load according to outdoor air temperature, and Equation 3.2 shows how to calculate the building cooling load according to outdoor air temperature:

$$Q_{BH} = \left(\frac{Q_{PH}}{t_{dh} - t_{bh}}\right) \bullet (t_{air} - t_{bh})$$ **Equation 3.1**

Where:

Q_{BH} = Building Heating Load at Specified Air Temperature (Btu/hr)
Q_{PH} = Building Peak Heating Load (Btu/hr)
t_{dh} = Outdoor Design Temperature for Heating (F)
t_{bh} = Building Balance Point Temperature for Heating (F)
t_{air} = Selected Outdoor Air Temperature (F)

$$Q_{BC} = \left(\frac{Q_{PC}}{t_{dc} - t_{bc}}\right) \bullet (t_{air} - t_{bc})$$ **Equation 3.2**

Where:

Q_{BC} = Building Sensible Cooling Load at Specified Air Temperature (Btu/hr)
Q_{PC} = Building Peak Sensible Cooling Load (Btu/hr)
t_{dc} = Outdoor Design Temperature for Cooling (F)
t_{bc} = Building Balance Point Temperature for Cooling (F)

Note: Use the sensible cooling load for this calculation (not total). The space thermostat operates based on sensible cooling load and sensible equipment cooling capacity.

Next, equipment capacity, as it varies throughout the year, needs to be determined. Because the capacity of a ground source heat pump varies with entering water temperature rather than being dependent on outside air temperature, the entering water temperature from the loopfield needs to be estimated. The entering water temperature for an open-loop system will remain fairly constant throughout the year and can be reasonably approximated by the deep earth temperature for the geographical location. The entering water temperature for a closed-loop ground source heat pump system is extremely difficult to quantify because it depends on a large number of factors such as time of year, type of soil at the installation site, the ground heat exchanger design as it relates to the size of the installed ground source heat pump versus the size of the heat exchanger itself, equipment full-load run hours, amount of formation water flow (long-term considerations), loopfield backfill material, etc. However, the ability to roughly approximate entering water temperature lies within the premise that a properly designed system is sized such that the entering water temperatures will not fall below a certain temperature during the heating season or go above a certain temperature during the cooling season. Additionally, the assumption can be made that the minimum entering temperature will occur during the coldest month with the highest level of equipment full-load run hours in heating, and the maximum entering water temperature will occur during the warmest month with the highest level of equipment full-load run hours in cooling. Assumptions similar to the occurrence of extreme entering water temperatures can be made to approximate when the mean entering water temperature will occur. Entering water temperatures will approach their mean almost concurrently with air temperatures during season changes. For practical purposes, the mean entering water temperature can be assumed to be the same as the mean entering air temperature for the geographical region. Figure 3.8 illustrates this principle:

Assuming the relationship demonstrated by Figure 3.8, the entering water temperatures throughout the

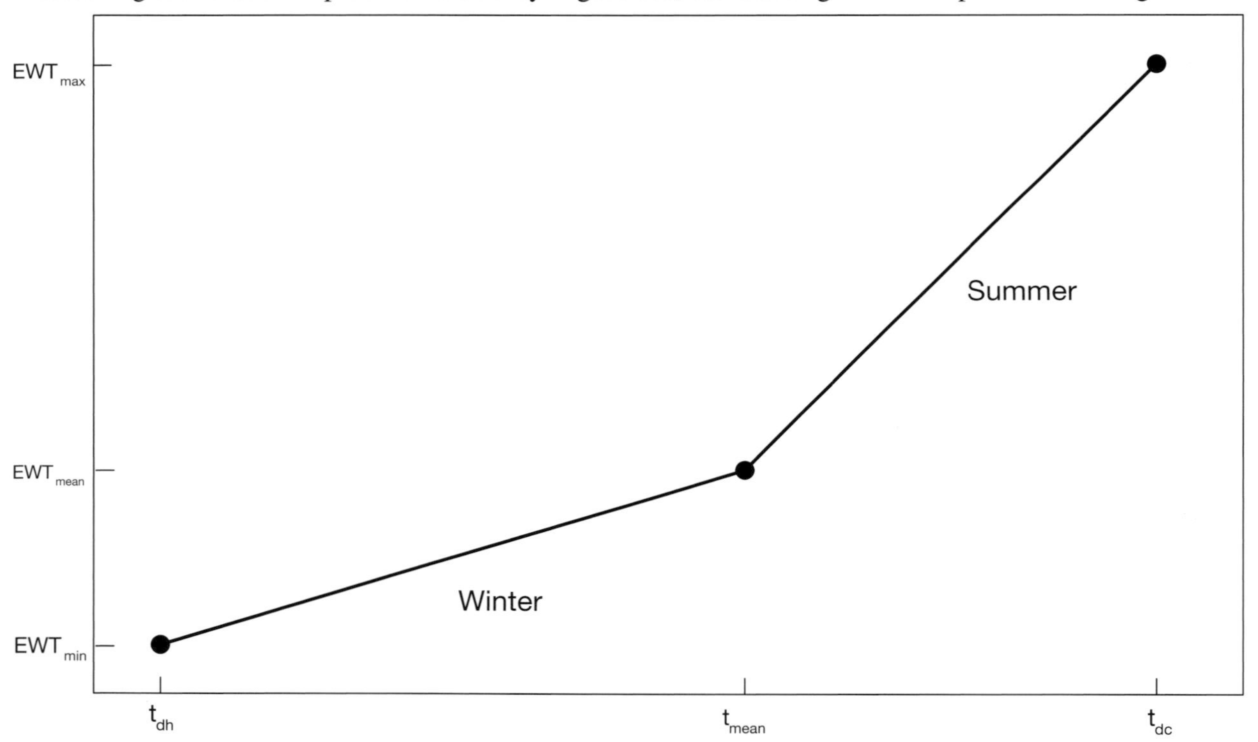

Figure 3.8. Entering Water Temperature versus Entering Air Temperature

year can be calculated according to Equation 3.3 in heating and Equation 3.4 in cooling.

$$EWT_h = EWT_{min} + \left(\frac{EWT_{mean} - EWT_{min}}{t_{mean} - t_{min}}\right) \cdot (t_{air} - t_{dh})$$ **Equation 3.3**

Where:

EWT_h = Entering Water Temperature in Heating at Specified Air Temperature (F)
EWT_{min} = Minimum Design Entering Water Temperature (F)
EWT_{mean} = Mean Entering Water Temperature (F)
t_{mean} = Mean Outside Air Temperature (F)

$$EWT_c = EWT_{mean} + \left(\frac{EWT_{max} - EWT_{mean}}{t_{dc} - t_{mean}}\right) \cdot (t_{air} - t_{mean})$$ **Equation 3.4**

Where:

EWT_c = Entering Water Temperature in Cooling at Specified Air Temperature (F)
EWT_{max} = Maximum Design Entering Water Temperature (F)

Now that the relationship between entering water temperature and outdoor air temperature has been defined, ground source heat pump capacity and demand, as they vary with entering water temperature, can be calculated. To do so, engineering performance data for the selected equipment is necessary. Equipment capacity and demand and their relationship with entering water temperature are shown by Figure 3.9. The figure displays the relationship between capacity and power input (demand) as they vary with EWT for a typical 2-ton unit.

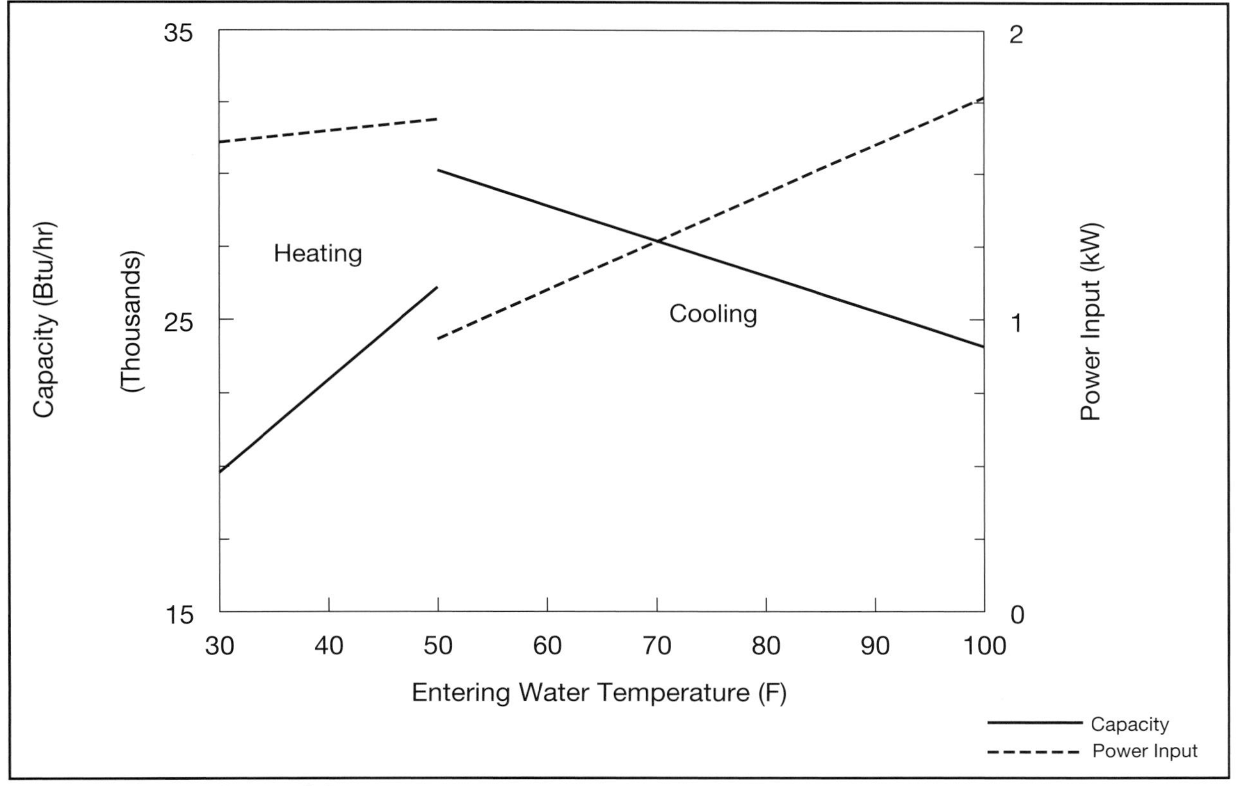

Figure 3.9. Capacity & Demand versus Entering Water Temperature

Development of the profile shown by Figure 3.9 is similar to the development of a building's heating and cooling load profile. Two points for capacity and demand in heating and cooling mode are necessary. A straight line relationship can be established after two points are identified for each mode of operation and the capacity and demand for a piece of equipment can then be approximated for any reasonable entering water temperature. Engineering performance data for a selected heat pump model can be used to find two points of operation as shown by Table 3.3:

Table 3.3. Engineering Data Operating Points

EWT (˚F)	Unit Capacity (Btu/hr)	Demand (kW)
EWT_{low}	$CAP_{lowtemp}$	$D_{lowtemp}$
EWT_{high}	$CAP_{hightemp}$	$D_{hightemp}$

From the engineering performance data, the capacity and demand profiles for heating and cooling can be defined by Equations 3.5-3.8:

$$HC = HC_{hth} + \left(\frac{HC_{hth} - HC_{lth}}{EWT_{hth} - EWT_{lth}}\right) \cdot (EWT_h - EWT_{hth}) \qquad \textbf{Equation 3.5}$$

Where:

- HC = Equipment Heating Capacity at Specified EWT (Btu/hr)
- EWT_{lth} = Low Entering Water Temp from Engineering Data for Heating (F)
- EWT_{hth} = High Entering Water Temp from Engineering Data for Heating (F)
- HC_{lth} = Equipment Rated Heating Capacity at EWT_{lth} (Btu/hr)
- HC_{hth} = Equipment Rated Heating Capacity at EWT_{hth} (Btu/hr)

$$SC = SC_{htc} + \left(\frac{SC_{htc} - SC_{ltc}}{EWT_{htc} - EWT_{ltc}}\right) \cdot (EWT_c - EWT_{htc}) \qquad \textbf{Equation 3.6}$$

Where:

- SC = Equipment Sensible Cooling Capacity at Specified EWT (Btu/hr)
- EWT_{ltc} = Low Entering Water Temp from Engineering Data for Cooling (F)
- EWT_{htc} = High Entering Water Temp from Engineering Data for Cooling (F)
- SC_{ltc} = Equipment Rated Sensible Cooling Capacity at EWT_{ltc} (Btu/hr)
- SC_{htc} = Equipment Rated Sensible Cooling Capacity at EWT_{htc} (Btu/hr)

$$DMD_h = DMD_{hth} + \left(\frac{DMD_{hth} - DMD_{lth}}{EWT_{hth} - EWT_{lth}}\right) \cdot (EWT_h - EWT_{hth}) \qquad \textbf{Equation 3.7}$$

Where:

- DMD_h = Equipment Demand in Heating at Specified EWT (kW)
- DMD_{lth} = Equipment Rated Demand in Heating at EWT_{lth} (kW)
- DMD_{hth} = Equipment Rated Demand in Heating at EWT_{hth} (kW)

$$DMD_c = DMD_{htc} + \left(\frac{DMD_{htc} - DMD_{ltc}}{EWT_{htc} - EWT_{ltc}}\right) \cdot (EWT_c - EWT_{htc})$$ **Equation 3.8**

Where:

DMD_c = Equipment Demand in Cooling at Specified EWT (kW)
DMD_{ltc} = Equipment Rated Demand in Cooling at EWT_{ltc} (kW)
DMD_{htc} = Equipment Rated Demand in Cooling at EWT_{htc} (kW)

Once the heat pump capacity in heating and cooling is calculated for every temperature bin, the theoretical run time fraction (RF_T) can be calculated according to the ratio of building load to equipment capacity as shown by Equation 3.9:

$$RF_T = \frac{Q_B}{CAP}$$ **Equation 3.9**

Where:

RF_T = Theoretical Run Fraction (Dimensionless)
Q_B = Building Load (Btu/hr)
CAP = Unit Capacity (Btu/hr)

Typically, manufacturers' performance data applies when equipment is operating under full-load, steady-state conditions. Realistically, the equipment will operate under partial load most of the time. To quantify the behavior of a heat pump when operating under part-load conditions, the part-load factor (PLF) needs to be computed according to Equation 3.10:

$$PLF = 1 - D_C \bullet \left(1 - \frac{Q_B}{CAP}\right)$$ **Equation 3.10**

Where:

PLF = Part-Load Factor (Dimensionless)
D_C = Degradation Coefficient (Dimensionless)

The degradation coefficient is a dimensionless number used to help quantify the amount of efficiency loss due to startup, shutdown, and other part-load operation. For practical purposes in the case of a ground source heat pump system, a value within the range of 0.10 - 0.15 is a reasonable approximation for this coefficient.

Partial-load factor is also defined to be the ratio of theoretical equipment run time to actual equipment run time. Because it is calculated according to this ratio, the actual run time fraction (RF_A) can be calculated according to Equation 3.11:

$$RF_A = \frac{RF_T}{PLF}$$ **Equation 3.11**

Where:

RF_A = Actual Run Time Fraction (Dimensionless)

Once the actual run time fraction has been calculated for all temperature bins, the actual amount of time a piece of equipment will run during the analysis period can be calculated according to Equation 3.12:

$$RT_A = h_b \cdot RF_A$$ **Equation 3.12**

Where:
RT_A = Actual Equipment Run Time (hrs)
h_b = Number of Hours in Each Bin for Analysis Period (hrs)

To calculate operating cost for a system during the analysis period, the rate of unit input (demand) needs to be calculated according to Equation 3.7 for heating and Equation 3.8 for cooling. Next, the amount of energy used during the analysis period needs to be calculated and can be done according to Equation 3.13:

$$E = DMD \cdot RT_A$$ **Equation 3.13**

Where:
E = Amount of Energy Used During Analysis Period (kWh)
DMD = Equipment Demand in Heating or Cooling at Specified EWT (kW)

Once the amount of energy use during the analysis period has been computed, the cost of operation can be calculated according to the local utility rate schedule in conjunction with Equation 3.14:

$$C_e = E \cdot r_e$$ **Equation 3.14**

Where:
C_e = Cost of Operation for Analysis Period ($)
r_e = Cost of Electricity According to Local Utility Rate ($/kWh)

Certain refinements to this procedure may be required in most cases. In some cases, ground source heat pumps are sized such that they cannot handle the entire peak load of a building, and supplemental resistance heat is required. In such cases, the PLF for the equipment will be assumed to be 1.0 because the unit will run continuously. In addition, when the unit does not have sufficient capacity dictated by the building load, supplemental resistance heat (COP=1.0) will be used and the additional cost to do so will need to be accounted for according to Equation 3.15.

$$DMD_{res} = (Q_{BH} - HC) / 3,412$$ **Equation 3.15**

Where:
DMD_{res} = Supplemental Resistance Heat Demand When $HC<Q_{BH}$ (kW)

In such cases where supplemental heat is required, the total system demand in heating is defined by Equation 3.16:

$$DMD_{htotal} = DMD_{h} + DMD_{res}$$ **Equation 3.16**

Where:
DMD_{htotal} = Total System Demand in Heating Mode (kW)

In the case of a ground source heat pump system, the amount of energy removed from the ground in heating mode or rejected to the ground in cooling mode during the analysis period can be calculated according to Equations 3.17 and 3.18, respectively. The amount of energy removed from the ground in heating mode is referred to as the heating ground load. The amount of energy rejected to the ground in cooling mode is referred to as the cooling ground load.

$$gl_{h} = (HC - DMD_{h} \bullet 3{,}412) \bullet RF \bullet h_{b}$$ **Equation 3.17**

Where:
gl_{h} = Heating Ground Load (Btu)

$$gl_{c} = (-TC - DMD_{c} \bullet 3{,}412) \bullet RF \bullet h_{b}$$ **Equation 3.18**

Where:
gl_{c} = Cooling Ground Load (Btu)
TC = Equipment Total Cooling Capacity at Specified EWT (Btu/hr)

Note: Use the total cooling capacity for this calculation (not sensible). While the operating pattern of the GSHP equipment will be determined by sensible cooling load and sensible equipment cooling capacity, the ground load will be determined by total cooling load and total equipment capacity.

By convention, ground load is considered positive into the building (extraction from the ground) and negative out of the building (rejection to the ground). By this convention, the net ground load for the analysis period is calculated using Equation 3.19.

$$gl_{net} = gl_{h} + gl_{c}$$ **Equation 3.19**

Where:
gl_{net} = Net Ground Load for Analysis Period (Btu)

The amount of heat removed from the ground/rejected to the ground via a ground source heat pump system will be dependent on the system's capacity, run hours in each temperature bin, and its efficiency. The net annual ground load is an important parameter in the design of a GHEX for a GSHP system. If the ground load is severely unbalanced (i.e. – much more heat extraction in heating than rejection in cooling annually or vice versa), bore design lengths will likely be affected and center-center bore spacing will need to be adjusted to include more ground heat capacity in the loopfield. The concept of unbalanced ground loads and their effect on ground-loop design is discussed in detail in Chapter 5.

Another parameter that can be calculated using the bin method is design-month equipment run fraction in each mode of operation. Unless otherwise noted, the design months for a given geographical location can be assumed to be January for heating and July for cooling. Run fraction is a parameter used to estimate annual ground loads and can greatly affect overall GHEX design lengths. This concept is also discussed in more detail in Chapter 5. Equipment run fraction can be calculated using Equation 3.20 for January and 3.21 for July:

$$F_{JAN} = \frac{RT_{JAN-HTG}}{h_{b-JAN}}$$ **Equation 3.20**

Where:

F_{JAN} = Run Fraction in Heating Mode during Heating Design Month (January – Dimensionless)
$RT_{JAN-HTG}$ = Actual Equipment Run Time in Heating Mode in January (hours)
h_{b-JAN} = Total Bin Hours in January (744 hours)

$$F_{JUL} = \frac{RT_{JUL-CLG}}{h_{b-JUL}}$$ **Equation 3.21**

Where:

F_{JUL} = Run Fraction in Cooling Mode during Cooling Design Month (July – Dimensionless)
$RT_{JUL-CLG}$ = Actual Equipment Run Time in Cooling Mode in July (hours)
h_{b-JUL} = Total Bin Hours in July (744 hours)

The actual run time in heating mode in January is found by totaling the number of run hours ($RT_{JAN-HTG}$) when the outdoor air temperature is lower than the balance point temperature for heating. The actual run time in cooling mode in July is found by totaling the number of run hours ($RT_{JUL-CLG}$) when the outdoor air temperature is higher than the balance point temperature for cooling. The total bin hours in the design month for the location are simply found by totaling the number of hours in the design month of interest (January for heating and July for cooling).

Another method of estimating run fraction is to use Figures 3.10 for heating and 3.11 for cooling according to heating and cooling degree days for the geographical location of the project and percent-sizing of the equipment compared to the building load. Heating and cooling degree days for various locations throughout the United States are shown in Tables 3.4a and 3.4b.

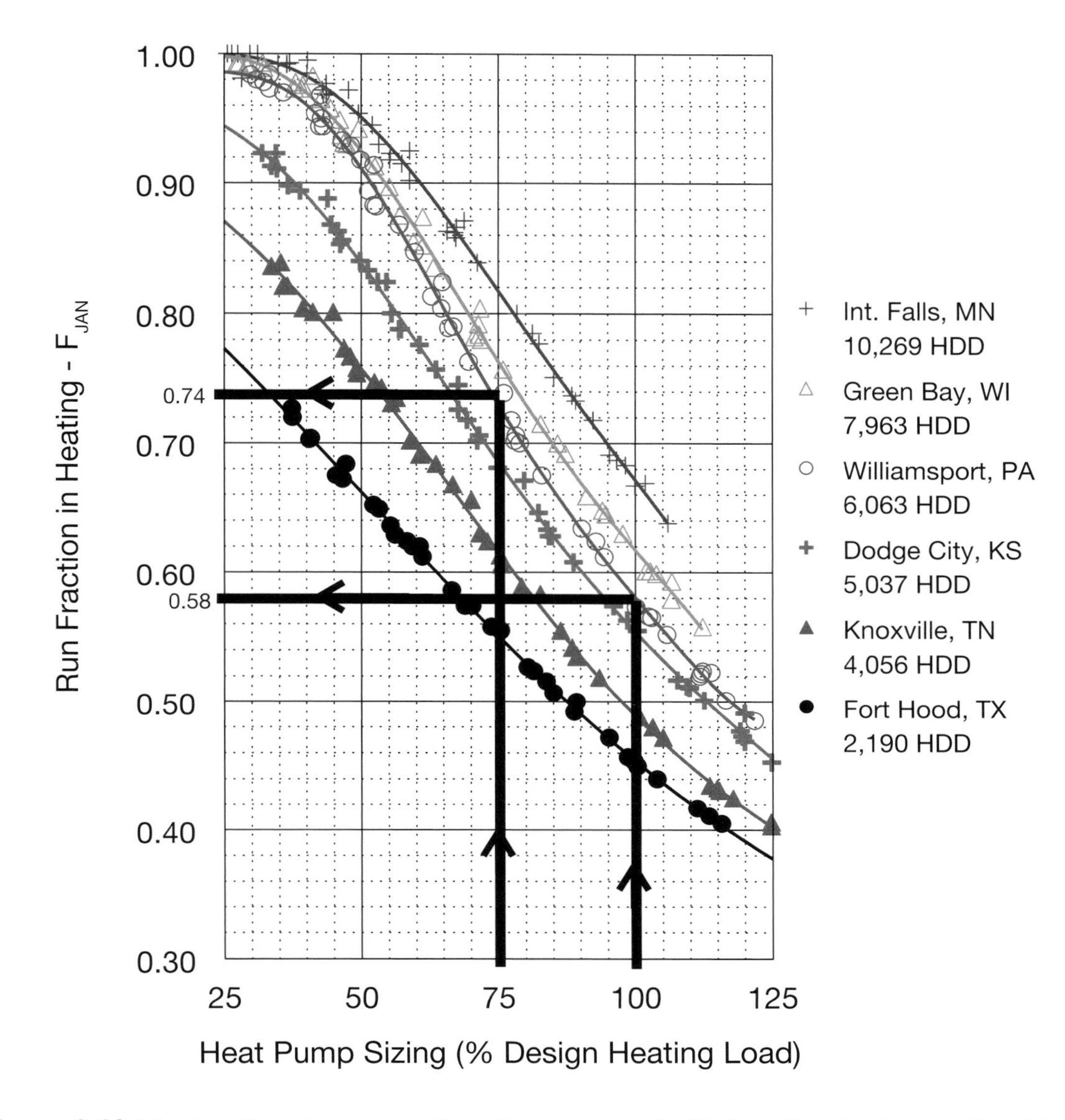

Figure 3.10. Heating Run Fraction vs. Heat Pump Sizing for Various Heating Degree Day Values

Run fraction in heating can be estimated by using Figure 3.10, which was developed using a bin analysis for locations with heating degree days (HDD) ranging from approximately 2,000 to 10,000. The curves were created using current heat pump performance data from every major heat pump manufacturer. As an example, assume that a GSHP system is to be installed in Dugway, UT. The number of heating degree days that occur in Dugway are HDD=5,963, which is very similar to Williamsport, PA (HDD=6,063 as shown by Figure 3.10). According to Figure 3.10, a GSHP system sized to handle 75% of the heating load for a home in Dugway will have a run fraction in heating of approximately F_{JAN}=0.74 (read from the Williamsport, PA curve) and a larger GSHP system sized to handle 100% of the heating load for the same home in Dugway will have a run fraction in heating of approximately F_{JAN}=0.58 (again, read from the same curve).

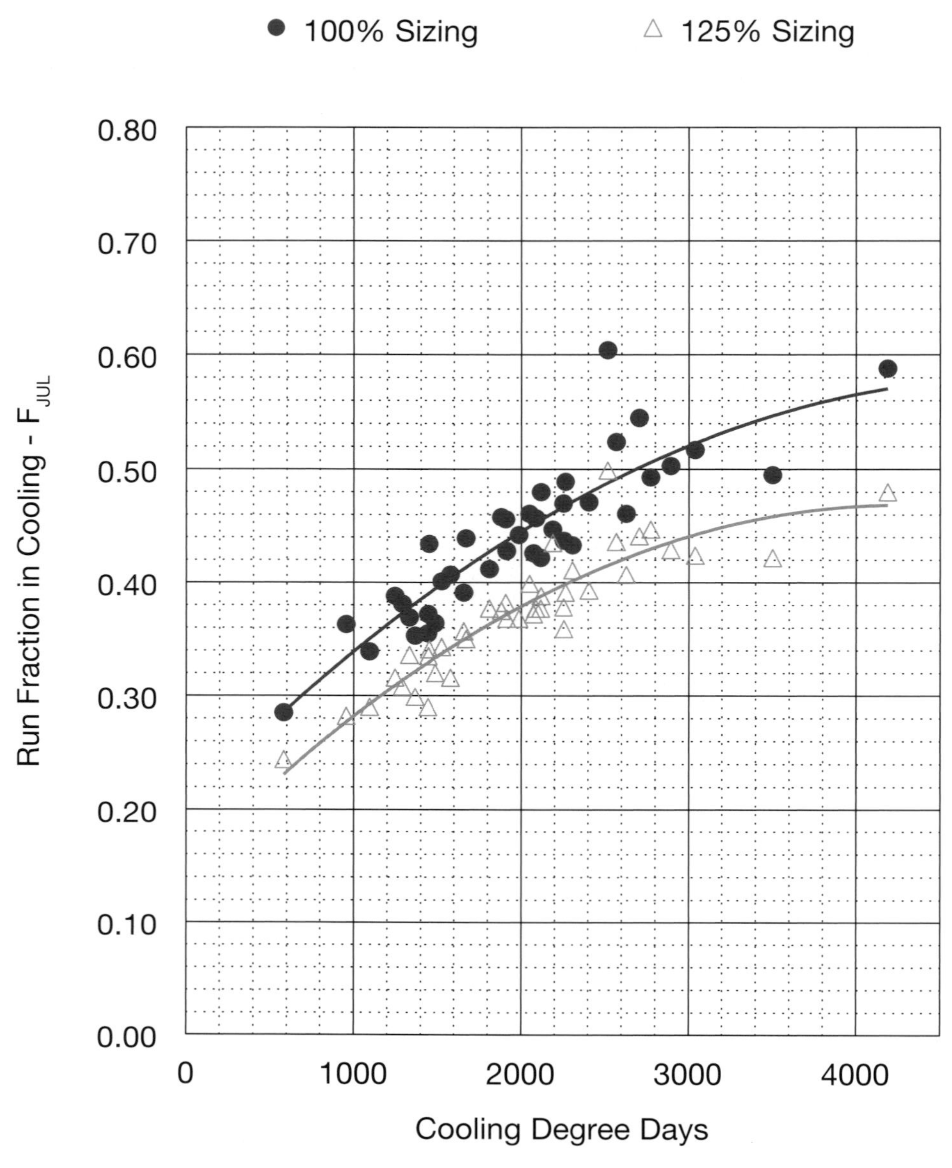

Figure 3.11. Cooling Run Fraction vs. Heat Pump Sizing for Various Cooling Degree Day Values

Run fraction in cooling can be estimated according to cooling degree days for a given location in the same manner as run fraction in heating is estimated from Figure 3.10. Additionally, the curves were generated in a manner similar to that of Figure 3.10. However, notice that the only curves given in Figure 3.11 are for 100% and 125% equipment sizing in cooling (based on sensible capacity). Generally, a system will not be sized such that it will handle less than 100% of the cooling load for a building. Additionally, if a system is oversized by a significant amount in cooling capacity, more than likely the design will be heating based thus explaining why 100%-sizing and 125%-sizing are the only curves given in the figure.

Table 3.4a. Heating and Cooling Degree Days for Various Locations in the U.S.

State		City	HDD	CDD
Alabama	1	Birmingham	2823	1881
	2	Fort Rucker	1947	2213
	3	Huntsville	3262	1671
	4	Mobile	1681	2539
	5	Montgomery	2194	2252
	6	Tuscaloosa	2371	2239
Alaska	7	Adak Navsta	8949	0
	8	Anchorage	10470	3
	9	Aniak	12769	1
	10	Barrow	19674	0
	11	Barter Island	19719	0
	12	Bettles Field	15357	38
	13	Fairbanks	13980	74
	14	Juneau	8574	0
	15	Kodiak	8862	0
	16	Nome	13674	2
Arizona	17	Flagstaff	8340	37
	18	Glendale	1125	4189
	19	Sierra Vista	2369	1739
	20	Tucson	1578	3017
	21	Winslow	4692	1104
	22	Yuma	852	4472
Arkansas	23	Blytheville	3544	1979
	24	El Dorado	2580	2127
	25	Fayetteville	4166	1439
	26	Fort Smith	3437	1929
	27	Little Rock	3084	2086
	28	Pine Bluff	2935	2099
	29	Texarkana	2421	2280
California	30	Alameda	2400	377
	31	Arcata	4403	7
	32	Bakersfield	2120	2286
	33	Barstow	2294	2566
	34	Bishop	4314	1003
	35	Crescent City	4687	6
	36	Fairfield	2649	975
	37	Fresno	2447	1963
	38	Imperial	1115	3764
	39	Lompoc	2241	322
	40	Long Beach	1211	1186
	41	Los Angeles	1274	679
	42	Marysville	2488	1687
	43	Merced	2602	1578
	44	Montague	5550	550
	45	Mountain View	4105	1494
	46	Oakland	2400	377
	47	Rosamond	3241	1733
	48	Sacramento	2666	1248
	49	San Bernardino	1599	1937
	50	San Diego	1063	866
	51	San Francisco	2862	142
	52	San Jose	2171	811
Colorado	53	Colorado Springs	6480	404
	54	Denver	6128	695
	55	Grand Junction	5700	1091
	56	Pueblo	5598	922
	57	Trinidad	5512	730
Connecticut	58	Bridgeport	5466	789
	59	Hartford	6121	654
Delaware	60	Dover	4212	1262
	61	Wilmington	4888	1125
Florida	62	Cape Canaveral	696	3289
	63	Cocoa Beach	696	3289
	64	Daytona Beach	815	2961
	65	Fort Lauderdale	178	4098
	66	Fort Myers	302	3957
	67	Gainesville	1256	2606
	68	Homestead AFB	228	3761
	69	Jacksonville	1354	2627
	70	Key West	62	4830
	71	Miami	149	4361
	72	Orlando	580	3428
	73	Panama City	1817	2174
	74	Pensacola	1498	2650
	75	Tallahassee	1604	2551
	76	Tampa	591	3482
	77	Valparaiso	1920	2271
	78	West Palm Beach	246	3999
Georgia	79	Albany	2108	2264
	80	Atlanta	2827	1810
	81	Augusta	2525	1986
	82	Brunswick	1545	2548
	83	Columbus	2154	2296
	84	Fort Benning	2154	2296
	85	Macon	2364	2115
	86	Rome	3510	1360
	87	Savannah	1799	2454
	88	Valdosta	2073	1999
Hawaii	89	Barbers Point	1	4388
	90	Hilo	0	3228
	91	Honolulu	0	3936
	92	Kahului	0	4561
Idaho	93	Boise	5727	807
	94	Idaho Falls	7917	322
	95	Lewiston	5220	792
	96	Mountain Home	6084	812
	97	Pocatello	7109	387
Illinois	98	Belleville	4612	1339
	99	Champaign	5916	979
	100	Chicago	6498	830
	101	Decatur	5458	1142
	102	Glenview	7149	624
	103	Moline	6415	969
	104	Peoria	6097	998
	105	Springfield	5596	1165
Indiana	106	Bunker Hill	6192	808
	107	Evansville	4617	1422
	108	Fort Wayne	6205	830
	109	Indianapolis	5521	1042
	110	South Bend	6294	812
	111	Terre Haute	5433	1107
Iowa	112	Cedar Rapids	6837	858
	113	Des Moines	6436	1052
	114	Fort Dodge	7513	746
	115	Mason City	7765	655
	116	Sioux City	6900	914
	117	Waterloo	7348	758
Kansas	118	Dodge City	5037	1481
	119	Fort Riley	5120	1465
	120	Goodland	6023	894
	121	Salina	4952	1600
	122	Topeka	5225	1357
	123	Wichita	4765	1658
Kentucky	124	Fort Campbell	4298	1433
	125	Fort Knox	4897	1091
	126	Lexington	4713	1154
	127	Louisville	4352	1443
Louisiana	128	Alexandria	1908	2602
	129	Baton Rouge	1689	2628
	130	Lake Charles	1546	2705
	131	Monroe	2190	2517
	132	New Orleans	1417	2773
	133	Shreveport	2251	2405
Maine	134	Augusta	7358	388
	135	Bangor	7676	313
	136	Brunswick	7353	297
	137	Limestone	9560	191
	138	Portland	7318	347
Maryland	139	Andrews AFB	4695	1162
	140	Baltimore	3807	1774
	141	Patuxent River	4077	1331
Massachusetts	142	Bedford	6370	485
	143	Boston	5630	777
	144	Chicopee	6831	371
	145	Falmouth	5734	740
	146	Worcester	6831	371
Michigan	147	Detroit	5898	920
	148	Grand Rapids	6896	613
	149	Lansing	7098	558
	150	Marquette	8272	260
	151	Mt. Clemens	6620	597
	152	Muskegon	6943	487
	153	Oscoda	7924	349
	154	Pellston	8392	267
	155	Sault Ste Marie	9224	145
	156	Traverse City	7550	432
Minnesota	157	Bemidji	9540	349
	158	Duluth	9724	189
	159	International Falls	10269	233
	160	Minneapolis - St. Paul	7876	699
	161	Rochester	8308	473
Mississippi	162	Biloxi	1645	2517
	163	Columbus	2740	2073
	164	Jackson	2401	2264
	165	McCain Field	2504	2085
	166	McComb	1932	2291
	167	Meridian	2504	2085
Missouri	168	Columbia	5177	1246
	169	Ft. Leonard Wood	4682	1255
	170	Grandview	4734	1676
	171	Joplin	4253	1555
	172	Springfield	4602	1366
	173	St. Louis	4758	1561

Table 3.4b. Heating and Cooling Degree Days for Various Locations in the U.S.

State		City	HDD	CDD
Montana	174	Billings	7006	583
	175	Butte	9399	127
	176	Glasgow	8560	494
	177	Great Falls	7828	288
	178	Havre	8250	377
	179	Helena	7975	277
	180	Kalispell	8193	142
	181	Miles City	7620	822
	182	Missoula	7622	256
Nebraska	183	Bellevue	6311	1095
	184	Grand Island	6385	1027
	185	Lincoln	6242	1154
	186	North Platte	6766	750
	187	Omaha	6311	1095
	188	Scottsbluff	6742	690
Nevada	189	Elko	7181	412
	190	Ely	7561	196
	191	Fallon	5156	912
	192	Las Vegas	2239	3214
	193	Mercury	3139	2132
	194	Reno	5600	493
	195	Tonopah	5549	773
	196	Winnemucca	6271	526
New Hampshire	197	Concord	7478	442
	198	Lebanon	7694	307
	199	Manchester	6834	445
	200	Portsmouth	7478	442
New Jersey	201	Atlantic City	5113	935
	202	Newark	4843	1220
	203	Trenton	4897	891
	204	Wrightstown	4897	891
New Mexico	273	Alamogordo	3108	1715
	274	Albuquerque	4281	1290
	275	Carlsbad	2935	2031
	276	Clovis	3955	1305
	277	Farmington	5508	805
	278	Gallup	6588	357
	279	Roswell	3332	1814
New York	280	Albany	6860	544
	281	Binghamton	7237	396
	282	Buffalo	6692	548
	283	Jamestown	7048	389
	284	New York City	4947	949
	285	Newburgh	6438	550
	286	Niagara Falls	6752	508
	287	Plattsburg	7817	387
	288	Rome	7164	444
	289	Syracuse	6803	551
	290	Warren	7542	254
	291	Watertown	7681	301
North Carolina	292	Asheville	4326	818
	293	Charlotte	3162	1681
	294	Cherry Point	2556	1857
	295	Elizabeth City	3073	1590
	296	Fort Bragg	3097	1721
	297	Goldsboro	2771	1922
	298	Greensboro	3848	1332
	299	Jacksonville	2656	1832
	300	Raleigh	3070	1572
North Dakota	301	Bismarck	8802	471
	302	Dickinson	8558	512
	303	Fargo	9092	533
	304	Grand Forks	9489	420
	305	Minot	8990	492
Ohio	306	Akron	6154	678
	307	Cincinnati	5228	942
	308	Cleveland	6121	702
	309	Columbus	5492	951
	310	Dayton	5690	935
	311	Mansfield	6528	548
	312	Toledo	6460	715
	313	Youngstown	6451	552
Oklahoma	314	Altus	3556	2162
	315	Enid	4269	1852
	316	Fort Sill	3326	2199
	317	Mc Alester	3199	2026
	318	Oklahoma City	3663	1907
	319	Tulsa	3642	2049
Oregon	320	Astoria	5056	22
	321	Burns	7785	218
	322	Eugene	4786	242
	323	Klamath Falls	6916	216
	324	Medford	4539	711
	325	North Bend	4464	12
	326	Pendleton	5321	644
	327	Portland	4400	390
	328	Redmond	6274	297
	329	Salem	4784	257
Pennsylvania	330	Allentown	5830	787
	331	Altoona	6055	546
	332	Dubois	6834	384
	333	Olmsted	7471	206
	334	Philadelphia	4759	1235
	335	Pittsburgh	5829	726
	336	Wilkes-Barre-Scranton	6234	611
	337	Williamsport	6063	709
Rhode Island	338	Providence	5754	714
South Carolina	339	Beaufort	1942	2395
	340	Charleston	2005	2306
	341	Columbia	2594	2074
	342	Florence	2524	2029
	343	Greenville-Spartanburg	3272	1526
	344	Myrtle Beach	2415	1987
	345	Sumter	2577	1913
South Dakota	205	Aberdeen	8348	626
	206	Huron	7834	741
	207	Pierre	7282	919
	208	Rapid City	7211	598
	209	Sioux Falls	7812	747
Tennessee	210	Bristol	4445	956
	211	Chattanooga	3427	1608
	212	Knoxville	4056	1297
	213	Memphis	3041	2187
	214	Nashville	3677	1652
Texas	215	Abilene	2659	2386
	216	Amarillo	4318	1344
	217	Austin	1648	2974
	218	Brownsville	644	3874
	219	Corpus Christi	786	3883
	220	Del Rio	1417	3226
	221	El Paso	2543	2254
	222	Fort Hood	2190	2477
	223	Fort Worth	2370	2568
	224	Galveston	1008	3268
	225	Houston	1525	2893
	226	Lubbock	3508	1769
	227	Lufkin	1900	2480
	228	Midland	2716	2139
	229	San Angelo	2396	2383
	230	San Antonio	1573	3038
	231	Sherman	2850	2137
	232	Tyler	1958	2521
	233	Waco	2164	2840
	234	Witchita Falls	3024	2396
Utah	235	Cedar City	5978	704
	236	Dugway	5963	1038
	237	Ogden	5868	980
	238	Salt Lake Ctiy	5631	1066
	239	Wendover	6002	1088
Vermont	240	Burlington	7665	489
Virginia	241	Charlottesville	4103	1212
	242	Fort Belvoir	4055	1531
	243	Hampton	3535	1432
	244	Norfolk	3368	1612
	245	Richmond	3919	1435
	246	Roanoke	4284	1134
Washington	247	Bellingham	5400	67
	248	Hanford	4731	909
	249	Longview	4900	148
	250	Olympia	5531	97
	251	Seattle	4615	192
	252	Spokane	6820	394
	253	Tacoma	4650	167
	254	Walla Walla	4882	957
	255	Wenatchee	5950	775
	256	Yakima	6104	431
West Virginia	257	Beckley	5427	529
	258	Charleston	4644	978
	259	Huntington	4583	1111
	260	Martinsburg	4807	1081
	261	Morgantown	5312	759
	262	Wheeling	5313	926
Wisconsin	263	Eau Claire	8196	554
	264	Green Bay	7963	463
	265	La Crosse	7340	775
	266	Madison	7493	582
	267	Milwaukee	7087	616
Wyoming	268	Casper	7571	428
	269	Cheyenne	7388	273
	270	Lander	7790	445
	271	Rock Springs	8670	230
	272	Sheridan	7721	398
District of Columbia	346	Washington D.C.	4925	1075

The numbers given for each location in Tables 3.4a-3.4b correspond to the numbered locations given in the map shown by Figure 3.12.

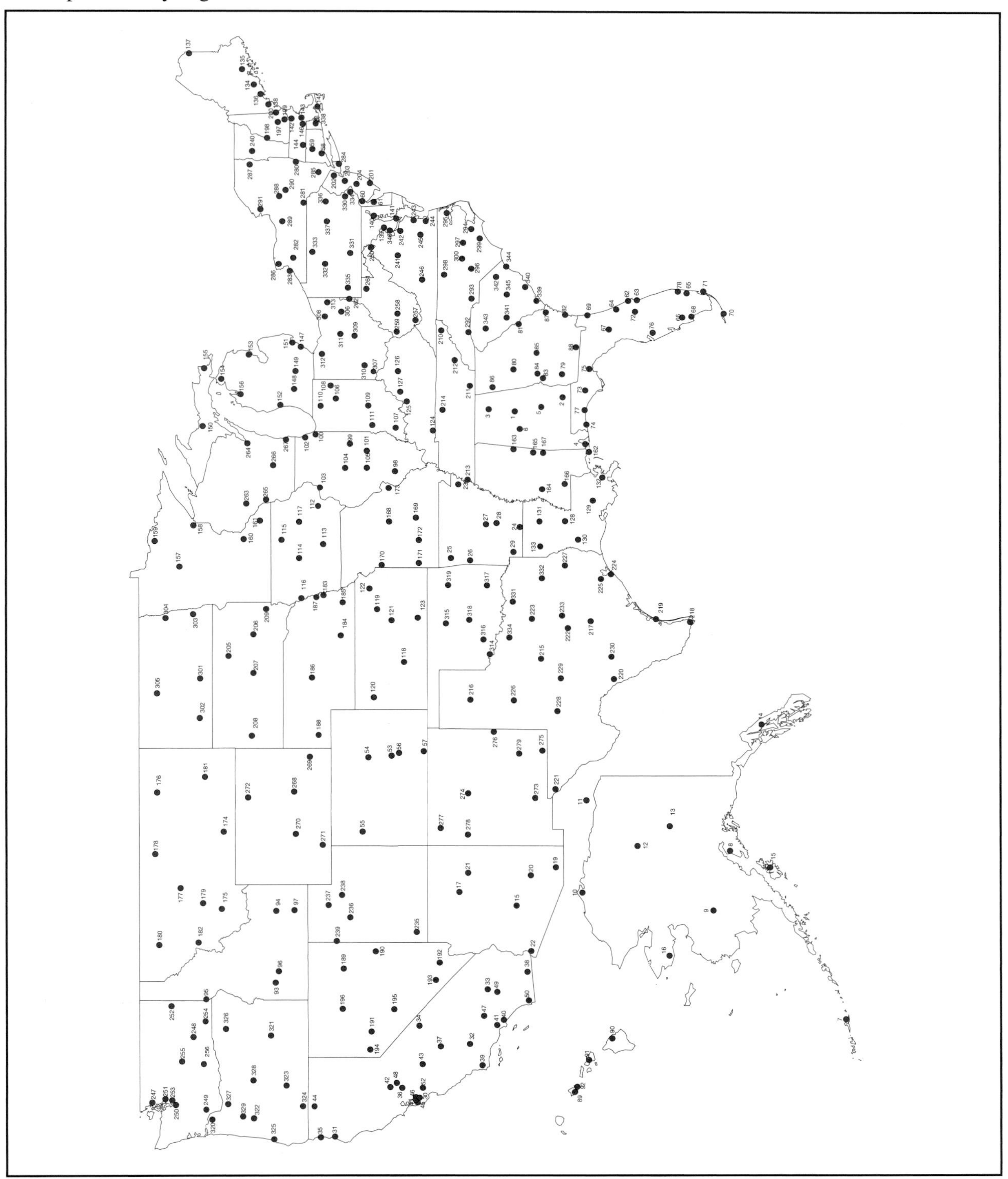

Figure 3.12. U.S. Bin Weather Data and H_{DD}/C_{DD} Data Location Map

To compare a ground source heat pump system to an alternative system in terms of operating cost, the simplest way is to calculate the total amount of energy supplied to a home in heating or rejected from the home in cooling and then calculate the cost to deliver/reject the same amount of energy with the alternative system type. The amount of heat actually delivered to a home in heating mode will be the sum of the energy removed from the ground, energy added by compressor work, and energy added via supplemental resistance heat, when needed. The amount of heat delivered to a home during the analysis period can be calculated according to Equation 3.22:

$$Q_{htotal} = gl_h + (DMD_{res} + DMD_h) \cdot h_b \cdot 3{,}412$$ **Equation 3.22**

Where:

Q_{htotal} = Total Energy Delivered to Residence in Heating Mode (Btu)

The amount of heat removed from a home in cooling mode will be the energy rejected to the ground minus the heat generated by the compressor. The amount of heat removed from a home during the analysis period can be calculated according to Equation 3.23:

$$Q_{ctotal} = gl_c - DMD_c \cdot h_b \cdot 3{,}412$$ **Equation 3.23**

Where:

Q_{ctotal} = Total Energy Removed from Residence in Cooling Mode (Btu)

Now, suppose a customer wants to know what the annual operating costs savings would be for a selected ground source heat pump system over a combustion-based furnace with central air-conditioning unit. Equation 3.24 displays how to calculate the cost of supplying the same amount of heat via combustion that would be supplied via a heat pump system:

$$C_{alth} = \frac{(Q_{htotal} / HV_f)}{(AFUE / 100)} \cdot r_f$$ **Equation 3.24**

Where:

C_{alth} = Cost of Heating Operation for Alternative System during Analysis Period ($)
HV_f = Fuel Heating Value (Btu/unit)
AFUE = Annual Fuel Utilization Efficiency Provided by Manufacturer (%)
r_f = Fuel Cost According to Local Rate ($/unit)

Typical heating values for several types of fuel are given by Table 3.5:

Table 3.5. Heating Values for Common Fuels

Fuel Type	HV_f	Units	r_f
Natural Gas	100,000	Btu/ccf	$/ccf
Propane	91,500	Btu/gal	$/gal
Heating Oil	138,000	Btu/gal	$/gal

Equation 3.25 displays how to calculate the cost of rejecting the same amount of heat via a central air-conditioning unit that would be rejected via a heat pump system:

$$C_{altc} = \frac{(Q_{ctotal} / SEER)}{1000} \bullet r_e$$ **Equation 3.25**

Where:

C_{altc} = Cost of Cooling Operation for Alternative System during Analysis Period ($)
SEER = Seasonal Energy Efficiency Ratio Provided by Manufacturer (dimensionless)

Other refinements to the bin method exist that are not explained in this section, such as accounting for pumping energy consumed by a circulating system for a ground source heat pump system, accounting for energy consumed by the fan for the alternative systems, etc. Such concepts can be easily incorporated into the procedure, but special care should be taken when doing so. All of the calculations necessary to perform the bin method can be easily executed by a spreadsheet program such as Microsoft Excel. As is the case for any procedure where an extensive amount of calculations are performed, software is recommended for increased accuracy and reliability in addition to minimizing time required to perform the calculation.

GSHP Selection Worksheets (Blank Copies) – See Appendix B

Figure 3.2. Single-Capacity, Water-to-Air Heat Pump Selection Worksheet
Figure 3.3. Two-Capacity, Water-to-Air Heat Pump Selection Worksheet
Figure 3.4. Single-Capacity, Water-to-Water Heat Pump Selection Worksheet

4 PIPING MATERIALS, PROPERTIES, AND FLOW CHARACTERISTICS

In This Section

4.1 Pipe Types and Uses

Many types of pipe exist in the HVAC marketplace. The intention of this section is to list the different types available and to point out which are most commonly utilized in the ground source heat pump (GSHP) industry.

4.1.1 STEEL PIPE

Steel pipe is mostly used to convey natural gas in homes, as well as in sprinkler systems because of its high resistance to damage caused by heat. Steel pipe is manufactured according to schedule or weight class to determine its wall thickness. The most common classes of steel pipe are Schedule 40 and Schedule 80 (Table 4.1). Steel pipe is most commonly joined by threading. It can also be joined via flared and compression fittings, welding, grooved joints, sleeves, etc. (*ASHRAE (2008) Handbook of Systems and Equipment – Chapter 45.*)

4.1.2 COPPER TUBING

Copper tubing is generally used for water supply lines in homes, as well as for refrigeration lines in the HVAC industry. Copper and copper alloys exhibit high resistance to corrosion, but are becoming costly. There are two main classes of copper tubing: ASTM Standard B88 (*ASTM (1996) Standard Specification for Seamless Copper Water Tube. Standard B88.*) (Types K, L, M, and DWV) used for water and drain service and ASTM Standard B280 (*ASTM (1997) Standard Specification for Seamless Copper Tube for Air Conditioning and Refrigeration Field Service. Standard B280.*) (ACR) used in refrigeration. The most commonly used types of copper tubing in the HVAC industry are Types L and M (Table 4.2). Copper tubing is most commonly joined by soldering or brazing socket end fittings. It can also be joined via flared and compression fittings, grooved joints, etc.

Table 4.1. Schedule 40 & 80 Pipe Data

Nominal Pipe Size (in)	SCH	ID (in)	OD (in)	Wall Thickness (in)
3/4	40	0.824	1.050	0.113
	80	0.742	1.050	0.154
1	40	1.049	1.315	0.133
	80	0.957	1.315	0.179
1-1/4	40	1.380	1.660	0.140
	80	1.278	1.660	0.191
1-1/2	40	1.610	1.900	0.145
	80	1.500	1.900	0.200
2	40	2.067	2.375	0.154
	80	1.939	2.375	0.218
2-1/2	40	2.469	2.875	0.203
	80	2.323	2.875	0.276
3	40	3.068	3.500	0.216
	80	2.900	3.500	0.300
4	40	4.026	4.500	0.237
	80	3.826	4.500	0.337
5	40	5.017	5.563	0.273
	80	4.767	5.563	0.398
6	40	6.065	6.625	0.280
	80	5.761	6.625	0.432
8	40	7.981	8.625	0.322
	80	7.625	8.625	0.500
10	40	10.020	10.750	0.365
	80	9.564	10.750	0.593

Table 4.2. Copper Tube Data

Nominal Pipe Size (in)	Type	ID (in)	OD (in)	Wall Thickness (in)
3/4	K	0.745	0.875	0.065
	L	0.785	0.875	0.045
	M	0.811	0.875	0.032
1	K	0.995	1.125	0.065
	L	1.025	1.125	0.050
	M	1.055	1.125	0.035
1-1/4	K	1.245	1.375	0.065
	L	1.265	1.375	0.055
	M	1.291	1.375	0.042
1-1/2	K	1.481	1.625	0.072
	L	1.505	1.625	0.060
	M	1.527	1.625	0.049
2	K	1.959	2.125	0.083
	L	1.985	2.125	0.070
	M	2.009	2.125	0.058
2-1/2	K	2.435	2.625	0.095
	L	2.465	2.625	0.080
	M	2.495	2.625	0.065
3	K	2.907	3.125	0.109
	L	2.945	3.125	0.090
	M	2.981	3.125	0.072
3-1/2	K	3.385	3.625	0.120
	L	3.425	3.625	0.100
	M	3.459	3.625	0.083
4	K	3.857	4.125	0.134
	L	3.905	4.125	0.110
	M	3.935	4.125	0.095
5	K	4.805	5.125	0.160
	L	4.875	5.125	0.125
	M	4.907	5.125	0.109
6	K	5.741	6.125	0.192
	L	5.845	6.125	0.140
	M	5.881	6.125	0.122

4.1.3 PLASTIC PIPING

Plastic piping is used extensively in plumbing and HVAC because it is lightweight, inexpensive, and corrosion resistant. Plastic piping is also very smooth, which leads to lower pumping power requirements compared to other types of piping. Many different types of plastic exist from which piping is made. Special care needs to be taken to ensure that the proper material is selected for its use. Table 4.3 lists applications pertinent to the HVAC industry and the associated compatibility of various materials with each listed application.

Of the many types of plastic piping available, those most commonly used for ground source heat pump applications are as follows (*ASHRAE (2008) Handbook of Systems and Equipment – Chapter 45.*):

PVC (Polyvinyl Chloride): The most widely used plastic because of its applicability in many different areas at the lowest cost. PVC is joined by solvent cementing, threading, or flanging. Larger sizes can be joined by gasketed push-on joints. The most common classes of PVC pipe are Schedule 40 and Schedule 80, which share the same dimensions as steel pipe (Table 4.1). **PVC piping is not recommended for the**

Table 4.3. Manufacturer's Recommendations[1] for Plastic Materials

	PVC	CPVC	PB	HDPE	PP	ABS	PVDF	RTRP
Cold Water Service	R	R	R	R	R	R	R	R
Hot (140˚F) Water	N	R	R	R	R	R	R	R
Potable Water Service	R	R	R	R	R	R	R	R
Drain, Waste, and Vent	R	R	N	--	R	R	--	--
Demineralized Water	R	R	--	--	R	R	R	--
Deionized Water	R	R	--	--	R	R	R	R
Salt Water	R	R	R	R	R	R	--	R
Heating (200F) Hot Water	N	N	N	N	N	N	--	R
Natural Gas	N	N	N	R	N	N	--	--
Compressed Air	N	N	--	R	N	R	--	--
Sunlight and Weather Resistance	N	N	N	R	--	R	R	R
Underground Service	R	R	R	R	R	R	--	R
Food Handling	R	R	--	--	R	R	R	R

1. ASHRAE Handbook – HVAC Systems & Equipment (Ch. 41)
R-Recommended N-Not Recommended -- - Insufficient Information

buried portion of the ground heat exchanger in a closed-loop, ground-coupled heat pump system. However, it has been used for the interior piping portion of the closed-loop, ground-coupled heat pump system. One drawback of using PVC is that it becomes brittle when cold and can crack with an accidental impact. Because of this, Schedule 80 PVC is used more commonly than Schedule 40 PVC. If accidental impact is an issue in the area where PVC will be used, other piping materials should be considered.

CPVC (Chlorinated Polyvinyl Chloride): Has the same properties as PVC, but is treated to resist loss of strength, which occurs in PVC at higher temperatures. CPVC is joined by the same methods as PVC. **CPVC piping is not recommended for the buried portion of the ground heat exchanger in a closed-loop, ground-coupled heat pump system. However, it has been used for the interior piping portion of the closed-loop, ground-coupled heat pump system. One drawback of using CPVC is that it becomes brittle when cold and can crack with an accidental impact. Because of this, Schedule 80 CPVC is used more commonly than Schedule 40 CPVC. If accidental impact is an issue in the area where CPVC will be used, other piping materials should be considered.**

PB (Polybutylene): A lightweight, flexible plastic that can be used at temperatures up to 210 F. **This material was once commonly used for the buried portion of the ground heat exchanger in a closed-loop, ground-coupled heat pump system, but is no longer available in the United States.** Polybutylene is joined by heat fusion or mechanical means.

PE (Polyethylene): Is classified into three main categories according to the density of the polyethylene material.

Low-density polyethylene (LDPE) is used in the food and beverage industry and for instrument tubing. Its use is recommended in low-temperature applications. LDPE is joined by mechanical means such as compression fittings or push-on connectors and clamps.

Medium-density polyethylene (MDPE) is used primarily in the gas distribution industry and is primarily joined by heat fusion for sizes equal to and larger than 3/4 inch, which are manufactured to Iron Pipe Size (IPS) dimensions. Smaller sizes are generally manufactured to Copper Tube Size (CTS) dimensions and usually referred to as tubing. Smaller tubing sizes can be joined via flare, compression, or insert fittings as well.

High-density polyethylene (HDPE) is one of the two materials suitable for use as the buried ground heat exchanger portion of a ground source heat pump system (the other less-commonly used material approved for use is cross-linked polyethylene, also referred to as PEX-A). HDPE is primarily joined by heat fusion for all Iron Pipe Size dimensions. Smaller tubing sizes (for non-GSHP applications) can be joined via flare, compression, or insert fittings as well. There may be multiple raw materials with different material designations that would be applicable to a category, such as PE3608 (prior to 2005, was designated as PE3408, per *ASTM D-3350 (2010)* or PE4710, which are both HDPE material suitable for pipe applications.

In order to maintain a consistent pressure rating over a range of dimensions, PE is manufactured according to dimension ratios to determine its wall thickness. Dimension ratio (DR) is the ratio of the pipe outside diameter to the wall thickness and relates to the pressure rating of the pipe, which will be discussed in Section 4.3. Table 4.4 provides pipe data for DRs of 7, 9, 11, 13.5, 15.5, and 17.

Occasionally, reference can be found to polyethylene being available or specified in schedule dimensions (e.g. Schedule 40). This is a holdover from the days when polyethylene was first manufactured to IPS dimensions, which already existed. In the schedule dimension category, pipe pressure ratings decrease as the pipe dimension increases because the pipe wall thickness does not increase proportionally to the pipe diameter.

Material designations are determined by *ASTM D-3350 (2010)*. Examining the material designation of PE3608 as an example, the first two characters (PE) represent the material, the third character (3) represents the density category, the fourth character (6) represents resistance to Slow Crack Growth, and the final two characters (08) represent Hydrostatic Design Stress in units of 100 psi based on a designated service factor per the appropriate ASTM standard.

Appropriate piping materials and the corresponding pressure ratings can be found in the current *IGSHPA (2009) Standards* document. IGSHPA will work on an ongoing basis to keep this document as updated as possible.

Table 4.4. DR-7, 9, 11, 13.5, 15.5, & 17 Pipe Data*

Dimension Ratio (DR)	Nominal Pipe Size (in)	ID (in)	OD (in)	Wall Thickness (in)
7	3/4	0.750	1.050	0.150
	1	0.939	1.315	0.188
	1-1/4	1.186	1.660	0.237
	1-1/2	1.357	1.900	0.271
	2	1.696	2.375	0.339
	3	2.500	3.500	0.500
9	3/4	0.817	1.050	0.117
	1	1.023	1.315	0.146
	1-1/4	1.291	1.660	0.184
	1-1/2	1.478	1.900	0.211
	2	1.847	2.375	0.264
	3	2.722	3.500	0.389
11	3/4	0.859	1.050	0.095
	1	1.076	1.315	0.120
	1-1/4	1.358	1.660	0.151
	1-1/2	1.555	1.900	0.173
	2	1.943	2.375	0.216
	3	2.864	3.500	0.318
13.5	1-1/2	1.619	1.900	0.141
	2	2.023	2.375	0.176
	3	2.981	3.500	0.259
	4	3.833	4.500	0.333
	6	5.644	6.625	0.491
15.5	1-1/2	1.655	1.900	0.123
	2	2.069	2.375	0.153
	3	3.048	3.500	0.226
	4	3.919	4.500	0.290
	6	5.770	6.625	0.427
17	1-1/2	1.676	1.900	0.112
	2	2.096	2.375	0.140
	3	3.088	3.500	0.206
	4	3.971	4.500	0.265
	6	5.846	6.625	0.390

*-Refer to HDPE manufacturer's data to determine which HDPE sizes are available in a given dimension ratio. Not all sizes are widely available.

4.1.4 REINFORCED RUBBER HOSE

In many instances, short lengths of reinforced rubber hose are used to connect flow centers to manifolds, heat pumps, etc. For most purposes, reinforced rubber hose pressure rated to 200 psi is adequate for use with residential GSHP systems. Table 4.5 provides reinforced rubber hose data (rated to 200 psi) for sizes commonly used in the GSHP industry.

Table 4.5. Reinforced Rubber Hose Data

Nominal Size (in)	ID (in)	OD (in)	Vol/100' (gal)	Wall Thickness (in)
1	1.00	1.44	4.08	0.22
1-1/4	1.25	1.73	6.37	0.24
1-1/2	1.50	1.98	9.18	0.24
2	2.00	2.50	16.32	0.25

4.2 Pipe Thermal Resistance

High-density polyethylene pipe is the only material available in the U.S. suitable for use in the underground portion of the ground-coupled heat exchanger portion of the ground-source heat pump system. For that reason, HDPE is the only piping material where heat transfer resistance is a major concern. The thermal conductivity of HDPE pipe is 0.225 Btu/hr-ft-F. The resistance to heat transfer of any pipe can be calculated by Equation 4.1. Table 4.6 provides thermal resistance values for various dimension ratios of HDPE pipe.

$$R_P = \frac{\ln(D_{p,o} / D_{p,i})}{2 \cdot \pi \cdot k_P}$$ **Equation 4.1**

Where:

R_P = Thermal Resistance of pipe (hr-ft-F/Btu)
$D_{p,o}$ = Outside Pipe Diameter (in)
$D_{p,i}$ = Inside Pipe Diameter (in)
k_P = Thermal Conductivity of pipe (Btu/hr-ft-F)

Table 4.6. Thermal Resistance for HDPE

DR	R_P (hr-ft-°F/Btu)
7	0.238
9	0.175
11	0.141
13.5	0.117
15.5	0.101
17	0.086

4.3 Pressure Ratings

Several factors affect the pressure rating of pipe, including material from which the pipe is made, wall thickness, and temperature. Pipe should not be used in situations where the system working pressure is higher than the manufacturer's specified maximum working pressure. Tables 4.7 through 4.10 list the pressure ratings for HDPE, PVC, steel, and copper pipe for various wall thicknesses.

Because vertically placed ground-loop piping is subjected to high working pressures over a wide range of temperatures, it is necessary to understand how the pressure ratings of HDPE pipe vary with temperature. Table 4.8 displays pressure ratings for various sizes and wall thicknesses of HDPE pipe at different temperatures.

Table 4.7. Pressure Ratings for HDPE 3408 Plastic Pipe at 73.4 F (23.5 C) – (2007)

Dimension Ratio	Pressure Rating (psi)
7	267
9	200
11	160
13.5	128
15.5	110
17	100

Source – Driscopipe 5300 Climate Guard System Brochure

Table 4.8. Pressure Ratings for HDPE 3408 Plastic Pipe at Various Temperatures (2007)

Temp (˚F)	Pipe DR					
	7	9	11	13.5	15.5	17
20	400	300	240	192	166	150
40	367	275	220	176	152	138
73.4	**267**	**200**	**160**	**128**	**110**	**100**
80	253	192	150	122	105	95
90	232	173	140	111	96	87
100	210	160	125	101	87	79
110	188	143	113	90	78	71
120	167	130	100	80	69	63
130	150	115	90	72	62	56
140	133	100	80	64	55	50

Table 4.9. Pressure Ratings for PVC/CPVC[1] and Steel[2] Pipe

	PVC & CPVC[3]		Steel Pipe[4]	
Nominal Pipe Size (in)	SCH	Maximum Working Pressure (psi)	SCH[5]	Maximum Working Pressure (psi)
3/4	40	289	40 ST	217
	80	413	80 XS	681
1	40	270	40 ST	226
	80	378	80 XS	642
1-1/4	40	221	40 ST	229
	80	312	80 XS	594
1-1/2	40	198	40 ST	231
	80	282	80 XS	576
2	40	166	40 ST	230
	80	243	80 XS	551
2-1/2	40	182	40 ST	533
	80	255	80 XS	835
3	40	158	40 ST	482
	80	225	80 XS	767
4	40	133	40 ST	430
	80	194	80 XS	695
5	40	117	--	--
	80	173	--	--
6	40	106	40 ST	696
	80	167	80 XS	1209
8	40	93	40 ST	643
	80	148	80 XS	1106
10	40	84	40 ST	606
	80	140	80	1081

1. ASHRAE Handbook – HVAC Systems & Equipment (Ch. 41) – PVC, CPVC Data
2. http://www.engineeringtoolbox.com (http://www.engineeringtoolbox.com/pvc-cpvc-pipes-pressures-d_796.html) – Steel Pipe Data
3. Rated at 73 F
4. Rated according to ASTM A53 to 400 F
5. XS = Extra Strong, ST = Standard Weight

Table 4.10. Working Pressure Ratings for Copper[1] Tubing

Nominal Pipe Size (in)	Type	Working Pressure[2] (psi) Annealed Tubing	Working Pressure[2] (psi) Drawn Tubing
3/4	K	677	1270
	L	469	879
	M	334	625
1	K	527	988
	L	405	760
	M	284	532
1-1/4	K	431	808
	L	365	684
	M	279	522
1-1/2	K	404	758
	L	337	631
	M	275	516
2	K	356	668
	L	300	573
	M	249	467
2-1/2	K	330	619
	L	278	521
	M	226	423
3	K	318	596
	L	263	492
	M	210	394
3-1/2	K	302	566
	L	252	472
	M	209	392
4	K	296	555
	L	243	456
	M	210	394
5	K	285	534
	L	222	417
	M	194	364
6	K	286	536
	L	208	391
	M	182	341

1. ASHRAE Handbook – HVAC Systems & Equipment (Ch. 41)
2. ASTM B88 to 250 F

As previously discussed, reinforced rubber hose pressure rated to 200 psi is sufficient for use in GSHP systems.

4.4 Head Loss

4.4.1 FLOW REGIMES

There are two basic flow regimes defined in fluid mechanics to describe the nature of fluid flow in any situation: laminar and turbulent. In the low-velocity laminar regime, fluid flow is streamlined and smooth. Once the fluid velocity has been increased above a critical value, flow becomes chaotic and disordered and enters into the turbulent regime. The evolution from laminar to turbulent flow does not occur at any one specific point, but occurs gradually. The gradual change from laminar to turbulent flow is described as transition flow.

The chaotic, intense mixing of fluid in turbulent flow maximizes the heat transfer capability of that fluid. Ground-loop piping should be sized such that fluid flow through each loop is turbulent to ensure maximum heat transfer. The ground heat exchanger loops are the heat transfer portion of the system. The supply-return piping is not considered in the heat transfer calculations. For this reason, turbulence is critical in the loops, but not in supply-return piping. The heat transfer enhancing effects of turbulent flow are, however, not without cost. Turbulent flow also produces higher levels of friction between the fluid and surface, which ultimately results in increased pumping power requirements that must be overcome by an electrically-driven pump.

The point at which fluid flow transitions from laminar to turbulent flow depends on many factors including: surface geometry and roughness, fluid velocity and temperature, and fluid type. Research performed by Osborn Reynolds in the 1880s demonstrated that the flow regime is dependent upon the ratio of inertia forces to viscous forces. This ratio is called the Reynolds number. It is a dimensionless number that can be calculated for internal pipe flow according to Equation 4.2.

$$Re = 124 \frac{V \bullet D_{p,i}}{(\mu / \rho)}$$ **Equation 4.2**

Where:

Re	= Reynolds Number (Dimensionless)
V	= Fluid Velocity (ft/sec)
$D_{p,i}$	= Inside Pipe Diameter (in)
μ	= Fluid Viscosity (cp)
ρ	= Fluid Density (lb_m/ft^3)

Fluid velocity in a pipe of given diameter can be calculated using Equation 4.3.

$$V = 0.4085 \frac{Q}{(D_{p,i})^2}$$ **Equation 4.3**

Where:

Q	= Fluid Flow Rate (gpm)

The Reynolds number at which flow becomes fully turbulent is called the critical Reynolds number. In most practical conditions, fluid flow through a circular pipe corresponding to Reynolds numbers less than 2,300 is considered to be laminar and greater than 10,000 is considered to be fully turbulent. In practice in the GSHP industry, Reynolds numbers greater than 2,500 are considered to correspond to turbulent flow.

4.4.2 MINIMUM FLOW RATES FOR TURBULENCE

From a design perspective, a Reynolds number of 2,500 is required to ensure turbulent flow through the loop of a ground heat exchanger. The minimum flow rate necessary to ensure turbulent flow for any specific type of circulating fluid can be calculated using Equation 4.4. The minimum flow rate for turbulence for water at 40 F is provided in Table 4.11 for common DR-11 PE pipe sizes used as heat exchanger loops.

$$Q_{min} = 49.35 \frac{D_{p,i} \bullet \mu}{\rho}$$ **Equation 4.4**

Where:

Q_{min} = Minimum Flow Rate for Turbulence (gpm)

Table 4.11. Minimum Flow Rates for Turbulence for Water at 40 F

Pipe Size (in)	ID[1] (in)	Q_{min} (gpm)
3/4	0.859	1.05
1	1.076	1.31
1-1/4	1.358	1.66
1-1/2	1.555	1.90
2	1.943	2.38

1. HDPE DR-11 pipe

Adding antifreeze to water to achieve a lower mixture freeze point is common for closed-loop ground heat exchangers in moderate and cold climates. The addition of antifreeze serves to increase the viscosity of the water-antifreeze mixture, especially at low temperatures. Table 4.12 lists the mixture freezing temperature for various volumetric concentrations of propylene glycol, methanol, and ethanol solutions in water. Figure 4.1 displays the freeze protection data graphically.

Table 4.12. Approximate Freeze Protection Levels for Various Types of Antifreeze

Concentration (by Volume)	Propylene Glycol	Methanol	Ethanol
	Freeze Point (°F)		
5.0%	29.3	26.2	29.5
7.5%	27.7	23.0	28.1
10.0%	26.1	19.7	26.4
12.5%	24.4	16.2	24.6
15.0%	22.5	12.6	22.6
17.5%	20.5	8.8	20.4
20.0%	18.4	4.9	18.1
22.5%	16.1	--	15.6
25.0%	13.8	--	12.9
27.5%	11.3	--	10.0
30.0%	8.8	--	7.0

*Note - Values are based on pure concentrations of propylene glycol, methanol, and ethanol. Antifreeze products have various levels of inhibitors, dyes, and other ingredients, which will affect all aspects of the antifreeze. Refer to manufacturer's product data for actual values.

Many states require the use of food grade propylene glycol as a freeze protecting agent for closed-loop GSHP systems. The question is how much propylene glycol is actually needed in a given system. The answer, in the case of a GSHP system, is to use at least enough propylene glycol (or any antifreeze) to freeze protect to 10 F below the average loop temperature under design conditions, typically measured on a percent by volume basis. For example, if the minimum design EWT for a loopfield is 30 F, the minimum design LWT will be approxi-

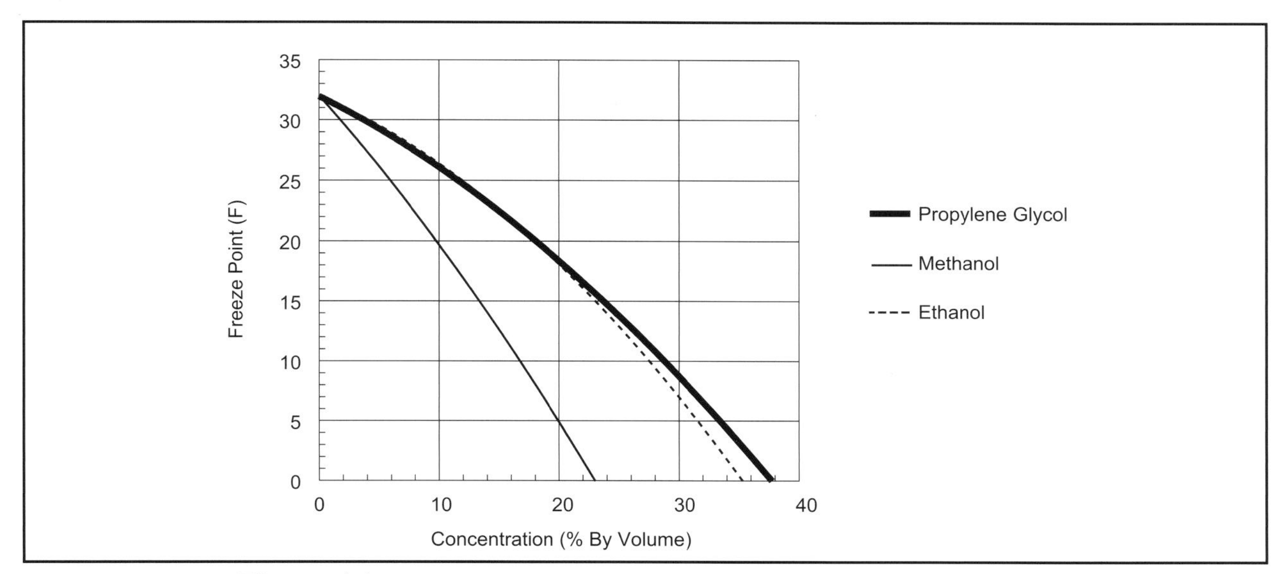

Figure 4.1. Freeze Protection Curves for Various Types of Antifreeze

mately 24-25 F. The average loop temperature under design conditions would be 27 F for this example [(30+24) / 2 = 27 F]. For a 27 F average loop temperature under design conditions, freeze protection to 17 F will be necessary. According to Table 4.12 and Figure 4.1, approximately 20%-22.5% propylene glycol by volume will be necessary for adequate freeze protection for the system. A loopfield designed such that the minimum entering water temperature will not fall below 45 F does not require freeze protection.

Another common question in the industry is whether or not inhibitor content should play a role in the determination of solution concentration level. Inhibitors in propylene glycol-based antifreeze solutions serve three important functions: bacterial growth prevention, corrosion protection, and fluid stability amplification. However, there are very few forms of microbial growth that can actually live without light, and they are relatively rare. The circulating fluid in a closed-loop GSHP system should not be exposed to light in any part of the system, making bacterial growth extremely unlikely and rare. In the event that bacterial growth is found in a system, common water treatment chemicals such as chlorine can be added to impede the growth. Secondly, high inhibitor content for corrosion protection is necessary in systems where the circulating fluid comes into contact with large amounts of corrosive metal. The majority of piping in a typical closed-loop GSHP system is either high-density polyethylene (HDPE) or polyvinyl chloride (PVC). The relatively small amount of metal in a GSHP system that comes into contact with the circulating fluid does not generally necessitate the use of high amounts of corrosion protection. Thirdly, propylene glycol-based solutions will break down when exposed to extreme temperatures greater than 250 F. Ground source heat pump systems operate well below the range of this acceptable temperature; fluid breakdown is a non-issue. For the reasons outlined above, inhibitor content should not play a substantial role in the determination of solution concentration level.

Turbulence for antifreeze-water solutions will occur at a flow rate that depends on the volumetric concentration of antifreeze in the water and the temperature of the mixture. Tables 4.13 through 4.15 provide the minimum flow rate to ensure turbulence (Re=2,500) in various HDPE loop diameters for common ranges of propylene glycol, methanol, and ethanol solutions. Fluid viscosities and densities for antifreeze will vary from manufacturer to manufacturer, depending on the type and amount of additives put into the antifreeze mixture to control corrosion, causing slight deviations from the values shown. Refer to manufacturer's data if available.

Table 4.13. Minimum Flow Rates (gpm) for Turbulence with Propylene Glycol

3/4" DR-11 HDPE (ID = 0.859")							
Concentration (By Volume)	Fluid Temperature (°F)						
	10 °F	15 °F	20 °F	25 °F	30 °F	35 °F	40 °F
5%	--	--	--	--	--	--	--
10%	--	--	--	--	1.87	1.70	1.53
15%	--	--	--	--	2.34	2.12	1.89
20%	--	--	3.54	3.17	2.79	2.53	2.26
25%	--	--	5.00	4.43	3.85	3.43	3.02
30%	8.76	7.61	6.47	5.67	4.88	4.33	3.77
1" DR-11 HDPE (ID = 1.076")							
Concentration (By Volume)	Fluid Temperature (°F)						
	10 °F	15 °F	20 °F	25 °F	30 °F	35 °F	40 °F
5%	--	--	--	--	--	--	--
10%	--	--	--	--	2.34	2.13	1.91
15%	--	--	--	--	2.92	2.65	2.37
20%	--	--	4.43	3.96	3.50	3.17	2.82
25%	--	--	6.26	5.54	4.81	4.29	3.78
30%	10.96	9.52	8.09	7.10	6.11	5.42	4.71
1-1/4" DR-11 HDPE (ID = 1.358")							
Concentration (By Volume)	Fluid Temperature (°F)						
	10 °F	15 °F	20 °F	25 °F	30 °F	35 °F	40 °F
5%	--	--	--	--	--	--	--
10%	--	--	--	--	2.96	2.69	2.41
15%	--	--	--	--	3.69	3.35	2.99
20%	--	--	5.59	5.00	4.41	4.00	3.56
25%	--	--	7.90	7.00	6.07	5.42	4.77
30%	13.83	12.01	10.21	8.96	7.71	6.84	5.95
1-1/2" DR-11 HDPE (ID = 1.555")							
Concentration (By Volume)	Fluid Temperature (°F)						
	10 °F	15 °F	20 °F	25 °F	30 °F	35 °F	40 °F
5%	--	--	--	--	--	--	--
10%	--	--	--	--	3.38	3.08	2.76
15%	--	--	--	--	4.22	3.83	3.42
20%	--	--	6.40	5.72	5.05	4.58	4.08
25%	--	--	9.05	8.01	6.96	6.20	5.46
30%	15.83	13.76	11.69	10.26	8.82	7.83	6.81
2" DR-11 HDPE (ID = 1.943")							
Concentration (By Volume)	Fluid Temperature (°F)						
	10 °F	15 °F	20 °F	25 °F	30 °F	35 °F	40 °F
5%	--	--	--	--	--	--	--
10%	--	--	--	--	4.23	3.85	3.45
15%	--	--	--	--	5.28	4.79	4.28
20%	--	--	7.99	7.15	6.31	5.72	5.10
25%	--	--	11.31	10.01	8.69	7.75	6.82
30%	19.79	17.19	14.61	12.82	11.03	9.78	8.51

-- Concentration level not recommended due to insufficient freeze protection

Table 4.14. Minimum Flow Rates (gpm) for Turbulence with Methanol

3/4" DR-11 HDPE (ID = 0.859")							
Concentration (By Volume)	Fluid Temperature (°F)						
	10 °F	15 °F	20 °F	25 °F	30 °F	35 °F	40 °F
5%	--	--	--	--	--	1.42	1.29
10%	--	--	--	2.09	1.87	1.68	1.53
12.5%	--	--	--	2.29	2.02	1.81	1.65
15%	--	3.10	2.79	2.49	2.18	1.94	1.76
17.5%	--	3.29	2.96	2.63	2.29	2.04	1.86
20%	--	3.49	3.12	2.77	2.41	2.15	1.95
1" DR-11 HDPE (ID = 1.076")							
Concentration (By Volume)	Fluid Temperature (°F)						
	10 °F	15 °F	20 °F	25 °F	30 °F	35 °F	40 °F
5%	--	--	--	--	--	1.77	1.62
10%	--	--	--	2.62	2.34	2.11	1.91
12.5%	--	--	--	2.86	2.53	2.27	2.06
15%	--	3.88	3.49	3.11	2.73	2.43	2.20
17.5%	--	4.12	3.70	3.29	2.87	2.56	2.32
20%	--	4.36	3.91	3.47	3.02	2.69	2.43
1-1/4" DR-11 HDPE (ID = 1.358")							
Concentration (By Volume)	Fluid Temperature (°F)						
	10 °F	15 °F	20 °F	25 °F	30 °F	35 °F	40 °F
5%	--	--	--	--	--	2.24	2.04
10%	--	--	--	3.31	2.95	2.66	2.42
12.5%	--	--	--	3.61	3.19	2.86	2.60
15%	--	4.89	4.40	3.93	3.44	3.07	2.78
17.5%	--	5.20	4.68	4.15	3.62	3.23	2.93
20%	--	5.50	4.93	4.38	3.81	3.40	3.07
1-1/2" DR-11 HDPE (ID = 1.555")							
Concentration (By Volume)	Fluid Temperature (°F)						
	10 °F	15 °F	20 °F	25 °F	30 °F	35 °F	40 °F
5%	--	--	--	--	--	2.56	2.34
10%	--	--	--	3.79	3.38	3.04	2.77
12.5%	--	--	--	4.13	3.65	3.28	2.98
15%	--	5.60	5.04	4.50	3.94	3.51	3.18
17.5%	--	5.95	5.35	4.76	4.14	3.69	3.36
20%	--	6.30	5.65	5.01	4.36	3.89	3.52
2" DR-11 HDPE (ID = 1.943")							
Concentration (By Volume)	Fluid Temperature (°F)						
	10 °F	15 °F	20 °F	25 °F	30 °F	35 °F	40 °F
5%	--	--	--	--	--	3.20	2.93
10%	--	--	--	4.73	4.22	3.80	3.46
12.5%	--	--	--	5.17	4.56	4.10	3.73
15%	--	7.00	6.30	5.62	4.93	4.39	3.97
17.5%	--	7.44	6.69	5.94	5.17	4.61	4.19
20%	--	7.88	7.06	6.27	5.45	4.86	4.40

-- Concentration level not recommended due to insufficient freeze protection

Table 4.15. Minimum Flow Rates (gpm) for Turbulence with Ethanol

3/4" DR-11 HDPE (ID = 0.859")							
Concentration (By Volume)	Fluid Temperature (°F)						
	10 °F	15 °F	20 °F	25 °F	30 °F	35 °F	40 °F
5%	--	--	--	--	--	--	--
10%	--	--	--	--	2.63	2.34	2.06
15%	--	--	--	3.42	3.08	2.73	2.39
20%	--	--	4.30	3.90	3.50	3.11	2.71
25%	--	5.27	4.82	4.37	3.92	3.46	3.01
30%	6.09	5.60	5.12	4.64	4.16	3.68	3.20
1" DR-11 HDPE (ID = 1.076")							
Concentration (By Volume)	Fluid Temperature (°F)						
	10 °F	15 °F	20 °F	25 °F	30 °F	35 °F	40 °F
5%	--	--	--	--	--	--	--
10%	--	--	--	--	3.29	2.93	2.58
15%	--	--	--	4.28	3.85	3.42	2.99
20%	--	--	5.38	4.89	4.38	3.89	3.39
25%	--	6.60	6.04	5.46	4.90	4.33	3.76
30%	7.61	7.01	6.41	5.81	5.21	4.60	4.00
1-1/4" DR-11 HDPE (ID = 1.358")							
Concentration (By Volume)	Fluid Temperature (°F)						
	10 °F	15 °F	20 °F	25 °F	30 °F	35 °F	40 °F
5%	--	--	--	--	--	--	--
10%	--	--	--	--	4.15	3.70	3.26
15%	--	--	--	5.40	4.86	4.31	3.78
20%	--	--	6.79	6.17	5.53	4.91	4.27
25%	--	8.33	7.62	6.90	6.18	5.46	4.75
30%	9.61	8.84	8.09	7.33	6.58	5.81	5.05
1-1/2" DR-11 HDPE (ID = 1.555")							
Concentration (By Volume)	Fluid Temperature (°F)						
	10 °F	15 °F	20 °F	25 °F	30 °F	35 °F	40 °F
5%	--	--	--	--	--	--	--
10%	--	--	--	--	4.75	4.24	3.73
15%	--	--	--	6.18	5.57	4.94	4.33
20%	--	--	7.77	7.06	6.33	5.62	4.89
25%	--	9.53	8.72	7.90	7.08	6.25	5.44
30%	11.00	10.13	9.26	8.40	7.53	6.65	5.78
2" DR-11 HDPE (ID = 1.943")							
Concentration (By Volume)	Fluid Temperature (°F)						
	10 °F	15 °F	20 °F	25 °F	30 °F	35 °F	40 °F
5%	--	--	--	--	--	--	--
10%	--	--	--	--	5.93	5.30	4.66
15%	--	--	--	7.72	6.95	6.17	5.41
20%	--	--	9.71	8.82	7.91	7.02	6.11
25%	--	11.91	10.90	9.87	8.85	7.81	6.79
30%	13.75	12.65	11.57	10.49	9.41	8.31	7.22

-- Concentration level not recommended due to insufficient freeze protection

4.4.3 HEAD LOSS CALCULATIONS

In order to produce flow, pumps must be large enough to generate sufficient pressure to overcome friction losses caused by contact between the circulating fluid and pipe walls. For a given flow rate, pressure drop due to friction (head loss) increases with fluid viscosity and flow rate and decreases with increasing fluid temperature and pipe size. To appropriately size circulating pumps for GSHP systems, a system head loss calculation is necessary. As in calculating ground-heat exchanger length requirements, head loss calculations must be performed for worst-case scenarios to ensure that circulating pumps can serve the system under all flow conditions. The worst-case scenario for systems utilizing water as the circulating fluid would be if that fluid reached a temperature of 40 F (for a minimum EWT=45 F, the minimum LWT will be approximately 40 F). If the system is designed to allow the circulating fluid to fall below 45 F entering the GSHP system (i.e. – if EWT_{min}<45 F), freeze protection to 10 F below the average loop temperature under design conditions is required. Head loss is calculated using the Darcy-Weisbach equation.

$$HL_f = f\left(\frac{L}{D_{p,i}/12}\right)\left(\frac{V^2}{2 \bullet g}\right)$$ **Equation 4.5**

Where:

HL_f	= Head Loss (ft)
f	= Pipe Friction Factor (Dimensionless)
L	= Length of Pipe (ft)
$D_{p,i}$	= Inside Pipe Diameter (in)
V	= Fluid Velocity (ft/sec)
g	= Acceleration Due to Gravity (32.2 ft/sec²)

Polyethylene pipe is considered to be smooth, so the friction factor for HDPE used in the Darcy-Weisbach equation can be calculated using the Blasius equation (Equation 4.6) for Reynolds numbers between 2,500 and 10,000. For Reynolds numbers greater than 10,000, the friction factor for HDPE can be calculated using the Nikoradze equation (Equation 4.7). In either case, Reynolds number is calculated according to Equation 4.2.

$$f = \frac{0.3164}{\sqrt[4]{Re}}$$ (2,500 < Re < 10,000) **Equation 4.6**

$$f = 0.0032 + 0.221\ Re^{-0.237}$$ (Re > 10,000) **Equation 4.7**

Table 4.16 lists head loss values (feet of head loss per 100 feet of pipe) in various sizes/dimension ratios for HDPE pipe and at various flow rates for water at 40 F. Tables 4.17-4.20 list head loss values (feet of head loss per 100 feet of pipe) in various sizes/pressure ratings for PVC, steel, copper tubing, and reinforced rubber heater hose at various flow rates for water at 40 F. To calculate the head loss for a circulating fluid other than water (when freeze protection is required), perform head loss calculations for water at 40 F then apply the appropriate antifreeze viscosity correction factor (refer to Section 4.4.4).

Table 4.16. Head Loss Values in HDPE (DR-7, 9, & 11) for Water at 40 F*

HDPE	DR-7					DR-9					DR-11				
	Nominal Pipe Size					Nominal Pipe Size					Nominal Pipe Size				
	3/4"	1"	1-1/4"	1-1/2"	2"	3/4"	1"	1-1/4"	1-1/2"	2"	3/4"	1"	1-1/4"	1-1/2"	2"
Flowrate (gpm)	HL/100' (ft)					HL/100' (ft)					HL/100' (ft)				
1.50	1.17					0.78					0.61				
1.75	1.53					1.02					0.80				
2.00	1.93					1.29					1.01				
2.25	2.37					1.58					1.24				
2.50	2.85					1.90					1.50				
2.75	3.37					2.25					1.77				
3.00	3.92	1.35				2.62	0.90				2.06	0.71			
3.25	4.51	1.55				3.01	1.03				2.37	0.81			
3.50	5.14	1.76				3.43	1.18				2.70	0.93			
3.75	5.16	1.99				3.87	1.33				3.04	1.04			
4.00	5.79	2.23				3.85	1.49				3.41	1.17			
4.25	6.45	2.48				4.29	1.65				3.37	1.30			
4.50	7.15	2.74				4.75	1.83				3.73	1.44			
4.75	7.87	2.68				5.23	2.01				4.11	1.58			
5.00	8.63	2.94				5.74	1.95				4.50	1.73			
5.50		3.48	1.29				2.32	0.86				1.82	0.67		
6.00		4.07	1.33				2.71	1.00				2.12	0.79		
6.50		4.70	1.54				3.12	1.02				2.45	0.90		
7.00		5.36	1.76				3.57	1.17				2.80	0.92		
7.50		6.07	1.99				4.04	1.32				3.17	1.04		
8.00		6.82	2.23				4.53	1.48				3.55	1.16		
8.50		7.60	2.49				5.05	1.65				3.96	1.30		
9.00		8.42	2.76				5.60	1.83				4.39	1.44		
10.00			3.33	1.74				2.21	1.16				1.74	0.91	
11.00			3.95	2.07				2.63	1.37				2.06	1.08	
12.00			4.62	2.42				3.07	1.61				2.41	1.26	
13.00			5.33	2.79				3.54	1.85				2.78	1.45	
14.00			6.09	3.19				4.05	2.12				3.17	1.66	
15.00			6.89	3.61	1.24			4.58	2.40	0.82			3.59	1.88	0.65
16.00			7.74	4.05	1.39			5.14	2.69	0.92			4.03	2.11	0.72
17.00			8.63	4.51	1.55			5.74	3.00	1.03			4.50	2.35	0.81
18.25				5.13	1.76				3.41	1.17				2.67	0.92
19.50				5.78	1.98				3.84	1.32				3.01	1.03
20.75				6.46	2.21				4.29	1.47				3.37	1.15
22.00				7.18	2.46				4.77	1.64				3.74	1.28
23.25				7.92	2.72				5.27	1.81				4.13	1.42
24.00					2.88					1.91					1.50
26.00					3.32					2.21					1.73
28.00					3.79					2.52					1.98
30.00					4.30					2.85					2.24
32.00					4.82					3.21					2.51
34.00					5.38					3.58					2.80
36.00					5.96					3.96					3.11
38.00					6.57					4.37					3.43
40.00					7.21					4.79					3.76

*-Refer to HDPE manufacturer's data to determine which HDPE sizes are available in a given dimension ratio. Not all sizes are widely available.

Table 4.16. (continued). Head Loss Values in HDPE (DR-13.5, 15.5, & 17) for Water at 40 F*

HDPE	DR-13.5					DR-15.5					DR-17				
	Nominal Pipe Size					Nominal Pipe Size					Nominal Pipe Size				
	3"	4"	6"	8"	10"	3"	4"	6"	8"	10"	3"	4"	6"	8"	10"
Flowrate (gpm)	HL/100' (ft)					HL/100' (ft)					HL/100' (ft)				
70	1.32					1.18					1.11				
75	1.49					1.34					1.26				
80	1.68					1.51					1.42				
85	1.87					1.68					1.58				
90	2.07					1.86					1.75				
95	2.28					2.05					1.93				
100	2.51	0.75				2.25	0.67				2.12	0.63			
105	2.74	0.82				2.46	0.74				2.31	0.69			
110	2.98	0.89				2.67	0.80				2.51	0.75			
115	3.22	0.96				2.90	0.87				2.72	0.81			
120	3.48	1.04				3.13	0.94				2.94	0.88			
125	3.75	1.12				3.37	1.01				3.16	0.95			
130	4.02	1.20				3.61	1.08				3.40	1.02			
140		1.37	0.21				1.24	0.19				1.16	0.18		
170		1.95	0.30				1.75	0.27				1.65	0.26		
185		2.27	0.35				2.04	0.32				1.92	0.30		
200		2.62	0.41				2.35	0.37				2.21	0.34		
215		2.98	0.46				2.68	0.42				2.52	0.39		
230		3.37	0.52				3.03	0.47				2.84	0.44		
245		3.78	0.59				3.39	0.53				3.19	0.50		
260		4.20	0.65				3.78	0.59				3.55	0.55		
275		4.65	0.72				4.18	0.65				3.93	0.61		
290		5.12	0.80				4.60	0.72				4.33	0.67		
300			0.85	0.24				0.76	0.21				0.72	0.20	
400			1.43	0.40				1.28	0.36				1.20	0.34	
450			1.77	0.50				1.59	0.45				1.49	0.42	
500			2.14	0.60				1.92	0.54				1.80	0.51	
550			2.54	0.71				2.28	0.64				2.14	0.60	
600			2.97	0.84				2.67	0.75				2.51	0.71	
650			3.44	0.97				3.09	0.87				2.90	0.82	
700			3.93	1.10				3.53	0.99				3.32	0.93	
750			4.46	1.25				4.01	1.12				3.76	1.06	
800			5.01	1.41				4.50	1.26				4.23	1.19	
900				1.74	0.60				1.57	0.54				1.47	0.51
1000				2.11	0.73				1.90	0.66				1.78	0.62
1100				2.51	0.87				2.25	0.78				2.12	0.73
1200				2.94	1.02				2.64	0.91				2.48	0.86
1300				3.40	1.18				3.05	1.06				2.87	0.99
1400				3.89	1.35				3.49	1.21				3.28	1.14
1500				4.41	1.53				3.96	1.37				3.72	1.29
1600				4.96	1.71				4.45	1.54				4.18	1.45
1850					2.23					2.01					1.88
2100					2.81					2.53					2.37
2350					3.45					3.10					2.91
2600					4.15					3.73					3.50

*-Refer to HDPE manufacturer's data to determine which HDPE sizes are available in a given dimension ratio. Not all sizes are widely available.

Table 4.17. Head Loss Values in PVC (Schedule 40 & 80) for Water at 40 F

PVC	Schedule 40							Schedule 80						
	Nominal Pipe Size							Nominal Pipe Size						
	3/4"	1"	1-1/4"	1-1/2"	2"	2-1/2"	3"	3/4"	1"	1-1/4"	1-1/2"	2"	2-1/2"	3"
Flowrate (gpm)	HL/100' (ft)							HL/100' (ft)						
1.00	0.37							0.60						
1.50	0.75							1.23						
2.00	1.23							2.03						
2.50	1.82							3.00						
3.00	2.51	0.80						4.13	1.23					
3.50	3.29	1.04						5.41	1.61					
4.00	4.15	1.32						6.09	2.04					
4.50	4.55	1.62						7.52	2.51					
5.00	5.50	1.95						9.09	2.68					
6.00		2.40	0.73						3.72	1.05				
7.00		3.16	0.85						4.90	1.23				
8.00		4.01	1.08	0.52					6.23	1.56	0.72			
9.00		4.96	1.33	0.64					7.70	1.92	0.89			
10.00		5.99	1.61	0.77					9.30	2.32	1.08			
12.00			2.23	1.07						3.22	1.50			
14.00			2.94	1.40						4.25	1.97			
16.00			3.74	1.78	0.54					5.40	2.51	0.73		
18.00			4.62	2.20	0.67					6.67	3.10	0.90		
20.00			5.58	2.66	0.80					8.07	3.74	1.09		
22.50				3.29	0.99						4.62	1.35		
25.00				3.98	1.20						5.59	1.63		
27.50				4.72	1.42						6.63	1.93		
30.00				5.52	1.66	0.71					7.75	2.26	0.95	
32.50				6.38	1.92	0.82					8.95	2.61	1.10	
35.00				7.29	2.20	0.94					10.23	2.98	1.25	
38.00					2.55	1.09						3.46	1.45	
41.00					2.92	1.24						3.97	1.67	
44.00					3.31	1.41	0.50					4.51	1.89	0.65
47.00					3.73	1.59	0.56					5.07	2.13	0.74
50.00					4.17	1.78	0.63					5.67	2.38	0.82
53.75						2.03	0.71						2.71	0.94
57.50						2.29	0.81						3.06	1.06
61.25						2.56	0.90						3.43	1.18
65.00						2.85	1.01						3.82	1.32
68.75						3.15	1.1						4.23	1.46
72.50						3.47	1.22						4.65	1.60
76.25						3.80	1.34						5.09	1.76
80.00						4.15	1.46						5.56	1.91
87.50							1.72							2.25
95.00							1.99							2.61
102.50							2.28							2.99
110.00							2.59							3.40
117.50							2.92							3.83
125.00							3.27							4.28
132.50							3.63							4.76
140.00							4.01							5.25

Table 4.18. Head Loss Values in Steel (Schedule 40 & 80) for Water at 40 F

Steel	Schedule 40							Schedule 80						
	Nominal Pipe Size							Nominal Pipe Size						
	3/4"	1"	1-1/4"	1-1/2"	2"	2-1/2"	3"	3/4"	1"	1-1/4"	1-1/2"	2"	2-1/2"	3"
Flowrate (gpm)	HL/100' (ft)							HL/100' (ft)						
1.00	0.39							0.63						
1.50	0.76							1.25						
2.00	1.24							2.03						
2.50	1.82							2.97						
3.00	2.48	0.80						4.06	1.23					
3.50	3.23	1.03						5.30	1.59					
4.00	4.06	1.30						6.67	2.01					
4.50	4.98	1.59						8.18	2.46					
5.00	5.98	1.91						9.82	2.95					
6.00		2.62	0.72						4.04	1.03				
7.00		3.42	0.93						5.28	1.34				
8.00		4.31	1.18	0.57					6.66	1.69	0.79			
9.00		5.29	1.44	0.70					8.18	2.07	0.97			
10.00		6.36	1.73	0.83					9.84	2.49	1.17			
12.00			2.38	1.15						3.43	1.60			
14.00			3.11	1.50						4.49	2.10			
16.00			3.94	1.89						5.67	2.65			
18.00			4.84	2.33						6.98	3.26			
20.00			5.83	2.80	0.85					8.40	3.92	1.16		
22.50				3.44	1.05						4.82	1.42		
25.00				4.14	1.26						5.80	1.71		
27.50				4.90	1.49						6.87	2.02		
30.00				5.72	1.74	0.75					8.01	2.36	1.00	
32.50				6.59	2.00	0.86					9.23	2.71	1.15	
35.00				7.51	2.28	0.98					10.53	3.09	1.31	
38.00					2.64	1.13						3.58	1.51	
41.00					3.02	1.29						4.09	1.73	
44.00					3.42	1.46						4.64	1.96	
47.00					3.84	1.65						5.21	2.20	
50.00					4.29	1.84	0.65					5.82	2.46	0.85
53.75						2.09	0.74						2.79	0.97
57.50						2.35	0.83						3.15	1.09
61.25						2.63	0.93						3.52	1.22
65.00						2.92	1.04						3.91	1.36
68.75						3.23	1.14						4.32	1.50
72.50						3.55	1.26						4.75	1.65
76.25						3.88	1.38						5.20	1.80
80.00						4.23	1.50						5.66	1.96
87.50							1.76							2.30
95.00							2.03							2.66
102.50							2.33							3.05
110.00							2.64							3.46
117.50							2.97							3.89
125.00							3.32							4.35
132.50							3.68							4.82
140.00							4.06							5.32

Table 4.19. Head Loss Values in Copper Tubing (Types L & M) for Water at 40 F

Copper Tubing	Type L							Type M						
	Nominal Pipe Size							Nominal Pipe Size						
	3/4"	1"	1-1/4"	1-1/2"	2"	2-1/2"	3"	3/4"	1"	1-1/4"	1-1/2"	2"	2-1/2"	3"
Flowrate (gpm)	HL/100' (ft)							HL/100' (ft)						
1.00	0.46							0.40						
1.50	0.94							0.80						
2.00	1.55							1.33						
2.50	2.30							1.97						
3.00	3.16	0.89						2.71	0.78					
3.50	4.14	1.17						3.54	1.02					
4.00	4.65	1.47						3.98	1.28					
4.50	5.74	1.81						4.91	1.58					
5.00	6.94	2.18	0.80					5.93	1.90	0.73				
6.00		2.68	1.10						2.33	1.00				
7.00		3.53	1.29						3.07	1.17				
8.00		4.48	1.64	0.71					3.91	1.48	0.66			
9.00		5.54	2.02	0.88					4.82	1.83	0.82			
10.00		6.69	2.44	1.06					5.83	2.21	0.99			
12.00			3.38	1.47						3.07	1.37			
14.00			4.46	1.94						4.05	1.81			
16.00			5.67	2.47						5.15	2.30			
18.00			7.01	3.05						6.36	2.84			
20.00			8.47	3.68	0.98					7.68	3.43	0.92		
22.50				4.55	1.21						4.24	1.14		
25.00				5.50	1.46						5.13	1.38		
27.50				6.53	1.73						6.09	1.63		
30.00				7.63	2.02	0.72					7.12	1.91	0.68	
32.50				8.81	2.33	0.83					8.22	2.20	0.78	
35.00				10.07	2.67	0.94					9.39	2.52	0.89	
38.00					3.09	1.09						2.92	1.03	
41.00					3.55	1.25						3.35	1.18	
44.00					4.03	1.42						3.80	1.34	
47.00					4.53	1.60						4.28	1.51	
50.00					5.07	1.79	0.76					4.78	1.69	0.72
53.75						2.04	0.87						1.93	0.82
57.50						2.30	0.98						2.17	0.93
61.25						2.58	1.10						2.44	1.04
65.00						2.87	1.22						2.71	1.15
68.75						3.18	1.35						3.00	1.28
72.50						3.50	1.49						3.30	1.40
76.25						3.83	1.63						3.62	1.54
80.00						4.18	1.78						3.94	1.68
87.50							2.09							1.97
95.00							2.42							2.29
102.50							2.78							2.62
110.00							3.16							2.98
117.50							3.56							3.35
125.00							3.98							3.75
132.50							4.42							4.17
140.00							4.88							4.60

Table 4.20. Head Loss Values in Rubber Heater Hose (200-psi) for Water at 40 F

200-psi	Rubber Heater Hose		
	Nominal Pipe Size		
	1"	1-1/4"	1-1/2"
Flowrate (gpm)	HL/100' (ft)		
3.00	1.01		
3.75	1.49		
4.50	2.04		
5.25	2.66		
6.00	3.37	1.16	
6.75	4.14	1.43	
7.50	4.98	1.71	
8.25	5.89	2.03	
9.00	6.87	2.36	
9.75	7.92	2.72	
10.50	9.04	3.10	
11.25	10.22	3.50	
12.00	11.47	3.92	
12.75	12.78	4.36	
13.50	14.15	4.83	
14.25	15.60	5.32	
15.00	17.10	5.82	
16.00		6.53	2.72
17.50		7.67	3.19
19.00		8.88	3.69
20.50		10.18	4.22
22.00		11.56	4.79
23.50		13.01	5.39
25.00		14.55	6.02
26.50		16.16	6.69
28.00		17.86	7.38
29.50		19.63	8.11
31.00		21.48	8.86
31.00			8.86
33.50			10.19
36.00			11.61
38.50			13.11
41.00			14.69
43.50			16.36
46.00			18.11
48.50			19.95
51.00			21.86

The data in Table 4.16 is graphed in Figures 4.2-4.4 for visual comparison. The data in Tables 4.17-4.20 is graphed in Figures 4.5-4.8 for visual comparison. Note that the figures display a "best range" of head loss at any given flow. The optimum range for head loss is 1-3 feet of head loss per 100 feet of pipe. The rationale behind this range is discussed in Section 4.5. Figures 4.2-4.8 display graphs of head loss calculations for HDPE (DR-7, 9, 11, 13.5, 15.5, and 17), PVC (SCH 40 and 80), copper tubing (Type K, L, and M), steel pipe (SCH 40 and 80), and reinforced rubber hose for water at 40 F.

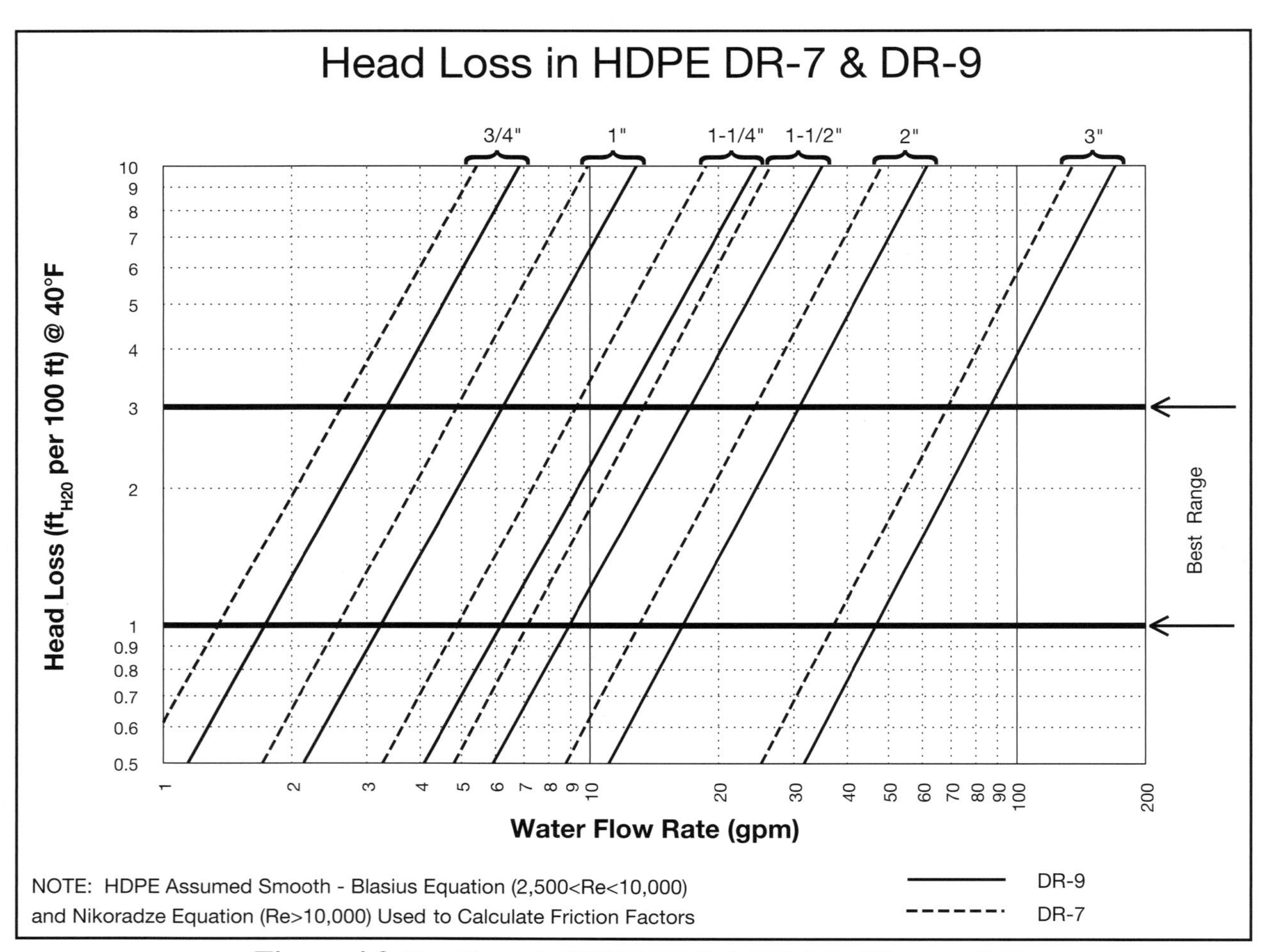

Figure 4.2. Head Loss in HDPE DR-7 & DR-9 for Water at 40 F

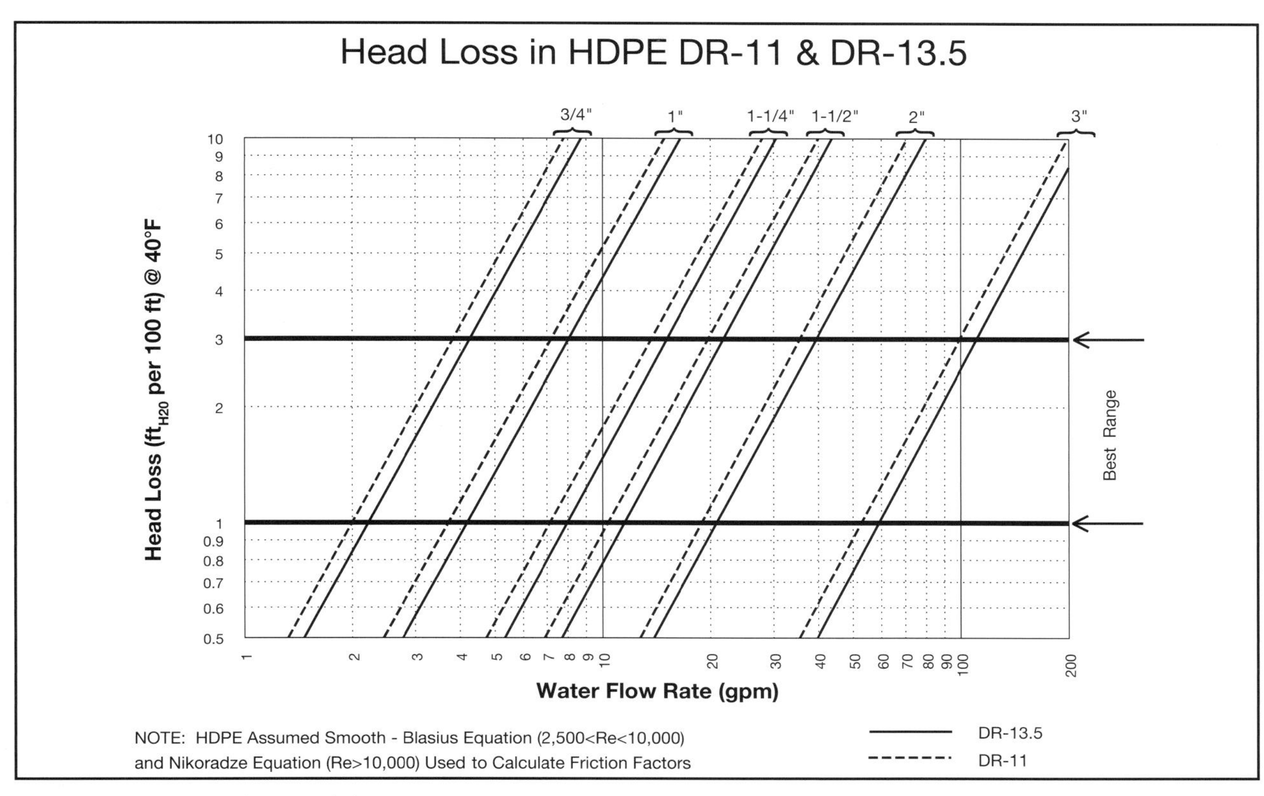

Figure 4.3. Head Loss in HDPE DR-11 & DR-13.5 for Water at 40 F

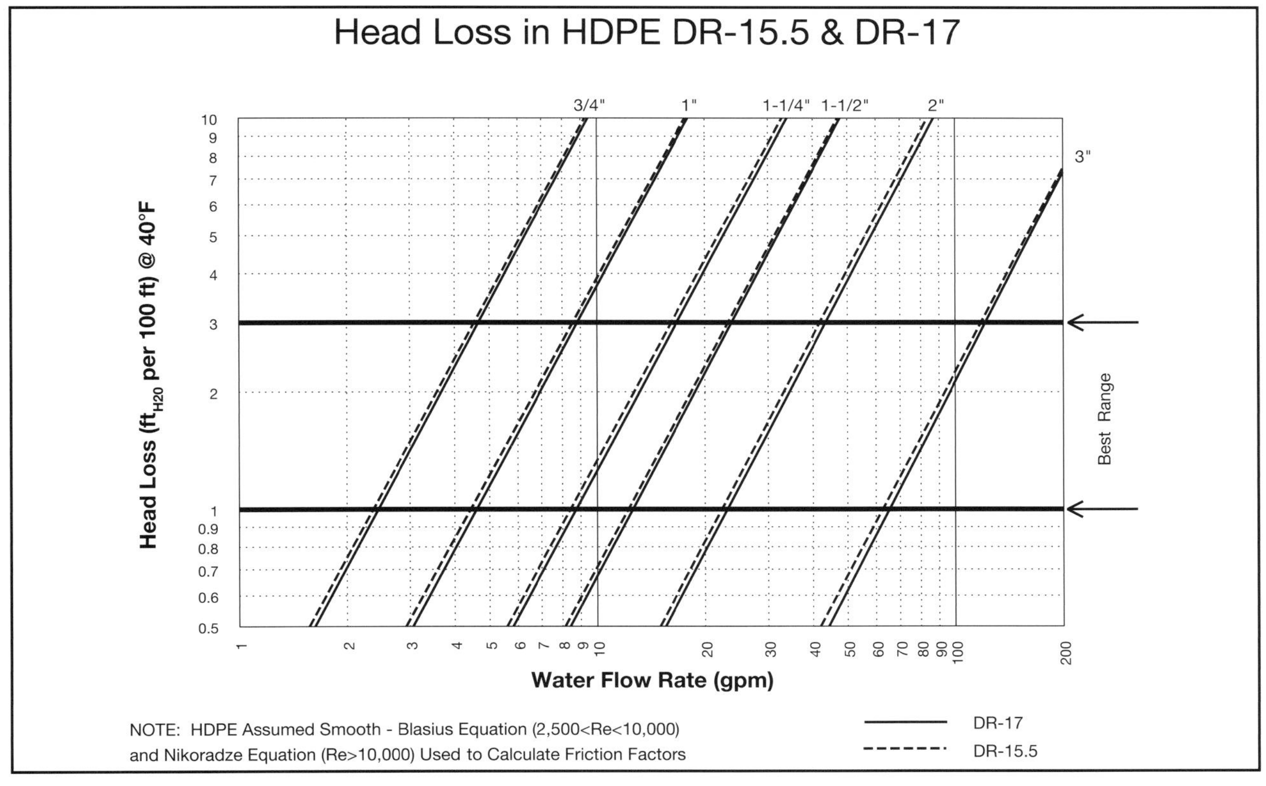

Figure 4.4. Head Loss in HDPE DR-15.5 & DR-17 for Water at 40 F

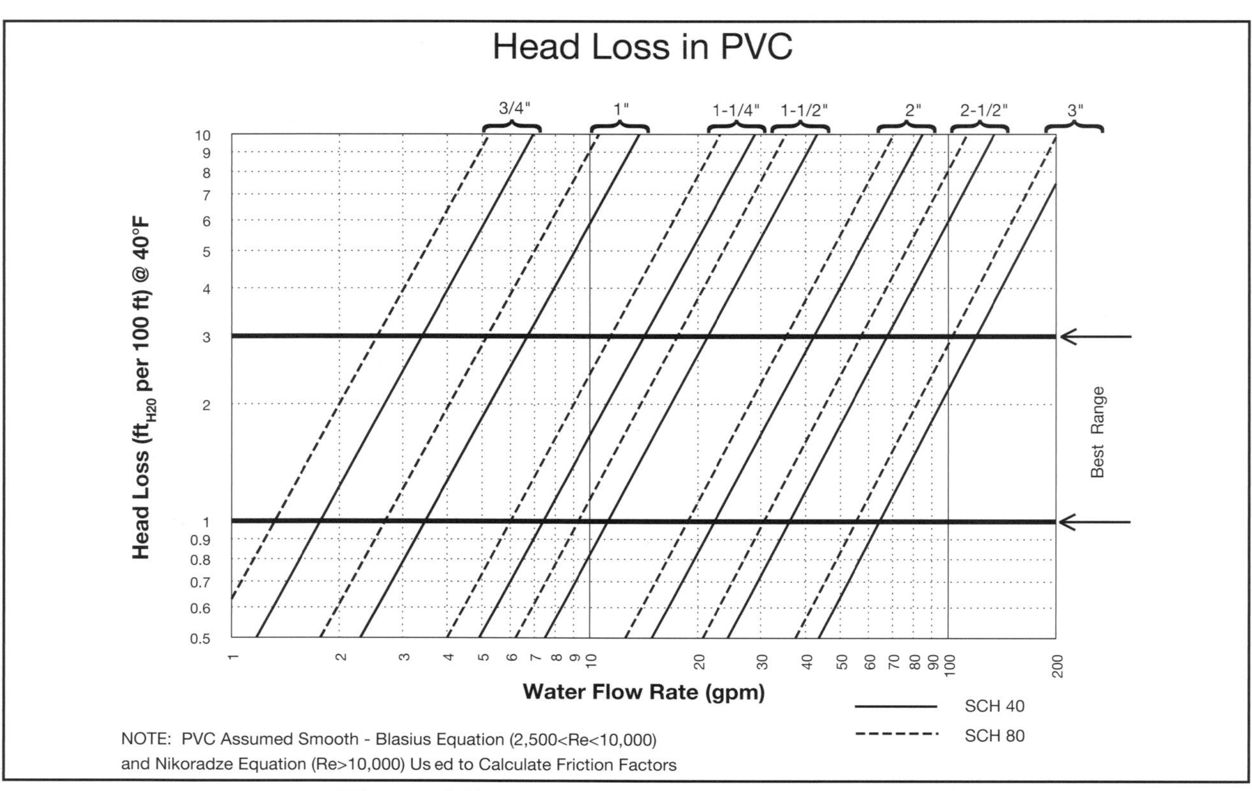

Figure 4.5. Head Loss in PVC for Water at 40 F

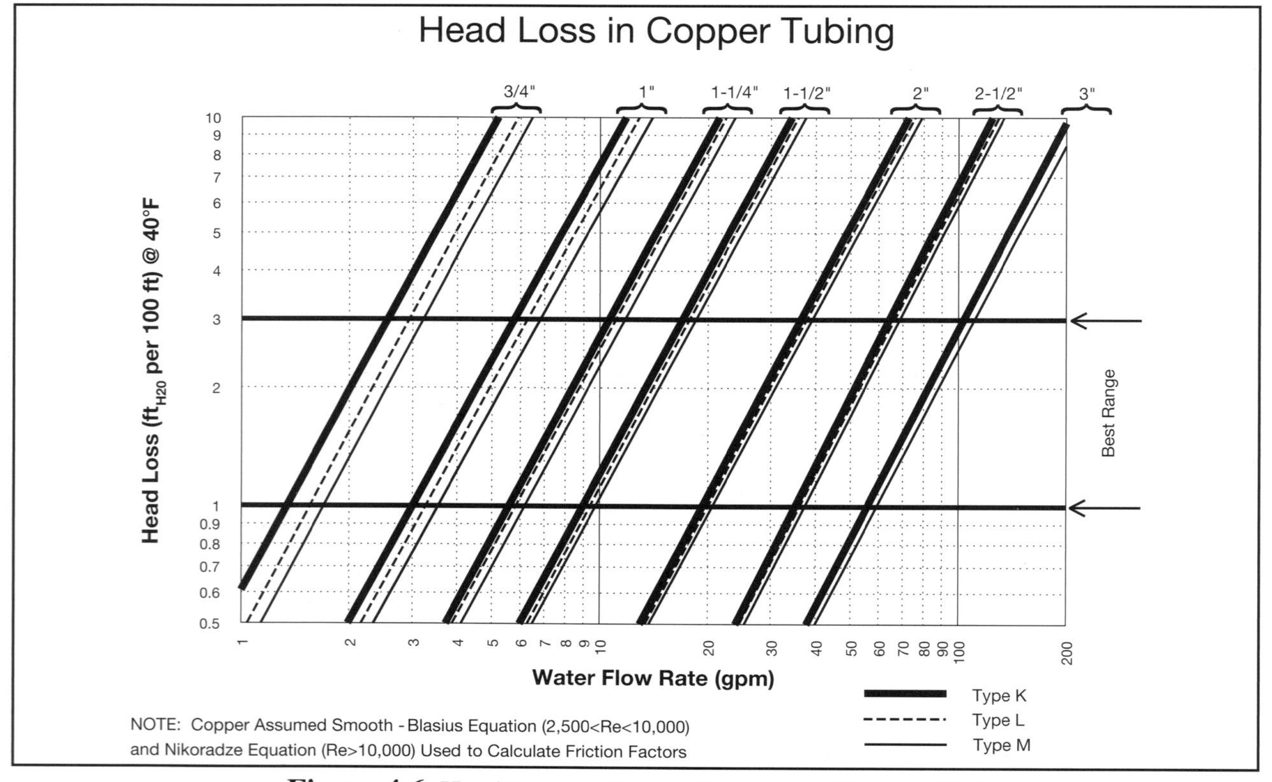

Figure 4.6. Head Loss in Copper Tubing for Water at 40 F

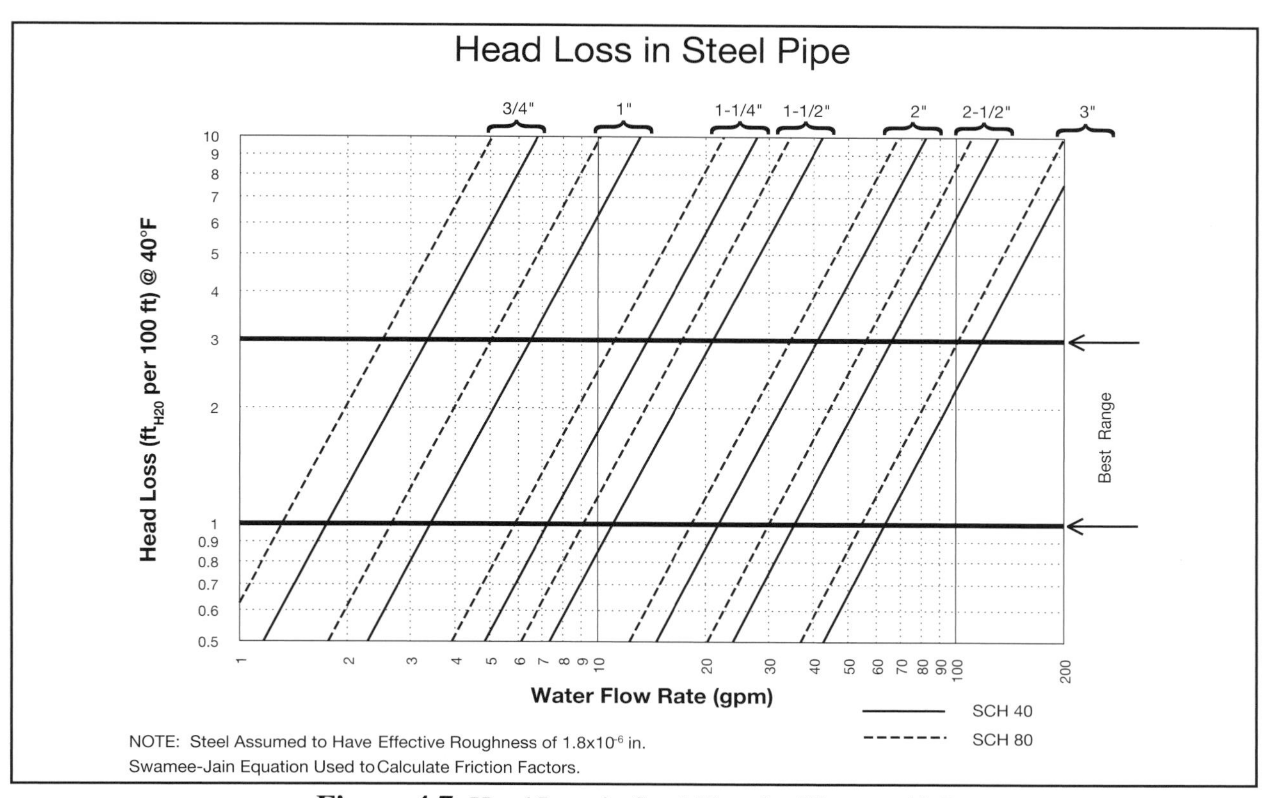

Figure 4.7. Head Loss in Steel Pipe for Water at 40 F

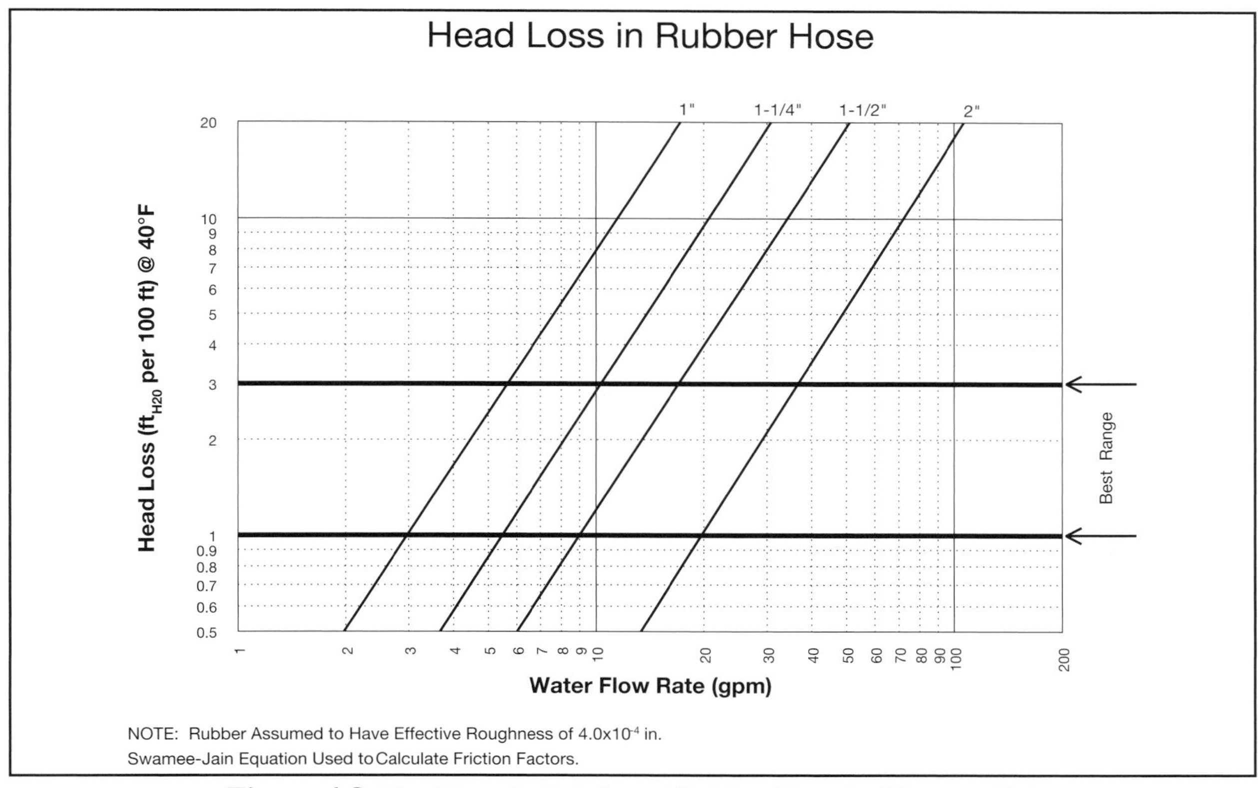

Figure 4.8. Head Loss in Reinforced Rubber Hose for Water at 40 F

4.4.4 ANTIFREEZE CORRECTION FACTORS

Because water-antifreeze solutions tend to be more viscous (and therefore harder to pump) than water itself, correction factors are needed to account for such viscosity increase for pump sizing purposes. In many cases, manufacturers will provide head loss tables based on water at 40 F as the circulating fluid. Antifreeze correction factors are calculated by dividing the head loss caused by the antifreeze at the temperature and concentration of interest by the head loss caused by water at 40 F. Again, keep in mind that fluid viscosities and densities for antifreeze will vary from manufacturer to manufacturer. Table 4.21 provides antifreeze viscosity correction factors for various types and concentrations of antifreeze at various temperatures.

Table 4.21. Antifreeze Viscosity Correction Factors (C_{VAF})

Propylene Glycol Viscosity Multipliers								
Concentration By Volume	Freeze Point (˚F)	Temperature (˚F)						
		40	35	30	25	20	15	10
10%	26.1	1.13	1.16	1.19	--	--	--	--
15%	22.5	1.19	1.22	1.25	--	--	--	--
20%	18.4	1.24	1.28	1.31	1.35	1.39	--	--
25%	13.8	1.34	1.38	1.42	1.47	1.52	--	--
30%	8.8	1.41	1.46	1.51	1.56	1.62	1.68	1.74
Methanol Viscosity Multipliers								
Concentration By Volume	Freeze Point (˚F)	Temperature (˚F)						
		40	35	30	25	20	15	10
5%	26.2	1.08	1.11	--	--	--	--	--
10%	19.7	1.13	1.15	1.18	1.22	--	--	--
12.5%	16.2	1.15	1.18	1.21	1.25	--	--	--
15%	12.6	1.17	1.20	1.23	1.27	1.31	1.34	--
17.5%	8.8	1.18	1.21	1.25	1.29	1.33	1.36	--
20%	4.9	1.20	1.23	1.26	1.31	1.35	1.38	--
Ethanol Viscosity Multipliers								
Concentration By Volume	Freeze Point (˚F)	Temperature (˚F)						
		40	35	30	25	20	15	10
10%	26.4	1.21	1.26	1.30	--	--	--	--
15%	22.6	1.26	1.31	1.36	1.40	--	--	--
20%	18.1	1.31	1.36	1.42	1.47	1.56	--	--
25%	12.9	1.35	1.41	1.47	1.57	1.64	1.67	--
30%	7.0	1.37	1.44	1.53	1.63	1.66	1.69	--

-- Concentration level not recommended due to insufficient freeze protection

Use Equation 4.8 to calculate head loss in a given type/size of pipe for antifreeze solutions of different temperatures/concentrations.

$$HL_{AF} = HL_{H2O} \cdot C_{VAF}$$ **Equation 4.8**

Where:

HL_{AF} = Head Loss for a Given Antifreeze Solution Type/Concentration/Temp (ft)

HL_{H20} = Head Loss for Water at 40 F (ft - refer to Tables 4.16-4.20 and Figures 4.2-4.8)

C_{VAF} = Viscosity Correction Factor for Given Antifreeze Solution Type/Concentration/Temp (refer to Table 4.21)

4.5 Flow Capacities

When selecting the size of pipe to be used as the ground heat exchanger portion of the GSHP system, several guidelines must be followed. First, pipe sizing needs to be large enough to keep pumping power requirements to a minimum. Secondly, pipe sizing needs to be small enough to ensure turbulent flow in the circulating fluid, thus maximizing heat transfer between the fluid and the pipe wall. A simple guideline to follow is to size the pipe such that the required flow will result in 1-3 feet of head loss per 100 feet of pipe for any type of circulating fluid. Designing the ground heat exchanger to stay within this range will ensure the proper balance between minimizing pumping power requirements and material costs and maximizing thermal performance. Pipe sizing that produces head loss significantly below this range will cause the pipe to be oversized and can dramatically increase project material and handling costs in addition to allowing fluid flow to fall out of the turbulent regime in cases where a water and antifreeze solution is used. Pipe sizing that produces head loss above this range can cause significantly higher system first cost and operating cost due to increased pump size and pumping power demand. Table 4.22 provides flow capacities for two common HDPE DRs and a range of pipe sizes to stay within the range of 1-3 feet of head loss per 100 feet of pipe for water at 40 F.

Table 4.22. Flow Capacities for Various Pipe Sizes for 1-3 feet of Head Loss / 100 feet of Pipe (Water at 40 F)

DR-11		DR-15.5	
Nom. Size (in)	Capacity (gpm)	Nom. Size (in)	Capacity (gpm)
3/4	2.0 - 3.7	2	22.5 - 41.7
1	3.7 - 7.3	3	63.5 - 117
1-1/4	7.3 - 13.6	4	124 - 229
1-1/2	10.5 - 19.5	6	348 - 640
2	19.1 - 35.3	8	700 - 1290
3	53.8 - 99.4	10	1260 - 2300

In some instances, it is not possible to stay within the recommended range of 1-3 feet of head loss per 100 feet of pipe length. The maximum recommended pressure drop for a closed-loop system is 4 feet of head loss per 100 feet of pipe. Tables 4.23 through 4.25 provide the maximum recommended fluid flow rates for various types and sizes of pipe. The maximum recommended flow rates given in the tables are based on 4 feet of head per 100 feet of pipe for water at 40 F, 20% propylene glycol by volume at 25 F, 12.5% methanol by volume at 25 F, and 20% ethanol by volume at 25 F.

Table 4.23. Maximum Recommended Flow Rates (gpm) for Circulating Fluids in HDPE*

Nom. Pipe	HDPE DR-11				HDPE DR-13.5				HDPE DR-15.5				HDPE DR-17			
Size (in)	Water	P.G.	Meth.	Eth.	Water	P.G.	Meth.	Eth.	Water	P.G.	Meth.	Eth	Water	P.G.	Meth.	Eth.
3/4	4.7	3.8	3.9	3.7	5.2	4.2	4.4	4.1	-	-	-	-	-	-	-	-
1	8.5	6.9	7.2	6.7	9.5	7.7	8.1	7.5	-	-	-	-	-	-	-	-
1-1/4	16	13	14	13	18	14	16	14	-	-	-	-	-	-	-	-
1-1/2	23	19	21	18	26	21	23	20	27	22	25	22	28	23	26	22
2	42	37	38	34	46	41	42	40	49	43	45	42	51	45	46	44
3	120	100	110	100	130	110	120	110	140	120	130	120	140	130	130	120

*Based on 4 ft/100 ft Head Loss

Table 4.24. Maximum Recommended Flow Rates (gpm) for Circulating Fluids in Steel & PVC*

Nom. Pipe	Steel Schedule 40				Steel Schedule 80				PVC Schedule 40				PVC Schedule 80			
Size (in)	Water	P.G.	Meth.	Eth.	Water	P.G.	Meth.	Eth.	Water	P.G.	Meth.	Eth.	Water	P.G.	Meth.	Eth.
3/4	4.0	3.3	3.5	3.3	3.0	2.4	2.6	2.4	3.9	3.3	3.5	3.3	2.9	2.5	2.6	2.5
1	7.7	6.4	6.8	6.1	6.0	4.9	5.3	4.8	8.0	6.4	6.8	6.2	6.2	5.0	5.3	4.9
1-1/4	16	14	14	13	13	11	12	10	17	13	15	13	14	11	12	11
1-1/2	25	21	22	20	20	17	18	16	25	21	23	20	21	17	19	17
2	48	41	43	40	41	35	36	33	49	43	45	42	41	36	38	33
2-1/2	78	67	70	65	66	57	59	55	78	69	72	68	67	59	61	58
3	140	120	130	120	120	100	110	100	140	120	130	120	120	110	110	100

*Based on 4 ft/100 ft Head Loss

Table 4.25. Maximum Recommended Flow Rates (gpm) for Circulating Fluids in Copper*

Nom. Pipe	Copper Type K				Copper Type L				Copper Type M			
Size (in)	Water	P.G.	Meth.	Eth.	Water	P.G.	Meth.	Eth.	Water	P.G.	Meth.	Eth.
3/4	3.0	2.5	2.7	2.5	3.4	2.9	3.1	2.9	3.8	3.2	3.4	3.2
1	6.9	5.6	5.8	5.4	7.5	6.1	6.3	5.9	8.1	6.5	6.9	6.4
1-1/4	13	10	11	10	13	11	11	10	14	11	12	11
1-1/2	20	16	18	16	21	17	19	17	22	18	20	17
2	42	37	39	34	44	39	40	38	45	40	42	39
2-1/2	76	67	69	65	78	69	72	68	81	71	74	70
3	120	110	110	100	120	110	110	110	130	110	120	110

*Based on 4 ft/100 ft Head Loss

As stated before, antifreeze properties will vary by manufacturer. When calculating minimum flow rates for turbulence, piping head loss, antifreeze correction factors, piping flow capacities, or freeze protection levels, always refer to manufacturer's data for fluid properties if available.

4.6 Flushing Flow Rates

After the ground heat exchanger is completely installed, it is necessary to purge the system of air and debris. Entrained air in circulating fluid is undesirable because it allows oxygen to corrode metallic components, causes flow noise, and can possibly prevent flow in entire sections of the system. Trapped air can be separated from water and therefore removed by power flushing the system at a minimum flow velocity of 2 feet per second (FPS). Debris in circulating fluid is undesirable because it can cause excessive wear on mechanical components of the system. Debris can be removed from the system by flushing through a filter, which is a component that can be integrated into the flushing unit itself. Typically, the flow rates required to generate the recommended fluid velocity exceed the capability of the system's circulating pumps. In order to satisfy the minimum velocity requirement, a flushing pump must be used that is large enough to supply flow at the necessary rate to attain the minimum flow velocity of 2 feet per second. Flushing flow rate for any type or size of pipe can be calculated using Equation 4.9.

$$Q_{flush} = 4.896(D_{p,i})^2$$

Equation 4.9

Where:

Q_{flush} = Flushing Flow Rate (gpm)
$D_{p,i}$ = Inside Pipe Diameter (in)

Table 4.26 provides required flushing flow rates to achieve the minimum flow velocity recommended for purging air in various sizes and DRs for HDPE pipe. Flushing flow rates for other types and sizes of pipe are provided in Tables 4.27 through 4.29.

Table 4.26. Flushing Flow Rates (gpm) to Achieve 2 FPS in HDPE Pipe*

Diameter (in)	DR-7	DR-9	DR-11	DR-13.5	DR-15.5	DR-17
3/4	2.8	3.3	3.6	3.9	4.1	4.2
1	4.3	5.1	5.7	6.1	6.4	6.6
1-1/4	6.9	8.2	9.0	9.8	10.2	10.5
1-1/2	9.0	10.7	11.8	12.8	13.4	13.8
2	14.1	16.7	18.5	20.0	20.9	21.5
3	30.6	36.3	40.1	43.5	45.5	46.7
4	50.6	59.9	66.3	71.9	75.2	77.1
6	110	130	144	156	163	167
8	186	220	244	264	276	283
10	289	342	379	410	429	440

*-Refer to HDPE manufacturer's data to determine which HDPE sizes are available in a given dimension ratio. Not all sizes are widely available.

Table 4.27. Flushing Flow Rates (gpm) to Achieve 2 FPS in Schedule 40 & Schedule 80 Pipe (PVC, CPVC, Steel)

Nominal Pipe Size (in)	Flushing Flowrate (gpm)	
	Pipe Class	
	SCH 40	SCH 80
3/4	3.3	2.7
1	5.4	4.5
1-1/4	9.3	8.0
1-1/2	12.7	11.0
2	20.9	18.4
2-1/2	29.8	26.4
3	46.1	41.2
4	79.4	71.7
5	123	111
6	180	162
8	312	285
10	492	448

Table 4.28. Flushing Flow Rates (gpm) to Achieve 2 FPS in Copper Tubing

Nominal Pipe Size (in)	Flushing Flowrate (gpm)		
	Copper Tube Type		
	Type K	Type L	Type M
3/4	2.7	3.0	3.2
1	4.8	5.1	5.4
1-1/4	7.6	7.8	8.2
1-1/2	10.7	11.1	11.4
2	18.8	19.3	19.8
2-1/2	29.0	29.7	30.5
3	41.4	42.5	43.5
3-1/2	56.1	57.4	58.6
4	72.8	74.7	75.8
5	113	116	118
6	161	167	169

Table 4.29. Flushing Flow Rates (gpm) to Achieve 2 FPS in Reinforced Rubber Hose

Rubber Hose	
Nominal Size (in)	Flushing Flowrate (gpm)
1	4.9
1-1/4	6.1
1-1/2	7.3

DESIGN OF CLOSED-LOOP GROUND HEAT EXCHANGERS

In This Section

The design of a closed-loop ground heat exchanger (GHEX) consists of selecting a ground heat exchanger configuration that will provide the required ground load demand of the heat pump for the least cost of installation. There is no "best" type or configuration of GHEX; so the designer must have some way of choosing between alternatives. There are three basic types of closed-loop ground heat exchangers: vertically-bored (Figure 5.1), horizontally-trenched (Figure 5.2), and horizontally-bored (Figure 5.3). The configuration of GHEX that is utilized will depend on available land area, drilling and trenching conditions, the relative cost of the various options, and the extent to which the installation site can be excavated without excessive re-landscaping costs.

The general design procedure for a closed-loop ground heat exchanger is:

1. Select a GHEX configuration based on a comprehensive site survey
2. Configure the GHEX for proper flow and ease of flushing and purging
3. Estimate the length of each flow path in the GHEX based on configuration
4. Calculate head loss to circulate fluid through the GHEX and size pump system
5. Evaluate the economics of design
6. Redesign, if necessary, or consider other configurations for comparison

Steps 1-3 will be considered in detail in this chapter. Head loss through the piping system and pumping system selection will be considered in Chapter 6.

The economics of any design are dependent on site-specific conditions along with the development of installation infrastructure in the region. With proper design and installation, each closed-loop GHEX can be made to work properly at design conditions, and the selection between alternatives generally comes down to installation cost and availability for the options that will work at the specific site. Prior to design of the GHEX, it would be well advised to contact several potential GHEX installers in your region to determine capabilities and relative costs of the various options. The actual cost to install a given design will generally not be known until bids are obtained and as experience is gained as more GHEX systems are installed, providing better knowledge of the relative costs of the various GHEX configurations.

5.1 Configuration and Layout of Closed-Loop Ground Heat Exchangers

5.1.1 LOCATION The configuration of the closed-loop GHEX is highly dependent on the size and layout of the installation site and whether the installation is new construction or retrofit. New construction is generally open to all GHEX configurations because, in most cases, landscaping has not been completed and the extent to which the site is excavated is less important. Retrofit to an existing structure with established lawn, shrubbery, trees, gardens, sprinkler systems, etc., presents more of an installation challenge to minimize damage and cost of repair to the landscape. In either case, a site plan must be developed that identifies all existing structures, utilities, lawn and garden obstacles, and planned future constructions to locate an acceptable location for the GHEX. This will normally require a site visit by the designer, which should include discussions with the owner as to future construction plans and other uses that may preclude specific areas on the site from GHEX installation. Limitations to the type of equipment that may be brought onto the site should be noted to avoid consideration of a GHEX configuration that requires installation equipment that is too large or heavy to bring onto the site. Ultimately, a detailed site plan should be developed which locates the GHEX relative to one or more permanent structures on the site. The site plan will provide the GHEX installer the required information for a successful installation. This site plan, modified as necessary to indicate any changes caused by unknown or unforeseen conditions, will serve as a document of record for the GHEX installer, the HVAC contractor, the home or business owner, and the local electric supplier (if required). The site plan will be discussed in more detail in Chapter 7.

5.1.2 CLOSED-LOOP CONFIGURATION As indicated earlier, there are three basic types of closed-loop ground heat exchangers: vertically-bored, horizontally-trenched, and horizontally-bored. The GHEX type that is utilized will be selected first based on available land area, with the vertically-bored configuration requiring the least surface area and the horizontally-trenched configuration requiring the most accessible surface area. The horizontally-bored configuration does require as much or more surface area than the horizontally-trenched, but all excavation is usually done from one end of the installation and access to the area above the GHEX is not required during installation (access is required at the far end of the installation where the U-bends are attached and a grouting rig may be located). Horizontally-bored GHEX systems offer other advantages as they can be placed under existing structures, lawn and garden obstacles, playfields, etc., without having to disturb any surface area above the GHEX.

5.1.2.1 VERTICALLY-BORED The vertically-bored GHEX (Figure 5.1) requires the least land area for installation, typically being installed with 15 to 25 foot spacing between boreholes to minimize thermal interference between boreholes. Each borehole is drilled to a design depth that depends on soil or rock type, required heating or cooling capacity of the borehole, and drilling conditions that may limit the maximum borehole depth for cost-effective drilling. There is no minimum or maximum drilling depth that a design must adhere to, and the designer should talk to vertical GHEX installers familiar with the region to determine drilling conditions and limitations that should be considered in the design phase. Vertically-bored GHEX systems encounter the largest variation in soil and rock conditions, resulting in a wide range of heat transfer capabilities that have, along with the grout or backfill thermal properties, a large effect on the required length of borehole to achieve the necessary heat transfer to or from the ground. A general rule-of-thumb for residential and light commercial vertical GHEX design is to install one vertical borehole for each nominal ton of heating or cooling capacity. This supports the use of a standardized-header arrangement (Section 5.1.3) and the use of ¾-inch U-bends in each borehole to accommodate the necessary 3 GPM of circulating fluid flow associated with one ton of heat pump capacity. This approach works best in unconsolidated formations (clays, silts, and sands and gravels), where the required GHEX lengths range from 150 to 250 feet, or more, of borehole per nominal ton of heat pump capacity. GHEX systems in rock formations with higher thermal conductivities, requiring 150 feet or less of borehole for each ton of heat pump capacity, generally utilize 1-inch or 1-1/4-inch U-bends to accommodate the required fluid flow rates for borehole depths up to approximately 600 feet that may represent 3 or more tons of heat pump capacity. Header arrangements for these types of layouts will be discussed in Section 5.1.3.

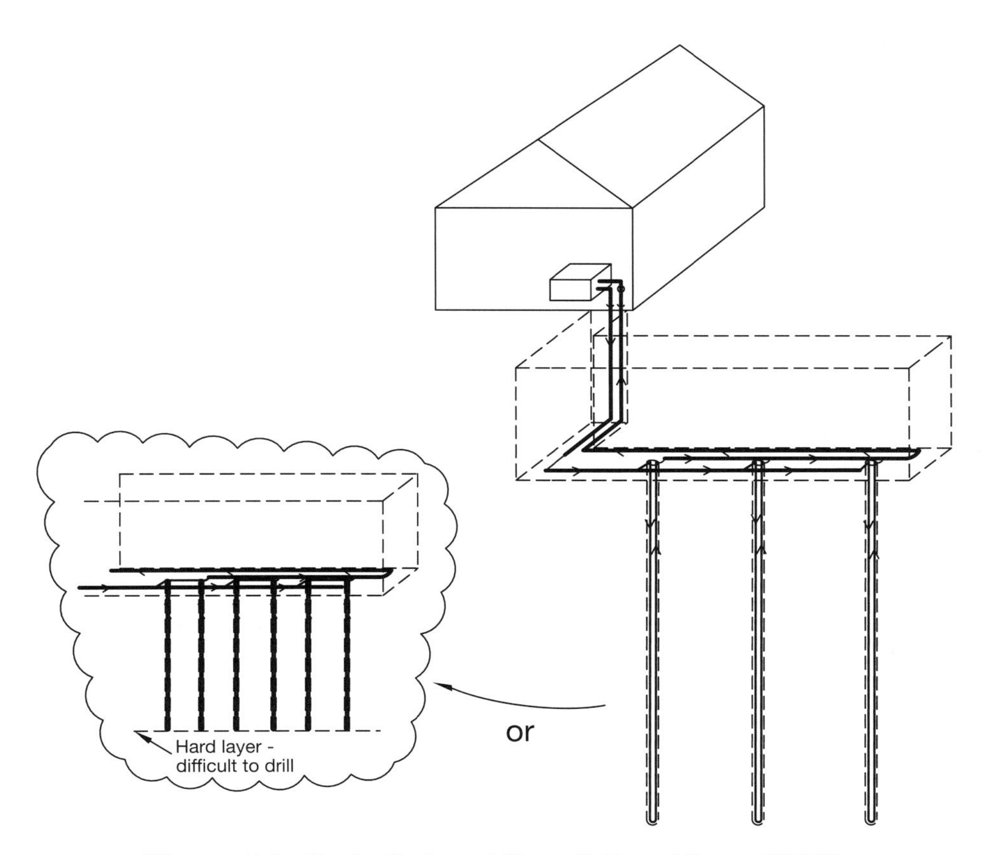

Figure 5.1. Vertically-bored "3-ton" Closed-Loop GHEX.

5.1.2.2 HORIZONTALLY-TRENCHED The horizontally-trenched GHEX (Figure 5.2) requires the most accessible land area for installation, with trenches commonly being installed approximately 10 feet apart and up to 150 feet in length per nominal ton of heat pump capacity. The trench width and depth along with the pipe configuration in the trench (bottom of Figure 5.2) directly affect the trench design length. Available trenching options may limit the trench width and depth and have an influence on the horizontal loop configurations that may be considered. Most often, horizontal GHEX systems are installed at depths ranging from 4 to 8 feet and the influence of ambient air temperature on design soil temperatures, as well as the potential for dry soil conditions that reduce the soil's ability to transfer heat, must be considered. The economics of horizontally-trenched GHEX systems are generally better than vertically or horizontally-bored systems, but land availability issues will often limit their application to rural acreages or very large lots. Soil conditions that may limit the use of the horizontally-trenched GHEX include caving, sandy soils, very rocky soils, or solid rock at a shallow depth. Large pit systems have been used in caving, sandy soils, but attention must be paid to the relationship between spacing between the ground loops and loop length to avoid extreme temperatures at design conditions, especially when soil conditions become dry and have reduced heat transfer capabilities.

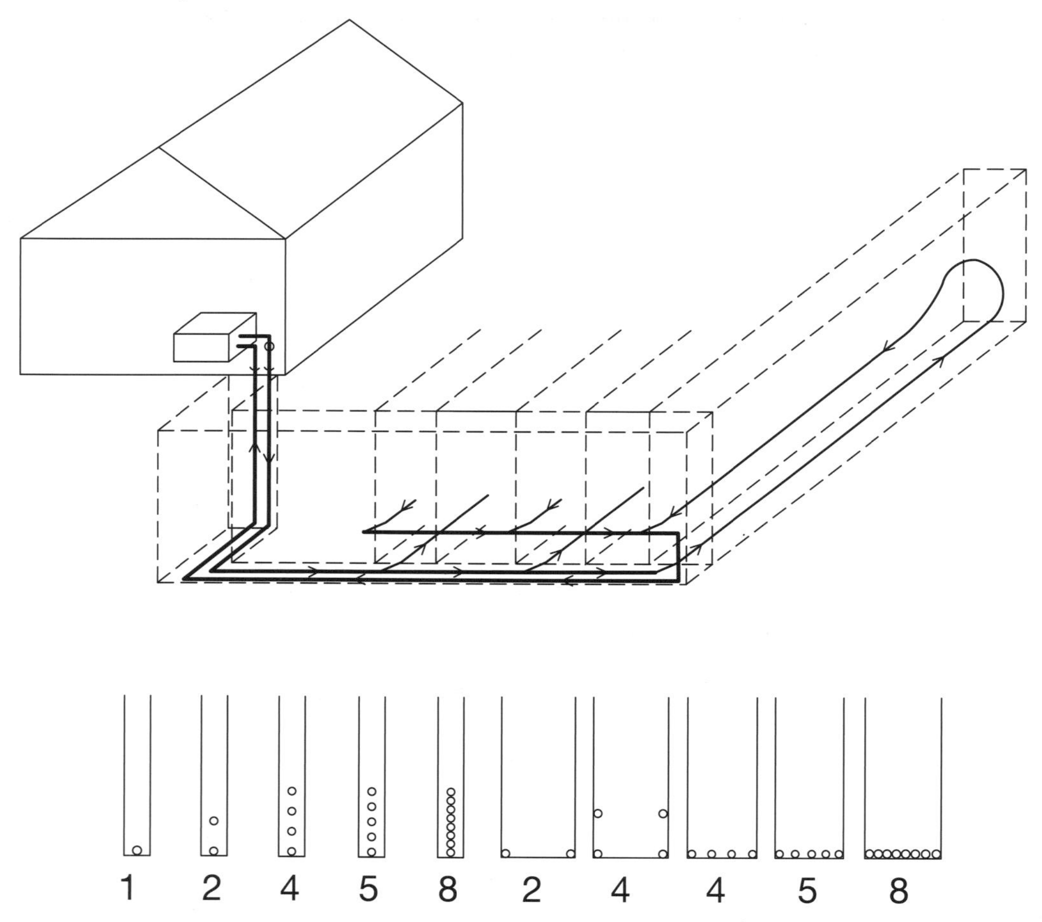

Figure 5.2. Horizontally-trenched "3-ton" Closed-Loop GHEX.

5.1.2.3 HORIZONTALLY-BORED The horizontally-bored GHEX (Figure 5.3) is a relatively new approach to installing a closed-loop GHEX in which a horizontal boring machine is used to create a small borehole at a minimum depth of about 15 feet into which a conventional U-bend is pulled and grouted. The length of the horizontal borehole depends on the site layout, but a minimum of about 225 feet of borehole per nominal ton of heat pump capacity is normally required in most unconsolidated formations (clays, silts, and sands and gravels), and the design length is also highly dependent on the grout or backfill thermal properties. Because the horizontal boring process requires additional borehole length to bore down to the required depth and to exit back to the surface at the far end of the bore field (approximately 30 feet on each end for a 15 foot boring depth), fitting a horizontally-bored GHEX on a small lot may require connecting two or more bores in series to create a single flow path and may also require stacking bores vertically (15, 30, and possibly 45 feet). Large sites may allow for horizontal bores of up to 600 feet or longer, representing approximately 3 tons of heat pump capacity per bore. Horizontally-bored GHEX systems can be installed under structures, lawn and garden obstacles, play fields, etc., from one end of the ground loop site, providing access to ground heat exchange areas that would not otherwise be available. These systems also offer an economic alternative to vertically-bored GHEX systems where depth to rock is shallow and the economics of installing well casing and drilling into the rock, or drilling shallow boreholes and increasing the headering cost of the installation, make vertically-bored GHEX installation prohibitive. The qualified infrastructure for horizontally-bored GHEX installation continues to grow as the design of these systems becomes more common.

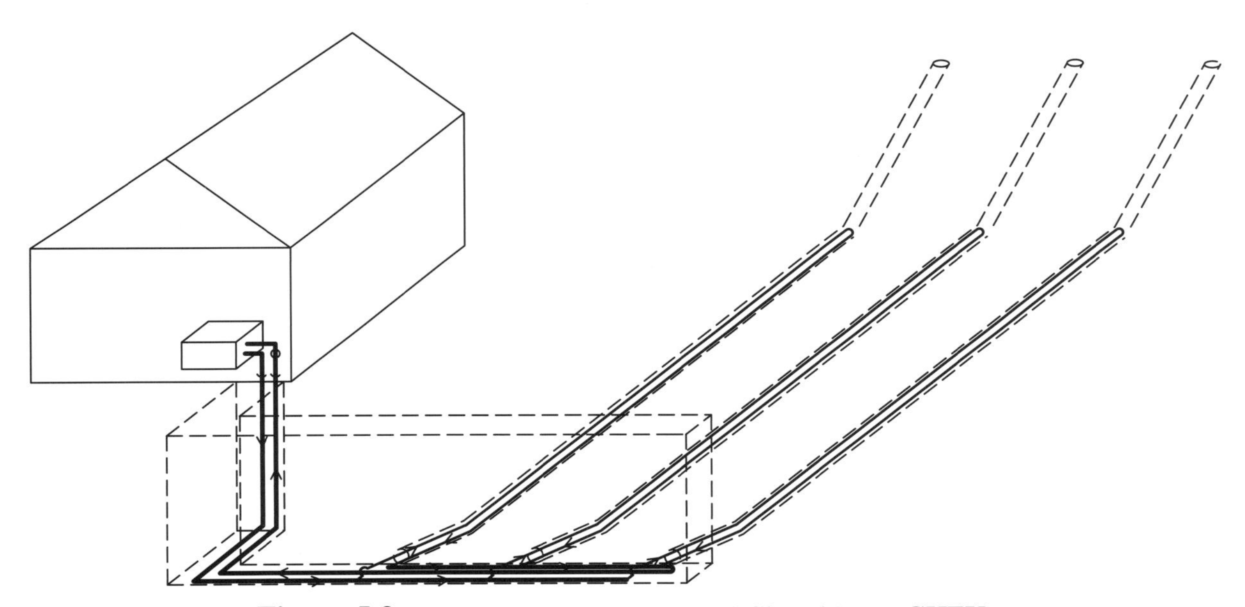

Figure 5.3. Horizontally-bored "3-ton" Closed-Loop GHEX.

5.1.3 LAYOUT The layout of the GHEX consists of individual loops (vertical boreholes, horizontal boreholes, horizontal trenches or individual pond loops) in parallel that act as primary heat exchange devices connected to supply and return lines via a headering system. The ideal GHEX configuration will be arranged such that all flow paths are of equal length (+/- 10%) and the piping is sized to promote proper heat transfer on the internal diameter of the pipe (turbulence) without excessive head loss for the pumping system to overcome. The individual parallel loops will be connected together in a manner that the circulating fluid will flow equally to each loop and that the piping system can be easily flushed and purged of air and debris using appropriately sized flushing equipment.

5.1.3.1 FLOW PATHS AND PIPE SIZE SELECTION Proper layout of a parallel-flow GHEX results in enough flow in the individual ground loops to create the turbulence required to provide proper heat transfer without producing excessive head loss that will consume unnecessary pumping power. That is, a minimum flow rate through each GHEX flow path is required for turbulent flow, and a maximum flow rate exists where the head loss becomes excessive (approximately 4 feet of head loss per 100 feet of pipe). For a given pipe size, the acceptable flow rate range depends on the type and temperature of the circulating fluid. Minimum flow rates for turbulence in DR-11 HDPE pipe are provided in Table 4.10 for water at 40 F and in Tables 4.12 through 4.14 for propylene glycol, methanol, and ethanol solutions, respectively, at various concentrations and temperatures. Maximum flow rates, resulting in 4 feet of head loss per 100 feet of pipe, can be estimated from Table 4.15 for water at 40 F and for 20% propylene glycol, 12.5% methanol, and 12.5% ethanol at 25 F. These percentages provide similar freeze points of 18.7 F, 16.2 F, and 18.5 F, respectively, for the three antifreeze solutions and represent the appropriate freeze protection level of about 10 F below the average circulating fluid temperature for a design minimum entering water temperature of 30 F, where the lowest temperature in the GHEX (where the fluid is most viscous) would be about 25 F.

Figure 5.4 summarizes the minimum and maximum flow rates for the four circulating fluids discussed above. Water allows the largest flow rate range for all pipe sizes, followed by methanol, propylene glycol, and ethanol. For all circulating fluids, ¾-inch DR-11 provides the narrowest acceptable flow rate range, with ethanol having virtually no difference between minimum and maximum acceptable flow rates. Minimum flow rates for turbulence and maximum flow rates for head loss can be determined for other antifreeze solutions and design temperatures from the data in Chapter 4.

Figure 5.4 provides a starting point for pipe size selection and the number of flow paths in the GHEX. In residential and light commercial design, the general rule is to design for one flow path per nominal ton of heat pump capacity. Because 1 ton of heat pump capacity requires approximately 3 gpm of flow, one flow path would accommodate that same 3 gpm of flow. Looking at Figure 5.4, only the flow rate range associated with ¾-inch pipe (approximately 1.00 - 4.75 for water, 2.25 - 4.00 for methanol, 3.25 - 3.75 for propylene glycol, and 3.65 for ethanol) would meet both the minimum and maximum flow rate constraints for all circulating fluids (very close for ethanol). To achieve turbulence at 3 gpm, both water and methanol would work for 1-inch pipe and only water for 1-1/4-inch pipe. In vertically and horizontally-bored GHEX systems, depending on drilling conditions and capabilities, the design bore length may frequently represent more than 1 nominal ton of heat pump capacity (Figure 5.5), and the required flow rate through each ground loop may require 1-inch or 1-1/4-inch pipe to control head loss. Looking at Figure 5.4, the 1-inch pipe should accommodate between 4.5 and 6 gpm for most circulating fluids and the 1-1/4-inch pipe between 6 and 12 gpm, representing 1.5 to 2.0 tons and 2.0 to 4.0 tons of heat pump capacity for the 1-inch and 1-1/4-inch pipes, respectively. Looking at Figure 5.5 and considering constant soil or rock thermal properties with depth, a 3-ton GHEX may be comprised of three bores to a given design depth (D_3) using ¾-inch pipe, two bores to a depth D_2 approximately 1.5 times deeper than D_3 using 1-inch pipe, or one bore approximately 3 times deeper ($D_1 \approx 3D_3$) using 1-1/4-inch pipe. Of course, if the soil or rock thermal properties change with depth, the relative design lengths of the three configurations in Figure 5.5 will change.

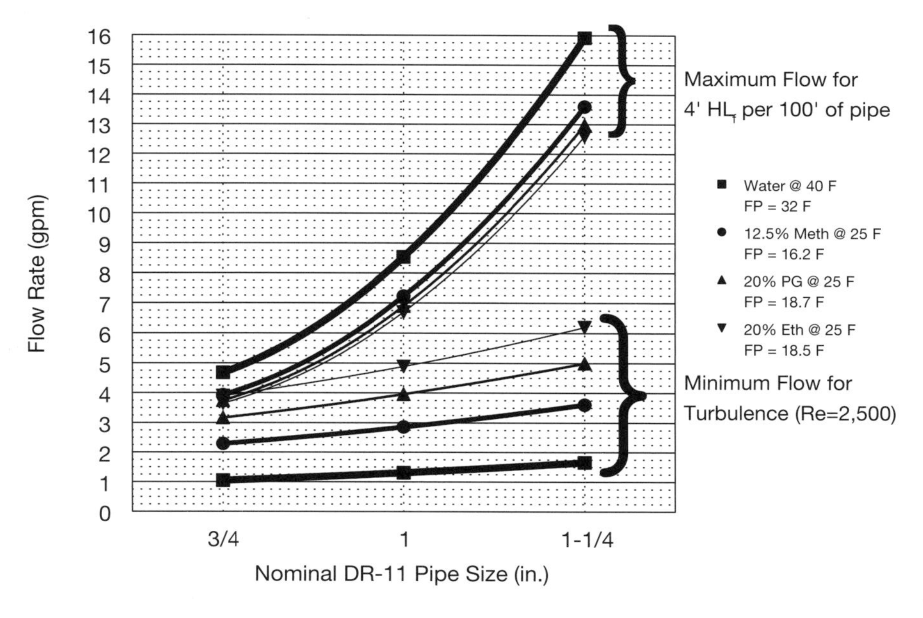

Figure 5.4. Flow rate range bands for selected circulating fluids.

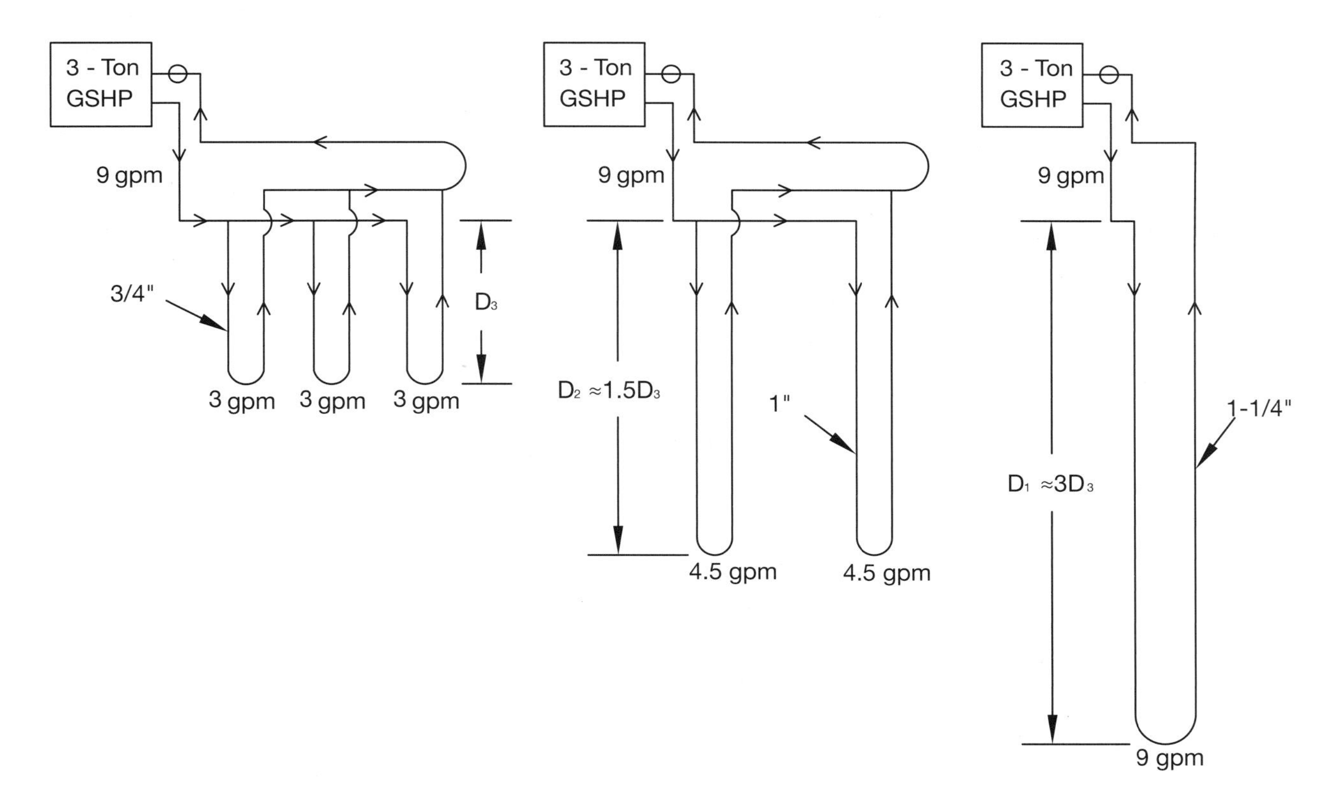

Figure 5.5. Optional ground loop layouts for a nominal "3-ton" GSHP.

The selection of pipe size for the ground loops is a balance between the flow constraints discussed previously and the design lengths that result from applying the appropriate design equations that are presented in Sections 5.2 through 5.4. The following pipe size selection and GHEX layout procedure is recommended to aid in the efficient design of the GHEX.

1. Divide the total system flow rate by the estimated flow rate per flow path depending on ground loop pipe size.
2. For the minimum design EWT, check for turbulence in the ground loop pipe.
3. For the minimum design EWT, check that head loss is approximately 4 feet per 100 feet of pipe or less.
4. Calculate the ground loop lengths and determine if the GHEX will fit properly on the site or if drilling depths are manageable.
5. If Step 4 is satisfied, then design is workable. Otherwise, rearrange the loopfield layout or consider a different ground loop pipe size and go back to Step 1.

5.1.3.2 HEADERING FOR FLUSHING AND PURGING The header is the portion of the piping system where the smaller diameter parallel-flow ground loops are connected to the larger diameter supply and return piping. The header is generally classified according to location and type. The header can be located inside or outside of the building, depending on the preferences of the installer and/or the building owner. If the header is located inside of the building, each individual ground loop's "pigtails" must have sufficient length to be run in a trench all the way back to a location inside of the building, where they are connected together into a single flow path to and from the pumping system. If the header is located in a header trench outside of the building, each individual ground loop's "pigtails" may be very short and connected to the headers in the trench and larger supply and return piping carries flow to and from the pumping system. Some of the advantages and disadvantages of each header location are listed below.

INSIDE OF BUILDING HEADER

Advantages

- Work in a controlled environment
- Fusion equipment not mandatory
- Able to isolate individual loops for flushing/purging using valves
- Easy to find and isolate problem loops

Disadvantages

- Long supply/return lines to/from each ground loop which limits distance between building and ground loop system
- Large wall penetration or several small ones
- Inside piping is cluttered and occupies more space
- More difficult to insulate inside piping system

OUTSIDE OF BUILDING HEADER

Advantages

- Larger supply/return lines to/from the ground loop system allows greater distance between building and ground loop system
- Two (or more) small wall penetrations for larger supply/return lines
- Inside piping is neater and simpler
- Inside piping is easier to insulate

Disadvantages

- Work in outdoor conditions that may be difficult and dirty
- Heat fusion must be done to connect supply/return lines to ground loops
- Time and space limitations relative to the header trench
- More difficult to find and isolate problem loops

The type of header refers to the configuration of the header relative to the location of the ground loop boreholes or trench ends. If all of the individual ground loop "pigtails" can be brought back to a central point without excessive run distances for one or more of the loops, then a close header arrangement (Figure 5.6) is preferred. The close header is most common for ground loop systems with six or less flow paths where the boreholes or trench ends are grouped closely and the individual ground loop "pigtails" can be brought back to a central location in an orderly manner. The close header is the only type used for inside the building headers. Examples of several pre-manufactured close header configurations are provided in Table 5.1. The close header is classified according to the supply-return line size (inlet) and the number and size of ground loop connections or flow paths (outlets). The selection of the appropriate close header depends on the number and size of individual ground loops in the ground loop systems, the size of supply and return lines that will be utilized, and the length of the supply and return lines (relative to head loss in the piping system).

If all of the individual ground loop "pigtails" cannot be brought back to a central point without excessive run distances, then a Step-Down Step-Up Reverse-Return Header (SDSURR) arrangement (Figure 5.7) is preferred. The SDSURR header is only used with outside the building headers and is most commonly used for ground loop systems with a higher number of flow paths, where the boreholes or trenches are spaced such that it would be difficult to get all of the individual ground loop "pigtails" back to a central location without excessively long runs. This header arrangement is used on commercial systems because it allows for orderly connection of all of the ground loops to the supply/return piping system and promotes equal flow to all ground loops due to its reverse-return construction. The SDSURR header is constructed using a combination of tees and elbows (Table 5.2) along with pipe sizes that depend on the flushing flow rates required to flush the ground loops at various locations along the header flow path. Several headers are presented in Figure 5.8 (a)-(h) for ¾-inch ground loops, in Figure 5.9 (a)-(f) for 1-inch ground loops, and in Figure 5.10 (a)-(c) for 1-1/4-inch ground loops. The flushing flow rates (identified in the boxes) and the tee and elbow fittings (identified in the circles) are included. Pipe sizes that can be flushed—by the flushing flow rate that would occur if the minimum flushing flow rate (Table 4.21) were provided to each individual ground loop—are identified for each section of pipe in the supply side header. Note the supply and return portions of the header are mirror images of one another.

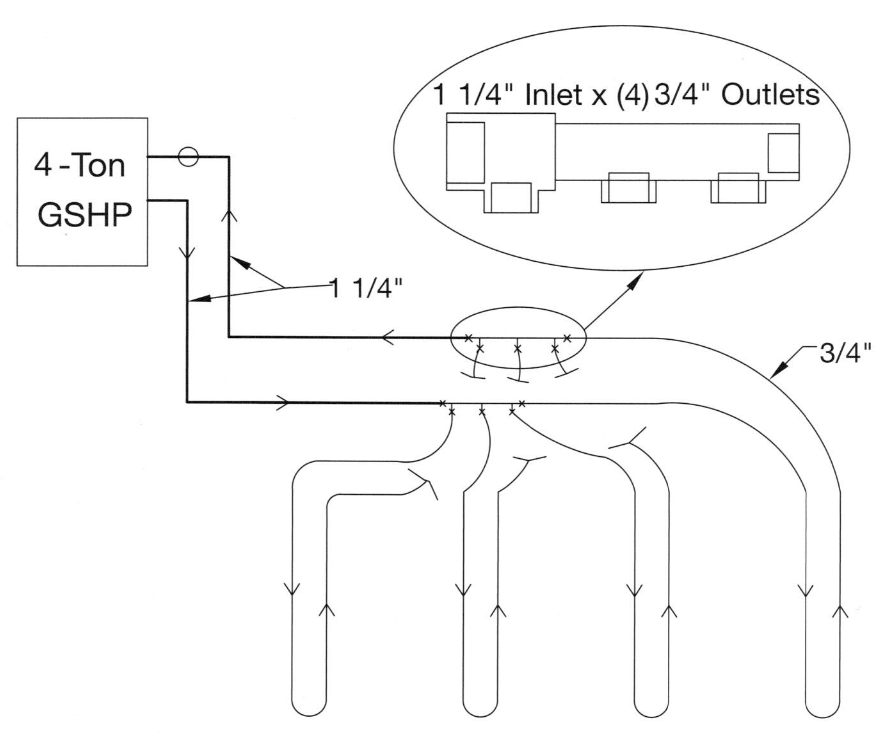

Figure 5.6. Nominal "4-ton" Close Header layout.

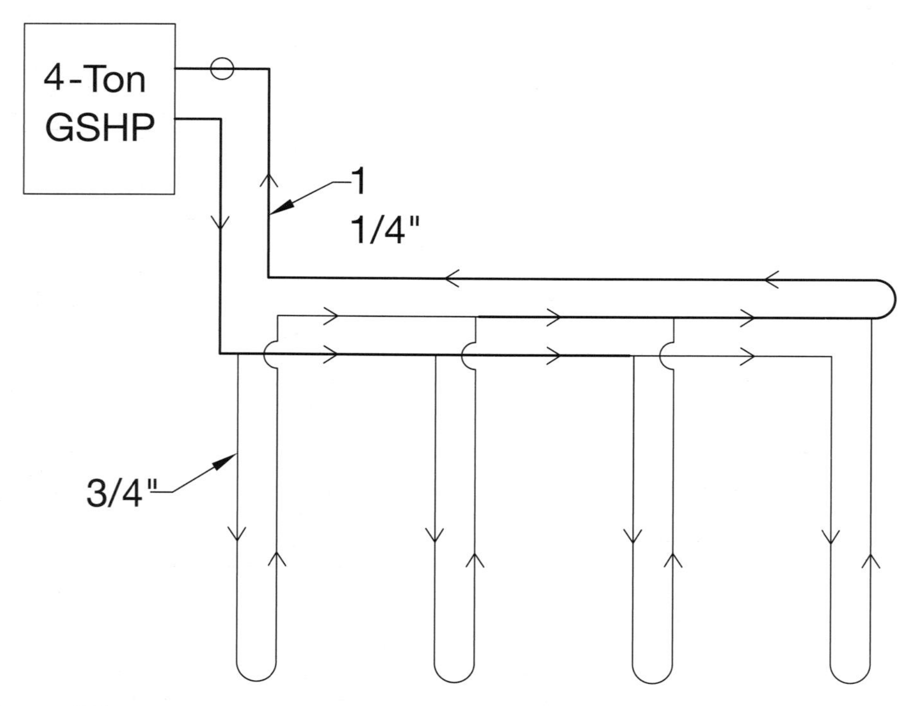

Figure 5.7. Nominal "4-ton" Step-Down Step-Up Reverse-Return Header layout.

Table 5.1. Common Pre-Manufactured Close Header Configurations.

Header	Supply/Return Size	Ground Loop Size	# Ground Loops[1]	Description
H1	1-1/4"	¾"	3	1-1/4" Inlet x (3) ¾" Outlets
H2	1-1/4"	¾"	4	1-1/4" Inlet x (4) ¾" Outlets
H3	1-1/4"	¾"	5	1-1/4" Inlet x (5) ¾" Outlets
H4	1-1/4"	¾"	6	1-1/4" Inlet x (6) ¾" Outlets
H5	1-1/4"	¾"	7	1-1/4" Inlet x (7) ¾" Outlets
H6	2"	¾"	8	2" Inlet x (8) ¾" Outlets
H7	2"	¾"	9	2" Inlet x (9) ¾" Outlets
H8	2"	¾"	10	2" Inlet x (10) ¾" Outlets
H9	1-1/4"	1"	3	1-1/4" Inlet x (3) 1" Outlets
H10	1-1/4"	1"	4	1-1/4" Inlet x (4) 1" Outlets
H11	1-1/4"	1"	5	1-1/4" Inlet x (5) 1" Outlets
H12	1-1/4"	1"	6	1-1/4" Inlet x (6) 1" Outlets
H13	1-1/4"	1"	7	1-1/4" Inlet x (7) 1" Outlets
H14	2"	1"	8	2" Inlet x (8) 1" Outlets
H15	2"	1"	9	2" Inlet x (9) 1" Outlets
H16	2"	1"	10	2" Inlet x (10) 1" Outlets
H17	2"	1-1/4"	3	2" Inlet x (3) 1-1/4" Outlets
H18	2"	1-1/4"	4	2" Inlet x (4) 1-1/4" Outlets

1. The number of ground loops = the number of flow paths to and from the headers.

Table 5.2. Common Pre-Manufactured Fusion Elbows and Tees.

Fusion Elbow		Fusion Tee	
Label	Description	Label	Description[1]
E1	¾" x ¾"	T1	¾" x ¾" x ¾"
E2	1" x ¾"	T2	1" x 1" x ¾"
E3	1" x 1"	T3	1" x 1" x 1"
E4	1-1/4" x ¾"	T4	1-1/4" x ¾" x ¾"
E5	1-1/4" x 1"	T5	1-1/4" x 1" x 1"
E6	1-1/4" x 1-1/4"	T6	1-1/4" x 1-1/4" x ¾"
E7	1-1/2" x 1-1/2"	T7	1-1/4" x 1-1/4" x 1"
E8	2" x 1-1/4"	T8	1-1/4" x 1-1/4" x 1-1/4"
E9	2" x 2"	T9	1-1/2" x ¾" x ¾"
		T10	1-1/2" x 1" x 1"
		T11	1-1/2" x 1-1/4" x ¾"
		T12	1-1/2" x 1-1/4" x 1"
		T13	1-1/2" x 1-1/4" x 1-1/4"
		T14	1-1/2" x 1-1/2" x ¾"
		T15	1-1/2" x 1-1/2" x 1"
		T16	1-1/2" x 1-1/2" x 1-1/4"
		T17	1-1/2" x 1-1/2" x 1-1/2"
		T18	2" x 1-1/4" x ¾"
		T19	2" x 1-1/4" x 1"
		T20	2" x 1-1/4" x 1-1/4"
		T21	2" x 1-1/2" x ¾"
		T22	2" x 1-1/2" x 1"
		T23	2" x 1-1/2" x 1-1/4"
		T24	2" x 1-1/2" x 1-1/2"
		T25	2" x 2" x ¾"
		T26	2" x 2" x 1"
		T27	2" x 2" x 1-1/4"
		T28	2" x 2" x 1-1/2"
		T29	2" x 2" x 2"

1. Last dimension in the listing is the Tee outlet size.

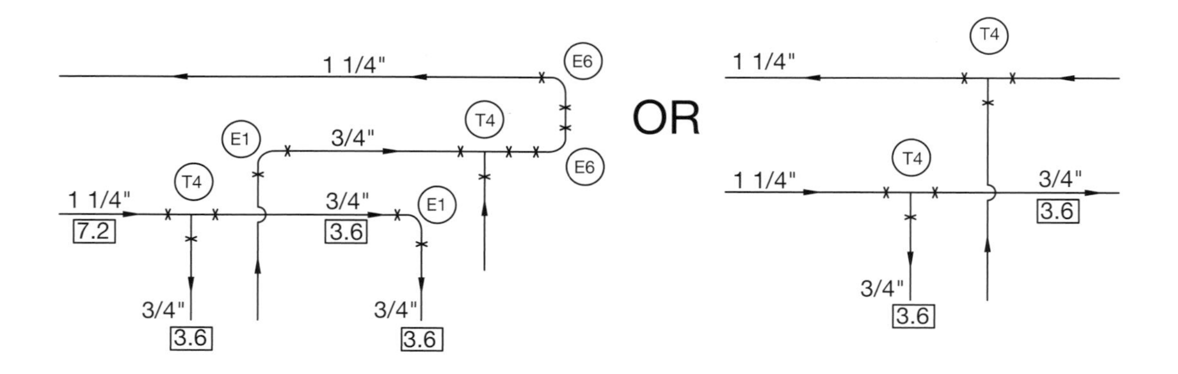

(a) "2 Ton" Step-Down Step-Up Reverse-Return Header for 3/4" Ground Loops

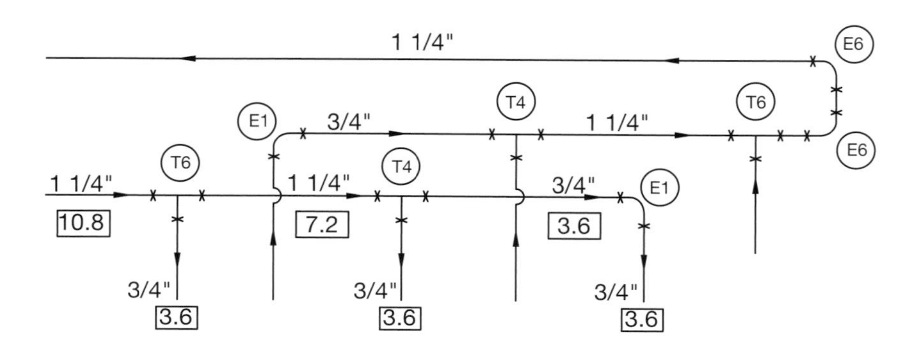

(b) "3 Ton" Step-Down Step-Up Reverse-Return Header for 3/4" Ground Loops

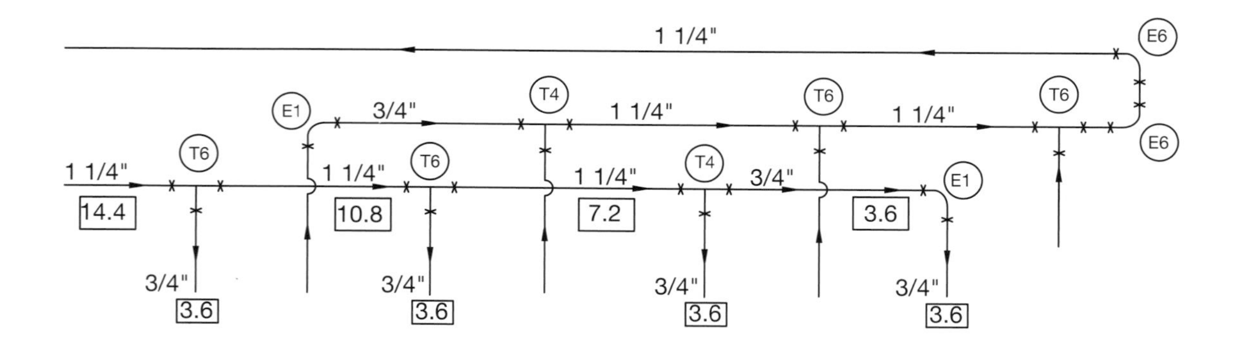

(c) "4 Ton" Step-Down Step-Up Reverse-Return Header for 3/4" Ground Loops

Figure 5.8. Step-Down Step-Up Reverse-Return Headers for ¾-inch Ground Loops.

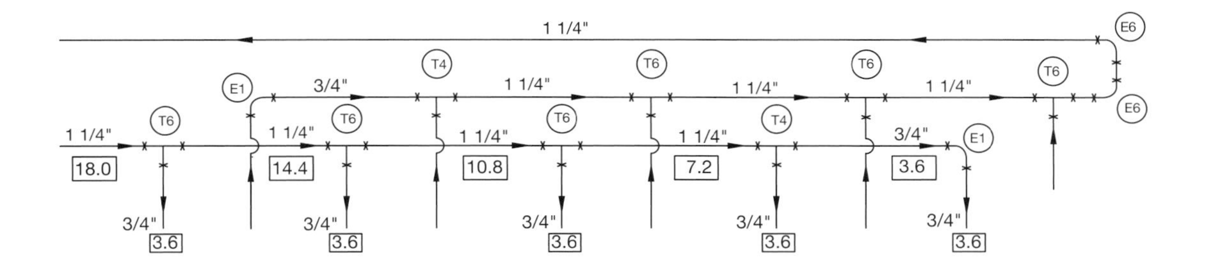

(d) "5 Ton" Step-Down Step-Up Reverse-Return Header for 3/4" Ground Loops

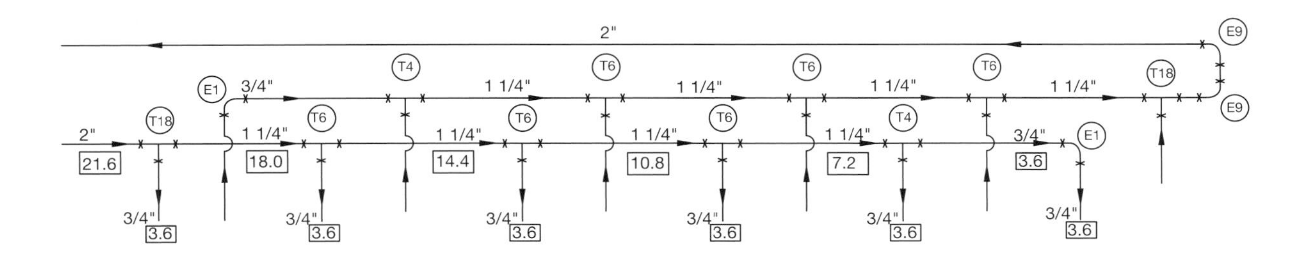

(e) "6 Ton" Step-Down Step-Up Reverse-Return Header for 3/4" Ground Loops

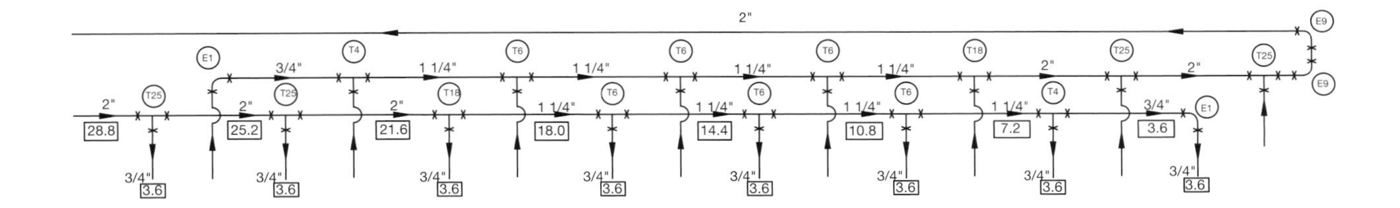

(f) "8 Ton" Step-Down Step-Up Reverse-Return Header for 3/4" Ground Loops

Figure 5.8. Step-Down Step-Up Reverse-Return Headers for ¾-inch Ground Loops. (Cont.)

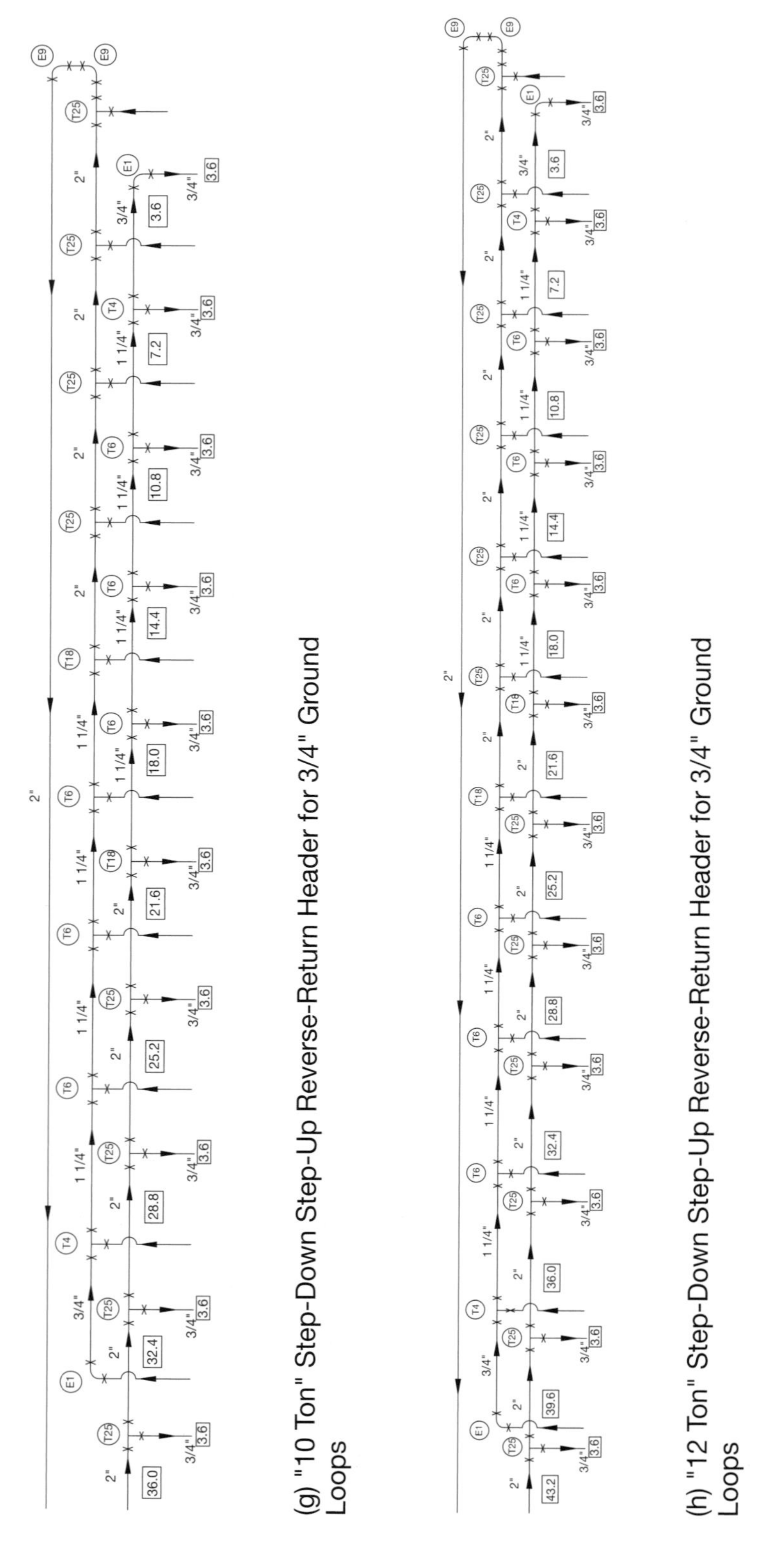

Figure 5.8. Step-Down Step-Up Reverse-Return Headers for ¾-inch Ground Loops. (Cont.)

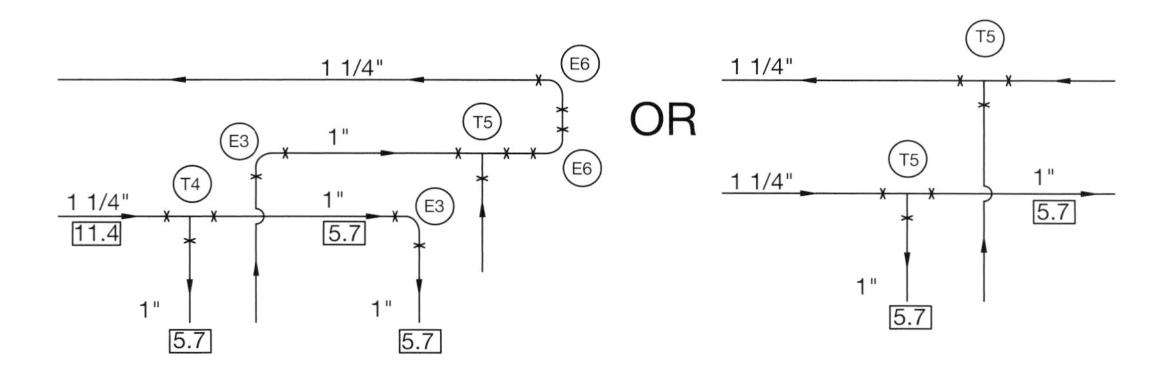

(a) "2.5 - 3.5 Ton" Step-Down Step-Up Reverse-Return Header for 1" Ground Loops

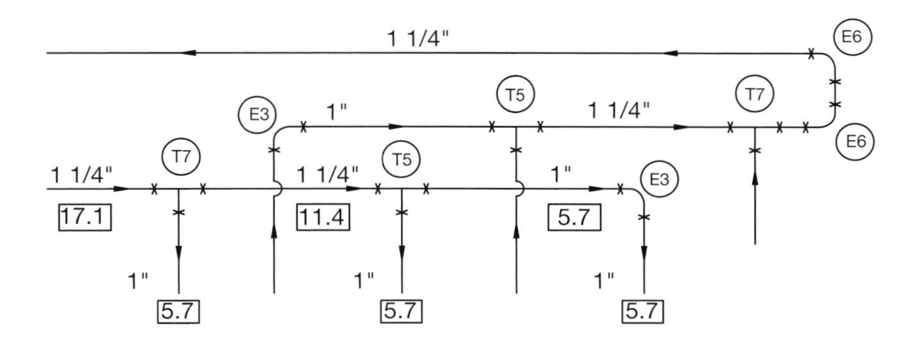

(b) "4-5 Ton" Step-Down Step-Up Reverse-Return Header for 1" Ground Loops

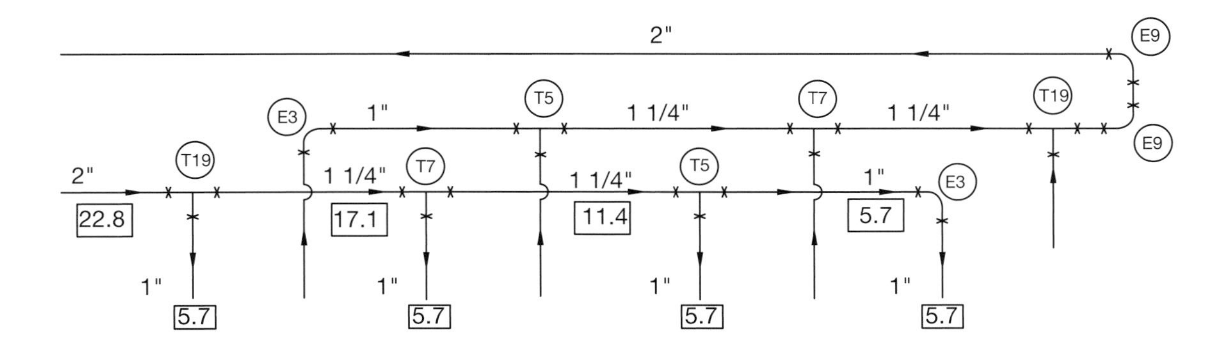

(c) "6-7 Ton" Step-Down Step-Up Reverse-Return Header for 1" Ground Loops

Figure 5.9. Step-Down Step-Up Reverse-Return Headers for 1-inch Ground Loops.

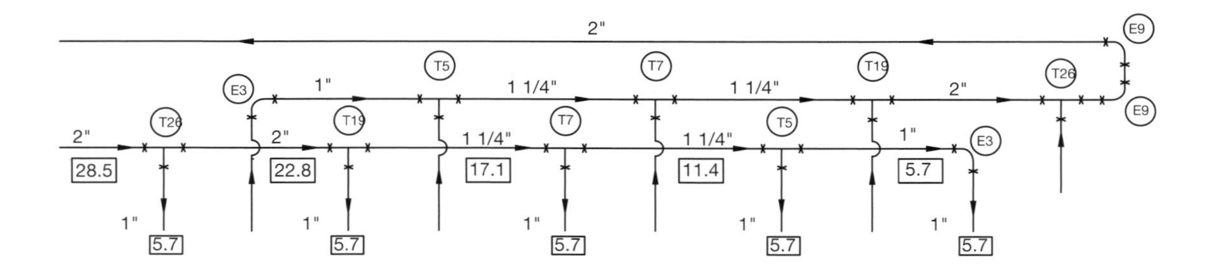

(d) "8-9 Ton" Step-Down Step-Up Reverse-Return Header for 1" Ground Loops

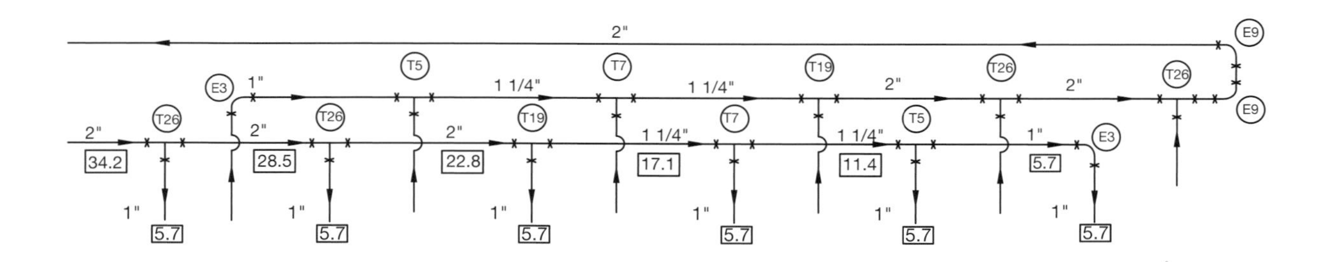

(e) "10-11 Ton" Step-Down Step-Up Reverse-Return Header for 1" Ground Loops

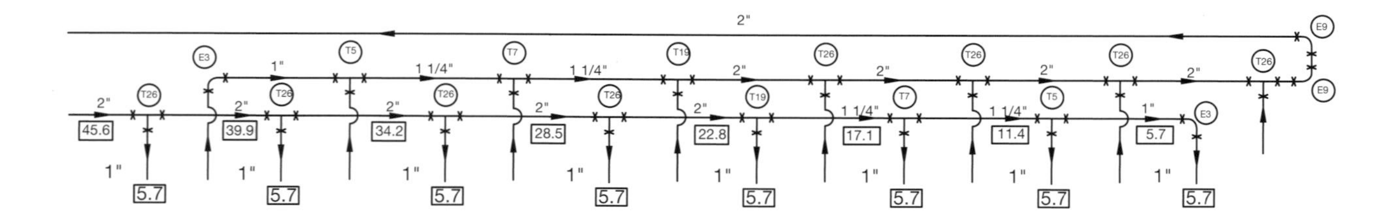

(f) "12 Ton" Step-Down Step-Up Reverse-Return Header for 1" Ground Loops

Figure 5.9. Step-Down Step-Up Reverse-Return Headers for 1-inch Ground Loops. (Cont.)

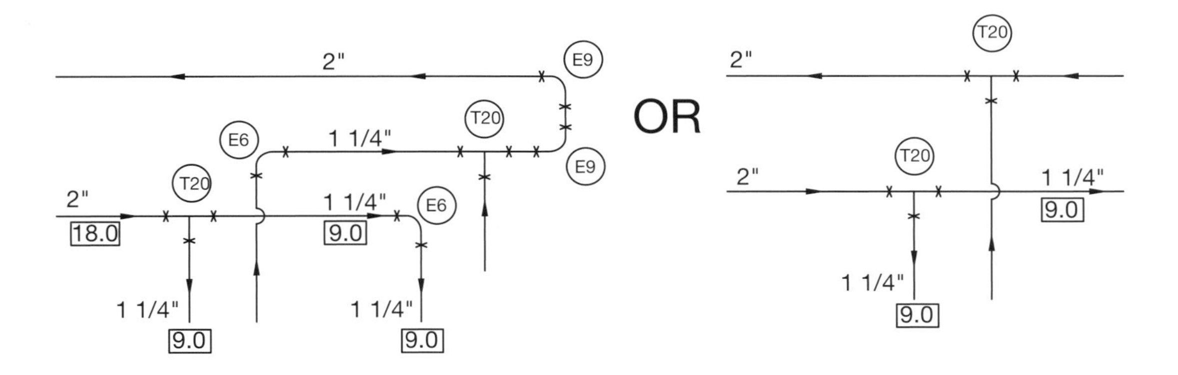

(a) "5-7 Ton" Step-Down Step-Up Reverse-Return Header for 1 1/4" Ground Loops

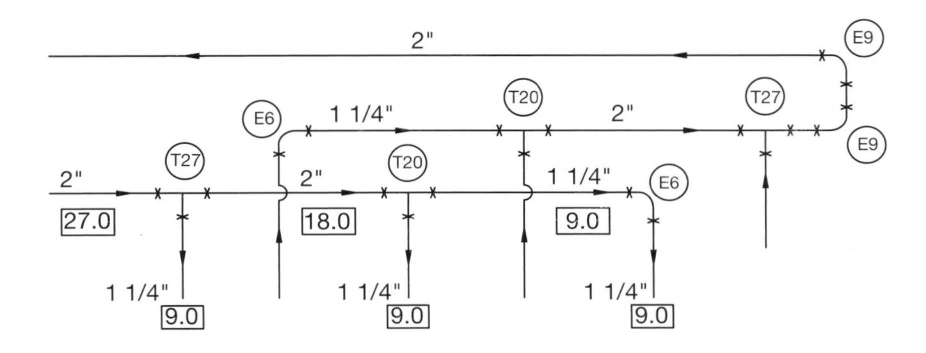

(b) "8-10 Ton" Step-Down Step-Up Reverse-Return Header for 1 1/4" Ground Loops

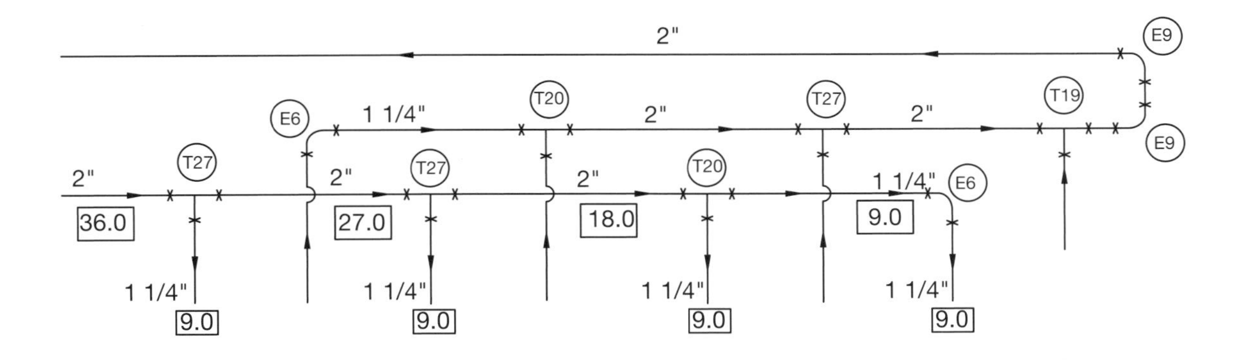

(c) "11-13 Ton" Step-Down Step-Up Reverse-Return Header for 1 1/4" Ground Loops

Figure 5.10. Step-Down Step-Up Reverse-Return Headers for 1-1/4-inch Ground Loops.

Whether a Close Header or a Step-Down Step-Up Reverse-Return Header is used for a specific ground loop, the actual number of ground loops or flow paths depends on:

- the flow rate of the system
- the pipe size selected for the ground loops
- the minimum flow rate for turbulence for the pipe size selected
- the maximum flow rate for head loss for the pipe size selected, and
- the required length of pipe in the trench or borehole to meet the required heating or cooling ground load of the heat pump system

There is no exact header configuration that depends solely on heat pump size or flow rate. The type of ground loop and the type of soil/rock in which the ground loop will be located must also be taken into account. In general, when a horizontally-trenched ground loop is buried in an unconsolidated formation (silt, sand, clay, glacial till, etc., with thermal conductivities less than about 1.2 Btu/hr ft F) the required length of pipe to accommodate about 1 nominal ton (3 gpm) of heat pump capacity is relatively long and an individual ground loop may contain 500 feet or more of pipe. In a heating dominated design where antifreeze is utilized, ¾-inch ground loop piping is commonly used and each ground loop represents about 1 ton of heat pump capacity. Horizontally-bored ground loops are always placed in unconsolidated formations, and the pipe size will depend on the bore length, with ¾-inch used for bores up to 250 feet long (1 ton), 1-inch used for bores up to 400 feet long (approximately 2 tons) and 1-1/4-inch pipe used for bores up to 600 feet long (approximately 3 tons). Vertically-bored ground loops, for which ¾-inch pipe is used, that are placed in unconsolidated formations generally require between 175 and 250 feet of borehole per nominal ton, depending on the grout thermal conductivity. For vertically-bored ground loops into consolidated formations (shale, sandstone, limestone, granite, etc., with thermal conductivities greater than 1.2 Btu/hr ft F) the required length of pipe to accommodate 1 ton (3 gpm) of heat pump capacity is much less than in unconsolidated formations. In addition, drillers tend to want to drill deeper into consolidated formations because of the efficiencies associated with drilling time versus setup time, and one borehole many times represents more than 1 ton of heat pump capacity. A borehole capacity of 1-1/2 to 2 ton requires 1-inch pipe to accommodate 4.5 to 6 gpm and a borehole capacity of 2 to 4 tons requires 1-1/4-inch pipe to accommodate 6 to 12 gpm.

Tables 5.3, 5.4, and 5.5 provide suggested header configurations for ¾-inch, 1-inch, and 1-1/4-inch ground loop pipe sizes, based on formation type, for nominal heat pump capacities ranging from 1 to 12 tons. Three-quarter-inch ground loops (Table 5.3) are normally used for horizontally-trenched GHEXs, with or without antifreeze, and vertically or horizontally-bored GHEXs in unconsolidated formations, where drilling lengths are limited to about 250 feet (500 foot flow path lengths). One-inch ground loops (Table 5.4) are sometimes used in horizontally-trenched GHEXs in cooling climates, in horizontally-bored GHEXs in unconsolidated formations up to 400 feet, or vertically-bored GHEXs in consolidated formations to a depth of up to about 400 feet (800 foot flow path lengths). One and 1-1/4-inch ground loops (Figure 5.4) are most often used in horizontally-bored GHEXs in unconsolidated formations to lengths of about 600 feet and vertically-bored GHEXs in consolidated formations to depths of about 600 feet (1,200 foot flow path lengths).

Table 5.3. Suggested Header Configurations for ¾-inch Ground Loops – Unconsolidated Formations.

		Horizontally-trenched (w/ or w/o antifreeze) Horizontally-bored (250' max) Vertical (175-250'/Ton-250' max)			
Nom HP Cap. (Ton)	Nom GPM (GPM)	# of GL's (GPM/3)	Close Header Fitting[2] (S/R)	Alt. Close Header (S/R)	SDSURR[3]
1	3	1	None		None
1-1/2	4.5	1.5 (2)[1]	T13 (1-1/4")		Fig. 5.8(a)
2	6	2	T13 (1-1/4")		Fig. 5.8(a)
2-1/2	7.5	2.5 (3)[1]	H1 (1-1/4")		Fig. 5.8(b)
3	9	3	H1 (1-1/4")		Fig. 5.8(b)
3-1/2	10.5	3.5 (4)[1]	H2 (1-1/4")		Fig. 5.8(c)
4	12	4	H2 (1-1/4")		Fig. 5.8(c)
5	15	5	H3 (1-1/4")		Fig. 5.8(d)
6	18	6	H4 (2")	2 x H1 (1-1/4")	Fig. 5.8(e)
8	24	8	H6 (2")	2 x H2 (1-1/4")	Fig. 5.8(f)
10	30	10	H8 (2")	2 x H3 (1-1/4")	Fig. 5.8(g)
12	36	12		2 x H4 (2")	Fig. 5.8(h)

1. Number of ground loops rounded up to accommodate flow without excessive head loss. Design lengths per ground loop will be shorter than normal, but design flow rate may need to be increased to ensure turbulence during heating design conditions, especially if antifreeze will be used. Rounding down would result in long design lengths per ground loop and excessive head loss in each.
2. Tees are listed in Table 5.2 and Close Headers are listed in Table 5.1.
3. Tees and elbows used in figures are listed in Table 5.2.

Table 5.4. Suggested Header Configurations for 1-inch Ground Loops.

		Horizontally-trenched (w/o antifreeze) in cooling climate Horizontally-bored in unconsolidated formation (400' max)			
Nom HP Size (Ton)	Nom GPM (GPM)	# of GL's (GPM/4.5)[1]	Close Header Fitting[2] (S/R)	Alt. Close Header (S/R)	SDSURR[3]
1	3	0.67 (1)	None		None
1-1/2	4.5	1	None		None
2	6	1.33 (1)	None		None
2-1/2	7.5	1.67 (2)	T14 (1-1/4")		Fig. 5.9(a)
3	9	2	T14 (1-1/4")		Fig. 5.9(a)
3-1/2	10.5	2.33 (2)	T14 (1-1/4")		Fig. 5.9(a)
4	12	2.67 (3)	H9 (1-1/4")		Fig. 5.9(b)
5	15	3.33 (3)	H9 (1-1/4")		Fig. 5.9(b)
6	18	4	H10 (2")	2 x T14 (1-1/4")	Fig. 5.9(c)
8	24	5.33 (5)	H11 (2")		Fig. 5.9(d)
10	30	6.67 (6)	H12 (2")	2 x H9 (1-1/4")	Fig. 5.9(e)
12	36	8	H14 (2")	2 x H10 (1-1/4")	Fig. 5.9(f)

		Vertical in consolidated formation (125-175'/Ton - 400' max)			
Nom HP Size (Ton)	Nom GPM (GPM)	# of GL's (GPM/6)[1]	Close Header Fitting[2] (S/R)	Alt. Close Header (S/R)	SDSURR[3]
1	3	0.5 (1)	None		None
1-1/2	4.5	0.75 (1)	None		None
2	6	1	None		None
2-1/2	7.5	1.25 (2)	T14 (1-1/4")		Fig. 5.9(a)
3	9	1.5 (2)	T14 (1-1/4")		Fig. 5.9(a)
3-1/2	10.5	1.75 (2)	T14 (1-1/4")		Fig. 5.9(a)
4	12	2	T14 (1-1/4")		Fig. 5.9(a)
5	15	2.5 (3)	H9 (1-1/4")		Fig. 5.9(b)
6	18	3	H9 (2")		Fig. 5.9(b)
8	24	4	H10 (2")	2 x T14 (1-1/4")	Fig. 5.9(c)
10	30	5	H11 (2")		Fig. 5.9(d)
12	36	6	H12 (2")	2 x H9 (1-1/4")	Fig. 5.9(e)

1. Acceptable flow rate range for 1-inch is between 4.5 and 6 GPM, with 4.5 GPM representing 1-1/2 tons of nom. heat pump capacity and 6 GPM representing 2 tons of nom. heat pump capacity.
2. Tees are listed in Table 5.2 and Close Headers are listed in Table 5.1.

Table 5.5. Suggested Header Configurations for 1-1/4-inch Ground Loops.

		Horizontally-bored in unconsolidated formation (600' max)			
Nom HP Size (Ton)	Nom GPM (GPM)	# of GL's (GPM/9)[1]	Close Header Fitting[2] (S/R)	Alt. Close Header (S/R)	SDSURR[3]
1	3				
1-1/2	4.5				
2	6				
2-1/2	7.5	0.83 (1)	None		
3	9	1	None		
3-1/2	10.5	1.17 (1)	None		
4	12	1.33 (2)	T17 (1-1/4")		Fig. 5.10(a)
5	15	1.67 (2)	T17 (1-1/4")		Fig. 5.10(a)
6	18	2	T31 (2")		Fig. 5.10(a)
8	24	2.67 (3)	H17 (2")		Fig. 5.10(b)
10	30	3.33 (3)	H17 (2")		Fig. 5.10(b)
12	36	4	H18 (2")		Fig. 5.10(c)
		Vertical in consolidated formation (125-175'/Ton - 600' max).			
Nom HP Size (Ton)	Nom GPM (GPM)	# of GL's (GPM/12)[1]	Close Header Fitting[2] (S/R)	Alt. Close Header (S/R)	SDSURR[3]
1	3				
1-1/2	4.5				
2	6				
2-1/2	7.5	0.63 (1)	None		
3	9	0.75 (1)	None		
3-1/2	10.5	0.88 (1)	None		
4	12	1	None		
5	15	1.25 (2)	T17 (1-1/4")		Fig. 5.10(a)
6	18	1.5 (2)	T31 (2")		Fig. 5.10(a)
8	24	2	T31 (2")		Fig. 5.10(a)
10	30	2.50 (3)	H17 (2")		Fig. 5.10(b)
12	36	3	H17 (2")		Fig. 5.10(b)

1. Acceptable flow rate range for 1-1/4-inch is between 6 and 12 GPM, with 6 GPM representing 2 tons of nom. heat pump capacity and 12 GPM representing 4 tons of nom. heat pump capacity.
2. Tees are listed in Table 5.2 and Close Headers are listed in Table 5.1.
3. Tees and elbows used in figures are listed in Table 5.2.

As indicated earlier each pipe size is associated with a flow range where both turbulence and head loss constraints are satisfied (Figure 5.4), and those flow rates are used to estimate the number of flow paths or ground loops that should be used in the layout of the GHEX. For ¾-inch pipe that flow rate was about 3 gpm (which was used in Table 5.3), for 1-inch pipe that flow rate range was from 4.5 to 6 gpm (4.5 gpm used in Table 5.4) and for 1-1/4-inch pipe that flow rate range was from 6 to 12 gpm (9 gpm used in Table 5.5). Based on a nominal flow rate of 3 gpm per ton of heat pump capacity, the number of ground loops and the recommended Close Header fitting (identified in Tables 5.1 and 5.2) or Reverse-Return Header layout (Figure 5.8 (a)-(h) for ¾-inch, Figure 5.9 (a)-(f) for 1-inch and Figure 5.10 (a)-(c) for 1-1/4-inch) are determined in each table. These header configurations can serve as a starting point for a GHEX layout and design, and the number of flow paths or the size of the supply-return lines may be adjusted depending on ground loop design lengths and head loss considerations.

Example 5.1. Utilize the data for the single-capacity, water-to-air GSHP in Example 3.1.1 to select a ground loop pipe size and header arrangement for the ground loop types and ground formations.

Example 3.1.1: Nominal 4-Ton GSHP, GPM = 12 , EWT_{min} = 30 F, LWT_{min} = 25.5 F

HORIZONTALLY-TRENCHED GROUND LOOP IN UNCONSOLIDATED SOIL: (3/4-INCH GROUND LOOPS)

Table 5.3: Nominal 4-Ton using 4 - ¾-inch Ground Loops
1. Close Header H2 w/ 1-1/4-inch S/Rs
2. SDSURR in Figure 5.8c

HORIZONTALLY-BORED GROUND LOOP IN UNCONSOLIDATED SOIL: (1-INCH GROUND LOOPS)

Table 5.4: Nominal 4-Ton using 3 – 1-inch Ground Loops
1. Close Header H9 w/ 1-1/4-inch S/Rs
2. SDSURR in Figure 5.9b

VERTICAL GROUND LOOP IN CONSOLIDATED FORMATION: (1-INCH GROUND LOOPS)

Table 5.4: Nominal 4-Ton using 2 – 1-inch Ground Loops
1. Close Header T14 w/ 1-1/4-inch S/Rs
2. SDSURR in Figure 5.9a

HORIZONTALLY-BORED GROUND LOOP IN UNCONSOLIDATED SOIL: (1-1/4-INCH GROUND LOOPS)

Table 5.5: Nominal 4-Ton using 2 - 1-1/4-inch Ground Loops
1. Close Header T17 w/ 1-1/4-inch S/Rs
2. SDSURR in Figure 5.10a

VERTICAL GROUND LOOP IN CONSOLIDATED FORMATION: (1-1/4-INCH GROUND LOOPS)

Table 5.5: Nominal 4-Ton using 1 - 1-1/4-inch Ground Loops
1. One ground loop – no header
2. One ground loop – no header

5.2 Vertically-Bored, Closed-Loop GHEX Design

Heating and cooling design equations and data for vertically-bored, closed-loop GHEX design are presented in this section. The design equations are upgrades to the previous IGSHPA design equations that now include the influence of borehole resistance (Grouting Manual) on the required length of borehole to meet the design heating or cooling ground load of the connected ground source heat pump equipment. The design data include deep-earth temperatures and thermal properties of various soil and rock types that will be encountered at drilling depths to 600 or more feet. Grout thermal properties, pipe diameter, borehole diameter, and pipe location in the borehole all influence borehole resistance. Ground loop resistance tables are provided, which combine the effects of ground thermal properties, pipe diameter, borehole diameter, grout thermal conductivity, and run fraction of the heat pump during the design month. The effects of unbalanced ground loads in heating and cooling are addressed utilizing a bore length multiplier to account for long-term cooling or warming of the ground near the vertically-bored ground loops. Vertically-bored GHEX design worksheets for both heating and cooling mode are provided to facilitate the design process and provide documentation of the design for easy record keeping.

The steps in designing the vertically-bored, closed-loop GHEX include:

1. Define the design heating or cooling load and select GSHP equipment (Section 3.4) based on the EWT_{min} or EWT_{max} and GPM that the GHEX will be designed for.
2. Lay out the GHEX according to Section 5.1.3 by selecting pipe sizes for the ground loops that meet turbulence and head loss constraints and using a header arrangement that balances flow between the ground loops and facilitates flushing and purging of the GHEX.
3. Estimate the design heating or cooling month run fraction and the annual heating and cooling ground loads using a bin analysis or graphed data.
4. Determine the deep-earth temperature where the vertically-bored GHEX will be installed.
5. Estimate the thermal properties (thermal conductivity and diffusivity) of the ground formation into which the vertically-bored GHEX will be installed.
6. Select borehole completion information (borehole diameter, U-bend pipe size, and grout thermal conductivity) that will be used in the installation.
7. Determine the ground loop resistance from tabled data for the selected U-bend pipe size accounting for design month run fraction, ground formation thermal conductivity, borehole diameter, and grout thermal conductivity.
8. Calculate the total design heating or cooling borehole length and divide by the number of boreholes to determine the active borehole length.
9. Adjust the heating or cooling design lengths to account for unbalanced ground loads, if necessary.
10. Revise the design, as necessary, to meet potential space limitations on the installation site by balancing the number and spacing of boreholes against drilling depths that can be achieved economically.

The design heating or cooling borehole lengths that result directly from the equations and data presented below are for residential and light commercial GSHP applications with a relatively small number of ground loops spaced such that they minimally interfere with each other. If the heating and cooling ground loads are relatively unbalanced, then borehole length multipliers are provided for one, two, or three-row borefields that account for borehole spacing and adjust borehole lengths for long-term energy depletion or buildup in the soil or rock in and around the vertical ground loopfield. For GSHP applications that require a large number of boreholes in a grid arrangement (not considered in this manual), and have heating and cooling ground loads that are extremely unbalanced, it is highly recommended that the designs lengths obtained with the methods presented in this section be checked against commercial ground loop design software tools (GCHPCalc, GLHEPRO, GLD, etc.) that are designed to account for unbalanced ground loads and the influence of borehole spacing. The equations presented here are not to be used for the design of commercial vertically-bored, closed-loop GHEX systems.

5.2.1 DESIGN EQUATIONS

5.2.1.1 HEATING DESIGN EQUATIONS The total borehole design length for heating can be estimated using Equation 5.1, which represents the active borehole length for heat transfer below the bottom of the header trench. The various parameters in Equation 5.1 include the heat of extraction rate from the ground, which is based on the heating capacity (HC_D) and the coefficient of performance (COP_D) of the heat pump system at design heating conditions, including corrections for operating conditions not directly provided in the engineering specifications (Section 3.4). The design minimum entering (EWT_{min}) and leaving (LWT_{min}) water temperatures are also determined as part of the heat pump selection (Section 3.4). Run fraction in heating mode during the heating design month (F_{Jan}) was discussed in Section 3.6 and is combined with hot water generation run fraction (F_{HWG}), if utilized, for a design month run fraction in heating (F_H). (This is a conservative approach and will ensure that the design length calculated using Equation 5.1 will be long enough to provide for the hot water generation load.) The remaining design parameters (T_G, R_G, and R_B) will be discussed in the following sections.

$$L_{H,T} = \frac{HC_D \cdot \left(\frac{COP_D - 1}{COP_D}\right) \cdot \left(R_B + R_G \cdot F_H\right)}{T_G - \left(\frac{EWT_{min} + LWT_{min}}{2}\right)}$$ **Equation 5.1**

Where:

$L_{H,T}$ = total borehole design length for heating (below header trench), ft
HC_D = heat pump heating capacity at design heating conditions, Btu/hr
COP_D = coefficient of performance at design heating conditions, dimensionless
R_B = borehole thermal resistance, hr ft F/Btu
R_G = steady-state thermal resistance of ground surrounding the borehole, hr ft F/Btu
F_H = run fraction in heating mode during heating design month (Jan.), dimensionless
T_G = average ground temperature along borehole length, F
EWT_{min} = minimum entering water temperature at heating design conditions, F
LWT_{min} = minimum leaving water temperature at heating design conditions, F

The individual borehole design length for heating can be calculated using Equation 5.2.

$$L_{H,B} = \frac{L_{H,T}}{N_B}$$ **Equation 5.2**

Where:

$L_{H,B}$ = individual borehole design length for heating (below header trench), ft
N_B = number of boreholes connected to the heat pump system

5.2.1.2 COOLING DESIGN EQUATIONS The total borehole design length for cooling can be estimated using Equation 5.3, which represents the active borehole length for heat transfer below the bottom of the header trench. The various parameters in Equation 5.3 include the heat of rejection rate to the ground, which is based on the total cooling capacity (TC_D) and the energy efficiency ratio (EER_D) of the heat pump system at design cooling conditions, including corrections for operating conditions not directly provided in the engineering specifications (Section 3.4). The design maximum entering (EWT_{max}) and leaving (LWT_{max}) water temperatures are also determined as part of the heat pump selection (Section 3.4). Run fraction in cooling mode during the cooling design month (F_{Jul}) was discussed in Section 3.6 and represents the design month run fraction in cooling (F_C). The effect of hot water generation on the design month run fraction in cooling has not been accounted for as a conservative measure. (Including that effect would decrease the design month run fraction in cooling, but if no hot water were generated, then the design length calculated using Equation 5.3 would be short.) The remaining design parameters (T_G, R_G, and R_B) will be discussed in the following sections.

$$L_{C,T} = \frac{TC_D \cdot \left(\frac{EER_D + 3.412}{EER_D}\right) \cdot \left(R_B + R_G \cdot F_C\right)}{\left(\frac{EWT_{max} + LWT_{max}}{2}\right) - T_G}$$ **Equation 5.3**

Where:

$L_{C,T}$ = total borehole design length for cooling (below header trench), ft
TC_D = heat pump total cooling capacity at design cooling conditions, Btu/hr
EER_D = energy efficiency ratio at design cooling conditions, dimensionless
R_B = borehole thermal resistance, hr ft F/Btu
R_G = steady-state thermal resistance of ground surrounding the borehole, hr ft F/Btu
F_C = run fraction in cooling mode during cooling design month (July), dimensionless
T_G = average ground temperature along borehole length, F
EWT_{max} = maximum entering water temperature at cooling design conditions, F
LWT_{max} = maximum leaving water temperature at cooling design conditions, F

The individual borehole design length for heating can be calculated using Equation 5.4.

$$L_{C,B} = \frac{L_{C,T}}{N_B}$$ **Equation 5.4**

Where:

$L_{C,B}$ = individual borehole design length for cooling (below header trench), ft
N_B = number of boreholes connected to the heat pump system

5.2.2 DESIGN PARAMETERS

5.2.2.1 DEEP-EARTH TEMPERATURE (T_G) The deep-earth temperature is the average temperature of the soil or rock into which the vertical ground heat exchanger will be installed and is generally constant at depths ranging from 20 to 200 feet. Between the surface and about a 20-foot depth the temperature of the soil varies with location, depth, type of soil, and time of year. This variation has little influence on the design of a vertically-bored GHEX because a very small percentage of the active borehole length is in the approximately 14-16 feet of ground between the 20-foot depth and the bottom of the header trench, which is usually 4-6 feet. These temperature variations are much more of an issue with horizontally-trenched and horizontally-bored GHEX systems and will be discussed in detail in Section 5.3.2.1. At depths greater than 200 feet, the earth temperature begins to increase at a rate that depends on location and the degree of heat flow toward the surface from high temperature geothermal sources deep below the surface. According to maps available from the *Southern Methodist University Geothermal Lab (2009)*, the deep earth temperature can increase from a low of about 0.25 F per 100 feet to a high of about 4.5 F per 100 feet. This is only an issue if the vertical GHEX is going to be installed to extreme depths in those areas of high thermal gradients, most of which exist in the northwest part of the United States where near-surface high temperature geothermal is common.

A good first estimate of deep-earth temperature can be obtained from Figure 5.11, which shows the approximate groundwater temperatures in the contiguous 48 states and can be considered accurate to within +/- 4 F, depending on location. Local water well logs are a very good source of temperature information, if recorded by the well driller and reported to the state geological survey. Contact your state geological survey to determine if state groundwater temperature maps are available. The deep-earth temperature can also be estimated to be 2 to 5 F above the local average annual air temperature (Figure 5.12), which can also be estimated from the bin data for a given location. The best source of deep-earth temperature is the direct measurement of water temperature obtained from a local water well. The temperature of the water from a well that is delivered via underground piping and through a conditioned space may not be a good estimate of the actual deep-earth temperature because the ground temperatures around the buried piping or the ambient temperatures surrounding the distribution piping may have a significant effect on the delivery temperature of the water at the faucet.

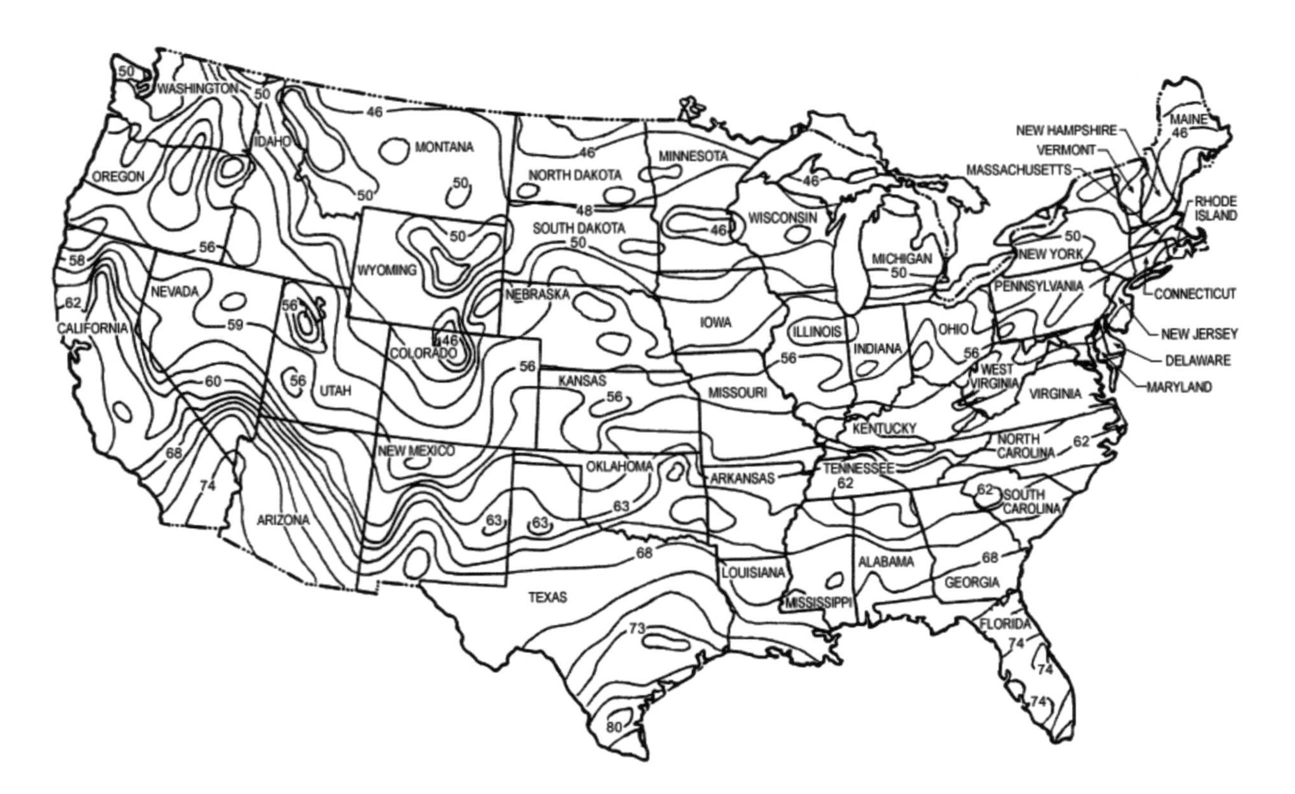

Figure 5.11. Approximate Groundwater Temperatures (F) *(NWWA, 1985)*

MEAN ANNUAL AIR TEMPERATURE (°F)

BASED ON NORMAL PERIOD 1961-1990

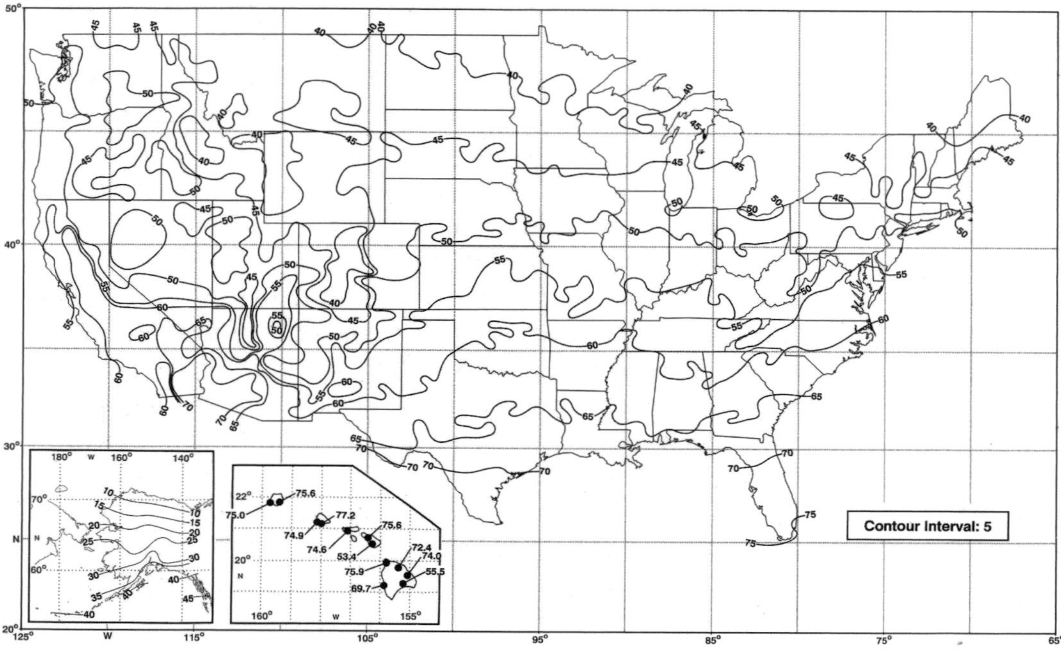

Figure 5.12. Mean Annual Air Temperature (F) *(NOAA, 2009)*

5.2.2.2 THERMAL PROPERTIES OF THE GROUND AND GROUND THERMAL RESISTANCE (R_G)

There are three thermal properties of soils and rocks that are important in the design of the closed-loop, vertically-bored GHEX. Thermal conductivity (k) is the property of a material that defines the rate at which heat energy is transported through a material due to a temperature difference across the material and is measured in units of Btu/hr ft F. Materials with high thermal conductivity will carry heat at a higher rate for a given temperature differential and allow for higher heat transfer rates for a given heat exchanger surface area. Heat capacity (C), the product of density (ρ), and specific heat (c), are the properties of a material that define the amount of energy stored in a material volume relative to its temperature change and are measured in units of Btu/ft^3 F. Materials with high heat capacity will store more heat for a given increase in temperature and make for more efficient energy storage for a given heat exchanger size. Thermal diffusivity (α), the ratio of the thermal conductivity to the heat capacity, is a measure of the relative rate of heat movement in the material to the material's heat storage capacity and is measured in units of ft^2/day. A low diffusivity material will have a relatively large heat storage capacity compared to its ability to transport heat, and temperature at various points in the material will change more slowly when a thermal disturbance occurs at some point in the material. High diffusivity, on the other hand, indicates that the material has a relatively low heat capacity compared to its ability to transport heat, and temperature at various points in the material will change more rapidly when a thermal disturbance occurs at some point in the material. For heat exchanger devices that are designed to both transport and store energy efficiently, a material with both high thermal conductivity and high heat capacity (resulting in a relatively low thermal diffusivity) is desirable to maximize both heat transfer rate and energy storage capacity.

The thermal conductivity of soils and rocks is highly dependent on the mineralogy, density, and water content of the soil or rock. The material that comprises the soil or rock formation defines the limits of thermal conductivity for that material, and density and water content affect the thermal conductivity within those limits. Table 5.6 provides thermal property ranges of selected soils and rocks *(ASHRAE, 2011)* for a range of water contents and densities that would likely be encountered in vertical ground loop applications (depths where soil moisture is relatively stable throughout the year). Soils are generally comprised of sand, silt, and clay in various percentages, with the percentages of sand and clay having the largest influence on the thermal performance of the mixture. Sands can have a thermal conductivity ranging between 0.5 and 2.2 Btu/hr ft F depending on mineralogy, density, and water content. Extremely dry, sandy soils may have a thermal conductivity lower than 0.5 Btu/hr ft F at very low water contents. Clays (and silts) have a greater affinity for water and thus exist at higher water contents under most conditions. Clay soils have a very high affinity for water and hold a large amount of water, even during periods of very dry weather conditions, and have a thermal conductivity ranging between 0.30 and 1.10 Btu/hr ft F. The thermal conductivity ranges shown for the various rock types included in Table 5.6 are also affected by density, water content, and mineralogy. Thermal conductivity ranges for rock types not included in Table 5.6 may be estimated from Figure 11.1 (1, 2, and 3) in the IGSHPA Soil and Rock Manual. It is very difficult to determine if the low or high end of the thermal conductivity range applies without direct measurement, and it is recommended (as a conservative approach) that an average to below average value be used for a known soil or rock type unless better information is available.

The thermal diffusivity for each soil and rock listing in Table 5.6 represents the ratio of thermal conductivity and heat capacity, as described above, and reflects the influence of dry density and water content on the heat capacity of the soil or rock type. Higher dry density and water content results in higher heat capacity, which lowers the thermal diffusivity for a given thermal conductivity. As with thermal conductivity, it is very difficult to determine if the low or high end of the thermal diffusivity range applies. It is recommended (as a conservative approach) that an average to above average value from the thermal diffusivity range be used for a known soil or rock type unless better information is available.

Table 5.6. Thermal Property Ranges of Selected Soils and Rocks – *ASHRAE (2011)*[1]

	Dry Density (lb/ft^3)	Thermal Conductivity (Btu/hr ft F)	Thermal Diffusivity (ft^2/day)
Soils			
Heavy Clay – 15% water	120	0.8 – 1.1	0.45 – 0.65
Heavy Clay – 5 % water	120	0.6 – 0.8	0.50 – 0.65
Light Clay – 15% water	80	0.4 – 0.6	0.35 – 0.50
Light Clay – 5 % water	80	0.3 – 0.5	0.35 – 0.60
Heavy Sand – 15% water	120	1.6 – 2.2	0.9 – 1.2
Heavy Sand – 5 % water	120	1.2 – 1.9	1.0 – 1.5
Light Sand – 15% water	80	0.6 – 1.2	0.5 – 1.0
Light Sand – 5 % water	80	0.5 – 1.1	0.6 – 1.3
Rocks			
Granite	165	1.3 – 2.1	0.9 – 1.4
Limestone	150 – 175	1.4 – 2.2	0.9 – 1.4
Sandstone	160 – 170	1.2 – 2.0	0.7 – 1.2
Shale – Wet	130 – 165	0.8 – 1.4	0.7 – 0.9
Shale – Dry	130 – 165	0.6 – 1.2	0.6 – 0.8

1. Kavanaugh and Rafferty (1997)

The thermal properties that are utilized for vertically-bored, closed-loop GHEX design have a large impact on the calculated design length (affected by thermal conductivity) and the adjustment of the design length for unbalanced ground loads (affected by thermal diffusivity). The closer the selected design values are to the average thermal conductivity and thermal diffusivity values that are encountered along the length of the vertically-bored GHEX, the more accurate the calculated design lengths will be. It is good design practice to utilize thermal property values that are as close to the actual values, but on the conservative side for each, meaning thermal conductivity should be slightly below the actual value and thermal diffusivity should be slightly above the actual value.

There are three ways to estimate the thermal conductivity and diffusivity of the soil or rock into which the vertically-bored GHEX is to be installed. The easiest, and least accurate, is to estimate the values based on a best guess of what the formation contains. In some areas, where the geology is very consistent with depth, this method may be acceptable. The next best method is to obtain information on the geology as it varies with depth, and then, use a weighted averaging process to estimate the average thermal conductivity and diffusivity along the vertically-bored GHEX length. Table 5.7 provides a tool for documenting the averaging process for the thermal conductivity and diffusivity based on soil and rock layer thickness. Information on the geology of the location may be obtained from geologic maps, drill logs from water well or other drilled excavations, or a test well drilled expressly for that purpose. In each case, the type of soil or rock is recorded as a function of depth, and thermal conductivity and thermal diffusivity values from Table 5.6 are assigned to each layer in the strata. Then, the layer thickness is multiplied times the thermal conductivity and diffusivity for each layer and the resulting products are summed over the entire length that the vertical borehole will be installed. The summed thermal conductivity and diffusivity products are then divided by the vertical borehole length to provide a weighted average for each.

A third, and much more accurate, method for determining the formation thermal conductivity and diffusivity is to perform a formation thermal conductivity (FTC) test. This procedure, which is documented in the *IGSHPA 2009 Standards* and *ASHRAE (2011)*, consists of drilling a test well to the desired depth and creating a drill log documenting the soil or rock types as they change with depth. A U-bend is installed and grouted into the test borehole and, after a sufficient waiting period, equipment designed to heat water at a constant rate is connected to the U-bend. The temperature of the water left standing in the U-bend during the waiting period is first measured to estimate the deep earth temperature (T_G). Then, the heating elements are energized, and the water is heated and circulated through the U-bend for 36 to 48 hours. Temperature data obtained from the test is analyzed to determine the average thermal conductivity along the test borehole length. Average thermal diffusivity is estimated from the measured average thermal conductivity and estimated average heat capacity for the soil or rock types along the test borehole length. *IGSHPA 2009 Standards* and *ASHRAE (2011)* provide recommended tests specifications for the procedure according to research conducted by *Kavanaugh (2000, 2001)*.

Table 5.7. Thermal Conductivity and Diffusivity Averaging Table

Layer	Start Depth	End Depth	t_{Layer}[1] (ft)	k_{Layer}[2] (Btu/hr ft F)	α_{Layer}[3] (ft^2/day)	$t_{Layer} \times k_{Layer}$ (Btu/hr ft^2 F)	$t_{Layer} \times \alpha_{Layer}$ (ft^3/day)
Sum							
k_{avg} (Btu/hr ft F)	= Sum ($t_{Layer} \times k_{Layer}$) / Sum ($t_{Layer}$)						
α_{avg} (ft^2/day)	= Sum ($t_{Layer} \times \alpha_{Layer}$) / Sum ($t_{Layer}$)						

1. Layer thickness = End Depth – Start Depth
2. Estimated thermal conductivity for the layer (Table 5.6 or other source)
3. Estimated thermal diffusivity for the layer (Table 5.6 or other source)

The ground thermal resistance (R_G) in Equations 5.1 and 5.3 accounts for the thermal conductivity of the soil or rock (k_G) surrounding the borehole and the radial nature of heat transfer into and out of the borehole during heat pump operation in heating or cooling. R_G is calculated using Equation 5.5 for steady-state heat transfer, a worst-case situation, which will only occur in normal heat pump operation during periods of extreme heating or cooling demand. In Equations 5.1 and 5.3, R_G is multiplied by the design month heating or cooling run fraction, which reduces the ground resistance in proportion to monthly run fraction to account for recovery time of the ground when the heat pump system is not operating. The borehole diameter (D_B) and the diameter of the ground away from the borehole ($D_{G,O}$) (where the deep earth temperature is assumed constant) define the soil or rock area around the borehole affected by the heat transfer. Borehole diameter normally ranges between 4 and 6 inches and depends on the bit diameter used by the driller. The outer soil or rock diameter has been selected for calculations at 15 feet, which is a standard spacing between boreholes. Using a larger value for $D_{G,O}$ will increase the ground thermal resistance only slightly and will have minimal influence on the ground loop lengths obtained from Equations 5.1 and 5.3.

Equation 5.5

$$R_G = \frac{\ln\left(\dfrac{D_{G,O}}{D_B}\right)}{2 \bullet \pi \bullet k_G}$$

Where:

k_G = thermal conductivity of ground around borehole, Btu/hr ft F
$D_{G,O}$ = diameter of ground surrounding borehole affected by heat transfer, ft
D_B = diameter of borehole, ft

5.2.2.3 BOREHOLE THERMAL RESISTANCE (R_B) Borehole thermal resistance (R_B) is a measure of the resistance to heat transfer between the circulating fluid inside the pipe and the borehole wall. The research that resulted in the definition of borehole thermal resistance is addressed in detail in IGSHPA's Grouting for Vertical Geothermal Heat Pump Systems (2001). Equation 5.6 defines borehole thermal resistance, which includes the effects of the pipe size and geometry in the borehole and the thermal conductivity of the grouting material filling the annular area between the outside pipe surface and the borehole wall.

$$R_B = R_{Grout} + R_{PP}$$ **Equation 5.6**

Where:

R_{Grout} = thermal resistance due to grout and pipe position in borehole, hr ft F/Btu
R_{PP} = thermal resistance of pipe walls for two pipes in parallel, hr ft F/Btu
= $R_P/2$

Grout resistance (R_{Grout}) accounts for the thermal conductivity of the grouting material (k_{Grout}) and the geometry of the two pipes inside the borehole represented by the borehole shape factor (S_B), as shown by Equation 5.7. The thermal conductivities of grouting materials can range from 0.40 to 1.20 Btu/hr ft F, with values in the lower half of that range being most common for residential and light commercial applications. The geometry of the two pipes inside the borehole includes the pipe outside diameter and the configuration of the pipes relative to each other and to the borehole wall. *IGSHPA (2000) Grouting for Vertical Geothermal Heat Pump Systems* details combinations of pipe diameter and configuration in the borehole that represents the worst-case (pipes touching in the center of the borehole-Configuration A), the intermediate case (equal distance between pipes and borehole walls-Configuration B), and the best-case (pipes diametrically opposite touching the borehole walls-Configuration C) situations for heat transfer. Without making special efforts to achieve best-case pipe configuration, the best approximation for the naturally occurring pipe configuration in the borehole can be expressed by Equation 5.8, which is an average of the Configuration B and Configuration C scenarios.

$$R_{Grout} = \frac{1}{S_B \bullet k_{Grout}}$$ **Equation 5.7**

Where:

k_{Grout} = grout thermal conductivity, Btu/hr ft F
S_B = borehole shape factor, dimensionless

$$S_B = \frac{17.44 \cdot \left(\frac{D_B}{D_{P,O}}\right)^{-0.6052} + 21.91 \cdot \left(\frac{D_B}{D_{P,O}}\right)^{-0.3796}}{2}$$ **Equation 5.8**

Where:

D_B = borehole diameter, inches
$D_{P,O}$ = pipe outside diameter, inches

The thermal resistance of the piping material (R_P) is addressed in Section 4.2 and calculated according to Equation 4.1, which accounts for the pipe wall thickness and the thermal conductivity of the piping material. Table 4.6 provides the thermal resistance of high-density polyethylene pipe for various DR ratios. For most residential and light commercial applications, DR 11 pipe is used in the borehole and R_P has a value of 0.141 hr ft F/Btu.

5.2.2.4 GROUND LOOP RESISTANCE Ground loop resistance represents the total resistance to heat transfer at design heating or cooling conditions between the circulating fluid in the pipe and the undisturbed deep earth temperature and has been defined in Equation 5.1 as ($R_B + R_G \bullet F_H$) and Equation 5.3 as ($R_B + R_G \bullet F_C$). The ground loop resistance includes the effects of borehole thermal resistance (pipe size and geometry in the borehole, the grout thermal conductivity, and borehole diameter), ground thermal resistance (soil or rock thermal conductivity and borehole diameter), and the run fraction for the design heating or cooling month. To simplify the determination of ground loop resistance, tables have been constructed which express ground loop resistance as a function of design month run fraction in heating (F_H) or cooling (F_C), soil or rock thermal conductivity (k_G), borehole diameter (D_B), and grout thermal conductivity (k_{Grout}) for nominal U-bend diameters of ¾-inch (Table 5.8), 1-inch (Table 5.9) and 1-1/4-inch (Table 5.10). Interpolation between tabled values is acceptable. Care must be used when extrapolating beyond tabled values, especially on the low end of the tabled variables, and it would be better practice to calculate the ground loop resistance using Equations 5.5 through 5.8.

Table 5.8. Vertical Borehole Ground Loop Resistance ($R_B + R_G \bullet F_H$ for Heating or $R_B + R_G \bullet F_C$ for Cooling) for ¾-inch Nominal U-bend (hr ft F/Btu)

F_H or F_C	k_G (Btu/hr ft F)	Borehole Diameter = 4.00 in. k_{Grout} (Btu/hr ft F) 0.40	0.57	0.69	0.79	0.88	1.00	1.07	1.14	1.20
0.30	0.70	0.569	0.498	0.469	0.451	0.439	0.426	0.419	0.414	0.410
	0.85	0.523	0.452	0.423	0.405	0.393	0.380	0.374	0.368	0.364
	1.00	0.491	0.420	0.391	0.373	0.361	0.348	0.342	0.336	0.332
	1.15	0.467	0.396	0.367	0.349	0.337	0.324	0.318	0.312	0.308
	1.30	0.449	0.378	0.349	0.331	0.319	0.306	0.300	0.294	0.290
	1.60	0.423	0.352	0.323	0.305	0.293	0.280	0.273	0.268	0.264
	1.90	0.405	0.334	0.305	0.287	0.275	0.262	0.255	0.250	0.246
	2.20	0.392	0.321	0.292	0.274	0.262	0.249	0.242	0.237	0.233
0.40	0.70	0.655	0.584	0.555	0.538	0.525	0.512	0.506	0.501	0.496
	0.85	0.594	0.523	0.494	0.477	0.464	0.451	0.445	0.439	0.435
	1.00	0.552	0.480	0.451	0.434	0.421	0.408	0.402	0.397	0.392
	1.15	0.520	0.449	0.420	0.402	0.390	0.377	0.371	0.365	0.361
	1.30	0.496	0.425	0.395	0.378	0.366	0.352	0.346	0.341	0.337
	1.60	0.461	0.390	0.360	0.343	0.331	0.318	0.311	0.306	0.302
	1.90	0.437	0.366	0.337	0.319	0.307	0.294	0.287	0.282	0.278
	2.20	0.419	0.348	0.319	0.302	0.289	0.276	0.270	0.264	0.260
0.50	0.70	0.742	0.671	0.642	0.624	0.612	0.599	0.593	0.587	0.583
	0.85	0.666	0.594	0.565	0.548	0.535	0.522	0.516	0.511	0.507
	1.00	0.612	0.541	0.512	0.494	0.482	0.469	0.463	0.457	0.453
	1.15	0.573	0.501	0.472	0.455	0.442	0.429	0.423	0.418	0.414
	1.30	0.542	0.471	0.442	0.424	0.412	0.399	0.393	0.387	0.383
	1.60	0.499	0.427	0.398	0.381	0.368	0.355	0.349	0.344	0.339
	1.90	0.469	0.398	0.368	0.351	0.339	0.326	0.319	0.314	0.310
	2.20	0.447	0.376	0.347	0.329	0.317	0.304	0.298	0.292	0.288
0.60	0.70	0.829	0.757	0.728	0.711	0.698	0.685	0.679	0.674	0.669
	0.85	0.737	0.666	0.637	0.619	0.607	0.594	0.587	0.582	0.578
	1.00	0.673	0.602	0.572	0.555	0.543	0.530	0.523	0.518	0.514
	1.15	0.625	0.554	0.525	0.508	0.495	0.482	0.476	0.470	0.466
	1.30	0.589	0.518	0.489	0.471	0.459	0.446	0.439	0.434	0.430
	1.60	0.536	0.465	0.436	0.419	0.406	0.393	0.387	0.382	0.377
	1.90	0.501	0.429	0.400	0.383	0.370	0.357	0.351	0.346	0.341
	2.20	0.474	0.403	0.374	0.357	0.344	0.331	0.325	0.320	0.315
0.70	0.70	0.915	0.844	0.815	0.797	0.785	0.772	0.766	0.760	0.756
	0.85	0.808	0.737	0.708	0.690	0.678	0.665	0.659	0.653	0.649
	1.00	0.733	0.662	0.633	0.616	0.603	0.590	0.584	0.578	0.574
	1.15	0.678	0.607	0.578	0.560	0.548	0.535	0.529	0.523	0.519
	1.30	0.635	0.564	0.535	0.518	0.505	0.492	0.486	0.481	0.476
	1.60	0.574	0.503	0.474	0.457	0.444	0.431	0.425	0.419	0.415
	1.90	0.532	0.461	0.432	0.415	0.402	0.389	0.383	0.378	0.373
	2.20	0.502	0.431	0.402	0.384	0.372	0.359	0.353	0.347	0.343
0.80	0.70	1.002	0.930	0.901	0.884	0.871	0.858	0.852	0.847	0.843
	0.85	0.879	0.808	0.779	0.762	0.749	0.736	0.730	0.725	0.720
	1.00	0.794	0.723	0.694	0.676	0.664	0.651	0.645	0.639	0.635
	1.15	0.731	0.660	0.630	0.613	0.601	0.588	0.581	0.576	0.572
	1.30	0.682	0.611	0.582	0.564	0.552	0.539	0.533	0.527	0.523
	1.60	0.612	0.541	0.512	0.494	0.482	0.469	0.463	0.457	0.453
	1.90	0.564	0.493	0.464	0.447	0.434	0.421	0.415	0.409	0.405
	2.20	0.530	0.458	0.429	0.412	0.399	0.386	0.380	0.375	0.370

F_H or F_C	k_G (Btu/hr ft F)	Borehole Diameter = 5.00 in. k_{Grout} (Btu/hr ft F) 0.40	0.57	0.69	0.79	0.88	1.00	1.07	1.14	1.20
0.30	0.70	0.580	0.501	0.468	0.449	0.435	0.421	0.414	0.408	0.403
	0.85	0.536	0.458	0.425	0.406	0.392	0.378	0.371	0.365	0.360
	1.00	0.506	0.427	0.395	0.376	0.362	0.348	0.341	0.335	0.330
	1.15	0.484	0.405	0.373	0.353	0.340	0.325	0.318	0.312	0.308
	1.30	0.467	0.388	0.356	0.336	0.322	0.308	0.301	0.295	0.290
	1.60	0.442	0.363	0.331	0.312	0.298	0.283	0.276	0.270	0.266
	1.90	0.425	0.346	0.314	0.295	0.281	0.266	0.260	0.253	0.249
	2.20	0.413	0.334	0.302	0.282	0.269	0.254	0.247	0.241	0.237
0.40	0.70	0.661	0.582	0.550	0.530	0.517	0.502	0.495	0.489	0.485
	0.85	0.604	0.525	0.492	0.473	0.459	0.445	0.438	0.432	0.427
	1.00	0.563	0.484	0.452	0.433	0.419	0.405	0.398	0.392	0.387
	1.15	0.534	0.455	0.422	0.403	0.389	0.375	0.368	0.362	0.357
	1.30	0.511	0.432	0.399	0.380	0.366	0.352	0.345	0.339	0.334
	1.60	0.478	0.399	0.367	0.347	0.333	0.319	0.312	0.306	0.301
	1.90	0.455	0.376	0.344	0.325	0.311	0.297	0.290	0.284	0.279
	2.20	0.439	0.360	0.328	0.308	0.295	0.280	0.273	0.267	0.262
0.50	0.70	0.743	0.664	0.631	0.612	0.598	0.584	0.577	0.571	0.566
	0.85	0.671	0.592	0.559	0.540	0.526	0.512	0.505	0.499	0.494
	1.00	0.620	0.541	0.509	0.490	0.476	0.462	0.455	0.449	0.444
	1.15	0.583	0.504	0.472	0.453	0.439	0.424	0.417	0.411	0.407
	1.30	0.555	0.476	0.443	0.424	0.410	0.396	0.389	0.383	0.378
	1.60	0.513	0.434	0.402	0.383	0.369	0.355	0.348	0.342	0.337
	1.90	0.485	0.406	0.374	0.355	0.341	0.327	0.320	0.314	0.309
	2.20	0.465	0.386	0.354	0.334	0.320	0.306	0.299	0.293	0.288
0.60	0.70	0.824	0.745	0.713	0.693	0.680	0.665	0.658	0.652	0.648
	0.85	0.738	0.659	0.627	0.607	0.593	0.579	0.572	0.566	0.561
	1.00	0.677	0.598	0.566	0.547	0.533	0.519	0.512	0.506	0.501
	1.15	0.633	0.554	0.522	0.502	0.488	0.474	0.467	0.461	0.456
	1.30	0.598	0.520	0.487	0.468	0.454	0.440	0.433	0.427	0.422
	1.60	0.549	0.470	0.438	0.418	0.405	0.390	0.383	0.377	0.373
	1.90	0.515	0.436	0.404	0.385	0.371	0.357	0.350	0.344	0.339
	2.20	0.491	0.412	0.380	0.360	0.346	0.332	0.325	0.319	0.314
0.70	0.70	0.906	0.827	0.794	0.775	0.761	0.747	0.740	0.734	0.729
	0.85	0.805	0.726	0.694	0.674	0.661	0.646	0.639	0.633	0.628
	1.00	0.734	0.656	0.623	0.604	0.590	0.576	0.569	0.563	0.558
	1.15	0.682	0.603	0.571	0.552	0.538	0.524	0.517	0.511	0.506
	1.30	0.642	0.563	0.531	0.512	0.498	0.484	0.477	0.471	0.466
	1.60	0.585	0.506	0.474	0.454	0.440	0.426	0.419	0.413	0.408
	1.90	0.545	0.466	0.434	0.415	0.401	0.387	0.380	0.374	0.369
	2.20	0.517	0.438	0.405	0.386	0.372	0.358	0.351	0.345	0.340
0.80	0.70	0.987	0.908	0.876	0.856	0.843	0.828	0.821	0.815	0.811
	0.85	0.872	0.793	0.761	0.741	0.728	0.713	0.706	0.700	0.696
	1.00	0.791	0.713	0.680	0.661	0.647	0.633	0.626	0.620	0.615
	1.15	0.732	0.653	0.621	0.601	0.588	0.573	0.566	0.560	0.556
	1.30	0.686	0.607	0.575	0.556	0.542	0.527	0.520	0.514	0.510
	1.60	0.620	0.541	0.509	0.490	0.476	0.462	0.455	0.449	0.444
	1.90	0.575	0.496	0.464	0.445	0.431	0.417	0.410	0.404	0.399
	2.20	0.543	0.464	0.431	0.412	0.398	0.384	0.377	0.371	0.366

F_H or F_C	k_G (Btu/hr ft F)	Borehole Diameter = 6.00 in. k_{Grout} (Btu/hr ft F) 0.40	0.57	0.69	0.79	0.88	1.00	1.07	1.14	1.20
0.30	0.70	0.590	0.504	0.469	0.448	0.433	0.418	0.410	0.404	0.398
	0.85	0.549	0.464	0.428	0.407	0.392	0.377	0.369	0.363	0.358
	1.00	0.521	0.435	0.400	0.379	0.364	0.348	0.341	0.334	0.329
	1.15	0.500	0.414	0.379	0.357	0.343	0.327	0.319	0.313	0.308
	1.30	0.483	0.397	0.362	0.341	0.326	0.311	0.303	0.296	0.291
	1.60	0.460	0.374	0.339	0.318	0.303	0.287	0.280	0.273	0.268
	1.90	0.444	0.358	0.323	0.302	0.287	0.271	0.264	0.257	0.252
	2.20	0.432	0.346	0.311	0.290	0.275	0.259	0.252	0.245	0.240
0.40	0.70	0.668	0.582	0.547	0.526	0.511	0.495	0.487	0.481	0.476
	0.85	0.613	0.527	0.492	0.471	0.456	0.440	0.433	0.426	0.421
	1.00	0.575	0.489	0.454	0.433	0.418	0.402	0.395	0.388	0.383
	1.15	0.547	0.461	0.426	0.405	0.390	0.374	0.366	0.360	0.355
	1.30	0.525	0.439	0.404	0.383	0.368	0.352	0.345	0.338	0.333
	1.60	0.494	0.408	0.373	0.352	0.337	0.321	0.313	0.307	0.302
	1.90	0.472	0.386	0.351	0.330	0.315	0.300	0.292	0.286	0.280
	2.20	0.457	0.371	0.336	0.315	0.300	0.284	0.277	0.270	0.265
0.50	0.70	0.745	0.659	0.624	0.603	0.588	0.572	0.565	0.558	0.553
	0.85	0.677	0.591	0.556	0.535	0.520	0.504	0.497	0.490	0.485
	1.00	0.629	0.543	0.508	0.487	0.472	0.456	0.449	0.442	0.437
	1.15	0.594	0.508	0.473	0.452	0.437	0.421	0.414	0.407	0.402
	1.30	0.566	0.481	0.446	0.424	0.410	0.394	0.386	0.380	0.375
	1.60	0.527	0.442	0.407	0.385	0.371	0.355	0.347	0.341	0.336
	1.90	0.501	0.415	0.380	0.359	0.344	0.328	0.321	0.314	0.309
	2.20	0.481	0.396	0.360	0.339	0.324	0.309	0.301	0.295	0.290
0.60	0.70	0.822	0.736	0.701	0.680	0.665	0.650	0.642	0.636	0.630
	0.85	0.740	0.655	0.619	0.598	0.583	0.568	0.560	0.554	0.549
	1.00	0.683	0.597	0.562	0.541	0.526	0.510	0.503	0.496	0.491
	1.15	0.641	0.555	0.520	0.499	0.484	0.468	0.461	0.454	0.449
	1.30	0.608	0.522	0.487	0.466	0.451	0.436	0.428	0.421	0.416
	1.60	0.561	0.475	0.440	0.419	0.404	0.389	0.381	0.375	0.369
	1.90	0.529	0.443	0.408	0.387	0.372	0.357	0.349	0.342	0.337
	2.20	0.506	0.420	0.385	0.364	0.349	0.333	0.326	0.319	0.314
0.70	0.70	0.900	0.814	0.779	0.758	0.743	0.727	0.719	0.713	0.708
	0.85	0.804	0.718	0.683	0.662	0.647	0.631	0.624	0.617	0.612
	1.00	0.737	0.651	0.616	0.595	0.580	0.565	0.557	0.550	0.545
	1.15	0.688	0.602	0.567	0.546	0.531	0.515	0.508	0.501	0.496
	1.30	0.650	0.564	0.529	0.508	0.493	0.477	0.470	0.463	0.458
	1.60	0.595	0.509	0.474	0.453	0.438	0.423	0.415	0.408	0.403
	1.90	0.558	0.472	0.437	0.416	0.401	0.385	0.378	0.371	0.366
	2.20	0.531	0.445	0.410	0.389	0.374	0.358	0.350	0.344	0.339
0.80	0.70	0.977	0.891	0.856	0.835	0.820	0.804	0.797	0.790	0.708
	0.85	0.868	0.782	0.747	0.726	0.711	0.695	0.688	0.681	0.612
	1.00	0.791	0.706	0.670	0.649	0.634	0.619	0.611	0.605	0.545
	1.15	0.735	0.649	0.614	0.593	0.578	0.562	0.555	0.548	0.496
	1.30	0.691	0.606	0.571	0.549	0.534	0.519	0.511	0.505	0.458
	1.60	0.629	0.543	0.508	0.487	0.472	0.456	0.449	0.442	0.403
	1.90	0.586	0.500	0.465	0.444	0.429	0.414	0.406	0.399	0.366
	2.20	0.555	0.469	0.434	0.413	0.398	0.383	0.375	0.368	0.339

1. Shape factor based on average of "B" and "C" configurations.

Table 5.9. Vertical Borehole Ground Loop Resistance ($R_B + R_G \cdot F_H$ for Heating or $R_B + R_G \cdot F_C$ for Cooling) for 1-inch Nominal U-bend (hr ft F/Btu).

F_H or F_C	k_G (Btu/hr ft F)	Borehole Diameter = 4.00 in. k_{Grout} (Btu/hr ft F)									Borehole Diameter = 5.00 in. k_{Grout} (Btu/hr ft F)									Borehole Diameter = 6.00 in. k_{Grout} (Btu/hr ft F)								
		0.40	0.57	0.69	0.79	0.88	1.00	1.07	1.14	1.20	0.40	0.57	0.69	0.79	0.88	1.00	1.07	1.14	1.20	0.40	0.57	0.69	0.79	0.88	1.00	1.07	1.14	1.20
0.30	0.70	0.545	0.481	0.455	0.439	0.428	0.416	0.411	0.406	0.402	0.554	0.482	0.453	0.436	0.423	0.410	0.404	0.399	0.395	0.562	0.485	0.453	0.434	0.421	0.406	0.400	0.394	0.389
	0.85	0.499	0.435	0.409	0.393	0.382	0.370	0.365	0.360	0.356	0.510	0.439	0.410	0.393	0.380	0.367	0.361	0.356	0.351	0.521	0.444	0.412	0.393	0.380	0.365	0.359	0.353	0.348
	1.00	0.467	0.403	0.377	0.361	0.350	0.338	0.333	0.328	0.324	0.480	0.409	0.380	0.362	0.350	0.337	0.331	0.325	0.321	0.492	0.415	0.383	0.364	0.351	0.337	0.330	0.324	0.319
	1.15	0.444	0.380	0.353	0.338	0.326	0.315	0.309	0.304	0.300	0.458	0.387	0.358	0.340	0.328	0.315	0.309	0.303	0.299	0.471	0.394	0.362	0.343	0.330	0.316	0.309	0.303	0.298
	1.30	0.425	0.361	0.335	0.319	0.308	0.296	0.291	0.286	0.282	0.441	0.370	0.340	0.323	0.311	0.298	0.291	0.286	0.282	0.455	0.378	0.346	0.327	0.313	0.299	0.293	0.287	0.282
	1.60	0.399	0.335	0.309	0.293	0.282	0.270	0.265	0.260	0.256	0.416	0.345	0.316	0.298	0.286	0.273	0.267	0.261	0.257	0.432	0.354	0.323	0.303	0.290	0.276	0.269	0.263	0.259
	1.90	0.381	0.317	0.291	0.275	0.264	0.252	0.247	0.242	0.238	0.399	0.328	0.299	0.281	0.269	0.256	0.250	0.244	0.240	0.416	0.338	0.306	0.287	0.274	0.260	0.253	0.247	0.243
	2.20	0.368	0.304	0.278	0.262	0.251	0.239	0.234	0.229	0.225	0.387	0.316	0.287	0.269	0.257	0.244	0.238	0.232	0.228	0.404	0.326	0.295	0.276	0.262	0.248	0.241	0.235	0.231
0.40	0.70	0.632	0.568	0.541	0.526	0.515	0.503	0.497	0.492	0.488	0.635	0.564	0.535	0.517	0.505	0.492	0.486	0.480	0.476	0.639	0.562	0.530	0.511	0.498	0.484	0.477	0.471	0.466
	0.85	0.571	0.507	0.480	0.465	0.453	0.442	0.436	0.431	0.427	0.577	0.506	0.477	0.460	0.447	0.434	0.428	0.423	0.418	0.585	0.507	0.476	0.457	0.443	0.429	0.422	0.416	0.412
	1.00	0.528	0.464	0.438	0.422	0.411	0.399	0.393	0.388	0.385	0.537	0.466	0.437	0.419	0.407	0.394	0.388	0.382	0.378	0.547	0.469	0.438	0.418	0.405	0.391	0.384	0.378	0.374
	1.15	0.496	0.432	0.406	0.390	0.379	0.367	0.362	0.357	0.353	0.507	0.436	0.407	0.390	0.377	0.364	0.358	0.353	0.348	0.518	0.441	0.409	0.390	0.377	0.363	0.356	0.350	0.345
	1.30	0.472	0.408	0.382	0.366	0.355	0.343	0.337	0.332	0.329	0.485	0.413	0.384	0.367	0.354	0.341	0.335	0.330	0.326	0.497	0.419	0.388	0.369	0.355	0.341	0.334	0.328	0.324
	1.60	0.437	0.373	0.347	0.331	0.320	0.308	0.302	0.298	0.294	0.452	0.381	0.351	0.334	0.322	0.309	0.302	0.297	0.293	0.465	0.388	0.356	0.337	0.324	0.310	0.303	0.297	0.292
	1.90	0.413	0.349	0.323	0.307	0.296	0.284	0.279	0.274	0.270	0.429	0.358	0.329	0.311	0.299	0.286	0.280	0.274	0.270	0.444	0.367	0.335	0.316	0.302	0.288	0.282	0.276	0.271
	2.20	0.396	0.332	0.305	0.290	0.278	0.267	0.261	0.256	0.252	0.413	0.342	0.313	0.295	0.283	0.270	0.263	0.258	0.254	0.428	0.351	0.319	0.300	0.287	0.273	0.266	0.260	0.256
0.50	0.70	0.718	0.654	0.628	0.612	0.601	0.589	0.584	0.579	0.575	0.716	0.645	0.616	0.599	0.586	0.573	0.567	0.562	0.557	0.717	0.639	0.608	0.589	0.575	0.561	0.554	0.548	0.544
	0.85	0.642	0.578	0.552	0.536	0.525	0.513	0.507	0.502	0.499	0.645	0.573	0.544	0.527	0.514	0.501	0.495	0.490	0.486	0.648	0.571	0.539	0.520	0.507	0.493	0.486	0.480	0.476
	1.00	0.589	0.524	0.498	0.482	0.471	0.460	0.454	0.449	0.445	0.594	0.523	0.494	0.477	0.464	0.451	0.445	0.439	0.435	0.601	0.523	0.492	0.473	0.459	0.445	0.438	0.432	0.428
	1.15	0.549	0.485	0.459	0.443	0.432	0.420	0.414	0.409	0.406	0.557	0.486	0.457	0.439	0.427	0.414	0.408	0.402	0.398	0.565	0.488	0.456	0.437	0.424	0.410	0.403	0.397	0.392
	1.30	0.519	0.454	0.428	0.412	0.401	0.390	0.384	0.379	0.375	0.528	0.457	0.428	0.411	0.398	0.385	0.379	0.374	0.369	0.538	0.461	0.429	0.410	0.397	0.383	0.376	0.370	0.365
	1.60	0.475	0.411	0.385	0.369	0.358	0.346	0.340	0.335	0.332	0.487	0.416	0.387	0.370	0.357	0.344	0.338	0.333	0.328	0.499	0.422	0.390	0.371	0.358	0.344	0.337	0.331	0.326
	1.90	0.445	0.381	0.355	0.339	0.328	0.316	0.310	0.305	0.302	0.459	0.388	0.359	0.341	0.329	0.316	0.310	0.304	0.300	0.472	0.395	0.363	0.344	0.331	0.317	0.310	0.304	0.300
	2.20	0.423	0.359	0.333	0.317	0.306	0.294	0.289	0.284	0.280	0.439	0.368	0.338	0.321	0.309	0.296	0.289	0.284	0.280	0.453	0.376	0.344	0.325	0.312	0.297	0.291	0.285	0.280
0.60	0.70	0.805	0.741	0.715	0.699	0.688	0.676	0.670	0.665	0.662	0.798	0.727	0.698	0.680	0.668	0.655	0.649	0.643	0.639	0.794	0.717	0.685	0.666	0.653	0.638	0.632	0.626	0.621
	0.85	0.713	0.649	0.623	0.607	0.596	0.584	0.579	0.574	0.570	0.712	0.641	0.611	0.594	0.582	0.569	0.562	0.557	0.553	0.712	0.635	0.603	0.584	0.571	0.556	0.550	0.544	0.539
	1.00	0.649	0.585	0.559	0.543	0.532	0.520	0.514	0.510	0.506	0.651	0.580	0.551	0.534	0.521	0.508	0.502	0.496	0.492	0.655	0.577	0.546	0.527	0.513	0.499	0.492	0.486	0.482
	1.15	0.602	0.538	0.511	0.496	0.484	0.473	0.467	0.462	0.458	0.607	0.536	0.506	0.489	0.477	0.464	0.457	0.452	0.448	0.612	0.535	0.503	0.484	0.471	0.457	0.450	0.444	0.440
	1.30	0.565	0.501	0.475	0.459	0.448	0.436	0.431	0.426	0.422	0.572	0.501	0.472	0.455	0.442	0.429	0.423	0.418	0.413	0.580	0.503	0.471	0.452	0.438	0.424	0.417	0.411	0.407
	1.60	0.513	0.449	0.422	0.407	0.396	0.384	0.378	0.373	0.369	0.523	0.452	0.423	0.405	0.393	0.380	0.374	0.368	0.364	0.533	0.456	0.424	0.405	0.392	0.377	0.371	0.365	0.360
	1.90	0.477	0.413	0.387	0.371	0.360	0.348	0.342	0.337	0.334	0.489	0.418	0.389	0.371	0.359	0.346	0.340	0.334	0.330	0.501	0.424	0.392	0.373	0.359	0.345	0.339	0.333	0.328
	2.20	0.451	0.387	0.360	0.345	0.334	0.322	0.316	0.311	0.308	0.465	0.394	0.364	0.347	0.335	0.322	0.315	0.310	0.306	0.478	0.400	0.369	0.350	0.336	0.322	0.315	0.309	0.305
0.70	0.70	0.891	0.827	0.801	0.785	0.774	0.762	0.757	0.752	0.748	0.879	0.808	0.779	0.762	0.749	0.736	0.730	0.725	0.720	0.871	0.794	0.762	0.743	0.730	0.716	0.709	0.703	0.698
	0.85	0.785	0.720	0.694	0.678	0.667	0.656	0.650	0.645	0.641	0.779	0.708	0.679	0.661	0.649	0.636	0.629	0.624	0.620	0.776	0.698	0.667	0.648	0.634	0.620	0.613	0.607	0.603
	1.00	0.710	0.646	0.619	0.604	0.592	0.581	0.575	0.570	0.566	0.708	0.637	0.608	0.591	0.578	0.565	0.559	0.554	0.549	0.709	0.632	0.600	0.581	0.567	0.553	0.547	0.541	0.536
	1.15	0.654	0.590	0.564	0.548	0.537	0.525	0.520	0.515	0.511	0.656	0.585	0.556	0.539	0.526	0.513	0.507	0.501	0.497	0.660	0.582	0.551	0.531	0.518	0.504	0.497	0.491	0.487
	1.30	0.612	0.548	0.521	0.506	0.495	0.483	0.477	0.472	0.469	0.616	0.545	0.516	0.498	0.486	0.473	0.467	0.461	0.457	0.622	0.544	0.512	0.493	0.480	0.466	0.459	0.453	0.449
	1.60	0.551	0.487	0.460	0.445	0.433	0.422	0.416	0.411	0.407	0.559	0.487	0.458	0.441	0.429	0.416	0.409	0.404	0.400	0.567	0.489	0.458	0.439	0.425	0.411	0.404	0.398	0.394
	1.90	0.509	0.445	0.418	0.403	0.392	0.380	0.374	0.369	0.365	0.519	0.448	0.419	0.401	0.389	0.376	0.370	0.364	0.360	0.529	0.452	0.420	0.401	0.388	0.374	0.367	0.361	0.357
	2.20	0.478	0.414	0.388	0.372	0.361	0.349	0.344	0.339	0.335	0.491	0.419	0.390	0.373	0.360	0.347	0.341	0.336	0.332	0.502	0.425	0.393	0.374	0.361	0.347	0.340	0.334	0.329
0.80	0.70	0.978	0.914	0.888	0.872	0.861	0.849	0.843	0.838	0.835	0.961	0.890	0.861	0.843	0.831	0.818	0.812	0.806	0.802	0.949	0.871	0.840	0.821	0.807	0.793	0.786	0.780	0.776
	0.85	0.856	0.792	0.765	0.750	0.739	0.727	0.721	0.716	0.712	0.846	0.775	0.746	0.728	0.716	0.703	0.697	0.691	0.687	0.840	0.762	0.730	0.711	0.698	0.684	0.677	0.671	0.667
	1.00	0.770	0.706	0.680	0.664	0.653	0.641	0.636	0.631	0.627	0.765	0.694	0.665	0.648	0.635	0.622	0.616	0.611	0.606	0.763	0.686	0.654	0.635	0.622	0.607	0.601	0.595	0.590
	1.15	0.707	0.643	0.617	0.601	0.590	0.578	0.572	0.568	0.564	0.706	0.635	0.606	0.588	0.576	0.563	0.557	0.551	0.547	0.707	0.629	0.598	0.579	0.565	0.551	0.544	0.538	0.534
	1.30	0.658	0.594	0.568	0.552	0.541	0.529	0.524	0.519	0.515	0.660	0.589	0.560	0.542	0.530	0.517	0.511	0.505	0.501	0.663	0.586	0.554	0.535	0.522	0.508	0.501	0.495	0.490
	1.60	0.589	0.524	0.498	0.482	0.471	0.460	0.454	0.449	0.445	0.594	0.523	0.494	0.477	0.464	0.451	0.445	0.439	0.435	0.601	0.523	0.492	0.473	0.459	0.445	0.438	0.432	0.428
	1.90	0.541	0.477	0.450	0.435	0.423	0.412	0.406	0.401	0.397	0.549	0.478	0.449	0.431	0.419	0.406	0.400	0.394	0.390	0.558	0.481	0.449	0.430	0.416	0.402	0.396	0.390	0.385
	2.20	0.506	0.442	0.416	0.400	0.389	0.377	0.371	0.366	0.363	0.516	0.445	0.416	0.399	0.386	0.373	0.367	0.362	0.357	0.527	0.450	0.418	0.399	0.385	0.371	0.364	0.358	0.354

1. Shape factor based on average of "B" and "C" configurations.

Table 5.10. Vertical Borehole Ground Loop Resistance ($R_B + R_G \bullet F_H$ for Heating or $R_B + R_G \bullet F_C$ for Cooling) for 1-1/4-inch Nominal U-bend (hr ft F/Btu).

F_H or F_C	k_G	Borehole Diameter = 4.00 in.								
		k_{Grout} (Btu/hr ft F)								
	(Btu/hr ft F)	0.40	0.57	0.69	0.79	0.88	1.00	1.07	1.14	1.20
0.30	0.70	0.523	0.466	0.442	0.428	0.418	0.408	0.403	0.398	0.395
	0.85	0.478	0.420	0.397	0.382	0.372	0.362	0.357	0.352	0.349
	1.00	0.446	0.388	0.365	0.350	0.340	0.330	0.325	0.320	0.317
	1.15	0.422	0.364	0.341	0.327	0.317	0.306	0.301	0.297	0.293
	1.30	0.404	0.346	0.323	0.308	0.298	0.288	0.283	0.278	0.275
	1.60	0.377	0.320	0.296	0.282	0.272	0.262	0.257	0.252	0.249
	1.90	0.359	0.302	0.278	0.264	0.254	0.244	0.239	0.234	0.231
	2.20	0.346	0.289	0.265	0.251	0.241	0.231	0.226	0.221	0.218
0.40	0.70	0.610	0.553	0.529	0.515	0.505	0.494	0.489	0.485	0.481
	0.85	0.549	0.491	0.468	0.454	0.444	0.433	0.428	0.424	0.420
	1.00	0.506	0.449	0.425	0.411	0.401	0.390	0.385	0.381	0.378
	1.15	0.475	0.417	0.394	0.379	0.369	0.359	0.354	0.349	0.346
	1.30	0.450	0.393	0.369	0.355	0.345	0.335	0.330	0.325	0.322
	1.60	0.415	0.358	0.334	0.320	0.310	0.300	0.295	0.290	0.287
	1.90	0.391	0.334	0.310	0.296	0.286	0.276	0.271	0.266	0.263
	2.20	0.374	0.316	0.293	0.279	0.269	0.258	0.253	0.249	0.245
0.50	0.70	0.697	0.639	0.616	0.601	0.591	0.581	0.576	0.571	0.568
	0.85	0.620	0.563	0.539	0.525	0.515	0.505	0.499	0.495	0.492
	1.00	0.567	0.509	0.486	0.472	0.462	0.451	0.446	0.442	0.438
	1.15	0.527	0.470	0.446	0.432	0.422	0.412	0.407	0.402	0.399
	1.30	0.497	0.439	0.416	0.402	0.392	0.381	0.376	0.372	0.368
	1.60	0.453	0.396	0.372	0.358	0.348	0.337	0.332	0.328	0.325
	1.90	0.423	0.366	0.342	0.328	0.318	0.308	0.303	0.298	0.295
	2.20	0.402	0.344	0.320	0.306	0.296	0.286	0.281	0.276	0.273
0.60	0.70	0.783	0.726	0.702	0.688	0.678	0.667	0.662	0.658	0.655
	0.85	0.691	0.634	0.610	0.596	0.586	0.576	0.571	0.566	0.563
	1.00	0.627	0.570	0.546	0.532	0.522	0.512	0.507	0.502	0.499
	1.15	0.580	0.522	0.499	0.485	0.475	0.464	0.459	0.455	0.451
	1.30	0.543	0.486	0.462	0.448	0.438	0.428	0.423	0.418	0.415
	1.60	0.491	0.434	0.410	0.396	0.386	0.375	0.370	0.366	0.362
	1.90	0.455	0.398	0.374	0.360	0.350	0.339	0.334	0.330	0.327
	2.20	0.429	0.372	0.348	0.334	0.324	0.313	0.308	0.304	0.301
0.70	0.70	0.870	0.812	0.789	0.774	0.765	0.754	0.749	0.745	0.741
	0.85	0.763	0.705	0.682	0.668	0.658	0.647	0.642	0.638	0.634
	1.00	0.688	0.630	0.607	0.593	0.583	0.572	0.567	0.563	0.559
	1.15	0.633	0.575	0.552	0.537	0.527	0.517	0.512	0.507	0.504
	1.30	0.590	0.533	0.509	0.495	0.485	0.474	0.469	0.465	0.462
	1.60	0.529	0.471	0.448	0.434	0.424	0.413	0.408	0.404	0.400
	1.90	0.487	0.430	0.406	0.392	0.382	0.371	0.366	0.362	0.358
	2.20	0.457	0.399	0.376	0.361	0.351	0.341	0.336	0.331	0.328
0.80	0.70	0.956	0.899	0.875	0.861	0.851	0.841	0.835	0.831	0.828
	0.85	0.834	0.777	0.753	0.739	0.729	0.718	0.713	0.709	0.705
	1.00	0.749	0.691	0.667	0.653	0.643	0.633	0.628	0.623	0.620
	1.15	0.685	0.628	0.604	0.590	0.580	0.570	0.565	0.560	0.557
	1.30	0.637	0.579	0.556	0.541	0.531	0.521	0.516	0.512	0.508
	1.60	0.567	0.509	0.486	0.472	0.462	0.451	0.446	0.442	0.438
	1.90	0.519	0.461	0.438	0.424	0.414	0.403	0.398	0.394	0.390
	2.20	0.484	0.427	0.403	0.389	0.379	0.368	0.363	0.359	0.356

F_H or F_C	k_G	Borehole Diameter = 5.00 in.								
		k_{Grout} (Btu/hr ft F)								
	(Btu/hr ft F)	0.40	0.57	0.69	0.79	0.88	1.00	1.07	1.14	1.20
0.30	0.70	0.529	0.466	0.440	0.424	0.413	0.401	0.395	0.391	0.387
	0.85	0.486	0.422	0.396	0.381	0.370	0.358	0.352	0.347	0.344
	1.00	0.456	0.392	0.366	0.350	0.339	0.328	0.322	0.317	0.313
	1.15	0.434	0.370	0.344	0.328	0.317	0.305	0.300	0.295	0.291
	1.30	0.417	0.353	0.327	0.311	0.300	0.288	0.283	0.278	0.274
	1.60	0.392	0.328	0.302	0.286	0.275	0.264	0.258	0.253	0.249
	1.90	0.375	0.311	0.285	0.269	0.258	0.247	0.241	0.236	0.232
	2.20	0.363	0.299	0.273	0.257	0.246	0.234	0.229	0.224	0.220
0.40	0.70	0.611	0.547	0.521	0.505	0.494	0.483	0.477	0.472	0.468
	0.85	0.553	0.490	0.463	0.448	0.437	0.425	0.419	0.414	0.411
	1.00	0.513	0.449	0.423	0.408	0.396	0.385	0.379	0.374	0.370
	1.15	0.483	0.420	0.393	0.378	0.367	0.355	0.349	0.344	0.341
	1.30	0.461	0.397	0.371	0.355	0.344	0.332	0.327	0.322	0.318
	1.60	0.428	0.364	0.338	0.322	0.311	0.299	0.294	0.289	0.285
	1.90	0.405	0.341	0.315	0.299	0.288	0.277	0.271	0.266	0.262
	2.20	0.389	0.325	0.299	0.283	0.272	0.260	0.255	0.250	0.246
0.50	0.70	0.692	0.629	0.602	0.587	0.576	0.564	0.558	0.553	0.550
	0.85	0.621	0.557	0.531	0.515	0.504	0.492	0.487	0.482	0.478
	1.00	0.570	0.506	0.480	0.465	0.453	0.442	0.436	0.431	0.428
	1.15	0.533	0.469	0.443	0.427	0.416	0.405	0.399	0.394	0.390
	1.30	0.504	0.441	0.414	0.399	0.388	0.376	0.370	0.365	0.362
	1.60	0.463	0.399	0.373	0.358	0.347	0.335	0.329	0.324	0.321
	1.90	0.435	0.371	0.345	0.329	0.318	0.307	0.301	0.296	0.292
	2.20	0.415	0.351	0.325	0.309	0.298	0.286	0.281	0.276	0.272
0.60	0.70	0.774	0.710	0.684	0.668	0.657	0.645	0.640	0.635	0.631
	0.85	0.688	0.624	0.598	0.582	0.571	0.559	0.554	0.549	0.545
	1.00	0.627	0.563	0.537	0.522	0.510	0.499	0.493	0.488	0.485
	1.15	0.583	0.519	0.493	0.477	0.466	0.454	0.449	0.444	0.440
	1.30	0.548	0.484	0.458	0.443	0.432	0.420	0.414	0.409	0.406
	1.60	0.499	0.435	0.409	0.393	0.382	0.370	0.365	0.360	0.356
	1.90	0.465	0.401	0.375	0.359	0.348	0.337	0.331	0.326	0.322
	2.20	0.441	0.377	0.351	0.335	0.324	0.312	0.307	0.302	0.298
0.70	0.70	0.855	0.792	0.765	0.750	0.739	0.727	0.721	0.716	0.713
	0.85	0.755	0.691	0.665	0.649	0.638	0.626	0.621	0.616	0.612
	1.00	0.684	0.620	0.594	0.579	0.568	0.556	0.550	0.545	0.542
	1.15	0.632	0.568	0.542	0.527	0.515	0.504	0.498	0.493	0.490
	1.30	0.592	0.528	0.502	0.486	0.475	0.464	0.458	0.453	0.449
	1.60	0.535	0.471	0.445	0.429	0.418	0.406	0.401	0.396	0.392
	1.90	0.495	0.431	0.405	0.389	0.378	0.367	0.361	0.356	0.352
	2.20	0.466	0.403	0.377	0.361	0.350	0.338	0.332	0.328	0.324
0.80	0.70	0.937	0.873	0.847	0.831	0.820	0.808	0.803	0.798	0.794
	0.85	0.822	0.758	0.732	0.716	0.705	0.693	0.688	0.683	0.679
	1.00	0.741	0.677	0.651	0.636	0.625	0.613	0.607	0.602	0.599
	1.15	0.682	0.618	0.592	0.576	0.565	0.553	0.548	0.543	0.539
	1.30	0.636	0.572	0.546	0.530	0.519	0.508	0.502	0.497	0.493
	1.60	0.570	0.506	0.480	0.465	0.453	0.442	0.436	0.431	0.428
	1.90	0.525	0.461	0.435	0.420	0.408	0.397	0.391	0.386	0.382
	2.20	0.492	0.429	0.402	0.387	0.376	0.364	0.358	0.354	0.350

F_H or F_C	k_G	Borehole Diameter = 6.00 in.								
		k_{Grout} (Btu/hr ft F)								
	(Btu/hr ft F)	0.40	0.57	0.69	0.79	0.88	1.00	1.07	1.14	1.20
0.30	0.70	0.536	0.466	0.438	0.421	0.409	0.396	0.390	0.385	0.381
	0.85	0.495	0.426	0.397	0.380	0.368	0.355	0.349	0.344	0.340
	1.00	0.466	0.397	0.368	0.351	0.339	0.327	0.320	0.315	0.311
	1.15	0.445	0.376	0.347	0.330	0.318	0.305	0.299	0.294	0.290
	1.30	0.429	0.359	0.331	0.314	0.302	0.289	0.283	0.278	0.274
	1.60	0.405	0.336	0.308	0.290	0.278	0.266	0.260	0.254	0.250
	1.90	0.389	0.320	0.292	0.274	0.262	0.250	0.244	0.238	0.234
	2.20	0.378	0.308	0.280	0.263	0.251	0.238	0.232	0.227	0.222
0.40	0.70	0.613	0.544	0.515	0.498	0.486	0.474	0.467	0.462	0.458
	0.85	0.559	0.489	0.461	0.444	0.432	0.419	0.413	0.407	0.403
	1.00	0.520	0.451	0.423	0.405	0.393	0.381	0.375	0.369	0.365
	1.15	0.492	0.423	0.394	0.377	0.365	0.352	0.346	0.341	0.337
	1.30	0.471	0.401	0.373	0.356	0.343	0.331	0.325	0.319	0.315
	1.60	0.439	0.370	0.341	0.324	0.312	0.300	0.293	0.288	0.284
	1.90	0.418	0.348	0.320	0.303	0.291	0.278	0.272	0.267	0.263
	2.20	0.402	0.333	0.304	0.287	0.275	0.263	0.257	0.251	0.247
0.50	0.70	0.691	0.621	0.593	0.576	0.564	0.551	0.545	0.539	0.535
	0.85	0.622	0.553	0.524	0.507	0.495	0.483	0.477	0.471	0.467
	1.00	0.575	0.505	0.477	0.460	0.448	0.435	0.429	0.423	0.419
	1.15	0.539	0.470	0.441	0.424	0.412	0.400	0.393	0.388	0.384
	1.30	0.512	0.443	0.414	0.397	0.385	0.372	0.366	0.361	0.357
	1.60	0.473	0.404	0.375	0.358	0.346	0.333	0.327	0.322	0.318
	1.90	0.446	0.377	0.349	0.331	0.319	0.307	0.301	0.295	0.291
	2.20	0.427	0.358	0.329	0.312	0.300	0.287	0.281	0.276	0.272
0.60	0.70	0.768	0.698	0.670	0.653	0.641	0.628	0.622	0.617	0.613
	0.85	0.686	0.617	0.588	0.571	0.559	0.546	0.540	0.535	0.531
	1.00	0.629	0.559	0.531	0.514	0.502	0.489	0.483	0.478	0.473
	1.15	0.586	0.517	0.488	0.471	0.459	0.447	0.441	0.435	0.431
	1.30	0.554	0.484	0.456	0.439	0.427	0.414	0.408	0.403	0.399
	1.60	0.507	0.437	0.409	0.392	0.380	0.367	0.361	0.356	0.352
	1.90	0.475	0.405	0.377	0.360	0.348	0.335	0.329	0.324	0.320
	2.20	0.452	0.382	0.354	0.337	0.325	0.312	0.306	0.300	0.296
0.70	0.70	0.845	0.776	0.747	0.730	0.718	0.706	0.699	0.694	0.690
	0.85	0.750	0.680	0.652	0.635	0.623	0.610	0.604	0.599	0.594
	1.00	0.683	0.613	0.585	0.568	0.556	0.543	0.537	0.532	0.528
	1.15	0.633	0.564	0.536	0.518	0.506	0.494	0.488	0.482	0.478
	1.30	0.595	0.526	0.498	0.480	0.468	0.456	0.450	0.444	0.440
	1.60	0.541	0.471	0.443	0.426	0.414	0.401	0.395	0.390	0.385
	1.90	0.503	0.434	0.405	0.388	0.376	0.364	0.358	0.352	0.348
	2.20	0.476	0.407	0.378	0.361	0.349	0.336	0.330	0.325	0.321
0.80	0.70	0.923	0.853	0.825	0.808	0.796	0.783	0.777	0.771	0.767
	0.85	0.813	0.744	0.716	0.698	0.686	0.674	0.668	0.662	0.658
	1.00	0.737	0.668	0.639	0.622	0.610	0.597	0.591	0.586	0.582
	1.15	0.681	0.611	0.583	0.566	0.553	0.541	0.535	0.529	0.525
	1.30	0.637	0.568	0.539	0.522	0.510	0.497	0.491	0.486	0.482
	1.60	0.575	0.505	0.477	0.460	0.448	0.435	0.429	0.423	0.419
	1.90	0.532	0.462	0.434	0.417	0.405	0.392	0.386	0.381	0.377
	2.20	0.501	0.431	0.403	0.386	0.374	0.361	0.355	0.350	0.346

1. Shape factor based on average of "B" and "C" configurations.

5.2.3 UNBALANCED GROUND LOAD CORRECTION FACTOR (B_M) The heating and cooling design lengths calculated using Equations 5.1 and 5.3, respectively, assume that the temperature of the ground, from and to which heat is being transferred, stays constant throughout the year and from year to year. This is a good assumption if the amount of energy extracted from the ground equals the amount of energy rejected into the ground each year, or the ground energy load is balanced. If more energy is extracted from than rejected to the ground each year, and the ground does not have the thermal capability to replace that energy without a long-term reduction in the temperature of the ground around the vertical GHEX, then the entering water temperature from the GHEX will decrease each year until the ground temperature around the vertical GHEX reaches a point where a net energy flow toward the GHEX will make up for the lack of annual heat rejection from the GHEX. The opposite is true if more energy is rejected to than absorbed from the ground each year. In either case, the degree of unbalanced ground energy load will require additional soil volume to transfer heat with, which can be obtained through a combination of increasing the spacing of the boreholes and/or increasing the GHEX design length. If that is not done, then it is likely that, after several years of operation, the entering water temperature will be lower than the design minimum entering water temperature for net annual heat extraction or will be higher than the design maximum entering water temperature for net annual heat rejection.

A correction factor has been defined to adjust the design heating or cooling borehole length for unbalanced annual ground energy load based on heat transfer theory used by commercial ground loop design programs including GCHPCalc, GLHEPRO, GLD, etc. The bore length multiplier (B_M) is based on the net annual ground energy adjusted for the design heating or cooling length and the temperature difference between the loop and the ground (the Normalized Net Annual Ground Load - NNAGL), the thermal properties of the soil and rock surrounding the GHEX, the spacing of the boreholes along a row, and the number of rows in the bore field. Figures 5.13 through 5.18 provide B_M for soil and rock thermal conductivity/diffusivity combinations of 0.7/0.45; 1.0/0.60; 1.3/0.75; 1.6/0.90; 1.9/1.05; and 2.2/1.20 (Btu/hr ft F/ft^2/day), respectively. In each figure, a series of curves represent various combinations of bore spacing (15, 20, and 25 feet) and number of rows in the bore field (1, 2, and 3). For a GHEX with more than three rows of bores, the use of a commercial design program should be considered.

The definition of NNAGL depends on whether heating or cooling design lengths are dominant, as expressed by the equations labeled **Heating** and **Cooling** in Equation 5.9a and 5.9b, respectively. In either case, an energy analysis (described in Section 3.6) provides the annual heating (AGL_{DH}) and cooling (AGL_{DC}) ground energy loads. The net annual energy load, which is the difference in AGL_{DH} and AGL_{DC}, is divided by both the design length calculated using Equation 5.1 or 5.3 and the temperature difference between the deep earth and the average circulating fluid, which normalizes the data. Using the figure (Figure 5.13 through 5.18) with the closest soil or rock thermal conductivity, read from the NNAGL value vertically upward to the curve representing the proper combination of bore spacing and number of bore rows. At the intersection of that curve read horizontally to the left axis to obtain an estimate for B_M, which is utilized in Equations 5.10 and 5.12 to adjust the total borehole design heating or cooling length and in Equations 5.11 and 5.13 to adjust the individual borehole design heating or cooling length. If, in an unusual situation, NNAGL has a negative value, then the adjustment to the design heating or cooling length is not required and the design length from either Equation 5.1 or 5.3 is utilized directly.

$$NNAGL = \frac{AGL_{DH} - AGL_{DC}}{L_{H,T} \cdot \left(T_G - \left(\frac{EWT_{min} + LWT_{min}}{2}\right)\right)}$$

(Heating) Equation 5.9a

$$NNAGL = \frac{AGL_{DC} - AGL_{DH}}{L_{C,T} \cdot \left(\left(\frac{EWT_{max} + LWT_{max}}{2}\right) - T_G\right)}$$

(Cooling) Equation 5.9b

Where:

$NNAGL$ = normalized net annual ground energy load, Btu/ft F
AGL_{DH} = annual design heating ground energy load, Btu
AGL_{DC} = annual design cooling ground energy load, Btu

$$L_{H,T,UGL} = B_M \cdot L_{H,T}$$

Equation 5.10

Where:

$L_{H,T,UGL}$ = total borehole design length for heating adjusted for unbalanced ground load, ft
B_M = bore length multiplier

$$L_{H,B,UGL} = \frac{L_{H,T,UGL}}{N_B}$$

Equation 5.11

Where:

$L_{H,B,UGL}$ = individual borehole design length for heating adjusted for unbalanced ground load, ft
N_B = number of boreholes connected to the heat pump system

$$L_{C,T,UGL} = B_M \cdot L_{C,T}$$

Equation 5.12

Where:

$L_{C,T,UGL}$ = total borehole design length for cooling adjusted for unbalanced ground load, ft
B_M = bore length multiplier

$$L_{C,B,UGL} = \frac{L_{C,T,UGL}}{N_B}$$

Equation 5.13

Where:

$L_{C,B,UGL}$ = individual borehole design length for cooling adjusted for unbalanced ground load, ft
N_B = number of boreholes connected to the heat pump system

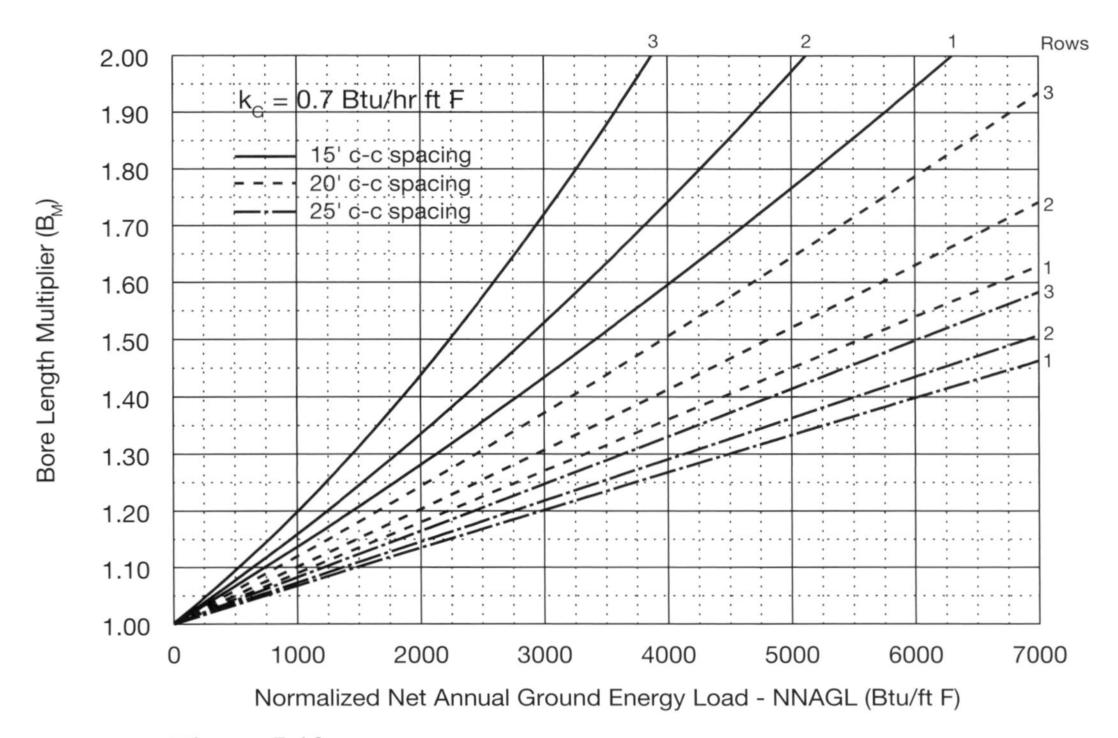

Figure 5.13. Bore Length Multiplier (B_M) for k_G = 0.7 Btu/hr ft F

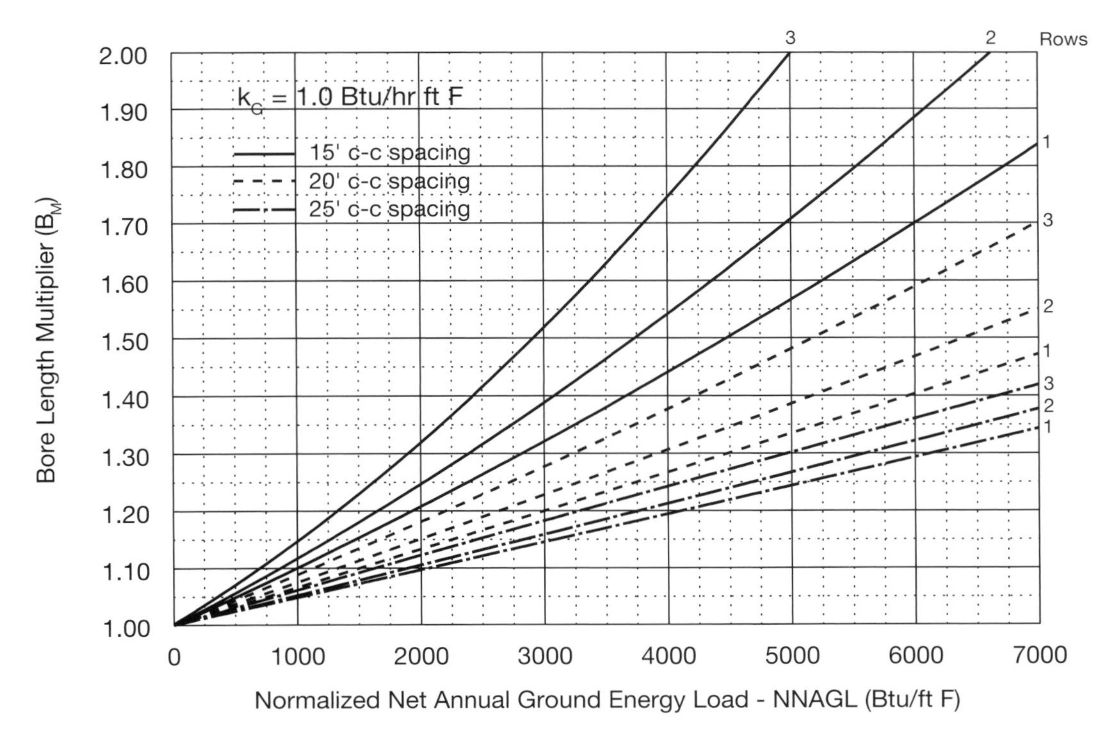

Figure 5.14. Bore Length Multiplier (B_M) for k_G = 1.0 Btu/hr ft F

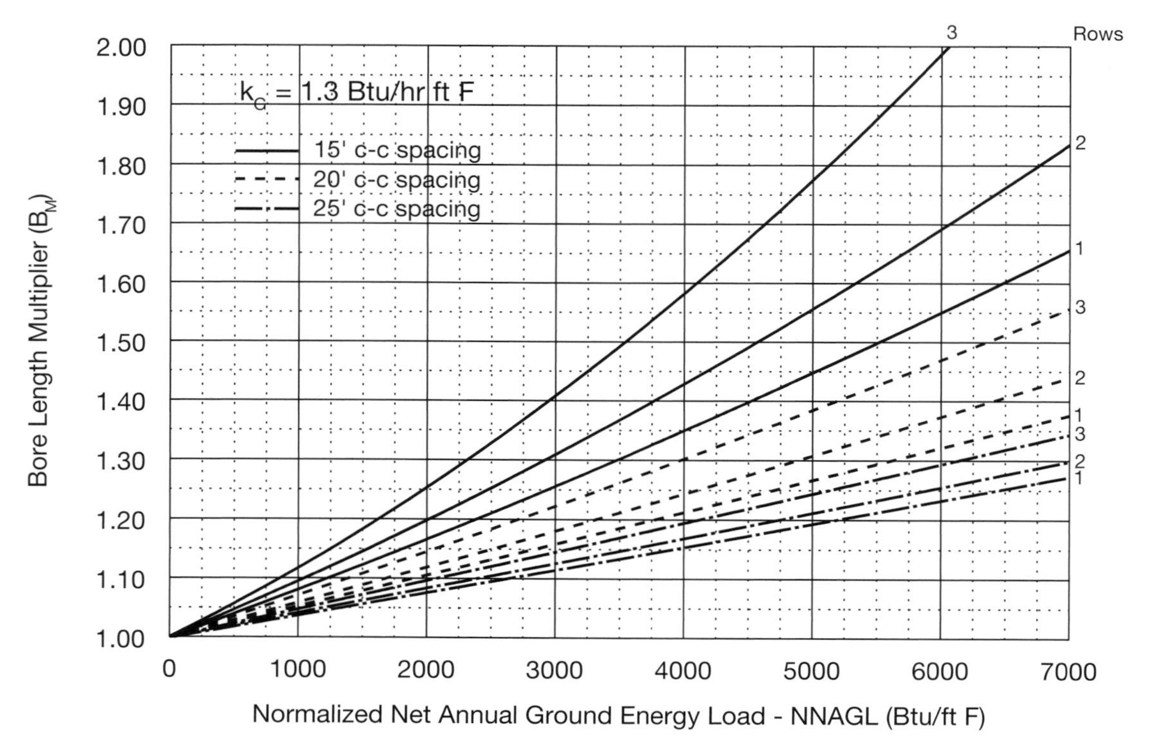

Figure 5.15. Bore Length Multiplier (B_M) for k_G = 1.3 Btu/hr ft F

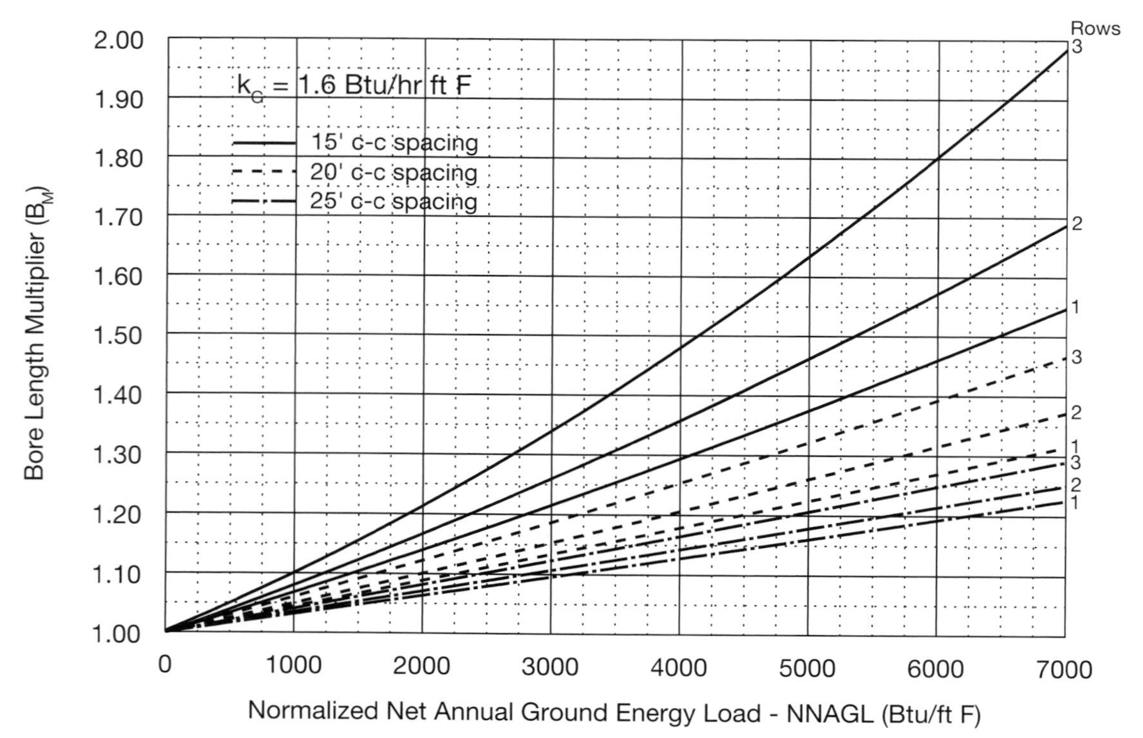

Figure 5.16. Bore Length Multiplier (B_M) for k_G = 1.6 Btu/hr ft F

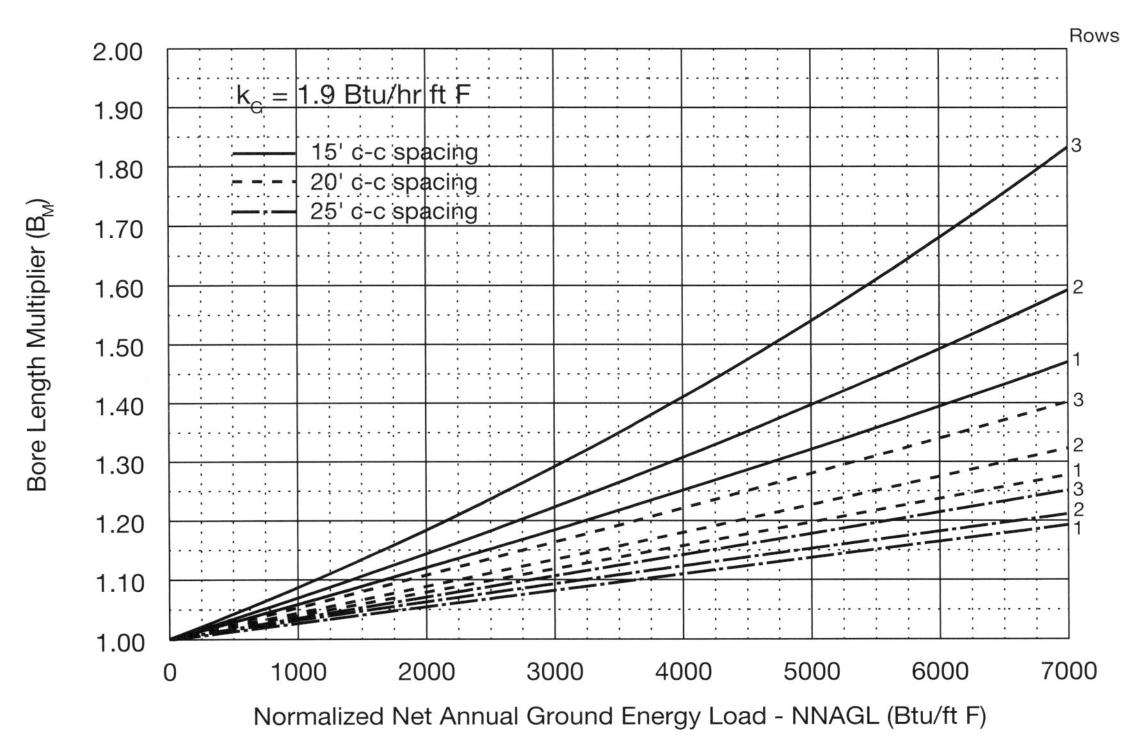

Figure 5.17. Bore Length Multiplier (B_M) for k_G = 1.9 Btu/hr ft F

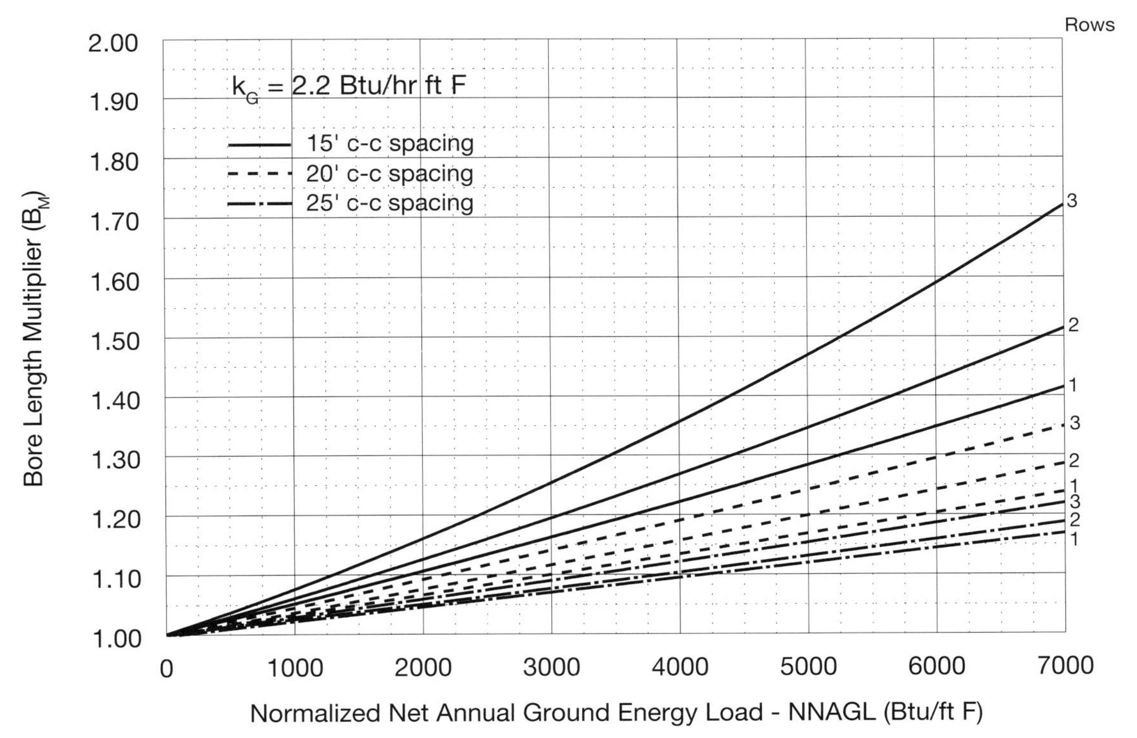

Figure 5.18. Bore Length Multiplier (B_M) for k_G = 2.2 Btu/hr ft F

5.2.4 VERTICALLY-BORED GHEX DESIGN WORKSHEETS The proper design of a vertically-bored GHEX requires reliable heating and cooling load design data and heat pump selection, consideration of domestic hot water generation, specification of vertical loopfield design parameters, proper layout of the vertical borefield, application of the heating or cooling design equations to determine the required borehole lengths, and adjustment for unbalanced annual ground loads to assure long-term GHEX performance. Heating and cooling load design data are determined and heat pumps selected according to procedures presented in Chapter 3, and worksheets were provided to document the selection of the water-to-air or water-to-water GSHP units for the design space heating or cooling loads. Part of the heat pump selection is based on specification of the minimum and maximum entering water temperatures from the vertical GHEX, along with the water flow rate that the GHEX must accommodate, each of which are critical parameters in the proper design of the vertical GHEX system. Hot water generation, by either a dedicated water-to-water GSHP or a desuperheater in the space heating and cooling GSHP, will affect the vertical loopfield design as discussed in Chapter 9 and procedures are provided to account for hot water generation in Tables 9.4 and 9.5. Layout of the vertical borefield, in terms of number of vertical bores, U-bend diameter, and maximum drilling depth are considered to accommodate required water flow rate to achieve proper turbulence for heat transfer, in a heating dominant design, and to control head loss to keep pumping power at an acceptable level at heating or cooling design conditions. Application of the appropriate borehole design length equation (Equation 5.1 or 5.3) using design data for the ground and borehole results in the required total length of vertical borehole below the header trench. Adjustment of that length may be required for an unbalanced annual ground energy load, which is accomplished using procedures described in Section 5.2.3.

In an effort to simplify the design of the vertical GHEX and provide a method to easily document the design, worksheets have been developed for both heating-mode dominant (Figure 5.19) and cooling-mode dominant (Figure 5.20) design. The worksheets are broken into sections to allow input for heat pump design and energy analysis data, domestic hot water generation, borefield layout, design data for the vertical GHEX, calculation of the total borehole design length, and adjustment for unbalanced annual ground energy load. Additional calculations can be made to refine the vertical GHEX design lengths in a table at the bottom of each worksheet. Each input blank in Figures 5.19 and 5.20 has been labeled with a number identifying a short input description, a listing of which follows each figure. A blank copy of each figure is provided in Appendix B.

Vertically-Bored GHEX Design Worksheet – Heating Mode

Space Heating GSHP Design Data

HC_C[1] = 1 Btu/hr
DMD_C[1] = 2 kW
GPM[1] = 3 gpm
AGL_H = 4 Btu
AGL_C = 5 Btu
Total Heat Loss[1] = 6 Btu/hr
Percent Sizing = 7 %
F_{Jan} = 8

HWG GSHP Design Data

HC_{HWG} = 9 Btu/hr
DMD_{HWG} = 10 kW
GPM_{HWG} = 11 gpm
AGL_{HWG} = 12 Btu
F_{HWG} = 13

Total Design Data

HC_D = 14 Btu/hr
DMD_D = 15 kW
GPM_D = 16 gpm
AGL_{HD} = 17 Btu
AGL_{CD} = 18 Btu

GHEX Design Data

COP_D[2] (= HC_D / (DMD_D x 3,412) = 19
HE_D[2] (= HC_D x (COP_D-1)/COP_D) = 20 Btu/hr
F_H (= (F_{Jan} x HC_C + F_{HWG} x HC_{HWG}) / HC_D) = 21
EWT_{min}[3] = 22 F
LWT_{min}[2] (= EWT_{min} – HE_D/(500xGPM_D)) = 23 F

1. For single heat pump installation use data directly from the appropriate GSHP Selection Worksheet. For multiple heat pumps in the installation sum all heat pump Total Heat Losses, HC_C's, DMD_C's and GPM's that will be connected to a single GHEX using the table provided below.
2. For single heat pump installation use data directly from appropriate GSHP Selection Worksheet. For multiple heat pumps use the equation provided.
3. EWT_{min} is obtained directly from the appropriate GSHP Selection Worksheet and must be the same for the selection of all heat pumps for a multiple heat pump installation connected to a single GHEX. **(24)**

	Zone 1	Zone 2	Zone 3	Zone 4	Zone 5	Zone 6	Zone 7	Zone 8	Zone 9	Zone 10	Total
Total Heat Loss											
HC_C											
DMD_C											
GPM											

Vertical Borefield Layout

U-bend D_P = 25 in
GPM_{FP} = 26 gpm/flowpath
N_{FP} = 27 flowpaths
N_{BIS} = 28 bores in series
N_B = 29 bores
Layout = 30 x 31 (N_{Rows} x $N_{Bores/Row}$)
Bore Spacing = 32 ft

Vertical Borefield Design Data

T_G = 33 F
k_G = 34 Btu/hr ft F (Table 5.6)
D_{bore} = 35 in
k_{Grout} = 36 Btu/hr ft F
$(R_B + R_G \cdot F_H)$ = 37 hr ft F/Btu (Table 5.8, 5.9 or 5.10)

Borehole Design Lengths (Equations 5.1 and 5.2) **(38)**

$$L_{H,T} = \frac{HC_D \cdot \left(\frac{COP_D - 1}{COP_D}\right) \cdot (R_B + R_G \cdot F_H)}{T_G - \left(\frac{EWT_{min} + LWT_{min}}{2}\right)} = \frac{\_\_\_ \cdot \left(\frac{\_\_\_ - 1}{\_\_\_}\right) \cdot (\_\_\_)}{\_\_\_ - \left(\frac{\_\_\_ + \_\_\_}{2}\right)} = \frac{\_\_\_ \cdot (\_\_\_) \cdot (\_\_\_)}{\_\_\_ - (\_\_\_)} = \_\_\_ \text{ ft of bore}$$

$$L_{H,B} = \frac{L_{H,T}}{N_B} = \frac{\_\_\_}{\_\_\_} = \_\_\_ \text{ ft per bore (Below Header Trench)}$$

Unbalanced Ground Load Borehole Design Lengths (Equations 5.9a, 5.10, and 5.11) **(39)**

$$NNAGL = \frac{AGL_{HD} - AGL_{CD}}{L_{H,T} \cdot \left(T_G - \left(\frac{EWT_{min} + LWT_{min}}{2}\right)\right)} = \frac{\_\_\_ - \_\_\_}{\_\_\_ \cdot \left(\_\_\_ - \left(\frac{\_\_\_ + \_\_\_}{2}\right)\right)} = \_\_\_ \text{ Btu / ftF}$$

B_M = ______ (Figures 5.13 through 5.18)

$$L_{H,T,UGL} = B_M \cdot L_{H,T} = \_\_\_ \cdot \_\_\_ = \_\_\_ \text{ ft of bore} \qquad L_{H,B,UGL} = \frac{L_{H,T,UGL}}{N_B} = \frac{\_\_\_}{\_\_\_} = \_\_\_ \text{ ft per bore (Below Header Trench)}$$

Heating Design Length Calculations Summary Table **(40)**

Layout for Flow and Number of Bores						Design Length Calculations				Unbalanced Ground Load Design Lengths					
D_P	GPM_{FP}	N_{FP}	N_{BIS}	N_B	Layout	k_{Grout}	R_B+$R_G \cdot F_H$	$L_{H,T}$	$L_{H,B}$	Layout	Spacing	NNAGL	B_M	$L_{H,T,UGL}$	$L_{H,B,UGL}$
					/										
					/										
					/										
					/										
					/										

Figure 5.19. Vertically-Bored GHEX Design Worksheet – Heating Mode.

5.2.4.1 HEATING MODE DESIGN WORKSHEET (FIGURE 5.19)

1. The corrected heating capacity (Figure 3.1, 3.2, or 3.3) of the heat pump at design heating conditions for the zone for which the ground loop is being designed. For multiple zones on a single ground loop, this is the sum of the corrected heating capacities for all of the zones connected to the ground loop. A table (24) is provided to sum the corrected heating capacities for all space heating zones connected to the ground loop.
2. The corrected electrical demand (Figure 3.1, 3.2, or 3.3) of the heat pump at design heating conditions for the zone for which the ground loop is being designed. For multiple zones on a single ground loop, this is the sum of the corrected electrical demands for all of the zones connected to the ground loop. A table (24) is provided to sum the corrected electrical demands for all space heating zones connected to the ground loop.
3. The circulating fluid flow rate (Figure 3.1, 3.2, or 3.3) through the heat pump at design heating conditions for the zone for which the ground loop is being designed. For multiple zones on a single ground loop, this is the sum of the circulating fluid flow rates for all of the zones connected to the ground loop. A table (24) is provided to sum the circulating fluid flow rates for all space heating zones connected to the ground loop.
4. Obtained from an annual bin analysis on the design total heat loss (6) at the installation location. Whether for single or multiple space heating zones, it is important that the GSHP heating capacity and coefficient of performance for the heat pumps used in the bin analysis reflect the corrected heating capacity (1) and corrected electrical demand (2) GSHP data used for the vertical GHEX design.
5. Obtained from an annual bin analysis on the design total heat gain (46) at the installation location. Whether for single or multiple space cooling zones, it is important that the GSHP cooling capacity and energy efficiency ratio for the heat pumps used in the bin analysis reflect the corrected total cooling capacity (41) and corrected electrical demand (42) GSHP data used for the vertical GHEX design.
6. The ACCA Manual J heat loss (Figure 3.1, 3.2, or 3.3) for the zone for which the ground loop is being designed. For multiple zones on a single ground loop, this is the sum of the heat losses for all of the zones connected to the ground loop. A table (24) is provided to sum the total heat losses for all space heating zones connected to the ground loop.
7. The percent of total heat loss (6) that the corrected heating capacity (1) of the heat pump at design heating conditions is capable of providing. If this is less than 100%, a supplementary heating source is required, or the temperature of the heated zone may drop below the set point on the thermostat at design heating conditions.
8. Space heating load run fraction during the design heating month of January from a January bin analysis on a single or multiple zone Total Heat Loss (6) at the installation location. An alternate method to estimate the space heating load run fraction in January is to use Figure 3.9 with Percent Sizing (7) and heating degree day (HDD) data for the installation location from Table 3.4.
9. If no HWG option is used, then $HC_{HWG} = 0$. If HWG is used, refer to Table 9.4 for input that is dependent on HWG system type.
10. If no HWG option is used, then $DMD_{HWG} = 0$. If HWG is used, refer to Table 9.4 for input that is dependent on HWG system type.
11. If no HWG option is used, then $GPM_{HWG} = 0$. If HWG is used, refer to Table 9.4 for input that is dependent on HWG system type.
12. If no HWG option is used, then $AGL_{HWG} = 0$. If HWG is used, refer to Table 9.4 for input that is dependent on HWG system type.
13. If no HWG option is used, then $F_{HWG} = 0$. If HWG is used, refer to Table 9.4 for input that is dependent on HWG system type.
14. If no HWG option is used, then $HC_D = HC_C$. If HWG is used, refer to Table 9.4 for input dependent on HWG system type.

15. If no HWG option is used, then DMD_D = DMD_C. If HWG is used, refer to Table 9.4 for input dependent on HWG system type.
16. If no HWG option is used, then GPM_D = GPM. If HWG is used, refer to Table 9.4 for input dependent on HWG system type.
17. If no HWG option is used, then AGL_{HD} = AGL_H. If HWG is used, refer to Table 9.4 for input dependent on HWG system type.
18. AGL_{CD}=AGL_C.
19. The design coefficient of performance (Figure 3.1, 3.2, or 3.3) of the heat pump at design heating conditions for the zone for which the ground loop is being designed. For multiple zones and hot water generation on a single ground loop, this is calculated based on the design heating capacity (14) and design electrical demand (15) for all heating loads connected to the ground loop.
20. The design heat of extraction (Figure 3.1, 3.2, or 3.3) from the ground loop at design heating conditions for the zone for which the ground loop is being designed. For multiple zones and hot water generation on a single ground loop, this is calculated based on the design heating capacity (14) and design coefficient of performance (19) for all heating loads connected to the ground loop.
21. The design heating month run fraction that accounts for both space heating load and hot water generation load during the month of January.
22. The minimum entering water temperature for which the heat pump heating capacities (Figure 3.1, 3.2, or 3.3) have been selected for design heating conditions.
23. The minimum leaving water temperature based on the minimum entering water temperature (22), the design heat of extraction (20), and the design circulating fluid flow rate (16) at design heating conditions.
24. For multiple space heating zones, utilize this table to sum the Total Heat Loss, HC_C, DMD_C, and GPM for each zone for input into Total Heat Loss (6), HC_C (1), DMD_C (2), and GPM (3).
25. The nominal diameter of the U-bend used in the vertical borehole. Common U-bend diameters are ¾, 1, and 1-1/4-inch, and the size that is used is based on the flow rate through the U-bend as discussed in Section 5.1.3.
26. The flow rate through each flow path must be high enough for turbulence (Section 4.4.2), but in a range that keeps head loss at a reasonable level (Section 4.4.3) at design heating conditions. For most circulating fluids, recommended flow rates of 3 gpm for ¾-inch, 4.5 to 6 gpm for 1-inch, and 6 to 12 gpm for 1-1/4-inch pipe will provide a good trade off between turbulence, head loss, and the resulting design length of the vertical borehole.
27. The number of flow paths for the circulating fluid is estimated by dividing the design system flow rate (16) by the flow rate through each flow path (26). If the result is not a whole number, then round to the nearest whole number and, for the case of rounding up to the next highest whole number, increase the total system flow rate accordingly when designing the pumping system.
28. The number of vertical boreholes that have U-bends hooked together in series. That is, flow through one U-bend is diverted directly down a second (and possibly a third, fourth, etc.) U-bend before the flow is connected back to the return header returning flow to the heat pump system. This isn't common, but there are times when drilling conditions do not allow for the desired depth needed to provide the required heat transfer for the flow rate through the U-bend to be achieved economically. As an example, if 3 gpm is being circulated through a ¾-inch U-bend, which is the flow rate associated with about 1 ton of heat pump heating capacity, and the calculated total borehole design length was 200 feet, then 200 feet of drilled borehole below the header trench would be required. If difficult drilling were encountered at about 100 feet, then it would be acceptable to drill two, 100-foot deep boreholes and connect the U-bends in each of the boreholes in series to achieve proper flow through each U-bend (3 gpm) while having sufficient borehole length (200 feet) to provide the required heat transfer for the 1 ton of connected heat pump heating capacity. Care must be taken to account for variations in the thermal

conductivity of the soil or rock with depth when designing these "series-parallel" vertical GHEX systems. This is discussed in Section 5.1.3.1.

29. The number of vertical boreholes is calculated by multiplying the number of flow paths (27) and the number of bores in series (28).
30. The number of rows of vertical boreholes in the GHEX. In many residential ground loop designs, the boreholes can be arranged in a single row, but if the length of the area being used for the vertical GHEX is limiting, then more than one row can be used.
31. The number of vertical boreholes in each row in the GHEX, which is the number of vertical bore holes (29) divided by the number of rows of vertical boreholes (30).
32. The average center-to-center spacing between the boreholes. A minimum spacing of 15 feet is recommended as a starting point in the design. Spacing may be increased to accommodate an unbalanced annual ground energy load. A spacing of less than 15 feet is not recommended unless area for the vertical GHEX is extremely limited, and then care must be taken to carefully evaluate the design length for unbalanced annual ground energy load.
33. The deep earth temperature into which the vertical GHEX is being installed. May be estimated from groundwater temperatures in Figure 5.11, from a measured well water temperature at or near the installation location, or by direct measurement using methods specified by formation thermal conductivity test protocol.
34. The average thermal conductivity of the soil or rock into which the vertical GHEX is being installed. May be estimated based on a best guess of the soil or rock type that is being drilled into, by analyzing a drill log obtained from a well drilled at or near the location, or by direct measurement using methods specified by formation thermal conductivity test protocol.
35. The borehole diameter, which usually ranges between 4 and 6 inches depending on the type of drilling being done, the type of soil or rock being drilled into, and the need for casing in the borehole.
36. The grout thermal conductivity, which is defined by the relative mixture of the grout base material (bentonite or cement) and any additive that may be used for thermal enhancement (silica sand or other thermal enhancement compound). Grout thermal conductivity ranges between 0.40 and 1.20 Btu/hr ft F, with values in the middle and lower half of that range being most common for residential and light commercial applications.
37. This is the ground loop resistance, which depends on many factors described in Sections 5.2.2.2 and 5.2.2.3 (run fraction during design heating month (21), soil or rock thermal conductivity (34), borehole diameter (35), and grout thermal conductivity (36)), presented in Tables 5.8, 5.9, and 5.10 for nominal U-bend diameters of ¾, 1, and 1-1/4-inch, respectively.
38. The total borehole design length for heating (Equation 5.1) and the individual borehole design length for heating (Equation 5.2) are calculated using the indicated design parameters from the design sheet.
39. If the annual ground energy load is unbalanced, then the total and individual borehole design lengths for heating must be adjusted using the calculations and data provided in Section 5.2.3. The NNAGL is calculated as shown and used with the bore length multiplier figure (Figures 5.13 through 5.18) for the closest k_G to estimate B_M for the appropriate borehole spacing and loopfield layout rows x bores/row combination. The unbalanced ground load borehole design lengths are obtained by multiplying the borehole design lengths from (38) by B_M.
40. This table has been included to summarize heating design length calculations for various combinations of design parameters.

Vertically-Bored GHEX Design Worksheet – Cooling Mode

Space Cooling GSHP Design Data

TC_C[1] = **41** Btu/hr
DMD_C[1] = **42** kW
GPM[1] = **43** gpm
AGL_H = **44** Btu
AGL_C = **45** Btu
Total Heat Gain[1] = **46** Btu/hr
Percent Sizing = **47** %
F_{Jul} = **48**

HWG GSHP Design Data

GPM_{HWG} = **49** gpm
AGL_{HWG} = **50** Btu

Total Design Data

TC_D = **51** Btu/hr
DMD_D = **52** kW
GPM_D = **53** gpm
AGL_{HD} = **54** Btu
AGL_{CD} = **55** Btu

GHEX Design Data

EER_D[2] (= TC_D / (DMD_D x 1000)) = **56**
HR_D[2] (= TC_D x (EER_D+3.412)/EER_D) = **57** Btu/hr
F_C (= F_{Jul}) = **58**
EWT_{max}[3] = **59** F
LWT_{max}[2] (= EWT_{max} + HR_D/(500xGPM_D) = **60** F

1. For single heat pump installation, use data directly from the appropriate GSHP Selection Worksheet. For multiple heat pumps in the installation, sum all heat pump Total Heat Gains, TC_C's, DMD_C's, and GPM's that will be connected to a single GHEX using the table provided below.
2. For single heat pump installation, use data directly from appropriate GSHP Selection Worksheet. For multiple heat pumps, use the equation provided.
3. EWT_{max} is obtained directly from the appropriate GSHP Selection Worksheet and must be the same for the selection of all heat pumps for a multiple heat pump installation connected to a single GHEX. **(61)**

	Zone 1	Zone 2	Zone 3	Zone 4	Zone 5	Zone 6	Zone 7	Zone 8	Zone 9	Zone 10	Total
Total Heat Gain											
TC_C											
DMD_C											
GPM											

Vertical Borefield Layout

U-bend D_P = **62** in
GPM_{FP} = **63** gpm/flowpath
N_{FP} = **64** flowpaths
N_{BIS} = **65** bores in series
N_B = **66** bores
Layout = **67** x **68** (N_{Rows} x $N_{Bores/Row}$)
Bore Spacing = **69** ft

Vertical Borefield Design Data

T_G = **70** F
k_G = **71** Btu/hr ft F (Table 5.6)
D_{bore} = **72** in
k_{Grout} = **73** Btu/hr ft F
($R_B + R_G \cdot F_C$) = **74** hr ft F/Btu (Table 5.8, 5.9 or 5.10)

Borehole Design Lengths (Equations 5.3 and 5.4) **(75)**

$$L_{C,T} = \frac{TC_D \cdot \left(\frac{EER_D + 3.412}{EER_D}\right) \cdot (R_B + R_G \cdot F_C)}{\left(\frac{EWT_{max} + LWT_{max}}{2}\right) - T_G} = \frac{\_\_\_ \cdot \left(\frac{\_\_\_ + 3.412}{\_\_\_}\right) \cdot (\_\_\_)}{\left(\frac{\_\_\_ + \_\_\_}{2}\right) - \_\_\_} = \frac{\_\_\_ \cdot (\_\_\_) \cdot (\_\_\_)}{(\_\_\_) - \_\_\_} = \_\_\_ \text{ ft of bore}$$

$$L_{C,B} = \frac{L_{C,T}}{N_B} = \frac{\_\_\_}{\_\_\_} = \_\_\_ \text{ ft per bore (Below Header Trench)}$$

Unbalanced Ground Load Borehole Design Lengths (Equations 5.9b, 5.12, and 5.13) **(76)**

$$NNAGL = \frac{AGL_{CD} - AGL_{HD}}{L_{C,T} \cdot \left(\left(\frac{EWT_{max} + LWT_{max}}{2}\right) - T_G\right)} = \frac{\_\_\_ - \_\_\_}{\_\_\_ \cdot \left(\left(\frac{\_\_\_ + \_\_\_}{2}\right) - \_\_\_\right)} = \_\_\_ \text{ Btu / ftF}$$

B_M = ___ (Figures 5.13 through 5.18)

$$L_{C,T,UGL} = B_M \cdot L_{C,T} = \_\_\_ \cdot \_\_\_ = \_\_\_ \text{ ft of bore} \qquad L_{C,B,UGL} = \frac{L_{C,T,UGL}}{N_B} = \frac{\_\_\_}{\_\_\_} = \_\_\_ \text{ ft per bore (Below Header Trench)}$$

Cooling Design Length Calculations Summary Table **(77)**

Layout for Flow and Number of Bores						Design Length Calculations				Unbalanced Ground Load Design Lengths					
D_P	GPM_{FP}	N_{FP}	N_{BIS}	N_B	Layout	k_{Grout}	R_B+R_G·F_C	$L_{C,T}$	$L_{C,B}$	Layout	Spacing	NNAGL	B_M	$L_{C,T,UGL}$	$L_{C,B,UGL}$
					/										
					/										
					/										
					/										
					/										

Figure 5.20. Vertically-Bored GHEX Design Worksheet – Cooling Mode.

5.2.4.2 COOLING MODE DESIGN WORKSHEET (FIGURE 5.20)

41. The corrected total cooling capacity (Figure 3.1, 3.2, or 3.3) of the heat pump at design cooling conditions for the zone for which the ground loop is being designed. For multiple zones on a single ground loop, this is the sum of the corrected total cooling capacities for all of the zones connected to the ground loop. A table (61) is provided to sum the corrected total cooling capacities for all space cooling zones connected to the ground loop.
42. The corrected electrical demand (Figure 3.1, 3.2, or 3.3) of the heat pump at design cooling conditions for the zone for which the ground loop is being designed. For multiple zones on a single ground loop, this is the sum of the corrected electrical demands for all of the zones connected to the ground loop. A table (61) is provided to sum the corrected electrical demands for all space cooling zones connected to the ground loop.
43. The circulating fluid flow rate (Figure 3.1, 3.2, or 3.3) through the heat pump at design cooling conditions for the zone for which the ground loop is being designed. For multiple zones on a single ground loop, this is the sum of the circulating fluid flow rates for all of the zones connected to the ground loop. A table (61) is provided to sum the circulating fluid flow rates for all space cooling zones connected to the ground loop.
44. Obtained from an annual bin analysis on the design total heat loss (6) at the installation location. Whether for single or multiple heating zones, it is important that the GSHP heating capacity and coefficient of performance for the heat pumps used in the bin analysis reflect the corrected heating capacity (1) and corrected electrical demand (2) GSHP data used for the vertical GHEX design.
45. Obtained from an annual bin analysis on the design total heat gain (46) at the installation location. Whether for single or multiple cooling zones, it is important that the GSHP cooling capacity and energy efficiency ratio for the heat pumps used in the bin analysis reflect the corrected total cooling capacity (41) and corrected electrical demand (42) GSHP data used for the vertical GHEX design.
46. The ACCA Manual J heat gain (Figure 3.1, 3.2, or 3.3) for the zone for which the ground loop is being designed. For multiple zones on a single ground loop, this is the sum of the heat gains for all of the zones connected to the ground loop. A table (61) is provided to sum the total heat gains for all space cooling zones connected to the ground loop.
47. The percent of total heat gain (46) that the corrected total cooling capacity (41) of the heat pump at design cooling conditions is capable of providing. If this is less than 100%, then the temperature of the cooled zone may rise above the set point on the thermostat at design cooling conditions.
48. Space cooling load run fraction during the design cooling month of July from a July bin analysis on a single or multiple zone Total Heat Gain (46) at the installation location. An alternate method to estimate the space cooling load run fraction in July is to use Figure 3.10 with Percent Sizing (47) and cooling degree days (CDD) data for the installation location from Table 3.4.
49. If no HWG option is used, then $GPM_{HWG} = 0$. If HWG is used, refer to Table 9.5 for input that is dependent on HWG system type.
50. If no HWG option is used, then $AGL_{HWG} = 0$. If HWG is used, refer to Table 9.5 for input that is dependent on HWG system type.
51. TC_D=TC_C.
52. DMD_D=DMD_C.
53. If no HWG option is used, then GPM_D = GPM. If HWG is used, refer to Table 9.5 for input dependent on HWG system type.
54. If no HWG option is used, then $AGL_{HD} = AGL_H$. If HWG is used, refer to Table 9.5 for input dependent on HWG system type.
55. AGL_{CD}=AGL_C.
56. The design energy efficiency ratio (Figure 3.1, 3.2, or 3.3) of the heat pump at design cooling conditions for

the zone for which the ground loop is being designed. For multiple zones on a single ground loop, this is calculated based on the design total cooling capacity (51) and design electrical demand (52) for all cooling loads connected to the ground loop.

57. The design heat of rejection (Figure 3.1, 3.2, or 3.3) to the ground loop at design cooling conditions for the zone for which the ground loop is being designed. For multiple zones on a single ground loop, this is calculated based on the design total cooling capacity (51) and design energy efficiency ratio (56) for all cooling loads connected to the ground loop.
58. The design cooling month run fraction that accounts for space cooling loads during the month of July.
59. The maximum entering water temperature for which the heat pump cooling capacities (Figure 3.1, 3.2, or 3.3) have been selected for design cooling conditions.
60. The maximum leaving water temperature based on the maximum entering water temperature (59), the design heat of rejection (57), and the design circulating fluid flow rate (53) at design cooling conditions.
61. For multiple space cooling zones, utilize this table to sum the Total Heat Gain, TC_C, DMD_C, and GPM for each zone for input into Total Heat Gain (46), TC_C (41), DMD_C (42), and GPM (43).
62. The nominal diameter of the U-bend used in the vertical borehole. Common U-bend diameters are ¾, 1, and 1-1/4-inch, and the size that is used is based on the flow rate through the U-bend as discussed in Section 5.1.3.
63. The flow rate through each flow path must be high enough for turbulence (Section 4.4.2), but in a range that keeps head loss at a reasonable level (Section 4.4.3) at design heating conditions. For most circulating fluids, the minimum recommended flow rates of 3 gpm for ¾-inch, 4.5 to 6 gpm for 1-inch, and 6 to 12 gpm for 1-1/4-inch pipe will provide a good trade off between turbulence, head loss, and the resulting design length of the vertical borehole.
64. The number of flow paths for the circulating fluid is estimated by dividing the design system flow rate (53) by the flow rate through each flow path (64). If the result is not a whole number, then round to the nearest whole number and, for the case of rounding up to the next highest whole number, increase the total system flow rate accordingly when designing the pumping system.
65. The number of vertical boreholes that have U-bends hooked together in series. That is, flow through one U-bend is diverted directly down a second (and possibly a third, fourth, etc.) U-bend before the flow is connected back to the return header returning flow to the heat pump system. This isn't common, but there are times when drilling conditions do not allow for the desired depth needed to provide the required heat transfer for the flow rate through the U-bend to be achieved economically. As an example, if 3 gpm is being circulated through a ¾-inch U-bend, which is the flow rate associated with about 1 ton of heat pump heating capacity, and the calculated total borehole design length were 200 feet, then 200 feet of drilled borehole below the header trench would be required. If difficult drilling were encountered at about 100 feet, then it would be acceptable to drill two, 100-foot deep boreholes and connect the U-bends in each of the boreholes in series to achieve proper flow through each U-bend (3 gpm) while having sufficient borehole length (200 feet) to provide the required heat transfer for the 1 ton of connected heat pump heating capacity. Care must be taken to account for variations in the thermal conductivity of the soil or rock with depth when designing these "series-parallel" vertical GHEX systems. This is discussed in Section 5.1.3.1.
66. The number of vertical boreholes is calculated by multiplying the number of flow paths (64) and the number of bores in series (65).
67. The number of rows of vertical boreholes in the GHEX. In many residential ground loop designs, the boreholes can be arranged in a single row, but if the length of the area being used for the vertical GHEX is limiting, then more than one row can be used.
68. The number of vertical boreholes in each row in the GHEX, which is the number of vertical bore holes (66) divided by the number of rows of vertical boreholes (67).
69. The average center-to-center spacing between the boreholes. A minimum spacing of 15 feet is recommended

as a starting point in the design. Spacing may be increased to accommodate an unbalanced annual ground energy load. A spacing of less than 15 feet is not recommended unless area for the vertical GHEX is extremely limited, and then, care must be taken to carefully evaluate the design length for unbalanced annual ground energy load.

70. The deep earth temperature into which the vertical GHEX is being installed. May be estimated from groundwater temperatures in Figure 5.11, from a measured well water temperature at or near the installation location, or by direct measurement using methods specified by formation thermal conductivity test protocol.
71. The average thermal conductivity of the soil or rock into which the vertical GHEX is being installed. May be estimated based on a best guess of the soil or rock type that is being drilled into, by analyzing a drill log obtained from a well drilled at or near the location, or by direct measurement using methods specified by formation thermal conductivity test protocol.
72. The borehole diameter, which usually ranges between 4 and 6 inches depending on the type of drilling being done, the type of soil or rock being drilled into, and the need for casing in the borehole.
73. The grout thermal conductivity, which is defined by the relative mixture of the grout base material (bentonite or cement) and any additive that may be used for thermal enhancement (silica sand or other thermal enhancement compound). Grout thermal conductivity ranges between 0.40 and 1.20 Btu/hr ft F, with values in the middle to lower half of that range being most common for residential and light commercial applications.
74. This is the ground loop resistance, which depends on many factors described in Sections 5.2.2.2 and 5.2.2.3 (run fraction during design cooling month (58), soil or rock thermal conductivity (71), borehole diameter (72), and grout thermal conductivity (73)), presented in Tables 5.8, 5.9, and 5.10 for nominal U-bend diameters of ¾, 1, and 1-1/4-inch, respectively.
75. The total borehole design length for cooling (Equation 5.3) and the individual borehole design length for cooling (Equation 5.4) are calculated using the indicated design parameters from the design sheet.
76. If the annual ground energy load is unbalanced, then the total and individual borehole design lengths for cooling should be adjusted using the calculations and data provided in Section 5.2.3. The NNAGL is calculated as shown and used with the bore length multiplier figure (Figures 5.13 through 5.18) for the closest k_G to estimate B_M for the appropriate borehole spacing and loopfield layout rows x bores/row combination. The unbalanced ground load borehole design lengths are obtained by multiplying the borehole design lengths from (75) by B_M.
77. This table has been included to summarize cooling design length calculations for various combinations of design parameters.

5.2.5 VERTICALLY-BORED GHEX DESIGN EXAMPLES

Appendix A contains examples of vertically-bored GHEX designs for a heating dominant installation (Appendix A.1), a cooling dominant installation (Appendix A.2), and a multiple heat pump installation (Appendix A.3).

5.3 Horizontally-Trenched, Closed-Loop GHEX Design

Heating and cooling design equations and data for horizontally-trenched, closed-loop GHEX design are presented in this section. The design equations are upgrades to the previous IGSHPA design equations and now include the influence of trench spacing on the required length of horizontal trench to meet the design heating or cooling ground load of the connected ground source heat pump equipment. The design data include earth temperature at the average pipe burial depth and thermal conductivity of the soil that will be encountered at depths encountered in horizontally-trenched GHEX systems. Soil thermal resistance tables are provided for several pipe configurations in the horizontal trench as a function of soil thermal conductivity. The effect of trench spacing is addressed by the use of a trench spacing multiplier (S_M), which accounts for interference between trenches during design heating or cooling mode operation. Horizontally-trenched GHEX design worksheets for both heating and cooling mode are provided to facilitate the design process and provide documentation of the design for easy record keeping.

The steps in designing the horizontally-trenched, closed-loop GHEX include:

1. Define the design heating or cooling load and select GSHP equipment (Section 3.4) based on the EWT_{min} or EWT_{max} and GPM that the GHEX will be designed for.
2. Lay out the GHEX according to Section 5.1.3 by selecting pipe sizes for the ground loops that meet turbulence and head loss constraints and using a header arrangement that balances flow between the ground loops and facilitates flushing and purging of the GHEX.
3. Estimate the design heating or cooling month run fraction using a bin analysis or graphed data.
4. Estimate the thermal properties (thermal conductivity and diffusivity) of the ground formation into which the horizontally-trenched GHEX will be installed.
5. Determine the soil temperature at the average depth that the horizontally-trenched GHEX will be installed at the end of the design month based on location and estimated thermal diffusivity of the soil.
6. Determine the soil resistance from tabled data for the selected pipe configuration in the trench accounting for estimated thermal conductivity of the soil.
7. For multiple trench installations, account for trench spacing on ground loop performance based on center-to-center trench spacing and the number of trenches in parallel.
8. Calculate the total design heating or cooling pipe length, and divide by the number of trenches to determine the length of pipe in each trench.
9. Calculate the length of each trench by dividing the pipe length per trench by the length of pipe per unit trench length based on the pipe configuration in the trench.
10. Revise the design, as necessary, to meet potential space limitations on the installation site by varying the depth and spacing of the trenches to adjust the length of the trenches, or, if necessary, consider another pipe configuration in the trench.

5.3.1 DESIGN EQUATIONS

5.3.1.1 HEATING DESIGN EQUATIONS The total pipe design length for heating can be estimated using Equation 5.14, which represents the required length of pipe arranged in a given configuration in a horizontal trench to provide the heat of extraction of the heat pump system at design heating conditions. The various parameters in Equation 5.14 include the heating capacity (HC_D) and the coefficient of performance (COP_D) of the heat pump at design heating conditions, including corrections for operating conditions not directly provided in the engineering specifications (Section 3.4). The design minimum entering (EWT_{min}) and leaving (LWT_{min}) water temperatures are also determined as part of the heat pump selection (Section 3.4). Run fraction in heating mode during the heating design month (F_{Jan}) was discussed in Section 3.6 and is combined with hot water generation run fraction (F_{HWG}), if utilized, for a design month run fraction in heating (F_H). (This is a conservative approach and will ensure that the design length calculated using Equation 5.1 will be long enough to provide for the hot water generation load.) The pipe thermal resistance (R_P) was discussed in Section 4.2 and values are provided in Table 4.6 for different DR ratios. IGSHPA standard specifies a minimum pressure rating of 160 psi for horizontal ground loop piping, which requires a maximum DR ratio of 11 for PE3408 pipe and has a thermal resistance of 0.141 hr ft F/Btu. The remaining design parameters ($T_{S,L}$, R_S, S_M, and P_M) will be discussed in the following sections.

$$L_{H,P} = \frac{HC_D \cdot \left(\frac{COP_D - 1}{COP_D}\right) \cdot \left(R_P + R_S \cdot P_M \cdot S_M \cdot F_H\right)}{T_{S,L} - \left(\frac{EWT_{min} + LWT_{min}}{2}\right)}$$ **Equation 5.14**

Where:

$L_{H,P}$ = total pipe design length for heating, ft
HC_D = heat pump heating capacity at design heating conditions, Btu/hr
COP_D = coefficient of performance at design heating conditions, dimensionless
R_P = pipe thermal resistance, hr ft F/Btu
R_S = soil thermal resistance at steady-state, hr ft F/Btu
P_M = multiplier to account for pipe diameter other than ¾ inch, dimensionless
S_M = multiplier to account for trench spacing, dimensionless
F_H = run fraction in heating mode during heating design month (January), dimensionless
$T_{S,L}$ = design soil temperature for heating at average horizontal GHEX pipe depth, F
EWT_{min} = minimum entering water temperature at design heating conditions, F
LWT_{min} = minimum leaving water temperature at design heating conditions, F

The per trench design length of pipe for heating can be calculated using Equation 5.15. The number of trenches is usually equal to the number of nominal tons of heat pump heating capacity, based on the use of nominal ¾-inch pipe in the trench. If larger diameter pipe is used in the trench, the relationship between flow rate, turbulence, and head loss must be balanced as discussed in Section 5.1.3.

$$L_{H,P/T} = \frac{L_{H,P}}{N_T}$$ **Equation 5.15**

Where:

$L_{H,P/T}$ = design length of pipe in each trench for heating, ft
N_T = number of trenches that the total pipe design length is divided among

The trench design length for heating can be calculated using Equation 5.16. The number of pipes in each trench is determined by the configuration of the pipe in the trench, which will be addressed in Section 5.3.2.

$$L_{H,T} = \frac{L_{H,P/T}}{N_P}$$ **Equation 5.16**

Where:

$L_{H,T}$ = design length of each trench for heating, ft
N_P = number of pipes in each trench, ft_{pipe}/ft_{trench}

5.3.1.2 COOLING DESIGN EQUATIONS The total pipe design length for cooling can be estimated using Equation 5.17, which represents the required length of pipe arranged in a given configuration in a horizontal trench to absorb the heat of rejection of the heat pump system at design cooling conditions. The various parameters in Equation 5.17 include the total cooling capacity (TC_D) and the energy efficiency ratio (EER_D) of the heat pump at design cooling conditions, including corrections for operating conditions not directly provided in the engineering specifications (Section 3.4). The design maximum entering (EWT_{max}) and leaving (LWT_{max}) water temperatures are also determined as part of the heat pump selection (Section 3.4). Run fraction in cooling mode during the cooling design month (F_{Jul}) was discussed in Section 3.6 and represents the design month run fraction in cooling (F_C). The pipe thermal resistance (R_P) was discussed in Section 4.2 and values are provided in Table 4.6 for different DR ratios. IGSHPA standard specifies a minimum pressure rating of 160 psi for horizontal ground loop piping, which requires a maximum DR ratio of 11 for PE3408 pipe and has a thermal resistance of 0.141 hr ft F/Btu. The remaining design parameters ($T_{S,H}$, R_S, S_M, and P_M) will be discussed in the following sections.

$$L_{C,P} = \frac{TC_D \cdot \left(\frac{EER_D + 3.412}{EER_D}\right) \cdot \left(R_P + R_S \cdot P_M \cdot S_M \cdot F_C\right)}{\left(\frac{EWT_{max} + LWT_{max}}{2}\right) - T_{S,H}}$$ **Equation 5.17**

Where:

$L_{C,P}$ = total pipe design length for cooling, ft
TC_D = heat pump total cooling capacity at design cooling conditions, Btu/hr
EER_D = energy efficiency ratio at design cooling conditions, dimensionless
R_P = pipe thermal resistance, hr ft F/Btu
R_S = soil thermal resistance at steady-state, hr ft F/Btu
P_M = multiplier to account for pipe diameter other than ¾-inch, dimensionless
S_M = multiplier to account for trench spacing, dimensionless
F_C = run fraction in cooling mode during cooling design month (July), dimensionless
$T_{S,H}$ = design soil temperature for cooling at average horizontal GHEX pipe depth, F
EWT_{max} = maximum entering water temperature at design cooling conditions, F
LWT_{max} = maximum leaving water temperature at design cooling conditions, F

The per trench design length of pipe for cooling can be calculated using Equation 5.18. The number of trenches is usually equal to the number of nominal tons of heat pump heating capacity, based on the use of nominal ¾-inch pipe in the trench. If larger diameter pipe is used in the trench, the relationship between flow rate, turbulence, and head loss must be balanced as discussed in Section 5.1.3.

$$L_{C,P/T} = \frac{L_{C,P}}{N_T}$$ **Equation 5.18**

Where:

$L_{C,P/T}$ = design length of pipe for cooling in each trench, ft
N_T = number of trenches that the total pipe design length is divided among

The trench design length for cooling can be calculated using Equation 5.19. The number of pipes in each trench is determined by the configuration of the pipe in the trench, which will be addressed in Section 5.3.2.

$$L_{C,T} = \frac{L_{C,P/T}}{N_P}$$ **Equation 5.19**

Where:

$L_{C,T}$ = design length of each trench for cooling, ft
N_P = number of pipes in each trench, ft_{pipe}/ft_{trench}

5.3.2 DESIGN PARAMETERS

5.3.2.1 DESIGN SOIL TEMPERATURES ($T_{S,L}$ AND $T_{S,H}$) The temperature of the soil in the top 20 feet of the earth's surface varies with location, time of year, and depth. The proper design of the horizontally-bored GHEX depends on good estimates for the minimum ($T_{S,L}$) and maximum ($T_{S,H}$) soil temperatures at the average GHEX burial depth that occur during the design heating and cooling periods of January and July, respectively. The temperature of the soil can be estimated using Equation 5.20 (*Kusuda and Achenbach, 1965*), which requires an estimate for the soil thermal diffusivity, which is a function of soil type and moisture content (discussed in Section 5.3.2.2), along with the parameters T_M, A_S, and T_O, which are functions of location. Table 5.11 provides values for T_M, A_S, and T_O for 111 locations in the continental U.S., and those locations are mapped in Figure 5.21 for visual reference when selecting values that are not listed directly in Table 5.11. The best estimate for T_M and A_S for non-tabled locations would be linear interpolation between the two closest locations on opposite sides of the design site. Use a value for T_O from the closest tabled location with a similar latitude.

$$T(d,t) = T_M - A_S \bullet \mathrm{Exp}\left[-d \bullet \left(\frac{\pi}{365\alpha}\right)^{½}\right] \bullet \cos\left[\frac{2\pi}{365}\left(t - T_O - \frac{d}{2} \bullet \left(\frac{365}{\pi\alpha}\right)^{½}\right)\right]$$ **Equation 5.20**

Where:

$T(d,t)$ = earth temperature at soil depth (d) after (t) days from January 1, F
T_M = mean earth temperature in top 10 feet, F
A_S = earth surface temperature annual swing above and below T_M, F
T_O = number of days after January 1 to minimum earth surface temperature, days
d = depth into soil from earth's surface, ft
t = number of days after January 1, days
α = soil thermal diffusivity, ft^2/day
π = numerical constant pi, 3.1415927….

The annual variation in temperature, at depths ranging from 0 to 16 feet, is shown in Figure 5.22 for Louisville, KY (data from Table 5.11) in a soil with thermal diffusivity equal to 0.60 ft^2/day. The minimum surface temperature of T_M - A_S (60 – 22 = 38 F) occurs at T_O (33 days), and the maximum surface temperature of T_M + A_S (60 + 22 = 82 F) occurs at T_O+180 days (213 days). As depth increases, the amplitude of the soil temperature change is less and the minimum and maximum temperatures for the year are delayed. Equations 5.21 and 5.22 can be used to calculate the minimum and maximum temperatures and the days that they occur for any depth in the soil. As an example, at a 6-foot depth the minimum soil temperature is 49.3 F (much higher than the minimum surface temperature of 38 F) and occurs 75 days after January 1 (42 days after T_O), as calculated using Equation 5.21. The maximum soil temperature at a 6-foot depth is 70.7 F (much lower than the maximum surface temperature of 82 F) and occurs 257 days after January 1 (44 days after T_O+180), as calculated using Equation 5.22. At progressively deeper depths, the soil minimum and maximum temperatures approach T_M and are delayed further into the year, although the reductions in minimum and maximum temperature get smaller with depth, as indicated by the curves getting closer together with increasing depth. This is an indication that increasing the average depth of the horizontally-bored GHEX has an ever decreasing advantage toward achieving higher temperatures for heating design and lower temperatures for cooling design.

The influence of average depth on minimum and maximum soil temperatures can be more clearly seen by constructing a soil temperature funnel graph using Equations 5.21 and 5.22, as shown in Figure 5.23. Thermal diffusivities of 0.96, 0.60, and 0.26 ft^2/day have been used to represent the expected range between a very heavy, wet soil (0.96) down to a light, dry soil (0.26). An average thermal diffusivity of 0.60 ft^2/day is a good estimate for most soils, unless extremely wet or extremely dry conditions are known to consistently exist. As can be seen in Figure 5.23, higher soil thermal diffusivity (which is generally associated with a higher thermal conductivity, as discussed in Section 5.3.2.2) results in larger annual temperature swings with depth. For an average soil (α=0.60 ft^2/day) at a depth of 6 feet, the temperature can be expected to swing about 28% of the annual swing value (A_S) for the particular location of interest. At a depth of 8 feet, the temperature swing of that same soil is reduced to about 18%, and in even the wettest soils (α=0.96 ft^2/day) would be no more than about 26% of the annual swing value. With this knowledge, it is recommended that a minimum average burial depth for the pipes in a horizontally-trenched GHEX be between 6 and 8 feet. Horizontally-trenched GHEX average burial depths beyond 8 feet are not normally justified by further changes in soil temperature, but may be justified when the effect of soil depth on the thermal properties of the soil are considered (Section 5.3.2.2).

$$T(d,t_{min}) = T_M - A_S \bullet \text{Exp}\left[- d \bullet \left(\frac{\pi}{365\alpha} \right)^{1/2} \right]$$

Equation 5.21

Where:

$T(d,t_{min})$= annual minimum earth temperature at soil depth (d), F

t_{min} = number of days after January 1 when $T(d,t_{min})$ occurs

$$= T_O + \frac{d}{2} \bullet \left(\frac{365}{\pi\alpha} \right)^{1/2}$$

$$T(d,t_{max}) = T_M + A_S \bullet \text{Exp}\left[-d \bullet \left(\frac{\pi}{365\alpha}\right)^{1/2}\right]$$

Equation 5.22

Where:

$T(d,t_{max})$ = annual maximum earth temperature at soil depth (d), F

t_{max} = number of days after January 1 when $T(d,t_{max})$ occurs

$$= T_O + \frac{d}{2} \bullet \left(\frac{365}{\pi\alpha}\right)^{1/2} + \frac{365}{2}$$

Table 5.11. Earth Temperature Data for Selected U.S. Cities.

STATE	Map #	CITY	T_M(°F)	A_s(°F)	T_o
Alabama	1	Birmingham	65	19	31
	2	Montgomery	67	18	31
Arizona	3	Phoenix	73	23	33
	4	Tucson	70	18	34
Arkansas	5	Little Rock	64	21	32
California	6	Los Angeles	64	7	54
	7	Merced	68	25	33
	8	San Diego	64	7	54
Colorado	9	Colorado Springs	51	21	36
	10	Denver	52	22	37
	11	Grand Junction	55	25	32
DC	12	Washington	57	22	36
Flordia	13	Appalachicola	70	15	36
	14	Jacksonville	71	14	32
Georgia	15	Atlanta	62	19	32
	16	Augusta	65	18	30
	17	Macon	67	17	33
Idaho	18	Boise	53	21	34
	19	Idaho Falls	46	23	34
Illinois	20	Chicago	51	25	37
	21	E. St. Louis	57	24	34
	22	Urbana	53	26	35
Indiana	23	Fort Wayne	53	24	35
	24	Indianapolis	55	24	34
	25	South Bend	52	25	37
Iowa	26	Des Moines	52	28	35
	27	Sioux City	51	29	34
Kansas	28	Dodge City	57	25	35
	29	Topeka	56	26	35
Kentucky	30	Louisville	60	22	33
Louisiana	31	Lake Charles	70	16	32
	32	New Orleans	70	15	32
	33	Shreveport	66	19	32
Maine	34	Portland	48	22	39
Massachusetts	35	Plymouth	51	21	43
Michigan	36	Battle Creek	50	24	35
	37	Detroit	50	25	39
	38	Sau St. Marie	42	26	40

STATE	Map #	CITY	T_M(°F)	A_s(°F)	T_o
Minnesota	39	Duluth	41	28	37
	40	International Falls	39	31	34
	41	Minneapolis	47	29	35
Mississippi	42	Biloxi	70	17	32
	43	Columbus	65	19	32
	44	Jackson	67	18	31
Missouri	45	Columbia	57	24	35
	46	Kansas City	58	26	35
	47	Springfield	58	23	34
Montana	48	Billings	49	23	37
	49	Great Falls	48	23	36
	50	Missoula	46	21	32
Nebraska	51	Grand Island	52	27	35
	52	Lincoln	53	28	34
	53	North Platte	51	26	35
Nevada	54	Ely	47	22	35
	55	Las Vegas	69	23	32
	56	Winnemucca	52	22	33
New Jersey	57	Trenton	55	22	38
New Mexico	58	Albuquerque	59	22	31
	59	Roswell	63	22	30
New York	60	Albany	50	25	38
	61	Binghamton	48	24	38
	62	Niagra Falls	50	24	24
	63	Syracuse	50	24	38
North Carolina	64	New Bern	65	17	35
	65	Greensboro	60	20	31
North Dakota	66	Bismark	44	31	33
	67	Grand Forks	42	33	35
	68	Williston	45	29	34
Ohio	69	Akron	52	23	37
	70	Columbus	55	22	34
	71	Dayton	56	24	35
	72	Toledo	51	25	36
Oklahoma	73	Altus	65	24	33
	74	Oklahoma City	62	23	34
	75	Tulsa	62	23	34

STATE	Map #	CITY	T_M(°F)	A_s(°F)	T_o
Oregon	76	Astoria	53	9	45
	77	Medford	55	17	34
	78	Portland	54	13	37
Pennsylvania	79	Philidelphia	55	22	34
	80	Pittsburg	52	23	36
	81	Wilkes-Barre	52	23	36
	82	Middletown	55	23	35
South Carolina	83	Charleston	66	16	32
	84	Greenville	62	19	33
	85	Sumpter	65	18	32
South Dakota	86	Huron	47	30	35
	87	Rapid City	50	25	38
Tennesee	88	Bristol	59	20	32
	89	Knoxville	61	21	31
	90	Memphis	63	21	32
	91	Nashville	60	21	32
Texas	92	El Paso	66	20	28
	93	Fort Worth	68	21	34
	94	Houston	71	16	33
	95	San Antonio	72	16	32
Utah	96	Salt Lake	53	24	35
Vermont	97	Burlington	46	26	37
Virginia	98	Norfolk	61	20	37
	99	Richmond	60	19	33
	100	Roanoke	59	20	33
Washington	101	Seattle	53	12	36
	102	Spokane	49	21	32
	103	Moses Lake	53	23	29
West Virginia	104	Charleston	58	20	33
	105	Elkins	52	20	32
Wisconsin	106	Green Bay	46	26	37
	107	Madison	49	27	36
Wyoming	108	Casper	49	24	37
	109	Cheyenne	48	21	39
	110	Lander	46	24	38
	111	Sheridan	48	24	35

Figure 5.21. Earth Temperature Data Locations.

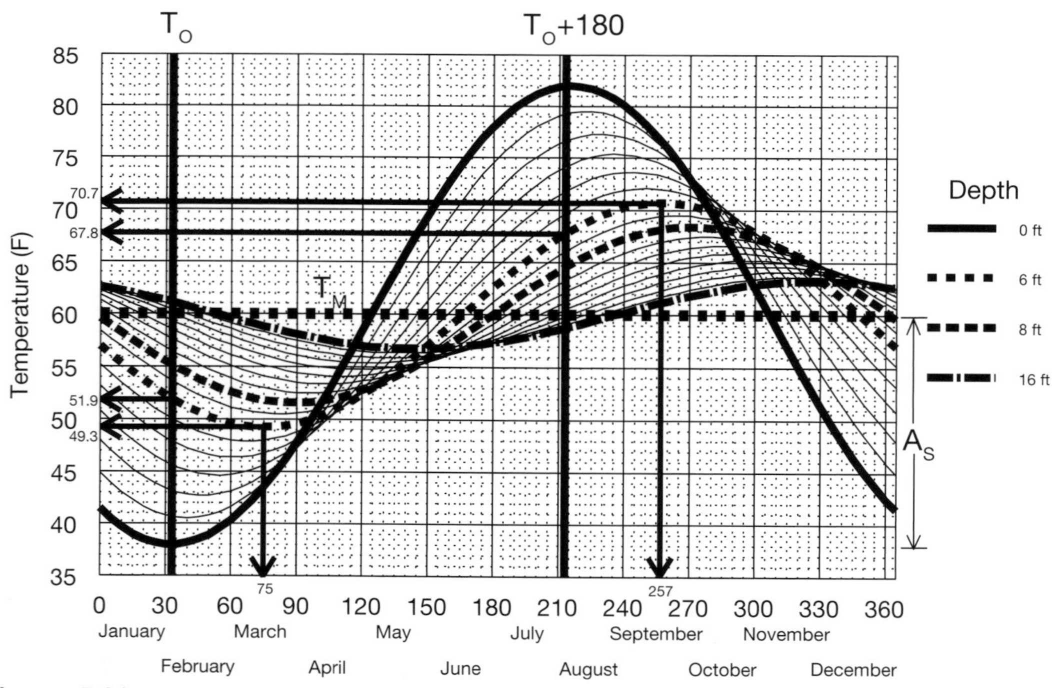

Figure 5.22. Soil Temperatures for Louisville, KY using Equation 5.20 with α=0.60 ft^2/day.

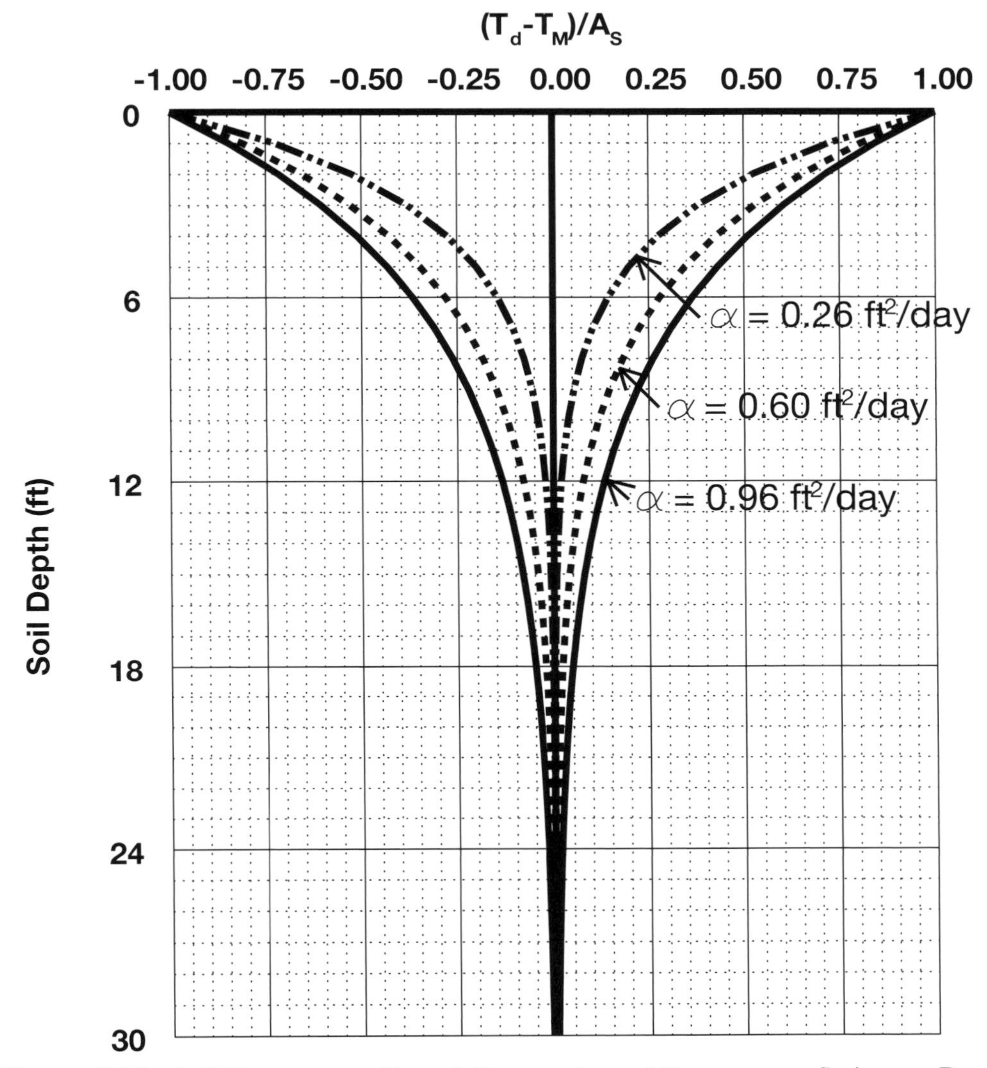

Figure 5.23. Soil Temperature Funnel Curve – Annual Temperature Swing vs. Depth

The minimum and maximum temperatures that are calculated by Equations 5.21 and 5.22 for a depth (d) in the soil occur at t_{min} and t_{max} days into the year which, as discussed above for the 6-foot depth in an average soil (α=0.60 ft²/day), will not occur until about 42 days after the day of minimum surface temperature (T_O) and 44 days after the day of maximum surface temperature (T_O+180). That would, for many locations, put the occurrence of minimum soil temperature at 6 feet about 75 days after January 1, or the middle to end of March, and put the occurrence of maximum soil temperature at 6 feet about 257 days after January 1, or the middle to end of September. However, as discussed in Section 3.2, the heating and cooling design months that result in the maximum heating and cooling loads that will be placed on the heat pump system, are January and July, respectively. That is, during the design heating month of January the horizontally-trenched GHEX must provide the maximum heat of extraction by the heat pump, and during the design cooling month of July the horizontally-trenched GHEX must absorb the maximum heat of rejection from the heat pump. The GHEX design length is therefore based on design parameters that are valid for January (heating length) and July (cooling length), and the minimum and maximum soil temperatures calculated using Equations 5.21 and 5.22 would be too low and high, respectively, because they occur much later in the year than the maximum heating and cooling design loads. Therefore, it would be valid to use soil temperatures at the average depth of pipe burial at or near the end of the design heating (January) and cooling (July) months, which for most locations would be closely predicted by the temperatures that occur at T_O and T_O+180, respectively.

Equations 5.23 and 5.24 express the minimum and maximum temperatures in the soil, in a dimensionless format, which are based on the assumption that the minimum soil temperature for heating design will occur at $t=T_O$ (early February for most locations), and the maximum soil temperature for cooling design will occur at $t=T_O+180$ (early August for most locations). The data in Tables 5.12 and 5.13 were generated from Equations 5.23 and 5.24 for a range of values for average pipe depth ($d=d_{avg}$) and soil thermal diffusivity (α) that would be expected in most horizontally-trenched GHEX applications.

$$T'(d_{avg},T_O) = -\,\text{Exp}\left[-d \bullet \left(\frac{\pi}{365\alpha}\right)^{1/2}\right] \bullet \cos\left[\frac{2\pi}{365}\left(-\frac{d}{2} \bullet \left(\frac{365}{\pi\alpha}\right)^{1/2}\right)\right] \qquad \textbf{Equation 5.23}$$

Where:

$T'(d_{avg},T_O)$ = dimensionless minimum soil temperature for heating design

$$= \frac{T(d,T_O) - T_M}{A_S}$$

$$T'(d_{avg}, T_O + 180) = -\,\text{Exp}\left[-d \bullet \left(\frac{\pi}{365\alpha}\right)^{1/2}\right] \bullet \cos\left[\frac{2\pi}{365}\left(180 - \frac{d}{2} \bullet \left(\frac{365}{\pi\alpha}\right)^{1/2}\right)\right] \qquad \textbf{Equation 5.24}$$

Where: $T'(d_{avg},T_O+180)$ = dimensionless maximum soil temperature for cooling design

$$= \frac{T(d,T_O + 180) - T_M}{A_S}$$

The minimum soil temperature for design of the horizontally-trenched GHEX for heating ($T_{S,L}$) can be determined at the average GHEX pipe depth (d_{avg}) in a soil with thermal diffusivity (α) by first obtaining the dimensionless minimum soil temperature ($T'(d_{avg},T_O)$) from Table 5.12 and plugging that into Equation 5.25 along with the A_S and T_M for the design site location. Likewise, the maximum soil temperature for design of the horizontally-trenched GHEX for cooling ($T_{S,H}$) can be determined at the average GHEX pipe depth (d_{avg}) in a soil with thermal diffusivity (α) by first obtaining the dimensionless maximum soil temperature ($T'(d_{avg},T_O+180)$) from Table 5.13 and plugging that into Equation 5.26 along with the A_S and T_M for the design site location.

$$T_{S,L} = T'(d_{avg},T_O) \bullet A_S + T_M \qquad \textbf{Equation 5.25}$$

$$T_{S,H} = T'(d_{avg},T_O + 180) \bullet A_S + T_M \qquad \textbf{Equation 5.26}$$

Table 5.12. Dimensionless Minimum Soil Temperature for Heating Design.[1]

α (ft²/day)	Soil Depth - d_{avg} (ft)							
	5	6	7	8	9	10	15	20
0.30	-0.2840	-0.1906	-0.1148	-0.0552	-0.0101	0.0226	0.0650	0.0328
0.35	-0.3232	-0.2299	-0.1520	-0.0887	-0.0387	-0.0005	0.0670	0.0434
0.40	-0.3568	-0.2642	-0.1854	-0.1197	-0.0662	-0.0239	0.0652	0.0521
0.45	-0.3858	-0.2944	-0.2153	-0.1481	-0.0923	-0.0468	0.0606	0.0585
0.50	-0.4112	-0.3212	-0.2423	-0.1743	-0.1167	-0.0689	0.0541	0.0630
0.55	-0.4337	-0.3452	-0.2668	-0.1984	-0.1396	-0.0900	0.0461	0.0657
0.60	-0.4538	-0.3669	-0.2891	-0.2206	-0.1610	-0.1100	0.0371	0.0669
0.65	-0.4719	-0.3865	-0.3096	-0.2411	-0.1810	-0.1290	0.0275	0.0668
0.70	-0.4884	-0.4045	-0.3284	-0.2602	-0.1998	-0.1470	0.0175	0.0656
0.75	-0.5033	-0.4209	-0.3457	-0.2779	-0.2174	-0.1641	0.0072	0.0635
0.80	-0.5171	-0.4361	-0.3618	-0.2944	-0.2339	-0.1803	-0.0031	0.0606
0.85	-0.5297	-0.4501	-0.3767	-0.3099	-0.2495	-0.1956	-0.0136	0.0571
0.90	-0.5414	-0.4631	-0.3907	-0.3244	-0.2642	-0.2101	-0.0239	0.0531
0.95	-0.5523	-0.4753	-0.4038	-0.3380	-0.2781	-0.2240	-0.0342	0.0487

1. $T'(d_{avg},T_O)$

Table 5.13. Dimensionless Maximum Soil Temperature for Cooling Design.[1]

α (ft²/day)	Soil Depth - d_{avg} (ft)							
	5	6	7	8	9	10	15	20
0.30	0.2699	0.1772	0.1025	0.0443	0.0007	-0.0304	-0.0669	-0.0324
0.35	0.3091	0.2161	0.1391	0.0770	0.0283	-0.0084	-0.0699	-0.0434
0.40	0.3426	0.2502	0.1720	0.1073	0.0550	0.0140	-0.0690	-0.0525
0.45	0.3717	0.2803	0.2016	0.1353	0.0804	0.0362	-0.0653	-0.0595
0.50	0.3972	0.3071	0.2284	0.1611	0.1044	0.0576	-0.0596	-0.0645
0.55	0.4198	0.3311	0.2528	0.1849	0.1269	0.0782	-0.0523	-0.0678
0.60	0.4401	0.3527	0.2750	0.2069	0.1480	0.0978	-0.0441	-0.0695
0.65	0.4583	0.3724	0.2954	0.2272	0.1677	0.1165	-0.0351	-0.0700
0.70	0.4749	0.3904	0.3142	0.2462	0.1863	0.1342	-0.0256	-0.0693
0.75	0.4900	0.4070	0.3315	0.2638	0.2037	0.1510	-0.0159	-0.0677
0.80	0.5039	0.4222	0.3476	0.2803	0.2201	0.1670	-0.0059	-0.0653
0.85	0.5167	0.4363	0.3626	0.2957	0.2356	0.1821	0.0041	-0.0623
0.90	0.5285	0.4494	0.3766	0.3102	0.2502	0.1965	0.0140	-0.0587
0.95	0.5395	0.4617	0.3897	0.3238	0.2640	0.2102	0.0239	-0.0547

1. $T'(d_{avg},T_O+180)$

5.3.2.2 SOIL THERMAL RESISTANCE (R_S) AND TRENCH SPACING MULTIPLIER (S_M)

The resistance to heat transfer in the soil around the buried piping of a horizontally-trenched GHEX, for an average depth of at least 5 feet, is a function of the thermal conductivity of the soil, the number, diameter, and configuration of the pipes in the trench, and the distance between horizontal trenches for multiple trench systems. Soil thermal conductivity and diffusivity will be discussed first, followed by presentation of soil resistance (R_S) as a function of soil thermal conductivity for common pipe configurations using ¾-inch pipe in the horizontally-trenched GHEX. A multiplier to account for thermal interference between adjacent trenches will be defined in terms of trench spacing distance and number of trenches in parallel. Pipe diameter will be considered in the next section.

The thermal conductivity of the soil depends on the type of soil (which is defined by the percentage of sand, silt, and clay), the density of the soil, and the water content of the soil. For a given soil type, the density does not change significantly, but the water content can change during the year, depending on the type of soil and the level of precipitation for that location. Sandy soils do not hold water well and will tend to exhibit large swings in water content and thermal conductivity depending on season and precipitation levels, while clay soils tend to hold water very well, and the water content and thermal conductivity will not vary as much throughout the year. For the design of the horizontally-trenched GHEX, it is important to select the lowest soil thermal conductivity that could be expected for the soil type and location of the installation. Because the moisture and thermal stability of soil generally increases with depth in the soil, higher design values for soil thermal conductivity can be used as the average depth of the horizontally-trenched GHEX piping system is increased.

ASHRAE (2009) provides apparent thermal conductivity values for different soil types (Table 5.14) based on work by *Salomone and Marlowe (1989)*. Included is the normal range that can be expected for each soil type, low values for the design of systems where ground heat exchange systems are being designed to meet worst-case conditions and high values for analysis of systems where maximum heat transfer rates are predicted. For the design of horizontally-trenched GHEX, the lower values shown in Table 5.14 are satisfactory unless better information is available.

Table 5.14. Typical Apparent Thermal Conductivity Values[1] for Soils – From *ASHRAE (2009)*.

Soil Type	Normal Range	Recommended Values for Design[2]	
		Low[3]	High[4]
Sands	0.35 – 1.45	0.45	1.30
Silts	0.50 – 1.45	0.95	1.30
Clays	0.50 – 0.95	0.65	0.90
Loams	0.50 – 1.45	0.55	1.30

1. Btu/hr ft F
2. Reasonable values for use when no site or soil-specific data are available.
3. Moderately conservative values for minimum heat loss through soil (use in soil heat exchanger or earth-contact cooling applications). Values are from Salomone and Marlowe (1989).
4. Moderately conservative values for maximum heat loss through soil (use in peak heat loss calculations). Values are from *Salomone and Marlowe (1989)*.

Remund (1998a,b) studied several different soil types in an effort to classify the thermal performance of common soils based on the percentage of sand, silt, and clay and the related densities and water contents that those soils would commonly exist at burial depths of horizontally-trenched GHEX. Soils containing fractions of sand, silt, and clay, ranging from nearly all clay to all sand, were tested. Data were gathered on each soil type to determine what densities and water contents each would naturally occur, including determination of the wilting point and field capacity water contents. According to *ASHRAE (2009)*, although thermal conductivity varies greatly over the complete range of possible water contents for a soil, that range can be narrowed if it is assumed that the water content of most field soils lie between the wilting point and the field capacity of the soil. Wilting point (W.P.) is the water content below which a plant cannot draw water from the soil through its root system and begins to wilt. The field capacity (F.C.) is the water content that the soil can maintain against drainage affected by gravity. Thermal conductivity was measured over the entire water content range at average dry density and one standard deviation either side of average.

Thermal conductivities at wilting point and field capacity for dry density were estimated from the data, as shown in Table 5.15. Average dry density, indicated by the "M" designation in the Field Density Class column, is considered the best estimate for which thermal conductivity and thermal diffusivity should be selected, unless better information is available. For soils that may be susceptible to drought, or extremely dry conditions, then the Wilting Point (W.P.) values for thermal diffusivity and thermal conductivity should be used for estimating soil design temperatures ($T_{S,L}$ and $T_{S,H}$) and soil resistance (R_S), respectively. For soils that will remain moist to wet throughout the year, the field capacity values for thermal diffusivity and thermal conductivity may be used for estimating soil design temperatures ($T_{S,L}$ and $T_{S,H}$) and soil resistance (R_S), respectively.

Table 5.15. Thermal Properties for Various Soils from *Remund (1998a,b)*.

Soil[1]	USDA System[2]			Field Density Class[3]	Soil Water Content		Thermal Conductivity[4] (Btu/hr ft F)		Thermal Diffusivity[5] (ft^2/day)	
	Sand (%)	Silt (%)	Clay (%)	(g/cm^3-lb/ft^3)	W.P.	F.C.	W.P.	F.C.	W.P.	F.C.
Benclare (Clay)	1.7	25.5	72.9	L (1.10-68.8)	30.1	39.9	0.42	0.55	0.29	0.33
				M (1.20-75.1)		37.2	0.51	0.62	0.33	0.36
				H (1.30-81.4)		35.1	0.63	0.70	0.37	0.38
Sharpsburg (Silty-Clay-Loam)	3.5	58.2	38.3	L (1.30-81.4)	16.5	31.3	0.39	0.77	0.32	0.45
				M (1.40-87.6)		28.6	0.53	0.79	0.41	0.46
				H (1.50-93.6)		26.5	0.64	0.83	0.46	0.47
Moody (Silty-Loam)	12.5	65.6	21.9	L (1.40-87.6)	9.1	23.5	0.40	0.75	0.39	0.49
				M (1.50-93.9)		21.7	0.48	0.83	0.43	0.52
				H (1.60-100)		20.1	0.60	0.92	0.50	0.58
Cecil (Clay)	35.0	17.8	47.2	L (1.40-87.6)	17.7	26.6	0.79	1.07	0.57	0.64
				M (1.50-93.9)		24.6	0.97	1.12	0.67	0.66
				H (1.60-100)		23.0	1.09	1.18	0.71	0.68
Kranzburg (Clay-Loam)	35.5	36.4	28.1	L (1.50-93.9)	11.5	21.7	0.60	0.90	0.50	0.57
				M (1.60-100)		20.2	0.73	1.00	0.58	0.62
				H (1.70-106)		18.7	0.88	1.05	0.65	0.58
Brookings (Sandy-Clay-Loam)	45.6	22.7	31.8	L (1.50-93.9)	13.1	21.1	0.64	1.03	0.50	0.65
				M (1.60-100)		19.6	0.80	1.05	0.60	0.66
				H (1.70-106)		18.1	0.93	1.05	0.65	0.65
Grovena (Sandy-Loam)	53.4	35.3	11.3	L (1.50-93.9)	4.9	13.6	0.40	0.81	0.42	0.64
				M (1.60-100)		13.4	0.51	0.92	0.51	0.68
				H (1.70-106)		12.6	0.62	1.04	0.58	0.75
Vienna (Sandy-Loam)	66.5	21.0	12.5	L (1.50-93.9)	5.9	15.6	0.47	0.92	0.48	0.68
				M (1.60-100)		15.4	0.62	0.97	0.60	0.68
				H (1.70-106)		14.6	0.74	1.05	0.67	0.72
Lamoure (Sandy-Loam) or (Loamy-Sand)	78.9	11.4	9.7	L (1.60-100)	4.2	12.1	0.47	0.87	0.48	0.68
				M (1.70-106)		11.9	0.65	1.10	0.63	0.80
				H (1.80-113)		11.1	0.80	1.21	0.74	0.87
Maddock (Loamy-Sand)	87.6	5.6	6.8	L (1.60-100)	3.3	11.7	0.42	0.97	0.45	0.77
				M (1.70-106)		11.5	0.49	1.07	0.50	0.80
				H (1.80-113)		10.7	0.50	0.20	0.48	0.87
Fordville (Sand)	100.0	0.0	0.0	L (1.60-100)	0.70	5.0	0.20	0.71	0.25	0.71
				M (1.70-106)		5.0	0.28	0.95	0.32	0.89
				H (1.80-113)		5.1	0.39	1.04	0.42	0.92

1. Soil series name - USDA textural class included in parenthesis.
2. According to USDA System sand, silt, and clay particle size limits.
3. M – average dry density for field soils below approximately 6 feet with indicated USDA particle size breakdown and should be used unless better information is available. L and H indicate one standard deviation below and above average, respectively.
4. Wilting point (W.P.) thermal conductivity is lowest expected value in the field. Field Capacity (F.C.) thermal conductivity is highest expected value in the field, except for saturated condition.
5. Wilting point (W.P.) thermal diffusivity is lowest expected value in the field. Field Capacity (F.C.) thermal diffusivity is highest expected value in the field, except for saturated condition.

Soil thermal resistance (R_S) represents the resistance to heat transfer between the outer wall of the buried pipe and the design heating or cooling soil temperature ($T_{S,L}$ or $T_{S,H}$), and depends on the diameter, number, and configuration of the pipe(s) in the trench, the thermal conductivity of the soil into which the GHEX is buried, and the number and spacing between trenches. The thermal conductivity of soils has been discussed above and commonly ranges from 0.5 to 1.1 Btu/hr ft F for soil conditions into which most horizontally-trenched GHEX systems are installed. However, thermal conductivity can be lower than 0.5 Btu/hr ft F in extremely dry sands and gravels or higher than 1.1 Btu/hr ft F in saturated sandy soils. Soil thermal resistance values have been calculated using CLGS (a program developed by IGSHPA to determine steady-state soil resistance for any pipe diameter, configuration in the trench, number and spacing of trenches, and soil thermal conductivity) for several commonly-used pipe configurations for a soil thermal conductivity range between 0.20 to 1.40 Btu/hr ft F. Tables 5.17 through 5.29 contain R_S values for several pipe configurations in "standing", "laying," and "rectangular" configurations that represent the most popular methods used for installing horizontally-trenched GHEX systems. Interpolation can be used to determine R_S for values of soil thermal conductivity between tabled values. All calculations for R_S are based on the use of ¾-inch pipe and a pipe multiplier (P_M), which will be discussed in Section 5.3.2.3, has been defined to adjust the soil resistance if 1 or 1-1/4-inch pipe is used. Included in Tables 5.17 through 5.29 is a trench spacing factor (S_M), which was developed to adjust soil resistance for thermal interference between parallel trenches and is expressed in terms of number of parallel trenches and center-to-center spacing (S_{cc}) between trenches. For more than six trenches in parallel, S_M does not increase significantly and can be assumed the same as for six trenches. Linear interpolation can be used to estimate S_M when either the number of trenches or S_{cc}, or both, fall between tabled values.

Table 5.16 summarizes the various piping configurations of Tables 5.17 through 5.29 and provides information helpful in the proper layout of these configurations based on pipe diameter and required flow rate for turbulence under most flow conditions. Also provided are recommended minimum piping depths in the trench and a brief description of the flow path in the trench. From a piping perspective, one, two, and four pipe configurations are usually achieved by unrolling the pipe and laying it out in the trench to achieve the number of pipe passes required. For five or more (sometimes four) pipes in the trench, the slinky is commonly used where the pipe is configured into multiple loops with proper offset spacing between loops to achieve the equivalent number of pipe passes in the trench. (IGSHPA's (1994) *Slinky Installation Guide* provides details on the layout and construction of the slinky to provide between 4 and 12 feet of pipe per foot of trench.) The "standing" configurations are generally installed in a trench constructed using a chain trencher, with a trench width usually between 6 and 12 inches. The "laying" and "rectangular" configurations would be installed in a trench constructed using a backhoe or excavator with a 24 to 42-inch wide bucket. Stacked "laying" configurations have been used when laying two slinkies in a single trench, where the trench has been excavated deeper than would be used for a single slinky, with a minimum of 3 feet of backfill placed between the slinky layers.

Table 5.16. Common Pipe and Trench Configurations for ¾, 1, and 1-¼-inch Pipe Diameters[1]

Table	Pipe Configuration[2]	# Flow Paths per Trench	D_P (in)	GPM per Flow Path	Single Trench Capacity (Ton)	Minimum Recommended Depths[3] (ft)			Common Application
						d_{avg}	d_{bp}	d_{tp}	
5.17	1-Pipe[4]	½	¾	3.0	½	6	6	6	Two trenches contain a single pipe, out in one trench and back in another, with one flow path.
			1	4.5	¾				
			1-¼	6.0	1				
5.18	2-Pipe Standing	1	¾	3.0	1	6	7	5	One trench contains two pipes, out-back, with one flow path.
			1	4.5	1-½				
			1-¼	6.0	2				
5.19	4-Pipe Standing[5]	1	¾	3.0	1	6	7.5	4.5	One trench contains 4 pipes, out-back-out-back or a 56" pitch slinky, with one flow path.
			1	4.5	1-½				
			1-¼	6.0	2				
5.20	5-Pipe Standing[5]	1	¾	3.0	1	6	7.5	4.5	One trench contains a 36" pitch slinky with one flow path.
			1	4.5	1-½				
			1-¼	6.0	2				
5.21	8-Pipe Standing[5]	1	¾	3.0	1	6	7.5	4.5	One trench contains an 18" pitch slinky with one flow path.
			1	4.5	1-½				
			1-¼	6.0	2				
5.22	2-Pipe Laying	1	¾	3.0	1	6	6	6	One trench contains two pipes, out-back, with one flow path.
			1	4.5	1-½				
			1-¼	6.0	2				
5.23	4-Pipe Laying[5]	1	¾	3.0	1	6	6	6	One trench contains 4 pipes, out-back-out-back or a 56" pitch slinky, with one flow path.
			1	4.5	1-½				
			1-¼	6.0	2				
5.24	5-Pipe Laying[5]	1	¾	3.0	1	6	6	6	One trench contains a 36" pitch slinky with one flow path.
			1	4.5	1-½				
			1-¼	6.0	2				
5.25	8-Pipe Laying[5]	1	¾	3.0	1	6	6	6	One trench contains an 18" pitch slinky with one flow path.
			1	4.5	1-½				
			1-¼	6.0	2				
5.26	4-Pipe Rectangular	1	¾	3.0	1	6	7	5	One trench contains 4 pipes, out-back-out-back, with one flow path.
			1	4.5	1-½				
			1-¼	6.0	2				
		2	¾	3.0	2				One trench contains 4 pipes, 2 out-2 back, with two parallel flow paths.
			1	4.5	3				
			1-¼	6.0	4				
5.27	2-Layer 4-Pipe Laying[5]	1	¾	3.0	1	7.5	9	6	One trench contains two-56" pitch slinkies, hooked in series with one flow path.
			1	4.5	1-½				
			1-¼	6.0	2				
		2	¾	3.0	2				One trench contains two-56" pitch slinkies, two parallel flow paths.
			1	4.5	3				
			1-¼	6.0	4				
5.28	2-Layer 5-Pipe Laying[5]	1	¾	3.0	1	7.5	9	6	One trench contains two-36" pitch slinkies, hooked in series with one flow path.
			1	4.5	1-½				
			1-¼	6.0	2				
		2	¾	3.0	2				One trench contains two-36" pitch slinkies, two parallel flow paths.
			1	4.5	3				
			1-¼	6.0	4				
5.29	2-Layer 8-Pipe Laying[5]	1	¾	3.0	1	7.5	9	6	One trench contains two-18" pitch slinkies, hooked in series with one flow path.
			1	4.5	1-½				
			1-¼	6.0	2				
		2	¾	3.0	2				One trench contains two-18" pitch slinkies, two parallel flow paths.
			1	4.5	3				
			1-¼	6.0	4				

1. Trench configurations assume header trench is located at one end of loopfield.
2. See indicated table for schematic of pipe configuration in the trench.
3. Minimum recommended average burial depths for the configuration (d_{avg}), for the bottom pipe (d_{bp}), and for top pipe (d_{tp}) and to reach thermally stable soil conditions. Burial depth (d_{avg} and associated d_{bp} and d_{tp}) should be adjusted downward (by up to 2 feet) in extremely cold or hot climates.
4. 1-Pipe configuration requires two trenches, one for the supply line away from the header trench and a second for the return line to the header trench, thus only ½ of the flow path exists in a single trench.
5. Configuration will be attained using a slinky with appropriate pitch to obtain equivalent feet of pipe per foot of trench.

Table 5.17. Soil Resistance (R_S) and Trench Spacing Multiplier (S_M) – 1-Pipe.

k_{soil}	Single Trench R_s	# Trenches	Trench Spacing Multiplier (S_M) Center-Center Trench Spacing (S_{cc}) 11	9	7	5
0.20	3.92	2	1.01	1.02	1.04	1.07
		4	1.01	1.02	1.05	1.12
		6	1.01	1.03	1.06	1.14
0.50	1.65	2	1.02	1.04	1.07	1.12
		4	1.03	1.06	1.11	1.20
		6	1.04	1.07	1.12	1.23
0.75	1.12	2	1.03	1.04	1.08	1.13
		4	1.04	1.07	1.13	1.23
		6	1.04	1.08	1.14	1.27
1.00	0.85	2	1.04	1.05	1.08	1.13
		4	1.05	1.08	1.14	1.25
		6	1.06	1.09	1.15	1.29
1.40	0.61	2	1.05	1.07	1.10	1.15
		4	1.07	1.10	1.16	1.28
		6	1.07	1.11	1.18	1.33

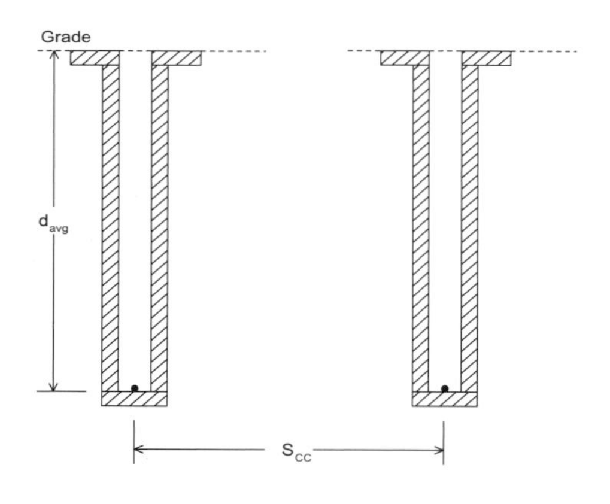

Table 5.18. Soil Resistance (R_S) and Trench Spacing Multiplier (S_M) – 2-Pipe Standing.

k_{soil}	Single Trench R_s	# Trenches	Trench Spacing Multiplier (S_M) Center-Center Trench Spacing (S_{cc}) 11	9	7	5
0.20	4.83	2	1.01	1.02	1.05	1.11
		4	1.01	1.04	1.08	1.18
		6	1.02	1.04	1.09	1.21
0.50	2.10	2	1.03	1.06	1.10	1.17
		4	1.05	1.09	1.16	1.30
		6	1.05	1.10	1.18	1.34
0.75	1.44	2	1.04	1.07	1.11	1.18
		4	1.06	1.10	1.19	1.33
		6	1.07	1.12	1.22	1.39
1.00	1.10	2	1.05	1.08	1.12	1.19
		4	1.07	1.13	1.21	1.36
		6	1.08	1.14	1.24	1.43
1.40	0.80	2	1.05	1.09	1.14	1.20
		4	1.09	1.14	1.23	1.39
		6	1.10	1.16	1.26	1.46

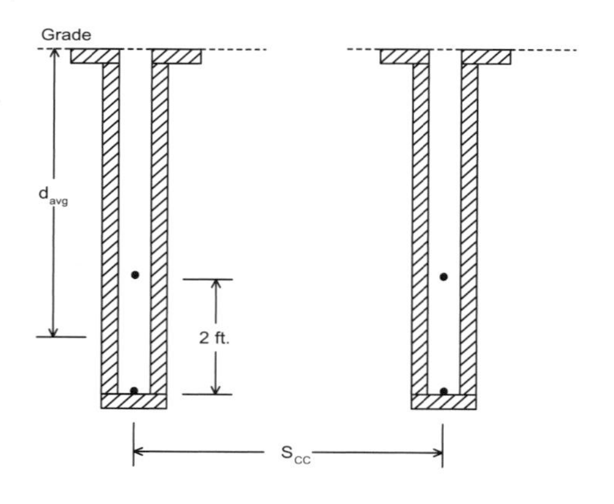

Table 5.19. Soil Resistance (R_S) and Trench Spacing Multiplier (S_M) – 4-Pipe[1] Standing.

k_{soil}	Single Trench R_s	# Trenches	Trench Spacing Multiplier (S_M) Center-Center Trench Spacing (S_{cc}) 11	9	7	5
0.20	7.30	2	1.01	1.03	1.07	1.15
		4	1.02	1.05	1.11	1.24
		6	1.02	1.05	1.12	1.27
0.50	3.27	2	1.04	1.07	1.12	1.21
		4	1.06	1.11	1.20	1.38
		6	1.06	1.12	1.23	1.44
0.75	2.26	2	1.05	1.09	1.15	1.23
		4	1.08	1.14	1.24	1.43
		6	1.09	1.15	1.27	1.50
1.00	1.73	2	1.06	1.10	1.16	1.25
		4	1.10	1.17	1.27	1.47
		6	1.12	1.18	1.31	1.55
1.40	1.27	2	1.07	1.11	1.17	1.26
		4	1.11	1.18	1.30	1.50
		6	1.13	1.20	1.35	1.60

1. Equivalent to a 56-inch pitch x 36-inch diameter slinky.

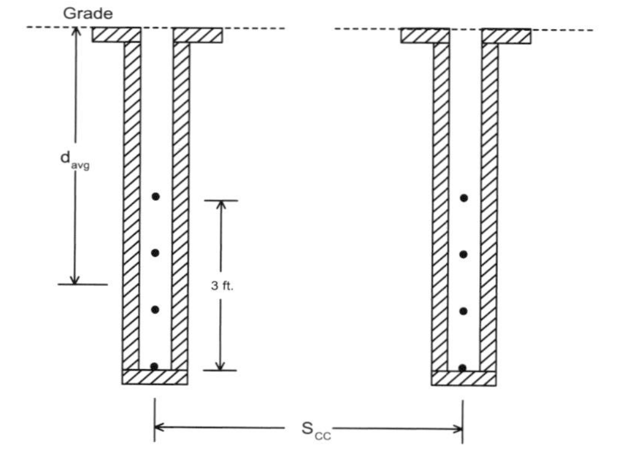

Table 5.20. Soil Resistance (R_S) and Trench Spacing Multiplier (S_M) – 5-Pipe[1] Standing.

k_{soil}	Single Trench R_s	# Trenches	Trench Spacing Multiplier (S_M) Center-Center Trench Spacing (S_{cc}) 11	9	7	5
0.20	8.84	2	1.01	1.03	1.07	1.15
		4	1.02	1.05	1.11	1.25
		6	1.02	1.04	1.13	1.29
0.50	3.98	2	1.04	1.07	1.13	1.22
		4	1.06	1.11	1.21	1.39
		6	1.07	1.11	1.23	1.45
0.75	2.75	2	1.05	1.09	1.15	1.24
		4	1.08	1.14	1.25	1.44
		6	1.09	1.15	1.28	1.52
1.00	2.11	2	1.07	1.10	1.16	1.26
		4	1.10	1.17	1.28	1.48
		6	1.12	1.18	1.32	1.57
1.40	1.55	2	1.07	1.12	1.17	1.27
		4	1.12	1.19	1.31	1.52
		6	1.13	1.20	1.35	1.61

1. Equivalent to a 36-inch pitch x 36-inch diameter slinky.

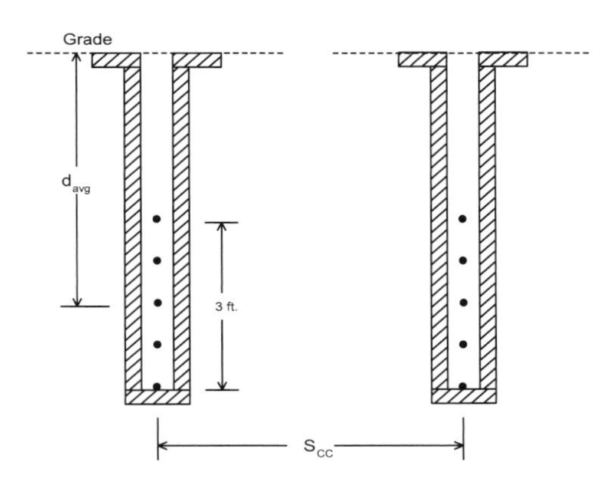

Table 5.21. Soil Resistance (R_S) and Trench Spacing Multiplier (S_M) – 8-Pipe[1] Standing.

k_{soil}	Single Trench R_s	# Trenches	Trench Spacing Multiplier (S_M) Center-Center Trench Spacing (S_{cc}) 11	9	7	5
0.20	13.70	2	1.01	1.03	1.08	1.16
		4	1.02	1.05	1.12	1.26
		6	1.02	1.06	1.13	1.30
0.50	6.19	2	1.04	1.08	1.13	1.23
		4	1.06	1.12	1.21	1.40
		6	1.07	1.13	1.24	1.31
0.75	4.29	2	1.06	1.09	1.15	1.25
		4	1.08	1.15	1.26	1.46
		6	1.09	1.17	1.29	1.53
1.00	3.28	2	1.07	1.11	1.17	1.27
		4	1.11	1.18	1.30	1.50
		6	1.13	1.20	1.34	1.59
1.40	2.41	2	1.08	1.12	1.19	1.29
		4	1.12	1.20	1.32	1.54
		6	1.14	1.22	1.37	1.64

1. Equivalent to a 18-inch pitch x 36-inch diameter slinky.

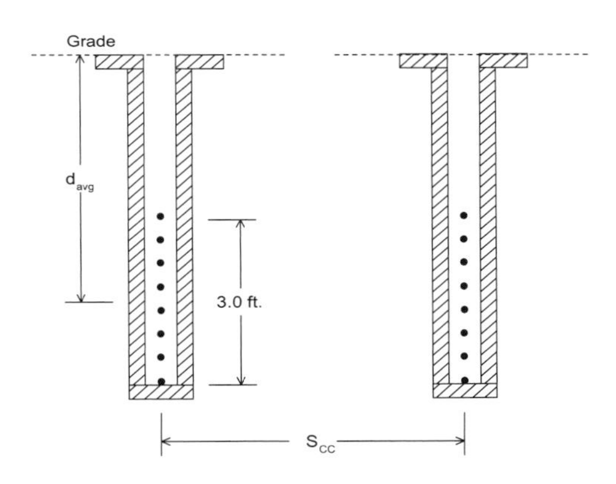

Table 5.22. Soil Resistance (R_S) and Trench Spacing Multiplier (S_M) – 2-Pipe Laying.

k_{soil}	S_H	Single Trench R_s	Trench Spacing Multiplier (S_M)				
			# Trenches	Center-Center Trench Spacing (S_{cc}) 11	9	7	5
0.20	1.0	5.36	2	1.01	1.02	1.05	1.11
			4	1.01	1.04	1.08	1.18
			6	1.02	1.04	1.09	1.21
	1.5	5.05	2	1.01	1.03	1.06	1.12
			4	1.02	1.04	1.09	1.20
			6	1.02	1.04	1.10	1.23
	2.0	4.83	2	1.01	1.03	1.06	1.14
			4	1.02	1.04	1.10	1.23
			6	1.02	1.04	1.11	1.26
0.50	1.0	2.32	2	1.03	1.05	1.09	1.16
			4	1.04	1.08	1.15	1.28
			6	1.05	1.09	1.17	1.33
	1.5	2.19	2	1.03	1.06	1.10	1.18
			4	1.05	1.09	1.16	1.31
			6	1.05	1.10	1.19	1.36
	2.0	2.10	2	1.03	1.06	1.11	1.19
			4	1.05	1.10	1.18	1.34
			6	1.06	1.10	1.20	1.39
0.75	1.0	1.58	2	1.04	1.07	1.11	1.18
			4	1.06	1.11	1.18	1.33
			6	1.07	1.12	1.21	1.39
	1.5	1.50	2	1.04	1.07	1.11	1.19
			4	1.07	1.11	1.19	1.35
			6	1.07	1.13	1.22	1.41
	2.0	1.44	2	1.04	1.08	1.13	1.21
			4	1.07	1.12	1.21	1.38
			6	1.08	1.13	1.03	1.44
1.00	1.0	1.21	2	1.04	1.07	1.12	1.18
			4	1.07	1.12	1.20	1.35
			6	1.08	1.13	1.23	1.41
	1.5	1.14	2	1.06	1.08	1.13	1.20
			4	1.09	1.13	1.22	1.39
			6	1.09	1.15	1.25	1.45
	2.0	1.10	2	1.05	1.08	1.14	1.21
			4	1.08	1.14	1.23	1.40
			6	1.09	1.15	1.26	1.47
1.40	1.0	0.88	2	1.05	1.08	1.13	1.19
			4	1.08	1.14	1.22	1.38
			6	1.09	1.15	1.25	1.44
	1.5	0.83	2	1.06	1.08	1.14	1.22
			4	1.10	1.14	1.24	1.41
			6	1.10	1.17	1.28	1.48
	2.0	0.80	2	1.06	1.09	1.15	1.23
			4	1.09	1.15	1.25	1.44
			6	1.10	1.16	1.29	1.51

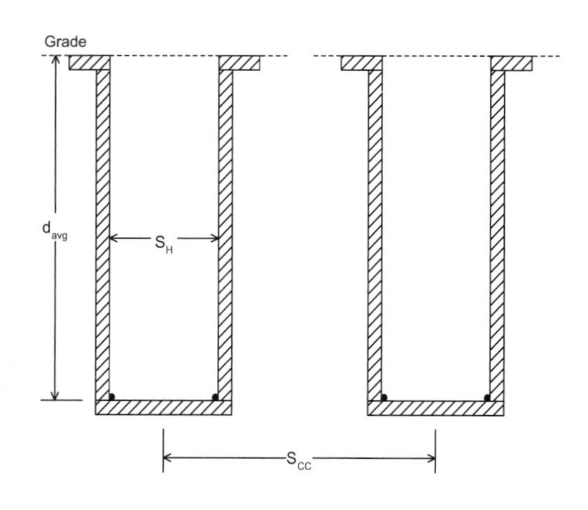

Table 5.23. Soil Resistance (R_S) and Trench Spacing Multiplier (S_M) – 4-Pipe[1] Laying.

k_{soil}	Single Trench R_s	Trench Spacing Multiplier (S_M)				
		# Trenches	Center-Center Trench Spacing (S_{cc}) 11	9	7	5
0.20	7.30	2	1.02	1.04	1.09	1.19
		4	1.02	1.06	1.14	1.31
		6	1.03	1.07	1.15	1.35
0.50	3.26	2	1.05	1.08	1.14	1.25
		4	1.07	1.13	1.23	1.44
		6	1.08	1.14	1.26	1.52
0.75	2.25	2	1.06	1.10	1.16	1.28
		4	1.09	1.16	1.27	1.49
		6	1.10	1.17	1.31	1.58
1.00	1.72	2	1.07	1.11	1.18	1.29
		4	1.11	1.18	1.30	1.53
		6	1.13	1.21	1.35	1.63
1.40	1.26	2	1.08	1.13	1.19	1.30
		4	1.13	1.20	1.33	1.56
		6	1.14	1.23	1.38	1.67

1. Equivalent to a 56-inch pitch x 36-inch diameter slinky.

Table 5.24. Soil Resistance (R_S) and Trench Spacing Multiplier (S_M) – 5-Pipe[1] Laying.

k_{soil}	Single Trench R_s	Trench Spacing Multiplier (S_M)				
		# Trenches	Center-Center Trench Spacing (S_{cc})			
			11	9	7	5
0.20	8.84	2	1.02	1.04	1.09	1.19
		4	1.03	1.06	1.14	1.32
		6	1.03	1.07	1.16	1.36
0.50	3.97	2	1.05	1.08	1.15	1.25
		4	1.07	1.13	1.23	1.45
		6	1.08	1.15	1.27	1.52
0.75	2.74	2	1.06	1.10	1.17	1.28
		4	1.09	1.16	1.28	1.50
		6	1.10	1.18	1.31	1.59
1.00	2.10	2	1.07	1.11	1.18	1.29
		4	1.11	1.18	1.30	1.54
		6	1.13	1.21	1.35	1.63
1.40	1.54	2	1.08	1.12	1.19	1.31
		4	1.12	1.20	1.33	1.57
		6	1.14	1.23	1.38	1.68

1. Equivalent to a 36-inch pitch x 36-inch diameter slinky.

Table 5.25. Soil Resistance (R_S) and Trench Spacing Multiplier (S_M) – 8-Pipe[1] Laying.

k_{soil}	Single Trench R_s	Trench Spacing Multiplier (S_M)				
		# Trenches	Center-Center Trench Spacing (S_{cc})			
			11	9	7	5
0.20	13.70	2	1.02	1.04	1.09	1.19
		4	1.03	1.06	1.18	1.32
		6	1.03	1.07	1.16	1.36
0.50	6.18	2	1.05	1.08	1.15	1.26
		4	1.07	1.13	1.27	1.46
		6	1.08	1.14	1.27	1.53
0.75	4.28	2	1.06	1.10	1.17	1.28
		4	1.09	1.16	1.31	1.51
		6	1.10	1.18	1.32	1.59
1.00	3.27	2	1.07	1.12	1.18	1.30
		4	1.12	1.19	1.34	1.54
		6	1.13	1.21	1.36	1.65
1.40	2.40	2	1.08	1.13	1.20	1.31
		4	1.13	1.21	1.37	1.58
		6	1.15	1.23	1.39	1.69

1. Equivalent to a 18-inch pitch x 36-inch diameter slinky.

Table 5.26. Soil Resistance (R_S) and Trench Spacing Multiplier (S_M) – 4-Pipe Rectangular.

k_{soil}	S_H	Single Trench R_s	# Trenches	Trench Spacing Multiplier (S_M) Center-Center Trench Spacing (S_{cc}) 11	9	7	5
0.20	1.0	7.09	2	1.01	1.03	1.08	1.16
			4	1.02	1.05	1.12	1.26
			6	1.02	1.06	1.13	1.30
	1.5	6.70	2	1.01	1.04	1.08	1.17
			4	1.02	1.06	1.13	1.29
			6	1.03	1.06	1.14	1.32
	2.0	6.39	2	1.02	1.04	1.09	1.19
			4	1.03	1.06	1.14	1.32
			6	1.03	1.07	1.16	1.36
0.50	1.0	3.18	2	1.04	1.07	1.13	1.22
			4	1.06	1.11	1.21	1.40
			6	1.07	1.13	1.24	1.46
	1.5	3.02	2	1.04	1.08	1.14	1.24
			4	1.07	1.12	1.23	1.42
			6	1.07	1.14	1.25	1.49
	2.0	2.89	2	1.05	1.09	1.15	1.26
			4	1.07	1.13	1.25	1.46
			6	1.08	1.15	1.28	1.53
0.75	1.0	2.20	2	1.05	1.09	1.15	1.24
			4	1.08	1.14	1.25	1.45
			6	1.09	1.16	1.28	1.52
	1.5	2.09	2	1.06	1.10	1.16	1.26
			4	1.09	1.15	1.27	1.48
			6	1.10	1.17	1.30	1.56
	2.0	2.00	2	1.07	1.11	1.18	1.29
			4	1.10	1.17	1.29	1.52
			6	1.11	1.19	1.33	1.60
1.00	1.0	1.68	2	1.07	1.08	1.17	1.26
			4	1.10	1.17	1.28	1.48
			6	1.11	1.18	1.32	1.57
	1.5	1.60	2	1.07	1.11	1.18	1.28
			4	1.11	1.18	1.30	1.51
			6	1.13	1.21	1.34	1.61
	2.0	1.53	2	1.08	1.12	1.19	1.30
			4	1.12	1.20	1.33	1.56
			6	1.13	1.22	1.37	1.65
1.40	1.0	1.23	2	1.07	1.11	1.18	1.28
			4	1.11	1.19	1.31	1.52
			6	1.13	1.21	1.36	1.62
	1.5	1.17	2	1.09	1.13	1.20	1.30
			4	1.13	1.21	1.33	1.56
			6	1.15	1.23	1.38	1.67
	2.0	1.13	2	1.08	1.13	1.19	1.31
			4	1.13	1.21	1.35	1.58
			6	1.14	1.24	1.40	1.70

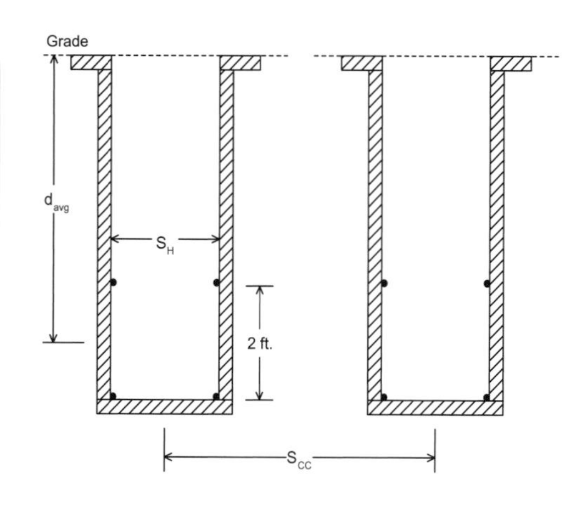

Table 5.27. Soil Resistance (R_S) and Trench Spacing Multiplier (S_M) – 2-Layer 4-Pipe[1] Laying.

k_{soil}	Single Trench R_s	# Trenches	Trench Spacing Multiplier (S_M) Center-Center Trench Spacing (S_{cc}) 11	9	7	5
0.20	9.46	2	1.02	1.05	1.12	1.25
		4	1.04	1.08	1.19	1.41
		6	1.04	1.09	1.21	1.46
0.50	4.45	2	1.06	1.11	1.19	1.33
		4	1.09	1.17	1.31	1.58
		6	1.10	1.19	1.35	1.67
0.75	3.12	2	1.08	1.13	1.21	1.36
		4	1.12	1.21	1.36	1.64
		6	1.13	1.23	1.41	1.75
1.00	2.41	2	1.10	1.15	1.23	1.37
		4	1.15	1.24	1.40	1.68
		6	1.17	1.27	1.46	1.81
1.40	1.78	2	1.11	1.17	1.25	1.39
		4	1.17	1.28	1.44	1.74
		6	1.19	1.31	1.51	1.88

1. Equivalent to a 56-inch pitch x 36-inch diameter slinky.

Table 5.28. Soil Resistance (R_S) and Trench Spacing Multiplier (S_M) – 2-Layer 5-Pipe[1] Laying.

k_{soil}	Single Trench R_s	# Trenches	Trench Spacing Multiplier (S_M) Center-Center Trench Spacing (S_{cc}) 11	9	7	5
0.20	11.58	2	1.02	1.05	1.12	1.25
		4	1.04	1.08	1.19	1.41
		6	1.04	1.09	1.21	1.46
0.50	5.46	2	1.06	1.11	1.19	1.33
		4	1.09	1.17	1.31	1.58
		6	1.10	1.19	1.35	1.67
0.75	3.83	2	1.08	1.13	1.21	1.36
		4	1.12	1.21	1.36	1.64
		6	1.13	1.23	1.41	1.75
1.00	2.96	2	1.10	1.15	1.23	1.37
		4	1.15	1.24	1.40	1.68
		6	1.17	1.27	1.46	1.81
1.40	2.20	2	1.11	1.17	1.25	1.39
		4	1.17	1.28	1.44	1.74
		6	1.19	1.31	1.51	1.88

Grade
d_{avg}
3 ft.
3 ft.
S_{CC}

1. Equivalent to a 36-inch pitch x 36-inch diameter slinky.

Table 5.29. Soil Resistance (R_S) and Trench Spacing Multiplier (S_M) – 2-Layer 8-Pipe[1] Laying.

k_{soil}	Single Trench R_s	# Trenches	Trench Spacing Multiplier (S_M) Center-Center Trench Spacing (S_{cc}) 11	9	7	5
0.20	18.14	2	1.02	1.05	1.12	1.25
		4	1.04	1.08	1.19	1.41
		6	1.04	1.09	1.21	1.46
0.50	8.59	2	1.06	1.11	1.19	1.33
		4	1.09	1.17	1.31	1.58
		6	1.10	1.19	1.35	1.67
0.75	6.04	2	1.08	1.13	1.21	1.36
		4	1.12	1.21	1.36	1.64
		6	1.13	1.23	1.41	1.75
1.00	4.67	2	1.10	1.15	1.23	1.37
		4	1.15	1.24	1.40	1.68
		6	1.17	1.27	1.46	1.81
1.40	3.46	2	1.11	1.17	1.25	1.39
		4	1.17	1.28	1.44	1.74
		6	1.19	1.31	1.51	1.88

Grade
d_{avg}
3 ft.
3 ft.
S_{CC}

1. Equivalent to a 18-inch pitch x 36-inch diameter slinky.

5.3.2.3 PIPE DIAMETER MULTIPLIER (P_M)

The soil thermal resistance (R_S) values provided in Tables 5.17 through 5.29 were developed using ¾-inch polyethylene pipe in the calculations. Using 1 or 1-¼-inch pipe is known to increase heat transfer (or reduce soil thermal resistance) because the pipe outside surface is located farther from the center of the pipe (a radial heat transfer issue) and has larger surface area. A simple correction for pipe diameter was developed by calculating soil thermal resistance for the 1 and 1-¼-inch pipes and taking those values and dividing them by the soil thermal resistance for the ¾-inch pipe for the corresponding soil thermal conductivity and pipe configuration, as defined by Equation 5.27. The resulting data were correlated by the number of pipes in the trench, averaged across all pipe configurations, and a curve-fit was fitted to the results, as shown in Figure 5.24. The curve-fit results for the corresponding range of number of pipes in the trench (N_P) are summarized in Table 5.30.

$$P_M = \frac{R_S(D_P,\, k_s,\, \text{configuration})}{R_S(\text{¾"},\, k_s,\, \text{configuration})}$$ **Equation 5.27**

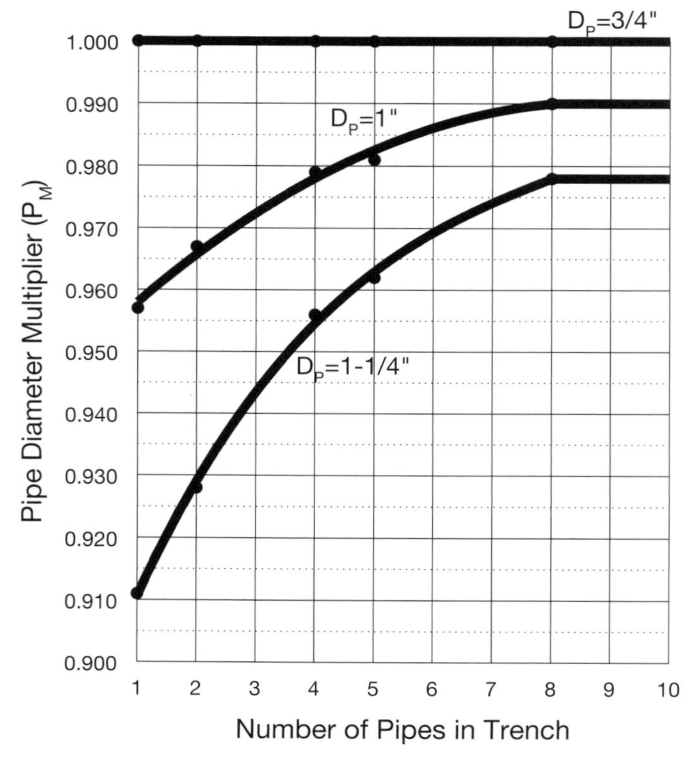

Figure 5.24. Pipe Diameter Multiplier (P_M) vs. Number of Pipes in Trench.

Table 5.30. Pipe Diameter Multiplier (P_M) for ¾, 1, and 1-¼-inch PE Pipe.

D_P (in)	Pipe Diameter Multiplier Equation	Notes[1]
¾	$P_M = 1.0$	R_S based on ¾" pipe
1	$P_M = 0.9492 + 0.0093N_P - 0.0005N_P^2$	for $1 \leq N_P < 8$
	$P_M = 0.990$	for $N_P \geq 8$
1-¼	$P_M = 0.8877 + 0.0254N_P - 0.0026N_P^2 + 0.0001N_P^3$	for $1 \leq N_P < 8$
	$P_M = 0.978$	for $N_P \geq 8$

1. N_P =number of pipes in the trench.

5.3.4 HORIZONTALLY-TRENCHED GHEX DESIGN WORKSHEETS Proper design of a horizontally-trenched GHEX requires reliable heating and cooling load design data and heat pump selection; consideration of domestic hot water generation; identification of installation location and soil identification; specification of horizontal trench design parameters and proper layout of the horizontal loopfield; and application of the heating or cooling design equations to determine the required pipe and trench lengths. Heating and cooling load design data are determined and heat pumps are selected according to procedures presented in Chapter 3. Worksheets were provided to document the selection of the water-to-air or water-to-water

GSHP units for the design space heating or cooling loads. Part of the heat pump selection is based on specification of the minimum and maximum entering water temperatures from the horizontally-trenched GHEX, along with the water flow rate that the GHEX must accommodate, each of which are critical parameters in the proper design of the horizontally-trenched GHEX system. Hot water generation, by either a dedicated water-to-water GSHP or a desuperheater in the space heating and cooling GSHP, will affect the horizontal loopfield design as discussed in Chapter 9, and procedures are provided to account for hot water generation in Tables 9.4 and 9.5. Installation location determines the soil temperatures at the average depth of pipe burial and soil type determines the thermal properties for design. Layout of the horizontally-trenched loopfield, in terms of number of flow paths, pipe diameter, and maximum flow distance are considered to accommodate required water flow rate to achieve proper turbulence for heat transfer, in a heating dominant design, and to control head loss to keep pumping power at an acceptable level at heating or cooling design conditions. Application of the appropriate design equation (Equation 5.14 or 5.17) using design data for the ground and trench configuration results in the required total length of pipe for heating or cooling. The piping configuration in the trench is then utilized to determine the length of the individual trenches.

In an effort to simplify the design of the horizontally-trenched GHEX and provide a method to easily document the design, worksheets have been developed for both heating mode dominant (Figure 5.25) and cooling mode dominant (Figure 5.26) design. The worksheets are broken into sections to allow input for heat pump design and energy analysis data, domestic hot water generation, location information and identification of soil thermal properties, loopfield layout, and design data for the horizontal GHEX for the calculation of the total length of pipe and trench. Additional calculations can be made to refine horizontally-trenched GHEX design lengths in a table at the bottom of each worksheet. Each input blank in Figures 5.25 and 5.26 has been labeled with a number identifying a short input description, a listing of which follows each figure. A blank copy of each figure is provided in Appendix B.

Horizontally-Trenched GHEX Design Worksheet – Heating Mode

Space Heating GSHP Design Data

HC_C[1] = 1 Btu/hr
DMD_C[1] = 2 kW
GPM[1] = 3 gpm
Total Heat Loss[1] = 4 Btu/hr
Percent Sizing = 5 %
F_{Jan} = 6

HWG GSHP Design Data

HC_{HWG} = 7 Btu/hr
DMD_{HWG} = 8 kW
GPM_{HWG} = 9 gpm
F_{HWG} = 10

Total Design Data

HC_D = 11 Btu/hr
DMD_D = 12 kW
GPM_D = 13 gpm

GHEX Design Data

COP_D[2] $(= HC_D / (DMD_D \times 3{,}412))$ = 14
HE_D[2] $(= HC_D \times (COP_D-1)/COP_D)$ = 15 Btu/hr
F_H $(= (F_{Jan} \times HC_C + F_{HWG} \times HC_{HWG}) / HC_D)$ = 16
EWT_{min}[3] = 17 F
LWT_{min}[2] $(= EWT_{min} - HE_D/(500 \times GPM_D))$ = 18 F

1. For single heat pump installation, use data directly from the appropriate GSHP Selection Worksheet. For multiple heat pumps in the installation, sum all heat pump Total Heat Losses, HC_C's, DMD_C's, and GPMs that will be connected to a single GHEX using the table provided below.
2. For single heat pump installation, use data directly from appropriate GSHP Selection Worksheet. For multiple heat pumps, use the equation provided.
3. EWT_{min} is obtained directly from the appropriate GSHP Selection Worksheet and must be the same for the selection of all heat pumps for a multiple heat pump installation connected to a single GHEX.

(19)

	Zone 1	Zone 2	Zone 3	Zone 4	Zone 5	Zone 6	Zone 7	Zone 8	Zone 9	Zone 10	Total
Total Heat Loss											
HC_C											
DMD_C											
GPM											

Horizontal Trench Design Data

Location 20 , ______ (City, State)
T_M = 21 F (Table 5.11)
A_s = 22 F (Table 5.11)
T_O = 23 Days (Table 5.11)

Soil Type 24 (Section 5.3.2.2)
k_{soil} = 25 Btu/hr ft F
α_{soil} = 26 ft²/day

Trench and Pipe Configuration ==> Table 27 (Select from Tables 5.17 – 5.29)
N_P = 28 ft_{pipe}/ft_{trench}
Conf. = 29 (Stand / Laying / Rect)
S_H = 30 ft
R_s = 31 hr ft F/Btu
S_{cc} = 32 ft
S_M = 33
d_{bp} = 34 ft
d_{avg} = 35 ft

D_P = 36 in (Nom.)
GPM_{FP} = 37 gpm/flowpath
N_{FP} = 38 flowpaths
$N_{FP/T}$ = 39 flowpaths/trench
N_T = 40 trenches

Design Soil Temperature for Heating

$T'(d_{avg},T_O)$ = 41 F (Table 5.12)
$T_{S,L} = T'(d_{avg},T_O)\bullet A_s+T_M$ = ______ • ______ + ______ = 42 F (Eq. 5.25)

Pipe Resistance and Pipe Multiplier

R_p = 43 hr ft F/Btu (Table 4.6)
P_M = 44 (Figure 5.24)

Horizontal Trench Design Lengths (Equations 5.14, 5.15, and 5.16) **(45)**

$$L_{H,P} = \frac{HC_D \cdot \left(\frac{COP_D - 1}{COP_D}\right)(R_P + R_S \cdot P_M \cdot S_M \cdot F_H)}{T_{S,L} - \left(\frac{EWT_{min} + LWT_{min}}{2}\right)} = \frac{\_\_\_ \cdot \left(\frac{\_\_\_ - 1}{\_\_\_}\right)(\_\_\_ + \_\_\_ \cdot \_\_\_ \cdot \_\_\_ \cdot \_\_\_)}{\_\_\_ - \left(\frac{\_\_\_ + \_\_\_}{2}\right)} = \frac{\_\_\_ \cdot (\_\_\_)(\_\_\_)}{\_\_\_ - (\_\_\_)} = \_\_\_ \text{ ft}$$

$$L_{H,P/T} = \frac{L_{H,P}}{N_T} = \frac{\_\_\_}{\_\_\_} = \_\_\_ \ ft_{pipe} / \text{trench} \qquad L_{H,T} = \frac{L_{H,P/T}}{N_P} = \frac{\_\_\_}{\_\_\_} = \_\_\_ \ ft_{trench}$$

Heating Design Length Calculations Summary Table **(46)**

Layout for Flow and Number of Trenches						Pipe and Trench Design Length Calculations								
D_P	GPM_{FP}	N_{FP}	Trench/N_P	$N_{FP/T}$	N_T	d_{avg}	$T'(d_{avg},T_o)$	$T_{S,L}$	R_s	P_M	S_{cc}/S_M	$L_{H,P}$	$L_{H,P/T}$	$L_{H,T}$
			/								/			
			/								/			
			/								/			
			/								/			
			/								/			

Figure 5.25. Horizontally-Trenched GHEX Design Worksheet – Heating Mode.

5.3.4.1 HEATING MODE DESIGN WORKSHEET (FIGURE 5.25)

1. The corrected heating capacity (Figure 3.1, 3.2, or 3.3) of the heat pump at design heating conditions for the zone for which the ground loop is being designed. For multiple zones on a single ground loop, this is the sum of the corrected heating capacities for all of the zones connected to the ground loop. A table (19) is provided to sum the corrected heating capacities for all space heating zones connected to the ground loop.
2. The corrected electrical demand (Figure 3.1, 3.2, or 3.3) of the heat pump at design heating conditions for the zone for which the ground loop is being designed. For multiple zones on a single ground loop, this is the sum of the corrected electrical demands for all of the zones connected to the ground loop. A table (19) is provided to sum the corrected electrical demands for all space heating zones connected to the ground loop.
3. The circulating fluid flow rate (Figure 3.1, 3.2, or 3.3) through the heat pump at design heating conditions for the zone for which the ground loop is being designed. For multiple zones on a single ground loop, this is the sum of the circulating fluid flow rates for all of the zones connected to the ground loop. A table (19) is provided to sum the circulating fluid flow rates for all space heating zones connected to the ground loop.
4. The ACCA Manual J heat loss (Figure 3.1, 3.2, or 3.3) for the zone for which the ground loop is being designed. For multiple zones on a single ground loop, this is the sum of the heat losses for all of the zones connected to the ground loop. A table (19) is provided to sum the total heat losses for all space heating zones connected to the ground loop.
5. The percent of Total Heat Loss (4) that the corrected heating capacity (1) of the heat pump at design heating conditions is capable of providing (=(1)/(4)x100%). If this is less than 100%, a supplementary heating source is required or the temperature of the heated zone may drop below the set point on the thermostat at design heating conditions.
6. Space heating load run fraction during the design heating month of January from a January bin analysis on a single or multiple zone total heat loss (4) at the installation location. An alternate method to estimate the space heating load run fraction in January is to use Figure 3.9 with Percent Sizing (5) and heating degree day (HDD) data for the installation location from Table 3.4.
7. If no HWG option is used, then $HC_{HWG} = 0$. If HWG is used, refer to Table 9.4 for input that is dependent on HWG system type.
8. If no HWG option is used, then $DMD_{HWG} = 0$. If HWG is used, refer to Table 9.4 for input that is dependent on HWG system type.
9. If no HWG option is used, then $GPM_{HWG} = 0$. If HWG is used, refer to Table 9.4 for input that is dependent on HWG system type.
10. If no HWG option is used, then $F_{HWG} = 0$. If HWG is used, refer to Table 9.4 for input that is dependent on HWG system type.
11. If no HWG option is used, then $HC_D = HC_C$. If HWG is used, refer to Table 9.4 for input dependent on HWG system type.
12. If no HWG option is used, then $DMD_D = DMD_C$. If HWG is used, refer to Table 9.4 for input dependent on HWG system type.
13. If no HWG option is used, then $GPM_D = GPM$. If HWG is used, refer to Table 9.4 for input dependent on HWG system type.
14. The design coefficient of performance (Figure 3.1, 3.2, or 3.3) of the heat pump at design heating conditions for the zone for which the ground loop is being designed. For multiple zones and hot water generation on a single ground loop, this is calculated based on the design heating capacity (11) and design electrical demand (12) for all heating loads connected to the ground loop.
15. The design heat of extraction (Figure 3.1, 3.2, or 3.3) from the ground loop at design heating conditions for the zone for which the ground loop is being designed. For multiple zones and hot water generation on a single ground loop, this is calculated based on the design heating capacity (11) and design coefficient of performance (14) for all heating loads connected to the ground loop.
16. The design heating month run fraction that accounts for both space heating load and hot water generation load during the month of January.

17. The minimum entering water temperature for which the heat pump heating capacities (Figure 3.1, 3.2, or 3.3) have been selected for design heating conditions.
18. The minimum leaving water temperature based on the minimum entering water temperature (17), the design heat of extraction (15), and the design circulating fluid flow rate (13) at design heating conditions.
19. For multiple space heating zones, utilize this table to sum the Total Heat Loss, HC_C, DMD_C, and GPM for each zone for input into Total Heat Loss (4), HC_C (1), DMD_C (2), and GPM (3).
20. The location of the horizontally-trenched GHEX determines the minimum soil temperature for design ($T_{S,L}$) and is calculated using the soil temperature data provided in Table 5.11. Figure 5.21 provides a map of locations for which earth temperature data are available. Parameters for the location nearest the actual installation site should be used, but care should be taken to make sure the selected location has similar geographic characteristics.
21. Obtained from Table 5.11 for the selected soil temperature data location.
22. Obtained from Table 5.11 for the selected soil temperature data location.
23. Obtained from Table 5.11 for the selected soil temperature data location.
24. The soil type to a depth of 6 to 8 feet at the installation location. The soil type affects the moisture holding capability of the soil and the resulting thermal conductivity that can be used for design of the horizontally-trenched GHEX, as discussed in Section 5.3.2.2.
25. The soil thermal conductivity, which is the lowest values that would be expected during heating mode operation of the heat pump system and is used to estimate the maximum soil thermal resistance (R_S) for use in calculating the total pipe design length for heating ($L_{H,T}$).
26. The soil thermal diffusivity, which is the highest value that would be expected during any year at the installation site and is used to predict minimum soil temperature for design ($T_{S,L}$).
27. The trench and pipe configurations for which R_S and S_M data have been determined are included in Tables 5.17 through 5.29. The trench dimensions include the depth and width required for the specified pipe configuration. All slinky pipe configurations assume a 3-foot slinky height, which requires an additional trench depth of 1.5 feet below the average trench depth for a standing configuration or a trench width of at least 3 feet for a laying configuration. Design data for pipe configurations not included in Tables 5.17 through 5.29 may be estimated using data from the most similar configuration with approximately the same pipe density (feet of pipe per foot of trench).
28. Number of pipes in the trench, as indicated by the Pipe Configuration name and shown in the schematics accompanying Tables 5.17 through 5.29. This is the feet of pipe per foot of trench and, except for extremely high soil thermal conductivity conditions, should not be more that 5 feet because thermal interference between pipes limits the effectiveness of the additional pipes for heat transfer under low thermal conductivity conditions. Therefore, in dry, low thermal conductivity soils, pipe configurations with lower pipe densities are a better choice to assure maximum thermal performance of the horizontally-trenched GHEX.
29. Pipe configuration in the trench is related to the geometry of the pipes relative to each other and the trench walls. A "Standing" configuration represents pipes that are stacked vertically atop one another and can be installed in a very narrow trench (usually chain trenched). A "Laying" configuration represents pipes that lie in the same horizontal plane and require a wider trench to achieve the desired spacing between pipes (usually trenched with a backhoe or excavator). The "Rectangular" configuration consists of 4 pipes, each representing one corner of a rectangular geometry, that may have equal width to depth dimension (Square) or either wider than tall or vice-versa (trenched with a backhoe or excavator).
30. The horizontal spacing between pipes for the "Rectangular" configuration.
31. The soil thermal resistance for the selected pipe configurations (27) and provided as a function of soil thermal conductivity in Tables 5.17 through 5.29. For design soil thermal conductivities between the tabled values, use linear interpolation to estimate (R_S).
32. The center-to-center spacing between trenches, or the center-to-center spacing between slinkies laid in a large pit, that should be as large as reasonable (10 feet is the recommended minimum). Putting the trenches or laying slinkies closer together results in thermal interference that reduces the thermal performance of the piping configuration (R_S), which was calculated assuming a single trench, and must be accounted for by using a correction factor (33) to increase the thermal resistance of the soil for design of the horizontally-trenched GHEX.
33. The trench spacing multiplier, which adjusts the soil thermal resistance to account for thermal interference

between adjacent trenches (or laying slinkies in a pit system). The trench spacing multiplier is defined, for each piping configuration, in terms of the number of trenches that are in parallel and the center-to-center spacing (32) between trenches (or laying slinkies).

34. The depth to the bottom pipe in the pipe configuration, which is also the actual trench depth and is the same as the average trench depth (35) for "Laying" configurations or equal to the average trench depth (35) plus 1/2 the heights of the "Standing" or "Rectangular" configuration distances.
35. The average depth of the pipe configuration in the trench, which is equal to the trench depth (34) for "Laying" configurations and is the actual trench depth (34) minus 1/2 the heights of the "Standing" or "Rectangular" configuration distances.
36. The nominal pipe diameter being used in the horizontally-trenched GHEX. Common nominal pipe diameters are ¾, 1, and 1-1/4-inch, and the size that is used is based on the flow rate through the pipe as discussed in Section 5.1.3.
37. The flow rate through each flow path must be high enough for turbulence (Section 4.4.2), but in a range that keeps head loss at a reasonable level (Section 4.4.3) at design heating conditions. For most circulating fluids recommended flow rates of 3 gpm for ¾-inch, 4.5 to 6 gpm for 1 inch, and 6 to 12 gpm for 1-1/4-inch pipe will provide a good trade off between turbulence, head loss, and the resulting design length of the horizontally-trenched GHEX.
38. The number of flow paths for the circulating fluid is estimated by dividing the design system flow rate (13) by the flow rate through each flow path (37). If the result is not a whole number, then round to the nearest whole number and, for the case of rounding up to the next highest whole number, increase the total system flow rate accordingly when designing the pumping system.
39. Generally the number of flow paths per trench is one, meaning that a single coil of pipe is placed in the selected pipe configuration with one supply and one return line connection. In that case, the flow rate through the selected pipe size required to satisfy the heating capacity provided by the design trench length results in a head loss within guidelines discussed in Section 4.5. Sometimes it may be desirable to have two flow paths in the trench, primarily when the area for installation allows for very long trenches and each trench may represent more heating or cooling capacity. In that case, the flow rate through the selected pipe size required to satisfy the heating capacity provided by the design trench length would result in excessive head loss, and the use of two flow paths through two coils of pipe arranged in the desired pipe configuration in the trench would alleviate that problem. The use of two flow paths per trench would most often be considered with the 4-Pipe Rectangular or any of the 2-Layer configurations. It would be very rare to lay out a piping system with more than two flow paths per trench.
40. The number of trenches is determined by dividing the number of flow paths (38) by the number of flow paths per trench (39).
41. The dimensionless minimum soil temperature for heating design, which is expressed as a function of soil thermal diffusivity (26) and average depth in the soil (35) in Table 5.12.
42. The minimum soil temperature for heating design (Equation 5.25) using the mean earth temperature in the top 10 feet (21) and annual swing (22) for the installation location and the dimensionless minimum soil temperature for heating design (41).
43. The pipe thermal resistance is provided for PE pipe as a function of DR in Table 4.6. Normally DR11 PE3408 pipe is used for the GHEX, but the introduction of PE4710 pipe will allow the use of thinner-wall DR13.5 to achieve the IGSHPA recommended 160 psi pressure rating.
44. The pipe multiplier, which adjusts the soil thermal resistance for a pipe diameter other than ¾-inch based on the number of pipes in the trench (feet of pipe for foot of trench). Figure 5.24 provides P_M for ¾-inch (always 1), 1-inch, and 1-1/4-inch PE pipe for the number of pipes in the trench up to 8, above which P_M is constant. Table 5.30 expresses P_M in an equation form for each pipe diameter.
45. The total design length for heating (Equation 5.14), the design length of pipe in each trench for heating (Equation 5.15), and the design length of each trench for heating (Equation 5.16) are calculated using the indicated design parameters from the design sheet.
46. This table has been included to summarize heating design length calculations for various combinations of design parameters.

Horizontally-Trenched GHEX Design Worksheet – Cooling Mode

Space Cooling GSHP Design Data

TC_C[1] = 47 Btu/hr
DMD_C[1] = 48 kW
GPM[1] = 49 gpm
Total Heat Gain[1] = 50 Btu/hr
Percent Sizing = 51 %
F_{Jul} = 52

HWG GSHP Design Data

GPM_{HWG} = 53 gpm

Total Design Data

TC_D = 54 Btu/hr
DMD_D = 55 kW
GPM_D = 56 gpm

GHEX Design Data

EER_D[2] (= TC_D / (DMD_D x 1000)) = 57
HR_D[2] (= TC_D x (EER_D+3.412)/EER_D) = 58 Btu/hr
F_C (= F_{Jul}) = 59
EWT_{max}[3] = 60 F
LWT_{max}[2] (= EWT_{max} + HR_D/(500xGPM_D)) = 61 F

1. For single heat pump installation, use data directly from the appropriate GSHP Selection Worksheet. For multiple heat pumps in the installation, sum all heat pump Total Heat Gains, TC_C's, DMD_C's, and GPM's that will be connected to a single GHEX using the table provided below.
2. For single heat pump installation, use data directly from appropriate GSHP Selection Worksheet. For multiple heat pumps, use the equation provided.
3. EWT_{max} is obtained directly from the appropriate GSHP Selection Worksheet and must be the same for the selection of all heat pumps for a multiple heat pump installation connected to a single GHEX. **(62)**

	Zone 1	Zone 2	Zone 3	Zone 4	Zone 5	Zone 6	Zone 7	Zone 8	Zone 9	Zone 10	Total
Total Heat Gain											
HC_C											
DMD_C											
GPM											

Horizontal Trench Design Data

Location 63 , ______ (City, State)
T_M = 64 F (Table 5.11)
A_s = 65 F (Table 5.11)
T_O = 66 Days (Table 5.11)

Soil Type 67 (Section 5.3.2.2)
k_{soil} = 68 Btu/hr ft F
α_{soil} = 69 ft²/day

Trench and Pipe Configuration ==> Table 70 (Select from Tables 5.17 – 5.29)
N_P = 71 ft_{pipe}/ft_{trench}
Conf. = 72 (Stand / Laying / Rect)
S_H = 73 ft
S_{cc} = 75 ft
d_{bp} = 77 ft
R_s = 74 hr ft F/Btu
S_M = 76
d_{avg} = 78 ft

D_P = 79 in (Nom.)
GPM_{FP} = 80 gpm/flowpath
N_{FP} = 81 flowpaths
$N_{FP/T}$ = 82 flowpaths/trench
N_T = 83 trenches

Design Soil Temperature for Cooling

$T'(d_{avg},T_O+180)$ = 84 F (Table 5.13)
$T_{S,H} = T'(d_{avg},T_O+180)\bullet A_s+T_M$ = ______ • ______ + ______ = 85 F (Eq. 5.26)

Pipe Resistance and Pipe Multiplier

R_p = 86 hr ft F/Btu (Table 4.6)
P_M = 87 (Figure 5.24)

Horizontal Trench Design Lengths (Equations 5.17, 5.18, and 5.19) **(88)**

$$L_{C,P} = \frac{TC_D \cdot \left(\frac{EER_D + 3.412}{EER_D}\right)(R_P + R_S \cdot P_M \cdot S_M \cdot F_C)}{\left(\frac{EWT_{max} + LWT_{max}}{2}\right) - T_{S,H}} = \frac{\_\_\_ \cdot \left(\frac{\_\_\_ + 3.412}{\_\_\_}\right)(\_\_\_ + \_\_\_ \cdot \_\_\_ \cdot \_\_\_ \cdot \_\_\_)}{\left(\frac{\_\_\_ + \_\_\_}{2}\right) - \_\_\_} = \frac{\_\_\_ \cdot (\_\_\_)(\_\_\_)}{(\_\_\_) - \_\_\_} = \_\_\_ \text{ ft of pipe}$$

$$L_{C,P/T} = \frac{L_{C,P}}{N_T} = \frac{\_\_\_}{\_\_\_} = \_\_\_ \; ft_{pipe} / \text{trench} \qquad L_{C,T} = \frac{L_{C,P/T}}{N_P} = \frac{\_\_\_}{\_\_\_} = \_\_\_ \; ft_{trench}$$

Cooling Design Length Calculations Summary Table **(89)**

Layout for Flow and Number of Trenches						Pipe and Trench Design Length Calculations								
D_P	GPM_{FP}	N_{FP}	Trench/N_P	$N_{FP/T}$	N_T	d_{avg}	$T'(d_{avg},T_o+180)$	$T_{S,H}$	R_s	P_M	S_{cc}/S_M	$L_{C,P}$	$L_{C,P/T}$	$L_{C,T}$
			/								/			
			/								/			
			/								/			
			/								/			
			/								/			

Figure 5.26. Horizontally-Trenched GHEX Design Worksheet – Cooling Mode.

5.3.4.2 COOLING MODE DESIGN WORKSHEET (FIGURE 5.26)

47. The corrected total cooling capacity (Figure 3.1, 3.2, or 3.3) of the heat pump at design cooling conditions for the zone for which the ground loop is being designed. For multiple zones on a single ground loop, this is the sum of the corrected total cooling capacities for all of the zones connected to the ground loop. A table (62) is provided to sum the corrected total cooling capacities for all space cooling zones connected to the ground loop.
48. The corrected electrical demand (Figure 3.1, 3.2, or 3.3) of the heat pump at design cooling conditions for the zone for which the ground loop is being designed. For multiple zones on a single ground loop, this is the sum of the corrected electrical demands for all of the zones connected to the ground loop. A table (62) is provided to sum the corrected electrical demands for all space cooling zones connected to the ground loop.
49. The circulating fluid flow rate (Figure 3.1, 3.2, or 3.3) through the heat pump at design cooling conditions for the zone for which the ground loop is being designed. For multiple zones on a single ground loop, this is the sum of the circulating fluid flow rates for all of the zones connected to the ground loop. A table (62) is provided to sum the circulating fluid flow rates for all space cooling zones connected to the ground loop.
50. The ACCA Manual J heat gain (Figure 3.1, 3.2, or 3.3) for the zone for which the ground loop is being designed. For multiple zones on a single ground loop, this is the sum of the heat gains for all of the zones connected to the ground loop. A table (62) is provided to sum the total heat gains for all space cooling zones connected to the ground loop.
51. The percent of Total Heat Gain (50) that the corrected total cooling capacity (47) of the heat pump at design cooling conditions is capable of providing (=(47)/(50)x100%). If this is less than 100%, then the temperature of the cooled zone may rise above the set point on the thermostat at design cooling conditions.
52. Space cooling load run fraction during the design cooling month of July from a July bin analysis on a single or multiple zone Total Heat Gain (50) at the installation location. An alternate method to estimate the space cooling load run fraction in July is to use Figure 3.11 with Percent Sizing (51) and cooling degree days (CDD) data for the installation location from Table 3.4.
53. If no HWG option is used, then $GPM_{HWG} = 0$. If HWG is used, refer to Table 9.5 for input that is dependent on HWG system type.
54. $TC_D=TC_C$.
55. $DMD_D=DMD_C$.
56. If no HWG option is used, then $GPM_D = GPM$. If HWG is used, refer to Table 9.5 for input dependent on HWG system type.
57. The design energy efficiency ratio (Figure 3.1, 3.2, or 3.3) of the heat pump at design cooling conditions for the zone for which the ground loop is being designed. For multiple zones on a single ground loop, this is calculated based on the design total cooling capacity (54) and design electrical demand (55) for all cooling loads connected to the ground loop.
58. The design heat of rejection (Figure 3.1, 3.2, or 3.3) to the ground loop at design cooling conditions for the zone for which the ground loop is being designed. For multiple zones on a single ground loop, this is calculated based on the design total cooling capacity (54) and design energy efficiency ratio (57) for all cooling loads connected to the ground loop.
59. The design cooling month run fraction that accounts for space cooling loads during the month of July.
60. The maximum entering water temperature for which the heat pump cooling capacities (Figure 3.1, 3.2, or 3.3) have been selected for design cooling conditions.
61. The maximum leaving water temperature based on the maximum entering water temperature (60), the design heat of rejection (58), and the design circulating fluid flow rate (55) at design cooling conditions.
62. For multiple space cooling zones, utilize this table to sum the Total Heat Gain, TC_C, DMD_C, and GPM for each zone for input into Total Heat Gain (50), TC_C (47), DMD_C (48), and GPM (49).
63. The location of the horizontally-trenched GHEX determines the maximum soil temperature for design ($T_{S,H}$) and is calculated using the soil temperature data provided in Table 5.11. Figure 5.21 provides a map of locations for

which earth temperature data are available. Parameters for the location nearest the actual installation site should be used, but care should be taken to make sure the selected location has similar geographic characteristics.

64. Obtained from Table 5.11 for the selected soil temperature data location.
65. Obtained from Table 5.11 for the selected soil temperature data location.
66. Obtained from Table 5.11 for the selected soil temperature data location.
67. The soil type to a depth of 6 to 8 feet at the installation location. The soil type affects the moisture holding capability of the soil and the resulting thermal conductivity that can be used for design of the horizontally-trenched GHEX, as discussed in Section 5.3.2.2.
68. The soil thermal conductivity, which is the lowest values that would be expected during cooling mode operation of the heat pump system and is used to estimate the maximum soil thermal resistance (R_S) for use in calculating the total pipe design length for cooling ($L_{C,T}$).
69. The soil thermal diffusivity, which is the highest value that would be expected during any year at the installation site and is used to predict maximum soil temperature for design ($T_{S,H}$).
70. The trench and pipe configurations for which R_S and S_M data have been determined are included in Tables 5.17 through 5.29. The trench dimensions include the depth and width required for the specified pipe configuration. All slinky pipe configurations assume a 3-foot slinky height, which requires an additional trench depth of 1.5 feet below the average trench depth for a standing configuration or a trench width of at least 3 feet for a laying configuration. Design data for pipe configurations not included in Tables 5.17 through 5.29 may be estimated by using data from the most similar configuration with approximately the same pipe density (feet of pipe per foot of trench).
71. Number of pipes in the trench, as indicated by the Pipe Configuration name and shown in the schematics accompanying Tables 5.17 through 5.29. This is the feet of pipe per foot of trench and, except for extremely high soil thermal conductivity conditions, should not be more that 5 feet because thermal interference between pipes limits the effectiveness of the additional pipes for heat transfer under low thermal conductivity conditions. Therefore, in dry, low thermal conductivity soils, pipe configurations with lower pipe densities are a better choice to assure maximum thermal performance of the horizontally-trenched GHEX.
72. Pipe configuration in the trench is related to the geometry of the pipes relative to each other and the trench walls. A "Standing" configuration represents pipes that are stacked vertically atop one another and can be installed in a very narrow trench (usually chain trenched). A "Laying" configuration represents pipes that lie in the same horizontal plane and require a wider trench to achieve the desired spacing between pipes (usually trenched with a backhoe or excavator). The "Rectangular" configuration consists of 4 pipes, each representing one corner of a rectangular geometry, that may have equal width to depth dimension (Square) or either wider than tall or vice-versa (trenched with a backhoe or excavator).
73. The horizontal spacing between pipes for the "Rectangular" configuration.
74. The soil thermal resistance for the selected pipe configurations (70) and provided as a function of soil thermal conductivity in Tables 5.17 through 5.29. For design soil thermal conductivities between the tabled values, use linear interpolation to estimate (R_S)
75. The center-to-center spacing between trenches, or the center-to-center spacing between slinkies laid in a large pit, that should be as large as reasonable (10 feet is the recommended minimum). Putting the trenches or laying slinkies closer together results in thermal interference that reduces the thermal performance of the piping configuration (R_S), which was calculated assuming a single trench, and must be accounted for by using a correction factor (76) to increase the thermal resistance of the soil for design of the horizontally-trenched GHEX.
76. The trench spacing multiplier, which adjusts the soil thermal resistance to account for thermal interference between adjacent trenches (or laying slinkies in a pit system). The trench spacing multiplier is defined, for each piping configuration, in terms of the number of trenches that are in parallel and the center-to-center spacing (75) between trenches (or laying slinkies).
77. The depth to the bottom pipe in the pipe configuration, which is also the actual trench depth and is the same as the average trench depth (78) for "Laying" configurations or equal to the average trench depth (78) plus 1/2 the

heights of the "Standing" or "Rectangular" configuration distances.

78. The average depth of the pipe configuration in the trench, which is equal to the trench depth (77) for "Laying" configurations and is the actual trench depth (77) minus 1/2 the heights of the "Standing" or "Rectangular" configuration distances.
79. The nominal pipe diameter being used in the horizontally-trenched GHEX. Common nominal pipe diameters are ¾, 1, and 1-1/4-inch, and the size that is used is based on the flow rate through the pipe as discussed in Section 5.1.3.
80. The flow rate through each flow path must be high enough for turbulence (Section 4.4.2), but in a range that keeps head loss at a reasonable level (Section 4.4.3) at design heating conditions. For most circulating fluids, recommended flow rates of 3 gpm for ¾-inch, 4.5 to 6 gpm for 1-inch, and 6 to 12 gpm for 1-1/4-inch pipe will provide a good trade off between turbulence, head loss, and the resulting design length of the horizontally-trenched GHEX.
81. The number of flow paths for the circulating fluid is estimated by dividing the design system flow rate (55) by the flow rate through each flow path (80). If the result is not a whole number, then round to the nearest whole number and, for the case of rounding up to the next highest whole number, increase the total system flow rate accordingly when designing the pumping system.
82. Generally the number of flow paths per trench is one, meaning that a single coil of pipe is placed in the selected pipe configuration with one supply and one return line connection. In that case, the flow rate through the selected pipe size required to satisfy the design cooling capacity provided by the design trench length results in a head loss within guidelines discussed in Section 4.5. Some times it may be desirable to have two flow paths in the trench, primarily when the area for installation allows for very long trenches and each trench may represent more heating or cooling capacity. In that case, the flow rate through the selected pipe size required to satisfy the design cooling capacity provided by the design trench length would result in excessive head loss, and the use of two flow paths through two coils of pipe arranged in the desired pipe configuration in the trench would alleviate that problem. The use of two flow paths per trench would most often be considered with the 4-Pipe Rectangular or any of the 2-Layer configurations. It would be very rare to lay out a piping system with more than two flow paths per trench.
83. The number of trenches is determined by dividing the number of flow paths (81) by the number of flow paths per trench (82).
84. The dimensionless maximum soil temperature for cooling design, which is expressed as a function of soil thermal diffusivity (69) and average depth in the soil (78) in Table 5.13.
85. The maximum soil temperature for cooling design (Equation 5.26) using the mean earth temperature in the top 10 feet (64) and annual swing (65) for the installation location and the dimensionless maximum soil temperature for cooling design (84).
86. The pipe thermal resistance is provided for PE pipe as a function of DR in Table 4.6. Normally DR11 PE3408 pipe is used for the GHEX, but the introduction of PE4710 pipe will allow the use of thinner-wall DR13.5 to achieve the IGSHPA recommended 160 psi pressure rating.
87. The pipe multiplier, which adjusts the soil thermal resistance for a pipe diameter other than ¾-inch based on the number of pipes in the trench (feet of pipe for foot of trench). Figure 5.24 provides P_M for ¾-inch (always 1), 1-inch, and 1-1/4-inch PE pipe for the number of pipes in the trench up to 8, above which P_M is constant. Table 5.30 expresses P_M in an equation form for each pipe diameter.
88. The total design length for cooling (Equation 5.17), the design length of pipe in each trench for cooling (Equation 5.18), and the design length of each trench for cooling (Equation 5.19) are calculated using the indicated design parameters from the design sheet.
89. This table has been included to summarize cooling design length calculations for various combinations of design parameters.

5.4 Horizontally-Bored, Closed-Loop GHEX Design

Heating and cooling design equations and data for horizontally-bored, closed-loop GHEX design are presented in this section. The design equations are similar to the previous vertically-bored GHEX design equations (Section 5.2) except that the deep-earth ground temperature is replaced by the soil temperature at the average GHEX pipe depth to account for annual variation in soil temperature present in the top 20 feet of the soil profile. The design data include thermal properties of soil types that will be encountered during the horizontal boring process and which will always be in unconsolidated formations. Ground loop resistance tables similar to those provided for vertically-bored GHEX design, which combined the effects of ground thermal properties, pipe diameter, borehole diameter, grout thermal conductivity, and run fraction of the heat pump during the design month, have been updated to reflect soil thermal conductivities in a lower range associated with unconsolidated formations. The effects of unbalanced ground loads in heating and cooling are again accounted for utilizing a bore length multiplier to account for long-term cooling or warming of the ground near the horizontally-bored ground loops. Corrections for unbalanced ground loads become more important as the average burial depth of the horizontal bores increases and at shallow depths the application of the multiplier will add a factor of safety to the design. Horizontally-bored GHEX design worksheets for both heating and cooling mode are provided to facilitate the design process and provide documentation of the design for easy record keeping.

The steps in designing the horizontally-bored, closed-loop GHEX include:

1. Define the design heating or cooling load and select GSHP equipment (Section 3.4) based on the EWT_{min} or EWT_{max} and GPM that the GHEX will be designed for.
2. Lay out the GHEX according to Section 5.1.3 by selecting pipe sizes for the ground loops that meet turbulence and head loss constraints and using a header arrangement that balances flow between the ground loops and facilitates flushing and purging of the GHEX.
3. Estimate the design heating or cooling month run fraction and the annual heating and cooling ground loads using a bin analysis or graphed data.
4. Determine the deep-earth temperature where the horizontally-bored GHEX will be installed.
5. Estimate the thermal properties (thermal conductivity and diffusivity) of the ground formation into which the horizontally-bored GHEX will be installed.
6. Select borehole completion information (borehole diameter, U-bend pipe size, and grout thermal conductivity) that will be used in the installation.
7. Determine the ground loop resistance from tabled data for the selected U-bend pipe size accounting for design month run fraction, ground formation thermal conductivity, borehole diameter, and grout thermal conductivity.
8. Calculate the total design heating or cooling borehole length and divide by the number of horizontal bores to determine the active borehole length.
9. Adjust the heating or cooling design lengths to account for unbalanced ground loads, if necessary.
10. Revise the design, as necessary, to meet potential space limitations on the installation site by balancing the number and spacing of horizontal bores against boring lengths that can be achieved on the site.

The design heating or cooling horizontal bore lengths that result directly from the equations and data presented below are for residential and light commercial GSHP applications with a relatively small number of ground loops spaced such that they minimally interfere with each other. If the heating and cooling ground loads are relatively unbalanced, then borehole length multipliers are utilized for one, two, or three-layer horizontal borefields that account for borehole spacing and adjust borehole lengths for long-term energy depletion or buildup in the soil or rock in and around the horizontally-bored GHEX. For GSHP applications that require a large number of horizontal bores in a grid arrangement (not considered in this manual) and have heating and cooling ground loads that are extremely unbalanced, it is highly recommended that the designs lengths obtained with the methods presented in this section be checked against commercial ground loop design software tools (GCHPCalc, GLHEPRO, GLD, etc.) that are designed to account for unbalanced ground loads and the influence of borehole spacing. The equations presented here are not to be used for the design of large commercial horizontally-bored, closed-loop GHEX systems.

5.4.1 DESIGN EQUATIONS

5.4.1.1 HEATING DESIGN EQUATIONS The total borehole design length for heating can be estimated using Equation 5.28, which represents the active borehole length for heat transfer. The various parameters in Equation 5.28 include the heat of extraction rate from the ground, which is based on the heating capacity (HC_D), and the coefficient of performance (COP_D) of the heat pump system at design heating conditions, including corrections for operating conditions not directly provided in the engineering specifications (Section 3.4). The design minimum entering (EWT_{min}) and leaving (LWT_{min}) water temperatures are also determined as part of the heat pump selection (Section 3.4). Run fraction in heating mode during the heating design month (F_{Jan}) was discussed in Section 3.6 and is combined with hot water generation run fraction (F_{HWG}), if utilized, for a design month run fraction in heating (F_H). (This is a conservative approach and will ensure that the design length calculated using Equation 5.28 will be long enough to provide for the hot water generation load.) The remaining design parameters ($T_{S,L}$, R_G, and R_B) were discussed in the previous sections ($T_{S,L}$ in Section 5.3.2.1, R_G in Section 5.2.2.2, and R_B in Section 5.2.2.3).

$$L_{H,T} = \frac{HC_D \cdot \left(\frac{COP_D - 1}{COP_D}\right) \cdot (R_B + R_G \cdot F_H)}{T_{S,L} - \left(\frac{EWT_{min} + LWT_{min}}{2}\right)} \qquad \textbf{Equation 5.28}$$

Where:

$L_{H,T}$ = total borehole design length for heating, ft
HC_D = heat pump heating capacity at design heating conditions, Btu/hr
COP_D = coefficient of performance at design heating conditions, dimensionless
R_B = borehole thermal resistance, hr ft F/Btu
R_G = steady-state thermal resistance of ground surrounding the borehole, hr ft F/Btu
F_H = run fraction in heating mode during heating design month (January), dimensionless
$T_{S,L}$ = design soil temperature for heating at average horizontally-bored GHEX pipe depth, F
EWT_{min} = minimum entering water temperature at heating design conditions, F
LWT_{min} = minimum leaving water temperature at heating design conditions, F

The individual borehole design length for heating can be calculated using Equation 5.29.

$$L_{H,B} = \frac{L_{H,T}}{N_B}$$ **Equation 5.29**

Where:

$L_{H,B}$ = individual borehole design length for heating, ft
N_B = number of boreholes connected to the heat pump system

5.4.1.2 COOLING DESIGN EQUATIONS The total borehole design length for cooling can be estimated using Equation 5.30, which represents the active borehole length for heat transfer. The various parameters in Equation 5.30 include the heat of rejection rate to the ground, which is based on the total cooling capacity (TC_D) and the energy efficiency ratio (EER_D) of the heat pump system at design cooling conditions, including corrections for operating conditions not directly provided in the engineering specifications (Section 3.4). The design maximum entering (EWT_{max}) and leaving (LWT_{max}) water temperatures are also determined as part of the heat pump selection (Section 3.4). Run fraction in cooling mode during the cooling design month (F_{Jul}) was discussed in Section 3.6 and represents the design month run fraction in cooling (F_C). The effect of hot water generation on the design month run fraction in cooling has not been accounted for as a conservative measure. (Including that effect would decrease the design month run fraction in cooling, but if no hot water were generated, then the design length calculated using Equation 5.30 would be short.) The remaining design parameters ($T_{S,H}$, R_G, and R_B) were discussed in the previous sections ($T_{S,H}$ in Section 5.3.2.1, R_G in Section 5.2.2.2, and R_B in Section 5.2.2.3).

$$L_{C,T} = \frac{TC_D \cdot \left(\frac{EER_D + 3.412}{EER_D}\right) \cdot \left(R_B + R_G \cdot F_C\right)}{\left(\frac{EWT_{max} + LWT_{max}}{2}\right) - T_{S,H}}$$ **Equation 5.30**

Where:

$L_{C,T}$ = total borehole design length for cooling, ft
TC_D = heat pump total cooling capacity at design cooling conditions, Btu/hr
EER_D = energy efficiency ratio at design cooling conditions, dimensionless
R_B = borehole thermal resistance, hr ft F/Btu
R_G = steady-state thermal resistance of ground surrounding the borehole, hr ft F/Btu
F_C = run fraction in cooling mode during cooling design month (July), dimensionless
$T_{S,H}$ = design soil temperature for cooling at average horizontally-bored GHEX pipe depth, F
EWT_{max} = maximum entering water temperature at cooling design conditions, F
LWT_{max} = maximum leaving water temperature at cooling design conditions, F

The individual borehole design length for heating can be calculated using Equation 5.31.

$$L_{C,B} = \frac{L_{C,T}}{N_B}$$ **Equation 5.31**

Where:

$L_{C,B}$ = individual borehole design length for cooling, ft
N_B = number of boreholes connected to the heat pump system

5.4.2 DESIGN PARAMETERS

5.4.2.1 DESIGN SOIL TEMPERATURES ($T_{S,L}$ AND $T_{S,H}$) Refer to Section 5.3.2.1 for discussion of design soil temperatures for horizontal GHEX applications and design data for the determination of $T_{S,L}$ or $T_{S,H}$ as a function of installation location, soil type, and average depth of loopfield burial. Horizontally-bored GHEX piping systems can be buried at any depth, but piping closest to the ground surface should be at a minimum depth of 6 to 8 feet, with 15 feet being a recommended minimum depth to achieve a depth where soil temperatures remain fairly constant throughout the year. Due to the horizontal nature of horizontally-bored GHEX piping systems, a large surface area is required to contain the GHEX footprint. Many times the required length of horizontally-bored GHEX cannot be fit on the property at one depth and a second (or third) layer of horizontal bores may be required to provide the necessary heating or cooling ground load for the installed heat pump capacity. If multiple-pipe layers are to be utilized, then it is recommended that the first pipe layer be located at a minimum 15-foot depth and a minimum 15-foot center-to-center spacing between loops (both vertically and horizontally) be utilized as a starting point for design to reduce the thermal interference between loops in the horizontal grid. If the average burial depth of two or three borehole layers is below 20 feet, then the design soil temperature ($T_{S,L}$ or $T_{S,H}$) can be estimated as the deep earth temperature (T_G) for the location (Section 5.2.2.1).

5.4.2.2 THERMAL PROPERTIES OF THE GROUND AND GROUND THERMAL RESISTANCE (R_G) The thermal properties of the soil that are important to the design of the closed-loop GHEX (thermal conductivity, heat capacity, and diffusivity) were discussed in Section 5.2.2.2. Table 5.14 provided thermal conductivity ranges for various soil types, and Table 5.15 provided thermal conductivity and diffusivity ranges for a several soils based on USDA textural class for commonly occurring densities and water contents (Remund, 1998a,b). Estimating thermal conductivity of soils for horizontally-trenched GHEX piping systems was discussed in Section 5.3.2.2, and that discussion applies equally well for horizontally-bored GHEX piping systems. Ground thermal resistance (R_G) can be calculated according to Equation 5.5 for a recommended soil diameter of 15 feet and has been incorporated into tables (Section 5.4.2.4) for the design parameters in Equation 5.28 (R_B+R_G•F_H) and Equation 5.30 (R_B+R_G•F_C) as a function of soil thermal conductivity.

5.4.2.3 BOREHOLE THERMAL RESISTANCE (R_B) Borehole thermal resistance was discussed in Section 5.2.2.3 for vertically-bored GHEX piping systems. Heat transfer from the pipe to the borehole wall does not depend on the orientation of the borehole, and the discussion for vertically-bored GHEX piping systems applies equally well to horizontally-bored GHEX piping systems.

5.4.2.4 GROUND LOOP RESISTANCE Ground loop resistance represents the total resistance to heat transfer at design heating or cooling conditions between the circulating fluid in the pipe and the design month earth temperature at the average depth of GHEX piping system burial, defined in Equation 5.28 as (R_B +R_G•F_H) and Equation 5.30 as (R_B +R_G•F_C). The ground loop resistance includes the effects of borehole thermal resistance (pipe size and geometry in the borehole, the grout thermal conductivity, and borehole diameter), ground thermal resistance (soil thermal conductivity and borehole diameter), and the run fraction for the design heating or cooling month. To simplify the determination of ground loop resistance, tables have been constructed, which express ground loop resistance as a function of design month run fraction in heating (F_H) or cooling (F_C), soil thermal conductivity (k_{soil}), borehole diameter (D_B), and grout thermal conductivity (k_{Grout}) for nominal U-bend diameters of ¾-inch (Table 5.31), 1-inch (Table 5.32), and 1-1/4-inch (Table 5.33). Interpolation between tabled values is acceptable. Care must be used when extrapolating beyond tabled values, especially on the low end of the tabled variables, and it would be better practice to calculate the ground loop resistance using Equations 5.5 through 5.8.

Table 5.31. Horizontal Borehole Ground Loop Resistance
($R_B + R_G \cdot F_H$ for Heating or $R_B + R_G \cdot F_C$ for Cooling) for ¾-inch Nominal U-bend (hr ft F/Btu)

F_H or F_C	k_{soil}	Borehole Diameter = 4.00 in.									Borehole Diameter = 5.00 in.									Borehole Diameter = 6.00 in.								
		k_{Grout} (Btu/hr ft F)									k_{Grout} (Btu/hr ft F)									k_{Grout} (Btu/hr ft F)								
	(Btu/hr ft F)	0.40	0.57	0.69	0.79	0.88	1.00	1.07	1.14	1.20	0.40	0.57	0.69	0.79	0.88	1.00	1.07	1.14	1.20	0.40	0.57	0.69	0.79	0.88	1.00	1.07	1.14	1.20
0.30	0.40	0.764	0.692	0.663	0.646	0.633	0.620	0.614	0.609	0.605	0.763	0.684	0.652	0.632	0.619	0.604	0.597	0.591	0.587	0.764	0.678	0.643	0.622	0.607	0.592	0.584	0.578	0.572
	0.55	0.640	0.569	0.539	0.522	0.510	0.497	0.490	0.485	0.481	0.646	0.567	0.535	0.516	0.502	0.488	0.481	0.475	0.470	0.654	0.568	0.533	0.512	0.497	0.481	0.473	0.467	0.462
	0.70	0.569	0.498	0.469	0.451	0.439	0.426	0.419	0.414	0.410	0.580	0.501	0.468	0.449	0.435	0.421	0.414	0.408	0.403	0.590	0.504	0.469	0.448	0.433	0.418	0.410	0.404	0.398
	0.85	0.523	0.452	0.423	0.405	0.393	0.380	0.374	0.368	0.364	0.536	0.458	0.425	0.406	0.392	0.378	0.371	0.365	0.360	0.549	0.464	0.428	0.407	0.392	0.377	0.369	0.363	0.358
	1.00	0.491	0.420	0.391	0.373	0.361	0.348	0.342	0.336	0.332	0.506	0.427	0.395	0.376	0.362	0.348	0.341	0.335	0.330	0.521	0.435	0.400	0.379	0.364	0.348	0.341	0.334	0.329
	1.15	0.467	0.396	0.367	0.349	0.337	0.324	0.318	0.312	0.308	0.484	0.405	0.373	0.353	0.340	0.325	0.318	0.312	0.308	0.500	0.414	0.379	0.357	0.343	0.327	0.319	0.313	0.308
	1.30	0.449	0.378	0.349	0.331	0.319	0.306	0.300	0.294	0.290	0.467	0.388	0.356	0.336	0.322	0.308	0.301	0.295	0.290	0.483	0.397	0.362	0.341	0.326	0.311	0.303	0.296	0.291
	1.45	0.435	0.363	0.334	0.317	0.304	0.291	0.285	0.280	0.276	0.453	0.374	0.342	0.323	0.309	0.294	0.288	0.281	0.277	0.470	0.384	0.349	0.328	0.313	0.298	0.290	0.284	0.278
0.40	0.40	0.915	0.844	0.815	0.797	0.785	0.772	0.766	0.760	0.756	0.906	0.827	0.794	0.775	0.761	0.747	0.740	0.734	0.729	0.900	0.814	0.779	0.758	0.743	0.727	0.719	0.713	0.708
	0.55	0.640	0.679	0.650	0.632	0.620	0.607	0.600	0.595	0.591	0.750	0.671	0.639	0.619	0.606	0.591	0.584	0.578	0.574	0.752	0.666	0.631	0.610	0.595	0.579	0.572	0.565	0.560
	0.70	0.655	0.584	0.555	0.538	0.525	0.512	0.506	0.501	0.496	0.661	0.582	0.550	0.530	0.517	0.502	0.495	0.489	0.485	0.668	0.582	0.547	0.526	0.511	0.495	0.487	0.481	0.476
	0.85	0.594	0.523	0.494	0.477	0.464	0.451	0.445	0.439	0.435	0.604	0.525	0.492	0.473	0.459	0.445	0.438	0.432	0.427	0.613	0.527	0.492	0.471	0.456	0.440	0.433	0.426	0.421
	1.00	0.552	0.480	0.451	0.434	0.421	0.408	0.402	0.397	0.392	0.563	0.484	0.452	0.433	0.419	0.405	0.398	0.392	0.387	0.575	0.489	0.454	0.433	0.418	0.402	0.395	0.388	0.383
	1.15	0.520	0.449	0.420	0.402	0.390	0.377	0.371	0.365	0.361	0.534	0.455	0.422	0.403	0.389	0.375	0.368	0.362	0.357	0.547	0.461	0.426	0.405	0.390	0.374	0.366	0.360	0.355
	1.30	0.496	0.425	0.395	0.378	0.366	0.352	0.346	0.341	0.337	0.511	0.432	0.399	0.380	0.366	0.352	0.345	0.339	0.334	0.525	0.439	0.404	0.383	0.368	0.352	0.345	0.338	0.333
	1.45	0.476	0.405	0.376	0.359	0.346	0.333	0.327	0.321	0.317	0.493	0.414	0.381	0.362	0.348	0.334	0.327	0.321	0.316	0.508	0.422	0.387	0.366	0.351	0.335	0.327	0.321	0.316
0.50	0.40	1.067	0.995	0.966	0.949	0.936	0.923	0.917	0.912	0.907	1.048	0.969	0.937	0.917	0.904	0.889	0.882	0.876	0.872	1.035	0.949	0.914	0.893	0.878	0.862	0.855	0.848	0.843
	0.55	0.860	0.789	0.760	0.742	0.730	0.717	0.711	0.705	0.701	0.854	0.775	0.742	0.723	0.709	0.695	0.688	0.682	0.677	0.850	0.765	0.729	0.708	0.693	0.678	0.670	0.664	0.659
	0.70	0.742	0.671	0.642	0.624	0.612	0.599	0.593	0.587	0.583	0.743	0.664	0.631	0.612	0.598	0.584	0.577	0.571	0.566	0.745	0.659	0.624	0.603	0.588	0.572	0.565	0.558	0.553
	0.85	0.666	0.594	0.565	0.548	0.535	0.522	0.516	0.511	0.507	0.671	0.592	0.559	0.540	0.526	0.512	0.505	0.499	0.494	0.677	0.591	0.556	0.535	0.520	0.504	0.497	0.490	0.485
	1.00	0.612	0.541	0.512	0.494	0.482	0.469	0.463	0.457	0.453	0.620	0.541	0.509	0.490	0.476	0.462	0.455	0.449	0.444	0.629	0.543	0.508	0.487	0.472	0.456	0.449	0.442	0.437
	1.15	0.573	0.501	0.472	0.455	0.442	0.429	0.423	0.418	0.414	0.583	0.504	0.472	0.453	0.439	0.424	0.417	0.411	0.407	0.594	0.508	0.473	0.452	0.437	0.421	0.414	0.407	0.402
	1.30	0.542	0.471	0.442	0.424	0.412	0.399	0.393	0.387	0.383	0.555	0.476	0.443	0.424	0.410	0.396	0.389	0.383	0.378	0.566	0.481	0.446	0.424	0.410	0.394	0.386	0.380	0.375
	1.45	0.518	0.447	0.418	0.400	0.388	0.375	0.369	0.363	0.359	0.532	0.453	0.421	0.401	0.388	0.373	0.366	0.360	0.355	0.545	0.459	0.424	0.403	0.388	0.372	0.365	0.358	0.353
0.60	0.40	1.218	1.147	1.118	1.100	1.088	1.075	1.069	1.063	1.059	1.191	1.112	1.079	1.060	1.046	1.032	1.025	1.019	1.014	1.170	1.084	1.049	1.028	1.013	0.998	0.990	0.984	0.978
	0.55	0.970	0.899	0.870	0.852	0.840	0.827	0.821	0.815	0.811	0.957	0.878	0.846	0.827	0.813	0.799	0.792	0.786	0.781	0.949	0.863	0.828	0.807	0.792	0.776	0.769	0.762	0.757
	0.70	0.829	0.757	0.728	0.711	0.698	0.685	0.679	0.674	0.669	0.824	0.745	0.713	0.693	0.680	0.665	0.658	0.652	0.648	0.822	0.736	0.701	0.680	0.665	0.650	0.642	0.636	0.630
	0.85	0.737	0.666	0.637	0.619	0.607	0.594	0.587	0.582	0.578	0.738	0.659	0.627	0.607	0.593	0.579	0.572	0.566	0.561	0.740	0.655	0.619	0.598	0.583	0.568	0.560	0.554	0.549
	1.00	0.673	0.602	0.572	0.555	0.543	0.530	0.523	0.518	0.514	0.677	0.598	0.566	0.547	0.533	0.519	0.512	0.506	0.501	0.683	0.597	0.562	0.541	0.526	0.510	0.503	0.496	0.491
	1.15	0.625	0.554	0.525	0.508	0.495	0.482	0.476	0.470	0.466	0.633	0.554	0.522	0.502	0.488	0.474	0.467	0.461	0.456	0.641	0.555	0.520	0.499	0.484	0.468	0.461	0.454	0.449
	1.30	0.589	0.518	0.489	0.471	0.459	0.446	0.439	0.434	0.430	0.598	0.520	0.487	0.468	0.454	0.440	0.433	0.427	0.422	0.608	0.522	0.487	0.466	0.451	0.436	0.428	0.421	0.416
	1.45	0.560	0.489	0.460	0.442	0.430	0.417	0.411	0.405	0.401	0.571	0.492	0.460	0.441	0.427	0.412	0.406	0.399	0.395	0.582	0.496	0.461	0.440	0.425	0.410	0.402	0.396	0.390
0.70	0.40	1.370	1.298	1.269	1.252	1.239	1.226	1.220	1.215	1.210	1.333	1.254	1.222	1.203	1.189	1.175	1.168	1.162	1.157	1.306	1.220	1.185	1.164	1.149	1.133	1.125	1.119	1.114
	0.55	1.080	1.009	0.980	0.963	0.950	0.937	0.931	0.925	0.921	1.061	0.982	0.950	0.930	0.917	0.902	0.895	0.889	0.885	1.047	0.961	0.926	0.905	0.890	0.875	0.867	0.860	0.855
	0.70	0.915	0.844	0.815	0.797	0.785	0.772	0.766	0.760	0.756	0.906	0.827	0.794	0.775	0.761	0.747	0.740	0.734	0.729	0.900	0.814	0.779	0.758	0.743	0.727	0.719	0.713	0.708
	0.85	0.808	0.737	0.708	0.690	0.678	0.665	0.659	0.653	0.649	0.805	0.726	0.694	0.674	0.661	0.646	0.639	0.633	0.628	0.804	0.718	0.683	0.662	0.647	0.631	0.624	0.617	0.612
	1.00	0.733	0.662	0.633	0.616	0.603	0.590	0.584	0.578	0.574	0.734	0.656	0.623	0.604	0.590	0.576	0.569	0.563	0.558	0.737	0.651	0.616	0.595	0.580	0.565	0.557	0.550	0.545
	1.15	0.678	0.607	0.578	0.560	0.548	0.535	0.529	0.523	0.519	0.682	0.603	0.571	0.552	0.538	0.524	0.517	0.511	0.506	0.688	0.602	0.567	0.546	0.531	0.515	0.508	0.501	0.496
	1.30	0.635	0.564	0.535	0.518	0.505	0.492	0.486	0.481	0.476	0.642	0.563	0.531	0.512	0.498	0.484	0.477	0.471	0.466	0.650	0.564	0.529	0.508	0.493	0.477	0.470	0.463	0.458
	1.45	0.602	0.531	0.501	0.484	0.472	0.459	0.452	0.447	0.443	0.611	0.532	0.499	0.480	0.466	0.452	0.445	0.439	0.434	0.620	0.534	0.499	0.478	0.463	0.447	0.439	0.433	0.428
0.80	0.40	1.521	1.450	1.421	1.403	1.391	1.378	1.372	1.366	1.362	1.476	1.397	1.365	1.345	1.332	1.317	1.310	1.304	1.299	1.441	1.355	1.320	1.299	1.284	1.268	1.261	1.254	1.114
	0.55	1.191	1.119	1.090	1.073	1.060	1.047	1.041	1.036	1.031	1.165	1.086	1.054	1.034	1.020	1.006	0.999	0.993	0.988	1.146	1.060	1.025	1.004	0.989	0.973	0.966	0.959	0.855
	0.70	1.002	0.930	0.901	0.884	0.871	0.858	0.852	0.847	0.843	0.987	0.908	0.876	0.856	0.843	0.828	0.821	0.815	0.811	0.977	0.891	0.856	0.835	0.820	0.804	0.797	0.790	0.708
	0.85	0.879	0.808	0.779	0.762	0.749	0.736	0.730	0.725	0.720	0.872	0.793	0.761	0.741	0.728	0.713	0.706	0.700	0.696	0.868	0.782	0.747	0.726	0.711	0.695	0.688	0.681	0.612
	1.00	0.794	0.723	0.694	0.676	0.664	0.651	0.645	0.639	0.635	0.791	0.713	0.680	0.661	0.647	0.633	0.626	0.620	0.615	0.791	0.706	0.670	0.649	0.634	0.619	0.611	0.605	0.545
	1.15	0.731	0.660	0.630	0.613	0.601	0.588	0.581	0.576	0.572	0.732	0.653	0.621	0.601	0.588	0.573	0.566	0.560	0.556	0.735	0.649	0.614	0.593	0.578	0.562	0.555	0.548	0.496
	1.30	0.682	0.611	0.582	0.564	0.552	0.539	0.533	0.527	0.523	0.686	0.607	0.575	0.556	0.542	0.527	0.520	0.514	0.510	0.691	0.606	0.571	0.549	0.534	0.519	0.511	0.505	0.458
	1.45	0.644	0.572	0.543	0.526	0.513	0.500	0.494	0.489	0.484	0.650	0.571	0.539	0.519	0.506	0.491	0.484	0.478	0.473	0.657	0.571	0.536	0.515	0.500	0.484	0.477	0.470	0.428

1. Shape factor based on average of "B" and "C" configurations.

Table 5.32. Horizontal Borehole Ground Loop Resistance ($R_B + R_G \cdot F_H$ for Heating or $R_B + R_G \cdot F_C$ for Cooling) for 1-inch Nominal U-bend (hr ft F/Btu)

F_H or F_C	k_{soil}	Borehole Diameter = 4.00 in.									Borehole Diameter = 5.00 in.									Borehole Diameter = 6.00 in.								
		k_{Grout} (Btu/hr ft F)									k_{Grout} (Btu/hr ft F)									k_{Grout} (Btu/hr ft F)								
	(Btu/hr ft F)	0.40	0.57	0.69	0.79	0.88	1.00	1.07	1.14	1.20	0.40	0.57	0.69	0.79	0.88	1.00	1.07	1.14	1.20	0.40	0.57	0.69	0.79	0.88	1.00	1.07	1.14	1.20
0.30	0.40	0.740	0.676	0.650	0.634	0.623	0.611	0.605	0.600	0.597	0.737	0.666	0.637	0.619	0.607	0.594	0.588	0.582	0.578	0.736	0.659	0.627	0.608	0.595	0.580	0.574	0.568	0.563
	0.55	0.616	0.552	0.526	0.510	0.499	0.487	0.481	0.477	0.473	0.620	0.549	0.520	0.502	0.490	0.477	0.471	0.465	0.461	0.625	0.548	0.516	0.497	0.484	0.470	0.463	0.457	0.452
	0.70	0.545	0.481	0.455	0.439	0.428	0.416	0.411	0.406	0.402	0.554	0.482	0.453	0.436	0.423	0.410	0.404	0.399	0.395	0.562	0.485	0.453	0.434	0.421	0.406	0.400	0.394	0.389
	0.85	0.499	0.435	0.409	0.393	0.382	0.370	0.365	0.360	0.356	0.510	0.439	0.410	0.393	0.380	0.367	0.361	0.356	0.351	0.521	0.444	0.412	0.393	0.380	0.365	0.359	0.353	0.348
	1.00	0.467	0.403	0.377	0.361	0.350	0.338	0.333	0.328	0.324	0.480	0.409	0.380	0.362	0.350	0.337	0.331	0.325	0.321	0.492	0.415	0.383	0.364	0.351	0.337	0.330	0.324	0.319
	1.15	0.444	0.380	0.353	0.338	0.326	0.315	0.309	0.304	0.300	0.458	0.387	0.358	0.340	0.328	0.315	0.309	0.303	0.299	0.471	0.394	0.362	0.343	0.330	0.316	0.309	0.303	0.298
	1.30	0.425	0.361	0.335	0.319	0.308	0.296	0.291	0.286	0.282	0.441	0.370	0.340	0.323	0.311	0.298	0.291	0.286	0.282	0.455	0.378	0.346	0.327	0.313	0.299	0.293	0.287	0.282
	1.45	0.411	0.347	0.321	0.305	0.294	0.282	0.276	0.271	0.268	0.427	0.356	0.327	0.309	0.297	0.284	0.278	0.272	0.268	0.442	0.365	0.333	0.314	0.301	0.286	0.280	0.274	0.269
0.40	0.40	0.891	0.827	0.801	0.785	0.774	0.762	0.757	0.752	0.748	0.879	0.808	0.779	0.762	0.749	0.736	0.730	0.725	0.720	0.871	0.794	0.762	0.743	0.730	0.716	0.709	0.703	0.698
	0.55	0.726	0.662	0.636	0.620	0.609	0.597	0.592	0.587	0.583	0.724	0.653	0.624	0.606	0.594	0.581	0.575	0.569	0.565	0.724	0.646	0.615	0.596	0.582	0.568	0.561	0.555	0.551
	0.70	0.632	0.568	0.541	0.526	0.515	0.503	0.497	0.492	0.488	0.635	0.564	0.535	0.517	0.505	0.492	0.486	0.480	0.476	0.639	0.562	0.530	0.511	0.498	0.484	0.477	0.471	0.466
	0.85	0.571	0.507	0.480	0.465	0.453	0.442	0.436	0.431	0.427	0.577	0.506	0.477	0.460	0.447	0.434	0.428	0.423	0.418	0.585	0.507	0.476	0.457	0.443	0.429	0.422	0.416	0.412
	1.00	0.528	0.464	0.438	0.422	0.411	0.399	0.393	0.388	0.385	0.537	0.466	0.437	0.419	0.407	0.394	0.388	0.382	0.378	0.547	0.469	0.438	0.418	0.405	0.391	0.384	0.378	0.374
	1.15	0.496	0.432	0.406	0.390	0.379	0.367	0.362	0.357	0.353	0.507	0.436	0.407	0.390	0.377	0.364	0.358	0.353	0.348	0.518	0.441	0.409	0.390	0.377	0.363	0.356	0.350	0.345
	1.30	0.472	0.408	0.382	0.366	0.355	0.343	0.337	0.332	0.329	0.485	0.413	0.384	0.367	0.354	0.341	0.335	0.330	0.326	0.497	0.419	0.388	0.369	0.355	0.341	0.334	0.328	0.324
	1.45	0.453	0.389	0.362	0.347	0.335	0.324	0.318	0.313	0.309	0.466	0.395	0.366	0.349	0.336	0.323	0.317	0.312	0.307	0.479	0.402	0.370	0.351	0.338	0.324	0.317	0.311	0.306
0.50	0.40	1.043	0.979	0.953	0.937	0.926	0.914	0.908	0.903	0.900	1.022	0.951	0.922	0.904	0.892	0.879	0.873	0.867	0.863	1.007	0.929	0.898	0.879	0.865	0.851	0.844	0.838	0.834
	0.55	0.836	0.772	0.746	0.730	0.719	0.707	0.702	0.697	0.693	0.828	0.756	0.727	0.710	0.697	0.684	0.678	0.673	0.669	0.822	0.745	0.713	0.694	0.681	0.666	0.660	0.654	0.649
	0.70	0.718	0.654	0.628	0.612	0.601	0.589	0.584	0.579	0.575	0.716	0.645	0.616	0.599	0.586	0.573	0.567	0.562	0.557	0.717	0.639	0.608	0.589	0.575	0.561	0.554	0.548	0.544
	0.85	0.642	0.578	0.552	0.536	0.525	0.513	0.507	0.502	0.499	0.645	0.573	0.544	0.527	0.514	0.501	0.495	0.490	0.486	0.648	0.571	0.539	0.520	0.507	0.493	0.486	0.480	0.476
	1.00	0.589	0.524	0.498	0.482	0.471	0.460	0.454	0.449	0.445	0.594	0.523	0.494	0.477	0.464	0.451	0.445	0.439	0.435	0.601	0.523	0.492	0.473	0.459	0.445	0.438	0.432	0.428
	1.15	0.549	0.485	0.459	0.443	0.432	0.420	0.414	0.409	0.406	0.557	0.486	0.457	0.439	0.427	0.414	0.408	0.402	0.398	0.565	0.488	0.456	0.437	0.424	0.410	0.403	0.397	0.392
	1.30	0.519	0.454	0.428	0.412	0.401	0.390	0.384	0.379	0.375	0.528	0.457	0.428	0.411	0.398	0.385	0.379	0.374	0.369	0.538	0.461	0.429	0.410	0.397	0.383	0.376	0.370	0.365
	1.45	0.495	0.430	0.404	0.388	0.377	0.366	0.360	0.355	0.351	0.506	0.435	0.406	0.388	0.376	0.363	0.356	0.351	0.347	0.517	0.439	0.408	0.389	0.375	0.361	0.354	0.348	0.344
0.60	0.40	1.194	1.130	1.104	1.088	1.077	1.065	1.060	1.055	1.051	1.165	1.093	1.064	1.047	1.035	1.022	1.015	1.010	1.006	1.142	1.065	1.033	1.014	1.001	0.986	0.980	0.974	0.969
	0.55	0.947	0.882	0.856	0.840	0.829	0.818	0.812	0.807	0.803	0.931	0.860	0.831	0.814	0.801	0.788	0.782	0.776	0.772	0.921	0.843	0.812	0.792	0.779	0.765	0.758	0.752	0.748
	0.70	0.805	0.741	0.715	0.699	0.688	0.676	0.670	0.665	0.662	0.798	0.727	0.698	0.680	0.668	0.655	0.649	0.643	0.639	0.794	0.717	0.685	0.666	0.653	0.638	0.632	0.626	0.621
	0.85	0.713	0.649	0.623	0.607	0.596	0.584	0.579	0.574	0.570	0.712	0.641	0.611	0.594	0.582	0.569	0.562	0.557	0.553	0.712	0.635	0.603	0.584	0.571	0.556	0.550	0.544	0.539
	1.00	0.649	0.585	0.559	0.543	0.532	0.520	0.514	0.510	0.506	0.651	0.580	0.551	0.534	0.521	0.508	0.502	0.496	0.492	0.655	0.577	0.546	0.527	0.513	0.499	0.492	0.486	0.482
	1.15	0.602	0.538	0.511	0.496	0.484	0.473	0.467	0.462	0.458	0.607	0.536	0.506	0.489	0.477	0.464	0.457	0.452	0.448	0.612	0.535	0.503	0.484	0.471	0.457	0.450	0.444	0.440
	1.30	0.565	0.501	0.475	0.459	0.448	0.436	0.431	0.426	0.422	0.572	0.501	0.472	0.455	0.442	0.429	0.423	0.418	0.413	0.580	0.503	0.471	0.452	0.438	0.424	0.417	0.411	0.407
	1.45	0.536	0.472	0.446	0.430	0.419	0.407	0.402	0.397	0.393	0.545	0.474	0.445	0.427	0.415	0.402	0.396	0.390	0.386	0.554	0.477	0.445	0.426	0.413	0.398	0.392	0.386	0.381
0.70	0.40	1.346	1.282	1.255	1.240	1.229	1.217	1.211	1.206	1.203	1.307	1.236	1.207	1.189	1.177	1.164	1.158	1.152	1.148	1.277	1.200	1.168	1.149	1.136	1.122	1.115	1.109	1.104
	0.55	1.057	0.993	0.966	0.951	0.939	0.928	0.922	0.917	0.913	1.035	0.964	0.935	0.917	0.905	0.892	0.886	0.880	0.876	1.019	0.942	0.910	0.891	0.877	0.863	0.857	0.851	0.846
	0.70	0.891	0.827	0.801	0.785	0.774	0.762	0.757	0.752	0.748	0.879	0.808	0.779	0.762	0.749	0.736	0.730	0.725	0.720	0.871	0.794	0.762	0.743	0.730	0.716	0.709	0.703	0.698
	0.85	0.785	0.720	0.694	0.678	0.667	0.656	0.650	0.645	0.641	0.779	0.708	0.679	0.661	0.649	0.636	0.629	0.624	0.620	0.776	0.698	0.667	0.648	0.634	0.620	0.613	0.607	0.603
	1.00	0.710	0.646	0.619	0.604	0.592	0.581	0.575	0.570	0.566	0.708	0.637	0.608	0.591	0.578	0.565	0.559	0.554	0.549	0.709	0.632	0.600	0.581	0.567	0.553	0.547	0.541	0.536
	1.15	0.654	0.590	0.564	0.548	0.537	0.525	0.520	0.515	0.511	0.656	0.585	0.556	0.539	0.526	0.513	0.507	0.501	0.497	0.660	0.582	0.551	0.531	0.518	0.504	0.497	0.491	0.487
	1.30	0.612	0.548	0.521	0.506	0.495	0.483	0.477	0.472	0.469	0.616	0.545	0.516	0.498	0.486	0.473	0.467	0.461	0.457	0.622	0.544	0.512	0.493	0.480	0.466	0.459	0.453	0.449
	1.45	0.578	0.514	0.488	0.472	0.461	0.449	0.443	0.439	0.435	0.584	0.513	0.484	0.467	0.454	0.441	0.435	0.430	0.425	0.591	0.514	0.482	0.463	0.450	0.436	0.429	0.423	0.418
0.80	0.40	1.497	1.433	1.407	1.391	1.380	1.368	1.363	1.358	1.354	1.450	1.379	1.350	1.332	1.320	1.307	1.300	1.295	1.291	1.413	1.335	1.304	1.285	1.271	1.257	1.250	1.244	1.240
	0.55	1.167	1.103	1.076	1.061	1.050	1.038	1.032	1.027	1.024	1.139	1.068	1.038	1.021	1.009	0.996	0.989	0.984	0.980	1.117	1.040	1.008	0.989	0.976	0.962	0.955	0.949	0.944
	0.70	0.978	0.914	0.888	0.872	0.861	0.849	0.843	0.838	0.835	0.961	0.890	0.861	0.843	0.831	0.818	0.812	0.806	0.802	0.949	0.871	0.840	0.821	0.807	0.793	0.786	0.780	0.776
	0.85	0.856	0.792	0.765	0.750	0.739	0.727	0.721	0.716	0.712	0.846	0.775	0.746	0.728	0.716	0.703	0.697	0.691	0.687	0.840	0.762	0.730	0.711	0.698	0.684	0.677	0.671	0.667
	1.00	0.770	0.706	0.680	0.664	0.653	0.641	0.636	0.631	0.627	0.765	0.694	0.665	0.648	0.635	0.622	0.616	0.611	0.606	0.763	0.686	0.654	0.635	0.622	0.607	0.601	0.595	0.590
	1.15	0.707	0.643	0.617	0.601	0.590	0.578	0.572	0.568	0.564	0.706	0.635	0.606	0.588	0.576	0.563	0.557	0.551	0.547	0.707	0.629	0.598	0.579	0.565	0.551	0.544	0.538	0.534
	1.30	0.658	0.594	0.568	0.552	0.541	0.529	0.524	0.519	0.515	0.660	0.589	0.560	0.542	0.530	0.517	0.511	0.505	0.501	0.663	0.586	0.554	0.535	0.522	0.508	0.501	0.495	0.490
	1.45	0.620	0.556	0.530	0.514	0.503	0.491	0.485	0.480	0.477	0.624	0.553	0.524	0.506	0.494	0.481	0.474	0.469	0.465	0.629	0.551	0.520	0.501	0.487	0.473	0.466	0.460	0.456

1. Shape factor based on average of "B" and "C" configurations.

Table 5.33. Horizontal Borehole Ground Loop Resistance ($R_B + R_G \bullet F_H$ for Heating or $R_B + R_G \bullet F_C$ for Cooling) for 1¼-inch Nominal U-bend (hr ft F/Btu)

F_H or F_C	k_{soil}	Borehole Diameter = 4.00 in. k_{Grout} (Btu/hr ft F)									Borehole Diameter = 5.00 in. k_{Grout} (Btu/hr ft F)									Borehole Diameter = 6.00 in. k_{Grout} (Btu/hr ft F)								
	(Btu/hr ft F)	0.40	0.57	0.69	0.79	0.88	1.00	1.07	1.14	1.20	0.40	0.57	0.69	0.79	0.88	1.00	1.07	1.14	1.20	0.40	0.57	0.69	0.79	0.88	1.00	1.07	1.14	1.20
0.30	0.40	0.718	0.661	0.637	0.623	0.613	0.603	0.597	0.593	0.590	0.713	0.649	0.623	0.607	0.596	0.584	0.579	0.574	0.570	0.710	0.640	0.612	0.595	0.583	0.570	0.564	0.559	0.555
	0.55	0.594	0.537	0.513	0.499	0.489	0.479	0.474	0.469	0.466	0.596	0.532	0.506	0.490	0.479	0.468	0.462	0.457	0.453	0.599	0.530	0.501	0.484	0.472	0.459	0.453	0.448	0.444
	0.70	0.523	0.466	0.442	0.428	0.418	0.408	0.403	0.398	0.395	0.529	0.466	0.440	0.424	0.413	0.401	0.395	0.391	0.387	0.536	0.466	0.438	0.421	0.409	0.396	0.390	0.385	0.381
	0.85	0.478	0.420	0.397	0.382	0.372	0.362	0.357	0.352	0.349	0.486	0.422	0.396	0.381	0.370	0.358	0.352	0.347	0.344	0.495	0.426	0.397	0.380	0.368	0.355	0.349	0.344	0.340
	1.00	0.446	0.388	0.365	0.350	0.340	0.330	0.325	0.320	0.317	0.456	0.392	0.366	0.350	0.339	0.328	0.322	0.317	0.313	0.466	0.397	0.368	0.351	0.339	0.327	0.320	0.315	0.311
	1.15	0.422	0.364	0.341	0.327	0.317	0.306	0.301	0.297	0.293	0.434	0.370	0.344	0.328	0.317	0.305	0.300	0.295	0.291	0.445	0.376	0.347	0.330	0.318	0.305	0.299	0.294	0.290
	1.30	0.404	0.346	0.323	0.308	0.298	0.288	0.283	0.278	0.275	0.417	0.353	0.327	0.311	0.300	0.288	0.283	0.278	0.274	0.429	0.359	0.331	0.314	0.302	0.289	0.283	0.278	0.274
	1.45	0.389	0.332	0.308	0.294	0.284	0.273	0.268	0.264	0.261	0.403	0.339	0.313	0.297	0.286	0.275	0.269	0.264	0.260	0.416	0.346	0.318	0.301	0.289	0.276	0.270	0.265	0.261
0.40	0.40	0.870	0.812	0.789	0.774	0.765	0.754	0.749	0.745	0.741	0.855	0.792	0.765	0.750	0.739	0.727	0.721	0.716	0.713	0.845	0.776	0.747	0.730	0.718	0.706	0.699	0.694	0.690
	0.55	0.704	0.647	0.623	0.609	0.599	0.589	0.584	0.579	0.576	0.700	0.636	0.610	0.594	0.583	0.571	0.566	0.561	0.557	0.698	0.628	0.600	0.583	0.571	0.558	0.552	0.546	0.542
	0.70	0.610	0.553	0.529	0.515	0.505	0.494	0.489	0.485	0.481	0.611	0.547	0.521	0.505	0.494	0.483	0.477	0.472	0.468	0.613	0.544	0.515	0.498	0.486	0.474	0.467	0.462	0.458
	0.85	0.549	0.491	0.468	0.454	0.444	0.433	0.428	0.424	0.420	0.553	0.490	0.463	0.448	0.437	0.425	0.419	0.414	0.411	0.559	0.489	0.461	0.444	0.432	0.419	0.413	0.407	0.403
	1.00	0.506	0.449	0.425	0.411	0.401	0.390	0.385	0.381	0.378	0.513	0.449	0.423	0.408	0.396	0.385	0.379	0.374	0.370	0.520	0.451	0.423	0.405	0.393	0.381	0.375	0.369	0.365
	1.15	0.475	0.417	0.394	0.379	0.369	0.359	0.354	0.349	0.346	0.483	0.420	0.393	0.378	0.367	0.355	0.349	0.344	0.341	0.492	0.423	0.394	0.377	0.365	0.352	0.346	0.341	0.337
	1.30	0.450	0.393	0.369	0.355	0.345	0.335	0.330	0.325	0.322	0.461	0.397	0.371	0.355	0.344	0.332	0.327	0.322	0.318	0.471	0.401	0.373	0.356	0.343	0.331	0.325	0.319	0.315
	1.45	0.431	0.373	0.350	0.336	0.326	0.315	0.310	0.306	0.302	0.442	0.379	0.352	0.337	0.326	0.314	0.308	0.303	0.300	0.453	0.384	0.355	0.338	0.326	0.314	0.307	0.302	0.298
0.50	0.40	1.021	0.964	0.940	0.926	0.916	0.905	0.900	0.896	0.893	0.998	0.934	0.908	0.892	0.881	0.870	0.864	0.859	0.855	0.981	0.911	0.883	0.866	0.854	0.841	0.835	0.829	0.825
	0.55	0.815	0.757	0.734	0.719	0.709	0.699	0.694	0.689	0.686	0.804	0.740	0.714	0.698	0.687	0.675	0.670	0.665	0.661	0.796	0.727	0.698	0.681	0.669	0.656	0.650	0.645	0.641
	0.70	0.697	0.639	0.616	0.601	0.591	0.581	0.576	0.571	0.568	0.692	0.629	0.602	0.587	0.576	0.564	0.558	0.553	0.550	0.691	0.621	0.593	0.576	0.564	0.551	0.545	0.539	0.535
	0.85	0.620	0.563	0.539	0.525	0.515	0.505	0.499	0.495	0.492	0.621	0.557	0.531	0.515	0.504	0.492	0.487	0.482	0.478	0.622	0.553	0.524	0.507	0.495	0.483	0.477	0.471	0.467
	1.00	0.567	0.509	0.486	0.472	0.462	0.451	0.446	0.442	0.438	0.570	0.506	0.480	0.465	0.453	0.442	0.436	0.431	0.428	0.575	0.505	0.477	0.460	0.448	0.435	0.429	0.423	0.419
	1.15	0.527	0.470	0.446	0.432	0.422	0.412	0.407	0.402	0.399	0.533	0.469	0.443	0.427	0.416	0.405	0.399	0.394	0.390	0.539	0.470	0.441	0.424	0.412	0.400	0.393	0.388	0.384
	1.30	0.497	0.439	0.416	0.402	0.392	0.381	0.376	0.372	0.368	0.504	0.441	0.414	0.399	0.388	0.376	0.370	0.365	0.362	0.512	0.443	0.414	0.397	0.385	0.372	0.366	0.361	0.357
	1.45	0.473	0.415	0.392	0.378	0.368	0.357	0.352	0.348	0.344	0.482	0.418	0.392	0.376	0.365	0.353	0.348	0.343	0.339	0.491	0.421	0.393	0.376	0.364	0.351	0.345	0.339	0.335
0.60	0.40	1.173	1.115	1.092	1.077	1.067	1.057	1.052	1.047	1.044	1.141	1.077	1.051	1.035	1.024	1.012	1.007	1.002	0.998	1.116	1.046	1.018	1.001	0.989	0.976	0.970	0.965	0.961
	0.55	0.925	0.867	0.844	0.830	0.820	0.809	0.804	0.800	0.796	0.907	0.843	0.817	0.802	0.790	0.779	0.773	0.768	0.765	0.894	0.825	0.797	0.779	0.767	0.755	0.749	0.743	0.739
	0.70	0.783	0.726	0.702	0.688	0.678	0.667	0.662	0.658	0.655	0.774	0.710	0.684	0.668	0.657	0.645	0.640	0.635	0.631	0.768	0.698	0.670	0.653	0.641	0.628	0.622	0.617	0.613
	0.85	0.691	0.634	0.610	0.596	0.586	0.576	0.571	0.566	0.563	0.688	0.624	0.598	0.582	0.571	0.559	0.554	0.549	0.545	0.686	0.617	0.588	0.571	0.559	0.546	0.540	0.535	0.531
	1.00	0.627	0.570	0.546	0.532	0.522	0.512	0.507	0.502	0.499	0.627	0.563	0.537	0.522	0.510	0.499	0.493	0.488	0.485	0.629	0.559	0.531	0.514	0.502	0.489	0.483	0.478	0.473
	1.15	0.580	0.522	0.499	0.485	0.475	0.464	0.459	0.455	0.451	0.583	0.519	0.493	0.477	0.466	0.454	0.449	0.444	0.440	0.586	0.517	0.488	0.471	0.459	0.447	0.441	0.435	0.431
	1.30	0.543	0.486	0.462	0.448	0.438	0.428	0.423	0.418	0.415	0.548	0.484	0.458	0.443	0.432	0.420	0.414	0.409	0.406	0.554	0.484	0.456	0.439	0.427	0.414	0.408	0.403	0.399
	1.45	0.515	0.457	0.433	0.419	0.409	0.399	0.394	0.389	0.386	0.521	0.457	0.431	0.415	0.404	0.393	0.387	0.382	0.378	0.528	0.458	0.430	0.413	0.401	0.388	0.382	0.377	0.373
0.70	0.40	1.324	1.267	1.243	1.229	1.219	1.208	1.203	1.199	1.196	1.283	1.219	1.193	1.177	1.166	1.155	1.149	1.144	1.140	1.251	1.182	1.153	1.136	1.124	1.111	1.105	1.100	1.096
	0.55	1.035	0.977	0.954	0.940	0.930	0.919	0.914	0.910	0.906	1.011	0.947	0.921	0.905	0.894	0.882	0.877	0.872	0.868	0.993	0.923	0.895	0.878	0.866	0.853	0.847	0.842	0.838
	0.70	0.870	0.812	0.789	0.774	0.765	0.754	0.749	0.745	0.741	0.855	0.792	0.765	0.750	0.739	0.727	0.721	0.716	0.713	0.845	0.776	0.747	0.730	0.718	0.706	0.699	0.694	0.690
	0.85	0.763	0.705	0.682	0.668	0.658	0.647	0.642	0.638	0.634	0.755	0.691	0.665	0.649	0.638	0.626	0.621	0.616	0.612	0.750	0.680	0.652	0.635	0.623	0.610	0.604	0.599	0.594
	1.00	0.688	0.630	0.607	0.593	0.583	0.572	0.567	0.563	0.559	0.684	0.620	0.594	0.579	0.568	0.556	0.550	0.545	0.542	0.683	0.613	0.585	0.568	0.556	0.543	0.537	0.532	0.528
	1.15	0.633	0.575	0.552	0.537	0.527	0.517	0.512	0.507	0.504	0.632	0.568	0.542	0.527	0.515	0.504	0.498	0.493	0.490	0.633	0.564	0.536	0.518	0.506	0.494	0.488	0.482	0.478
	1.30	0.590	0.533	0.509	0.495	0.485	0.474	0.469	0.465	0.462	0.592	0.528	0.502	0.486	0.475	0.464	0.458	0.453	0.449	0.595	0.526	0.498	0.480	0.468	0.456	0.450	0.444	0.440
	1.45	0.556	0.499	0.475	0.461	0.451	0.441	0.436	0.431	0.428	0.560	0.497	0.470	0.455	0.444	0.432	0.426	0.421	0.418	0.565	0.496	0.467	0.450	0.438	0.426	0.419	0.414	0.410
0.80	0.40	1.476	1.418	1.394	1.380	1.370	1.360	1.355	1.350	1.347	1.426	1.362	1.336	1.320	1.309	1.297	1.292	1.287	1.283	1.387	1.317	1.289	1.272	1.260	1.247	1.241	1.235	1.231
	0.55	1.145	1.088	1.064	1.050	1.040	1.029	1.024	1.020	1.017	1.115	1.051	1.025	1.009	0.998	0.986	0.981	0.976	0.972	1.091	1.022	0.993	0.976	0.964	0.952	0.945	0.940	0.936
	0.70	0.956	0.899	0.875	0.861	0.851	0.841	0.835	0.831	0.828	0.937	0.873	0.847	0.831	0.820	0.808	0.803	0.798	0.794	0.923	0.853	0.825	0.808	0.796	0.783	0.777	0.771	0.767
	0.85	0.834	0.777	0.753	0.739	0.729	0.718	0.713	0.709	0.705	0.822	0.758	0.732	0.716	0.705	0.693	0.688	0.683	0.679	0.813	0.744	0.716	0.698	0.686	0.674	0.668	0.662	0.658
	1.00	0.749	0.691	0.667	0.653	0.643	0.633	0.628	0.623	0.620	0.741	0.677	0.651	0.636	0.625	0.613	0.607	0.602	0.599	0.737	0.668	0.639	0.622	0.610	0.597	0.591	0.586	0.582
	1.15	0.685	0.628	0.604	0.590	0.580	0.570	0.565	0.560	0.557	0.682	0.618	0.592	0.576	0.565	0.553	0.548	0.543	0.539	0.681	0.611	0.583	0.566	0.553	0.541	0.535	0.529	0.525
	1.30	0.637	0.579	0.556	0.541	0.531	0.521	0.516	0.512	0.508	0.636	0.572	0.546	0.530	0.519	0.508	0.502	0.497	0.493	0.637	0.568	0.539	0.522	0.510	0.497	0.491	0.486	0.482
	1.45	0.598	0.541	0.517	0.503	0.493	0.482	0.477	0.473	0.470	0.600	0.536	0.510	0.494	0.483	0.471	0.466	0.461	0.457	0.603	0.533	0.505	0.488	0.476	0.463	0.457	0.451	0.447

1. Shape factor based on average of "B" and "C" configurations.

5.4.3 UNBALANCED GROUND LOAD CORRECTION FACTOR (B_M) The effect of an unbalanced annual ground load on the design of vertically-bored, closed-loop GHEX piping systems was discussed in Section 5.2.3. A correction factor (B_M) was presented based on the normalized net annual ground load (NNAGL) that adjusted the calculated borehole lengths for heating (Equation 5.28) or for cooling (Equation 5.30) based on ground thermal conductivity, borehole spacing, and loopfield geometry. Figures 5.13 through 5.18 provided correction factors for a range of soil thermal conductivity values. The same procedures that were utilized to adjust vertically-bored loopfield design lengths can be used to adjust the design length of the horizontally-bored loopfield. If the average depth of the horizontally-bored piping system is deeper than about 8 feet (below which ambient weather conditions at the ground surface will have little effect on the energy balance in the ground surrounding the buried GHEX piping) then application of the unbalanced ground load correction factor is highly recommended and becomes more important as the average depth of the piping system increases.

5.4.4 HORIZONTALLY-BORED GHEX DESIGN WORKSHEETS Proper design of a horizontally-bored GHEX requires reliable heating and cooling load design data and heat pump selection; consideration of domestic hot water generation; specification of horizontal loopfield design parameters; proper layout of the horizontal borefield; application of the heating or cooling design equations to determine the required borehole lengths; and adjustment for unbalanced annual ground loads to assure long-term GHEX performance. Heating and cooling load design data are determined and heat pumps are selected according to procedures presented in Chapter 3. Worksheets were provided to document the selection of the water-to-air or water-to-water GSHP units for the design space heating or cooling loads. Part of the heat pump selection is based on specification of the minimum and maximum entering water temperatures from the vertically-bored GHEX, along with the water flow rate that the GHEX must accommodate, each of which are critical parameters in the proper design of the horizontal GHEX system. Hot water generation, by either a dedicated water-to-water GSHP or a desuperheater in the space heating and cooling GSHP, will affect the horizontal loopfield design as discussed in Chapter 9, and procedures are provided to account for hot water generation in Tables 9.4 and 9.5. Layout of the horizontal borefield, in terms of number of horizontal bores, U-bend diameter, and maximum boring length are considered to accommodate required water flow rate to achieve proper turbulence for heat transfer in a heating dominant design and to control head loss to keep pumping power at an acceptable level at heating or cooling design conditions. Application of the appropriate borehole design length equation (Equation 5.28 or 5.30) using design data for the ground and borehole results in the required total length of horizontal borehole for heating or cooling. Adjustment of that length may be required for an unbalanced annual ground energy load, which is accomplished using procedures described in Section 5.2.3.

In an effort to simplify the design of the horizontal GHEX and provide a method to easily document the design, worksheets have been developed for both heating mode dominant (Figure 5.27) and cooling mode dominant (Figure 5.28) design. The worksheets are broken into sections to allow input for heat pump design and energy analysis data, domestic hot water generation, borefield layout, design data for the horizontal GHEX, calculation of the total borehole design length, and adjustment for unbalanced annual ground energy load. Additional calculations can be made to refine the horizontal GHEX design lengths in a table at the bottom of each worksheet. Each input blank in Figures 5.27 and 5.28 has been labeled with a number identifying a short input description, a listing of which follows each figure. A blank copy of each figure is provided in Appendix B.

Horizontally-Bored GHEX Design Worksheet – Heating Mode

Space Heating GSHP Design Data			HWG GSHP Design Data			Total Design Data		
HC_C [1]	= 1	Btu/hr	HC_{HWG}	= 9	Btu/hr	HC_D	= 14	Btu/hr
DMD_C [1]	= 2	kW	DMD_{HWG}	= 10	kW	DMD_D	= 15	kW
GPM [1]	= 3	gpm	GPM_{HWG}	= 11	gpm	GPM_D	= 16	gpm
AGL_H	= 4	Btu	AGL_{HWG}	= 12	Btu	AGL_{HD}	= 17	Btu
AGL_C	= 5	Btu				AGL_{CD}	= 18	Btu
Total Heat Loss [1]	= 6	Btu/hr						
Percent Sizing	= 7	%						
F_{Jan}	= 8		F_{HWG}	= 13				

GHEX Design Data

COP_D [2]	(= HC_D / (DMD_D x 3,412)	= 19		EWT_{min} [3]	= 22	F
HE_D [2]	(= HC_D x (COP_D-1)/COP_D)	= 20	Btu/hr	LWT_{min} [2] (= EWT_{min} – HE_D/(500xGPM_D))	= 23	F
F_H	(= (F_{Jan} x HC_C + F_{HWG} x HC_{HWG}) / HC_D)	= 21				

1. For single heat pump installation, use data directly from the appropriate GSHP Selection Worksheet. For multiple heat pumps in the installation, sum all heat pump Total Heat Losses, HC_C's, DMD_C's, and GPM's that will be connected to a single GHEX using the table provided below.
2. For single heat pump installation, use data directly from appropriate GSHP Selection Worksheet. For multiple heat pumps, use the equation provided.
3. EWT_{min} is obtained directly from the appropriate GSHP Selection Worksheet and must be the same for the selection of all heat pumps for a multiple heat pump installation connected to a single GHEX. (24)

	Zone 1	Zone 2	Zone 3	Zone 4	Zone 5	Zone 6	Zone 7	Zone 8	Zone 9	Zone 10	Total
Total Heat Loss											
HC_C											
DMD_C											
GPM											

Soil Design Data

Location 25 , ______ (City, State)

T_M	= 26	F	(Table 5.11)
A_s	= 27	F	(Table 5.11)
T_O	= 28	Days	(Table 5.11)
Soil Type	29		(Section 5.3.2.2)
k_{soil}	= 30	Btu/hr ft F	
α_{soil}	= 31	ft^2/day	

Horizontally-Bored GHEX Layout

U-bend D_P	= 32	in
GPM_{FP}	= 33	gpm/flowpath
N_{FP}	= 34	flowpaths
N_{BIS}	= 35	bores in series
N_B	= 36	bores
Layout	= 37 x 38	(N_{Layers} x $N_{Bores/Layer}$)
Bore Spacing	= 39	ft

Horizontally-Bored GHEX Design Data

D_{bore}	= 40	in	d_{avg}	= 43	ft
k_{Grout}	= 41	Btu/hr ft F	$T'(d_{avg},T_O)$	= 44	F (Table 5.12)
$(R_B + R_G \cdot F_H)$	= 42	hr ft F/Btu	$T_{S,L} = T'(d_{avg},T_O) \cdot A_s + T_M$ = ____ • ____ + ____	= 45	F

Borehole Design Lengths (Equations 5.28 and 5.29) (46)

$$L_{H,T} = \frac{HC_D \cdot \left(\frac{COP_D - 1}{COP_D}\right) \cdot (R_B + R_G \cdot F_H)}{T_{S,L} - \left(\frac{EWT_{min} + LWT_{min}}{2}\right)} = \frac{\_\_\_\_ \cdot \left(\frac{\_\_\_\_ - 1}{\_\_\_\_}\right) \cdot (\_\_\_\_)}{\_\_\_\_ - \left(\frac{\_\_\_\_ + \_\_\_\_}{2}\right)} = \frac{\_\_\_\_ \cdot (\_\_\_\_) \cdot (\_\_\_\_)}{\_\_\_\_ - (\_\_\_\_)} = \_\_\_\_ \text{ ft of bore}$$

$$L_{H,B} = \frac{L_{H,T}}{N_B} = \frac{\_\_\_\_}{\_\_\_\_} = \_\_\_\_ \text{ ft per bore}$$

Unbalanced Ground Load Borehole Design Lengths (Equations 5.9a, 5.10, and 5.11) (47)

$$NNAGL = \frac{AGL_{HD} - AGL_{CD}}{L_{H,T} \cdot \left(T_{S,L} - \left(\frac{EWT_{min} + LWT_{min}}{2}\right)\right)} = \frac{\_\_\_\_ - \_\_\_\_}{\_\_\_\_ \cdot \left(\_\_\_\_ - \left(\frac{\_\_\_\_ + \_\_\_\_}{2}\right)\right)} = \_\_\_\_ \text{ Btu / ftF}$$

B_M = ________ (Figures 5.13 through 5.18)

$$L_{H,T,UGL} = B_M \cdot L_{H,T} = \_\_\_\_ \cdot \_\_\_\_ = \_\_\_\_ \text{ ft of bore} \qquad L_{H,B,UGL} = \frac{L_{H,T,UGL}}{N_B} = \frac{\_\_\_\_}{\_\_\_\_} = \_\_\_\_ \text{ ft per bore}$$

Heating Design Length Calculations Summary Table (48)

Layout for Flow and Number of Bores						Design Length Calculations				Unbalanced Ground Load Design Lengths					
D_P	GPM_{FP}	N_{FP}	N_{BIS}	N_B	Layout	k_{Grout}	$R_B+R_G \cdot F_H$	$L_{H,T}$	$L_{H,B}$	Layout	Spacing	NNAGL	B_M	$L_{H,T,UGL}$	$L_{H,B,UGL}$
					/										
					/										
					/										
					/										

Figure 5.27. Horizontally-Bored GHEX Design Worksheet – Heating Mode.

5.4.4.1 HEATING MODE DESIGN WORKSHEET (FIGURE 5.27)

1. The corrected heating capacity (Figure 3.1, 3.2, or 3.3) of the heat pump at design heating conditions for the zone for which the ground loop is being designed. For multiple zones on a single ground loop, this is the sum of the corrected heating capacities for all of the zones connected to the ground loop. A table (24) is provided to sum the corrected heating capacities for all space heating zones connected to the ground loop.
2. The corrected electrical demand (Figure 3.1, 3.2, or 3.3) of the heat pump at design heating conditions for the zone for which the ground loop is being designed. For multiple zones on a single ground loop, this is the sum of the corrected electrical demands for all of the zones connected to the ground loop. A table (24) is provided to sum the corrected electrical demands for all space heating zones connected to the ground loop.
3. The circulating fluid flow rate (Figure 3.1, 3.2, or 3.3) through the heat pump at design heating conditions for the zone for which the ground loop is being designed. For multiple zones on a single ground loop, this is the sum of the circulating fluid flow rates for all of the zones connected to the ground loop. A table (24) is provided to sum the circulating fluid flow rates for all space heating zones connected to the ground loop.
4. Obtained from an annual bin analysis on the design total heat loss (6) at the installation location. Whether for single or multiple space heating zones, it is important that the GSHP heating capacity and coefficient of performance for the heat pumps used in the bin analysis reflect the corrected heating capacity (1) and corrected electrical demand (2) GSHP data used for the horizontally-bored GHEX design.
5. Obtained from an annual bin analysis on the design total heat gain (54) at the installation location. Whether for single or multiple space cooling zones, it is important that the GSHP cooling capacity and energy efficiency ratio for the heat pumps used in the bin analysis reflect the corrected total cooling capacity (49) and corrected electrical demand (50) GSHP data used for the horizontally-bored GHEX design.
6. The ACCA Manual J heat loss (Figure 3.1, 3.2, or 3.3) for the zone for which the ground loop is being designed. For multiple zones on a single ground loop, this is the sum of the heat losses for all of the zones connected to the ground loop. A table (24) is provided to sum the total heat losses for all space heating zones connected to the ground loop.
7. The percent of total heat loss (6) that the corrected heating capacity (1) of the heat pump at design heating conditions is capable of providing. If this is less than 100%, a supplementary heating source is required or the temperature of the heated zone may drop below the set point on the thermostat at design heating conditions.
8. Space heating load run fraction during the design heating month of January from a January bin analysis on a single or multiple zone Total Heat Loss (6) at the installation location. An alternate method to estimate the space heating load run fraction in January is to use Figure 3.9 with Percent Sizing (7) and heating degree day (HDD) data for the installation location from Table 3.4.
9. If no HWG option is used, then $HC_{HWG} = 0$. If HWG is used, refer to Table 9.4 for input that is dependent on HWG system type.
10. If no HWG option is used, then $DMD_{HWG} = 0$. If HWG is used, refer to Table 9.4 for input that is dependent on HWG system type.
11. If no HWG option is used, then $GPM_{HWG} = 0$. If HWG is used, refer to Table 9.4 for input that is dependent on HWG system type.
12. If no HWG option is used, then $AGL_{HWG} = 0$. If HWG is used, refer to Table 9.4 for input that is dependent on HWG system type.
13. If no HWG option is used, then $F_{HWG} = 0$. If HWG is used, refer to Table 9.4 for input that is dependent on HWG system type.
14. If no HWG option is used, then $HC_D = HC_C$. If HWG is used, refer to Table 9.4 for input dependent on HWG system type.
15. If no HWG option is used, then $DMD_D = DMD_C$. If HWG is used, refer to Table 9.4 for input dependent on HWG system type.
16. If no HWG option is used, then $GPM_D = GPM$. If HWG is used, refer to Table 9.4 for input dependent on HWG system type.
17. If no HWG option is used, then $AGL_{HD} = AGL_H$. If HWG is used, refer to Table 9.4 for input dependent on

HWG system type.

18. $AGL_{CD}=AGL_{C}$.
19. The design coefficient of performance (Figure 3.1, 3.2, or 3.3) of the heat pump at design heating conditions for the zone for which the ground loop is being designed. For multiple zones and hot water generation on a single ground loop, this is calculated based on the design heating capacity (14) and design electrical demand (15) for all heating loads connected to the ground loop.
20. The design heat of extraction (Figure 3.1, 3.2, or 3.3) from the ground loop at design heating conditions for the zone for which the ground loop is being designed. For multiple zones and hot water generation on a single ground loop, this is calculated based on the design heating capacity (14) and design coefficient of performance (19) for all heating loads connected to the ground loop.
21. The design heating month run fraction that accounts for both space heating load and hot water generation load during the month of January.
22. The minimum entering water temperature for which the heat pump heating capacities (Figure 3.1, 3 .2, or 3.3) have been selected for design heating conditions.
23. The minimum leaving water temperature based on the minimum entering water temperature (22), the design heat of extraction (20), and the design circulating fluid flow rate (16) at design heating conditions.
24. For multiple space heating zones, utilize this table to sum the Total Heat Loss, HC_C, DMD_C, and GPM for each zone for input into Total Heat Loss (6), HC_C (1), DMD_C (2), and GPM (3).
25. The location of the horizontally-trenched GHEX determines the minimum soil temperature for design ($T_{S,L}$) and is calculated using the soil temperature data provided in Table 5.11. Figure 5.21 provides a map of locations for which earth temperature data are available. Parameters for the location nearest the actual installation site should be used, but care should be taken to make sure the selected location has similar geographic characteristics.
26. Obtained from Table 5.11 for the selected soil temperature data location.
27. Obtained from Table 5.11 for the selected soil temperature data location.
28. Obtained from Table 5.11 for the selected soil temperature data location.
29. The soil type to the depth of the deepest horizontal bore at the installation location. The soil type affects the moisture holding capability of the soil and the resulting thermal conductivity that can be used for design of the horizontally-trenched GHEX, as discussed in Section 5.3.2.2. If more than one soil type layer exists to the depth of the deepest horizontal bore, then use the soil type with the lowest thermal conductivity value (conservative approach) or do a weighted average of thermal conductivity in similar fashion to what was done with vertical bores in Table 5.7.
30. The soil thermal conductivity, which is the lowest value that would be expected during heating mode operation of the heat pump system and is used to estimate the soil thermal resistance (R_S) for use in calculating the total pipe design length for heating ($L_{H,T}$).
31. The soil thermal diffusivity, which is the highest value that would be expected during any year at the installation site and is used to predict minimum soil temperature for design ($T_{S,L}$).
32. The nominal diameter of the U-bend used in the horizontal borehole. Common U-bend diameters are ¾, 1, and 1-1/4-inch, and the size that is used is based on the flow rate through the U-bend as discussed in Section 5.1.3.
33. The flow rate through each flow path must be high enough for turbulence (Section 4.4.2), but in a range that keeps head loss at a reasonable level (Section 4.4.3) at design heating conditions. For most circulating fluids recommended flow rates of 3 gpm for ¾-inch, 4.5 to 6 gpm for 1-inch, and 6 to 12 gpm for 1-1/4-inch pipe will provide a good trade off between turbulence, head loss, and the resulting design length of the horizontal borehole.
34. The number of flow paths for the circulating fluid is estimated by dividing the design system flow rate (16) by the flow rate through each flow path (33). If the result is not a whole number, then round to the nearest whole number and, for the case of rounding up to the next highest whole number, increase the total system flow rate accordingly when designing the pumping system.
35. The number of horizontal boreholes that have U-bends hooked together in series. That is, flow through one U-bend is diverted directly through a second (and possibly a third, fourth, etc.) before the flow is connected back to the return header returning flow to the heat pump system. This isn't common, but there are times when

site or boring conditions do not allow for a continuous bore length that is needed to provide the required heat transfer for the flow rate through the U-bend. As an example, if 3 gpm is being circulated through a ¾-inch U-bend, which is the flow rate associated with about 1 ton of heat pump heating capacity, and the calculated total borehole design length was 250 feet, then 250 feet of active borehole would be required. If a boring length limit were encountered at about 125 feet (short lot, for example), then it would be acceptable to bore two, 125-foot long boreholes and connect the U-bends in each of the boreholes in series to achieve proper flow through each U-bend (3 gpm) while having sufficient borehole length (250 feet) to provide the required heat transfer for the 1 ton of connected heat pump heating capacity.

36. The number of horizontal boreholes is calculated by multiplying the number of flow paths (34) and the number of bores in series (35).
37. The number of layers of horizontal boreholes in the GHEX. In many residential ground loop designs, the boreholes can be arranged in a single layer at the selected design depth, but if the area being used for the horizontally-bored GHEX is limiting then more than one layer can be used. The layers should be separated by a minimum of 15 feet of vertical distance. There are practical limits to the maximum boring depth that can be reached economically, and qualified horizontal boring contractors should be consulted for this information. Refer to Section 5.4.2.1 for further discussion concerning layout of the horizontally-bored GHEX.
38. The number of horizontal boreholes in each layer in the GHEX, which is the total number of horizontal boreholes (36) divided by the number of layers of horizontal boreholes (37).
39. The average center-to-center spacing between the horizontal boreholes. A minimum spacing of 15 feet is recommended as a starting point in the design. Spacing may be increased to accommodate an unbalanced annual ground energy load. A spacing of less than 15 feet is not recommended unless area for the horizontal GHEX is extremely limited, and then care must be taken to carefully evaluate the design length for unbalanced annual ground energy load.
40. The borehole diameter, which usually ranges between 4 and 6 inches for horizontal boring machines.
41. The grout thermal conductivity, which is defined by the relative mixture of the grout base material (bentonite or cement) and any additive that may be used for thermal enhancement (silica sand or other thermal enhancement compound). Grout thermal conductivity ranges between 0.40 and 1.20 Btu/hr ft F, with values in the middle and lower half of that range being most common for residential and light commercial applications.
42. This is the ground loop resistance, which depends on many factors described in Sections 5.4.2.2 and 5.4.2.3 (run fraction during design heating month (21), soil thermal conductivity (30), borehole diameter (40), and grout thermal conductivity (41)), and is presented in Tables 5.31, 5.32, and 5.33 for nominal U-bend diameters of ¾, 1, and 1-1/4 inch, respectively.
43. The average depth of the horizontal borefield, which would be the depth of the horizontal bores for a single-layer layout or the numerical average of the depths of each of the layers of horizontal bores for a multiple-layer layout.
44. The dimensionless minimum soil temperature for heating design, which is expressed as a function of soil thermal diffusivity (31) and average depth in the soil (43) in Table 5.12 or obtained from Equation 5.23.
45. The minimum soil temperature for heating design (Equation 5.25) using the mean earth temperature in the top 10 feet (26) and annual swing (27) for the installation location and the dimensionless minimum soil temperature for heating design (44). If the average depth of the horizontal borefield is greater than 20 feet, then it is acceptable to use deep earth temperature (T_G) from Section 5.2.2.1.
46. The total borehole design length for heating (Equation 5.28) and the individual borehole design length for heating (Equation 5.29) are calculated using the indicated design parameters from the design sheet.
47. If the annual ground energy load is unbalanced, then the total and individual borehole design lengths for heating must be adjusted using the calculations and data provided in Section 5.2.3. The NNAGL is calculated as shown and used with the bore length multiplier figure (Figures 5.13 through 5.18) for the closest k_{soil} to estimate B_M for the appropriate borehole spacing and loop field layout layers x bores/layer combination. The unbalanced ground load borehole design lengths are obtained by multiplying the borehole design lengths from (46) by B_M.
48. This table has been included to summarize heating design length calculations for various combinations of design parameters.

Horizontally-Bored GHEX Design Worksheet – Cooling Mode

Space Cooling GSHP Design Data

TC_C[1] = **49** Btu/hr
DMD_C[1] = **50** kW
GPM[1] = **51** gpm
AGL_H = **52** Btu
AGL_C = **53** Btu
Total Heat Gain[1] = **54** Btu/hr
Percent Sizing = **55** %
F_{Jul} = **56**

HWG GSHP Design Data

GPM_{HWG} = **57** gpm
AGL_{HWG} = **58** Btu

Total Design Data

TC_D = **59** Btu/hr
DMD_D = **60** kW
GPM_D = **61** gpm
AGL_{HD} = **62** Btu
AGL_{CD} = **63** Btu

GHEX Design Data

EER_D[2] (= TC_D / (DMD_D x 1000) = **64**
HR_D[2] (= TC_D x (EER_D+3.412)/EER_D) = **65** Btu/hr
F_C (= F_{Jul}) = **66**
EWT_{max}[3] = **67** F
LWT_{max}[2] (= EWT_{max} + HR_D/(500xGPM_D) = **68** F

1. For single heat pump installation, use data directly from the appropriate GSHP Selection Worksheet. For multiple heat pumps in the installation, sum all heat pump Total Heat Gains, TC_C's, DMD_C's, and GPM's that will be connected to a single GHEX using the table provided below.
2. For single heat pump installation, use data directly from appropriate GSHP Selection Worksheet. For multiple heat pumps, use the equation provided.
3. EWT_{max} is obtained directly from the appropriate GSHP Selection Worksheet and must be the same for the selection of all heat pumps for a multiple heat pump installation connected to a single GHEX.

(69)

	Zone 1	Zone 2	Zone 3	Zone 4	Zone 5	Zone 6	Zone 7	Zone 8	Zone 9	Zone 10	Total
Total Heat Gain											
TC_C											
DMD_C											
GPM											

Soil Design Data

Location ___**70**___ , ______ (City, State)
T_M = **71** F (Table 5.11)
A_s = **72** F (Table 5.11)
T_O = **73** Days (Table 5.11)
Soil Type ___**74**___ (Section 5.3.2.2)
k_{soil} = **75** Btu/hr ft F
α_{soil} = **76** ft²/day

Horizontally-Bored GHEX Layout

U-bend D_P = **77** in
GPM_{FP} = **78** gpm/flowpath
N_{FP} = **79** flowpaths
N_{BIS} = **80** bores in series
N_B = **81** bores
Layout = **82** x **83** (N_{Layers} x $N_{Bores/Layer}$)
Bore Spacing = **84** ft

Horizontally-Bored GHEX Design Data

D_{bore} = **85** in
k_{Grout} = **86** Btu/hr ft F
(R_B + $R_G \cdot F_H$) = **87** hr ft F/Btu
d_{avg} = **88** ft
$T'(d_{avg}, T_O+180)$ = **89** F (Table 5.13)
$T_{S,H} = T'(d_{avg}, T_O+180) \cdot A_s + T_M$ = ____ • ____ + ____ = **90** F

Borehole Design Lengths (Equations 5.30 and 5.31) (91)

$$L_{C,T} = \frac{TC_D \cdot \left(\frac{EER_D + 3.412}{EER_D}\right) \cdot (R_B + R_G \cdot F_C)}{\left(\frac{EWT_{max} + LWT_{max}}{2}\right) - T_{S,H}} = \frac{\_\_\_ \cdot \left(\frac{\_\_\_ + 3.412}{\_\_\_}\right) \cdot (\_\_\_)}{\left(\frac{\_\_\_ + \_\_\_}{2}\right) - \_\_\_} = \frac{\_\_\_ \cdot (\_\_\_) \cdot (\_\_\_)}{(\_\_\_) - \_\_\_} = \_\_\_ \text{ ft of bore}$$

$$L_{C,B} = \frac{L_{C,T}}{N_B} = \frac{\_\_\_}{\_\_\_} = \_\_\_ \text{ ft per bore}$$

Unbalanced Ground Load Borehole Design Lengths (Equations 5.9b, 5.12, and 5.13) (92)

$$NNAGL = \frac{AGL_{CD} - AGL_{HD}}{L_{C,T} \cdot \left(\left(\frac{EWT_{max} + LWT_{max}}{2}\right) - T_{S,H}\right)} = \frac{\_\_\_ - \_\_\_}{\_\_\_ \cdot \left(\left(\frac{\_\_\_ + \_\_\_}{2}\right) - \_\_\_\right)} = \_\_\_ \text{ Btu / ftF}$$

B_M = ________ (Figures 5.13 through 5.18)

$$L_{C,T,UGL} = B_M \cdot L_{C,T} = \_\_\_ \cdot \_\_\_ = \_\_\_ \text{ ft of bore} \qquad L_{C,B,UGL} = \frac{L_{C,T,UGL}}{N_B} = \frac{\_\_\_}{\_\_\_} = \_\_\_ \text{ ft per bore}$$

Cooling Design Length Calculations Summary Table (93)

Layout for Flow and Number of Bores						Design Length Calculations				Unbalanced Ground Load Design Lengths					
D_P	GPM_{FP}	N_{FP}	N_{BIS}	N_B	Layout	k_{Grout}	R_B+$R_G \cdot F_C$	$L_{C,T}$	$L_{C,B}$	Layout	Spacing	NNAGL	B_M	$L_{C,T,UGL}$	$L_{C,B,UGL}$
					/										
					/										
					/										
					/										

Figure 5.28. Horizontally-Bored GHEX Design Worksheet – Cooling Mode.

5.4.4.2 COOLING MODE DESIGN WORKSHEET (FIGURE 5.28)

49. The corrected total cooling capacity (Figure 3.1, 3.2, or 3.3) of the heat pump at design cooling conditions for the zone for which the ground loop is being designed. For multiple zones on a single ground loop, this is the sum of the corrected total cooling capacities for all of the zones connected to the ground loop. A table (69) is provided to sum the corrected total cooling capacities for all space cooling zones connected to the ground loop.
50. The corrected electrical demand (Figure 3.1, 3.2, or 3.3) of the heat pump at design cooling conditions for the zone for which the ground loop is being designed. For multiple zones on a single ground loop, this is the sum of the corrected electrical demands for all of the zones connected to the ground loop. A table (69) is provided to sum the corrected electrical demands for all space cooling zones connected to the ground loop.
51. The circulating fluid flow rate (Figure 3.1, 3.2, or 3.3) through the heat pump at design cooling conditions for the zone for which the ground loop is being designed. For multiple zones on a single ground loop, this is the sum of the circulating fluid flow rates for all of the zones connected to the ground loop. A table (69) is provided to sum the circulating fluid flow rates for all space cooling zones connected to the ground loop.
52. Obtained from an annual bin analysis on the design total heat loss (6) at the installation location. Whether for single or multiple heating zones, it is important that the GSHP heating capacity and coefficient of performance for the heat pumps used in the bin analysis reflect the corrected heating capacity (1) and corrected electrical demand (2) GSHP data used for the horizontally-bored GHEX design.
53. Obtained from an annual bin analysis on the design total heat gain (54) at the installation location. Whether for single or multiple cooling zones, it is important that the GSHP cooling capacity and energy efficiency ratio for the heat pumps used in the bin analysis reflect the corrected total cooling capacity (49) and corrected electrical demand (50) GSHP data used for the horizontally-bored GHEX design.
54. The ACCA Manual J heat gain (Figure 3.1, 3.2, or 3.3) for the zone for which the ground loop is being designed. For multiple zones on a single ground loop, this is the sum of the heat gains for all of the zones connected to the ground loop. A table (69) is provided to sum the total heat gains for all space cooling zones connected to the ground loop.
55. The percent of total heat gain (54) that the corrected total cooling capacity (49) of the heat pump at design cooling conditions is capable of providing. If this is less than 100%, then the temperature of the cooled zone may rise above the set point on the thermostat at design cooling conditions.
56. Space cooling load run fraction during the design cooling month of July from a July bin analysis on a single or multiple zone Total Heat Gain (54) at the installation location. An alternate method to estimate the space cooling load run fraction in July is to use Figure 3.10 with Percent Sizing (55) and cooling degree days (CDD) data for the installation location from Table 3.4.
57. If no HWG option is used, then $GPM_{HWG} = 0$. If HWG is used, refer to Table 9.5 for input that is dependent on HWG system type.
58. If no HWG option is used, then $AGL_{HWG} = 0$. If HWG is used, refer to Table 9.5 for input that is dependent on HWG system type.
59. $TC_D = TC_C$.
60. $DMD_D = DMD_C$.
61. If no HWG option is used, then $GPM_D = GPM$. If HWG is used, refer to Table 9.5 for input dependent on HWG system type.
62. If no HWG option is used, then $AGL_{HD} = AGL_H$. If HWG is used, refer to Table 9.5 for input dependent on HWG system type.
63. $AGL_{CD} = AGL_C$.
64. The design energy efficiency ratio (Figure 3.1, 3.2, or 3.3) of the heat pump at design cooling conditions for the zone for which the ground loop is being designed. For multiple zones on a single ground loop, this is calculated based on the design total cooling capacity (59) and design electrical demand (60) for all cooling loads connected to the ground loop.
65. The design heat of rejection (Figure 3.1, 3.2, or 3.3) to the ground loop at design cooling conditions for the zone for which the ground loop is being designed. For multiple zones on a single ground loop, this is calculated based on the design total cooling capacity (59) and design energy efficiency ratio (64) for all cooling loads connected to the ground loop.

66. The design cooling month run fraction that accounts for space cooling loads during the month of July.
67. The maximum entering water temperature for which the heat pump cooling capacities (Figure 3.1, 3.2, or 3.3) have been selected for design cooling conditions.
68. The maximum leaving water temperature based on the maximum entering water temperature (67), the design heat of rejection (65) and the design circulating fluid flow rate (61) at design cooling conditions.
69. For multiple space cooling zones utilize this table to sum the Total Heat Gain, TC_C, DMD_C, and GPM for each zone for input into Total Heat Gain (54), TC_C (49), DMD_C (50), and GPM (51).
70. The location of the horizontally-trenched GHEX determines the maximum soil temperature for design (TS,H) and is calculated using the soil temperature data provided in Table 5.11. Figure 5.21 provides a map of locations for which earth temperature data are available. Parameters for the location nearest the actual installation site should be used, but care should be taken to make sure the selected location has similar geographic characteristics.
71. Obtained from Table 5.11 for the selected soil temperature data location.
72. Obtained from Table 5.11 for the selected soil temperature data location.
73. Obtained from Table 5.11 for the selected soil temperature data location.
74. The soil type to the depth of the deepest horizontal bore at the installation location. The soil type affects the moisture holding capability of the soil and the resulting thermal conductivity that can be used for design of the horizontally-trenched GHEX, as discussed in Section 5.3.2.2. If more than one soil type layer exists to the depth of the deepest horizontal bore, then use the soil type with the lowest thermal conductivity value (conservative approach) or do a weighted average of thermal conductivity in similar fashion to what was done with vertical bores in Table 5.7.
75. The soil thermal conductivity, which is the lowest value that would be expected during cooling mode operation of the heat pump system and is used to estimate the soil thermal resistance (R_S) for use in calculating the total pipe design length for cooling ($L_{C,T}$).
76. The soil thermal diffusivity, which is the highest value that would be expected during any year at the installation site and is used to predict maximum soil temperature for design ($T_{S,H}$).
77. The nominal diameter of the U-bend used in the horizontal borehole. Common U-bend diameters are ¾, 1, and 1-1/4-inch, and the size that is used is based on the flow rate through the U-bend as discussed in Section 5.1.3.
78. The flow rate through each flow path must be high enough for turbulence (Section 4.4.2), but in a range that keeps head loss at a reasonable level (Section 4.4.3) at design heating conditions. For most circulating fluids recommended flow rates of 3 gpm for ¾-inch, 4.5 to 6 gpm for 1-inch, and 6 to 12 gpm for 1-1/4-inch pipe will provide a good trade-off between turbulence, head loss, and the resulting design length of the horizontal borehole.
79. The number of flow paths for the circulating fluid is estimated by dividing the design system flow rate (61) by the flow rate through each flow path (78). If the result is not a whole number, then round to the nearest whole number and, for the case of rounding up to the next highest whole number, increase the total system flow rate accordingly when designing the pumping system.
80. The number of horizontal boreholes that have U-bends hooked together in series. That is, flow through one U-bend is diverted directly through a second (and possibly a third, fourth, etc.) before the flow is connected back to the return header returning flow to the heat pump system. This isn't common, but there are times when site or boring conditions do not allow for a continuous bore length that is needed to provide the required heat transfer for the flow rate through the U-bend. As an example, if 3 gpm is being circulated through a ¾-inch U-bend, which is the flow rate associated with about 1 ton of heat pump heating capacity, and the calculated total borehole design length was 250 feet, then 250 feet of active borehole would be required. If a boring length limit were encountered at about 125 feet (short lot, for example), then it would be acceptable to bore two, 125-foot long boreholes and connect the U-bends in each of the boreholes in series to achieve proper flow through each U-bend (3 gpm) while having sufficient borehole length (250 feet) to provide the required heat transfer for the 1 ton of connected heat pump cooling capacity.
81. The number of horizontal boreholes is calculated by multiplying the number of flow paths (79) and the number of bores in series (80).
82. The number of layers of horizontal boreholes in the GHEX. In many residential ground loop designs, the boreholes can be arranged in a single layer at the selected design depth, but if the area being used for the horizontally-bored GHEX is limiting then more than one layer can be used. The layers should be separated by a minimum of 15 feet of vertical distance. There are practical limits to the maximum boring depth that can be reached economically, and qualified horizontal boring contractors should be consulted for this information. Refer to Section 5.4.2.1 for further

discussion concerning layout of the horizontally-bored GHEX.

83. The number of horizontal boreholes in each layer in the GHEX, which is the total number of horizontal boreholes (81) divided by the number of layers of horizontal boreholes (82).
84. The average center-to-center spacing between the horizontal boreholes. A minimum spacing of 15 feet is recommended as a starting point in the design. Spacing may be increased to accommodate an unbalanced annual ground energy load. A spacing of less than 15 feet is not recommended unless area for the horizontal GHEX is extremely limited, and then care must be taken to carefully evaluate the design length for unbalanced annual ground energy load.
85. The borehole diameter, which usually ranges between 4 and 6 inches for horizontal boring machines.
86. The grout thermal conductivity, which is defined by the relative mixture of the grout base material (bentonite or cement) and any additive that may be used for thermal enhancement (silica sand or other thermal enhancement compound). Grout thermal conductivity ranges between 0.40 and 1.20 Btu/hr ft F, with values in the middle to lower half of that range being most common for residential and light commercial applications.
87. This is the ground loop resistance, which depends on many factors described in Sections 5.4.2.2 and 5.4.2.3 (run fraction during design cooling month (66), soil thermal conductivity (75), borehole diameter (85), and grout thermal conductivity (86)), and is presented in Tables 5.31, 5.32, and 5.33 for nominal U-bend diameters of ¾, 1, and 1-1/4 inch, respectively.
88. The average depth of the horizontal borefield, which would be the depth of the horizontal bores for a single-layer layout or the numerical average of the depths of each of the layers of horizontal bores for a multiple-layer layout.
89. The dimensionless maximum soil temperature for cooling design, which is expressed as a function of soil thermal diffusivity (76) and average depth in the soil (88) in Table 5.13 or obtained from Equation 5.24.
90. The maximum soil temperature for cooling design (Equation 5.26) using the mean earth temperature in the top 10 feet (71) and annual swing (72) for the installation location and the dimensionless maximum soil temperature for cooling design (89). If the average depth horizontal borefield is greater than 20 feet, then it is acceptable to use deep earth temperature (T_G) from Section 5.2.2.1.
91. The total borehole design length for cooling (Equation 5.30) and the individual borehole design length for cooling (Equation 5.31) are calculated using the indicated design parameters from the design sheet.
92. If the annual ground energy load is unbalanced, then the total and individual borehole design lengths for cooling must be adjusted using the calculations and data provided in Section 5.2.3. The NNAGL is calculated as shown and used with the bore length multiplier figure (Figures 5.13 through 5.18) for the closest k_{soil} to estimate B_M for the appropriate borehole spacing and loop field layout layers x bores/layer combination. The unbalanced ground load borehole design lengths are obtained by multiplying the borehole design lengths from (91) by B_M.
93. This table has been included to summarize cooling design length calculations for various combinations of design parameters.

5.4.5 HORIZONTALLY-BORED GHEX DESIGN EXAMPLES Appendix A contains examples of horizontally-bored GHEX designs for a heating dominant installation (Appendix A.1), a cooling dominant installation (Appendix A.2), and a multiple heat pump installation (Appendix A.3).

GSHP Design Worksheets (Blank Copies) – See Appendix B

Figure 5.19. Vertically-Bored GHEX Design Worksheet – Heating Mode
Figure 5.20. Vertically-Bored GHEX Design Worksheet – Cooling Mode
Figure 5.25. Horizontally-Trenched GHEX Design Worksheet – Heating Mode
Figure 5.26. Horizontally-Trenched GHEX Design Worksheet – Cooling Mode
Figure 5.27. Horizontally-Bored GHEX Design Worksheet – Heating Mode
Figure 5.28. Horizontally-Bored GHEX Design Worksheet – Cooling Mode

6 INTERIOR CLOSED-LOOP PIPING DESIGN

In This Section

6.1 Overall Interior Piping Design Procedure

This chapter discusses how to design an interior piping system by following the procedure outlined below. Similar to designing the exterior portion of a GSHP system, following a set procedure during the design phase of the interior piping system layout not only minimizes overall process complication, but also allows for exploration of alternative designs. The recommended procedure for designing an interior piping system is as follows:

1. Determine the GSHP placement inside the building in addition to the loopfield placement outside of the building. According to the location of the two, determine the best method to penetrate the building and connect GHEX piping to interior piping.
2. Determine whether a pressurized or non-pressurized pumping system will be used.
3. Determine the interior piping layout according to the type of pumping system to be used. Include the placement of valves, P/T ports, standing column tanks (non-pressurized flow centers only), optional expansion tanks (pressurized flow centers only), and circulating pumps wherever they are necessary in the system.
4. Specify the type of pipe to be used for the interior piping portion of the system.
5. Perform system head loss calculations.
6. Size and select the circulating pumps according the results of the system head loss calculations.

While many options exist regarding pumping system types, piping materials, layouts, controls systems, etc., every situation will be unique and should be considered individually and tailored to a building owner's needs for what is most logical in terms of efficiency, performance, durability, and cost effectiveness.

6.2 Piping Materials

As stated in Chapter 4, high-density polyethylene (HDPE) and cross-linked polyethylene (PEX-A) are the only piping materials suitable for the buried portion of a ground source heat pump system. However, many options exist in regards to pipe type and joining methods for any piping not buried underground (e.g. - supply-return piping located inside of the building). Fluid flow in the turbulent regime is only critical in the buried portion of the GHEX and does not need to be considered for supply-return piping with respect to heat transfer. Compliance with code requirements, along with the ability to attain flushing flow, head loss, and cost and ease of installation are the primary factors to be considered when determining what type of pipe will be used for the interior portion of the system. While installers are not limited to any specific type of pipe for the interior portion of a system, the most commonly selected types are usually a combination of reinforced rubber heater hose, HDPE, and PVC because of their relatively low cost and ease of installation. Refer to Chapter 4 for a list of the types of pipe common in the HVAC industry in addition to relevant properties, descriptions, and joining methods for each type.

6.3 Pressurized & Non-Pressurized Flow Centers

Two basic types of flow centers exist that are used in conjunction with ground source heat pump systems: pressurized and non-pressurized (standing column). Systems that utilize pressurized flow centers must maintain positive pressure on the loopfield (to ensure positive suction-side pressure on the pumps) to produce flow. Positive pressure on the loopfield is attained via pressurization of the lines (discussed in Chapter 9). Non-pressurized flow centers maintain positive suction pressure on the pumps via a standing column of water and do not require that the loopfield piping be pressurized to ensure proper operation. Each type of flow center has its own advantages and disadvantages, and the type of system used for a GSHP is a matter of preference. The advantages and disadvantages for each type of flow center are as follows:

PRESSURIZED FLOW CENTERS

- **Advantages**
    - Pressurized flow centers are manufactured with incorporated flush/fill ports, eliminating the need to fabricate an external means to fill/flush the system
    - No possibility of fluid exposure to particulates or contaminants during operation
    - Pumps have adequate spacing (in two-pump flow centers) to ensure uniform flow patterns through each, thus maximizing pump performance

- **Disadvantages**
    - Thermal expansion of ground loop piping may cause large enough reduction in system pressure to cause pump cavitation (negative pump suction-side pressure), necessitating a service call to re-pressurize the system
    - Entrapped air not completely removed from the system during purging will not remove itself over time, increasing the possibility of pump cavitation
    - Check valves and isolation valves are not typically incorporated in the design of pressurized systems and must be field installed for multiple heat pump applications

NON-PRESSURIZED (STANDING COLUMN) FLOW CENTERS

- **Advantages**
    - Entrapped air not completely removed from the system during the purging process will remove itself over time
    - If circulating fluid levels lower over time, water/antifreeze can simply be added to the system from the top of the standing column tank via removable cap without necessitating reflushing/purging of the system
    - Antifreeze freeze protection levels can easily be tested from the top of the standing column tank
    - System flow can easily be measured (via a flowmeter tool) from the top of the standing column tank
    - Virtually maintenance free operation after initial installation
    - Does not require field charging, minimizing the need for service calls later in the system's life
    - Check valves and pump isolation valves are incorporated in the design of standing column systems, eliminating the need for field installed valves (unless zone valves are necessary, to be discussed later)
    - If an inside-the-building header system is utilized, system flushing and purging can be performed without the use of a flush cart (refer to Chapter 9)

- **Disadvantages**
    - Removable cap allows for the possibility of particulates and contaminants to enter the system
    - Flush/fill ports not incorporated with non-pressurized flow centers and must be fabricated and installed in the field

Figure 6.1a displays a detailed view of a typical pressurized flow center and Figure 6.1b displays a detailed view of a typical non-pressurized flow center.

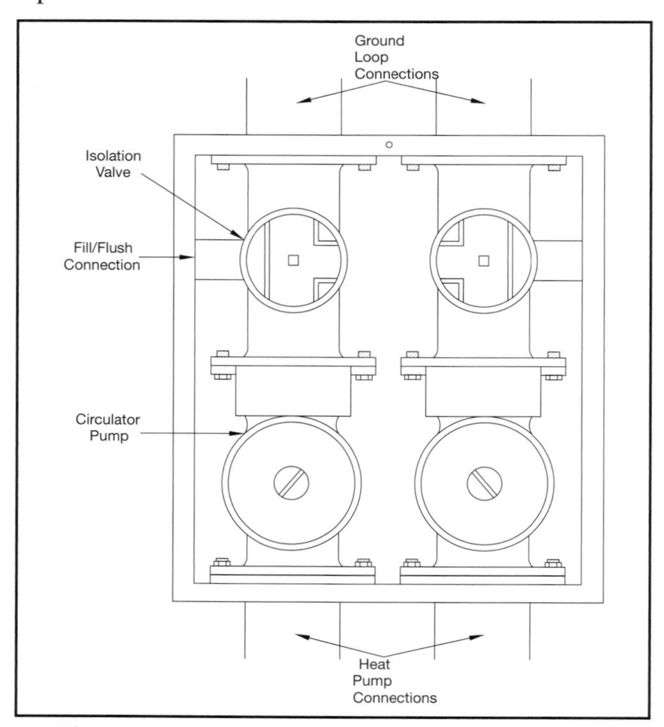

Figure 6.1a. Pressurized Flow Center Details

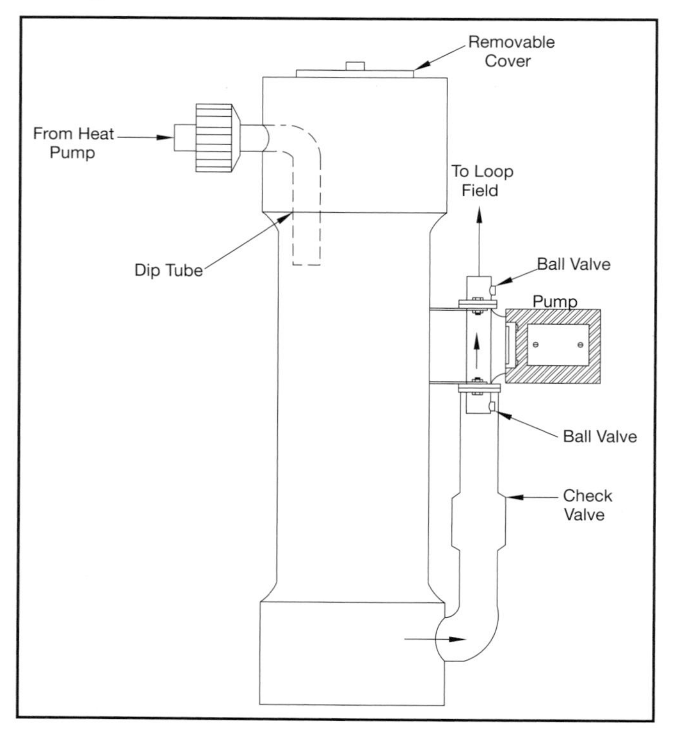

Figure 6.1b. Non-Pressurized (Standing Column) Flow Center Details (Patented)

6.4 Building Penetration Methods

Single heat pump systems are typical to smaller residential installations and are generally the least complicated type to install. The location of the GHEX outside in addition to the location of heat pump in the building will govern where the penetrations will need to be made to introduce the ground loop piping into the home. A minimum separation distance of 10 feet is recommended between the ground source loopfield and the footings for the building. Multiple heat pump systems tend to be more complicated, but the methods used to connect them to the GHEX are essentially the same. Generally, if a multi-circuit flow center is used, the location of that flow center will govern where the penetrations will need to be made to introduce the ground loop piping into the home. If smaller flow centers are hooked up to each individual heat pump unit, an interior piping manifold will need to be constructed (in a central location, more than likely in a mechanical room) and the ground loop piping will need to be brought to that manifold. Individual pairs of supply-return piping will then need to be run from the manifold to each heat pump unit. Interior piping will be discussed in more detail in Section 6.5.

The location of the heat pump(s)/flow center(s)/manifold in a building will determine the best penetration method to bring the ground loop piping into the home. If the heat pump/flow center is to be located in a basement near an exterior wall, it may be easiest to bring the lines in directly through that wall. The wall penetrations should be made approximately 1 foot above the basement floor and 1 foot apart in such situations. See Figure 6.2a for details. If the heat pump is to be located in the center of a building or if the building is without a basement (slab-on-grade or crawl space), it may be easiest to bring the supply-return lines into the building through the floor (such will always be the case for slab-on-grade homes). See Figure 6.2b for details for floor slab penetration and Figure 6.2c for crawl-space penetration. In any installation, the exterior portion of the ground loop piping should always remain buried 4-6 feet below the ground surface (which will be the depth of the header trench). Additionally, proper sealing materials should be used to prevent water from entering the building through the penetrations formed to introduce the ground loop piping to the building's interior.

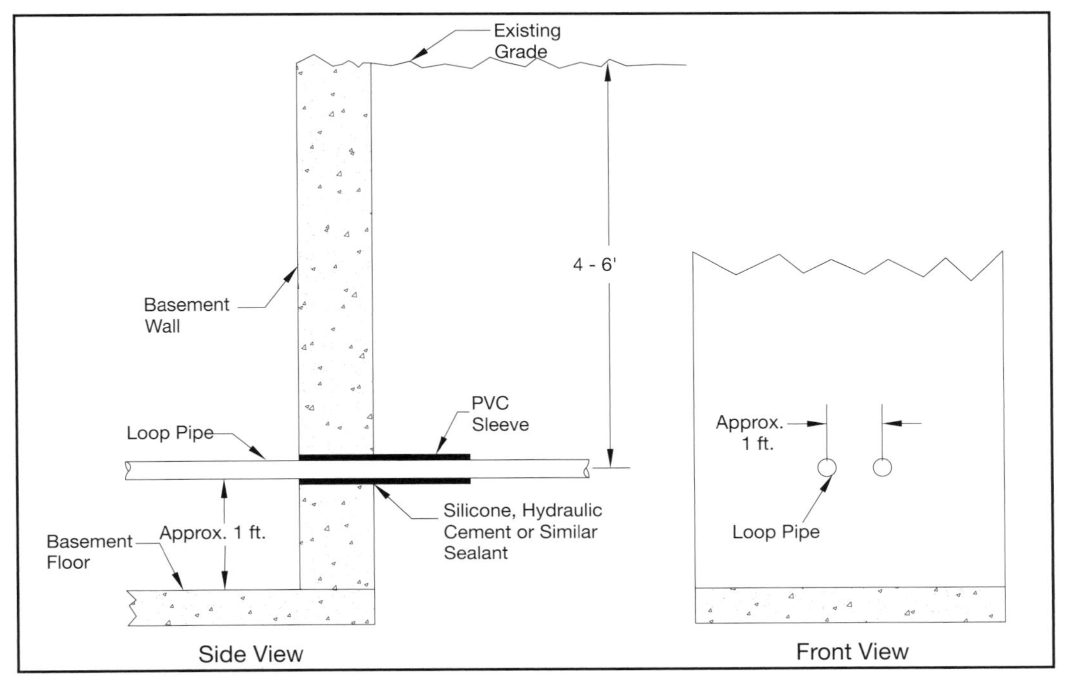

Figure 6.2a. Wall Penetration Details

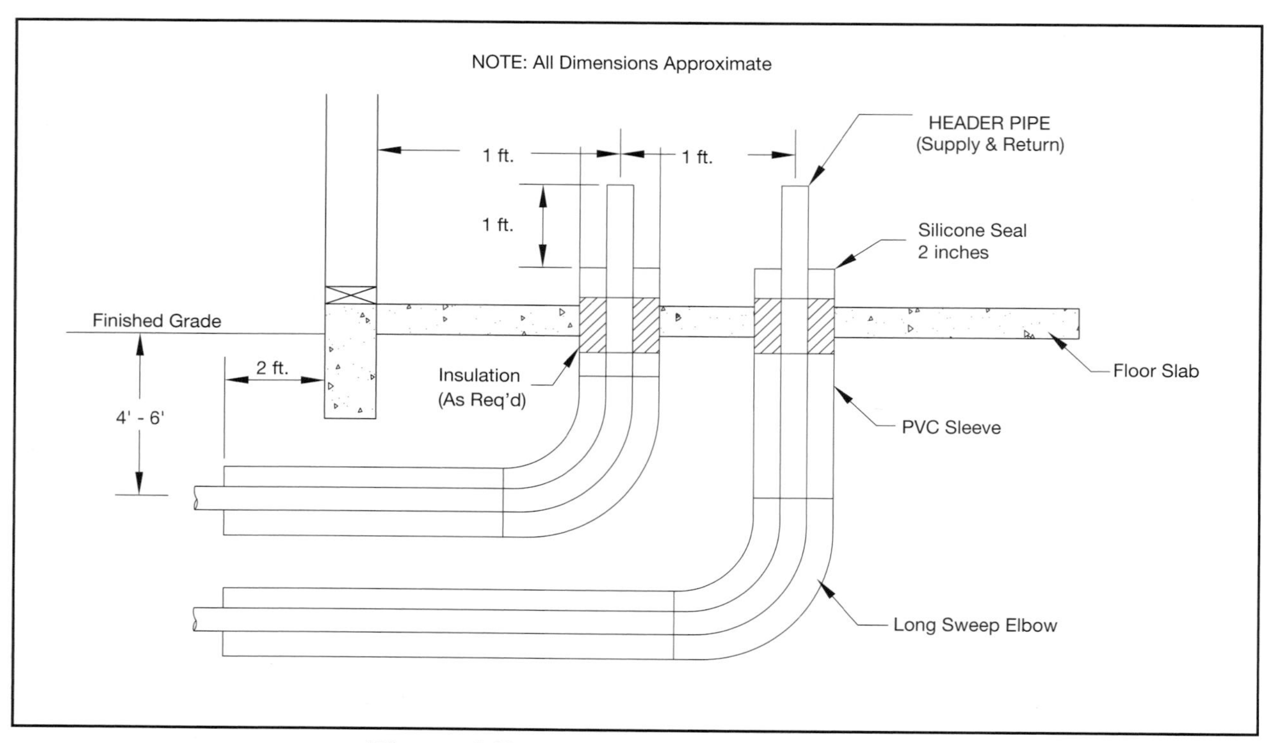

Figure 6.2b. Floor Slab Penetration Details

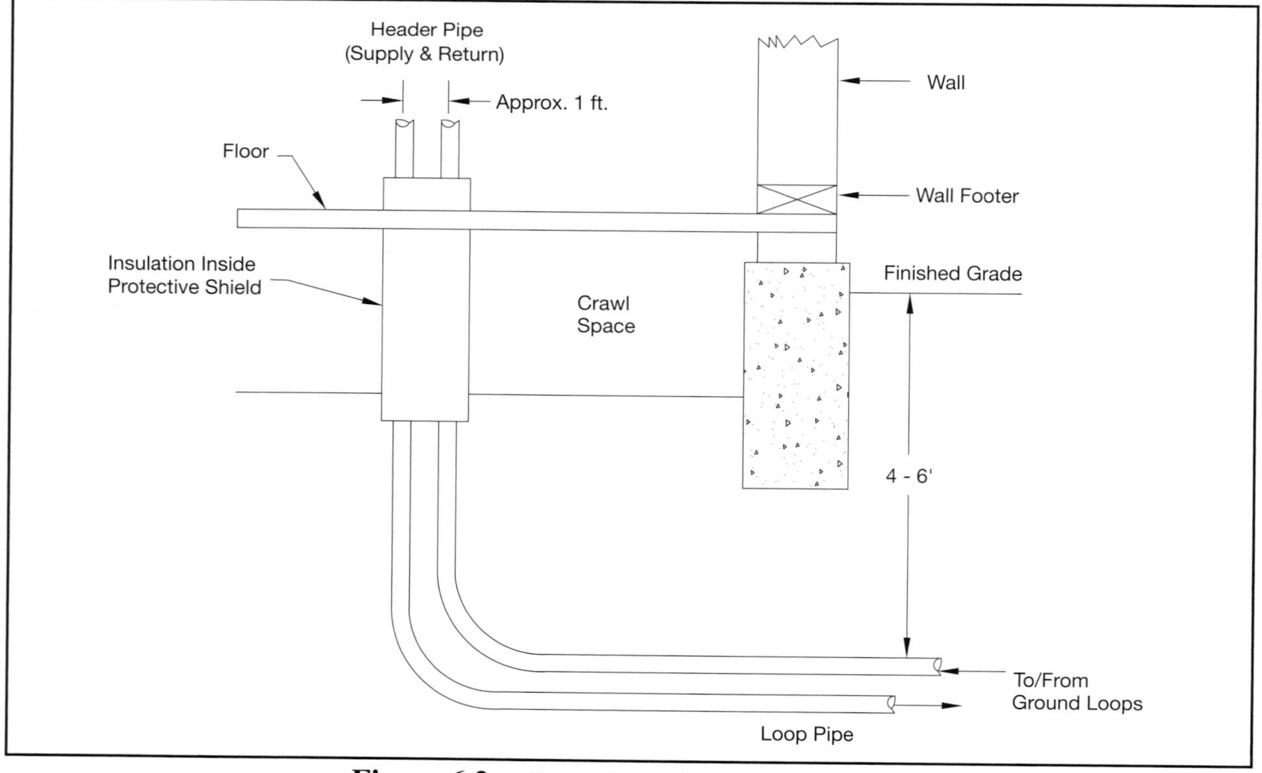

Figure 6.2c. Crawl Space Penetration Details

6.5 Interior Piping Layout

Once the building penetration has been made, the ground loop piping needs to be taken into the mechanical/heat pump room and connected. To properly connect the closed-loop GHEX to the system, three connections will need to be made:

1) Connect the supply line from the GHEX manifold to the pumping system(s) (flow center)
2) Connect the supply line from the flow center(s) to the supply line on the heat pump(s)
3) Connect the return line from the heat pump(s) to the return line on the GHEX manifold

Similar to ground loop piping design, piping sized such that it causes 1-3 feet of head loss per 100 feet of pipe is optimum. Some situations exist where slight pipe over sizing is justified, such as for multiple heat pump systems where flow-balancing issues may occur. Pipe over sizing is sometimes justified for multiple heat pump systems to limit the majority of the head loss due to fluid flow to the GHEX and the water-to-refrigerant coils in the heat pump units. Such situations will be discussed in Section 6.5.2.

6.5.1 SINGLE HEAT PUMP SYSTEMS

For single heat pump systems, the flow center should be located as close as possible to the heat pump unit that it serves. Typically, reinforced rubber heater hose is utilized for the connections from the flow center to the heat pump and from the heat pump to the GHEX return line. Use of reinforced rubber heater hose is actually preferred for those connections because condensation does not readily form on the piping (higher thermal resistance compared to other types of piping), plus it serves as a vibration break between the pumping station and the rest of the system, effectively reducing noise when the system is in operation. Additionally, reinforced rubber hose is flexible, simplifying field installation. When PVC or similar piping is used for all interior connections, inlets and outlets must be perfectly aligned with the piping runs because such piping is not flexible. However, it is best to keep the length of reinforced rubber hose used in the system to a minimum because the head loss through such piping tends to be relatively high.

Typical connections between a GSHP unit, flow center, and GHEX for a single heat pump system are illustrated by Figures 6.3a-6.3d. Figure 6.3a displays a typical system that utilizes a pressurized flow center for fluid circulation. Figure 6.3c displays a system that uses a non-pressurized flow center for fluid circulation with the GHEX manifolded outside. The connection method shown by Figure 6.3c for non-pressurized systems manifolded outside is one method given by the manufacturer of standing column flow centers (B&D MFG). Figure 6.3b displays an alternative method to connect a non-pressurized system with the GHEX manifolded outside, which utilizes a compact, PVC manifold that allows for simultaneous flushing of the interior and exterior piping. Figure 6.3d displays the connection details recommended for a non-pressurized system with the GHEX manifolded inside the building, allowing for flushing of each loop individually with the flow center, thus eliminating the need for a flush cart. The methods used in Figure 6.3d could also be used with a pressurized flow center. However, use of a flush cart during the flushing and purging process cannot be eliminated no matter how a pressurized system is manifolded.

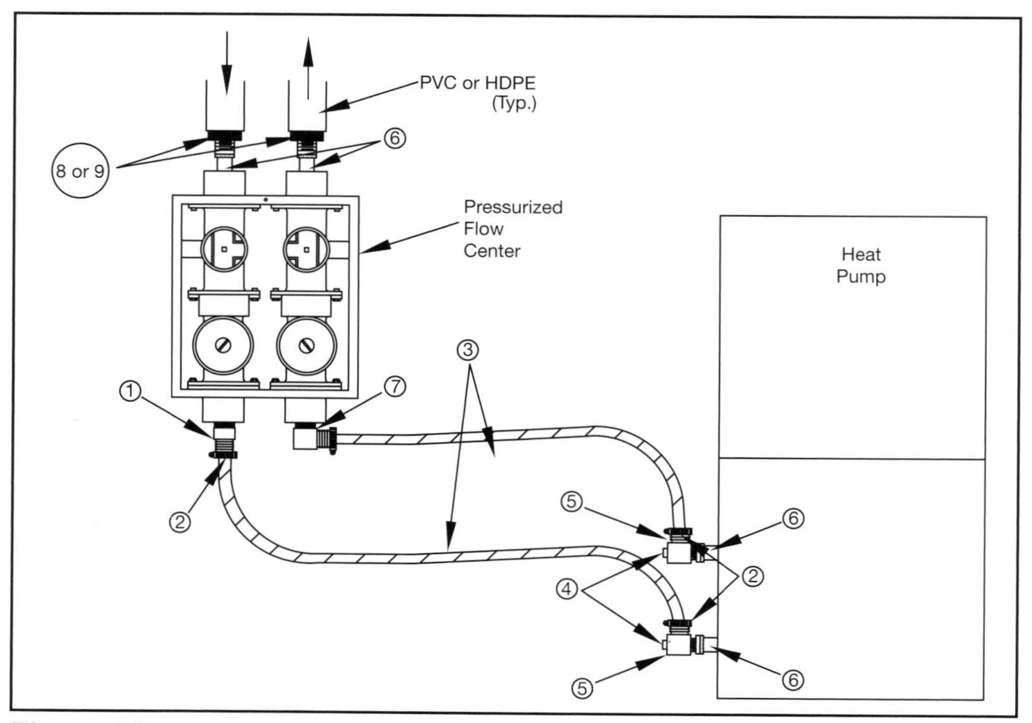

Figure 6.3a. Closed-Loop Connection to Heat Pump w/ Pressurized Flow Center

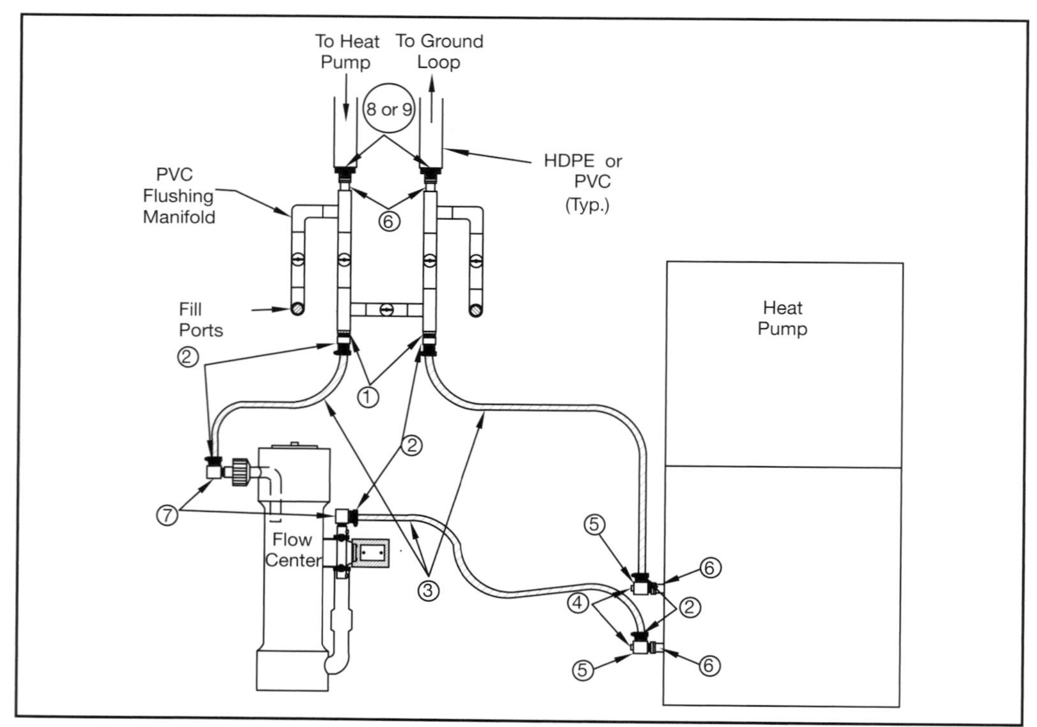

Figure 6.3b. Closed-Loop Connection to Heat Pump
w/ Standing Column Flow Center (Patented) – Outside Header

The numbered fittings shown in Figures 6.3a and 6.3b refer to Table 6.2 and Figure 6.5. See Table 6.2 and Figure 6.5 for more detail for each type of connector shown in Figures 6.3a-6.3b. The flushing manifold displayed in Figure 6.3b is shown in detail by Figure 6.4.

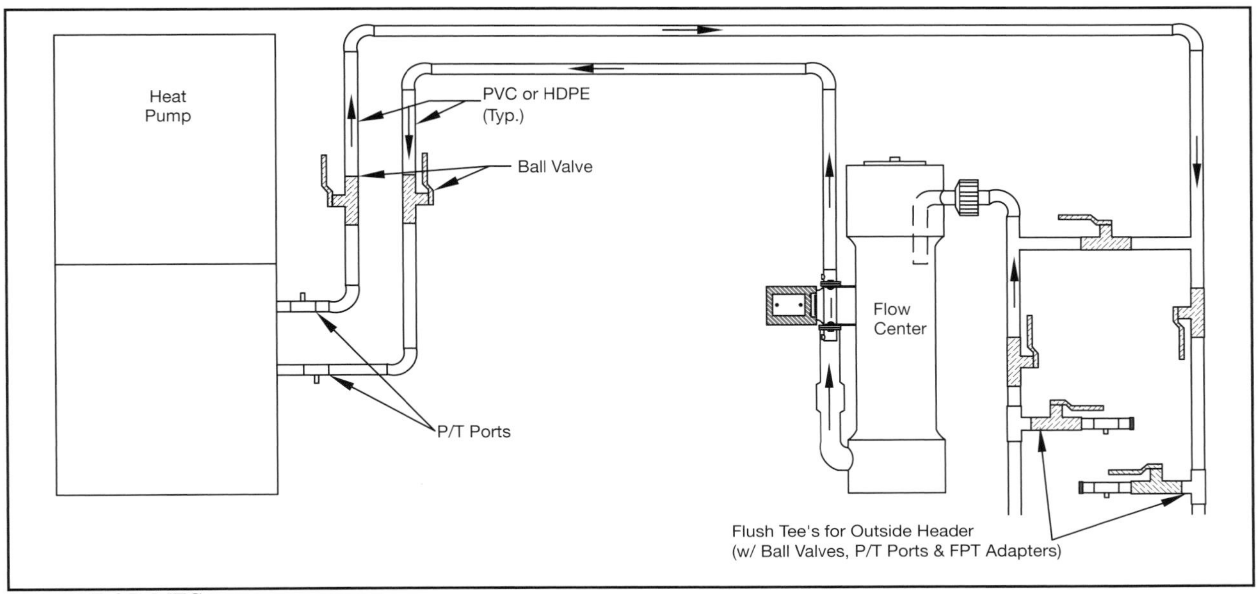

Source: B&D MFG

Figure 6.3c. Closed-Loop Connection to Heat Pump w/ Standing Column Flow Center (Patented) – Outside Header (Alternative Method)

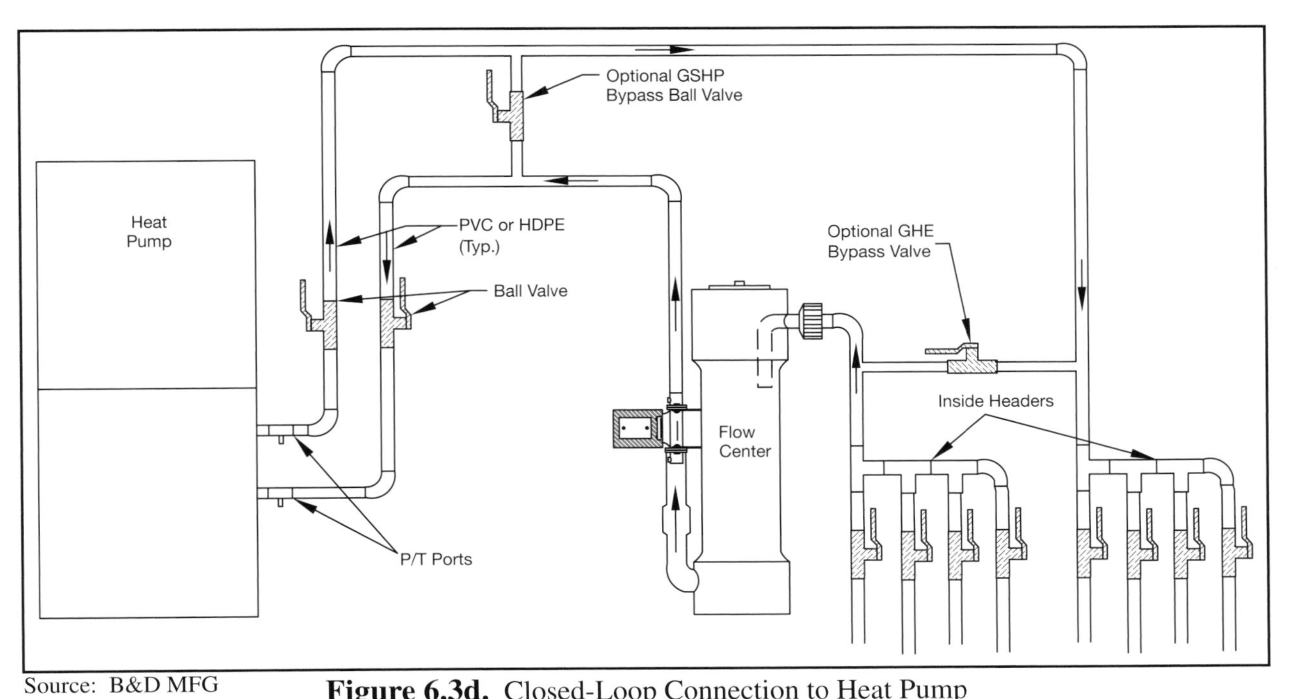

Source: B&D MFG

Figure 6.3d. Closed-Loop Connection to Heat Pump w/ Standing Column Flow Center (Patented) – Inside Header

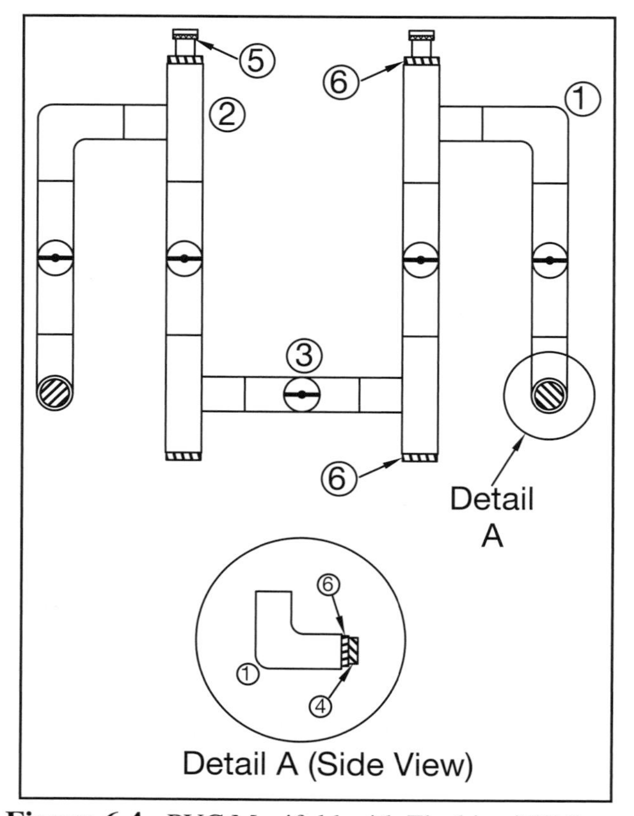

Figure 6.4. PVC Manifold with Flushing/Fill Ports

Using 1-1/4-inch components according to Table 6.1 is sufficient for up to 5-ton heat pumps (nominal capacity). For heat pumps with 6 to 8-ton (nominal) capacity, the recommended manifold size would be 1-1/2-inch. For heat pumps with 10 to 12-ton (nominal) capacity, the recommended manifold size would be 2-inch. Keep in mind that this PVC manifold is for single heat pump systems with the GHEX headered outside.

Part No.	Description	Quantity
1	1 1/4" 90° PVC Elbow	4
2	1 1/4" PVC Tee Section	4
3	1 1/4" PVC Ball Valve	5
4	1" PVC End Cap	2
5	1 1/4" MPT X 1" Swivel Nut	2
6	1 1/4" to 1" Reduction	6

Table 6.1. Parts List for PVC Manifold

Most distributors sell the appropriate hose and fittings in packages called "hose kits," which can be used to form the connections displayed in Figure 6.3a and Figure 6.3b. The parts list for a typical hose kit is given by items 1-5 in Table 6.2. The connections (with associated fittings) between the flow center, PVC manifold (if used), and GHEX shown in Figure 6.3a and Figure 6.3b are given in detail in Figure 6.5.

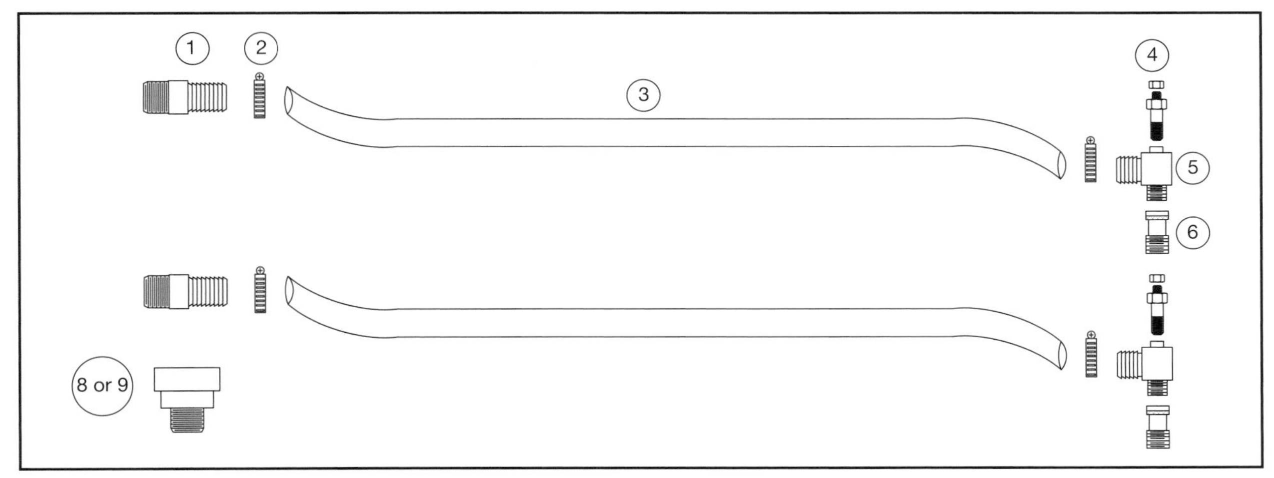

Figure 6.5. Typical GHEX-Flow Center-GSHP Connections

Part No.	Description	Quantity	
1	1" IPS x 1" Barb Straight	2	Hose Kit Parts
2	Size 20, 3/4" - 1 3/4" Clamp	4	
3	1" Blank 150 PD Hose 10 Ft.	10 Ft.	
4	1/4" MPT P/T Plug	2	
5	1" IPS x 1" Barb Elbow w/ P/T	2	
6	MPT x Swivel Nut Adapter	2	Misc. Parts
7	1" IPS x 1" Barb Elbow	2	
8	Fusion x MPT Adapter	2	
9	PVC x MPT Adapter	2	

Table 6.2. Typical GHEX-Flow Center-GSHP Connections Parts List

Most often, the parts provided in a typical hose kit will be sufficient for all of the connections necessary to get a system to its operational state. However, the connection method given by Figure 6.3b dictates the use of extra fittings (which again will allow for simultaneous flushing of the inside and outside loop in a single-unit GSHP system). The extra fittings needed for such a connection are Items 6-9 listed in Table 6.2. Always check with the GSHP manufacturer to ensure that the proper connector sizes are ordered for a specific project. Typically, heat pump units with 3-tons nominal capacity or less will require the use of ¾-inch IPS hose kits. Heat pump units with 4 to 6-ton nominal capacity will generally require the use of 1-inch IPS hose kits, and heat pump units larger than 6-ton nominal capacity will generally require the use of 1-1/4-inch IPS hose kits.

For performance checking, it is common for pressure/temperature measurement ports (P/T ports, Pete's ports) to be installed in the system. Many fittings are manufactured with the capability for P/T ports to be field installed. The best location for P/T ports to be installed is as close to the water inlet and outlet sides of the heat pump as possible. It is best to have the measurement ports as close to the inlets/outlets as possible to minimize the effects of heat loss/gain to/from the atmosphere when taking temperature readings and to minimize the effects of pressure loss in the piping when taking pressure readings. When performance checking a system, minimizing the losses described will aid in obtaining readings that agree with published manufacturer performance data, if the system is operating properly. Figure 6.6 displays common pressure and temperature sensing components.

Other pressure and temperature sensing adapter options exist. Contacting a HVAC or plumbing distributor is the best method to determine which options are actually available and which are best suited for a specific application.

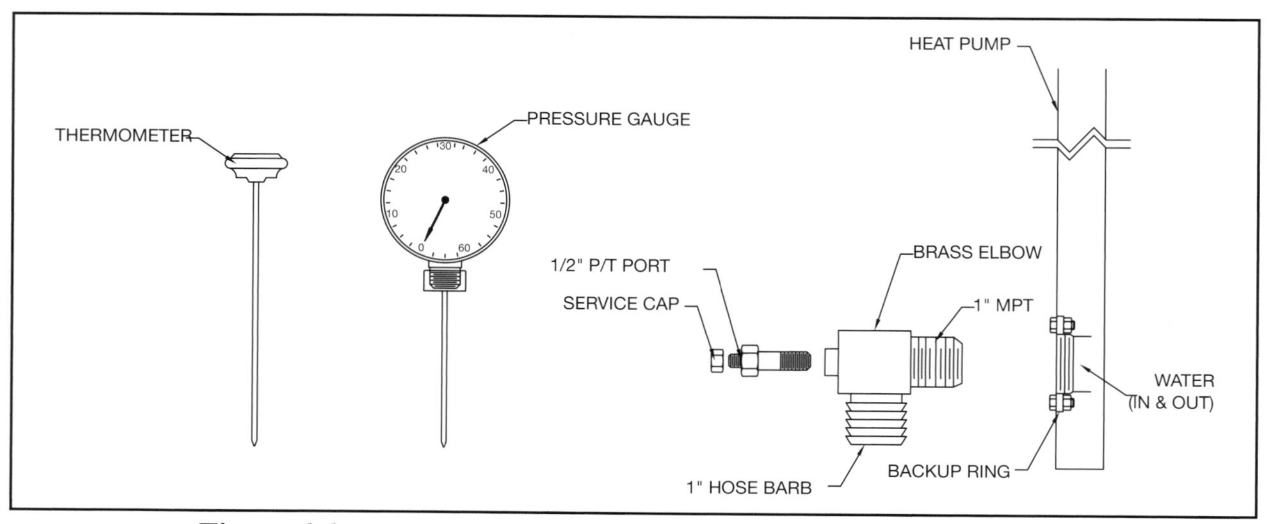

Figure 6.6. Pressure and Temperature Sensing Adapter and Components

In addition to incorporating P/T ports into the installation of a GSHP system, all ground loop and heat pump supply-return piping located indoors should be insulated to prevent condensation on the lines during periods when temperature of the fluid in the loops is relatively cold and the air is relatively warm and humid (occurring most commonly in the spring season). Condensation on the lines could cause problems such as mold, sheetrock decay due to moisture, etc. Closed-cell (moisture resistant) pipe insulation is recommended for such applications.

6.5.2 MULTIPLE HEAT PUMP SYSTEMS

While multiple heat pump systems tend to be more complicated, the methods used to connect them to ground-loop piping are similar to single heat pump systems. Several options exist in regards to hooking multiple heat pump systems to the circulation system, which fall into two basic categories: centralized pumping and distributed pumping. Centralized pumping systems are flow centers located centrally in a ground source system that produce flow to all heat pump units in that system. Distributed pumping systems are made up of smaller, individual pumping stations (one flow center for each heat pump) each controlled individually by the operation of the specific heat pump unit that they serve. The connections made to single-unit flow centers in a distributed pumping system could be made to each heat pump in the system using the methods discussed in Section 6.5.1.

6.5.2.1 MULTIPLE HEAT PUMP SYSTEMS WITH PRESSURIZED FLOW CENTERS

If a distributed pressurized flow system is to be utilized for a multiple heat pump system, one flow center will serve one GSHP unit. The number of flow centers used in the system will be determined by the number of heat pump units in the system. A supply manifold will need to be fabricated from the ground-loop supply line to each flow center. Additionally, a return manifold will need to be fabricated from each heat pump to the ground-loop return line. The only provisions that must be made to ensure proper system operation are accommodations for backflow prevention and loop isolation. Check valves must be installed at each flow center to prevent “short circuiting” or backflow in any part of the system. Backflow prevention is necessary to prevent backflow through GSHP units not in operation when the remainder of the system is in operation. Whenever they are not incorporated in the packaged flow center, isolation valves (ball valves) must also be installed at each flow center to allow for each interior flow path to be isolated for flushing in addition to allow for system maintenance when needed. Figure 6.7 displays an example of a 2-heat pump system that utilizes a distributed pressurized flow system.

A more generalized distributed pumping system is displayed by Figure 6.8.

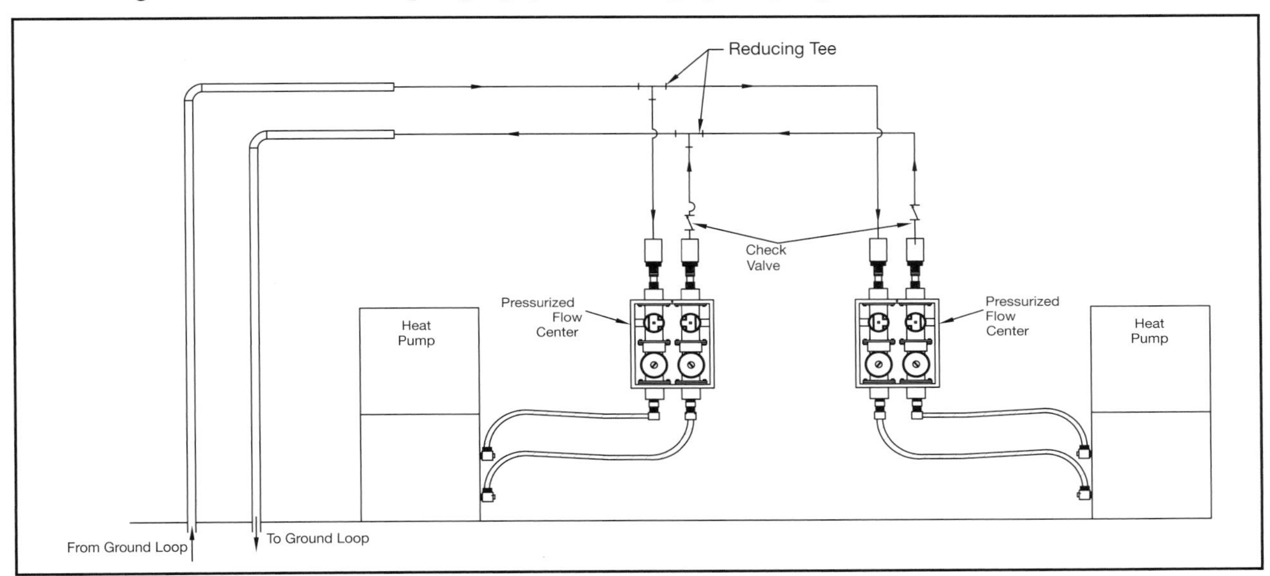

Figure 6.7. Distributed Pressurized Flow Center Example (2-Heat Pump)

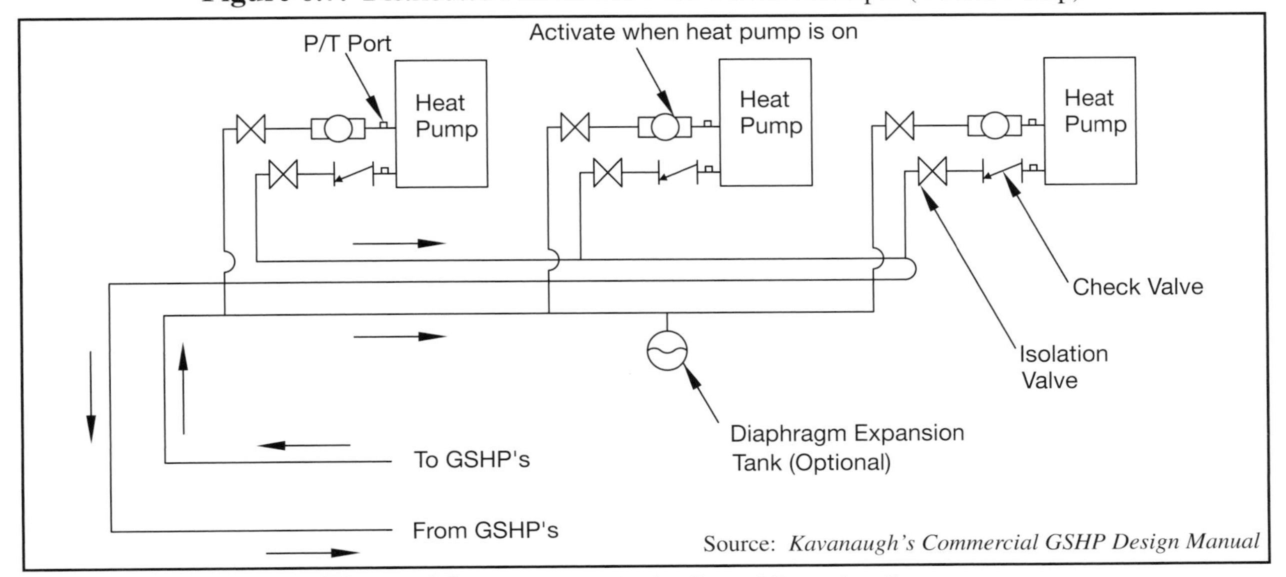

Figure 6.8. Generalized Distributed Pumping System

Distributed pumping systems are not generally recommended for systems with a large number of GSHP units. The pumps in a distributed pumping system shown by Figure 6.8 would need to be sized such that they would produce 2.5-3.0 gpm per ton of installed capacity when all pumps are operating (*Kavanaugh and Rafferty, 1997*).

If a centralized pressurized flow system is to be utilized for a multiple heat pump system, a single pumping system will serve every heat pump unit. There are two basic options regarding centralized pressurized flow systems: 1) producing full flow at all times regardless of how many GSHP units in the system are actually in operation and 2) using a system capable of producing variable flow to produce only the amount of flow necessary as determined by the number of GSHP units in operation. Producing full flow at all times regardless of the number of heat pump units in operation in a system (referred to as constant volume pumping) is the simplest type of system to operate and control, but is the least efficient type of system in terms of pumping power consumption. Figure 6.9 displays a pumping system that would produce full system flow at all times, regardless of how many pumps are actually in operation.

Notice the optional expansion tanks displayed in Figures 6.8-6.10. While expansion tanks are optional pieces of equipment for pressurized systems, they are useful because they can eliminate the detrimental effects of pressure swings due to thermal expansion and possible pump cavitation (refer to Section 6.3). Expansion tanks should usually be located in a central location in a building (mechanical room). To work properly, expansion tanks must be the single point of the system where no pressure change occurs with thermal expansion or contraction. Multiple expansion tanks should not be used in a system because such use may cause erratic (and possibly harmful) movement of air through system piping *(ASHRAE (2008) HVAC Systems and Equipment* – Chapter 11). Refer to the *ASHRAE (2008) Handbook of HVAC Systems and Equipment* for more information on expansion tank sizing and design. The most common type of expansion tank used in HVAC systems is the diaphragm tank.

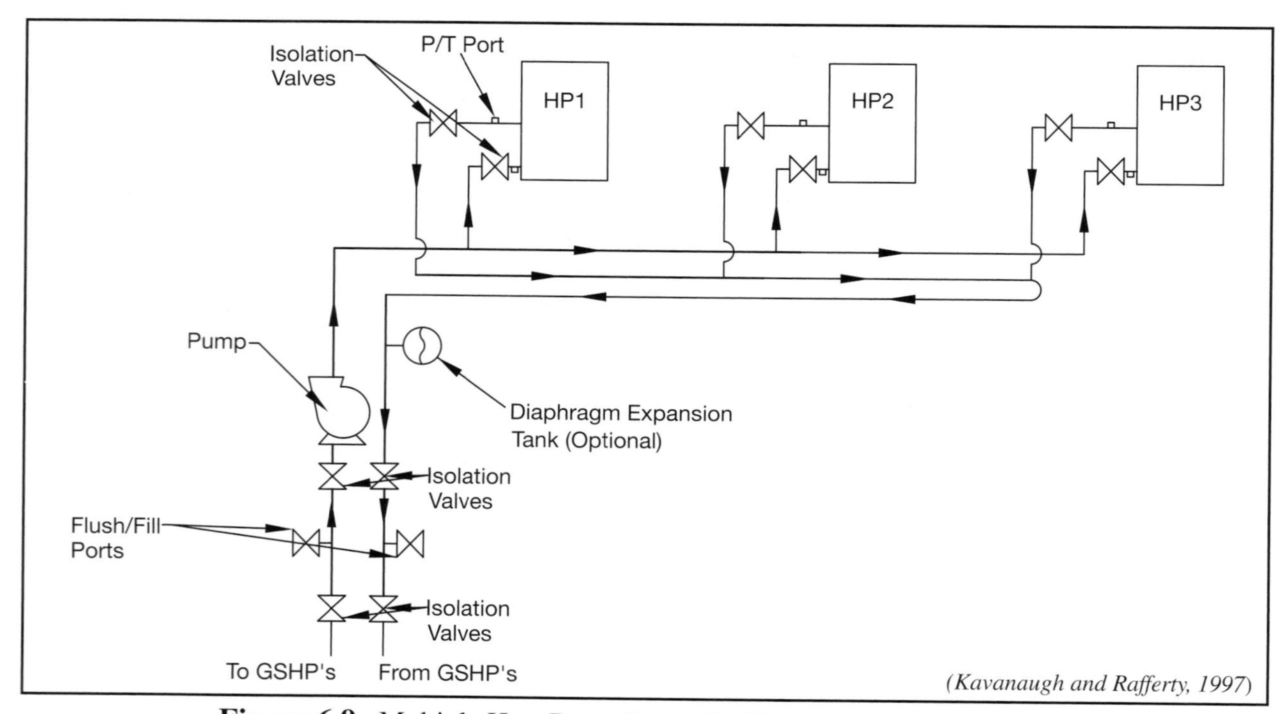

Figure 6.9. Multiple Heat Pump Example (Constant Volume Pumping)

The constant volume pumping scheme displayed by Figure 6.9 is generally only recommended for heat pump units with similar operating patterns. In other words, producing full system flow at all times regardless of number of heat pumps in operation would not be as wasteful in terms of pumping power consumption, if the heat pump units in the system tend to run at the same time. Multiple relays would need to be run in parallel with one another from each heat pump to the pumping station, all of which having capability to activate the constant volume pump. Interior pipe over sizing is recommended for applications such as constant volume pumping systems to reduce the system head loss to the GHEX and water-to-refrigerant coils in the GSHP units.

Two different methods exist of varying system flow through a system as determined by the number of GSHP units in operation: pump staging using multiple smaller pumps (usually 1-2 pumps for every heat pump in the system) or use of a single variable speed (variable frequency drive, VFD) pump. Staged pumping systems use smaller pumps on a centralized system dedicated to the operation of an individual GSHP. A staged pumping system would be controlled to have each pump or set of pumps properly sized and hard wired via relays to the GSHP that they serve. The pumps would be controlled such that they would operate only when their dedicated GSHP unit is in operation. A centralized staged pumping system is similar to a distributed pumping system in that individual pumps are controlled by a single heat pump. The difference is that all of the pumps in a centralized staged pumping system are centrally located in that system. Use of zone valves for such a system would be recommended to prevent flow through a heat pump when it is not in operation.

Using variable speed pumps is much more complicated in terms of controls and operation, but is much more efficient in that such systems will only produce the amount of flow necessary to satisfy the heat pumps in operation. The variable speed pump would be controlled by zone valves and pressure sensor. The zone valves would need to be placed at each heat pump in the system. They would need to be controlled such that they would be closed (thus preventing any flow through the GSHP that they serve), until that heat pump's thermostat calls for heating or cooling. The variable speed pump would be programmed to maintain a specific pressure in the system. If a single heat pump in a multiple heat pump system is operating, the pump would produce enough pressure to generate the amount of flow necessary to satisfy that individual GSHP unit. When an additional GSHP unit kicks on in the system, its zone valve will open, thus allowing flow through that heat pump. The variable speed pump will recognize a pressure drop in the system caused by the fluid flow being split through two parallel flow paths (to the two GSHP units in operation). The pump will then "ramp up" its speed, until its operation is within the pressure range it has been programmed to maintain. The pumping system will repeat the same process as more heat pumps in the system either turn on or turn off. It can easily be seen that such a system would be the most efficient in terms of pumping power consumption, but controls would be complicated. The assistance of an engineer is usually recommended for more complicated systems such as variable speed pumping systems. Figure 6.10 displays a centralized variable volume pumping system, which utilizes a VFD pump to produce system flow.

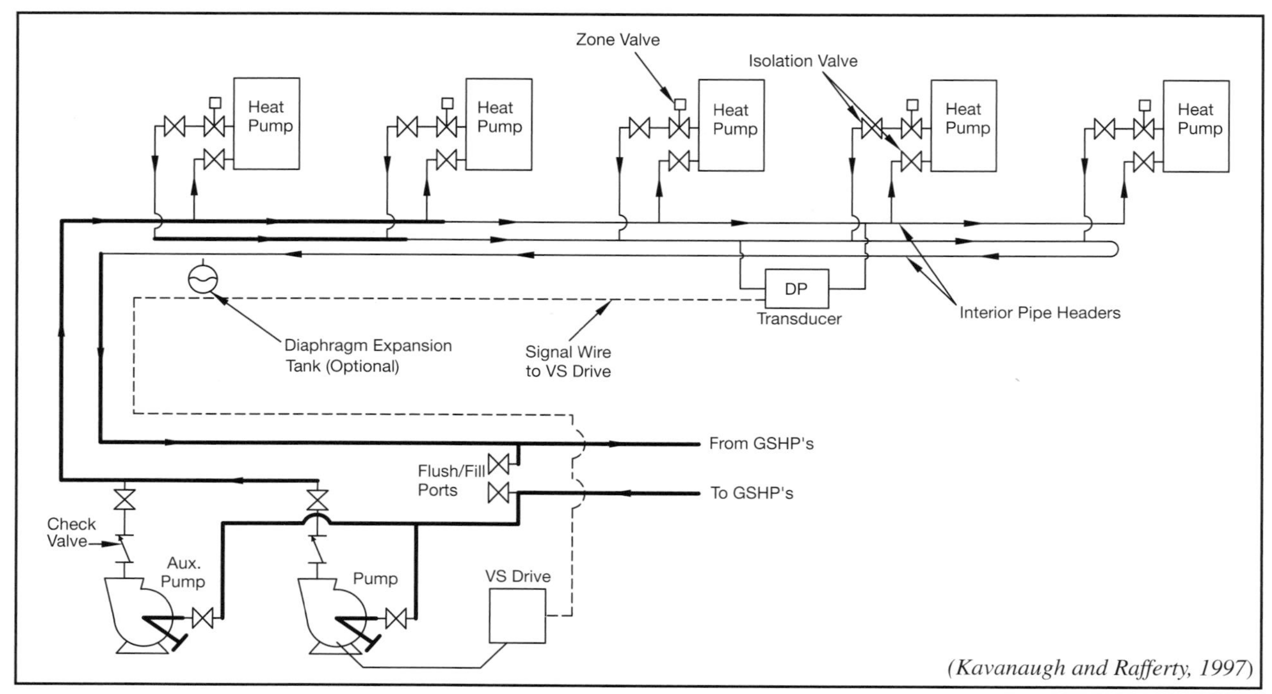

Figure 6.10. Centralized Pumping System with VFD Pumps

As shown in Figure 6.10, each heat pump in a variable volume pumping system needs to be equipped with normally closed, two-way zone valves. Each valve must have sufficient actuator torque to remain closed against available pumping pressure when the heat pump it serves is not calling for heating or cooling. The sensor (transducer connected to the variable speed drive) should normally be located near the end of the most remote supply-return header *(Kavanaugh and Rafferty, 1997*). However, the pressure sensor can be located anywhere in the system if reverse-return headers are used to provide balanced flow much like reverse-return headers used in GHEX piping. The auxiliary pump shown in Figure 6.10 is installed for redundancy in the event of primary pump failure. Pump redundancy is usually recommended for very large (commercial) systems.

6.5.2.2 MULTIPLE HEAT PUMP SYSTEMS WITH NON-PRESSURIZED FLOW CENTERS

If a distributed non-pressurized (standing column) flow system is to be utilized for a multiple heat pump system, an individual loopfield must be used for each flow center to ensure proper system operation as shown by Figure 6.3b. The connection methods for such a system would be exactly the same as described in Section 6.5.1. Individual distributed standing column flow centers cannot be tied together into a single loopfield. In general, distributed standing column pumping systems are not recommended except for situations where multiple heat pump systems have units that are very remote from one another.

For standing column flow centers, the recommended means to connect multiple heat pump systems to the circulation system is to use a centralized multi-circuit flow center to produce all of the flow necessary for every heat pump in the system. Figure 6.11 displays an example of a centralized multi-circuit standing column flow center with staged pumping that serves four heat pump units.

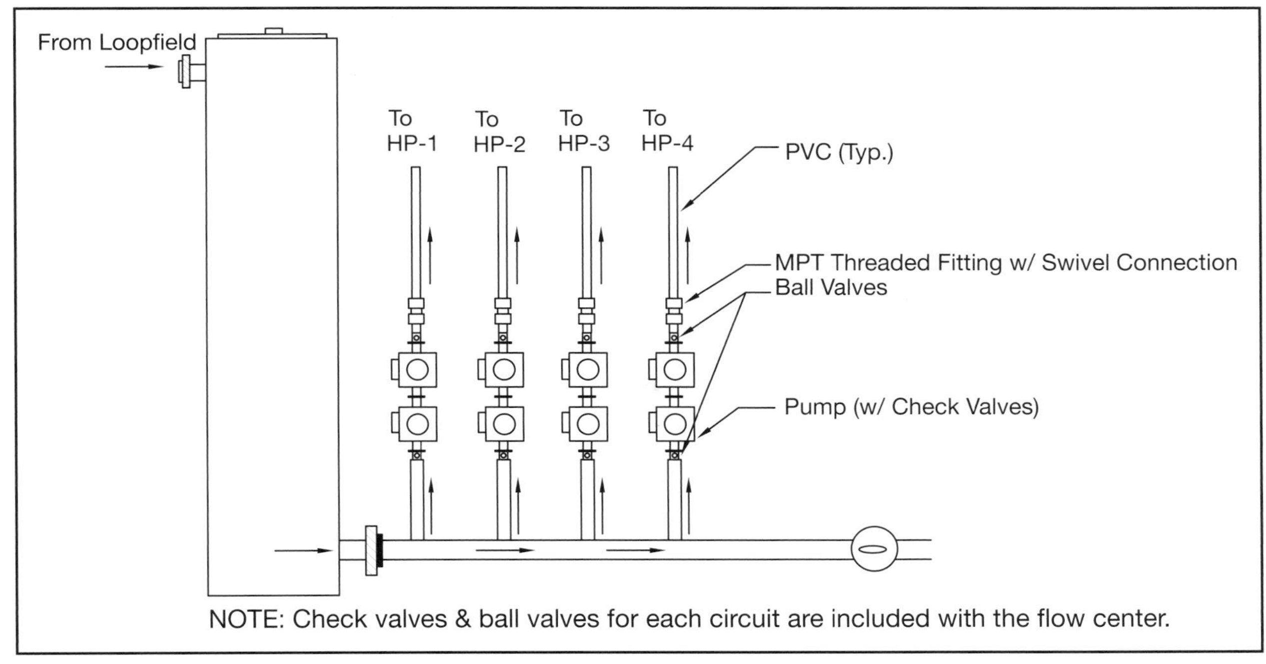

Figure 6.11. Multi-Circuit Standing Column Flow Center (for Multiple Heat Pump Systems - Patented)

The system displayed in Figure 6.11 displays individual supply lines run from the flow center to each heat pump. Each set of pumps is controlled by the heat pump unit they are directly connected to. For example, the set of pumps labeled "To HP-1" will only energize and produce flow when the thermostat for the heat pump labeled "HP-1" is calling for heating or cooling. Every set of pumps is controlled in a similar manner, creating an optimum pumping scheme. The staged pumping setup shown by Figure 6.11 ensures optimum pump performance in terms of energy consumption without the use of a complicated variable speed pump or zone valves. This system would not waste pumping energy by producing flow through heat pump units when they are not in operation. Each set of pumps in the figure would need their own set of check valves for backflow prevention. The particular flow center model shown in Figure 6.11 is actually manufactured with incorporated check valves, eliminating the need for field installation of such valves. Additionally, running individual supply lines from the multi-circuit flow center to each heat pump is the preferred method of connecting such a system. It is the simplest method in terms of controls and piping layout because zone valves are unnecessary for such a system. To minimize the amount of pipe necessary to complete the installation, a single return line can be brought from the farthest heat pump in the system to the flow center, "picking up" the flow from every heat pump as the line passes them in the building. An example of a 4-heat pump system served by the multi-circuit flow center in Figure 6.11 is discussed in detail in Example A.3 in Appendix A.

Centralized variable speed pumping systems can also be utilized for standing column systems. Such a system would need to be controlled with zone valves and pressure sensors in the same manner as described for centralized pressurized flow centers with variable speed pumps (Section 6.5.2.1). From a design perspective, the only difference between non-pressurized systems and pressurized systems is the use of expansion tanks in pressurized systems, which are optional, and the use of standing column tanks in non-pressurized systems, which are not optional. The systems shown in Figures 6.9-6.10 can be converted to non-pressurized systems simply by replacing the optional expansion tank with a standing column tank.

6.6 Pump Sizing

Regardless of the type of circulation system utilized in an installation, the pumping system needs to be sized such that it will generate the pressure head necessary to create the proper amount of system flow as dictated by installed GSHP capacity. The system head loss needs to be calculated for the longest supply-return path in the loopfield and the longest supply-return path for the interior circuit at the system's maximum flow rate.

6.6.1 PUMP PERFORMANCE CURVES

A GSHP system's circulating pump(s) must have sufficient power to overcome the total back pressure from the flow through the ground heat exchanger, the heat pump water-to-refrigerant source side heat exchanger, and all interior piping, fittings, manifolds, valves, hoses, etc. Figures 6.12-6.15 display pump performance curves for several different circulating pumps commonly used in the GSHP industry. The curves display the amount of back pressure (head loss, ft) that each type of circulating pump would be able to overcome at a given flow rate.

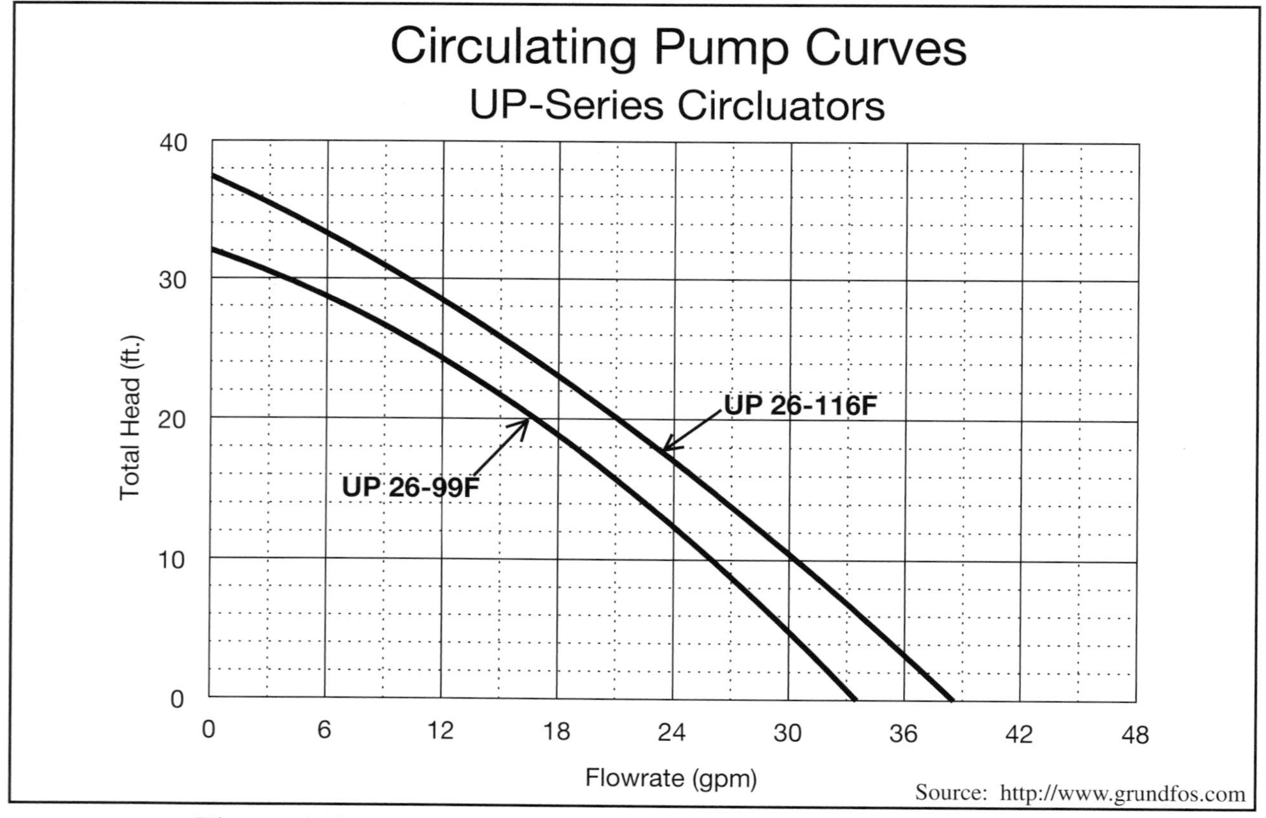

Figure 6.12. Grundfos Series UP Circulating Pump Performance Curves

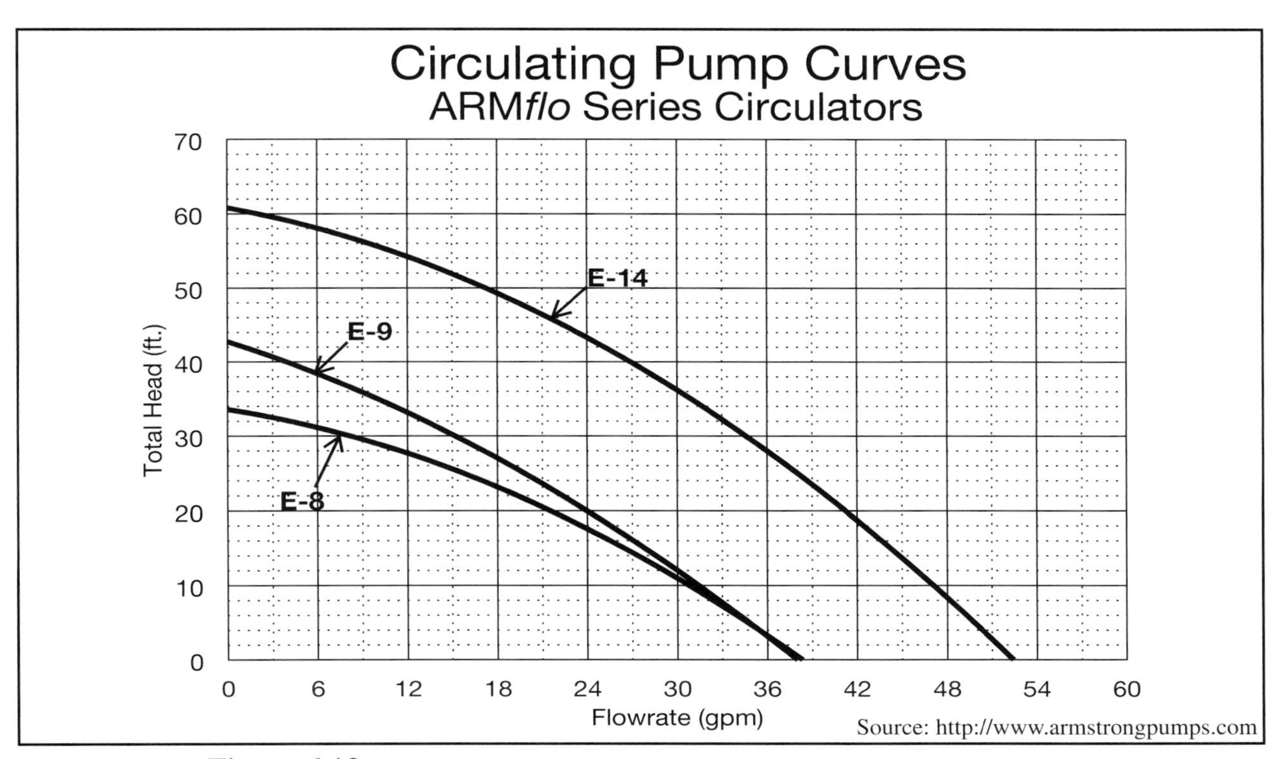

Figure 6.13. Armstrong ARMflo Circulating Pump Performance Curves

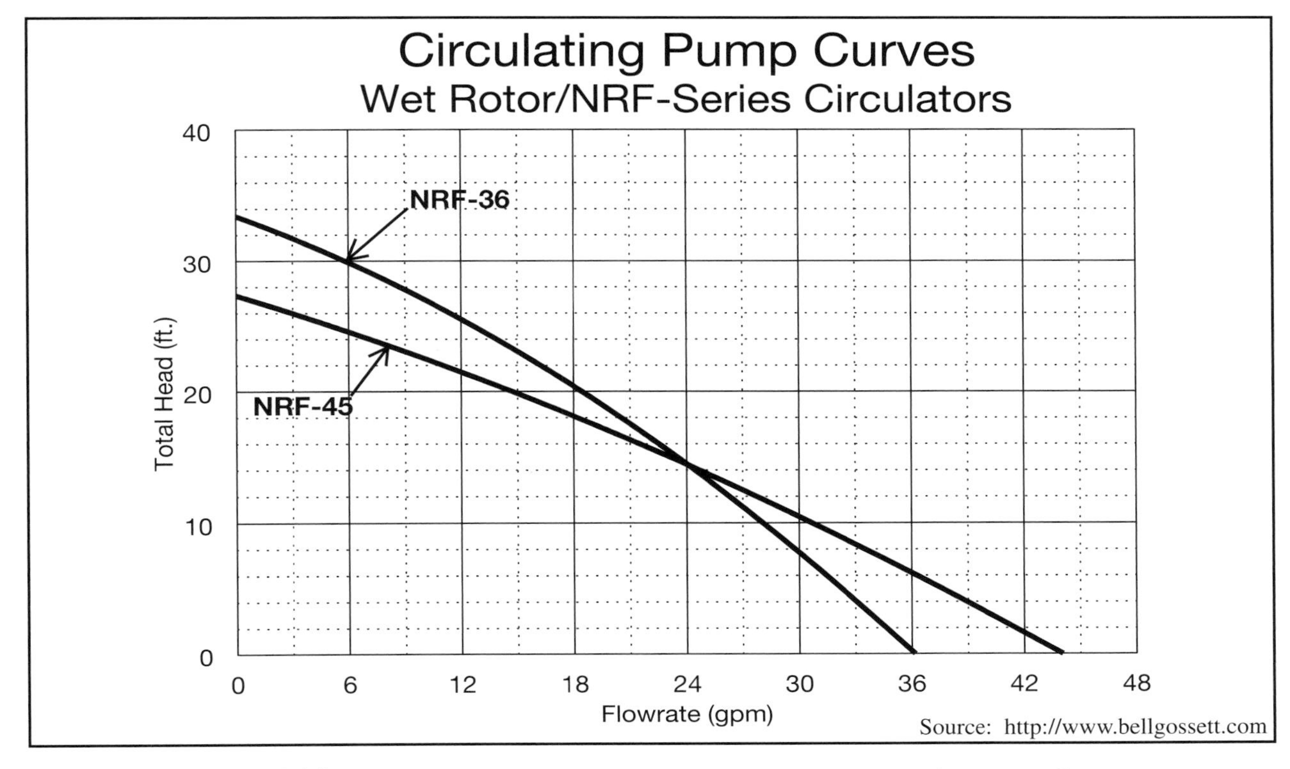

Figure 6.14. Bell & Gossett NRF-Series Circulating Pump Performance Curves

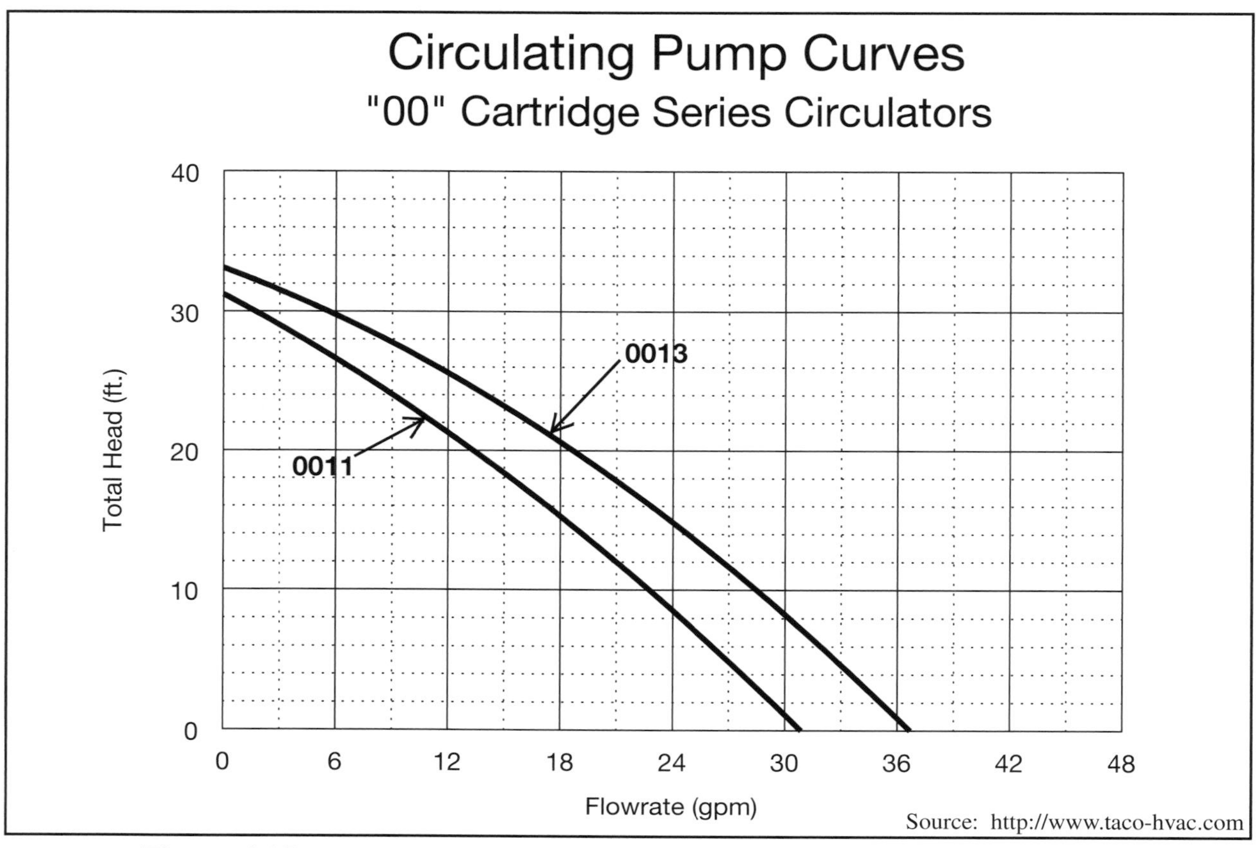

Figure 6.15. TACO "00" Cartridge Series Circulating Pump Performance Curves

All of the pump performance curves shown by Figures 6.12-6.15 are single pump performance curves. Many options are available regarding utilization of such circulating pumps (number of pumps used, series or parallel connection, etc). For example, when connected in series, the pressures that a pair of circulating pumps can produce at a given flow are added together. When connected in parallel, the flows that a pair of circulating pumps can produce are added together. Figures 6.16a-6.16b exemplify this basic principle of fluid dynamics. The curves shown in Figures 6.16a-6.16b are for the Grundfos Series-UP 26-99F model circulating pump.

As shown in Figure 6.16a, a single 26-99F circulator is capable of producing 24 feet of head at 12 gpm. Connecting two of the 26-99F circulators in series would yield a pumping system capable of producing 48 feet of head at 12 gpm (pressures add). Connecting two of the 26-99F circulators in parallel would yield a pumping system capable of producing 24 feet of head at 24 gpm (flows add).

For any type of mechanical system, always refer to current circulating pump performance data provided directly from a manufacturer as available models, performance curves, etc., may change with time.

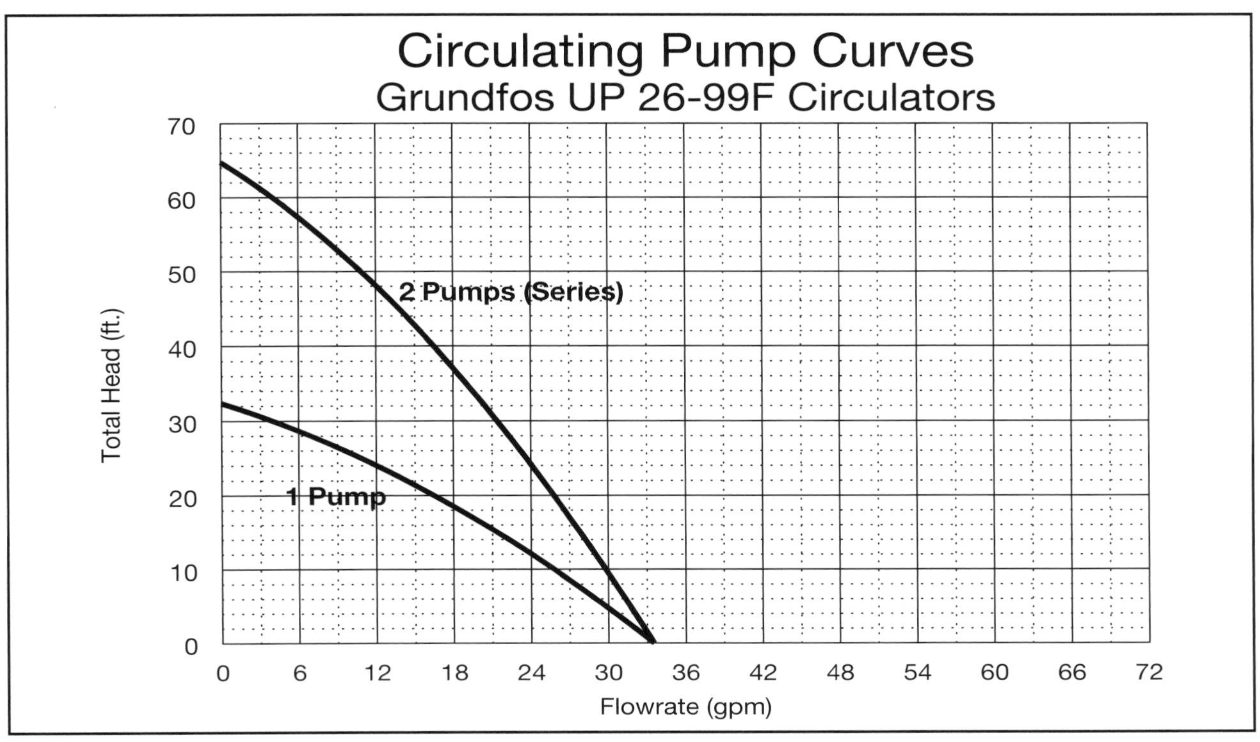

Figure 6.16a. Single Pump vs. 2-Pump (Series Connected) Performance Curves

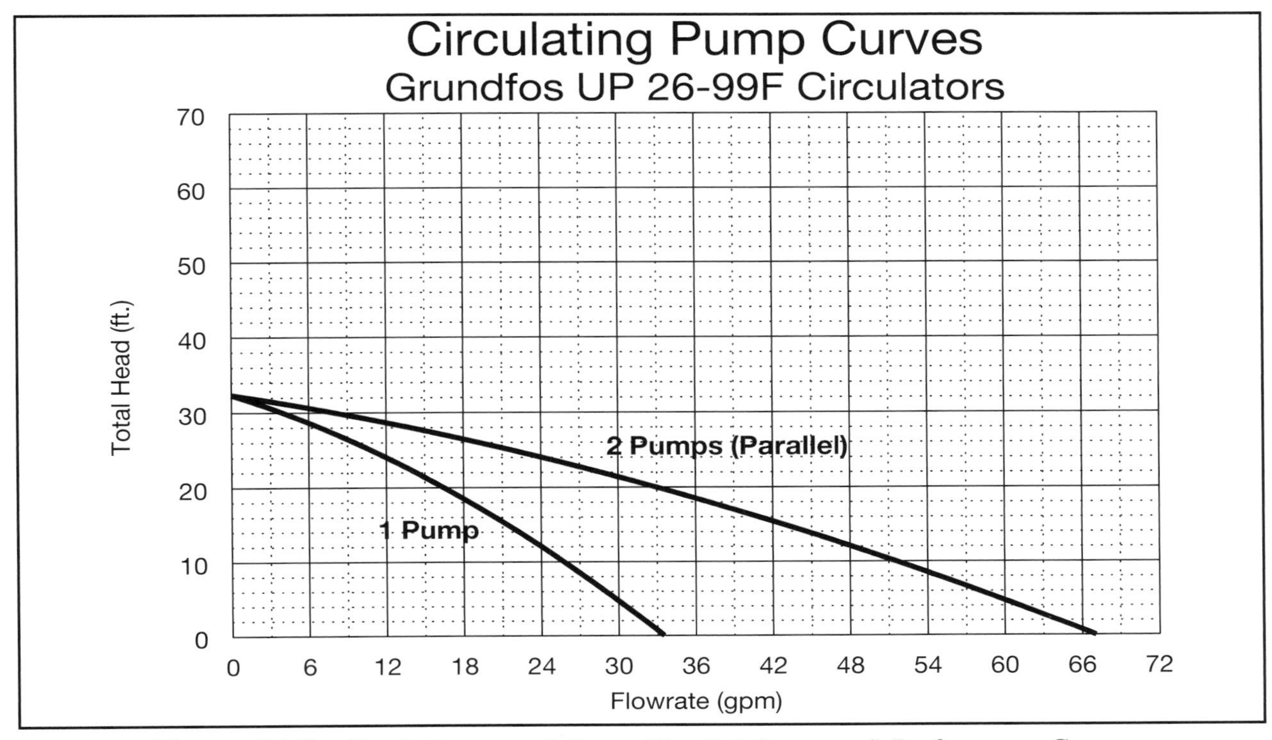

Figure 6.16b. Single Pump vs. 2-Pump (Parallel Connected) Performance Curves

6.6.2 HEAD LOSS CALCULATIONS

To determine which circulating pump model is best suited for a specific system, head loss calculations must be performed. To determine the pressure drop due to the flow of a specific fluid, at a given temperature through a specific type of pipe, follow the procedure as described in Section 4.4.3. Refer to Chapter 4 for graphs of head loss calculations for HDPE (DR-9, 11, 13.5, 15.5, and 17), PVC (SCH 40 and 80), copper tubing (Type K, L, and M), steel pipe (SCH 40 and 80), and reinforced rubber heater hose for water at 40 F. Refer to Section 4.4.4 for antifreeze correction factors to be applied to head loss calculations for water at 40 F.

As previously stated, design head loss calculations need to be performed for the longest flow path in the interior piping system, the longest flow path through the GHEX at the lowest fluid temperature (EWT_{min} for GSHP systems), and at the highest fluid flow rate (when every heat pump in a multiple heat pump system is in operation) experienced by the system. A properly designed GHEX will have the ground loops connected in a parallel or series-parallel configuration to minimize pumping power consumption. Additionally, a properly designed interior piping system will have all heat pumps (in a multiple heat pump configuration) connect to the pumping system in a parallel configuration, also minimizing pumping power consumption.

When calculating pressure drop in a system, pressure drop due to fittings in the longest flow path must also be accounted for. The simplest method to do so is to utilize an equivalent length associated with each type of fitting and add that length to the actual pipe length used in the flow path of interest. Table 6.3 lists equivalent pipe lengths for common fitting types and sizes used in the HVAC industry. In practice, an equivalent length of 3 foot of pipe per fitting is used unless better information is available.

Type Fitting & Application	Pipe & Fitting Material	Equivalent Length in Feet / Nominal Size of Fitting & Pipe						
		1/2"	3/4"	1"	1-1/4"	1-1/2"	2"	2-1/2"
Insert Coupling	Plastic	3'	3'	3'	3'	3'	3'	3'
Threaded Adapter	Copper	1'	1'	1'	1'	1'	1'	1'
Plastic or Copper Thread	Plastic	3'	3'	3'	3'	3'	3'	3'
90° Standard Elbow	Steel	2'	3'	3'	4'	4'	5'	6'
	Copper	2'	3'	3'	4'	4'	5'	6'
	Plastic	4'	5'	6'	7'	8'	9'	10'
Standard Tee, Straight	Steel	1'	2'	2'	3'	3'	4'	5'
Flow Through Run	Copper	1'	2'	2'	3'	3'	4'	5'
	Plastic	4'	4'	4'	5'	6'	7'	8'
Standard Tee, Turn	Steel	4'	5'	6'	8'	9'	11'	14'
Flow Through	Copper	4'	5'	6'	8'	9'	11'	14'
	Plastic	7'	8'	9'	12'	13'	17'	20'
Gate or Ball Valve	Steel	2'	3'	4'	5'	6'	7'	8'
Swing Check Valve	Steel	4'	5'	7'	9'	11'	13'	16'
Globe Valve	Steel	15'	20'	25'	35'	45'	55'	65'

LeClaire and Lafferty, 1996

Table 6.3. Equivalent Pipe Lengths for Common Fittings & HVAC Applications

Once the equivalent lengths for all of the fittings in a system are known, the next step is to calculate the actual length of each size and type of pipe in the longest flow path and add those lengths to the values of equivalent length of pipe for fittings of the same size/type. In addition, the fluid flow rate through every section of the flow path is needed. The system's design head loss would be the pressure drop (feet of head per 100 feet of pipe for the given fluid type/temperature/flow rate) multiplied by the total equivalent length of each type and size of pipe in the flow path. The circulating pump worksheet shown by Figure 6.17 can be used when performing system head loss calculations.

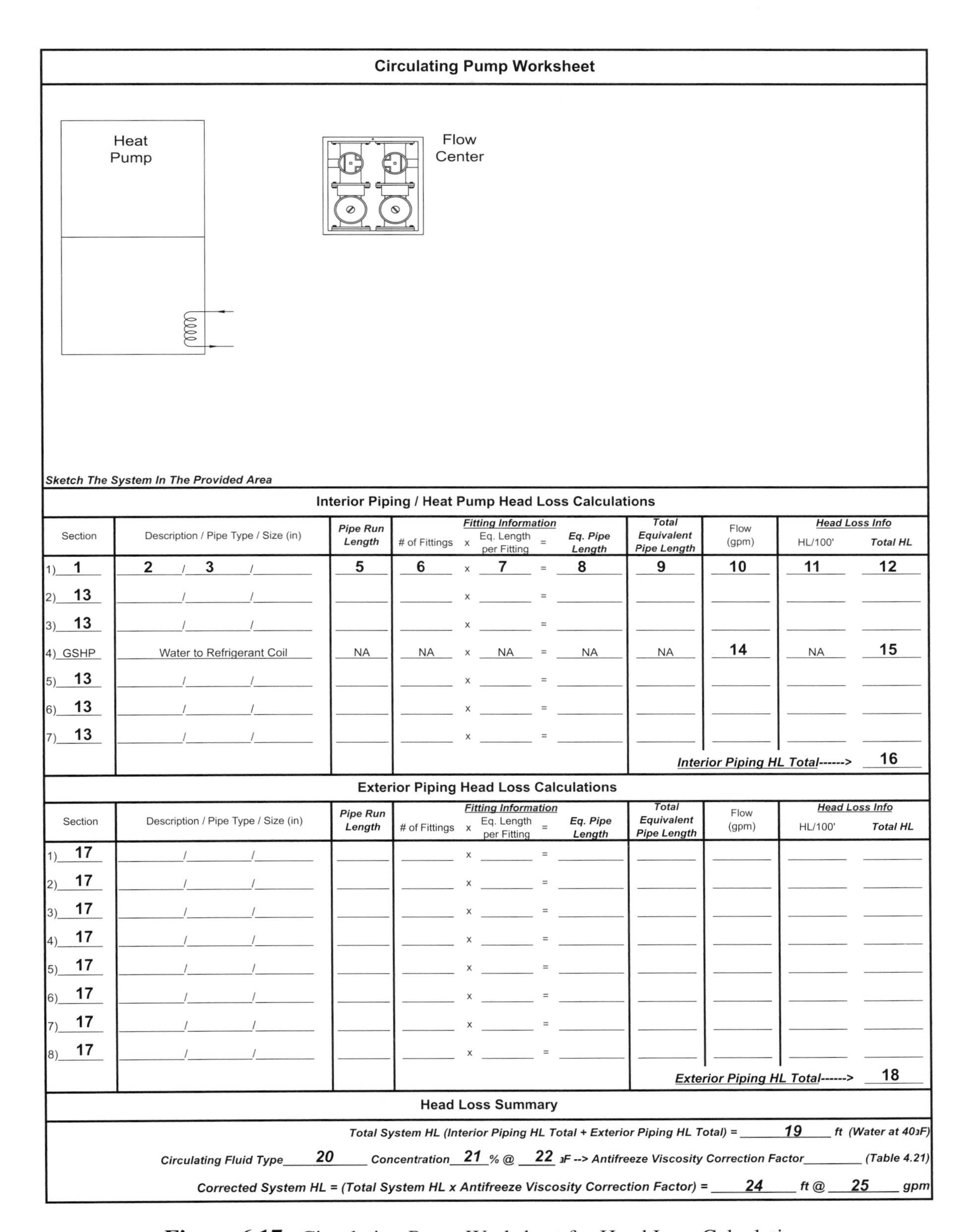

Circulating Pump Worksheet

Heat Pump

Flow Center

Sketch The System In The Provided Area

Interior Piping / Heat Pump Head Loss Calculations

Section	Description / Pipe Type / Size (in)	Pipe Run Length	Fitting Information: # of Fittings	x	Fitting Information: Eq. Length per Fitting	=	Fitting Information: Eq. Pipe Length	Total Equivalent Pipe Length	Flow (gpm)	Head Loss Info: HL/100'	Head Loss Info: Total HL
1) 1	2 / 3 /	5	6	x	7	=	8	9	10	11	12
2) 13	/ /			x		=					
3) 13	/ /			x		=					
4) GSHP	Water to Refrigerant Coil	NA	NA	x	NA	=	NA	NA	14	NA	15
5) 13	/ /			x		=					
6) 13	/ /			x		=					
7) 13	/ /			x		=					
										Interior Piping HL Total------>	16

Exterior Piping Head Loss Calculations

Section	Description / Pipe Type / Size (in)	Pipe Run Length	Fitting Information: # of Fittings	x	Fitting Information: Eq. Length per Fitting	=	Fitting Information: Eq. Pipe Length	Total Equivalent Pipe Length	Flow (gpm)	Head Loss Info: HL/100'	Head Loss Info: Total HL
1) 17	/ /			x		=					
2) 17	/ /			x		=					
3) 17	/ /			x		=					
4) 17	/ /			x		=					
5) 17	/ /			x		=					
6) 17	/ /			x		=					
7) 17	/ /			x		=					
8) 17	/ /			x		=					
										Exterior Piping HL Total------>	18

Head Loss Summary

Total System HL (Interior Piping HL Total + Exterior Piping HL Total) = 19 *ft (Water at 40°F)*

Circulating Fluid Type 20 *Concentration* 21 *% @* 22 *°F --> Antifreeze Viscosity Correction Factor*________ *(Table 4.21)*

Corrected System HL = (Total System HL x Antifreeze Viscosity Correction Factor) = 24 *ft @* 25 *gpm*

Figure 6.17. Circulating Pump Worksheet for Head Loss Calculations

To fill out the Circulating Pump Worksheet, sketch the piping system in the area provided at the top of the worksheet. Input data for each section of the piping system as described by the label directions provided below. Follow a single flow path through the entire piping system starting at the flow center to the GSHP, through the GSHP water-to-refrigerant coil, back to the flow center, out through a single loop in the loop-field, and back to the flow center.

1) Insert the pipe section for which the head loss (HL) will be calculated starting with the pipe section from the flow center to the heat pump.
2) Insert the description of the pipe section being analyzed.
3) Insert the pipe type for the pipe section.
4) Insert the pipe size for the pipe section.
5) Insert the actual pipe run length for the pipe section.
6) Insert the number of fittings in the pipe section.
7) Insert the equivalent pipe length per fitting in the pipe section. Refer to Table 6.3 for equivalent pipe lengths for various fitting types and for various materials. Use 3 feet of equivalent pipe length per fitting if no information is available.
8) Calculate the total equivalent pipe length for the fittings in the section by multiplying the number of fittings (6) by the equivalent pipe length per fitting (7)
9) Calculate the total equivalent pipe length for the section by adding the actual pipe run length (5) to the total equivalent pipe length for the fittings (8).
10) Insert the flow rate through the pipe section.
11) Insert the value for HL/100 feet of pipe length. Refer to Chapter 4, Section 4.4.3 for tables and graphs of head loss values for HDPE (DR-9, 11, 13.5, 15.5, and 17), PVC (SCH 40 and 80), copper tubing (Type K, L, and M), steel pipe (SCH 40 and 80), and reinforced rubber heater hose for water at 40 F. Refer to Section 4.4.4 for antifreeze correction factors to be applied to head loss calculations for water at 40 F.
12) Calculate the head loss through the pipe section by dividing the total equivalent pipe length (9) by 100 and multiplying that number by the value for HL/100 feet of pipe length (11).
13) Repeat this process through each section in the interior piping system until the flow path has been completely analyzed from the flow center, through the piping to the heat pump, through the GSHP water-to-refrigerant coil (14-15), and through the piping to the flow center.
14) Enter the system flow through the GSHP unit.
15) Enter the head loss at the system flow rate through the water-to-refrigerant coil in the GSHP, provided in manufacturer's performance data.
16) Calculate the total head loss through the interior piping system by adding all calculated head loss values in the Total HL column on the interior piping section of the worksheet.
17) Repeat this process for the exterior piping system. Start at the flow center, follow the flow path through a single loop out in the loopfield, through the step-down, step-up reverse-return header (or through the close header), and back to the starting point at the flow center.
18) Calculate the total head loss through the exterior piping system by adding all calculated head loss values in the Total HL column on the exterior piping section of the worksheet.
19) Calculate the total system head loss by adding the total head loss through the interior piping system (16) to the total head loss through the exterior piping system (18). The head loss value calculated here will be at the system flow rate with water at 40 F as the circulating fluid.
20) Insert the actual circulating fluid type to be used during system operation. If no antifreeze is to be used, no further calculations are necessary.
21) Insert the concentration of antifreeze to be used, if used at all.
22) Insert the minimum entering water temperature at which the ground loop was designed.

23) Insert the antifreeze viscosity correction factor for the actual circulation fluid to be used. Refer to Section 4.4.4 for a complete list of antifreeze viscosity correction factors. If antifreeze is not used in a given system, the correction factor entered here will be 1.0.
24) Calculate the actual system operating head loss by multiplying the total system HL (19) for water at 40 F by the antifreeze viscosity correction factor (23).
25) For recordkeeping, enter the flow rate at which the system head loss calculations were performed.

A copy of Figure 6.7 without labels in the input blanks is provided in Appendix B. To illustrate how the Circulating Pump Worksheet is to be used for head loss calculations, refer to Example 6.1.

Example 6.1. Calculate the head loss through the 4-ton ClimateMaster Tranquility 20™ Model 048 GSHP unit, for which performance data is given in Section 2.4.1. Assume that the heat pump is coupled to a ground loop consisting of 4 x ¾-inch HDPE U-bends. Also, assume that all bores are 200 feet deep and are connected by the step-down, step-up reverse-return piping system shown in Figure 5.9(c) in Section 5.1.3.2. Perform the head loss calculations using a 12 gpm system flow rate and 20% propylene glycol by volume as the circulating fluid. Figure 6.18 shows the completed Circulation Pump Worksheet. The center to center bore spacing is 15 feet and the supply-return pipe run from the flow center to the loopfield can be assumed to be 50 feet.

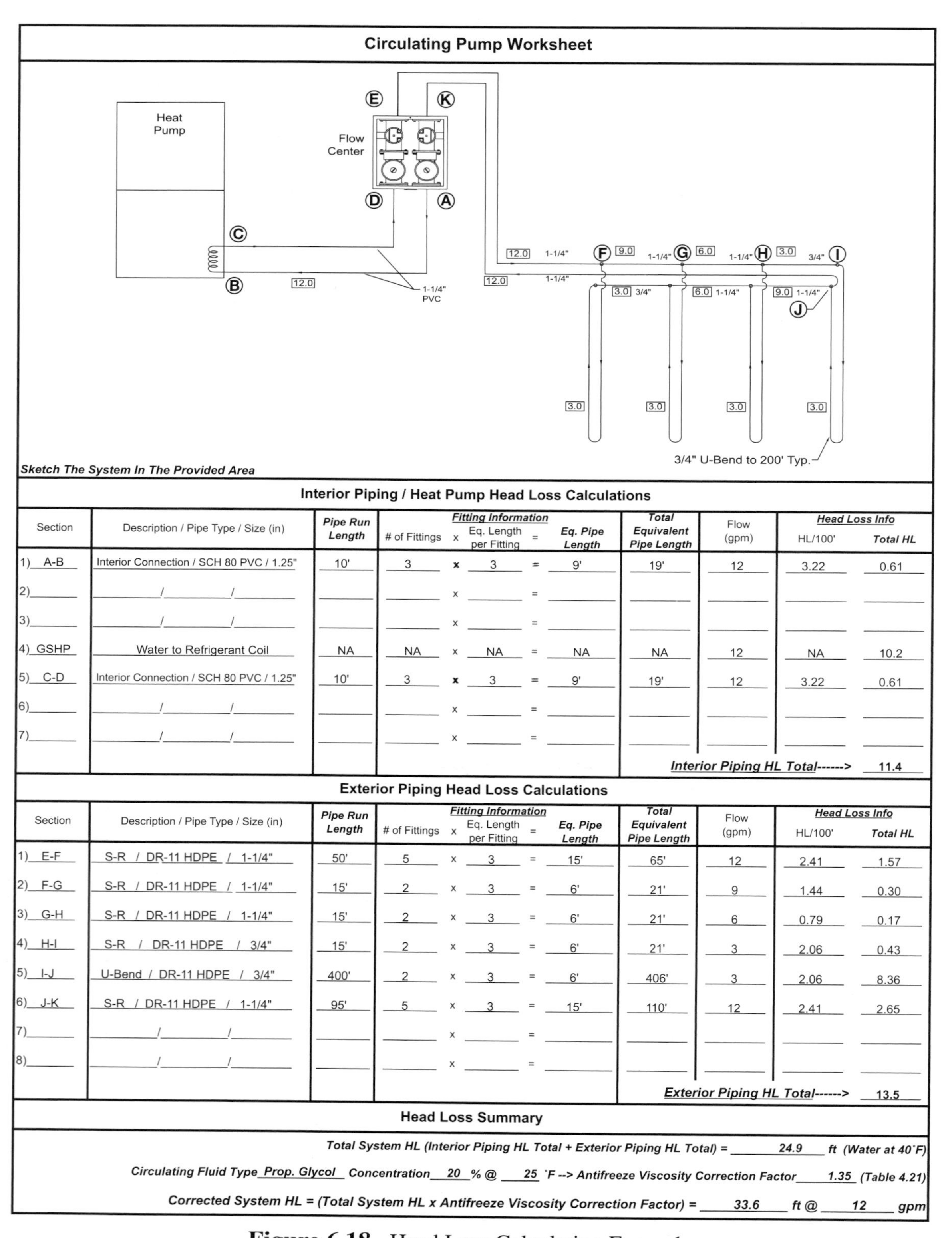

Circulating Pump Worksheet

Sketch The System In The Provided Area

Interior Piping / Heat Pump Head Loss Calculations

Section	Description / Pipe Type / Size (in)	Pipe Run Length	# of Fittings	x	Eq. Length per Fitting (Fitting Information)	=	Eq. Pipe Length	Total Equivalent Pipe Length	Flow (gpm)	HL/100' (Head Loss Info)	Total HL
1) A-B	Interior Connection / SCH 80 PVC / 1.25"	10'	3	x	3	=	9'	19'	12	3.22	0.61
2)	/ /			x		=					
3)	/ /			x		=					
4) GSHP	Water to Refrigerant Coil	NA	NA	x	NA	=	NA	NA	12	NA	10.2
5) C-D	Interior Connection / SCH 80 PVC / 1.25"	10'	3	x	3	=	9'	19'	12	3.22	0.61
6)	/ /			x		=					
7)	/ /			x		=					
									Interior Piping HL Total------>		11.4

Exterior Piping Head Loss Calculations

Section	Description / Pipe Type / Size (in)	Pipe Run Length	# of Fittings	x	Eq. Length per Fitting (Fitting Information)	=	Eq. Pipe Length	Total Equivalent Pipe Length	Flow (gpm)	HL/100' (Head Loss Info)	Total HL
1) E-F	S-R / DR-11 HDPE / 1-1/4"	50'	5	x	3	=	15'	65'	12	2.41	1.57
2) F-G	S-R / DR-11 HDPE / 1-1/4"	15'	2	x	3	=	6'	21'	9	1.44	0.30
3) G-H	S-R / DR-11 HDPE / 1-1/4"	15'	2	x	3	=	6'	21'	6	0.79	0.17
4) H-I	S-R / DR-11 HDPE / 3/4"	15'	2	x	3	=	6'	21'	3	2.06	0.43
5) I-J	U-Bend / DR-11 HDPE / 3/4"	400'	2	x	3	=	6'	406'	3	2.06	8.36
6) J-K	S-R / DR-11 HDPE / 1-1/4"	95'	5	x	3	=	15'	110'	12	2.41	2.65
7)	/ /			x		=					
8)	/ /			x		=					
									Exterior Piping HL Total------>		13.5

Head Loss Summary

Total System HL (Interior Piping HL Total + Exterior Piping HL Total) = 24.9 *ft (Water at 40°F)*

Circulating Fluid Type Prop. Glycol *Concentration* 20 *% @* 25 *°F --> Antifreeze Viscosity Correction Factor* 1.35 *(Table 4.21)*

Corrected System HL = (Total System HL x Antifreeze Viscosity Correction Factor) = 33.6 *ft @* 12 *gpm*

Figure 6.18. Head Loss Calculation Example

6.6.3 CIRCULATING PUMP SELECTION

The ultimate goal in performing head loss calculations is to match the circulating pump capabilities to the system head loss characteristics. A set of circulating pumps should be able to produce the required amount of system flow at the system's design head loss. The system operating point will be where the system head loss curve intersects the circulating pump curve. The operating point should lie at or above the recommended system flow rate as dictated by installed capacity. Figure 6.19 demonstrates the relationship between a system's head loss curve, circulating pump curve, and operating point. To generate a system head loss curve, perform head loss calculations through a given piping system [through a single flow path (parallel configuration) including interior piping, GHEX piping, and the GSHP water-to-refrigerant coil] at several different flow rates. Sketch a curve through the calculated data points or utilize software to generate a 2nd order polynomial curve fit through the head loss values calculated at different (minimum 3) flow rates (0 ft of head at 0 can be used as one operating point). For a demonstration illustrating how to generate a system head loss curve to determine the system operating point, refer to Examples A.1 and A.2 in Appendix A.

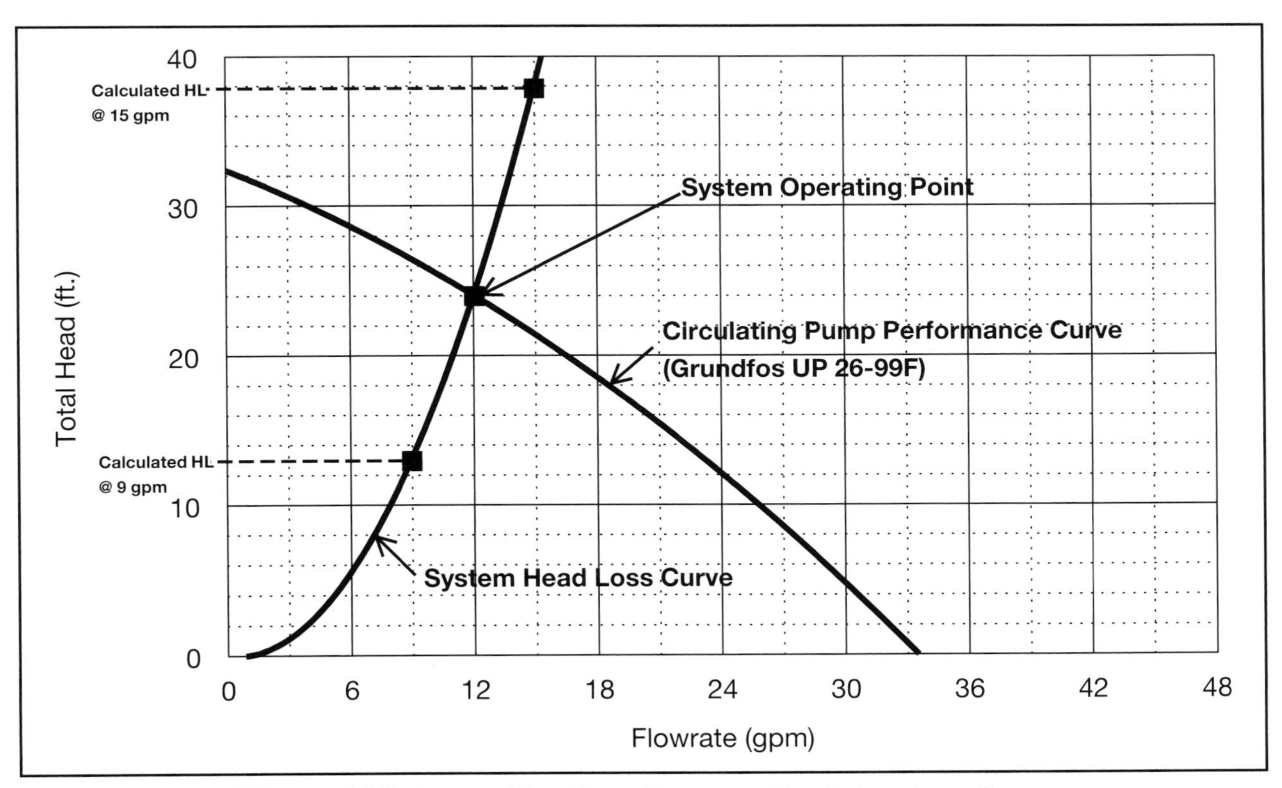

Figure 6.19. System Head Loss Curve vs. Circulating Pump Curve

INSTALLING THE CLOSED-LOOP GHEX PIPING SYSTEM

In This Section

7.1 Site evaluation
7.2 Designing for the site
7.3 Forming a site plan
7.4 Project preparation & installation guidelines
7.5 Vertically-bored, closed-loop GHEX installation
7.6 Horizontally-trenched, closed-loop GHEX installation
7.7 Horizontally-bored, closed-loop GHEX installation
7.8 Installing the Loopfield Header
7.9 As-built drawings
7.10 Site restoration

7.1 Site Evaluation

Installing a GHEX utilizes skill sets such as drilling, trenching, heat fusion pipe joining, and site restoration combined with lesser known specialty requirements for testing and verification of the completed GHEX. The GHEX installation may involve multiple subcontractors such as drillers, trenching contractors, heat fusion contractors, site grading contractors, landscapers, and others. The definition of a completed GHEX is as specified by project design requirements, specifications, and/or contract. The responsibility for completion of the GHEX in compliance with the design, specifications, and/or contract must rest with a single company – the GHEX contractor.

Performing the initial site evaluation should take place even before the design process begins. Previous chapters have covered the design of all types of closed-loop GHEX systems. However, prior to beginning design, the job site must be investigated to determine if it is suitable for a closed-loop ground heat exchange application and to identify the appropriate GHEX technology to be used.

Once the building peak heating and cooling loads have been determined and an appropriately sized ground source heat pump unit or units have been selected (Chapter 3), it is time to select the method of closed-loop ground heat exchange that best suits the site. The first step in designing the closed-loop GHEX is determining the type of GHEX appropriate for the site's conditions. This determination should be based on all information that can be gathered, including, but not limited to:

1. **SITE CONDITIONS:**
   - Develop an approximate layout of buried obstacles such as utilities, fuel tanks, septic systems, water wells, cisterns, etc. This preliminary layout is for initial reference only and is not a substitute for actually locating these or other items using utility locator services, as-built drawings, and/or any other available location resources to define unseen obstacles prior to beginning GHEX construction.
   - Inspect the site to determine if the space available is adequate for the type of GHEX selected and that the site is accessible for installation equipment necessary for GHEX installation. Check for the presence of any items that can create safety/health issues such as steep inclines, overhead power lines, unstable ground, possible site contamination, or other potential impediments to system installation.
   - Evaluate the ground conditions for moisture content, soil type, rocks, obstacles, and other benefits or challenges to installation. Review soil test boring data (if available) to identify trenching conditions to help determine the type of trenching equipment best suited for the ground conditions. If a vertical application is being considered, check with local well drillers for relevant information such as static water level, groundwater availability, and formations. Also check for any well logs available for wells within approximately ¼ mile of the job site.

2. **CODES AND REGULATIONS:**
   - Check with the applicable state agencies as well as county and city departments with similar responsibilities for code sets. It is necessary to identify the code/regulation authority that has jurisdiction and the most stringent constraints and to abide by their requirements. Key issues will be drilling prohibitions, if any, and drilling setbacks from structures, utilities, and property lines for vertical applications. Both horizontal and vertical applications will have to accommodate easements and set back requirements.
   - Obtain all of the necessary permits and/or licenses required to perform work in the area prior to beginning the installation process.
   - If the home or building has any indication of historical standing, be extremely cautious and determine what approvals and/or permits are required to perform work on the building and/or site. If the building carries historical identification plaques, check with the issuing agency for your obligations on such a property, regardless of what the general contractor or owner may say.

3. **ENVIRONMENTAL CONSIDERATIONS:**
   - Check if there is a protected wildlife habitat in the area where the installation will take place. Also, check for any endangered species populating the proposed work area or areas nearby.
   - Check for protected natural areas, such as wetlands or sources of flowing water on or adjoining the project site. This is particularly important if creeks, rivers, or streams flow through or near the site, which may be affected by groundwater or rain run off carrying drilling or trenching spoils residue (muddy water). This can be a significant issue. Run off prevention/control requirements need to be defined and proper preparations made prior to beginning installation.
   - Site contamination can be a very significant problem. If there is potential for contamination on the site, identify who has responsibility for clean up of any contaminated site materials that must be controlled and/or disposed of in the proper manner, possibly to the point of full remediation. Also be aware of ASBESTOS and/or LEAD PAINT present on the site that may become an issue at the point of building entry or during any inside work.

4. CLIENT INPUT:

- Check for the potential for future expansion of the building into the proposed GHEX installation area.
- Check for plans for additional improvements on the property such as an in-ground swimming pool or additional structures.
- Check for possible plans for division of the property that might separate all or a portion of the loopfield area from the building site at a later date.
- On retrofit applications, check for the presence of existing plantings, landscaping, or other surface improvements in the proposed GHEX area that are not to be disturbed.
- Check for a lawn sprinkler system on the property.

Unless prior knowledge indicates otherwise, all GHEX options should be evaluated to ensure the client receives the best heat exchanger performance possible for the least cost.

7.2 Designing for the Site

Closed-loop GHEX applications offer design flexibility. Vertically-bored, horizontally-trenched, and horizontally-bored GHEX systems can be applied as site conditions dictate. Refer to Chapter 5 for vertically-bored, horizontally-trenched, and horizontally-bored GHEX design procedures and guidelines.

7.2.1 VERTICALLY-BORED SYSTEMS Vertically-bored ground heat exchange systems require the least amount of surface area for buried closed-loop GHEX applications. Vertically-bored systems are typically the most expensive of all of the closed-loop options, but are sometimes the only feasible option because of space limitations on a job site. Depending on ground conditions, space availability, and design requirements, a single borehole can accommodate a partial nominal ton to multiple nominal tons of ground source heat pump capacity. When determining if a vertically-bored GHEX is a suitable choice for a given installation, key questions to answer before the design process can begin are:

- Are vertical GHEX installations allowed in the given area?
- Do ground conditions exist that will prevent the installation of a vertical GHEX such as caves, voids, sinkhole potential, or any other known natural impediments?
- Is a vertically-bored installation the most cost effective option as compared to other usable closed-loop alternatives?
- After considering code setback requirements and easements, does enough green space exist for the installation? If not, is the additional space under existing or planned driveways, sidewalks, etc., enough to accommodate installation space requirements?

7.2.2 HORIZONTALLY-TRENCHED SYSTEMS Horizontally-trenched ground heat exchange systems require the most surface area for buried closed-loop GHEX applications. Actual space requirements vary with pipe density in the trench. A system with a higher pipe density (5-pipe versus 2-pipe, refer to Chapter 5) will generally require less surface area for heat exchange. However, a higher pipe density design might not be the most economical option in terms of material and installation costs or system operating cost (pumping power consumption). Horizontally-trenched systems are generally a more economical first-cost option than vertically or horizontally-bored systems. Available space will generally be the limiting factor when deciding whether or not to utilize this ground heat exchange method, although the owner may not find the extensive excavation necessary for a horizontally-trenched GHEX acceptable and will vie for another option. When determining if a horizontally-trenched GHEX is a suitable choice for a given installation, key questions to answer before the design process can begin are:

- Does shallow rock exist on the site that would prevent cost effective trenching?
- Is a horizontally-trenched GHEX installation the most cost effective option as compared to other usable closed-loop alternative?
- After considering code setback requirements and easements, does enough green space exist for the installation? If not, is the additional space under existing or planned driveways, sidewalks, etc., enough to accommodate installation space requirements?

7.2.3 HORIZONTALLY-BORED SYSTEMS Horizontally-bored ground heat exchange systems generally require more surface area than horizontally-trenched systems and much more surface area than vertically-bored systems. Similar to vertically-bored systems, depending on ground conditions, space availability, and design requirements, a single horizontal borehole can accommodate a partial nominal ton to multiple nominal tons of ground source heat pump capacity. One benefit of horizontally-bored systems over other alternatives is that they can be installed under structures, lawn and garden obstacles, etc., without disturbing those existing structures, providing access to ground heat exchange areas that would not otherwise be available. Horizontally-bored systems can also be installed at lower cost in areas where depth to rock is shallow and the economics of drilling into the rock or drilling shallow boreholes to stay above the rock make a vertically-bored system cost prohibitive. A horizontally-bored system will also have lower headering costs compared to a vertically-bored system because a header trench will only need to be dug on one end of the field to connect all of the bores in the system.

When determining whether a horizontally-bored GHEX is a suitable choice for a given installation, key questions to answer before the design process can begin are:

- Are horizontally-bored GHEX installations allowed in the given area?
- Is a horizontally-bored installation the most cost effective option as compared to other usable closed-loop alternatives?
- After considering code setback requirements and easements, does enough green space exist for the installation? If not, is the additional space under existing or planned driveways, sidewalks, buildings, etc., enough to accommodate installation space requirements?

7.3 Forming a Site Plan

Initially, the site needs to be evaluated considering all issues addressed in Sections 7.1 and 7.2 along with anything else that may be discovered. This evaluation varies between new construction and retrofit applications. Key points to consider are:

- Check for existing walls, berms, retaining walls, permanent fencing, landscape architectural improvements, steep inclines, or other impediments that would restrict or prevent access or be safety issues on the site for drilling, trenching, and related equipment in the GHEX installation area.
- Construction sites often include significant changes in grade. Do not begin loopfield work until the loopfield area site work has been done or final grade elevations are established, documented, and understood by all parties involved.
- Perform a site visit. Such a visit is mandatory. Check for trees, rock outcroppings, creeks, or any other natural occurrence or obstruction that might not be shown on site drawings, but may create issues during GHEX installation.
- Identify and prepare for any other project site work that requires excavation in the GHEX installation area after loopfield installation, such as utility trenching or utility pole installation, fence post installation, and/or retaining wall or berm construction.
- Check with the OWNER, not just the general contractor for possible/planned future improvements that require excavation, such as existing building expansion or other improvements such as adding new buildings, an in-ground swimming pool, in-ground storm shelter, or similar items.
- On retrofit applications, identify special areas not to be disturbed with the OWNER, not just the general contractor, such as landscaping, gardens, etc.
- Identify the building's mechanical room location(s).

For new construction, a GHEX can be installed under various paved surfaces, specialty landscaping, or other surface improvements before they are constructed or installed. On retrofit applications these same items may prevent installation equipment access, GHEX installation, or just encumber site area that could have been used for all or a portion of the GHEX. Once the site evaluation is completed, the appropriate GHEX technology can be selected and the preliminary GHEX design completed.

The preliminary loopfield location should then be incorporated into the site plan project drawing as illustrated in Figure 7.1.

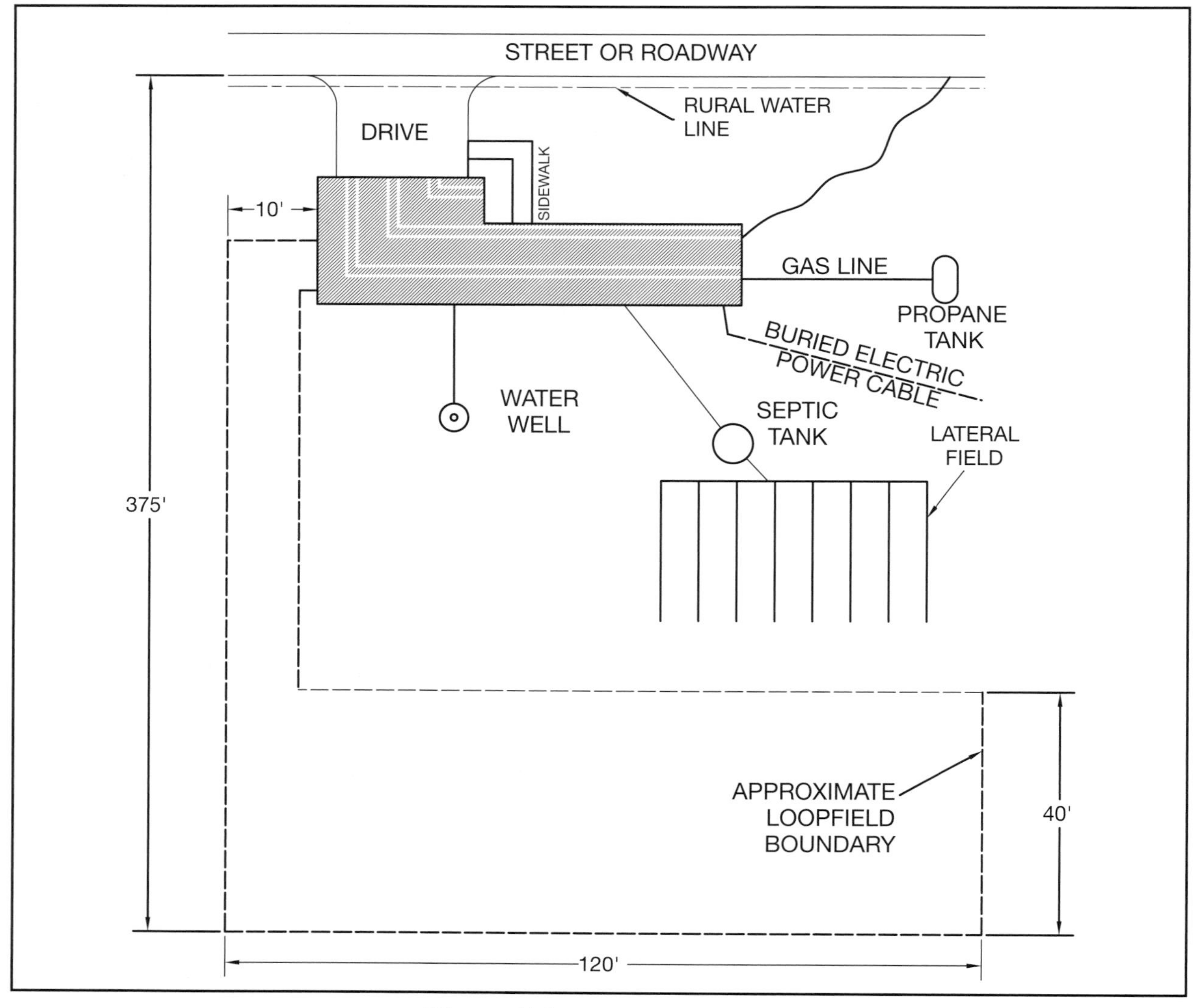

Figure 7.1. Site Plan Example

If the project is a retrofit application and there are no drawings, the GHEX installation area (including permanent improvements and/or landmark items that can be used as points of reference when triangulating the GHEX location) should be sketched and the GHEX installation area located. In either case, initial GHEX location and layout MUST be approved and accepted by the OWNER and/or GENERAL CONTRACTOR. The site plan drawing or sketch should include, but not be limited to:

- Dimensioned drawing of the planned GHEX layout and location on the property.
- Approximate location of all known utilities in or within 10 feet of the GHEX installation area.
- Approximate location of all known installation obstacles in or within 10 feet of the GHEX installation area.
- Approximate location of all easements within 5 feet of the GHEX installation area. (WARNING: Installing all or a portion of a GHEX in an easement is strongly discouraged).

Once the GHEX has been designed and the site plan layout accepted, permits (if required) can be applied for and acquired, and work can begin. Before beginning work on a project with existing improvements, it is strongly recommended that the entire site be inspected for damage and any damage be photographed. Typical examples are damaged fencing, cracked, or broken sidewalks, driveways, and/or patios, or damage to structures. This can verify that any existing damage was not caused by GHEX installation equipment or installation activity.

7.4 Project Preparation & Installation Guidelines

When preparing to start a project, do not assume that the requested services have been performed and resources necessary for job completion are available. Verify that everything is ready for use and that all necessary resources are on hand before mobilizing to the job site.

- Before beginning work of any kind, verify that a utility locate has been performed on the site and that the area is either clear or that all utilities are clearly marked.
- Electrical power will be required for the heat fusion process and to power the compressors, pumps, and other tools necessary for job completion. If power is not available on the site, a generator will be needed.
- Significant quantities of water will be required for drilling (if mud rotary drilling is the drilling method to be used), grouting the completed boreholes (if either vertically or horizontally-bored systems are used), and for filling the GHEX piping with water. If water is not available on the site, an alternative water supply will be needed. If the water is from a tank or water truck, a pump will be needed.
- An air compressor will be required for pressure testing GHEX piping and completed fusion joints and assemblies.
- Test caps with pressure gauges or test fitting assemblies will be required to pressure test the GHEX piping and completed fusion joints and assemblies.
- Ensure that the materials ordered for the project have been delivered in usable condition. If some of the materials have been back ordered, check with the supplier for a timeline as to when the undelivered materials will be available.

Before moving on site, ensure that the equipment to be used:

- is appropriate for the given project
- is the right machinery for the ground conditions expected
- can work within the space available
- has been properly maintained and serviced and is ready for several hours, days, or weeks of heavy duty operation.

7.4.1 GENERAL GHEX LOOP/HEADER CHECKS & INSTALLATION GUIDELINES Either high-density (HDPE) or cross-linked polyethylene (PEX-A) piping are used for the entire portion of the buried GSHP piping system. Pipe manufacturers extrude high quality pipe products. However, from the time the pipe leaves the extruder coiling machine or rack to the time it is used in the field, the opportunity for damage is always present. The following are typical checks that should be performed on all piping used on the job site both before and during the installation procedure.

- Visually inspect each U-bend assembly, roll of pipe, or bundle of straight pipe prior to removing the packing straps for shipping and handling damage. Coils or bundles can be mishandled when being loaded, can be dropped against sharp objects, and can be hit with fork lift blades. Nails can back out of pallet surfaces into the pipe and other causes of damage can occur. It is important to inspect the pipe while the integrity of the coil or bundle is intact. This ensures that surfaces, which were exposed during shipping and handling, can be thoroughly inspected for damage. Such damage is typically easily repaired and material is easier to replace prior to being installed. Replacement or repair is more difficult if damage is discovered after installation.
- Prior to installation, fill the ground-loop piping in the active legs of the GHEX with water and pressure test them for a minimum of 30 minutes, in accordance with the IGSHPA Design and Installation Standards.
- Some piping manufacturers sell U-bend assemblies and piping coils that are sealed at the factory. The hot pipe is sealed immediately after extrusion. When it cools and shrinks, the pipe interior becomes pressurized. Cutting the pipe end of the U-bend assembly causes the pipe to vent the pressure applied by the manufacturer to the atmosphere. Hearing noise caused by the release of pressure indicates that the pipe was not breached. However, the pipe could be severely gouged, but not penetrated and still hold this limited pressure. The sealed piping coils must still be inspected and tested as previously described.
- After testing and installation, seal the U-bend and/or coil pipe ends to prevent dirt, trash, or other sources of system contamination from entering the pipe.
- If side wall or saddle fusion fittings are used for loop connection to header piping, fuse the saddles onto the header. Saddle fittings have a natural curvature to their base that matches the exterior surface of the pipe. It is imperative that the curvature of the pipe wall outside diameter and fitting base are correctly seated together so proper fusion can occur. After the fusion has adequately cooled, completely drill the saddle out. Mark every drilled out saddle with a sticker or spray paint to indicate the saddle has been drilled.
- Before loops are attached to headers, remove the sealed ends and flow check each loop with water to ensure it flows freely and that the loop did not become blocked or restricted during installation. On saddle fitting applications, the fusion technician responsible for fusing the loop onto the header saddle fitting should perform a second check to ensure the saddle has been drilled out.
- After all loops are flow checked and fused to the headers, fill the entire GHEX piping system with water and pressure test it for a minimum of 30 minutes, in accordance with IGSHPA Design and Installation Standards.
- If possible, each loop that comprises a GHEX piping system should be of equal pipe length, with no greater variation in length than ± 5-10%.

7.4.2 SEALING LOOPS AND HEADERS DURING GHEX CONSTRUCTION As a GHEX is being constructed, completed bores often sit unconnected to the headers for extended periods of time. This can be due to the time frame required for GHEX construction, overnight delays, weather delays, or other activities that interrupt or delay construction. Loop pipe ends need to be sealed prior to being tied into the headers to prevent any foreign materials from inadvertently entering the piping system.

U-bend assemblies, coils, and straight pipe lengths of all diameters often come from the factory with caps on the pipe ends. A good short-term seal can be formed by placing (or replacing it if it has fallen off) the cap on the pipe end and duct taping it in place. If the caps are not available, duct tape several layers thick can be an acceptable temporary seal. However, duct tape is a poor seal in adverse conditions and is unacceptable for long-term sealing. Always be aware that if a cold U-bend assembly is inserted into a borehole with warmer surroundings, the pipe and water it is filled with will warm and expand. This expansion could cause improperly sealed pipe ends to suck whatever surrounds them into the U-bend assembly. Furthermore, duct tape is an indoor material and not intended to be exposed to outdoor temperature extremes combined with sunlight and/or weather extremes.

When long-term sealing is required for loop pipe ends, there are three alternatives:

- Fusing on caps
- Heating and closing pipe ends
- Barbed-plug fitting

Fusing on end caps is the most expensive and time-consuming approach. This requires fusion caps, fusion equipment, and a trained fusion technician. Heating and closing pipe ends is simpler than fusing on caps for short-term protection. To do so, use a torch to heat the pipe ends till they are hot and pliable, then mash them closed using welder's flat blade vice grips. The flattened pipe ends will be permanently closed, but will not form a seal. The ends will typically seep small amounts of water and can allow limited evaporation of the water from the loop. The best choice for long-term pipe end sealing is barbed-fitting plugs. For installation, only a plug and a hammer are required. Additionally, the plugs are permanent once driven in place. Finally, they are readily available, inexpensive, and disposable. A typical plug is shown in Figure 7.2.

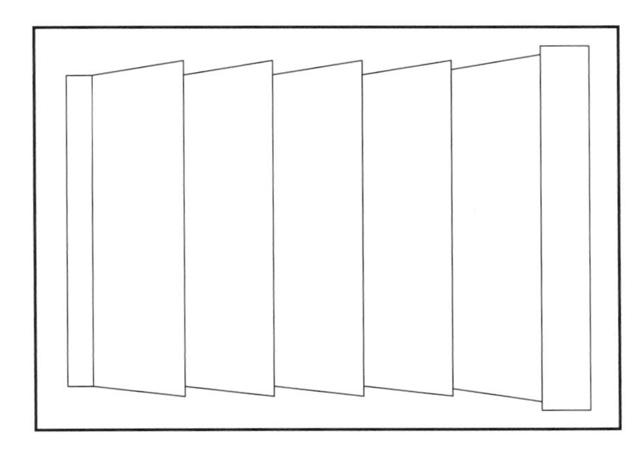

Figure 7.2. Barbed-Plug Fitting for Sealing Pipe Ends Prior to Headering

7.4.3 GENERAL TRENCHING REQUIREMENTS & GUIDELINES All closed-loop GHEX applications will require some degree of horizontal trenching. The following are universal trenching requirements that apply to all GHEX applications.

- Trenches for ground source loop piping and headers are not required to meet utility trenching specifications. They do not require graded bottoms, load bearing bottoms, specific dimensional pipe spacing, specific dimensional surrounding sand, or loose soil bedding (except as described below).
- Using sand for bedding is not required, unless there is no loose soil in the trenching spoils or available on the site.
- To facilitate GHEX piping location in the event that service is necessary, route #10 jacketed copper wire with the loopfield supply-return piping as shown in Figure 7.3. Loop the wire around header or individual loop piping throughout the entire trench length approximately once every 20 feet. Looping the wire intermittently will keep it relatively close to the pipe. Use one strand of wire per trench when there are multiple pipes in the trench. Route the wire into the mechanical room and coil around the header pipe at the room entry location for easy access.
- Remove rocks left in the bottom of the trench to minimize the possibility of damaging the pipe during installation. Use long handled tongs or a shovel.
- If rock with sharp or jagged edges is present in the bottom of the trench, the trench should be dug approximately 6 inches deeper than pipe placement design requires, allowing the trench bottom to be bedded with approximately 6 inches of loose soil or sand. If no rock with sharp or jagged edges is present, the pipe can rest on the trench bottom without bedding.
- If rock with sharp or jagged edges is present in the sidewalls of the trench, the pipe should be placed at least 6 inches away from the side wall and backfilled with loose soil or sand. If no rock with sharp or jagged edges is present in the sidewall, the pipe can rest anywhere on the bottom of the trench, including near or against the sidewall, but still needs to be backfilled as described previously.
- High density polyethylene (HDPE) expands and contracts with changes in loop temperature over an annual cycle – loop and header pipe should be "snaked" in the trench to accommodate movement associated with this expansion and contraction (refer to Figure 7.3).
- Trench bottoms can be wet/muddy or have a small amount of standing water. If the trench is flooded to the point the pipe floats, the trench should be pumped out, and the pipe must be secured in position during backfilling to ensure it is buried at the appropriate depth and location.
- Header trenches are often work areas. If trenches are deeper than the maximum OSHA recommended safe depth, make sure trenches are beveled, stepped, or shorn in accordance with OSHA requirements for the soil conditions present. If there is water in the trenches, do not use any electrically powered device while standing in the water in the trench.
- When backfilling around the pipe, use approximately 6 inches of loose soil or sand to ensure there are no voids around the pipe and that the pipe has good earth contact throughout.
- When backfilling, remove all rocks with sharp edges or rocks bigger than a golf ball from the loose soil used to bed around the pipe. After the pipe is bedded with at least 6 inches of loose soil or sand, trenching spoils can be used to backfill the trench.
- In areas with heavy clay soils, the trenching spoils are often large clumps of clay. Do not drop large clumps into the trench or allow them to fall directly on the pipe before bedding it with loose soil. The clumps typically will not break up and as a result, leave air gaps around the pipe.

- After bedding around the loop and header piping, the backfill should be watered in (if sufficient water is available) to settle the loose soil around the pipe to ensure there are no air gaps along the length of the pipe and to aid in soil compaction.
- When the trench has been backfilled to approximately 1 foot from finished grade, place warning tape in the trench, and then, finish backfilling as shown in Figure 7.3. On new construction sites, where there have been no previous structures, metallic warning/locator tape can be used. In areas where buildings have been demolished, there is often significant metal residue in the ground (nails, screws, metal pipe scrap, pieces of flashing, etc.) so the metallic tape may not be easily located, or the metallic residue may give a false location. Plastic warning tape that is non-metallic is recommended for these types of applications.
- For a system with multiple supply and return headers, the circulating fluid temperature in all supply piping will essentially be the same and the circulating fluid temperature in all return piping will essentially be the same. Multiple supply headers can be installed together, touching, or in very close proximity without affecting loopfield performance. Similarly, multiple return headers can be installed in close proximity to one another. It is good practice to try to separate the supply lines from the return lines in the header trench by 12 inches or more whenever possible. However, no unnecessary effort (insulating between the supply and return lines, digging a dedicated supply line header trench, or a dedicated return line header trench, etc.) needs to be made above and beyond spacing the lines out in the trench. Doing so would unnecessarily drive up installation costs for minimal benefit. The effects of thermal interference between supply and return piping is generally very minimal.

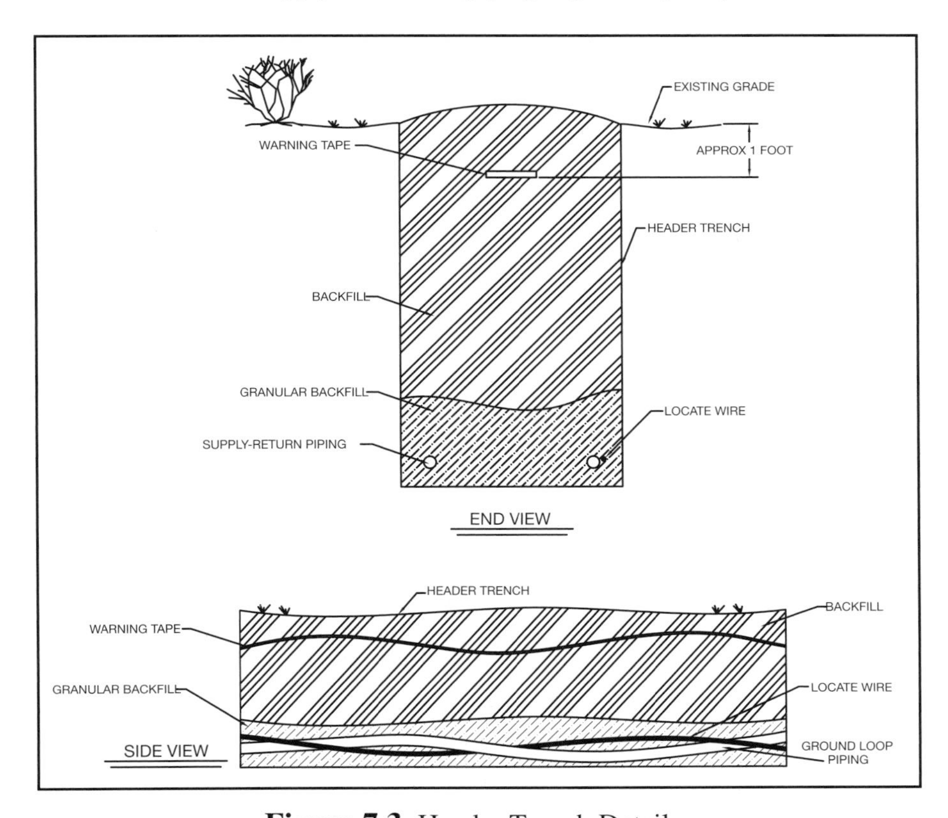

Figure 7.3. Header Trench Details

7.5 Vertically-Bored, Closed-Loop GHEX Installation

Vertically-bored U-bend piping is to be installed by an appropriately licensed driller. When the driller arrives at the site, issues such as site accessibility, drill rig mast overhead clearance, terrain, and other impediments to safe and productive drilling should already have been addressed and resolved. At this point, the site plan should have been completed, utilities located, and boreholes marked per the site plan. There should be no delays when the driller arrives onto the site. Typically, the driller will perform the following tasks:

- Drill the boreholes at the predetermined locations to a depth adequate to ensure proper U-bend insertion to the design depth
- Test U-bend assemblies and insert them to the design depth
- Finish by grouting the borehole from bottom to top

Four major drilling technologies are used in vertical GHEX drilling. The technology to be selected will be dictated by local ground conditions, the driller's experience, and by the infrastructure in place for the area. The technologies are mud or wet rotary, air rotary, air hammer, and sonic. In some soil conditions, hollow-stem auger drilling can be used, but the applications are limited and will not be addressed. Borehole sizes typically range from 3 ½-6 inches, but will vary based on pipe size used, ground conditions encountered, and state regulations.

Regardless of the drilling technology used, the objective of the driller is to install a GHEX with heat transfer capability that matches the peak rates of heat of rejection in cooling and heat of extraction in heating, accounting for unbalanced ground loads over the long term. In the past, statements have been made that the main objective when installing a vertical GHEX is to install a specific length of piping, not to reach a given depth. This statement is only true when the formation in which the GHEX is installed does not vary with depth. In most cases, however, the formation does change with depth and the statement is inaccurate. For example, if a vertical GHEX were designed to be installed to depths of 300 feet into a formation that consists of 100 feet of overburden (unconsolidated soil above the rock formation) at the top and 200 feet of rock at the bottom, in general, the bottom 200 feet of rock in each hole would be less resistant to heat transfer and therefore have better performance (refer to Section 5.2 for tables of the thermal properties of various types of soils and rock). If for some reason the drilling contractor decided to stay above the rock formation and install the entire loopfield with three times the number of bores to a depth of 100 feet, the performance of the loopfield will suffer due to both reduced thermal performance and long-term effects caused by unbalanced ground loads and an increased number of bores completely surrounded by neighboring bores. If the drilling contractor wanted to stay out of the rock portion of the formation and install a loopfield with the same heat transfer capability, the system design must be re-evaluated and additional bores installed to account for the reduced heat transfer capability of the upper portion of the formation. Refer to Section 5.2 for a complete discussion on vertically-bored GHEX design and the parameters that will affect design lengths.

Ground temperatures and soil moisture content naturally fluctuate near the earth's surface (at depths less than 20 feet from the surface), so boreholes of less than 50 feet are generally not recommended. The ultimate decision pertaining to optimum design bore depth should consider the cost of drilling (including set-up time to move the drill rig from hole to hole), borehole completion material cost (pipe, antifreeze, grout), and headering costs (material costs, trenching costs, and labor costs for time spent headering the system). A loopfield that uses a large number of shallow bores will have higher trenching and headering costs and will require more drill rig setup time. A loopfield that uses deeper bores will generally have lower trenching and headering costs and less drill rig setup time, but will have higher material costs because larger pipe is more expensive and requires more antifreeze volume. Also, a loopfield that uses deeper bores could have higher drilling costs because of the increased potential to have to drill through troublesome portions of a formation. If utilizing shallow bores seems to be the only feasible way to install a vertically-bored GHEX, an alternative closed-loop GHEX design should probably be considered (such as a horizontally-bored GHEX).

All borehole U-bend insertion depths on a project should be equal within ± 5%, if possible. However, if some boreholes are drilled to the appropriate depth during installation, but one or more are short because of a drilling problem or natural impediment, the hole(s) may still be usable. If adequate space is available to drill an additional borehole at the proper spacing, a companion bore can be drilled and the two can be tied together in series to form a single flow path with characteristics similar to the full-depth bores installed throughout the rest of the loopfield. This concept is illustrated in Figure 7.4. Remember, if a series-parallel configuration is used to stay out of a troublesome formation, refer to design documentation to ensure that the GHEX designer did not intend for the U-bend piping to be installed in that portion of the formation.

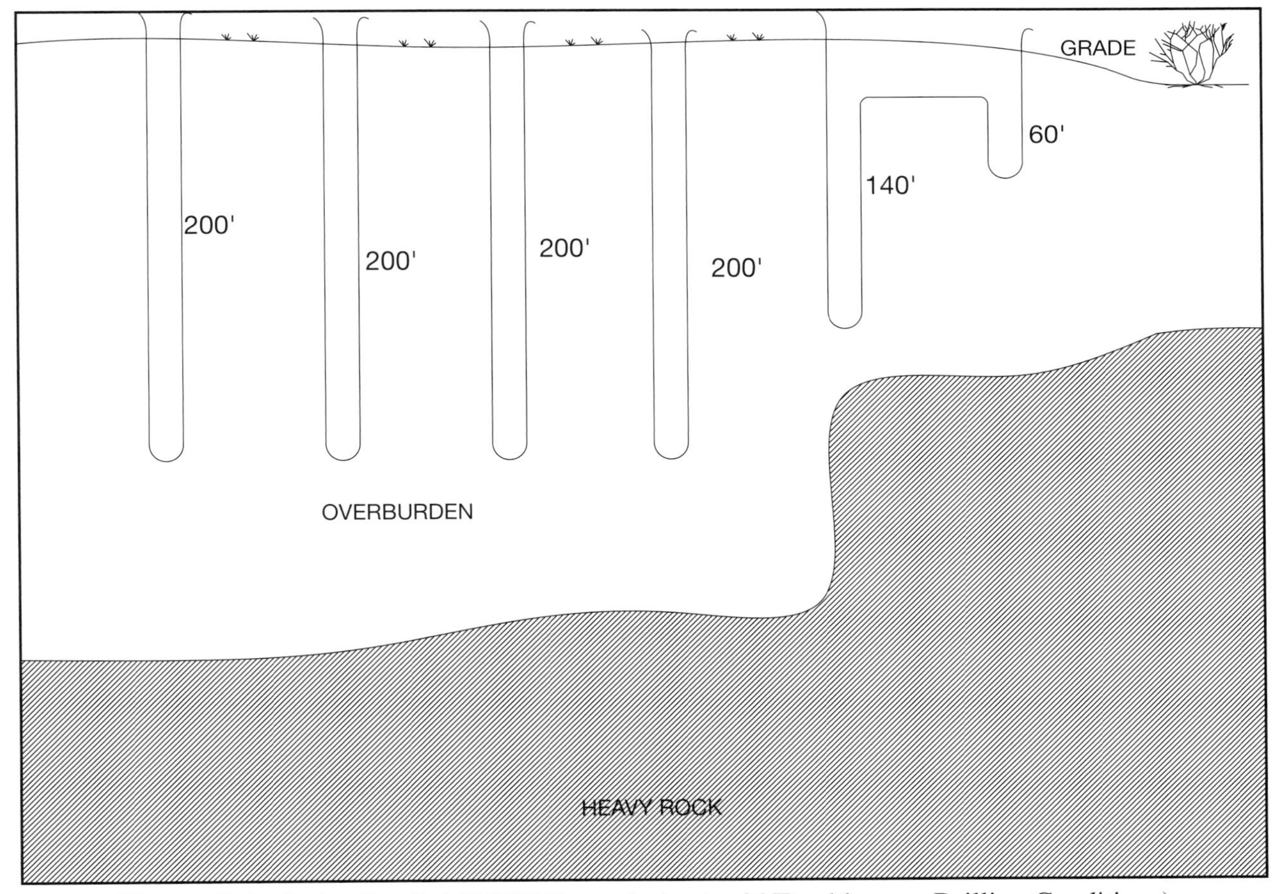

Figure 7.4. Series-Parallel GHEX Example (to Avoid Troublesome Drilling Conditions)

Using an appropriately licensed driller on residential and light commercial projects is very important for many reasons. Since formation thermal conductivity tests are rarely run on projects smaller than 25 tons (and often not on jobs of 40 tons or less) the well driller's knowledge of the formations in the area can be very helpful when estimating the site thermal conductivity to be used during the design process described in Chapter 5. The estimated design length information, combined with the site evaluation results, will determine the preliminary location, number, depth, and spacing of the project's boreholes, as well as U-bend diameter and grout requirements.

7.5.1 DRILLING METHODS As previously stated, there are four primary technologies used in vertically-bored GHEX installation. The technologies are mud or wet rotary, air rotary, air hammer, and sonic and will be discussed in detail in this section.

7.5.1.1. WET/MUD ROTARY DRILLING Wet/mud rotary drilling is primarily used in unconsolidated or semi-consolidated formations. This method uses a rotating drill bit at the end of a drill string. The drill string is hollow so that a water-based (or water and bentonite-based) drilling fluid can be pumped through the pipe to the bottom of the borehole. The drilling fluid carries the cuttings from the bottom of the hole up to the top and into the mud collection/settling pit. The drilling fluid is reused by pumping it from the top of the mud collection/settling pit and re-circulating it into the borehole. A wet/mud rotary drilling system is illustrated in Figure 7.5.

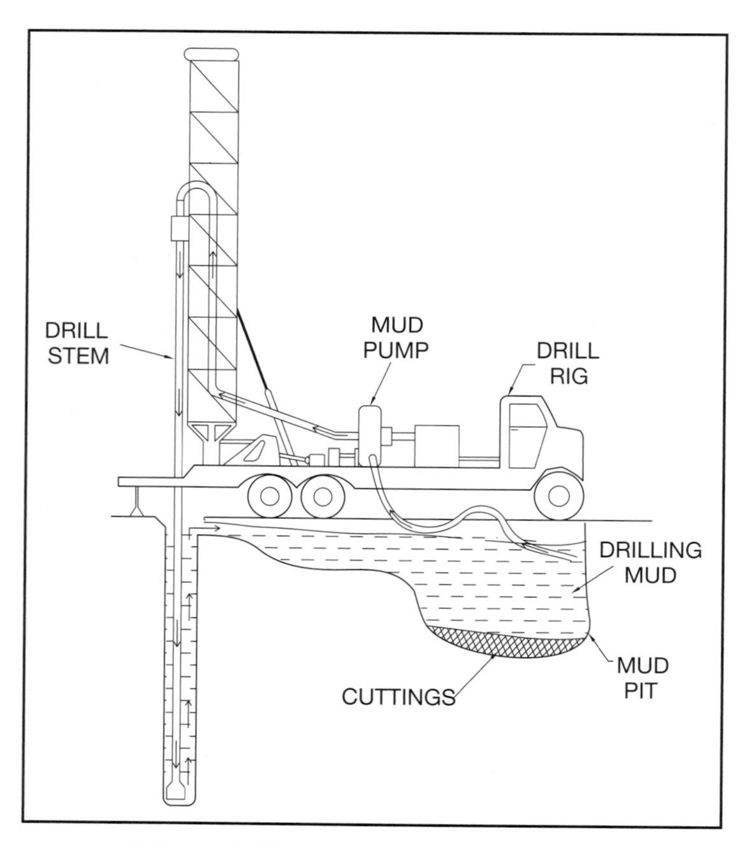

Figure 7.5. Wet/Mud Rotary Drill Rig

Drilling mud (water and drilling fluid) serves four main purposes in mud-rotary drilling. The primary functions of drilling mud in the wet rotary drilling process are as follows:

- Suspend and carry the cuttings out of the borehole and into the mud collection or settling pit
- Form a filter cake on the borehole walls to minimize the effects of penetration and possible contamination into permeable zones of the formation
- Cool and lubricate the drill bit during operation
- Prevent borehole wall collapse

One major drawback of using drilling mud is that it can be extremely messy on the job site if care is not taken to contain it. Two methods that are used to contain drilling mud on a job site are:

- Manually dig an earthen mud settling pit at the site with a backhoe (Figure 7.5)
- Utilize a portable surface mud collection pit (preferred method)

When an earthen mud pit is dug on the job site, a shallow trench will need to be dug to connect the borehole to the pit. The trench will need to be dug with enough slope to allow the drilling mud to freely flow from the borehole to the pit. The drawback to manually digging an earthen mud pit on the site is that the mess will likely be left behind after the drilling is completed. Using a portable surface mud collection system will generally do a better job of containing the mess and also facilitates easy removal of cuttings and drill mud from the site for disposal per project specifications or requirements.

7.5.1.2. AIR DRILLING Air drilling is used in unconsolidated or consolidated formations (rock). Air drilling is similar to wet/mud rotary drilling in that both methods use a circulating fluid (in this case air) to carry drill cuttings up to the surface and out of the hole. As discussed in Section 7.5.1.1, drilling mud is necessary for drilling in unstable unconsolidated formations to prevent borehole wall collapse and to prevent possible environmental contamination when drilling through a permeable zone. Borehole collapse is generally not an issue in consolidated formations. Also, consolidated formations are not permeable, unless a fracture is encountered in which case drilling mud will not prevent loss to the formation. Because of these two factors, air is a suitable circulating fluid when drilling through consolidated formations.

There are two types of air drilling, air rotary and air hammer. Air rotary drilling uses a drill bit at the end of a drill string similar to wet rotary. Air is pumped down through the drill string which escapes through holes in the bit. As the bit turns, it grinds the rock cuttings into small pieces, which are then carried up and out of the hole by the circulating air. Air rotary drilling is normally used in softer consolidated formations. Air hammer drilling is similar to air rotary except that the bit on the end of the drill string is a pneumatic bit driven by compressed air, which rapidly strikes the rock to break it up. After the rock is pulverized by the hammer bit, the air that is pumped down through the pneumatic bit escapes out the bottom and carries the cuttings up and out of the hole.

The air drilling process can discharge significant amounts of dust into the immediate surroundings. To control this dust, most air rigs utilize a shroud around the borehole that sprays water mist toward the hole. This mist will keep airborne dust to a minimum and will produce a limited amount of water run off, cuttings, and dust slurry. Air rigs use high pressure air and, when they encounter groundwater, there can be a significant amount of water discharge to the surface. If run off or water discharge is significant enough, it can be allowed to run into a settling area (similar to a mud settling pit as described in Section 7.5.1.1) to allow solids to settle out. Then, the water can be allowed to freely flow through grass, gravel, and/or riprap (heavy cobbles used at the waterfront of oceans, lakes, rivers, storm water discharge outflows, and drainage ditches, etc. to prevent erosion), which will act as an additional filter, and drain into low lands, flowing water, or storm drains. In areas where allowing the run off to flow off of the settling area is not possible or allowed, filter bags or other containment measures must be used to catch and control water and drill cutting run off.

Typically, when air drilling is used and the vertical GHEX installation must penetrate a layer of overburden, and then, continue drilling into the heavy rock formation, the portion of the hole that is in the overburden must be cased with PVC or steel to prevent borehole collapse caused by the air drilling process. The casing is installed with the intention of removing it from the borehole once it has been drilled and the U-bend has been inserted. The borehole can be grouted before or after casing removal.

For applications where the hole needs to be cased, in most instances, the casing can be removed without problem. However, if the formation has a history of not allowing casing to be easily removed, either use steel casing to minimize the effect on heat transfer capability of the hole or revisit the loopfield design and include the additional borehole resistance to accommodate the poor heat transfer characteristics of PVC. Refer to IGSHPA's (2000) *Grouting for Vertical Geothermal Heat Pump Systems* for borehole resistance calculation procedures, which account for PVC casing when it is left in a hole. A drill rig operator should always use caution when casing a borehole to avoid leaving gaps between the outside of the casing and the surrounding formation. Such gaps will increase the thermal resistance of the borehole and reduce its effectiveness with respect to heat transfer.

When the casing becomes seized in the borehole and cannot be removed, the borehole must be excavated to approximately 1 foot deeper than the planned trench depth to ensure that the U-bend pipes can be positioned as necessary to connect to the header. After the header trench has been dug to the appropriate depth, casing cutters will need to be used to score the casing at the bottom of the trench so that it can be snapped off. After scoring the casing with the cutters, use a backhoe or front-end loader to pull up on the casing, until it snaps off. Use caution not to damage the U-bend piping during this process. Additionally, the break point of the snapped off casing will have a jagged and very sharp edge. Either smooth off that sharp edge or place something in between the edge of the casing and the U-bend piping to ensure that they will not come in direct contact with one another. The U-bend piping can be blocked away from the edge with scrap pipe or lumber, or a piece of scrap HDPE split lengthways and notched so it can be bent in a tight circle and set in place over the casing end. Any method that will permanently separate the piping from the jagged casing edge is acceptable.

7.5.1.3. SONIC DRILLING Sonic drilling is a new form of drilling technology, which uses a combination of mechanical vibrations and rotary power to drill a borehole. The drill head consists of two counter-rotating, out-of-balance rollers. When the rollers rotate in the drill head, they cause a mechanical vibration, which is transferred through the drill string to the drill bit. The frequency of the vibration caused by the rotating unbalance can be adjusted to suit the geological conditions at the installation site. The goal is to vibrate the drill bit at the resonant frequency of the formation to maximize the cutting ability of the drill bit. The resonant vibration and weight of the drill string along with the downward thrust of the drill head drive the drill bit into the formation. Similar to wet rotary drilling, high pressure water is pumped down through the center of the drill string and out of the drill bit to carry the cuttings from the bottom of the hole to the top and into a fluid collection system. A major benefit of sonic drilling is that borehole casing is attached to the drill string and is sent simultaneously down the hole with the drill string and drill bit. Therefore, borehole collapse and drilling fluid penetration into permeable zones is not a concern and drilling mud is not needed. Additionally, because the casing is part of the drilling apparatus, it is removed along with the drill string and will not get stuck in the borehole.

The drill string and drill bit are hollow. Because of this, a U-bend and tremie line can be installed through the center of the string, and the hole can be grouted before the drill string and bit are removed. After grouting, the drill string is vibrated as it is being pulled out to allow for ease of removal from the completed borehole. Grouting is discussed in more detail in Section 7.5.3.

Sonic drilling works very well in all formations, except hard consolidated formations. However, a typical sonic rig is very versatile in that it can be used to drill through softer unconsolidated overburden. Once the hard rock portion of the formation is reached, the sonic drill string can be left in the hole and used as casing (similar to the ODEX system described in Section 7.5.1.2). Because the drill string and bit for a sonic rig are hollow, an air hammer bit can be inserted down through the sonic drill string and into the consolidated formation. From that point onward, standard air hammer drilling methods are used.

7.5.2 VERTICAL U-BEND INSTALLATION Once the boreholes have been drilled, the U-bend assembly will be inserted to the design depth. In the early years, loops were completely unrolled and laid out on the ground prior to insertion. Now, many drillers use loop reels. An example of a loop reel is shown in Figure 7.6. Loop reels can be mounted on drill rigs, backhoes, or trenchers (for use in horizontal applications) or wherever is most convenient during the loop installation process.

During the mud rotary drilling process, while the hole is being drilled, one person will set the U-bend on the loop reel, fill it with water and, using test caps, pressure test the U-bend assembly prior to installing it in the borehole. After testing is completed, the pressure should be relieved slowly, the test caps should be removed, and the loop pipe ends should be sealed as discussed in Section 7.4.2. The U-bend will then be ready for installation.

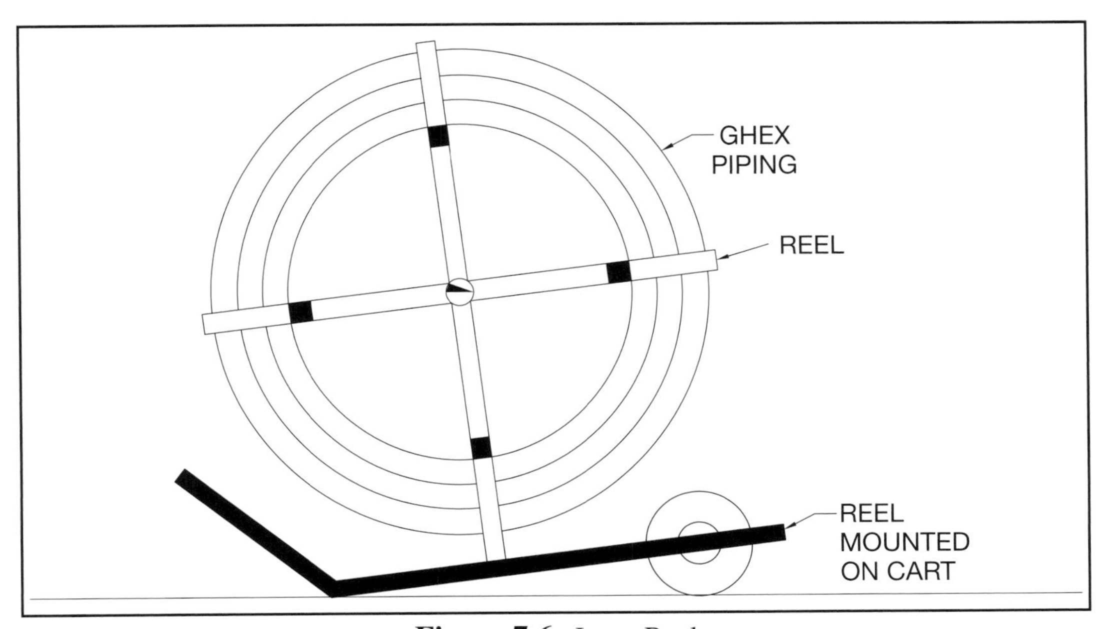

Figure 7.6. Loop Reel

There are two basic ways to install a U-bend in a drilling-fluid filled hole:

- **Manual Installation** To manually force a U-bend into a hole, a weight/stiffener will be attached to the end of the U-bend to: (1) take away the natural memory curvature of the pipe to ensure that it doesn't get stuck along the borehole wall on the way down and (2) add weight. A typical stiffener/weight would be a piece of pipe, rebar, or a fence post. Tape a piece of pipe or rebar (3-6 feet long) to the end of the pipe. Use extreme care to check for jagged or sharp edges on the ends of the stiffener and turn these sharp edges away from the pipe. Using duct tape, tape the stiffener securely to the U-bend. Also, tape over the jagged or sharp edges of the stiffener if they exist.

After securing a stiffener/weight to the end of the U-bend, several men will be needed to push the pipe into the hole. Because the pipe will tend to be buoyant in the drilling-fluid filled hole, the more weight that can be added to the end of the pipe, the easier U-bend insertion will be.

- **Mechanical Installation** To force the U-bend in the hole mechanically is simpler, faster, and easier than doing it manually. Before taking the U-bend assembly off the reel, cut approximately 1 foot off of one of the pipe legs and securely duct tape it to the end of the U-bend or attach a length of pipe or rebar to the end of the U-bend. A heavy 1-½-inch diameter, 20-foot long steel rod (commonly referred to as a "stinger") can be suspended from one of the rig's winch cables and inserted into or over the length of pipe or rebar secured at the end of the U-bend. The stinger will be heavy enough to offset the buoyancy forces on the U-bend and force it down to the bottom of the hole. If this method is used, care must be taken to ensure that the winch cable is in proper working order. After each insertion, the driller should wrap a rag or old cotton glove around the winch cable to squeegee the water and mud off of the cable as the stinger is withdrawn from the hole, while simultaneously checking for any broken cable strands protruding from the cable. These simple checks will identify any potential problem with cable fraying and cable strand ends that may damage the pipe. Figure 7.7 displays a U-bend being mechanically inserted into a borehole with a stinger bar.

Before installing the U-bend in the borehole, it is highly recommended that the tremie be attached to the stiffener bar and sent down to the bottom of the hole with the U-bend. This is done so that the hole can be grouted from bottom to top immediately following the installation of the U-bend. IGSHPA recommends that all boreholes be grouted within a relatively short amount of time after U-bend installation. It is not recommended practice to allow the boreholes to remain open and un-grouted for extended periods of time (12-hours maximum). Refer to *IGSHPA (2000) Grouting for Vertical Geothermal Heat Pump Systems* for an extensive discussion of practices commonly used to insert the tremie line into the borehole for grouting it from bottom to top.

Once the U-bend has been completely inserted, it must be secured in place because the loop will tend to float out of the drilling-fluid filled borehole. The U-bend can be secured by taping the pipe ends to a stake driven into the ground or by placing a heavy weight (such as a couple of bags of sand, grout, or a pallet) on the loop ends. Once the borehole is grouted and the grout has been allowed sufficient time to completely set up (typically 4-5 hours after placement), the stake or weights can be removed.

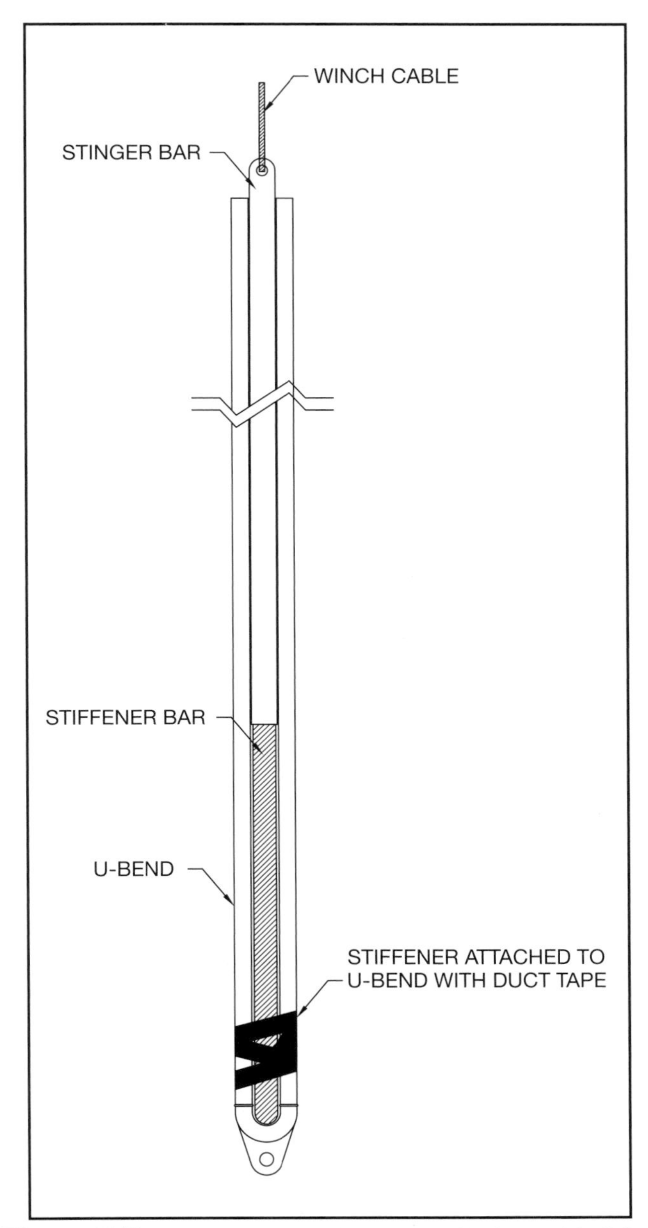

Figure 7.7. Mechanical U-bend Installation with Stinger Bar

Installing a U-bend in a borehole, where a drilling method that will not leave fluid in the hole was used (such as air drilling), is slightly different than the procedure used for a drilling-fluid filled hole. The loop is not to be filled with water on the reel before being inserted into the hole. Because the borehole will not be filled with any fluid, there will be no buoyancy forces in the hole to force the U-bend back out. The weight of the water in the U-bend is not necessary. In fact, adding excessive weight to a U-bend before inserting into an empty hole can be dangerous because the added weight will be too heavy, making it difficult to slow the pipe down to a controlled rate as it is lowered down into the hole. A braking mechanism will need to be added to a loop reel used for this method to slow the U-bend descent into the hole. A stiffener will need to be attached to the end of the U-bend to take away the natural memory curvature of the pipe and ensure that it doesn't get stuck along the borehole wall on the way down. In some instances, the bottom of the borehole will fill with water from the formation. When this occurs, weight will need to be added to the U-bend to force it to go through the water-filled portion of the borehole. Simply filling the U-bend with water, after reaching the point where it will no longer go down the hole because of the buoyant forces of the water, will be enough to force the U-bend down the hole to the bottom.

After the U-bend has been installed to the design depth in the dry hole, then it will need to be completely filled with water and pressure tested as previously described. Because the borehole may be partially or completely filled with water over time and because grouting the hole may tend to cause the U-bend to float out of the hole, the U-bends should be secured in place with a stake or a weight (as previously described), until the hole has been grouted, and the grout has been allowed to completely set up.

7.5.3 GROUTING THE VERTICAL GHEX Grouting is described in great detail in IGSHPA's *Grouting for Vertical Geothermal Heat Pump Systems*. As a standard procedure, IGSHPA recommends that all vertically-bored holes be grouted from bottom to top and that all horizontally-bored holes be grouted from end to end via pressure grouting with a tremie. There are two reasons to grout:

1. Protecting the integrity of all aspects of the deep earth environment.
2. Ensuring good heat transfer between the GHEX piping and the earth.

Considering the first point, it is imperative that the boreholes be sealed in the manner dictated by governing codes and regulations. If the governing regulations do not have geothermal-specific information, groundwater quality, and/or water well, codes for abandoning a well are to be followed. It is wise to research codes any time you move into a new area; regulations vary from jurisdiction to jurisdiction. It is also wise to check the code sets for changes or revisions. Finally, check for codes and regulations at the state, county, and municipal levels; code requirements and permitting is becoming more localized in many areas. Actual code and regulation information is purposely omitted from this section because of their wide diversity and potential for revision.

The second point, good earth contact, directly relates to the ability of the GHEX to transfer heat. There are two basic types of grouting materials that can be placed via the pressure grouting method: bentonite-based and cement-based. Of each of the two commercially available grouting materials, there are two types: high-solids and thermally-enhanced.

7.5.3.1 BENTONITE-BASED GROUTS High-solids, bentonite-based grouts typically contain approximately 20% solids and have a very poor thermal conductivity value (approximately 0.38-0.42 Btu/hr-ft-F). A borehole grouted with high-solids, bentonite-based grout will not perform as well as it would if that same hole were filled with a thermally-enhanced, bentonite-based grout. Thermally-enhanced, bentonite-based grouts were developed to overcome the poor thermal conductivity of high-solids, bentonite-based grouts by lowering overall borehole resistance to heat transfer, while still maintaining the required characteristics to seal the borehole and prevent penetration of contaminants down the borehole or interaction between aquifers. Commercially available thermally-enhanced grouts can be mixed to attain thermal conductivity values ranging from 0.50 to 1.40 Btu/hr-ft-F or better (the range varies from manufacturer to manufacturer). However, research has shown that the law of diminishing returns applies to grout thermal conductivity and bore length reduction in that increasing grout thermal conductivity above a value of approximately 1.00 Btu/hr-ft-F will lead to diminishing bore length reductions, while still driving up grout material, shipping, and labor costs. Exceptions exist where grout thermal conductivity values above 1.00 Btu/hr-ft-F are necessary (limited space, for example), but in general, using grout thermal conductivity values between 0.79 and 1.07 Btu/hr-ft-F will be most economically feasible.

7.5.3.2 CEMENT-BASED GROUTS Cementitious grouts were developed to compete with bentonite-based grouting materials and have been tested and proven to be comparable. However, the problems associated with cementitious grout have limited its use. Problems associated with cementitious grout are:

- It is cement based. If pumping is interrupted for any reason, the cement can set up and require extensive clean up for the grout pump assembly, as well as replacement of the tremie tube.
- If the boreholes are overfilled, the grout will cover the ground around the boreholes that were overfilled. Once the grout cures, it must be broken up and removed. Also, if the boreholes are overfilled, the grout will need to be broken up from the U-bend pipes to the bottom of the header trench to facilitate connection into the header. This will present an unnecessary opportunity to damage the pipe.
- Unless the flow of grout into the borehole is stopped a foot or so below the header trench bottom, the fixed location of each pipe creates an increased potential for kinks during header connection and backfilling.
- Cementitious grout does not bond to the pipe like a bentonite grout. It is solid and inflexible and can form voids between the grout and pipe wall when the pipe contracts during low temperature operation in the heating mode. Such voids will both decrease the thermal performance of the GHEX and provide an avenue for contaminants to flow down the borehole or for cross-contamination between aquifers

These and other negatives have significantly reduced the use of cementitious grouts. Refer to *IGSHPA (2000) Grouting for Vertical Geothermal Heat Pump Systems* for in-depth discussion of reasons for grouting, grouting products, grout placement methods, etc.

7.6 Horizontally-Trenched, Closed-Loop GHEX Installation

A horizontally-trenched, closed-loop GHEX will typically be installed by the one of the various pieces of excavation equipment shown in Figure 7.8, a chain trencher, a backhoe, a bulldozer, or an excavator. As has been described in Chapter 5, horizontal applications can be installed using a number of piping configurations and variations in pipe and trench spacing. Conventional horizontal systems can be installed with pipe densities in the trench ranging from 1 to 8 feet of pipe length per foot of trench (refer to Section 5.3 for common horizontally-trenched GHEX configurations), and the various configurations can either stand vertically in a narrow trench or lie horizontally in a wider trench. Horizontal slinky systems can be designed with various pipe densities to adjust trench length based on space availability. Refer to Section 5.3 for more details on horizontally-trenched configurations and GHEX design procedures.

METHOD A. EXCAVATION WITH CHAIN EXCAVATOR (TRENCHER)

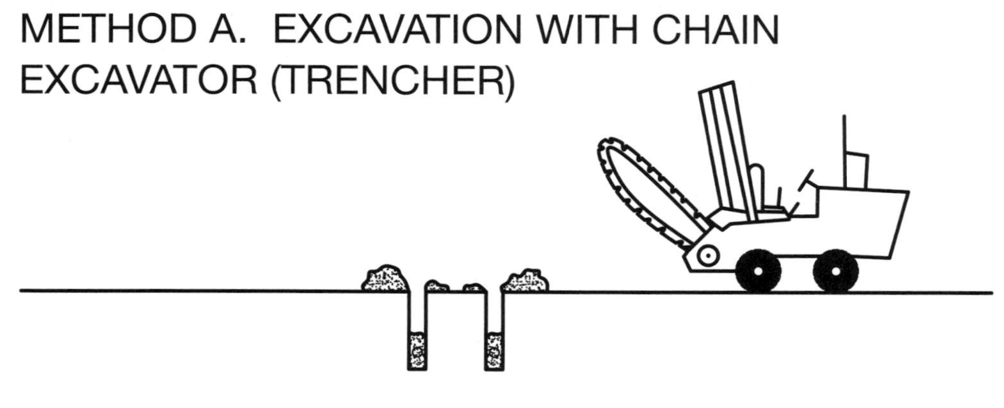

METHOD B. EXCAVATION WITH BACKHOE

METHOD C. EXCAVATION OF WHOLE SITE WITH BULLDOZER

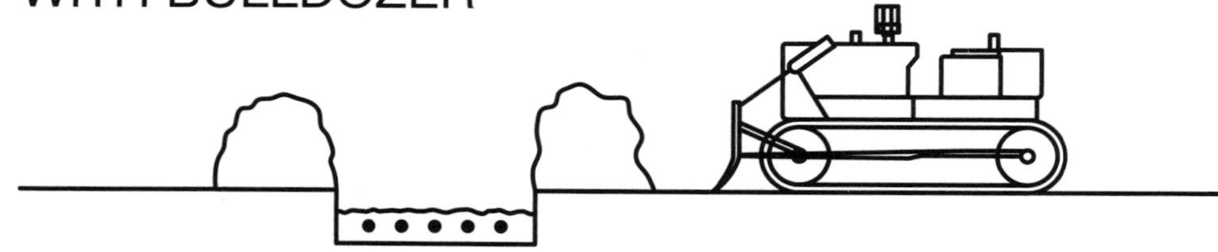

METHOD D. EXCAVATION WITH EXCAVATOR

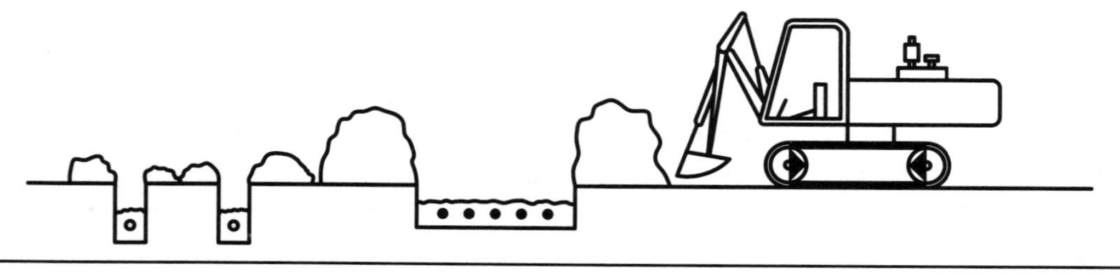

Figure 7.8. Excavation Equipment

7.6.1 EXCAVATION EQUIPMENT As previously stated, there are four main types of equipment used to install horizontally-trenched, closed-loop systems, a chain trencher, a backhoe, a bulldozer, or an excavator, all of which will be discussed in detail in this section. The choice of equipment to use depends on local site conditions and cost. In general, the machine that moves the least amount of soil will be the most cost effective. However, installations need to be evaluated on a case by case basis, as the actual cost to install a horizontally-trenched system will depend on:

- Availability of the excavation equipment
- Mobilization cost as it relates to traveling distance to get the excavation equipment to the job site
- Cost to rent/sub-contract/operate the equipment versus its productivity
    - Larger equipment may be more expensive to rent, but can typically excavate a large amount of soil in a short amount of time
- Access to the job site to begin excavation
- Ease of maneuverability during excavation

7.6.1.1 CHAIN TRENCHER A chain trencher will be used for applications that require narrow, but deep trenches, such as for the 1, 2, 4, 5, or 8-pipe standing configurations (refer to Section 5.3). Typical trench depths for chain-trench applications will be 4-6 feet with extended trench booms having the ability to go 7-8 feet deep. In most cases, chain trenchers will be the most economical option because the amount of soil or dirt removed during trenching is minimal when compared to the other available options. Also, trenching productivity will typically be higher than the other options, if the right soil conditions exist. Chain trenchers will not work well in areas where troublesome trenching conditions caused by loose, sandy soil, rocks, cobbles, or boulders exist. The soil removed by a chain trencher tends to be granular and will serve as excellent backfill material (with the exception of wet clay soils). Typically, when a chain trencher is used, the existing conditions will tend to have minimal rocks, cobbles, or boulders, so measures to protect the pipe against jagged or sharp rock should not be required. If rocks do happen to fall into the bottom of the trench, remove them with long-handled tongs or a shovel.

Caution should be used when backfilling chain-trenched, multiple-pipe GHEX applications in clay soils. The damp clay particles will stick to one another, forming clumps and creating the potential for air pockets to form around the pipe, which can greatly reduce heat transfer capability. The clay can also stick to the pipe and trench walls, significantly increasing the potential for bridging in the trench. Either break the clumps before placing them in the trench or use caution to only use the granular soil removed from the trench as backfill around the pipes. Refer to Section 7.4.3 for trench backfilling recommendations.

7.6.1.2 BACKHOE A backhoe will be used where there are rocks, cobbles, or boulders, which are too large or too difficult to remove with a chain trencher. When a backhoe is used, the horizontal loop installation will be a wide trench application, such as for the 1, 2, 4, 5, or 8-pipe laying configurations (refer to Section 5.3). Typical trench widths are 2-3 feet, with depths of 4-6 feet or more. A backhoe may also be used during the headering process for any type of GHEX installation to bring the supply-return piping from the GHEX to the building. The soil removed by a backhoe tends to be lumpy. If large clumps of soil are used for backfill material, air pockets can form around the pipe, which can greatly reduce the heat transfer capability of the loops. Care should be taken to use the granular soil removed from the trench (not the large clumps) to backfill around the piping for the first 6 inches or so. Refer to Section 7.4.3 for trench backfilling recommendations.

7.6.1.3 BULLDOZER A bulldozer will be used where a large amount of soil had been removed from the site for another purpose or if excavation of a large header pit were required. A bulldozer may also be required where a trench will not stay open because of granular, caving soils that exist at the site.

7.6.1.4 EXCAVATOR An excavator can be used in place of a backhoe for wide trench applications. An excavator can also be used in place of a bulldozer for large pit applications.

7.6.2 INSTALLING THE HORIZONTALLY-TRENCHED GHEX PIPING

Regardless of the excavation equipment used for a given horizontally-trenched GHEX installation, the length of pipe, depth of pipe, pipe spacing, trench spacing, or slinky configuration specified in the GHEX design must be adhered to during installation. However, trenches can change direction as necessary to avoid obstacles or be wrapped around the structure or other existing site features, if necessary. Straight trenches and square corners are not mandatory. During horizontally-trenched GHEX installation, use the following guidelines:

- Inspect the trench to ensure that all rocks are removed.
- Inspect the pipe as it is laid out beside the trench for cuts, kinks, or other damage.
- Pressure test the piping in accordance with the IGSHPA Design and Installation Standards.

For a multiple-pipe application, the first layer of pipe should be laid on the bottom of the trench, and then, the trench should be backfilled to the necessary level. After backfilling to the proper level, the soil should be watered in and compacted to ensure that the next layer of pipe is placed with proper separation from the first. This process should be repeated on each additional layer of pipe until all of the pipe has been installed according to the horizontally-trenched GHEX design. Additionally, for a standing multiple-pipe application (chain-trenched application), the fluid returning from the ground source heat pump to the GHEX should flow through the shallowest pipe and return to the heat pump from the deepest pipe. This takes the last run of the fluid supplying the ground source heat pump through the coolest soil in summer and the warmest soil in winter. When there are multiple GHEX loop pipes laying in a wide trench application, or for a slinky application, the connection order and flow direction of supply and return is irrelevant.

If the soil at the site is a heavy clay in a dry climate, it may shrink away from the pipe as the soil dries during the summer cooling cycle. In such cases, use a sand fill around the pipe to maintain contact with the pipe. The sand fill will not shrink away from the pipe when it dries. Use caution when using sand as a backfill material because sand has extremely poor heat transfer characteristics when it is dry. Another option would be to install a drip line whenever this situation exists to help prevent the soil from drying out. However, such practice is to be used as a last resort because most homeowners will not properly operate the drip line system. Figure 7.9 displays how a drip line may be used on a horizontally-trenched GHEX installation.

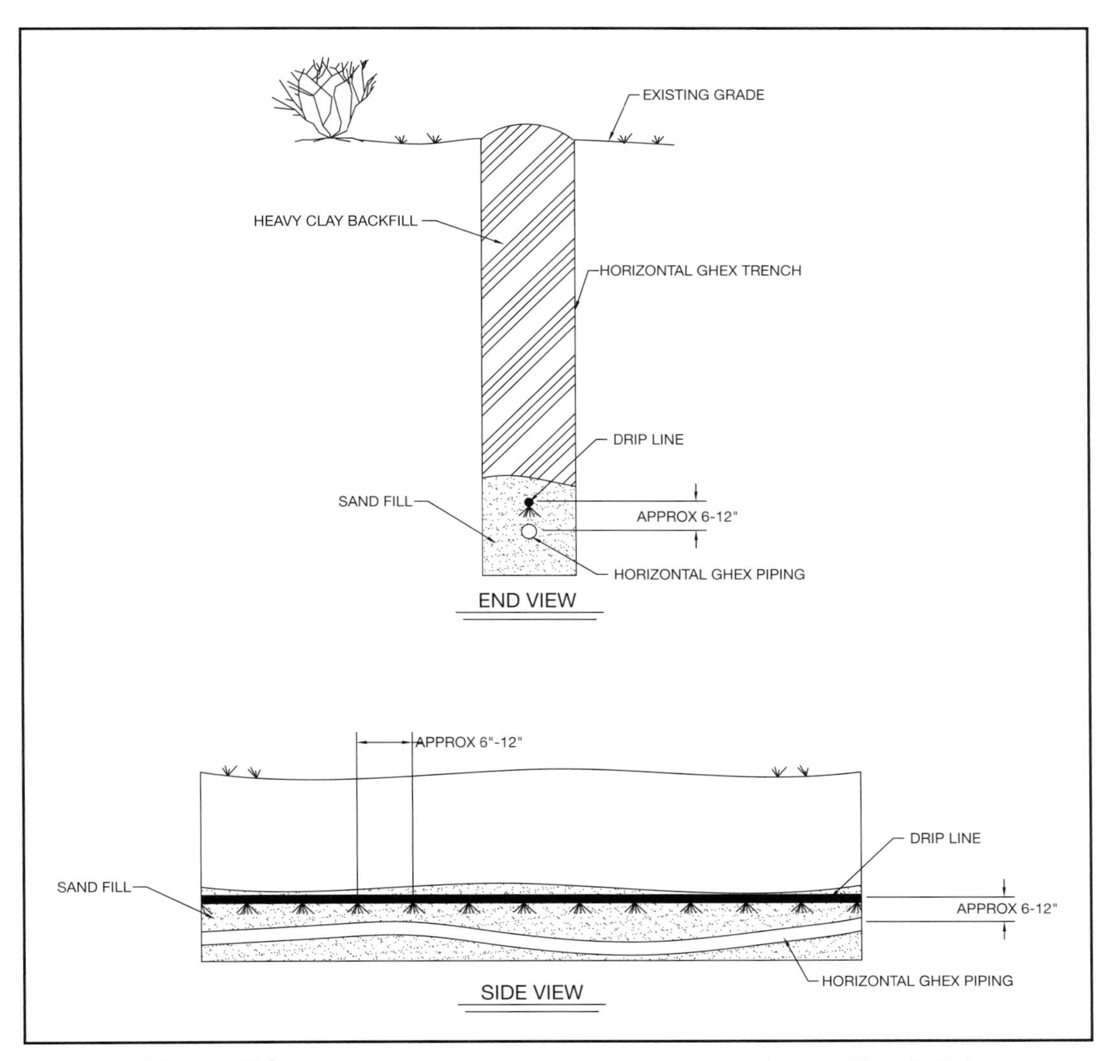

Figure 7.9. Drip Line Details (for Dry Climates & Heavy Clay Soils)

7.6.2.1 PIPE BEND RADIUS AND KINK PREVENTION When a standing multiple-pipe, horizontally-trenched (non-slinky) GHEX layout is installed, the piping will need to be brought back on top of itself when the pipe run reaches the end of the trench. For example, a 2-pipe standing configuration will be installed by placing a pipe run at the bottom of the trench. When the end of the trench is reached, proper backfilling procedures will need to be followed to ensure proper spacing between the two runs, and then the pipe will be brought back on top of itself to complete the piping run in the trench. Care needs to be taken to ensure that the pipe does not kink at the return bend at the end of the trench. Figure 7.10 shows the return bend options for any multi-pipe GHEX configuration. Extreme care should be taken if the large radius with support option is used (Figure 7.10). If this option is selected for the installation, hand backfilling around the bend is mandatory. The large radius with no support option must not be used as the potential to kink the pipe is too great.

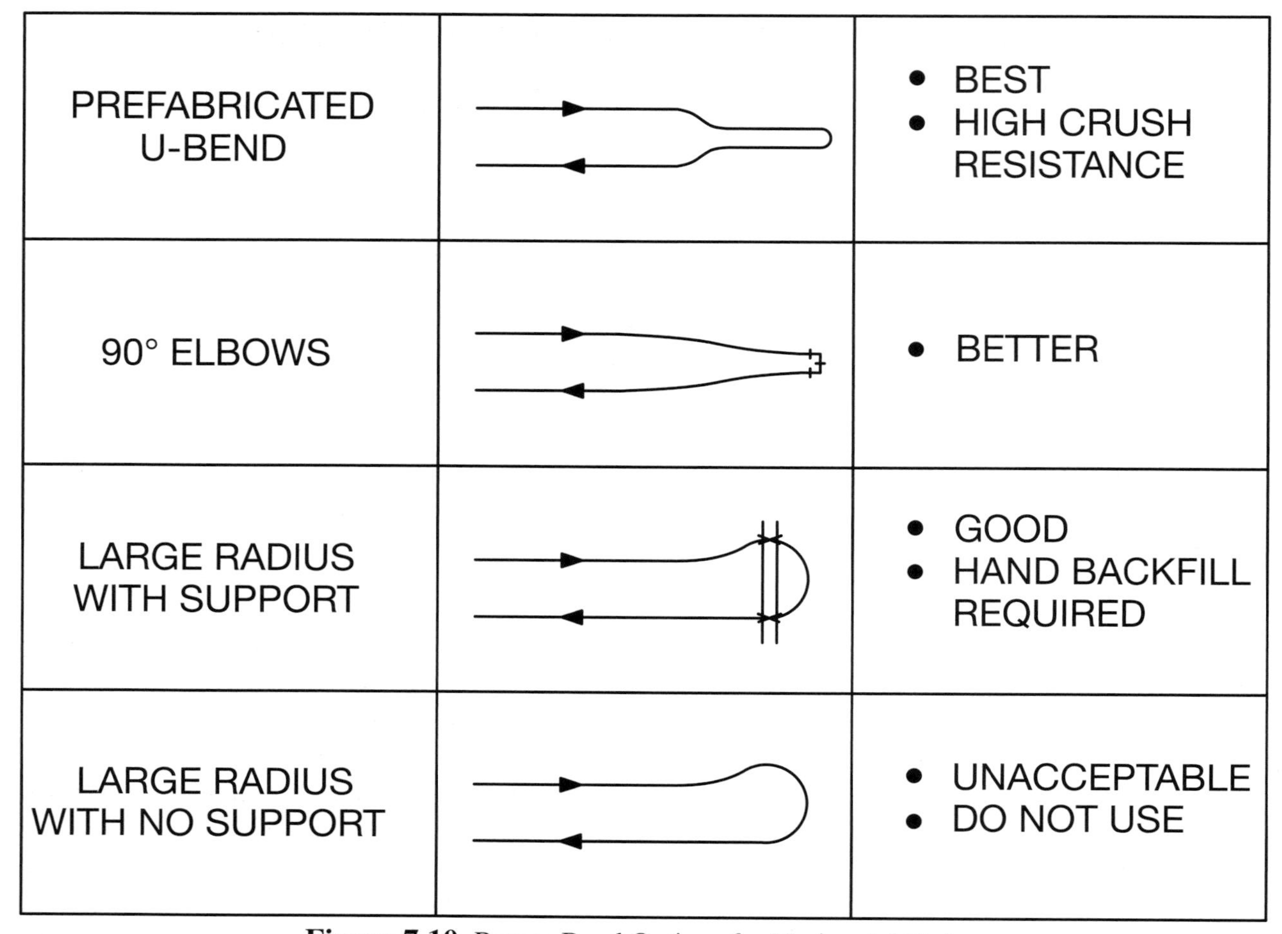

Figure 7.10. Return Bend Options for Horizontal Piping

When bending a pipe using a large radius, always hand backfill around the bend to prevent kinks, and always follow the manufacturer's recommendations for minimum pipe bend radius (regardless of whether the pipe is laying horizontally in the bottom of the trench or standing vertically in the trench). The minimum bend radius for HDPE piping is 25 times the diameter of the pipe.

7.7 Horizontally-Bored, Closed-Loop GHEX Installation

Horizontally-bored GHEX installations are performed using horizontal boring (directional boring, horizontal directional drilling, HDD) machines. Installation of a horizontally-bored U-bend is typically done in three steps:

- Bore the hole from behind the header trench, down through the header trench to the end of the loop run
- Attach the U-bend and tremie line to the hole reamer bit (which is attached to the end of the drill string) at the end of the borehole, and pull everything back to the header trench
- Detach the U-bend and tremie line from the drill string, and pull the tremie back out of the hole to the end of the field, while simultaneously grouting the entire borehole

Typically the HDD machine operator will bore down at an angle to a depth slightly below the depth specified by the GHEX design, level off in the horizontal plain at that depth for the predetermined distance, and then, angle upward to emerge at the surface at the end of the loopfield. At that point, the U-bend and tremie are attached to the end of the drill string. As the drill string is pulled back out of the hole, the U-bend and tremie line are pulled into place. Once the drill string is pulled completely back to the beginning of the hole, the U-bend and tremie line are disconnected in the header trench, and the tremie is pulled back to the end of the hole as it is grouted from beginning to end. The horizontal boring process is displayed in detail in Figure 7.11a. The details referred to in Figure 7.11a are illustrated in Figure 7.11b.

As shown on Figure 7.11a, the maximum U-bend installation depth is not the same as the design depth. The horizontally-bored piping should be installed at a lower depth than the design depth to account for the fact that a portion of the active GHEX piping will be installed above the maximum depth in the angled portion of the borehole. The AVERAGE pipe installation depth should be equal to the design depth.

Hole collapse has been used as a reason for not grouting horizontal boreholes. However, there is no way to guarantee that a horizontal borehole will collapse completely and uniformly around GHEX piping. IGSHPA recommends that all horizontally-bored holes be grouted from end to end to ensure good contact between the earth and the GHEX piping and to protect the integrity of environmental groundwater supply. Refer to Section 7.5.3 and to *IGSHPA (2000) Grouting for Vertical Geothermal Heat Pump Systems* for in-depth discussion of reasons for grouting, grouting products, grout placement methods, etc.

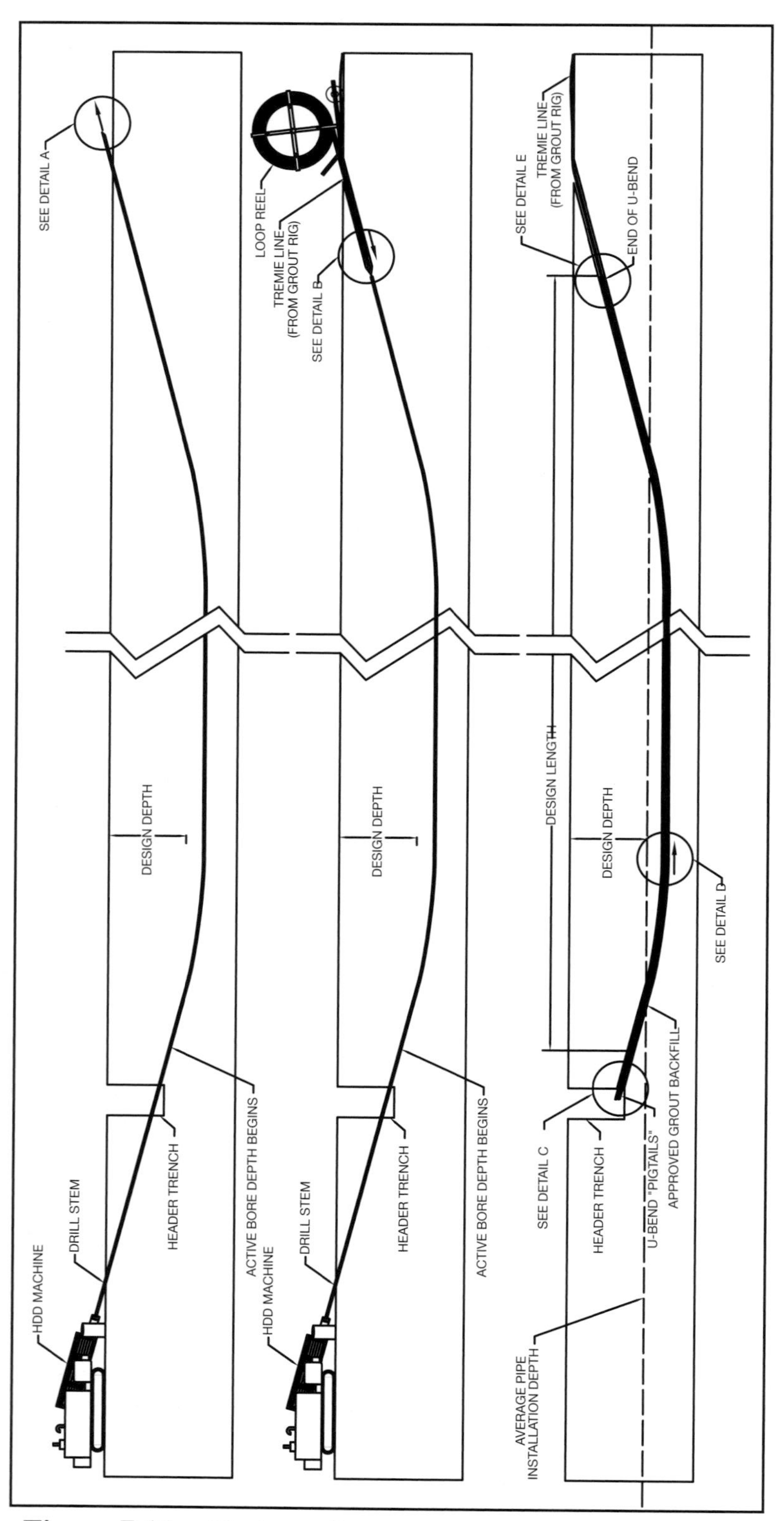

Figure 7.11a. Horizontally-Bored Loop Installation Process

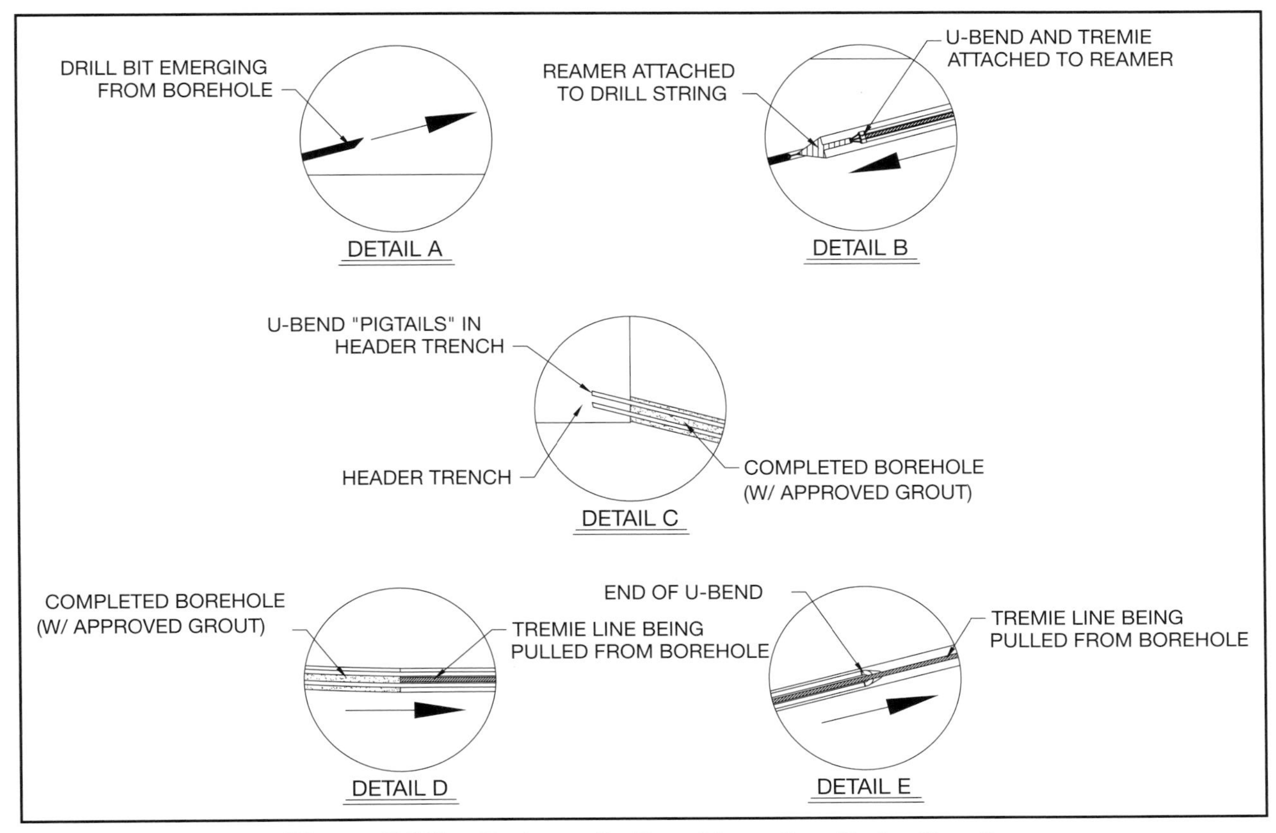

Figure 7.11b. Horizontally-Bored Loop Installation Details

7.8 Installing the Loopfield Header

The majority of the information for header installation has been provided in Section 7.4.1 and Section 7.4.3. Header connection and completion are discussed in the following sections.

7.8.1 HEADER CONNECTION AND COMPLETION Loopfield headers perform only one task: connecting the active legs of the GHEX (U-bends, slinky coils, etc.) to the interior pumping system. The GHEX should be designed such that all the necessary heat transfer capability is in the active legs of the GHEX. Do not account for the heat transfer capability of the supply-return header piping. Refer to Chapter 5 for step-down, step-up reverse-return (SDSURR) layout and pipe sizing guidelines and procedures. A SDSURR layout will facilitate flushing and will promote equal flow through all legs of the GHEX. However, the SDSURR does not need to have the third piping leg if site space allows. Figure 7.12 illustrates a conventional 6-ton (with 3/4-inch U-bends) straight-line SDSURR pattern and the "U" shaped pattern that will add some trenching, but eliminates the labor and material of the third piping leg.

During installation of the header piping, be prepared for what the site has to offer:

- Be prepared to cross utilities at some point of the trenching process. Use extreme care and hand dig around the lines to prevent damage.

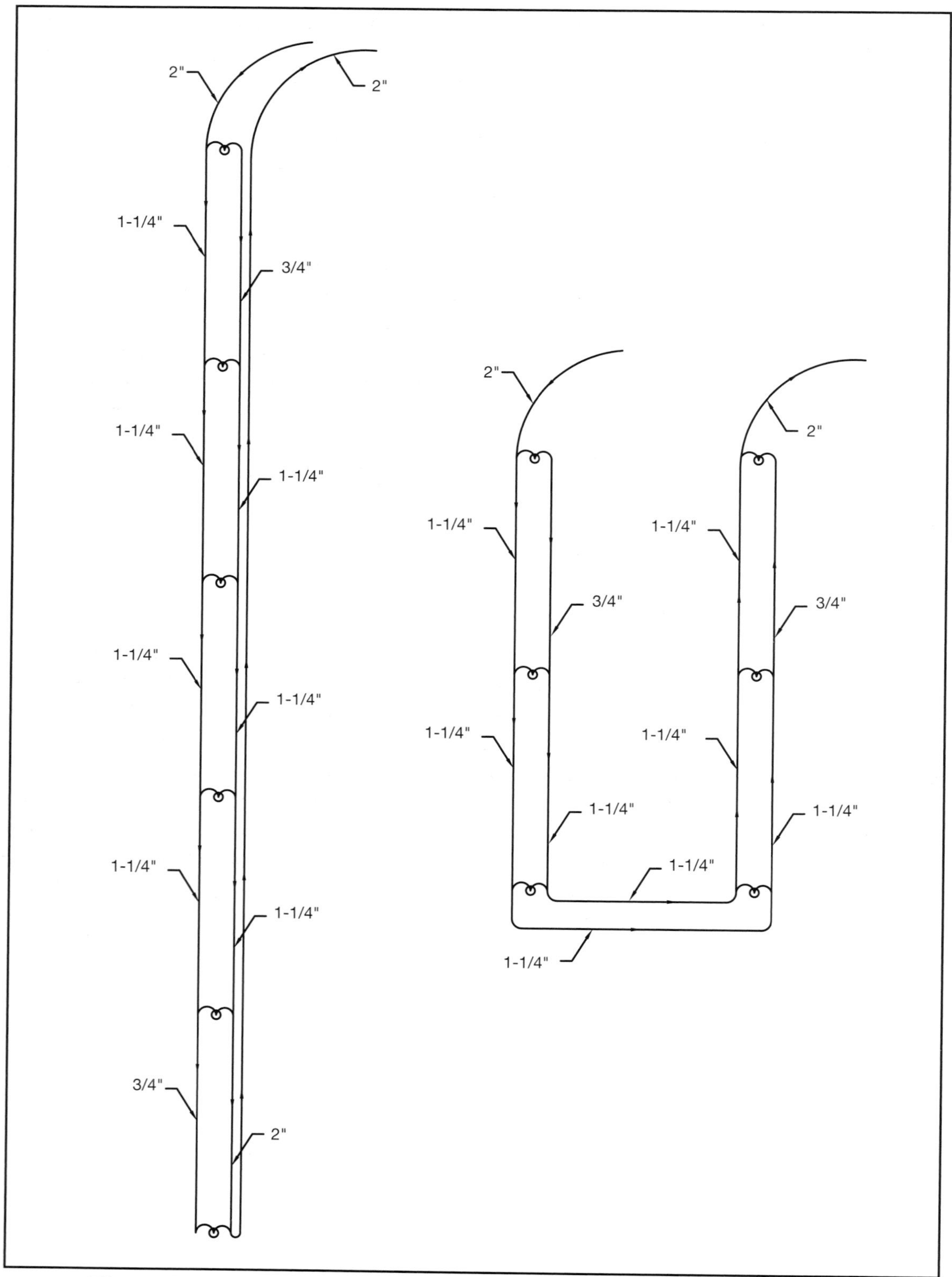

Figure 7.12. Step-Down, Step-Up Reverse-Return Options (6-ton Example)

- Use extreme care if working on a historical site. A contract on such a project may contain provisions to stop work and notify the appropriate authority if something of possible interest is encountered during excavation. Also, historic site improvements may exist that cannot be disturbed.
- Keep the loops under pressure while other trades are working in the same general area. If someone cuts into the loop, the loss of the pressure in the system will make that fact known as soon as the gauges are checked.

As the headers are installed, run the supply-return piping into the building or stop close to it until other work is completed. Seal the pipe ends using duct tape or a barbed-plug fitting (Figure 7.2), and identify matching headers if there are multiple header sets in the trench or at the building entry point.

Building entry is the last step in header construction and installation. Refer to Section 6.4 for building penetration and supply-return piping entry methods. When using PVC chases, always use an appropriately sized long sweep elbow to bring the header piping into the building. Standard sweep elbows or a pair of 45-degree elbows with a short straight length in between are not acceptable methods to use in place of a long sweep elbow. For slab-on-grade structures, instruct the plumber to rough in a chase as he performs the wastewater rough-in.

7.9 As-Built Drawings

As the GHEX is constructed, the site drawing must be revised as necessary to reflect the installation as it was actually installed. Boreholes, horizontal loops, or header trenches may be moved to accommodate some site condition or owner preference that was unknown until installation began. Each borehole and horizontal loop should be located via triangulation off permanent location points or by GPS location coordinates. The outside boundaries of the loopfield and header trench to the structure should be located in the same manner. Finally, note on the drawing if metallic or plastic warning tape was used in the trenches. A thorough, accurate as-built drawing, along with locate wire and warning tape, should protect the loopfield from damage if and when future work is done in the area. Always keep a copy of the as-built drawing for future reference, give a copy to the general contractor, and one to the owner. Figure 7.13 displays the as-built drawing for the site plan example given in Figure 7.1.

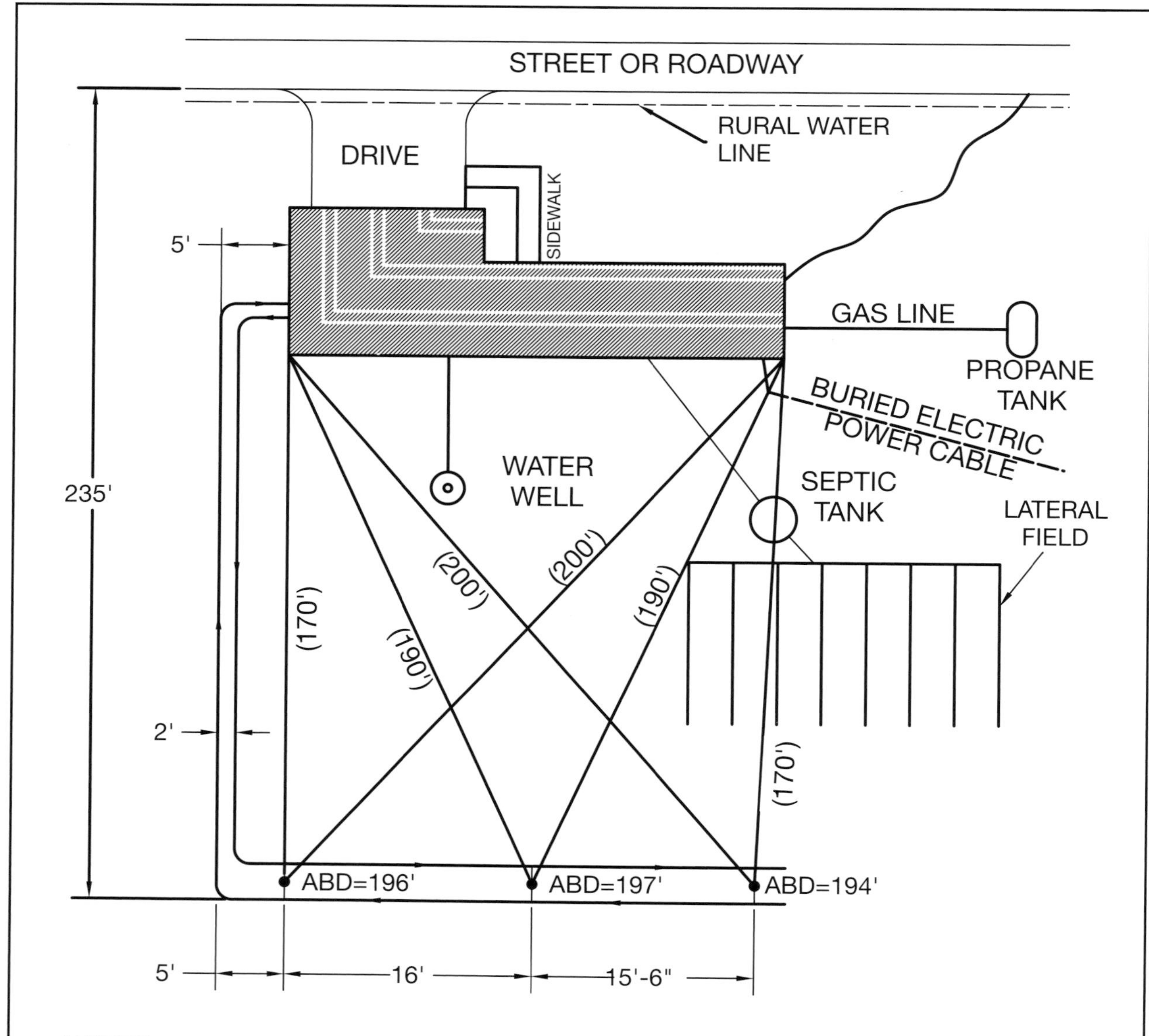

NOTES:

1) U-BEND PIPE DIAMETER = 3/4" & SUPPLY-RETURN PIPE DIAMETER = 1-1/4"
2) ABD = ACTUAL DEPTH OF VERTICAL BORE (MEASURED FROM BOTTOM OF HEADER TRENCH)
3) ALL BORES GROUTED WITH THERMALLY-ENHANCED GROUT, THERMAL CONDUCTIVITY VALUE = 0.88 BTU/HR-FT-F
4) HEADER TRENCH DEPTH APPROX. 6' BELOW GRADE

Figure 7.13. As-Built Drawing Example

7.10 Site Restoration

A contract for a given project should define who is responsible for site restoration and the protection of any surface features (landscaping, structures, etc.) that may exist. Always make sure that responsibility for site restoration is specifically detailed and understood. Site restoration could include, but is not limited to:

1. Restoration of existing grade after work and compaction is completed.
2. Restoration of existing grade with mulch, fertilizer, and grass seed. Clearly define what type of grass seed is to be used, who is responsible for watering the grass seed and for how long. When watering is done by the GHEX contractor, this duration is typically until the grass sprouts. After that point, future care reverts back to the client or building owner.
3. Restoration of existing grade with sod. Clearly define what type of sod is to be used.

Always ensure that the client or building owner fully understands how the job site will be left once work has been completed.

SYSTEM FLUSHING & PURGING

In This Section

8.1 System Pre-Startup

During construction of the GHEX, excavation, pipe assembly, and fusion work create the greatest opportunity for dirt and construction residue to enter the system. The GHEX must be flushed and purged in a manner that will remove any dirt, construction residue, or other contaminants, while avoiding contact with the pump impellers, the GSHP's water-side piping and water-to-refrigerant heat exchanger, or any other system interior piping components. Trash, metal or plastic pipe shavings, joint sealing materials, dust, dirt, etc., can find their way into the interior piping system and also must be removed. It is critical to remove small particles of dirt, sand, etc., in residential systems because these small particulates can damage the bearing surfaces of water-lubricated circulators commonly used for residential-class pumping systems or pumping modules. Before starting the completed system, the water side (GHEX, inside piping, pumping, and GSHPs) should be filled with potable water and flushed, purged, and pressurized as follows:

- Flush debris, contaminants, dirt, etc., from the GHEX
- Purge air from the GHEX
- Verify GHEX performance (pressure/flow)
- Check for possible blockage or restriction, if design verification does not reasonably agree with theoretical values
- Flush debris, contaminants, dirt, etc., from inside piping, pumping, and GSHP unit(s)
- Purge air from inside piping, pumping, and GSHP unit(s)
- Position or open valves to combine the GHEX and inside piping, pumping, and GSHP unit(s) into one continuous flushing operation
- Charge the system with antifreeze (if used)
- Pressurize the water side (if a pressurized flow center is used)

As a precautionary measure, pipe ends should be sealed as described in Chapter 7 immediately after installation to keep system contamination to a minimum. After pressure testing, individual loops or supply-return piping circuits should be sealed to ensure nothing inadvertently enters the pipe before being attached to the headers.

8.2 Flushing & Purging Basics

Before commissioning the GSHP system, the water side must be flushed with water to remove any dirt, trash, pipe shavings, or other debris after fabrication is completed. A properly designed GSHP system will allow for the interior piping and exterior GHEX piping to be isolated from one another (via valves) so the GHEX piping can first be flushed, followed by the interior piping and components. Only then can the combined outside and inside piping and system components be flushed in one continuous operation. Care must be taken to remove all debris and contaminants from the system to prevent blockage or restrictions in pump impellers and other system components and to prevent possible damage to water-side system components over time.

In addition to flushing debris from the system, it is also necessary to purge the system of any air trapped in the piping or components at the same time. Failure to properly purge all air from the system can result in:

- Corrosion of the system's metallic components, leading to eventual failure
- Air in pump volutes/impellers, causing cavitation and potentially interrupting system fluid flow
- Air trapped in one of several loops, possibly restricting or blocking flow in that loop

A minimum purging flow rate of 2 feet per second (FPS) throughout a piping system is required to completely remove any trapped air. This velocity must be maintained in each branch during air purging. Refer to Section 4.6, Tables 4.25-4.28 for flushing flow rates necessary to achieve 2 FPS for various types/sizes of piping commonly used in GSHP installations. Also, refer to Equation 4.9 to calculate flushing flow rate for any type/size of piping. The flushing flow rate for a supply-return (S-R) circuit will be determined by the number of loops on that S-R piping pair. Equation 8.1 should be used to calculate the flushing flow rate for a S-R circuit.

$$Q_{GLflush} = Q_i \cdot n$$ **Equation 8.1a**

-or-

$$Q_{GLflush} = Q_{S-R}$$ **Equation 8.1b**

Where:

$Q_{GLflush}$ = Flushing flow rate required for a ground-loop S-R circuit (gpm)
Q_i = Flushing flow rate required for an individual flow path in the GHEX (gpm)
n = Total number of flow paths in S-R piping circuit
Q_{S-R} = Minimum flushing flow rate required for 2 FPS in the S-R piping, determined from Equation 4.8 or Table 4.25 (gpm)

Use the largest value produced by Equation 8.1a and 8.1b. Examples 8.1 and 8.2 demonstrate how to use the equations.

Example 8.1. Calculate the flushing flow rate for an S-R circuit which has a single pair of 1-1/4-inch supply-return piping connected to two, ¾-inch U-bends:

Q_{Flush} = (3.6 gpm/U-bend for ¾-inch DR-11 HDPE)•(2 U-bends)=7.2 gpm
(from Equation 8.1a)

-or-

Q_{Flush} = 9.0 gpm for the 1-1/4-inch DR-11 HDPE supply-return piping
(from Equation 8.1b)

As shown in Example 8.1, the 1-1/4-inch DR-11 HDPE supply-return piping requires more flow for proper flushing than the two, parallel ¾-inch U-bends. The flushing flow rate for this example is 9.0 gpm, which will be evenly distributed through both U-bends (4.5 gpm each > 3.6 gpm required flushing flow rate for each ¾-inch DR-11 HDPE U-bend).

Example 8.2. Calculate the flushing flow rate for an S-R circuit, which has a single pair of 1-1/4-inch supply-return piping connected to four, ¾-inch U-bends:

Q_{Flush} = (3.6 gpm/U-bend for ¾-inch DR-11 HDPE)•(4 U-bends)=14.4 gpm
(from Equation 8.1a)

-or-

Q_{Flush} = 9.0 gpm for the 1-1/4-inch DR-11 HDPE supply-return piping
(from Equation 8.1b)

As shown in Example 8.2, the four, parallel ¾-inch DR-11 HDPE U-bends require more flow for proper flushing than the single 1-1/4-inch supply-return piping pair. The flushing flow rate for this example is 14.4 gpm, which will be evenly distributed through both U-bends (3.6 gpm each x 4 U-bends = 14.4 gpm > 9.0 gpm required flushing flow rate for 1-1/4-inch DR-11 HDPE).

A loopfield should be headered according to the guidelines given in Section 5.1.3. Doing so will ensure that the circulating fluid in the GHEX will flow equally through each loop and that the piping system can be easily flushed and purged of air and debris using appropriately sized flushing equipment. Figure 8.1 displays a step-down, step-up, reverse-return (SDSURR) layout for a typical 6-ton system (assuming 3 gpm per loop, 1 loop per ton). Table 8.1 gives the pipe size and operating/flushing flow rate details for the system displayed in Figure 8.1. Figure 8.1 and Table 8.1 were generated according to the guidelines given in Section 5.1.3 and specifically Figure 5.9(e).

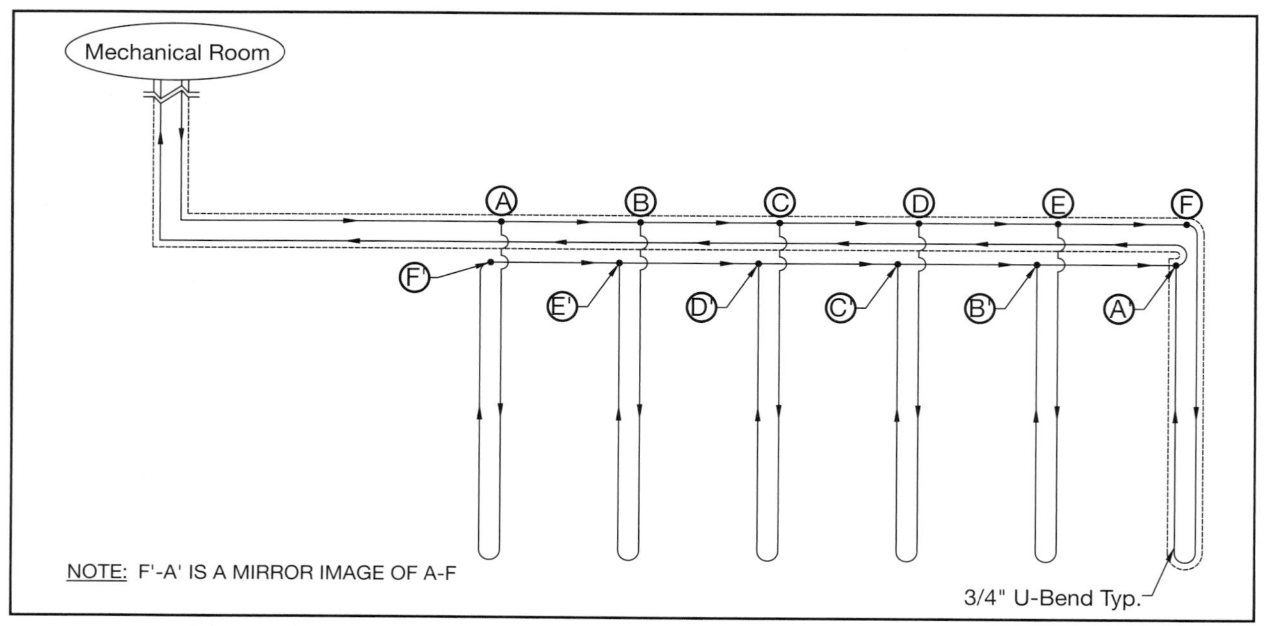

Figure 8.1. Step-Down, Step-Up Reverse-Return Layout for 6-ton System

Table 8.1. Step-Down, Step-Up Reverse-Return Header Details for 6-ton System

Pipe Section	Nom. Size (in)	Operating Flow Rate (gpm)	Flushing Flow Rate (gpm)
Mech-A	2	18.0	21.6
A-B	1-1/4	15.0	18.0
B-C	1-1/4	12.0	14.4
C-D	1-1/4	9.0	10.8
D-E	1-1/4	6.0	7.2
E-F	3/4	3.0	3.6

The ground-loop flushing flow rate will be determined by Equation 8.1, unless the system is headered similar to the method used in Figure 6.3d. If the inside of building header method displayed in Figure 6.3d is used, then the loopfield flushing flow rate will be the flow rate required to cause 2 FPS flow velocity in the individual loops (3.6 gpm for ¾-inch U-bends, 5.7 gpm for 1-inch U-bends, etc) as each loop will be manifolded such that it can be flushed individually. For the 6-ton example shown, each ¾-inch U-bend will require 3.6 gpm for flushing. Because six U-bends are connected to the single S-R piping pair, the required flushing flow rate will be 21.6 gpm (using Equation 8.1: Q_i=3.6 gpm/U-bend, n=6; Q_{Flush}=3.6 x 6 = 21.6 gpm), which is greater than the 18.5 gpm required for the 2-inch S-R piping.

The required minimum purging flow rate of 2 FPS fluid flow is only for air. This minimum requirement for the removal of air from a closed-loop heat exchanger will not guarantee the removal of solid, dense contaminants, such as coarse sand or gravel. Experience has shown that these materials can find their way to the circulation pumps and/or the GSHP water-side piping and water-to-refrigerant heat exchanger. The result of such contamination can be a failed circulation pump, blocked GSHP water-side piping, and/or blocked heat exchanger resulting in either low system fluid flow or no flow at all. Checks for solid contamination in a closed-loop system include, but are not limited to:

1. Check for noise, such as rattling or grinding, in the circulation pump volute or exposed piping. Using a long screwdriver as a stethoscope, place the flat point end against the pump volute or piping and the handle end against your ear. This check will easily identify solids moving in the volute or piping.
2. Visually inspect the pumps by removing the pump head from the volute and search for solid contaminants. Be sure to check for contaminants wedged within the impeller orifices or at orifice outlet openings.
3. Take note of an overheated pump. This can be caused by large contaminants that have locked the impeller or water lubricated bearings damaged by small particulate contamination. Such issues could cause a broken impeller shaft. USE EXTREME CARE under these conditions. Always check the motor and volute for warmth before removing the pump head. In some cases, small circulators can continue running without flow causing the fluid inside the circulator to get hot enough to flash to steam when the volute is opened.
4. Take note of an excessive pressure differential and temperature differential across the heat pump water-to-refrigerant heat exchanger (as compared to manufacturer's data). A large pressure and temperature differential across the coil will indicate that a low flow situation exists. If the circulation pump is in good working order, the low flow situation will likely be caused by a complete or partial blockage due to contaminants in the system. Having a record of the original start-up performance test, where pressure differential and temperature differential were recorded, is a good benchmark for comparison of pressure drop and temperature rise or drop. Heat pumps with brazed-plate, water-to-refrigerant heat exchangers are particularly sensitive to blockage due to their small orifices for fluid circulation within the heat exchanger.

8.2.1 HEAD LOSS CALCULATIONS & PUMP SIZING FOR FLUSHING/PURGING Head loss during flushing should be determined according to the procedures given in Section 6.6.2 for the minimum required flushing flow rate for each supply-return (S-R) circuit using water as the circulating fluid. The head loss calculations should then be used to size the pumping system used for flushing/purging. A purge pump will typically need to exhibit high-volume, high-head characteristics in order to have sufficient capacity to flush multiple loops simultaneously as they are connected in a single S-R parallel piping circuit. Table 8.2 gives an example of flushing flow rate head loss calculations for the 6-ton example displayed in Figure 8.1.

Table 8.2. Flushing Flow Head Loss Calculations for 6-ton System

Head Loss for Water at 40˚F						
Pipe Run	Flushing Flow Rate (gpm)	Pipe	Length (ft)	Nom Size (in)	hl/100'	hl (ft)
Mech-A	21.6	HDPE DR-11	50	2	1.24	0.62
A-B	18.0	HDPE DR-11	15	1-1/4	4.98	0.75
B-C	14.4	HDPE DR-11	15	1-1/4	3.34	0.50
C-D	10.8	HDPE DR-11	15	1-1/4	1.99	0.30
D-E	7.2	HDPE DR-11	15	1-1/4	0.96	0.14
E-F	3.6	HDPE DR-11	15	3/4	2.83	0.42
F-A'	3.6	HDPE DR-11	410	3/4	2.83	11.60
A'-Mech	21.6	HDPE DR-11	125	2	1.24	1.55
					Total	**15.89**

As shown in Table 8.2, a single flow path was selected for head loss calculations according to the procedures given in Section 6.6.2 (from the mechanical room to point A, through the U-bend to point F', and through the reverse-return header back to the mechanical room, refer to Figure 8.1). A single flow path is sufficient for the 6-ton loopfield head loss calculations because all of the loops on the S-R circuit are connected in parallel. According to the table, to flush the 6-ton GHEX in Figure 8.1, the purge pump will need to be capable of producing 21.6 gpm at 16 feet of head (the calculations exclude the minor effect of fittings in the piping system, such as elbows and tees). Similar calculations will need to be performed for the interior piping system and the interior and exterior piping systems combined (as flushed in one continuous operation). The values in the hl/100 feet column in Table 8.2 can be taken directly from Figure 4.3, Table 4.15, or can be calculated according to the procedures given in Section 4.4.3.

8.2.2 PURGE PUMP SIZING & FLUSH CART CONFIGURATIONS Flush carts fabricated for residential or light commercial use will typically utilize high-head, high-volume purge pumps from 1-1/2 hp to 3 hp in size. In general, recommended purge pump sizes for a given loopfield circuit size are as follows:

- 1-1/2-hp purge pump – up to 6-ton S-R piping circuits
- 2-hp purge pump – up to10-ton S-R piping circuits

A residential or light-commercial class purge pump should be capable of flushing a large loopfield assuming it is manifolded properly. For example, if a 100-ton loopfield utilizing ¾-inch U-bends were manifolded with 10 U-bends on each S-R circuit, then each 10-ton circuit could be isolated from the rest of the field (via valving) for flushing and a 2-hp purge pump would be sufficient. A pump curve for a typical 2-hp purge pump is displayed in Figure 8.2.

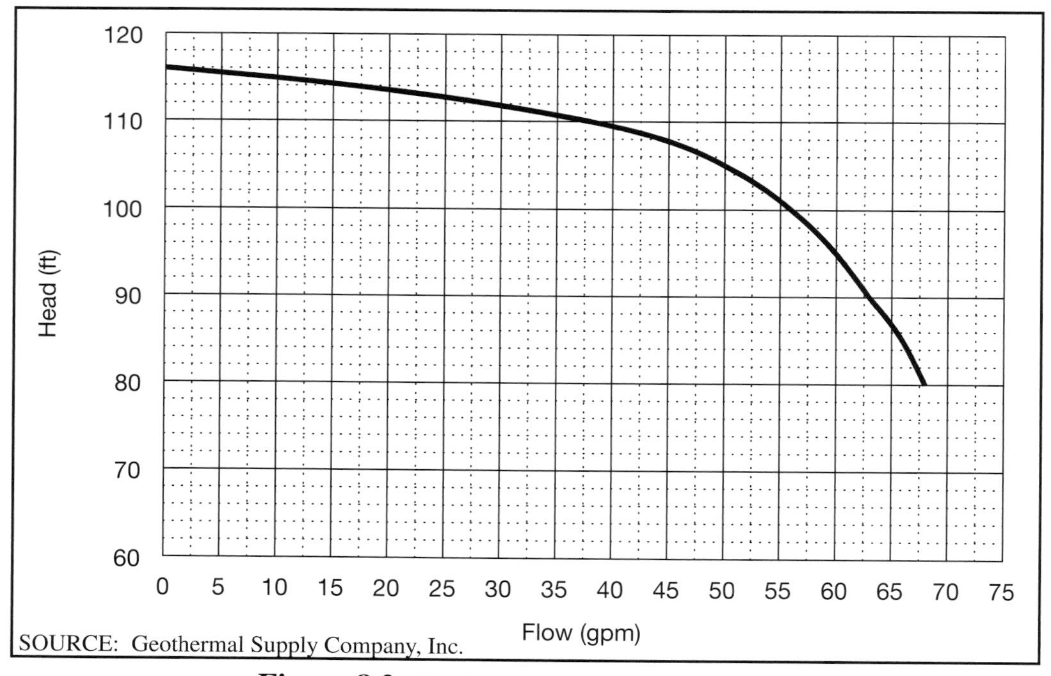

Figure 8.2. Typical 2-hp Purge Pump Curve

Several manufacturers and geothermal suppliers fabricate flush carts for purchase. Most commonly available flush carts integrate the purge pump with the valving, hose connections, electrical connections, filtration, and reservoir tank on a hand cart for maximum portability and ease of use during operation. A typical flush cart configuration is displayed in Figure 8.3.

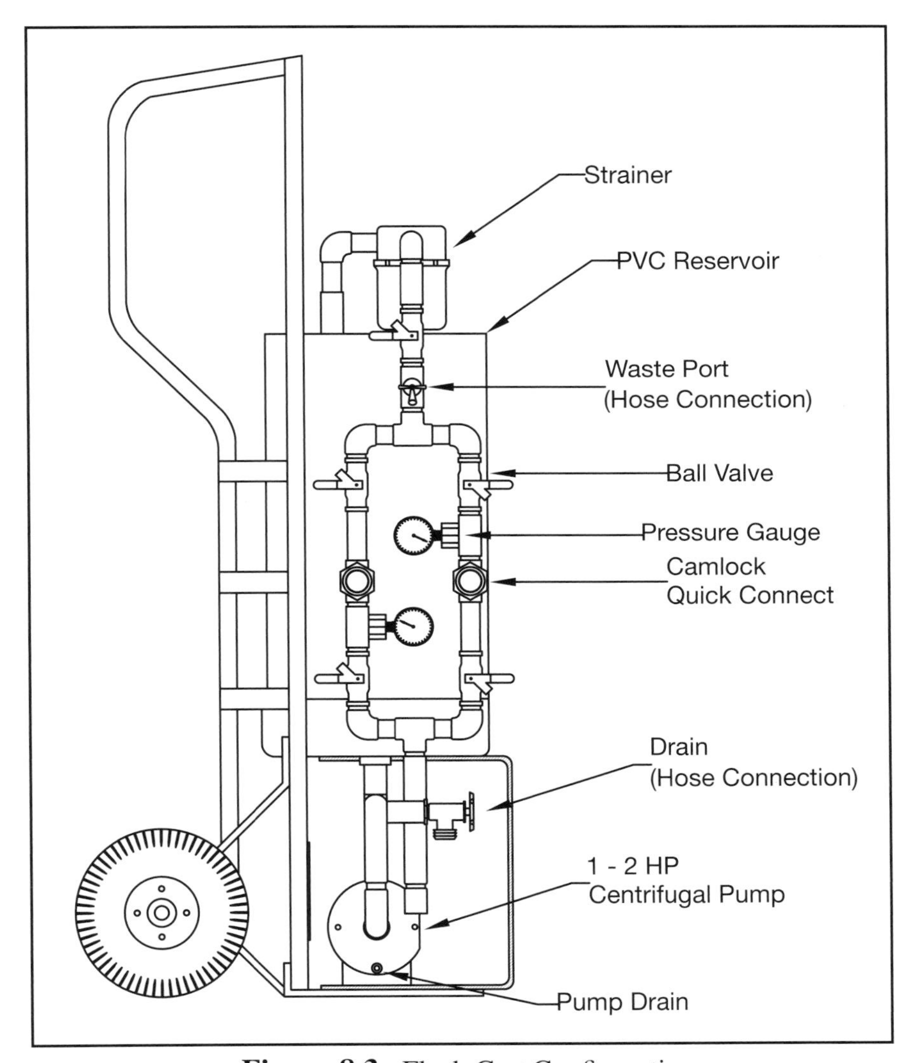

Figure 8.3. Flush Cart Configuration

The flush cart shown in Figure 8.3 has ball valves to allow for the fluid flow direction to be reversed during the flushing process (assuming there are no system components that prevent reversing flow, i.e. – backflow prevention, flowmeters, etc.). Such configurations are optional, but useful in the event that difficulties are encountered during the purging process. Reversing the flow direction may easily remove air trapped in the system that would take much longer if only one flow direction were used.

8.3 Verifying GHEX Pressure/Flow Design

Proper pressure/flow performance of the GHEX must be verified by testing and observation at the time of construction. As discussed in Chapter 7, even the most minor details must be addressed, inspected, verified, and documented. Verification of properly constructed saddle fusion joints and prevention of kinked lines are issues that require extreme attention. It is recommended that loopfield performance checking be done before trenches are backfilled. Otherwise, installation errors may necessitate that the GHEX be dug up later because of unsatisfactory system performance.

8.3.1 LOOPFIELD INSTALLATION ERROR DETECTION TECHNIQUES To test for blockages or restrictions in a GHEX circuit, use pressure gauges with the S-R circuit P/T ports. The pressure gauges should be used to monitor the pressure drop due to fluid flow through each S-R circuit in a system. A higher than expected pressure drop (as compared to theoretical head loss calculations, refer to Section 4.4.3 and Section 6.6.2) at a given flow will indicate that a problem might exist. Always keep in mind that there will be some error associated with both theoretical calculations and field measurements, and small differences between the calculations and measurements will usually be attributed to error and system variability.

Testing for blockages or restrictions is relatively easy for a small GHEX. A larger GHEX with multiple loops will be more difficult to diagnose. For example, if a 3-loop GHEX designed to flow at 3 gpm per loop has one loop blocked, the design flow rate through the remaining two loops will change to 4.5 gpm each. The resulting increase in flow through each loop will lead to an easily detectable increase in system pressure drop (from 3.6 psi at 3 gpm to 6.6 psi at 4.5 gpm for water at 40 F, assumed 200-ft x ¾-inch U-bends). However, if a 10-loop GHEX designed to flow at 3 gpm per loop has one loop blocked, the design flow rate through the remaining 9 loops will increase from 3 gpm to 3.33 gpm each (from 3.6 psi at 3 gpm to 4.4 psi at 3.33 gpm for water at 40 F, assumed 200-ft x ¾-inch U-bends). This minimal increase in flow in the individual loops will be harder to identify via the pressure drop test and will more than likely be attributed to inherent variability in the system.

Increasing the circuit flow rate during pressure/flow verification will increase the likelihood of detecting a problem. To do so, a purge pump will need to be used, rather than a system flow center, during the verification process. If the flow rate through the 10-loop system were increased to 4.5 gpm per loop with one loop blocked, flow rate in the remaining 9 loops will increase from 4.5 gpm to 5.0 gpm each (from 6.6 psi at 4.5 gpm to 8.0 psi at 5.0 gpm for water at 40 F, assumed 200-ft x ¾-inch U-bends).

8.3.2 LOOPFIELD INSTALLATION ERROR LOCATION TECHNIQUES Once detected, location of blocked or kinked lines is a fairly straightforward process. If a blocked or restricted loop has been detected, first ensure that all air has been properly removed from the system (refer to Section 8.2). If the system has been properly purged, a pinch-off tool should then be used to determine exactly which line is the problematic one. With fluid circulating through the lines, use the pinch-off tool to restrict each line in the circuit containing the blocked/restricted line. While doing so, note the pressure drop change in the S-R circuit associated with pinching off each line one by one. When the blocked line is squeezed off with the pinch-off tool, no pressure drop change will be detected in the circuit because a no-flow condition already exists in the loop. A pinch-off tool is shown in detail in Figure 8.4.

Fully examine the problematic loop for places where the installation errors could have occurred. Blocked lines due to failure to drill out saddle fittings can only occur at sidewall fusion joints. Completely or partially blocked lines due to kinks can occur in various places. Search for kinked lines at:

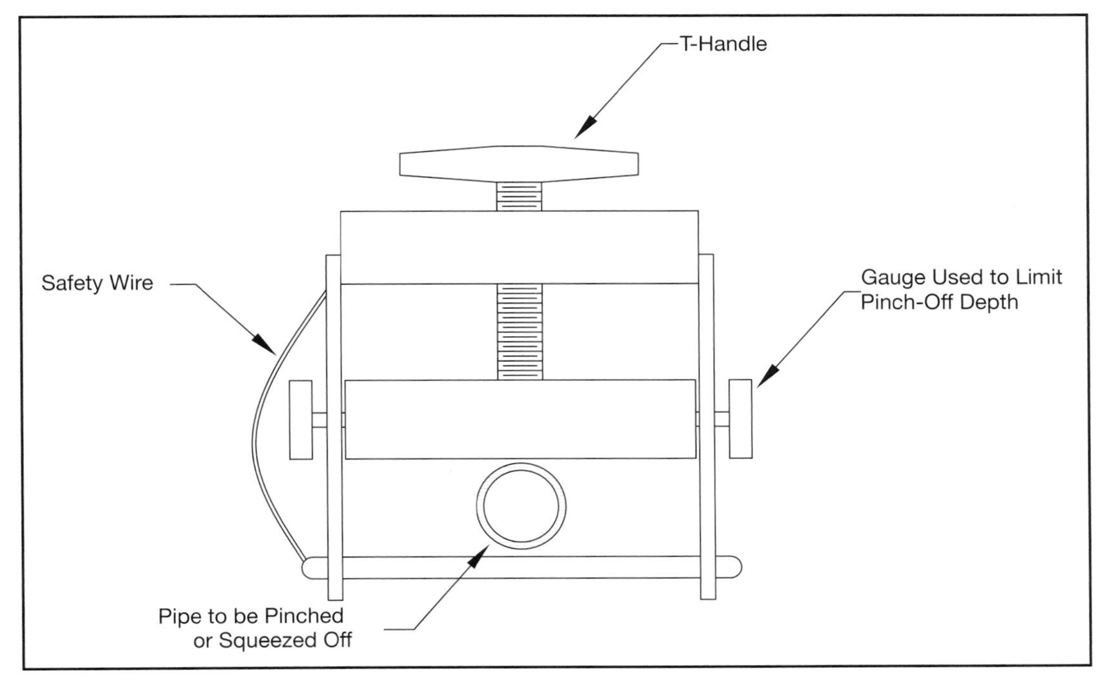

Figure 8.4. Pinch-Off Tool

- Locations where vertical GHEX lines have been bent over to tie into the headers.
- Locations where a horizontally-trenched loop has been bent back on itself to transition from the deepest piping path to a shallower piping path (for example, a 2-pipe standing configuration, refer to Section 5.3, Table 5.18).
- Locations where a vertical U-bend was kinked in a formation void during insertion into the borehole. This issue will be difficult to diagnose, but if found to be the case, the U-bend will need to be replaced.
- Locations where trenches intersect or turn and pipe manufacturer's bend radius recommendations are not followed.
- Locations where trenches intersect or turn and piping is pulled against the sidewall of the intersection causing a kink.

Kinked lines can usually be avoided by using either two, 90° elbows or a close-return U-bend fitting in place of a large-radius turn in a horizontally trenched GHEX or reverse-return header piping application. If kinked lines are discovered, do not attempt to straighten the line and leave it in service. Cut the kinked section out and replace that section with new, undamaged piping. After the cause of the blockage or restriction has been found and corrected, the entire flushing/purging process will need to be repeated. This is necessary to flush any dirt, pipe shavings, or other contaminants that may have entered the pipe during repair work as well as to remove any air that has been allowed to enter the system.

Blocked or restricted lines in a GHEX installation will generally be very difficult to detect. Loopfield installers must take extreme care during the installation process to ensure that no such issues occur. Review Chapter 7 for assembly and installation guidelines that will help prevent line blockages or restrictions.

8.4 Charging the System with an Antifreeze Solution

After the exterior GHEX piping and interior piping system have been flushed, purged, and pressure/flow verified, antifreeze can be added to the system (if used). Refer to system design documentation to ensure that the proper antifreeze type and concentration is used. Antifreeze solution freeze protection levels are calculated on a "by volume" basis. For example, to freeze protect to 18 F with a propylene glycol solution, 20% propylene glycol by volume will need to be added to the system (add 20 gallons of propylene glycol per 100 gallons of total pipe volume). Refer to Section 4.4.2 for freeze protection profiles for various types of antifreeze in addition to freeze protection level guidelines.

8.4.1 PIPE VOLUME CALCULATIONS To determine how much antifreeze is necessary for proper freeze protection levels, total system pipe volume calculations must first be performed. Tables 8.3-8.6 provide pipe volume values for various types/sizes of pipe commonly used in GSHP installations (gallons/100 feet of pipe length).

Table 8.3. Pipe Volume for HDPE

Dimension Ratio (DR)	Nominal Pipe Size (in)	ID (in)	OD (in)	Vol/100' (gal)
7	3/4	0.750	1.050	2.29
	1	0.939	1.315	3.60
	1-1/4	1.186	1.660	5.74
	1-1/2	1.357	1.900	7.51
	2	1.696	2.375	11.74
	3	2.500	3.500	25.50
9	3/4	0.817	1.050	2.72
	1	1.023	1.315	4.27
	1-1/4	1.291	1.660	6.80
	1-1/2	1.478	1.900	8.91
	2	1.847	2.375	13.92
	3	2.722	3.500	30.23
11	3/4	0.859	1.050	3.01
	1	1.076	1.315	4.72
	1-1/4	1.358	1.660	7.53
	1-1/2	1.555	1.900	9.86
	2	1.943	2.375	15.41
	3	2.864	3.500	33.46
13.5	1-1/2	1.619	1.900	10.69
	2	2.023	2.375	16.70
	3	2.981	3.500	36.27
	4	3.833	4.500	59.95
	6	5.644	6.625	129.94
15.5	1-1/2	1.655	1.900	11.17
	2	2.069	2.375	17.46
	3	3.048	3.500	37.91
	4	3.919	4.500	62.67
	6	5.770	6.625	135.84
17	1-1/2	1.676	1.900	11.47
	2	2.096	2.375	17.92
	3	3.088	3.500	38.91
	4	3.971	4.500	64.32
	6	5.846	6.625	139.42

Table 8.4. Pipe Volume for Schedule 40 & Schedule 80 (Steel & PVC)

Nominal Pipe Size (in)	SCH	ID (in)	OD (in)	Vol/100' (gal)
3/4	40	0.824	1.050	2.77
	80	0.742	1.050	2.25
1	40	1.049	1.315	4.49
	80	0.957	1.315	3.74
1-1/4	40	1.380	1.660	7.77
	80	1.278	1.660	6.66
1-1/2	40	1.610	1.900	10.58
	80	1.500	1.900	9.18
2	40	2.067	2.375	17.43
	80	1.939	2.375	15.34
2-1/2	40	2.469	2.875	24.87
	80	2.323	2.875	22.02
3	40	3.068	3.500	38.40
	80	2.900	3.500	34.31
4	40	4.026	4.500	66.13
	80	3.826	4.500	59.72
5	40	5.017	5.563	102.69
	80	4.767	5.563	92.71
6	40	6.065	6.625	150.08
	80	5.761	6.625	135.41
8	40	7.981	8.625	259.88
	80	7.625	8.625	237.21
10	40	10.020	10.750	409.63
	80	9.564	10.750	373.20

Table 8.5. Pipe Volume for Copper Tubing (Schedule K, L, M)

Nominal Pipe Size (in)	Type	ID (in)	OD (in)	Vol/100' (gal)
3/4	K	0.745	0.875	2.26
	L	0.785	0.875	2.51
	M	0.811	0.875	2.68
1	K	0.995	1.125	4.04
	L	1.025	1.125	4.29
	M	1.055	1.125	4.54
1-1/4	K	1.245	1.375	6.32
	L	1.265	1.375	6.53
	M	1.291	1.375	6.80
1-1/2	K	1.481	1.625	8.95
	L	1.505	1.625	9.24
	M	1.527	1.625	9.51
2	K	1.959	2.125	15.66
	L	1.985	2.125	16.08
	M	2.009	2.125	16.47
2-1/2	K	2.435	2.625	24.19
	L	2.465	2.625	24.79
	M	2.495	2.625	25.40
3	K	2.907	3.125	34.48
	L	2.945	3.125	35.39
	M	2.981	3.125	36.26
3-1/2	K	3.385	3.625	46.75
	L	3.425	3.625	47.86
	M	3.459	3.625	48.82
4	K	3.857	4.125	60.70
	L	3.905	4.125	62.22
	M	3.935	4.125	63.18
5	K	4.805	5.125	94.20
	L	4.875	5.125	96.96
	M	4.907	5.125	98.24
6	K	5.741	6.125	134.47
	L	5.845	6.125	139.39
	M	5.881	6.125	141.11

Table 8.6. Pipe Volume for Reinforced Rubber Hose

Nominal Pipe Size (in)	ID (in)	OD (in)	Vol/100' (gal)
1	1.00	1.44	4.08
1-1/4	1.25	1.73	6.37
1-1/2	1.50	1.98	9.18
2	2.00	2.50	16.32

With the values in the Vol/100' column in Tables 8.3-8.6, use the installed pipe lengths and Equation 8.2 to calculate the amount of antifreeze necessary for proper freeze protection.

$$V_{AF} = \sum_{i=1}^{n} V_i \bullet \frac{L_i}{100} \bullet \frac{C}{100}$$

Equation 8.2

Where:

V_{AF} = Required antifreeze volume (gal)
n = Number of pipe types used in a system
V_i = Volume per 100 feet of length of individual pipe type in the system (gal)
L_i = Length of individual pipe type in the system (ft)
C = Antifreeze concentration (%)

Example 8.3. Use Tables 8.3-8.6 and Equation 8.2 to calculate the antifreeze volume necessary for proper freeze protection for a system with 3 x 410 foot uni-coil ¾-inch DR-11 HDPE U-bends (200 feet bores); 50 feet of 1-1/4-inch DR-11 HDPE supply-return piping; 30 feet of 1-1/4-inch Schedule 40 PVC interior piping; and 15 feet of 1-inch 200-psi rubber heater hose. Assume that freeze protection to 18 F with 20% propylene glycol by volume is to be used.

For this example:

- n = 4 (4 types of pipe in the system, #1 = ¾-inch DR-11 HDPE, #2 = 1-1/4-inch DR-11 HDPE, #3 = 1-1/4-inch SCH 40 PVC, and #4 = 1-inch Rubber Heater Hose)
  - for n = 1 (3/4-inch DR-11 HDPE): V_1 = 3.01 gal/100 feet (Table 8.3), L_1 = 410 feet x 3 = 1230 feet
  - for n = 2 (1-1/4-inch DR-11 HDPE): V_2 = 7.53 gal/100 feet (Table 8.3), L_2 = 50 feet
  - for n = 3 (1-1/4-inch SCH 40 PVC): V_3 = 7.77 gal/100 feet (Table 8.4), L_3 = 30 feet
  - for n = 4 (1-inch 200 psi Heater Hose): V_4 = 4.08 gal/100 feet (Table 8.6), L_4 = 15 feet
- C = 20 (20% propylene glycol by volume needed)
- V_{AF} = [(3.01 x 1230/100) + (7.53 x 50/100) + (7.77 x 30/100) + 4.08 x 15/100)] x (20/100) = 8.7 gallons, refer to Table 8.7.

The amount of antifreeze needed (V_{AF}) for this example is calculated in Table 8.7.

Table 8.7. Antifreeze Volume Calculation Example

n	Pipe Type	V_i (gal/100')	L_i (ft)	$V_i \cdot L_i/100$	V_{AFi}*
1	3/4" DR-11 HDPE	3.01	1230	37.0	7.4
2	1-1/4" DR-11 HDPE	7.53	50	3.8	0.8
3	1-1/4" SCH 40 PVC	7.77	30	2.3	0.5
4	1" 200-psi Heater Hose	4.08	15	0.6	0.1
				V_{AF} =	8.7

* - $V_{AFi} = (V_i \times L_i/100 \times C/100), C=20$

Table 8.8 is a simple tool to be used to calculate total pipe volume and necessary antifreeze volume for a given GSHP system.

Table 8.8. Antifreeze Volume Calculation Worksheet

Antifreeze Volume Worksheet					
Job/Client Name______________		GSHP Technician ______________			
Loop Description ______________					
Design EWT_{min}______	Antifreeze Type______________	Concentration (By Vol.) ______			
Piping System Information					
Pipe Section Description	Pipe Type & Size	V_i (gal/100ft)	L_i (ft)		
1) Loop Volume	______________	______	______		
2) Exterior GHEX S-R Piping Volume	______________	______	______		
3) Interior S-R Piping Volume	______________	______	______		
4) Header Piping Volume	______________	______	______		
5) Piping Volume from Flow Center to GSHP	______________	______	______		
Total Piping System Volume Calculations					
Pipe Section Description	(V_i)	x	(L_i / 100)	=	Pipe Volume
1) Loop Volume	(__________)	x	(__________)	=	______
2) Exterior GHEX S-R Piping Volume	(__________)	x	(__________)	=	______
3) Interior S-R Piping Volume	(__________)	x	(__________)	=	______
4) Header Piping Volume	(__________)	x	(__________)	=	______
5) Piping Volume from Flow Center to GSHP	(__________)	x	(__________)	=	______
			Total Pipe Volume	=	______
(Total Pipe Volume) · (Concentration / 100) = Volume of Antifreeze Needed				=	______

Table 8.9 has been filled out according to Example 8.3.

Table 8.9. Antifreeze Volume Calculation Worksheet for Example 8.3

Antifreeze Volume Worksheet

Job/Client Name 3-ton Residence GSHP Technician John Doe

Loopfield Description 3 vertically installed 200' bores, 3/4" u-bends, 1-1/4" S-R piping

Design EWT_{min} 30˚F Antifreeze Type Propylene Glycol Concentration (By Vol.) 20%

Piping System Information

Pipe Section Description	Pipe Type & Size	V_i (gal/100')	L_i (ft)
1) Loop Volume	3/4" DR-11 HDPE	3.01	3 x 410 = 1230
2) Exterior GHEX S-R Piping Volume	1-1/4" DR-11 HDPE	7.53	50
3) Interior S-R Piping Volume	1-1/4" SCH-40 PVC	7.77	30
4) Header Piping Volume	N/A - Negligible (Close Header Used)	0	0
5) Piping Volume from Flow Center to GSHP	1" Rubber Heater Hose (200 psi)	4.08	15

Total Piping System Volume Calculations

Pipe Section Description	(V_i)	x	(L_i / 100)	=	Pipe Volume
1) Loop Volume	3.01	x	12.3	=	37.0
2) Exterior GHEX S-R Piping Volume	7.53	x	0.50	=	3.8
3) Interior S-R Piping Volume	7.77	x	0.30	=	2.3
4) Header Piping Volume		x		=	
5) Piping Volume from Flow Center to GSHP	4.08	x	0.15	=	0.6
			Total Pipe Volume	=	43.7
(Total Pipe Volume) · (Concentration / 100) = Volume of Antifreeze Needed				=	8.7

8.4.2 ANTIFREEZE ADDITION GUIDELINES & PROCEDURES The proper antifreeze concentration can be obtained by draining off the quantity of system water equal to the quantity of full strength antifreeze necessary per the total pipe volume calculations. During antifreeze addition to the system, several issues must be addressed:

- Obtain a manufacturer's Material Safety Data Sheet (MSDS) for the antifreeze and observe the recommended safety guidelines for handling, mixing, and use
- Use the appropriate personal protective gear (goggles, gloves, etc.) and follow the manufacturer's safety and use information during handling, installation, and mixing
- Have the appropriate safety equipment for fire, spill clean up, etc. available on site
- Use care to not introduce air into the system via aeration as the antifreeze is poured into the flush/purge reservoir or canister of a GT Flow Center
- During antifreeze addition, do not allow antifreeze levels to get low enough in the reservoir or canister to allow the pump to suck air into the system
- Include the volume of the reservoir of any flush/purge unit with the volume of the piping system (interior and exterior) to ensure the proper antifreeze strength is introduced

- Tag the installation with a permanent label or tag that identifies the type of antifreeze, strength of the antifreeze solution, and date of installation
- Dispose of any unused antifreeze solution remaining in the reservoir and/or hoses in accordance with governing codes and regulations

Addition of antifreeze will be the last step in commissioning the water side of a system that utilizes a non-pressurized flow center.

8.5 Pressurizing the GSHP System Water Side

The final step in commissioning the water side of a system with a pressurized flow center is to pressurize it. The system must be pressurized to a level that will prevent cavitation of the circulating pump(s) and maintain positive suction-side pumping pressure throughout the annual operating cycle. Pressurization of the system can only take place if it has already been fully flushed and purged of air and debris (Section 8.2), pressure/flow verified (Section 8.3), and if the proper concentration of antifreeze has been added (if necessary - Section 8.4). The water side of the system is ready to be placed into service as soon as it is pressurized. Pressurization of the water side of the ground source heat pump system will need to be performed with a purge pump. A complete discussion of how to pressurize the water side of a pressurized system can be found in Section 8.6.

Changes in system pressure due to temperature can be overcome by pressurizing the system to a minimum of 20-30 psi if commissioning takes place in the summer months and 40-50 psi if commissioning takes place in the winter months. Refer to *IGSHPA (2009) Closed-Loop/Geothermal Heat Pump Systems Design and Installation Standards,* Section 3A for updates to these recommendations.

8.6 Pressurized Flow Center Startup Procedure

The typical residential pressurized flow center, shown in Figure 6.1a, includes two, 3-way valves to facilitate system flushing and purging (located above the pumps in this illustration). The flush/purge ports are on each side of the pumping module's 3-way valves. Pressurized flow centers are available with multiple pumping options, offering flow rates up to approximately 40 gpm. The typical circulator pump(s) used on these modules do not have the capacity to adequately flush/purge the system. Flushing must be performed using a flush cart similar to the example shown in Figure 8.3.

The 3-way valves on the packaged pressurized flow center allow the GHEX to be flushed/purged first while isolated from the interior piping system. The interior piping, components, and the heat pump unit(s) are flushed/purged next (while isolated from the GHEX). Then, the combined interior and GHEX piping systems are flushed in one continuous operation. If a packaged pressurized flow center (as shown in Figure 6.1a) is not used, it is recommended that a valving configuration like that in Figure 8.5 be assembled and installed in the field. Using this valving configuration will give the same capabilities to isolate the interior and exterior piping systems from one another during flushing.

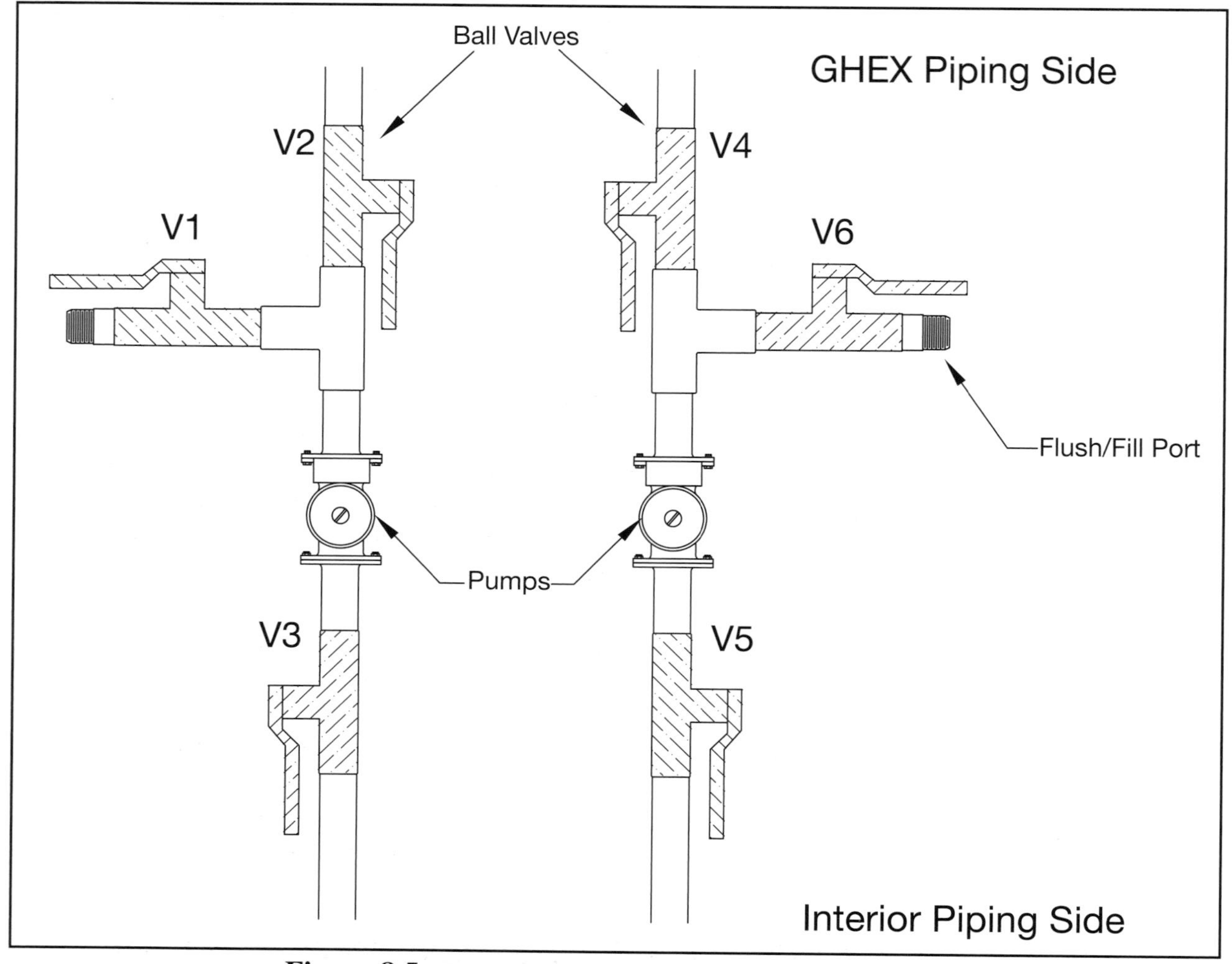

Figure 8.5. Pressurized System Servicing Configuration

8.6.1 FLUSHING/PURGING THE PRESSURIZED SYSTEM Flush and purge the pressurized system according to the following procedure:

1. Set the pump module 3-way valves or valve set to isolate the GHEX from the remainder of the system, as illustrated in Figure 8.6, Step 1 (or Figure 8.5 with valves V1, V2, V4, and V6 open and V5 and V3 closed).
2. Fill the flush/purge cart reservoir with potable water from the domestic water supply or from a clean water tank if no domestic water supply is available. Have additional potable water readily available as a make-up source to maintain the reservoir water at acceptable levels when flushing/ purging begins. This level will vary by flush/purge cart manufacturer.
3. Energize the pump and begin pumping reservoir water into the GHEX. As any remaining air in the GHEX piping is either displaced or compressed, the water level in the reservoir will drop. The water level must be maintained above the flush/purge cart return discharge into the reservoir to prevent

introducing air back into the GHEX. If significant amounts of air exist in the GHEX piping, it may be necessary to stop the purge pump completely and refill the reservoir.

4. Carefully monitor flushing return water quality. If it obviously contains dirt, sand, etc., it may be necessary to disconnect the return and discharge to a drain until the return flows clear water. Once the water flows clear, reconnect the return line and begin the flush/purge process again. If pipe shavings and other lighter debris are floating at the top of the flush cart reservoir, scoop them out of the water with a small mesh net or screen.
5. Once the reservoir water level is reasonably stable or begins to increase, turn the make-up water down or off as conditions dictate. Observe the discharge water as it re-enters the reservoir. Watch for air bubbles being discharged into the reservoir. The purging process should continue for a minimum of 15 minutes after no bubbles have been observed entering the reservoir.
6. With the pump on, close the flush/purge cart return valve and watch the reservoir water level. If it falls more than 1 inch, there is still air in the system that is being compressed by the power of the pump. Reverse the flow path or open and close the purge cart discharge valve to "shock" the flow and continue flushing/purging activity. If the flush/purge cart has no valving to reverse flow direction, but reversing is necessary, simply switch hose connections to manually reverse system flushing flow. Repeat the reservoir level check as necessary until the drop in reservoir water level is less than 1 inch.
7. Verify pressure/flow design (as described in Section 8.3)
8. Turn off the flush/purge pump and change the pump module or built-up pumping assembly valves to isolate the GHEX and access the inside piping for flushing/purging as shown in Figure 8.6, Step 2 (or Figure 8.5 with valves V1, V3, V5, and V6 open and V2 and V4 closed). Add water as required and flush the interior piping until all air and debris are removed.
9. Once the GHEX and interior piping and components have been flushed and purged, change the pump module or field-built pumping assembly valves to combine both into one continuous operation, as shown in Figure 8.6, Step 3 (or Figure 8.5 with all valves open).

8.6.2 CHARGING THE PRESSURIZED SYSTEM WITH ANTIFREEZE (IF USED)

If no antifreeze is required for a given installation, skip ahead to Section 8.6.3. Charge the pressurized system with antifreeze as follows:

1. Set the pump module 3-way valves or valve set to isolate the GHEX from the remainder of the system, as illustrated in Figure 8.6, Step 1 (or Figure 8.5 with valves V1, V2, V4, and V6 open and V5 and V3 closed).
2. Add the pre-calculated amount of antifreeze to the flush/purge unit reservoir with the unit circulating. As the antifreeze is added, disconnect the return line and allow an equal amount of water to be discarded. Use care to not allow air to enter the system through aeration, pump suction, or improper handling of the return line while draining off unneeded water. In most cases, the flush cart reservoir will not be large enough to hold all of the antifreeze required for a given installation. In such instances, antifreeze will need to be added in batches. For example, if the flush cart reservoir will hold 10 gallons of fluid, fill it with 10 gallons of antifreeze. Pump the antifreeze into the system and drain off an equal amount of water. Shut down the purge pump before completely emptying the reservoir to prevent air from entering the piping system. Refill the reservoir with

antifreeze, pump it into the system and drain off an equal amount of water. Repeat this process until the proper amount of antifreeze has been added to the system. Refer to Section 8.4 for guidelines in calculating the required amount of antifreeze for a given piping system.

3. After the pre-calculated amount of antifreeze has been added, reconnect the return line of the flush cart to the system to prepare it for pressurization. Normal circulator pump operation will adequately mix the antifreeze in the system.

8.6.3 PRESSURIZATION OF THE SYSTEM Pressurize the system as follows:

1. With the flush/purge cart pump running, close the 3-way valve or valve set that is allowing water to return to the flush/purge cart as shown in Figure 8.6, Step 4 (or Figure 8.5 with valves V1-V5 open and V6 closed).
2. Watch the pressure gauge on the flush/purge cart or a probe type pressure gauge installed in one of the ground source heat pump water-side piping P/T ports and allow the water-side pressure to build.
3. When the pressure reaches proper levels (20-30 psi in the summer months, 40-50 psi in the winter months, refer to Section 8.5), close the 3-way valve that is allowing water to be pumped into the ground source heat pump water side as shown in Figure 8.6, Step 5 (or Figure 8.5 with valves V2, V3, V4, and V5 open and V1 and V6 closed), then turn off the purge cart.
4. Observe the pressure gauge to ensure GHEX pipe expansion does not relieve system pressure to a level less than the target pressure. If pressure drops below the target pressure, repeat the procedure and ensure final system pressure does not fall below the prescribed level.
5. If a pressure gauge was used in a test port, remove the gauge and cap the test port.
6. Disconnect flush/purge cart hoses, remove any connectors or other flush/purge related fittings from the 3-way valves or valve sets, and cap.
7. If antifreeze was used, recover and dispose of the antifreeze solution remaining in the hoses and flush/purge cart reservoir in accordance with applicable codes and regulations.

After the system has been completely flushed and purged, flow verified, charged with antifreeze, and pressurized, the GSHP water side will be ready for use.

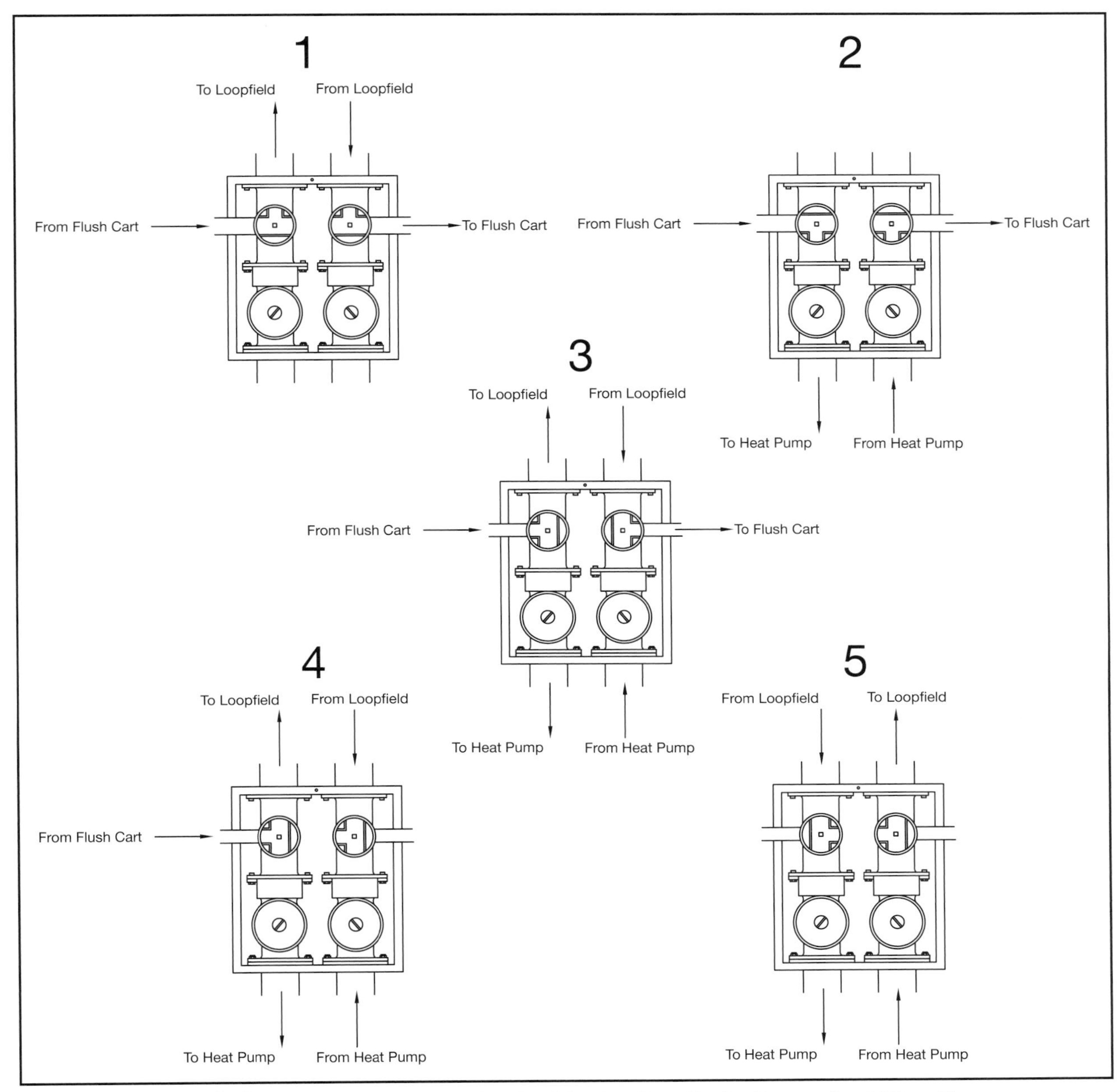

Figure 8.6. Flushing/Purging Valve Positioning Details for Pressurized Flow Center

8.7 Non-Pressurized (Standing Column) Flow Center Startup Procedure

The non-pressurized flow center shown in Figure 6.1b offers an alternative pumping option for closed-loop ground source heat pump systems. Figures 6.3b and 6.3c include provisions for conventional flushing/purging of a GHEX with outside-the-building buried headers. Figure 6.3d includes provisions for purging air from a GHEX with inside-the-building accessible headers that require shut-off valves on each leg of the GHEX loops.

Non-pressurized flow centers are available with multiple pumping options, offering flow rates up to approximately 200 gpm. The typical circulator pump(s) used on these modules do not have the capacity to adequately flush/purge an outside-the-building buried header system, but do have sufficient power to flush the individually isolated loops on parallel plumbed inside-the-building header systems. Flushing must be performed using a flush cart similar to the example shown in Figure 8.3, if an outside-the-building header system is used. The valving configurations shown in Figures 6.3b, 6.3c, and 6.3d allow the GHEX to be flushed/purged while being isolated from the interior piping system. When using a flush cart (as shown in Figure 8.3) to flush and purge a system with a non-pressurized flow center, never allow the flow caused by the use of the purge pump to travel through the non-pressurized flow center itself. The high flow velocity caused by use of the purge pump will damage the non-pressurized flow center. The pumps on the non-pressurized flow center itself should be used to flush the interior piping from the flow center to the heat pump, not the purge pump. One major benefit of using a non-pressurized flow center is that if all of the air is not completely removed from the system during the purging process, that air will work its way through the system and enter the canister of the flow center, which acts as an expansion tank for the system. Trapping of the air in the top of the canister will not detrimentally affect the performance of the circulator pumps. On the other hand, air trapped in a pressurized system will cause pump cavitation and will require a service call to re-purge the system before proper operation can take place.

8.7.1 FLUSHING/PURGING THE NON-PRESSURIZED SYSTEM (WITH OUTSIDE, BURIED HEADERS)

If outside-the-building buried headers are used in conjunction with a non-pressurized flow center, a flush cart will be necessary to properly flush and purge the system. Flush and purge the non-pressurized system with outside-the-building buried headers according to the following procedure:

1. Set the PVC manifold ball valves or valve set to isolate the GHEX from the remainder of the system, as illustrated in Figure 8.7, Step 1 (or Figure 8.8 with valves V4 and V5 closed).
2. Fill the flush/purge cart reservoir with potable water from the domestic water supply or from a clean water tank if no domestic water supply is available. Have additional potable water readily available as a make-up source to maintain the reservoir water at acceptable levels when flushing/purging begins. This level will vary by flush/purge cart manufacturer.
3. Energize the pump and begin pumping reservoir water into the GHEX. As any remaining air in the GHEX piping is either displaced or compressed, the water level in the reservoir will drop. The water level must be maintained above the flush/purge cart return discharge into the reservoir to prevent introducing air back into the GHEX. If significant amounts of air exist in the GHEX piping, it may be necessary to stop the purge pump completely and refill the reservoir.
4. Carefully monitor flushing return water quality. If it obviously contains dirt, sand, etc., it may be necessary to

disconnect the return and discharge to a drain until the return flows clear water. Once the water flows clear, reconnect the return line and begin the flush/purge process again. If pipe shavings and other lighter debris are floating at the top of the flush cart reservoir, scoop them out of the water with a small mesh net or screen.

5. Once the reservoir water level is reasonably stable or begins to increase, turn the make-up water down or off as conditions dictate. Observe the discharge water as it re-enters the reservoir. Watch for air bubbles being discharged into the reservoir. The purging process should continue for a minimum of 15 minutes after no bubbles have been observed entering the reservoir.
6. If a manifold system similar to the ones shown in Figure 6.4, Figure 8.7, or Figure 8.8 is used, the interior piping can be simultaneously flushed while the GHEX piping is flushed. To do so, first open the top of the canister and fill it with clean water. Check to ensure that the ball valves on both sides of the circulator pump(s) are open. Next, position the ball valves on the PVC manifold as shown in Figure 8.7, Step 2 (or Figure 8.8 with valves V1, V2, and V3 open). For a non-pressurized flow center equipped with Grundfos pumps, remove the screw on the end of the motor and rotate the motor shaft with a small screwdriver. This will vent air out of the pumping chamber and lubricate the motor bearings. When water appears at the screw opening, replace the screw. On pumps that do not have water lubricated bearings this venting procedure will not be necessary.
7. Energize the circulator pumps on the non-pressurized flow center (NOTE: Do not energize the circulator pumps with the thermostat without disconnecting the compressor power wiring. Failure to disconnect the power wiring will energize the compressor on the GSHP unit. The compressor should not be energized until the water side of the system has been completely flushed and purged and is prepared for operation). The flow center circulators will have adequate capacity to flush the interior piping system. As any remaining air in the interior piping is either displaced or compressed during flushing, the water level in the canister will drop. The water level must be maintained above the circulator pump suction side piping to prevent introducing air back into the piping system. If significant amounts of air are in the piping, it may be necessary to stop the circulator pumps completely and refill the canister.
8. With the pump on, close the flush/purge cart return valve and watch the reservoir water level. If it falls more than 1 inch, there is still air in the system that is being compressed by the power of the pump. Reverse the flow path or open and close the purge cart discharge valve to "shock" the flow and continue flushing/purging activity. If the flush/purge cart has no valving to reverse flow direction, but reversing is necessary, simply switch hose connections to manually reverse system flushing flow. Repeat the reservoir level check as necessary until the drop in reservoir water level is less than 1 inch. Make sure to release the pressure on the loopfield before disconnecting the flush cart by reopening the valve and allowing the system pressure to equalize. If pressure remains on the loopfield, the canister will overflow once the exterior piping and interior piping are combined into one continuous operation.
9. Verify pressure/flow design (as described in Section 8.3). Because a non-pressurized flow center has been utilized, a flowmeter tool can be installed in the system flow path to directly measure system flow. Refer to Figure 11.7 for flowmeter tool details.
10. Once the GHEX and interior piping and components have been flushed and purged, change the PVC manifold valve positioning to combine the interior piping and GHEX piping into one continuous operation, as shown in Figure 8.7, Step 3 (or Figure 8.8 with valves V1, V2, V4, and V5 open and valve V3 closed).
11. If antifreeze is required for a given installation, proceed to Section 8.7.3. Otherwise the filled, flushed, purged, and flow verified system will be ready for operation.

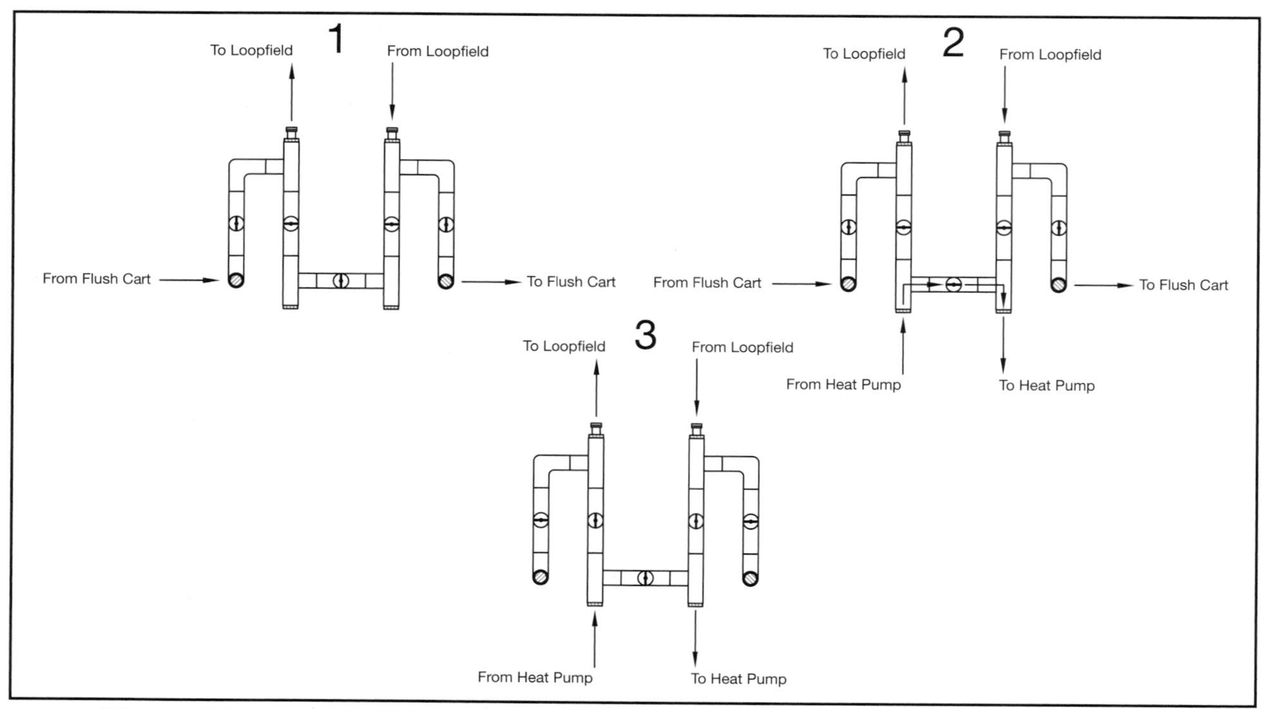

Figure 8.7. Flushing/Purging Valve Positioning Details for Non-Pressurized Flow Center w/ PVC Manifold

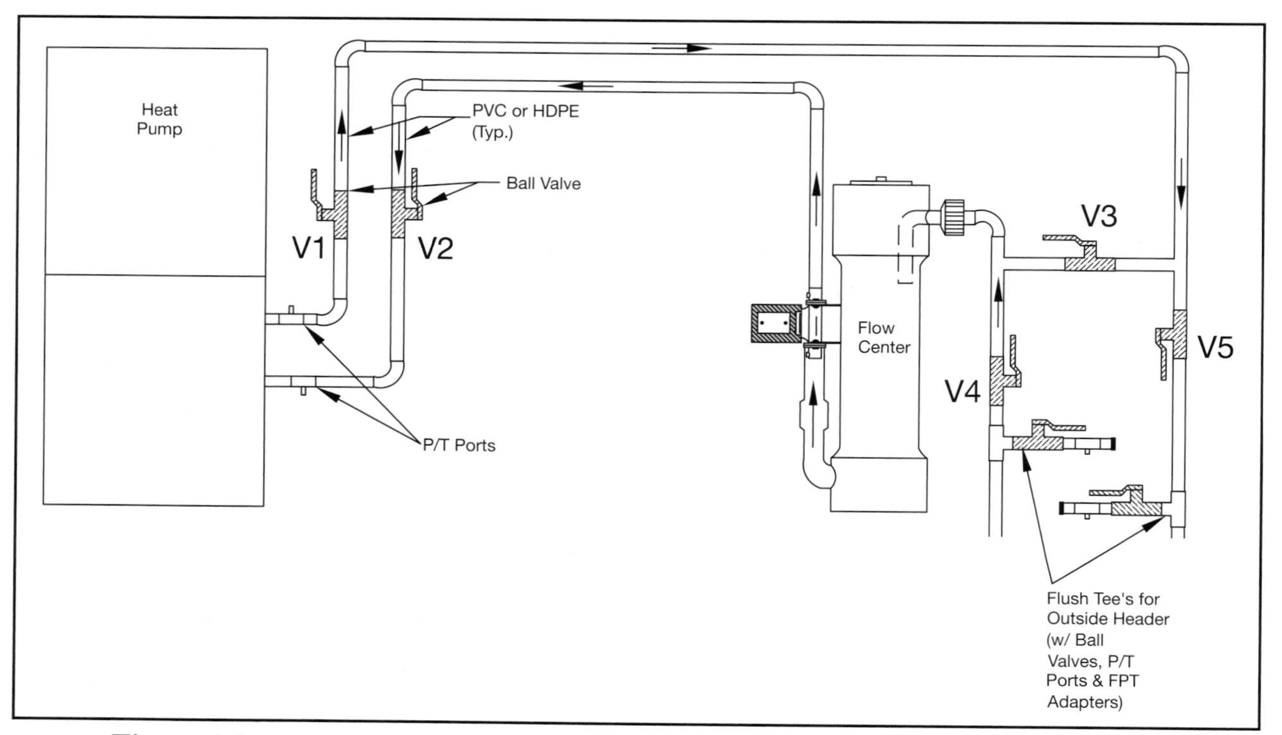

Figure 8.8. Flushing/Purging Valve Details for Non-Pressurized Flow Center (Patented) & Outside Headers

8.7.2 CHARGING A NON-PRESSURIZED SYSTEM WITH ANTIFREEZE (WITH OUTSIDE, BURIED HEADERS)

If an inside-the-building header configuration is used, charge the system with antifreeze according to Section 8.7.4. Charge the non-pressurized system with outside-the-building buried headers with antifreeze as follows:

1. Set the PVC manifold ball valves or valve set to isolate the GHEX from the remainder of the system, as illustrated in Figure 8.7, Step 1 (or Figure 8.8 with valves V4 and V5 closed).
2. Add the pre-calculated amount of antifreeze to the flush/purge unit reservoir with the unit circulating. As the antifreeze is added, disconnect the return line and allow an equal amount of water to be discarded. Use care to not allow air to enter the system through aeration, pump suction, or improper handling of the return line while draining off unneeded water. In most cases, the flush cart reservoir will not be large enough to hold all of the antifreeze required for a given installation. In such instances, antifreeze will need to be added in batches. For example, if the flush cart reservoir will hold 10 gallons of fluid, fill it with 10 gallons of antifreeze. Pump the antifreeze into the system and drain off an equal amount of water. Shut down the purge pump before completely emptying the reservoir to prevent air from entering the piping system. Refill the reservoir with antifreeze, pump it into the system and drain off an equal amount of water. Repeat this process until the proper amount of antifreeze has been added to the system. Refer to Section 8.4 for guidelines in calculating the required amount of antifreeze for a given piping system.
3. After the pre-calculated amount of antifreeze has been added, reconnect the return line and set the pump module 3-way valves or valve set to combine the GHEX with the remainder of the system, as illustrated in Figure 8.7, Step 3 (or Figure 8.8 with valves V1, V2, V4, and V5 open and valve V3 closed). The circulator pumps on the flow center will adequately mix the antifreeze solution during normal heat pump operation.

After the system has been completely flushed and purged, flow verified, and charged with antifreeze, the GSHP water side will be ready for use.

8.7.3 FLUSHING/PURGING THE NON-PRESSURIZED SYSTEM (WITH INSIDE HEADERS)

If inside-the-building accessible headers are used in conjunction with a non-pressurized flow center, a flush cart will not be necessary to properly flush and purge the system. Flush and purge the non-pressurized system with inside-the-building headers according to the following procedure:

1. Set the ball valves or valve set to isolate the GHEX from the remainder of the system by closing valves V1 and V2 and opening valve V3 as shown in Figure 8.9. Also, open the valves on a single pair of ground-loop header piping for flushing. For example, start with the valves open on the first pair (valves S1 and R1 on Figure 8.9 with valves S2, S3, S4, R2, R3, and R4 closed). If the optional ball valve (V3, Figure 8.9) is not used, the system will need to be flushed through the water-to-refrigerant coil on the GSHP unit.
2. For a non-pressurized flow center equipped with Grundfos pumps, remove the screw on the end of the motor and rotate the motor shaft with a small screwdriver. This will vent air out of the pumping chamber and lubricate the motor bearings. When water appears at the screw opening, replace the screw.

On pumps that do not have water lubricated bearings this venting procedure will not be necessary.

3. Open the top of the canister and fill it with clean water. Check to ensure that the ball valves on both sides of the circulator pump(s) are open.
4. Start the pumps and add water through the open top of the canister until a full stream of water is coming out of the dip tube and all air has been eliminated. Watch for air bubbles being ejected back into the canister. The purging process should continue after no bubbles have been observed entering the canister for a minimum of 15 minutes.
5. When the first loop is full and completely flushed and purged of debris and air, close the valves on that header piping pair and open the next pair of valves to flush the second loop. For example, close valves S1 and R1 and open valves S2 and R2 to flush the second pair of ground-loop header piping. Follow the same procedure to fill and flush/purge all ground-loop header piping pairs. If pipe shavings and other lighter debris are floating at the top of the flow center canister, scoop them out of the water with a small mesh net or screen. Heavier contaminants (sand, gravel, etc.) will settle to the bottom of the flow center canister in the sediment collector, which can be removed later with a long-handled small mesh net or screen.
6. When every loop has been filled, flushed, and purged, open all valves except valve V3 (Figure 8.9) to combine the interior piping and GHEX piping into one continuous operation.
7. If antifreeze is required for a given installation, proceed to Section 8.7.4. Otherwise the filled, flushed, purged, and flow verified system will be ready for operation.

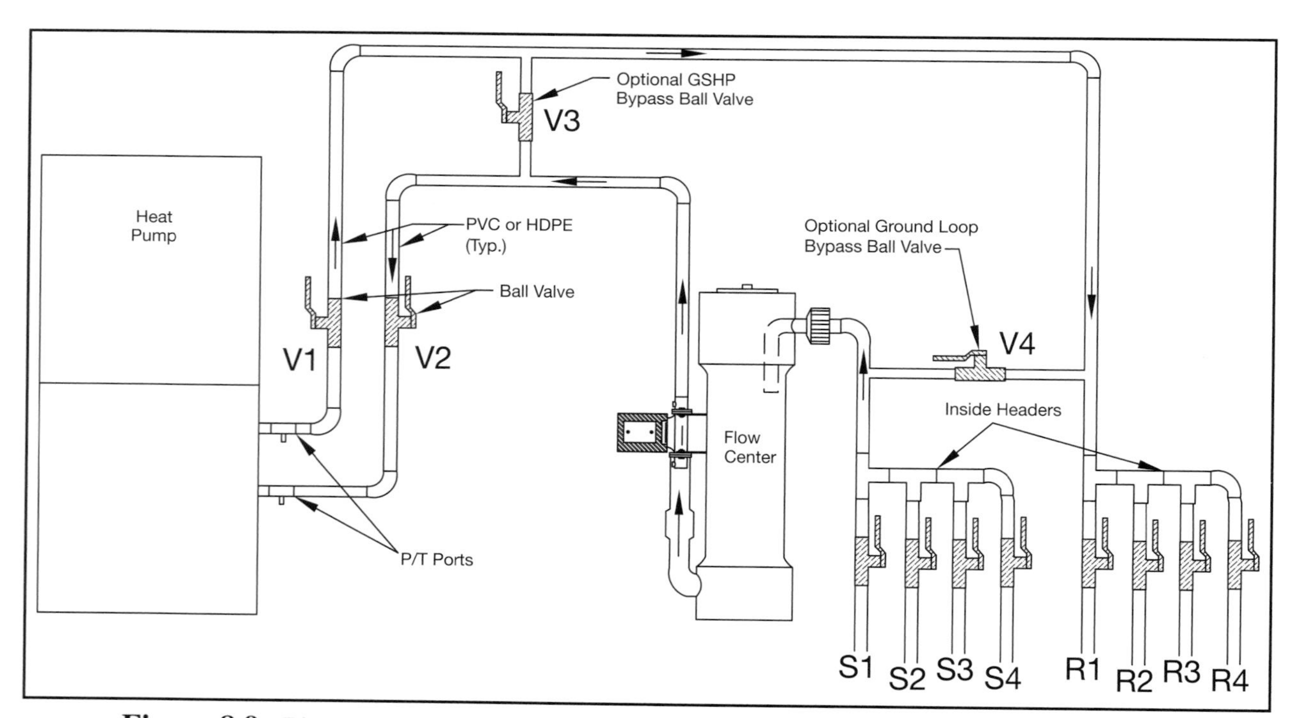

Figure 8.9. Flushing/Purging Valve Details for Non-Pressurized Flow Center (Patented) & Inside Header

8.7.4 CHARGING A NON-PRESSURIZED SYSTEM WITH ANTIFREEZE (WITH INSIDE HEADERS) If an outside-the-building header configuration is used, charge the system with antifreeze according to Section 8.7.2. If an inside-the-building header configuration is used, a means to divert the flow from the dip tube in the flow center to a discharge line will be necessary. A flowmeter tool can be used for this purpose. Connect the end of the flowmeter tool to the dip tube and place the hose at the end of the tool outside of the flow center into a drain or bucket for disposal of return fluid from the ground loop as shown in Figure 8.10. Charge the non-pressurized system with inside the building headers with antifreeze as follows:

1. Set the ball valves or valve set to isolate the GHEX from the remainder of the system by closing valves V1 and V2 and opening valve V3 as shown in Figure 8.9. Also, open the valves on all ground-loop header piping pairs (Figure 8.9 with valves S1, S2, S3, S4, R1, R2, R3, and R4 open).
2. Add the pre-calculated amount of antifreeze to the flow center canister with the unit circulating. After the canister is full, energize the flow center circulator pumps. As the antifreeze is added, place the return line from the flowmeter tool into a drain or bucket for disposal to allow an equal amount of water to be discarded. Use care to not allow air to enter the system through aeration, pump suction, or improper handling of the return line while draining off unneeded water. In most cases, the flow center canister will not be large enough to hold all of the antifreeze required for a given installation. In such instances, antifreeze will need to be added in batches. For example, if the flow center canister will hold 5 gallons of fluid, fill it with 5 gallons of antifreeze. Pump the antifreeze into the system and drain off an equal amount of water. Shut down the circulator pumps before completely emptying the canister to prevent air from entering the piping system. Refill the canister with antifreeze, pump it into the system and drain off an equal amount of water. Repeat this process until the proper amount of antifreeze has been added to the system. Refer to Section 8.4 for guidelines in calculating the required amount of antifreeze for a given system.
3. After the pre-calculated amount of antifreeze has been added, disconnect the flowmeter tool from the system and open all valves except valve V3 (Figure 8.9) to combine the interior piping and GHEX piping into one continuous operation. The circulator pumps on the flow center will adequately mix the antifreeze solution during normal heat pump operation.

After the system has been completely flushed and purged, flow verified, and charged with antifreeze, the GSHP water side will be ready for use.

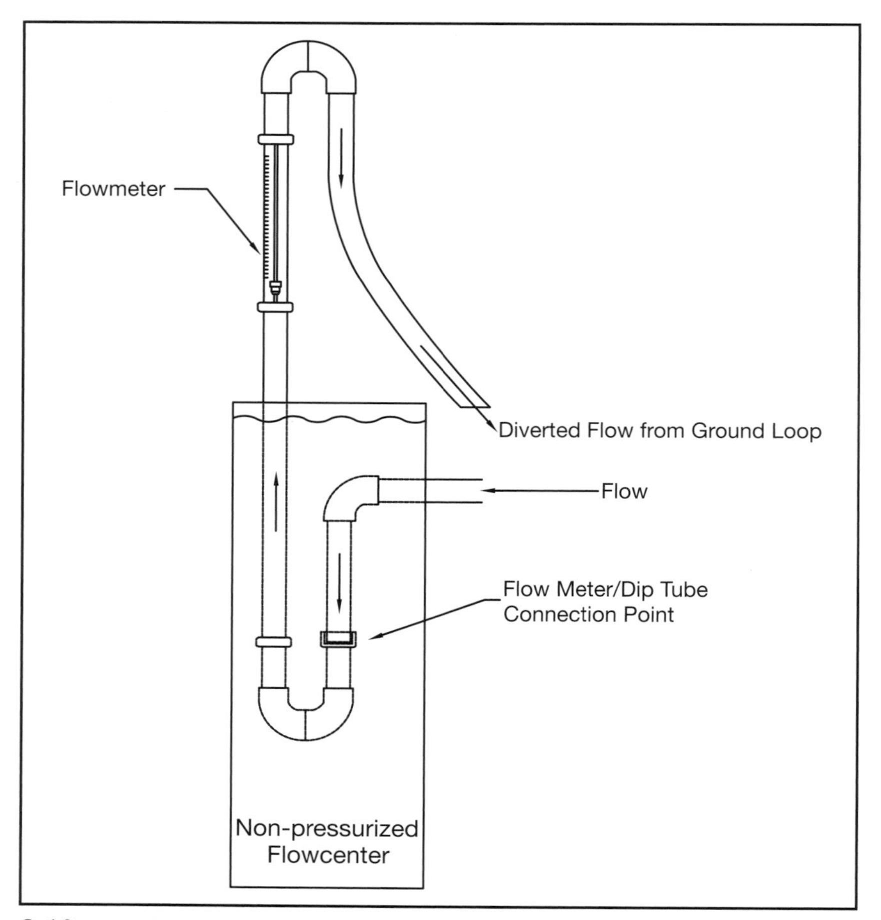

Figure 8.10. Antifreeze Charging Details with Flowmeter & Inside Headers

8.8 Flushing/Purging Guidelines for Multiple Heat Pump Systems

The same principles that apply to single heat pump systems apply to multiple heat pump systems. A minimum of 2 FPS flow velocity must be maintained in all parts of the system for the duration of the flushing/purging process. The major difference when flushing multiple heat pump systems comes when flushing the interior piping system. Provisions must be made because each heat pump must be isolated from the rest of the interior piping system and flushed individually. As shown in Figure 8.11, each heat pump in a multiple heat pump system should be installed with isolation valves both for flushing and for servicing.

8.8.1 FLUSHING/PURGING THE INTERIOR PIPING OF A MULTIPLE HEAT PUMP SYSTEM

Exterior GHEX piping will be flushed in the same manner as described in Section 8.6 and Section 8.7. Flush and purge the interior piping of a multiple heat pump system according to the following procedure:

1. Isolate the interior piping from the GHEX for flushing/purging by closing valves V3 and V4 and opening valves V1 and V2 as shown in Figure 8.11.
2. Isolate the heat pump unit farthest from the flush/fill ports for flushing/purging. In Figure 8.11, the

farthest unit from the flush/fill ports is HP3. Close the isolation valves on HP1 and HP2 and open the isolation valves on HP3. Flush and purge the interior piping circuit that runs through HP3 until all of the air is completely removed. Flushing the heat pump farthest from the flush/fill ports will minimize the possibility of trapping air in the interior supply-return piping.

3. Once the heat pump farthest from the flush/fill ports has been flushed, close the isolation valves for that unit and open the isolation valves for the next unit in the line (HP2 in Figure 8.11). Repeat the flushing process until all heat pump units have been completely purged of air.
4. When every heat pump has been flushed and purged, open all valves except the valves on the flush/fill ports (Figure 8.11) to combine the interior piping and GHEX piping into one continuous operation.

After the interior piping system has been completely flushed and purged, the GHEX can be charged with antifreeze. If no antifreeze is necessary, the GSHP water side will be ready for use.

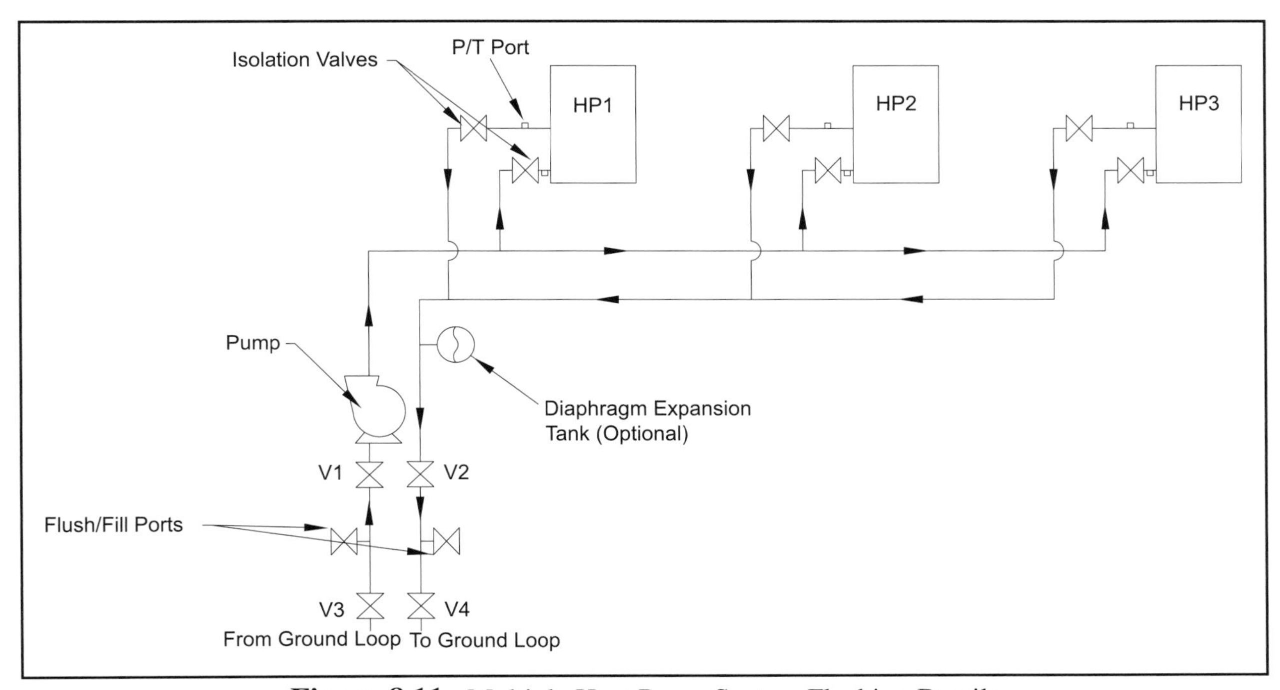

Figure 8.11. Multiple Heat Pump System Flushing Details

DOMESTIC HOT WATER HEATING

In This Section

9.1 Domestic Hot Water Options

There are three basic options regarding domestic hot water generation (HWG) with GSHP systems.

- Dedicated HWG is performed with a water-to-water GSHP or a GSHP capable of both air heating and cooling or water heating that includes controls to provide hot water when needed. In either case, operation of the GSHP in water heating mode is initiated whenever the hot water tank's high water temperature control/safety shut-off (aquastat) calls for heating.
- Supplemental HWG is performed with a desuperheater. The desuperheater generates hot water only when the GSHP unit is operating. Because the unit's operation is controlled by the space thermostat, HWG will only occur when the water temperature in the hot water storage tank is lower than the superheated refrigerant temperature, lower than the shut-off switch temperature setpoint inside the GSHP, and the GSHP happens to be running to meet the space heating or cooling requirements of the zone.
- Domestic HWG is an optional feature for a ground source heat pump system, and a customer may opt not to utilize the available HWG capabilities of GSHP equipment.

9.1.1 DEDICATED HOT WATER GENERATION If a dedicated HWG system is used, a separate water-to-water GSHP unit or a combination unit capable of heating water will be installed with the primary function of creating hot water. A typical configuration for a dedicated water-to-water GSHP for HWG is shown in Figure 9.1. The water-to-water GSHP load side is plumbed directly to the domestic hot water storage tank and, by code, the refrigerant-to-water heat exchanger must be double-walled and vented to prevent potential potable water contamination. If a single-walled refrigerant-to-water heat exchanger is used, then an intermediate water-to-water heat exchanger must be placed between the GSHP and the domestic hot water storage tank. An aquastat is placed in the domestic hot water storage tank to control the GSHP to maintain the desired hot water temperature. Emergency electric heat may be installed in the storage tank to maintain tank temperature if a problem develops with the GSHP unit.

A configuration using a combination GSHP to provide HWG is provided in Figure 9.2. The refrigerant-to-water heat exchanger on the GSHP load side is plumbed directly to the domestic hot water storage tank and, by code, the refrigerant-to-water heat exchanger must be double-walled and vented to prevent potential potable water contamination. If a single-walled refrigerant-to-water heat exchanger is used, then an intermediate water-to-water heat exchanger must be placed between the GSHP and the domestic hot water storage tank. An aquastat is placed in the domestic hot water storage tank to control the GSHP to maintain the desired hot water temperature. Priority must be given to the aquastat if 100% of the domestic hot water load is to be obtained with the combination GSHP. This will ensure that the GSHP provides water heating when called for and the space heating or cooling load will be satisfied by the water-to-air capability of the GSHP after the aquastat is satisfied. Emergency electric heat may be installed in the storage tank to maintain tank temperature if a problem develops with the GSHP unit.

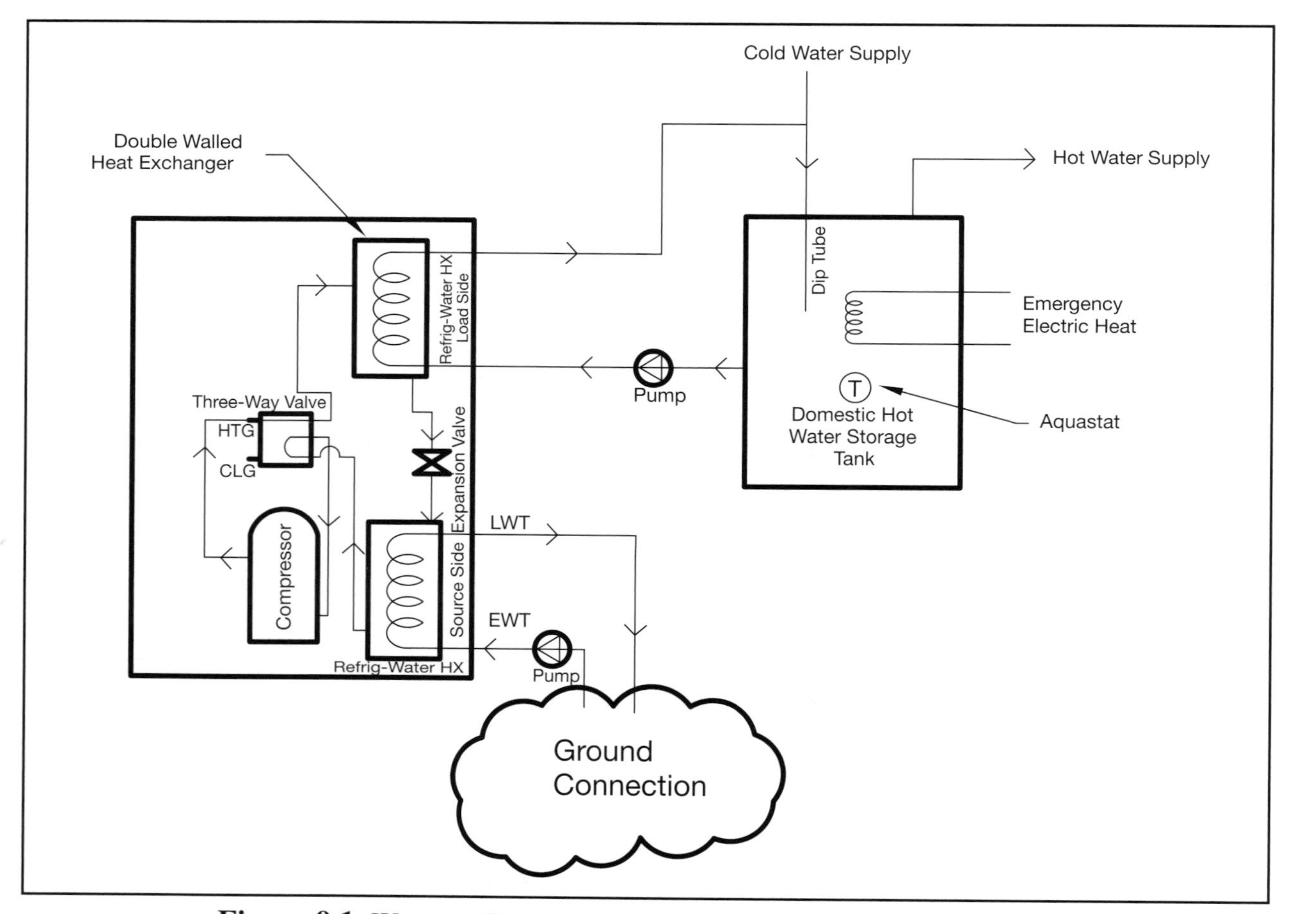

Figure 9.1. Water-to-Water GSHP for Dedicated Hot Water Generation.

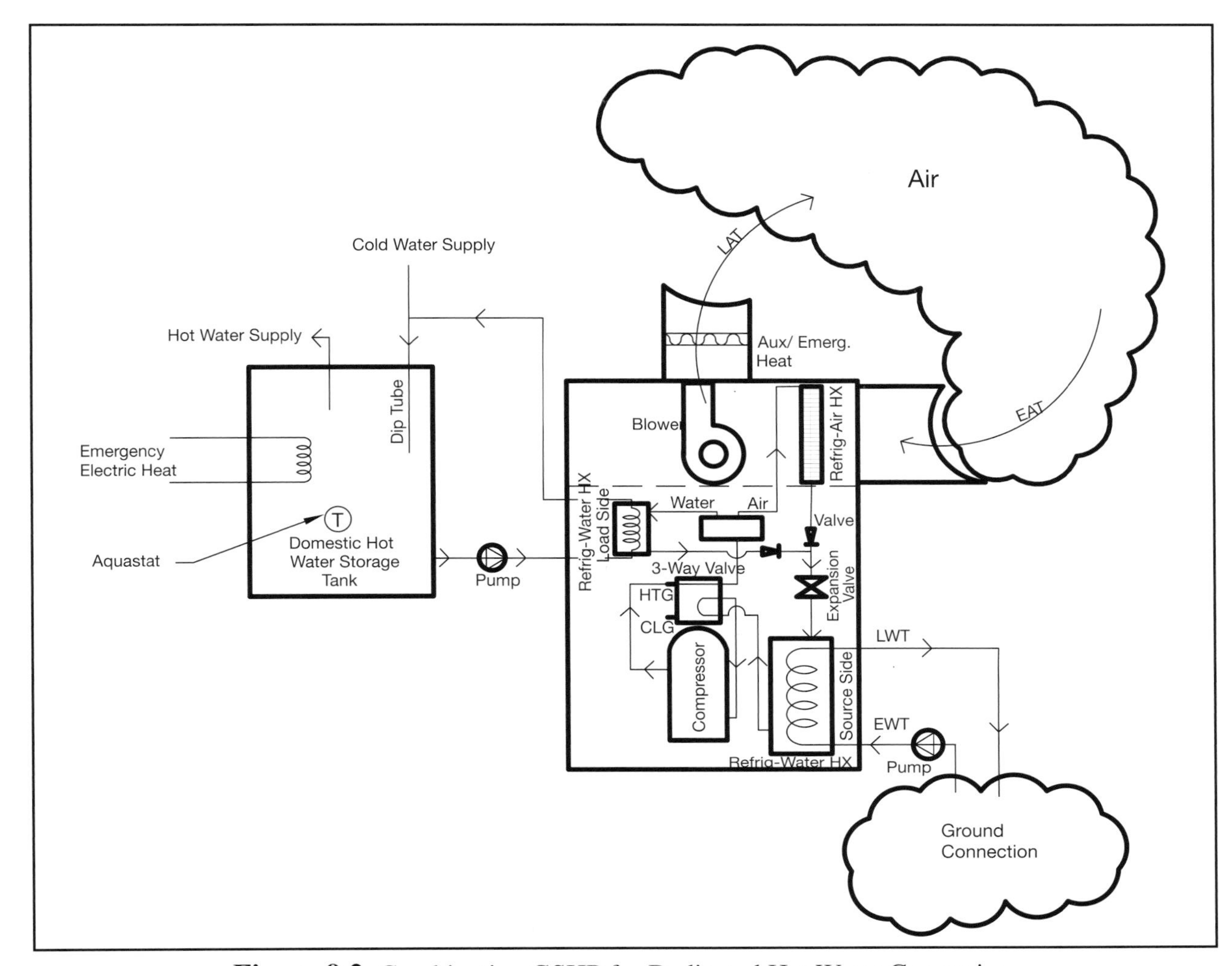

Figure 9.2. Combination GSHP for Dedicated Hot Water Generation.

9.1.2 SUPPLEMENTAL DOMESTIC HOT WATER GENERATION Supplemental domestic HWG is obtained using a desuperheater, which is comprised of a double-walled, vented water-to-refrigerant heat exchanger. This heat exchanger is located between the compressor discharge and the reversing valve to transfer heat from the superheated refrigerant vapor discharged by the compressor to heat domestic water as illustrated in Figure 9.3 (a revision of Figure 2.1). The desuperheater heat exchanger is sized to limit its heat exchange capacity to 10 to 15% of the GSHP's capacity so as not to condense the superheated vapor to liquid. Figure 2.2 illustrates a common configuration using a water-to-water GSHP with a desuperheater option, and Figure 2.3 illustrates a common configuration using a combination GSHP with a desuperheater option.

9.1.2.1 DESUPERHEATER BASICS Domestic hot water generation using the optional desuperheater feature on a GSHP system can significantly reduce water heating energy costs during both the heating and cooling seasons. However, a desuperheater only heats water when the GSHP is running. With a desuperheater, HWG is driven by the space thermostat and is independent of the need for domestic hot water. During GSHP operation

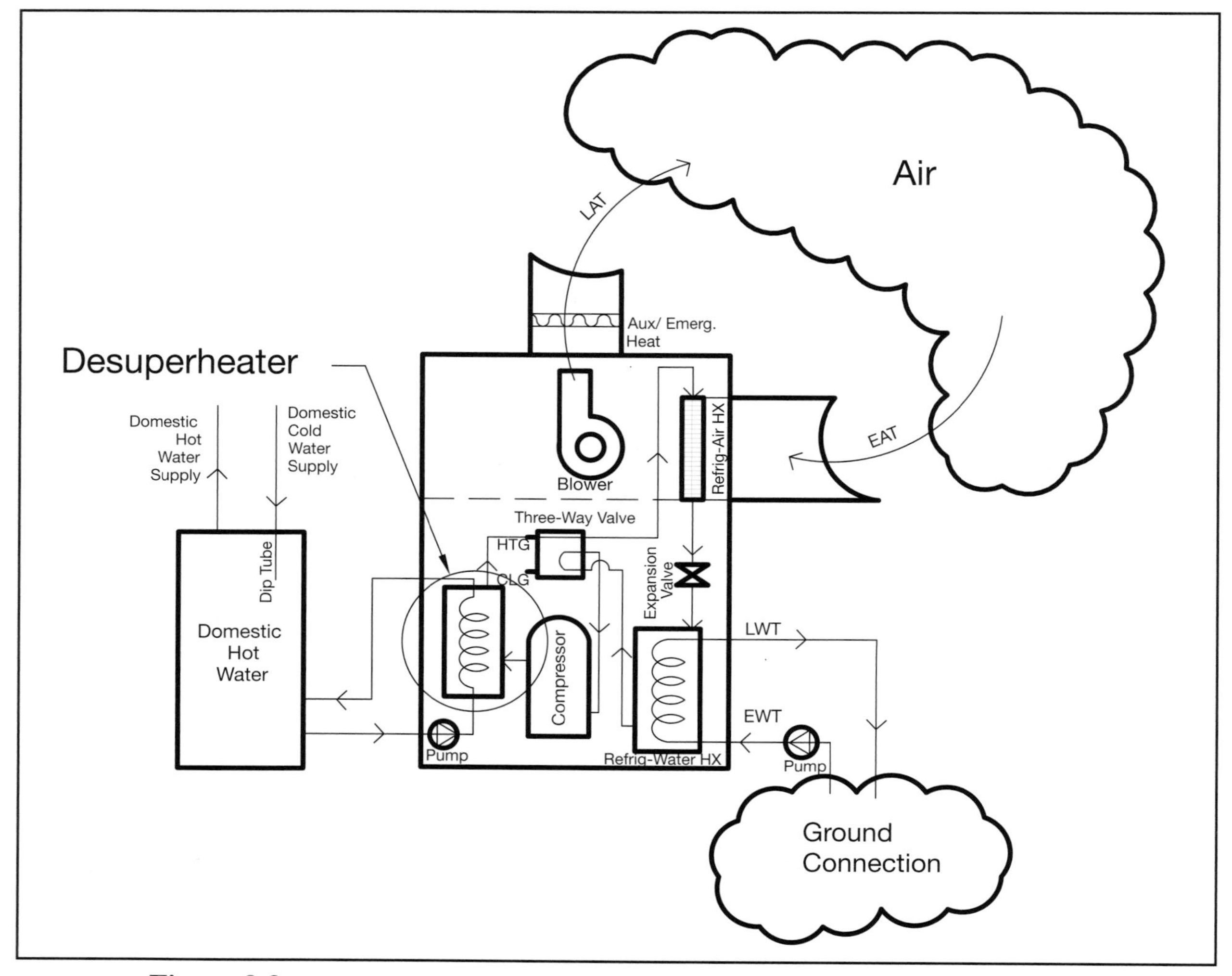

Figure 9.3. Water-to-Air GSHP with Desuperheater (Displayed in Heating Mode)

the superheated refrigerant vapor from the compressor is sent through the desuperheater heat exchanger, where it rejects some of the heat energy it contains to the domestic water being circulated. The domestic water from the hot water tank is circulated through the desuperheater circuit by a low power pump. As shown in Figure 9.4, the pump circuit includes an aquastat that stops pump operation when the hot water tank reaches the preset shut off temperature or if the tank temperature requirement is already satisfied. The high-temperature limit should never be removed from a desuperheater control circuit. During periods when the heat pump runs frequently, its operation could drive hot water tank temperatures above safe levels. Desuperheater "ON-OFF" operation is controlled by the GSHP control circuit. Desuperheater pump and control circuit power is provided by the GSHP.

During operation in the heating mode, desuperheater generated domestic hot water is not free. Instead, the desuperheater is a parasitic device that effectively reduces the total space heating capacity of the equipment by taking part of the heat energy for HWG that would otherwise be used to condition the heated space. In colder climates, space heating capacity reduction can sometimes be problematic when heat pump capacity is just enough to handle building space heating requirements, or when the heat pump and supplemental

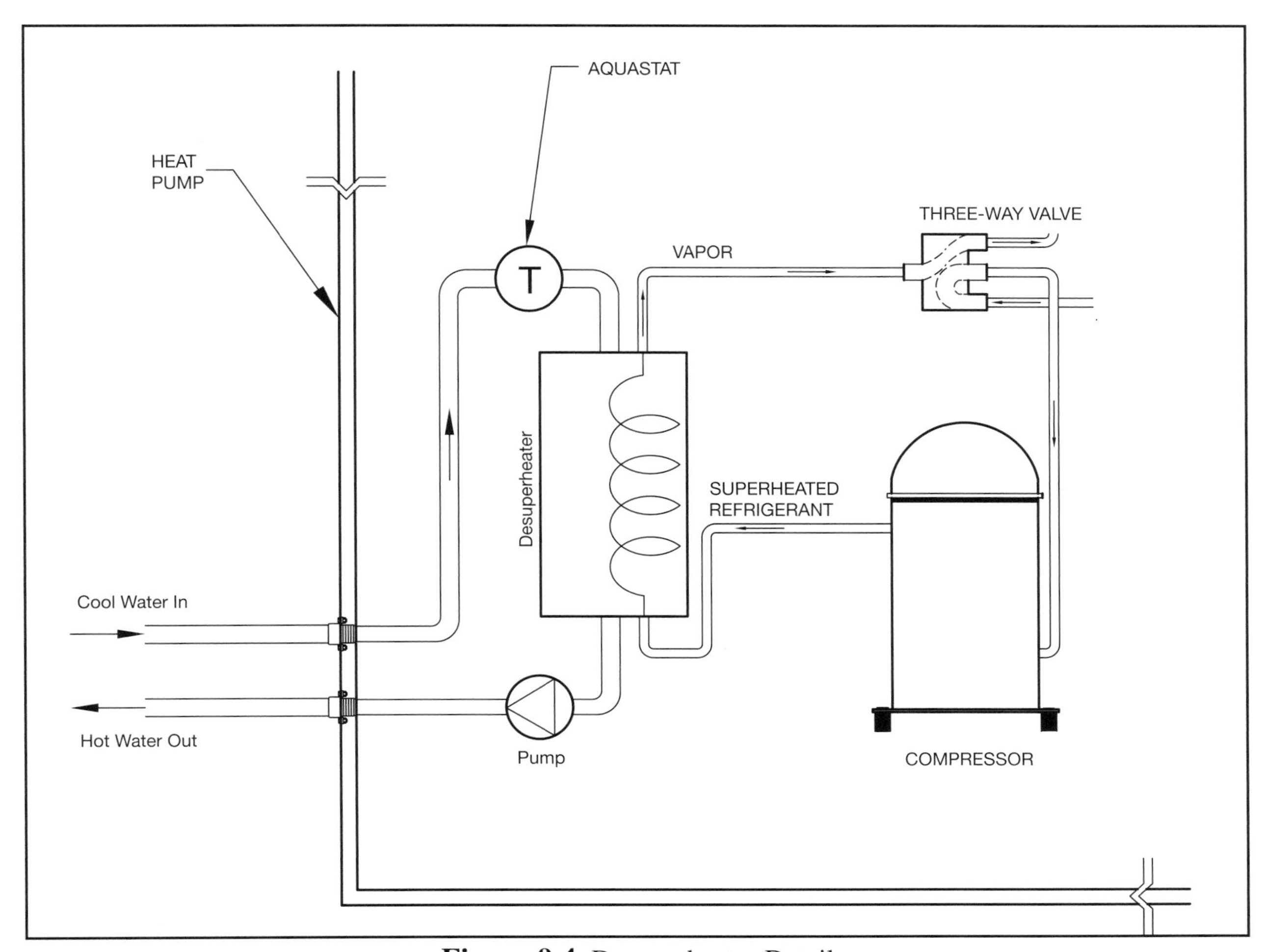

Figure 9.4. Desuperheater Detail

heating are both operating. Depending on sizing and supplemental heat use (if any), operation of the desuperheater during peak heating periods may not be advantageous. However, HWG by the GSHP is performed at a much higher efficiency than electric or gas-fired water heating.

A key cold-climate factor is low closed-loop entering water temperatures from the ground heat exchanger. During the peak winter heating season, closed-loop GSHP EWTs will be cold, possibly falling into the low 30s to mid 20s. When this occurs, compressor discharge vapor temperatures may be cooler than the water from the domestic hot water tank, which may reverse the operation of the desuperheater. Under these conditions, the cooler vapor coming from the compressor may absorb heat from the circulating water. Desuperheaters should not be allowed to operate under these conditions. Consult the manufacturer's control circuit data to see if it includes means to automatically interrupt desuperheater operation under these conditions, if it includes a desuperheater ON-OFF dip switch on the solid state control board, or if it includes a switch in the desuperheater pump circuit to manually turn off the desuperheater. If there is no factory-installed means to disable the desuperheater, a field installed manual ON-OFF switch must be placed in the desuperheater pump circuit, as shown in Figure 9.5. Be sure to contact the equipment manufacturer to determine 1) if the manufacturer allows this system modification, 2) if the manufacturer has a kit to facilitate this modification, and 3) if the manufacturer does not have a kit, do they have a recommended switch. If the desuperheater is manually controlled, it must be turned back on after peak winter heating and the resulting low ground heat exchanger temperatures moderate.

During operation in the cooling mode, the energy removed from the cooled space along with the electrical energy used to operate the GSHP must be rejected to the ground loop. A portion of that heat can be reclaimed by the desuperheater as long as the refrigerant temperature leaving the compressor is warmer than the supply water from the domestic hot water storage tank entering the desuperheater. If the domestic hot water from the storage tank is warmer than the superheated refrigerant leaving the compressor, then it is possible that the hot water will be cooled by the refrigerant and the heat will be lost from the storage tank. Otherwise, domestic HWG by the desuperheater is free in the cooling mode because heat absorbed by the circulated domestic hot water would otherwise be rejected to the GHEX. The only electrical energy required for desuperheater operation is that used by a low wattage domestic hot water circulating pump. Use of a desuperheater in cooling-dominant climates can also minimally reduce the long-term unbalanced ground load effects that significantly increase ground-loop sizing.

During the moderate heating months of fall and early winter, or late winter and early spring, the heat pump is somewhat oversized for the building loads and has excess capacity to use for heating domestic hot water. However, because the heat pump does not run as often, the desuperheater will not be as effective for hot water generation because the chances of needing hot water when the GSHP is operating are reduced. If significant hot water requirements exist, using the desuperheater for HWG during such times will result in extended heat pump run time and increased GSHP efficiency due to a small reduction in equipment cycling.

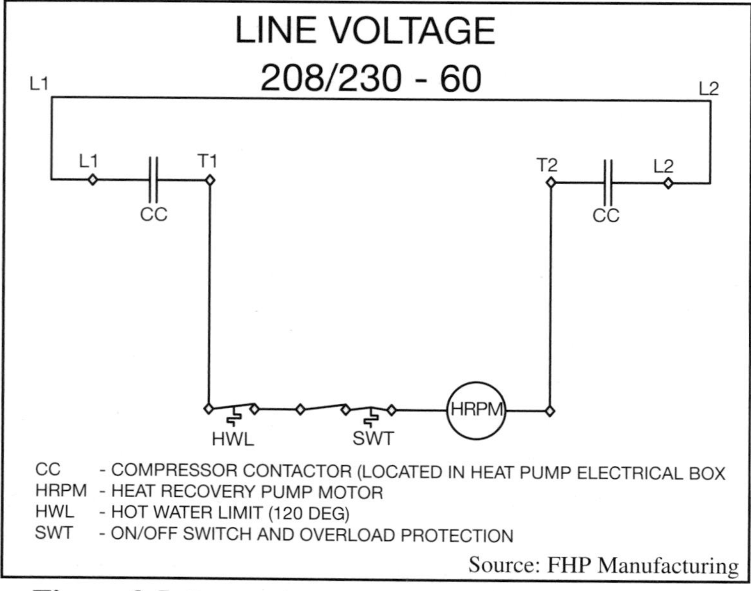

Figure 9.5. Desuperheater Line Voltage Control Diagram

9.1.2.2 DESUPERHEATER OPTIONS & CONFIGURATIONS Factory installed desuperheaters are an option available from all GSHP manufacturers. The desuperheater option is widely sold, but optional because its installation is dictated by the number of hot water tanks a structure may have or by the use of alternative water heating means. For example, if desuperheaters were to be used in a the building with four GSHP units and two hot water tanks, two of the heat pumps would have desuperheaters and two would not. The vast majority of optional desuperheaters have the water-to-refrigerant heat exchanger, water circulating pump, controls, and power connections built into the GSHP cabinet, as illustrated in Figure 9.3. A minority of desuperheater options are add-on packages that are in a small cabinet, but attached to the exterior of the GSHP unit and incorporated into the refrigerant, power, and control circuits by the manufacturer at the factory. An externally-mounted desuperheater is shown in Figure 9.6.

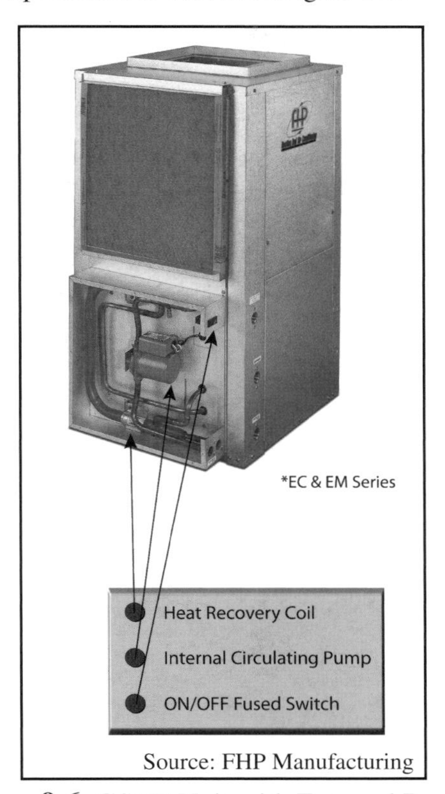

Source: FHP Manufacturing

Figure 9.6. GSHP Unit with External Desuperheater

Field installed desuperheaters are not recommended. In the unlikely event that a GSHP unit needs to have a desuperheater added in the field, contact the manufacturer for possible impact on warranty, control circuit, refrigerant circuit, etc. Field addition of an aftermarket packaged desuperheater may not be recommended by the GSHP manufacturer, may void the ground source heat pump warranty, and/or may not properly interface with solid state control circuits. If an aftermarket desuperheater will be installed, follow the desuperheater manufacturer's installation and refrigerant charge adjustment instructions, then evacuate and recharge the ground source heat pump/desuperheater in accordance with the GSHP manufacturer's evacuation and refrigerant charging procedures. Key concerns will be accurate refrigerant charge for optimum GSHP unit performance and efficiency and proper evacuation of the refrigerant circuit before charging to ensure all contaminants and moisture have been removed from the refrigerant circuit and compressor oil.

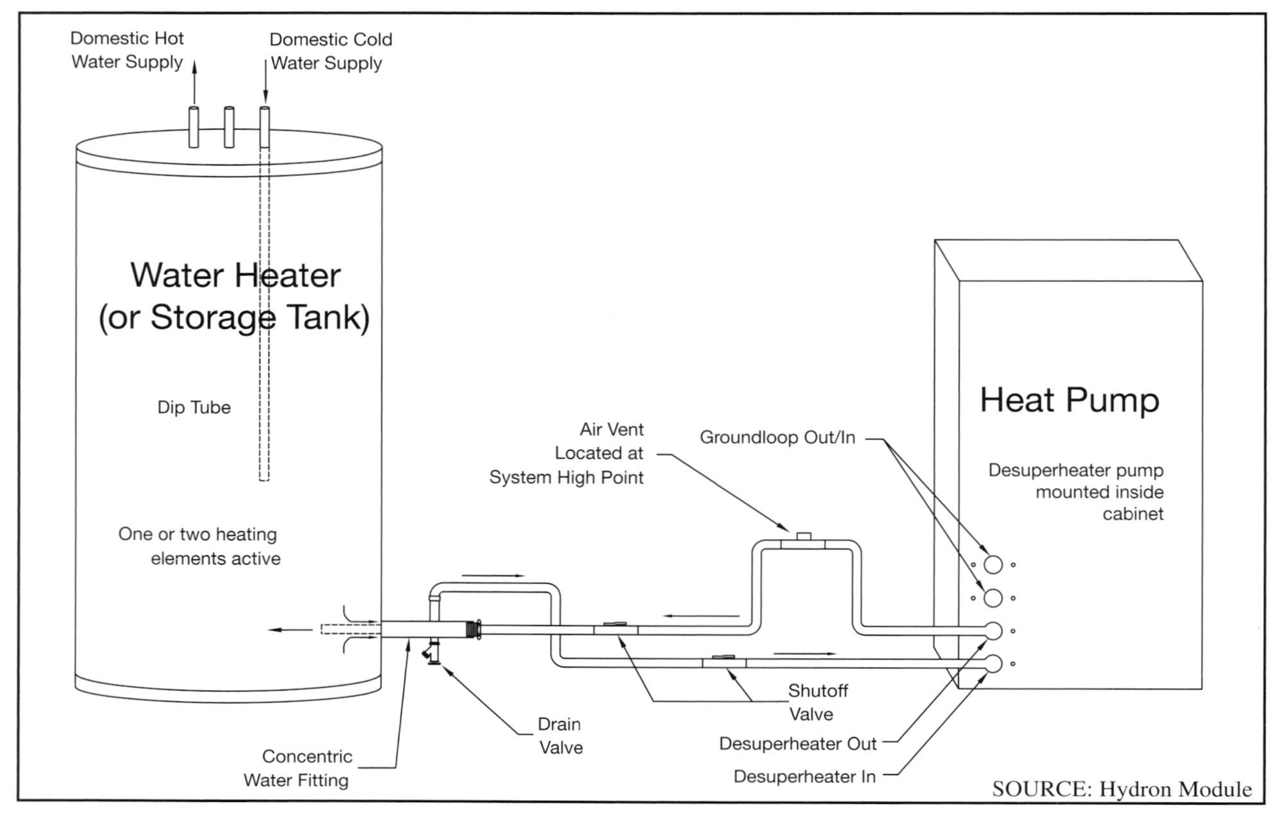

Figure 9.7. Single-Tank Desuperheater System (Concentric Connection)

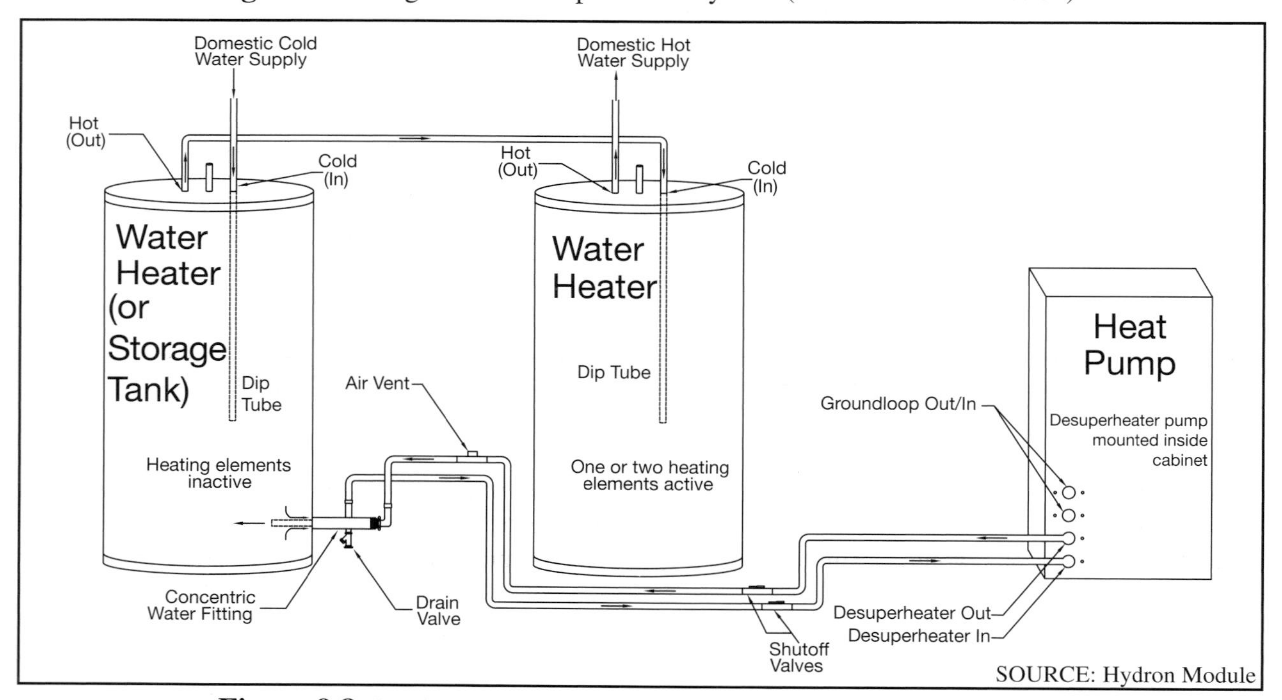

Figure 9.8. Dual-Tank Desuperheater System (Concentric Connection)

9.1.2.3 WATER HEATER TO DESUPERHEATER CONFIGURATIONS Several different water heater-to-desuperheater configurations exist. A single-tank desuperheater configuration is displayed in Figure 9.7. The advantage of using this configuration is minimal capital cost for equipment compared to alternative configurations. The single-tank desuperheater configuration is the most common configuration, but is the least efficient of the three typical configurations.

A dual-tank desuperheater configuration is displayed in Figure 9.8. If a dual-tank desuperheater configuration is utilized, hot water used in the space will be taken from the second storage tank. The desuperheater will generate hot water to be stored in the first tank, also known as the preheat tank. The first storage tank will feed the second whenever hot water is pulled from the second tank for domestic use. Using a dual-tank system will maximize desuperheater performance because water temperatures entering the desuperheater circuit from the first storage tank will, on average, be significantly lower than they would coming from a single-tank configuration. The dual-tank desuperheater configuration is more efficient than the single-tank configuration, but is more expensive because two tanks must be purchased.

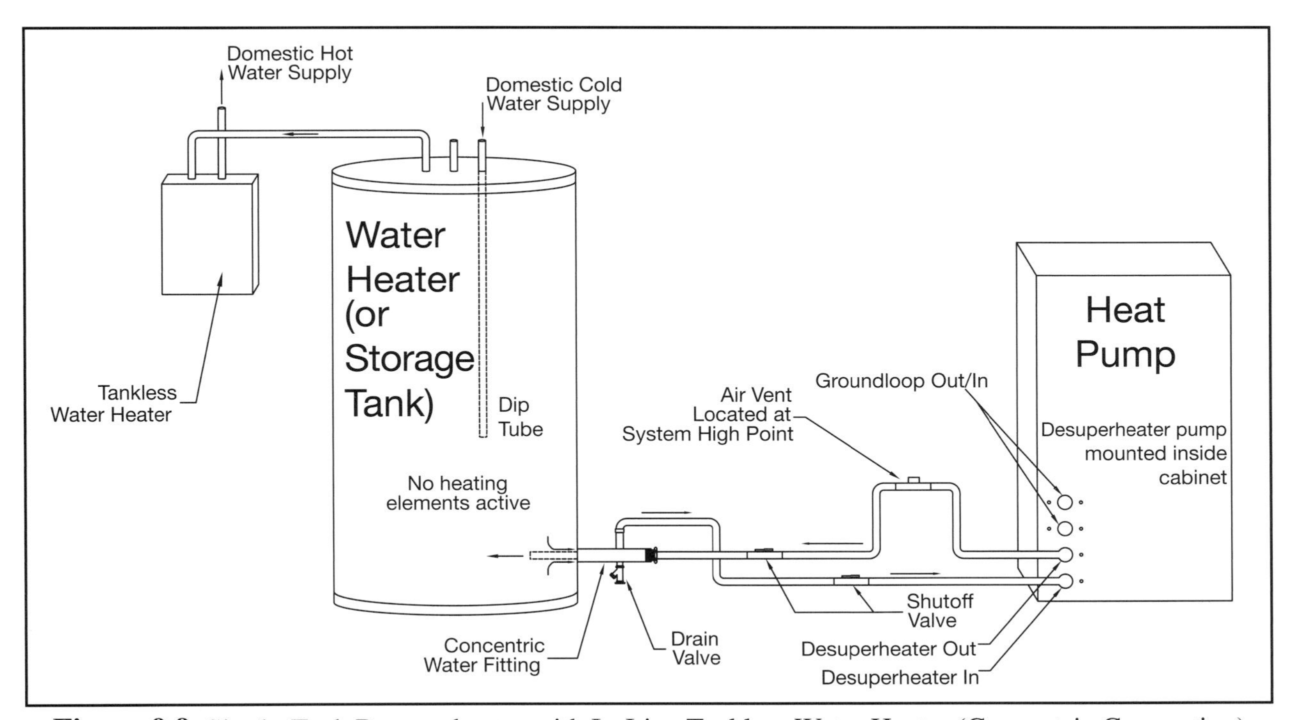

Figure 9.9. Single-Tank Desuperheater with In-Line Tankless Water Heater (Concentric Connection)

A third possible supplemental HWG configuration would be a single-tank configuration with an in-line, tankless water heater as shown in Figure 9.9. The benefits of using a single-tank configuration with an in-line tankless water heater would be similar to the dual-tank configuration in that desuperheater capacity and efficiency is maximized due to lower water temperatures entering the desuperheater circuit. There is the additional benefit with the tankless water heater in that there will be minimal energy losses to the surroundings from the HWG system. The single-tank configuration with in-line tankless HWG is the most efficient configuration of the three.

Figures 9.7-9.9 displayed various desuperheater connections using the concentric fitting. Figure 9.10 displays the concentric fitting in more detail.

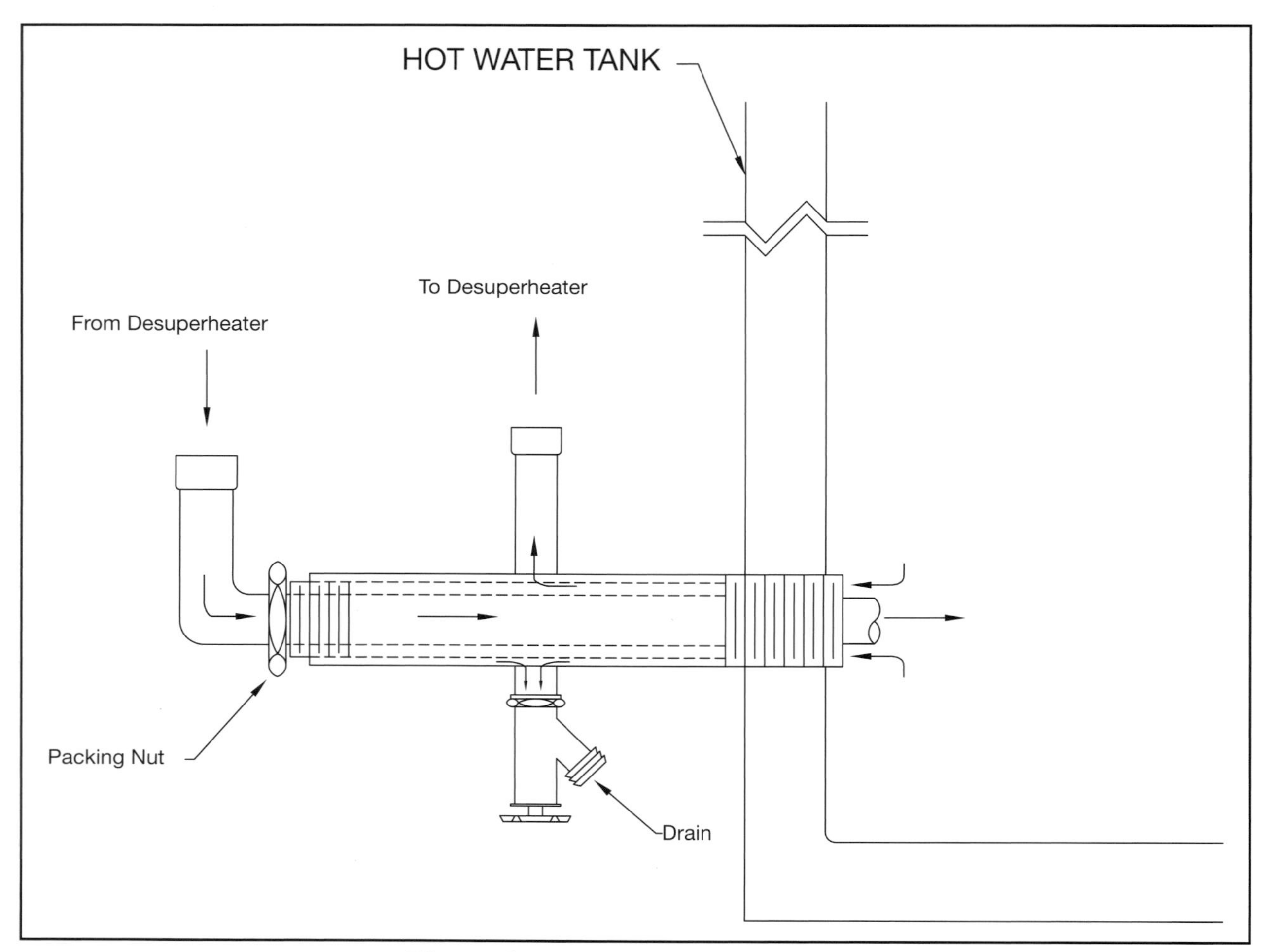

Figure 9.10. Concentric Fitting Detail

Electric water heaters are recommended for use with a GSHP desuperheater when the concentric connection method is used. If the concentric connection method is used in conjunction with a gas, propane, or fuel oil water heater, the dual-tank desuperheater configuration must be used, with the second tank being the combustion-based HWG tank. Additionally, if an electric water heater that has a single heating element is utilized, the dual-tank desuperheater configuration is recommended to ensure water temperatures low enough to promote HWG via the desuperheater circuit. The GSHP unit, water piping, pump, and hot water tank(s) should not be located in a space where the temperature falls below 50 F (Source: ClimateMaster Residential Installation Manual).

9.1.2.4 DESUPERHEATER PERFORMANCE Desuperheater performance will depend on the flow rate through the desuperheater circuit, the water temperature entering the desuperheater from the hot water tank, and the refrigerant temperature from the discharge of the compressor. Refer to Chapter 2 for further discussion on the GSHP refrigeration cycle and factors that affect its performance.

Most manufacturers' published data includes desuperheater performance data, though most is not extensive. If no desuperheater performance data is provided by the manufacturer, desuperheater capacity can be estimated to be 10% of the total heating capacity in the heating mode or 10% of the heat of rejection in the cooling mode. Figure 9.11 displays desuperheater performance data for several different GSHP models manufactured by Hydron Module:

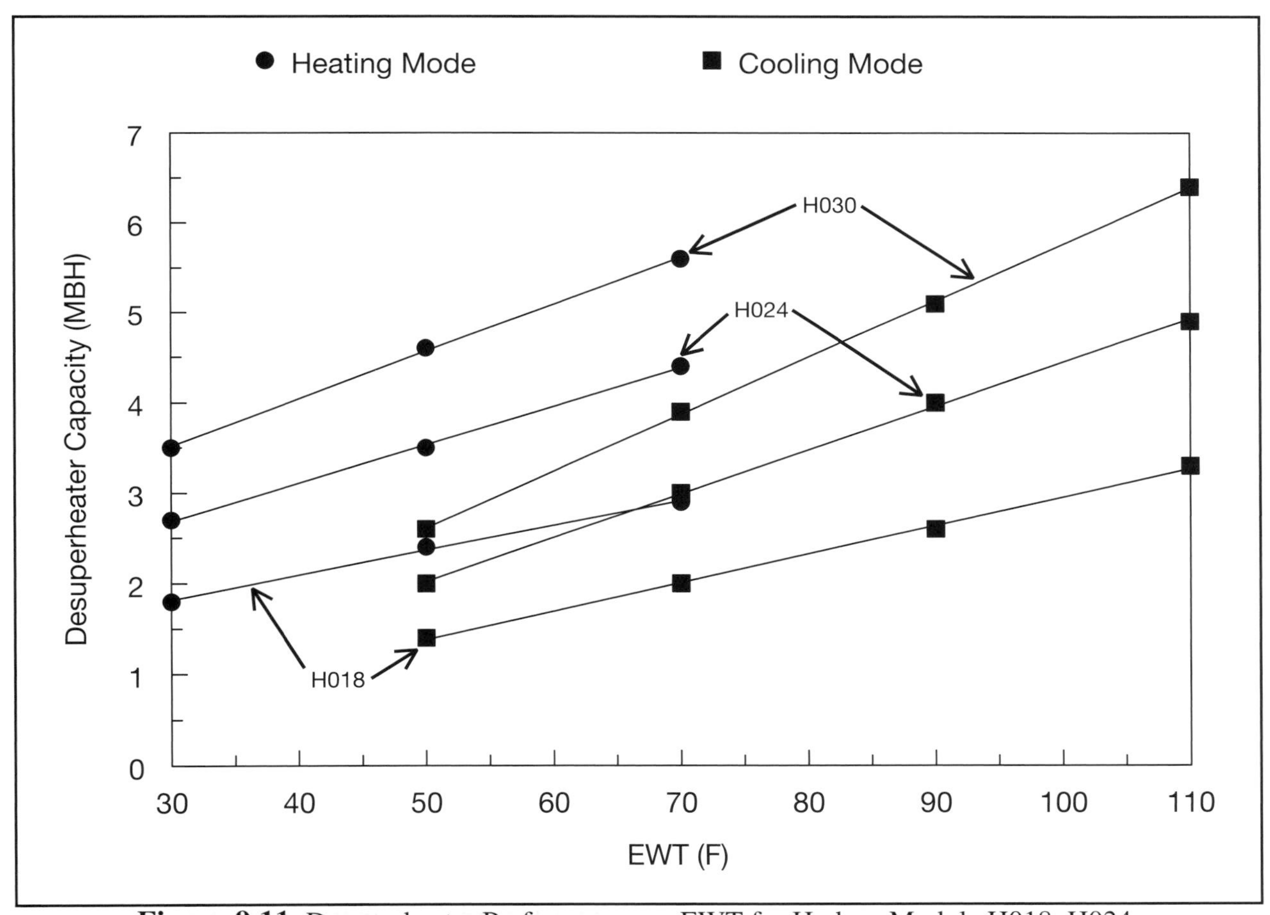

Figure 9.11. Desuperheater Performance vs. EWT for Hydron Module H018, H024, and H030 GSHP Units

The operating conditions at which the data in Figure 9.11 was recorded were 70 F EAT in heating and 80/67 F (db/wb) in cooling mode for all models, 600 CFM and 5.0 gpm (source side) for the H018, 800 CFM and 6.0 gpm (source side) for the H024, and 1000 CFM and 7.0 gpm (source side) for the H030. The manufacturer did not specify the flow rate or the temperature of the water being circulated through the desuperheater itself.

Equation 9.1 can be used to estimate the temperature rise of the domestic hot water through the desuperheater circuit.

$$\Delta T = (LWT_d - EWT_d) = \frac{CAP_d}{500 \bullet Q_d} \qquad \textbf{Equation 9.1}$$

Where:

ΔT = Domestic water temperature rise through desuperheater (F)
LWT_d = Temperature of the domestic hot water leaving the desuperheater (F)
EWT_d = Temperature of the domestic hot water entering the desuperheater (F)
CAP_d = Capacity of the desuperheater according to manufacturer's data (Btu/hr)
Q_d = Flow rate through the desuperheater circuit (gpm)

Contact the GSHP manufacturer to estimate flow rate through the desuperheater circuit given the actual installed diameter/run lengths of the desuperheater piping.

9.1.2.5 WATER HEATER PREPARATION & CONNECTION Desuperheaters can be connected to existing or new fossil fuel or electric hot water heaters. In either case, the hot water tank must be prepared for connection to the desuperheater. The necessary steps to prepare a hot water tank for desuperheater connection are as follows:

1) Turn off the electrical power or fossil fuel supply.
2) If a newly purchased hot water heater is to be used, skip ahead to Step 7. If an existing hot water heater is to be used, inspect the hot water tank. Check connection points, case condition, tank condition, elements/burner condition, temperature controls, etc., to determine if the hot water tank is in suitable condition to remain in service for an acceptable period of time. If not, replace the hot water tank.
3) Connect a hose to the drain valve on the water tank.
4) Turn off the domestic water supply to the tank, open a hot water faucet or the tank pressure relief valve, and use the tank drain valve to drain the hot water tank. Place the drain hose discharge where the outflow can be observed for sediment.
5) When the tank is drained, turn on the domestic water supply and flush the tank until the drain hose outflow is clear.
6) Close all valves and remove the drain line and drain valve in preparation for desuperheater connection.
7) Install HWG piping.

Connections on the ground source heat pump desuperheater supply and return should be swivel fittings. Some, but not all manufacturers, provide their units with swivel connectors. Using swivel connectors facilitates connection to the ground source heat pump unit and eliminates the need to add a union in the desuperheater supply and return piping. In general, desuperheater "one way" piping length between the hot water tank and the GSHP unit should be no more than the manufacturer specifies and is typically no more than 50 feet for either of the following connection methods. Either copper or PEX piping should be used for desuperheater connections and should be at least ½-inch in diameter (O.D.). Additionally, all HWG water piping and components should be insulated with no less than 3/8-inch wall closed-cell insulation. Refer to manufacturer's specifications for copper pipe sizing and insulation wall thickness recommendations.

9.1.2.5.1 CONCENTRIC WATER HEATER CONNECTION Using a concentric hot water tank connection fitting eliminates the need to tie into existing cold water piping. However, using this method can be problematic in areas where water quality is a concern. Connection to the desuperheater using the concentric hook up kit is the simplest and most widely used connection method. Figures 9.7-9.9 displayed desuperheater configurations, which used the concentric hot water tank connection method. Figure 9.10 displays the concentric fitting in detail.

1) Install the concentric connection fitting in the drain valve port of the hot water tank.
2) Install the drain valve in the concentric connection fitting.
3) Connect the desuperheater supply on the concentric fitting to the ground source heat pump desuperheater water in connection as illustrated in Figures 9.7 through 9.9. Install a shut off valve in this line near the concentric connection fitting and a "stop and waste" manual air vent in the high point of the line.
4) Connect the desuperheater return on the concentric fitting to the ground source heat pump desuperheater water out connection as shown in Figures 9.7 through 9.9. Install a shut off valve in this line near the concentric connection fitting.

Using the concentric water heater connection method can be problematic because it is more prone to scaling. In areas where water quality creates sediment in the bottom of the hot water tank, this installation method should not be used. If material builds up in the bottom of the tank it can be sucked into the desuperheater circuit, restricting or clogging piping and/or the pump impeller. This will impede or prevent desuperheater function, reducing or preventing hot water production, and the expected energy savings delivered by desuperheater operation. The concentric water heater connection method is recommended in areas with low scaling potential or where a water softener is used. In fact, use of water softening equipment is recommended on domestic water systems with supplemental HWG to reduce the probability of scaling and thus lengthening equipment life (Source: ClimateMaster Residential Installation Manual).

9.1.2.5.2 STANDARD WATER HEATER CONNECTION Using a standard water heater connection without the concentric fitting is also simple and widely used and avoids the sediment buildup problems that can occur with the concentric connection. The standard water heater connection method is shown in more detail by Figure 9.13.

1) Install a nipple and tee as shown in Figure 9.12.
2) Install the drain valve in one branch of the tee as shown in Figure 9.12.
3) Connect the desuperheater return to the other branch of the tee as shown in Figure 9.12 and to the ground source heat pump desuperheater water out connection as shown in Figures 9.13-9.15. Install a shut off valve in this line near the drain tee.
4) Install the cold water branch in the hot water tank cold water supply line immediately above the dielectric union. This branch is composed of a tee in the cold water line with a shut-off valve on the branch.
5) Connect the desuperheater supply from the cold water supply branch tee as shown in Figures 9.13-9.15 to the ground source heat pump desuperheater water in connection.

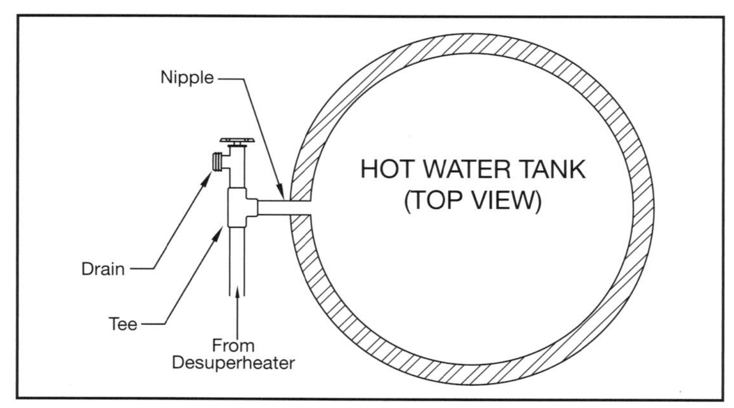

Figure 9.12. Standard Water Heater Connection Method Detail

This connection method either sucks water from the hot water tank (through the dip tube) to be circulated through the desuperheater, or circulates make up water (to replace used hot water) through the desuperheater before it enters the hot water tank. The desuperheater return to the hot water tank flows into the bottom of the tank through the drain port, ensuring that no sediment is sucked into the desuperheater piping or pump impeller. If the heat pump is running, make-up water will be preheated before it enters the hot water tank. The desuperheater water-to-refrigerant heat exchanger will regularly see both high-temperature water and cold-temperature makeup water, chill shocking the heat exchanger. This causes thermal expansion and contraction of the heat exchanger, which will keep scale from building up on the heat exchanger walls. If the standard water heater connection method is used, always check the dip tube in the water heater for a check valve. If a check valve is present, it will prevent water flow through the dip tube to the desuperheater. The absence of water flow can damage the desuperheater circulator. The standard water heater connection method is shown with the three desuperheater configurations in Figures 9.13-9.15.

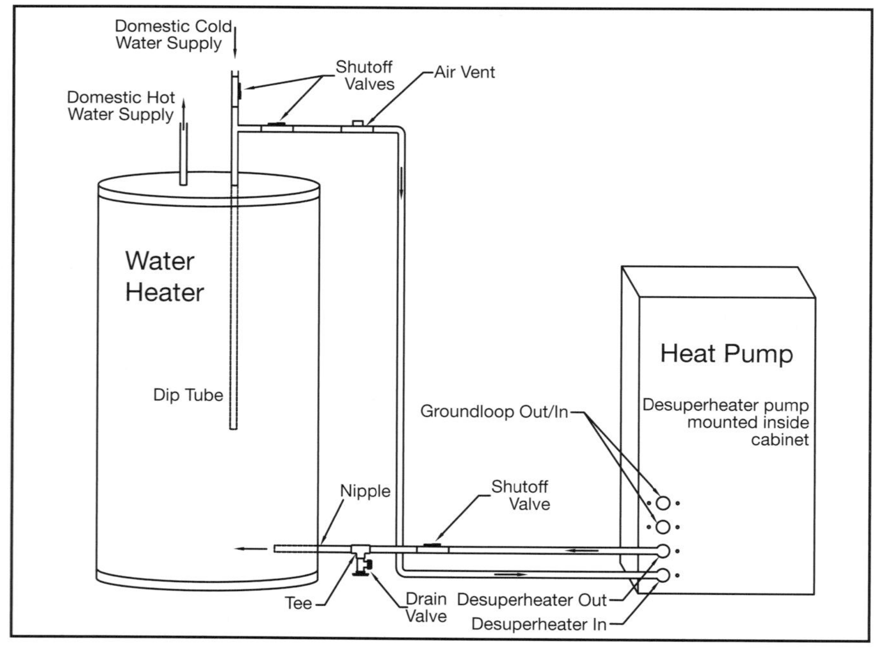

Figure 9.13. Single-Tank Desuperheater System (Standard Connection)

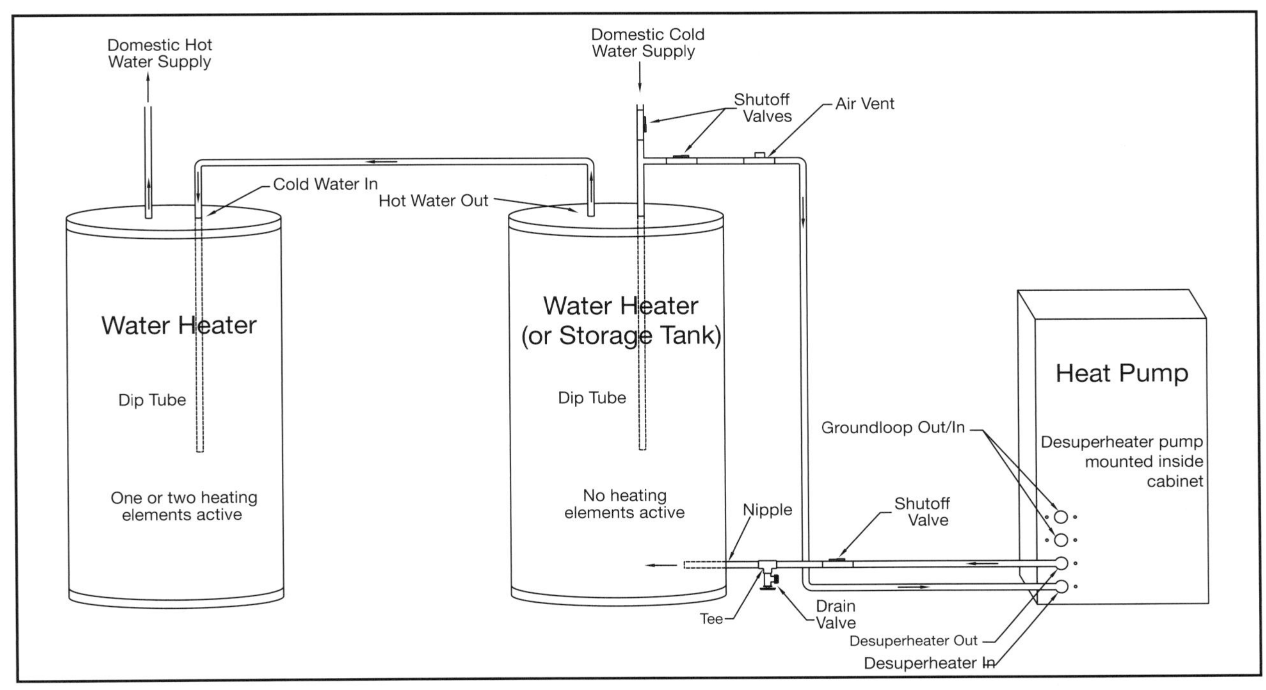

Figure 9.14. Dual-Tank Desuperheater System (Standard Connection)

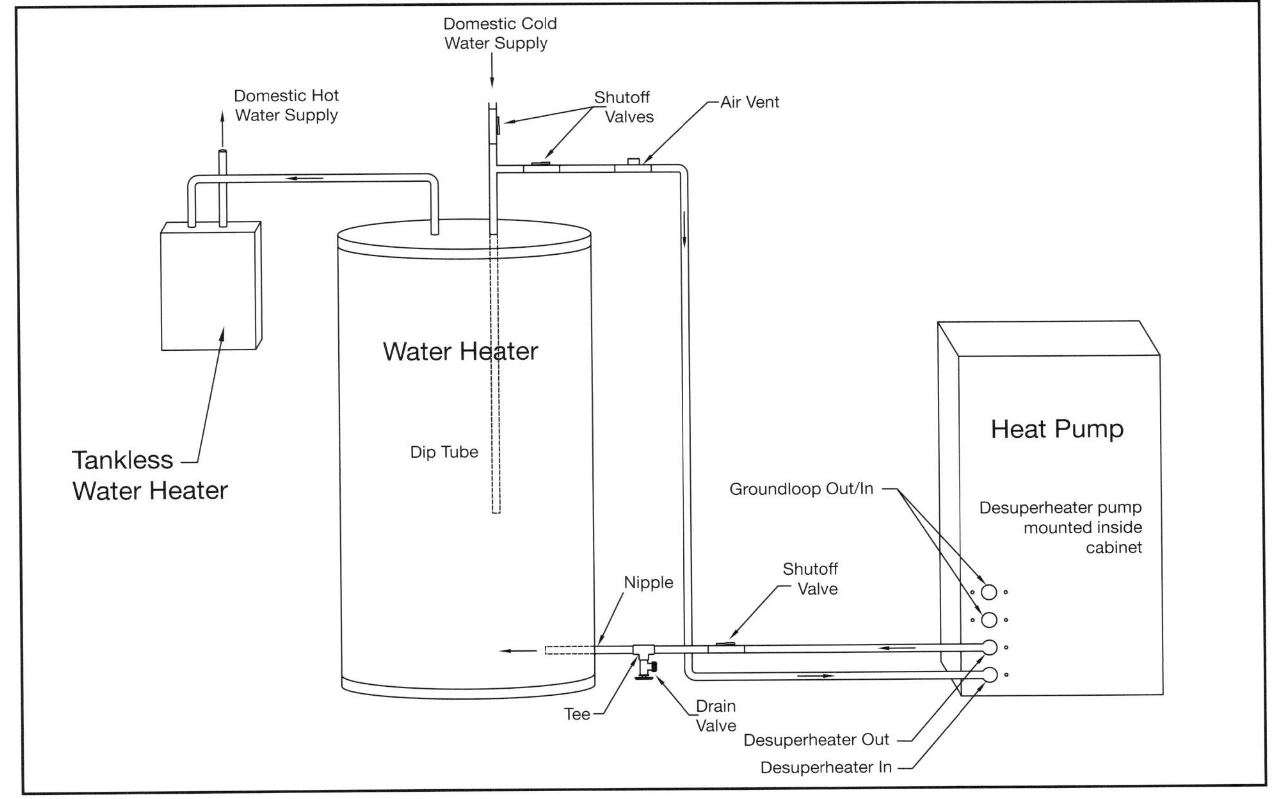

Figure 9.15. Single-Tank Desuperheater with In-Line Tankless Water Heater (Standard Connection)

9.1.2.6 DESUPERHEATER START-UP PROCEDURE Start-up and operation is the same for any connection methodology (concentric versus standard). First, the hot water tank must be set up for desuperheater operation:

1) Close the hot water tank drain valve, open the cold water supply, and fill the tank. While the tank is filling, open a hot and cold water faucet to vent any air in the domestic water piping.
2) On concentric hook up kit or dual tank applications, use the manual air vent to clear air from the desuperheater piping system. This manual vent is not required on sediment avoiding connection systems.
3) Reset hot water tank thermostats:
    a. Hot water tanks with two elements should be reset from factory settings. The lower element should be set to approximately 100 F. The upper element should be set to approximately 120 F.
    b. Hot water tanks with a single element or burner should be reset from factory settings. The element or burner should be set to approximately 120 F or the "low" setting. These settings will maximize the utilization of heat energy available from the refrigerant circuit.
4) Replace the access cover(s) and restore the electricity or fossil fuel supply to the hot water tank.
5) Check desuperheater connection line valves to be sure they are in the open position.
6) Check the desuperheater pump to make sure it is properly lubricated. Some of the water lubricated pumps have a large screw in the head of the pump. With the tank filled and under pressure, loosen/remove this screw. Leave it loose/open until water weeps from the screw hole. This weeping indicates that water has moved into the back bearings on the pump shaft and the pump bearings are properly lubricated.
7) Turn on the ground source heat pump using the thermostat. When the heat pump starts, the desuperheater pump should start, or start shortly thereafter.
8) On concentric hook up kit applications, check to confirm that the system is not air locked by using the manual air vent to clear any remaining air from the desuperheater piping system.
9) Check the supply and return water temperatures of the desuperheater after about 10-15 minutes of ground source heat pump run time. The temperature differential (rise) through the desuperheater should be 5-15 F. This differential can not be further quantified because it is site specific and depends on water temperature entering the desuperheater circuit, flow rate, etc.

9.2 Domestic Hot Water Generation Costs and GHEX Design Parameters

9.2.1 DOMESTIC HOT WATER GENERATION ENERGY AND COST CALCULATIONS

The annual energy requirements and cost to generate hot water for domestic use depends on many factors, including:

1. The annual volume of hot water consumption
2. The average temperature of the cold water supply
3. The hot water storage tank setpoint temperature
4. The heat loss from the hot water storage tank
5. The energy source cost, and
6. The efficiency of the water heating device

The annual energy required to generate hot water depends directly on the annual volume of hot water consumption, which is directly related to the type of facility and the occupancy. A commercial food service facility, for example, will have a large hot water load that must be determined on a case-by-case basis. A residential structure will have hot water use that depends on the number of occupants and their style of living (showers per day, volume of clothes to be washed, etc.). Table 9.1 provides average overall and peak hot water uses on an hourly, daily, weekly, and monthly basis for 'All Families' and 'Typical Families' *(ASHRAE 2011)*. The overall or peak monthly hot water use values in Table 9.1 can be used to estimate the monthly hot water use requirements for residential applications, and the annual consumption of hot water can be calculated according to Equation 9.2.

$$AHWU = MHWU \cdot 12$$ **Equation 9.2**

Where:

AHWU = annual hot water usage (gal/mo)
MHWU= monthly hot water usage (gal/yr)

Table 9.1. Overall (OVL) and Peak Average Hot-Water Use[1]

	Average Hot Water Use, gal							
	Hourly		Daily		Weekly		Monthly	
Group	OVL	Peak	OVL	Peak	OVL	Peak	OVL	Peak
All Families	2.6	4.6	62.4	67.1	436	495	1897	2034
"Typical Families"	2.6	5.8	63.1	66.6	442	528	1921	2078

1 - ASHRAE 2011 Handbook - HVAC Applications (Chapter 50)

The temperature of the cold water supply into the water heating system depends primarily on location and, without better information, can be estimated on an annual basis to be approximately equal to the average temperature of the earth in the top 10 feet or T_M from Table 5.11.

Table 9.2 provides representative hot water temperatures typical for various applications. Many uses for domestic hot water can be served with a hot water storage tank setpoint of approximately 120 F, a temperature which GSHP units can easily generate with a desuperheater or efficiently generate with dedicated water-to-water equipment. Some manufacturers have equipment with capabilities to generate hot water up to 145 F. Always check manufacturers' specifications to determine the capabilities of their water-to-water GSHP equipment. If water temperatures greater than 120 F are required, and the water-to-water GSHP that is being used can only produce hot water temperatures up to 120 F, a second-stage water heater (tank with heating capability or a tankless water heater) may be necessary to "boost" the temperatures. Such a heater would only be necessary to provide high-temperature water for fixtures that require higher water temperatures, such as dishwashing stations and other light commercial applications.

Table 9.2. Representative Hot-Water Temperatures for Various Uses[1]

Use	Temp (˚F)
Lavatory - Hand Washing	105
Lavatory - Shaving	115
Showers & Tubs	110
Therapeutic Baths	95
Commercial or Institutional Laundry	up to 180
Residential Dish Washing & Laundry	140

1 - ASHRAE 2011 Handbook - HVAC Applications (Chapter 50)

To determine the necessary installed HWG capacity to accommodate a hot water load, the hot water generation rate and storage capacity that is required must be estimated for the size of the residence. Table 9.3 gives minimum water heater capacities and storage tank sizes for one and two-family living units. Table 9.3 should be used to determine storage capacity and water-to-water GSHP equipment heating capacity for dedicated domestic HWG. For example, a home with 2.5 baths and 4 bedrooms would require a 50-gallon hot water storage tank (minimum) coupled with a GSHP unit with at least 18,766 Btu/hr heating capacity at loopfield design conditions (EWT_{min}). A GSHP unit used for dedicated domestic HWG must have a vented, double-walled water-to-refrigerant heat exchanger to minimize the risk of a refrigerant leak into the domestic water supply.

Table 9.3. HUD-FHA Minimum Water Heater Capacities for One & Two-Family Living Units[1]

Number of Baths	1 to 1.5			2 to 2.5				3 to 3.5			
Number of Bedrooms	1	2	3	2	3	4	5	3	4	5	6
Storage, gal	20	30	40	40	50	50	66	50	66	66	80
kW Input	2.5	3.5	4.5	4.5	5.5	5.5	5.5	5.5	5.5	5.5	5.5
Heating Cap (Btu/hr)	8,530	11,942	15,354	15,354	18,766	18,766	18,766	18,766	18,766	18,766	18,766
1 hr draw, gal	30	44	58	58	72	72	88	72	88	88	102
Recovery, gph	10	14	18	18	22	22	22	22	22	22	22

1 - ASHRAE 2011 Handbook - HVAC Applications (Chapter 50)

According to Table 9.3, most hot water heating systems consist of a storage tank with heating elements that maintain the temperature of the water in the tank at a preset level. The location of the domestic hot water storage tank in a conditioned space (50 to 80 F) will lead to heat loss from the tank (referred to as standing tank loss) that can be estimated to be between 10 and 20% of the total water heating energy usage.

Equation 9.3 can be used to estimate the annual energy (Btu) that must be delivered to the annual hot water usage volume (gal) to raise its temperature from T_{supply} to T_{tank} and accounts for standing tank losses (STL).

$$AHWE = \frac{AHWU \cdot 62.3 \cdot 1}{7.48} \cdot \left(T_{tank} - T_{supply}\right) \cdot \left(1 + \frac{STL}{100}\right) \qquad \textbf{Equation 9.3}$$

Where:

AHWE = Annual energy to heat water (Btu/yr)
T_{tank} = Hot water storage tank setpoint temperature (F)
T_{supply} = Cold water supply temperature (F)
STL = Standing tank losses (%)

The type and efficiency of the water heating system will determine the volume of primary fuel required to deliver the AHWE. The most common domestic hot water systems include electric resistance heating, propane heating, and natural gas heating. Heating water with a GSHP system utilizes electricity to move a majority of the energy to the water from a ground energy source with a coefficient of performance that results in an electric energy consumption of 1/4th to 1/3rd that of an electric resistance water heating system. Equations 9.4 through 9.7 can be used to calculate the kWhs for an electric resistance water heating system, the gallons of propane for a propane water heating system, the therms of natural gas for a natural gas water heating system, and the kWhs for a GSHP water heating system, respectively.

$$kWh_{elec-A} = \frac{AHWE}{\frac{\eta_{elec}}{100} \cdot 3{,}412}$$

Equation 9.4

Where:

kWh_{elec-A} = Annual resistance heat electric energy to heat water (kWh/yr)
η_{elec} = Efficiency of electric water heating system (%)

$$Gal_{Prop-A} = \frac{AHWE}{\frac{\eta_{propane}}{100} \cdot 92{,}000}$$

Equation 9.5

Where:

Gal_{Prop-A} = Annual gallons of propane to heat water (gal/yr)
η_{Prop} = Efficiency of propane water heating system (%)

$$Therm_{NG-A} = \frac{AHWE}{\frac{\eta_{NG}}{100} \cdot 100{,}000}$$

Equation 9.6

Where:

$Therm_{NG-A}$ = Annual therms of natural gas to heat water (therms/yr)
η_{NG} = Efficiency of natural gas water heating system (%)

$$kWh_{GSHP-A} = \frac{AHWE}{COP_{avg} \cdot 3{,}412}$$

Equation 9.7

Where:

kWh_{GSHP-A} = Annual GSHP electric energy to heat water (kWh/yr)
COP_{Avg} = Average annual efficiency of GSHP water heating system (%)

The annual cost for each system is the number of fuel units multiplied by the cost per fuel unit, as shown by Equations 9.8 through 9.11. For a domestic water heating system that combines a GSHP water heating system with one of the other water heating systems, the annual cost of producing hot water will depend on the percentage of the annual hot water requirement produced by each (Equation 9.12).

$$Cost_{elec-A} = kWh_{elec-A} \bullet ElectricEnergyCost$$ **Equation 9.8**

Where:

$Cost_{elec-A}$ = Annual cost to heat water using electric resistance water heating system ($/yr)
Electric Energy Cost = Unit cost of electric energy ($/kWh)

$$Cost_{Prop-A} = Gal_{Prop-A} \bullet PropaneEnergyCost$$ **Equation 9.9**

Where:

$Cost_{Prop-A}$= Annual cost to heat water using propane water heating system ($/yr)
Propane Energy Cost = Unit cost of propane energy ($/gal)

$$Cost_{NG-A} = Therm_{NG-A} \bullet NaturalGasEnergyCost$$ **Equation 9.10**

Where:

$Cost_{NG-A}$= Annual cost to heat water using natural gas water heating system ($/yr)
Natural Gas Energy Cost = Unit cost of natural gas energy ($/therm)

$$Cost_{GSHP-A} = kWh_{GSHP-A} \bullet ElectricEnergyCost$$ **Equation 9.11**

Where:

$Cost_{GSHP-A}$= Annual cost to heat water using GSHP water heating system ($/yr)
Electric Energy Cost = Unit cost of electric energy ($/kWh)

$$Cost_{Combined} = \frac{\%HWG_{GSHP-A}}{100} \bullet Cost_{GSHP-A} + \left(1 - \frac{\%HWG_{GSHP-A}}{100}\right) \bullet Cost_{OtherFuel-A}$$ **Equation 9.12**

Where:

$Cost_{Combined}$ = Annual cost to heat water using a combined GSHP/other fuel water heating system ($/yr)
$\%HWG_{GSHP-A}$ = Percentage of annual hot water use generated by GSHP water heating system (%)
$Cost_{OtherFuel-A}$ = Annual cost of electric resistance, propane, or natural gas water heating system ($/yr)

9.2.2 DETERMINATION OF HOT WATER GENERATION GHEX DESIGN PARAMETERS

When the GSHP is utilized to make hot water, the kWhs of electrical energy calculated using Equation 9.7 represents only a portion of the energy required to heat the water. The majority of the energy comes from the ground connection, which adds to the total energy that is extracted from the GHEX annually. For annual hot water generation, the annual energy ground load can be calculated using Equation 9.13.

$$AGL_{HWG} = AHWE \bullet \left(\frac{COP_{Avg} - 1}{COP_{Avg}} \right) \qquad \textbf{Equation 9.13}$$

For design of the vertically and horizontally-bored GHEX, the additional annual ground load for hot water generation will further unbalance the annual ground energy load and result in a larger adjustment to the GHEX lengths if the design is heating dominant. If the hot water generation capability of the GSHP system is disabled for some reason the resulting GHEX design will be slightly conservative. For design of the vertically and horizontally-bored GHEX in the cooling mode, the additional annual ground load for hot water generation will help to balance the annual ground energy load and result in a smaller adjustment to the GHEX lengths. However, if the hot water generation capability of the GSHP system is disabled for some reason, the resulting GHEX design may result in an increase in entering water temperatures above the design value over the long term. For that reason, it would be conservative to ignore the annual ground load for hot water generation for cooling dominant designs, unless the hot water generation capabilities of the GSHP system will be consistently utilized. For horizontally-trenched GHEX design, there is no adjustment for unbalanced ground loads and the additional annual ground energy load for hot water generation is not an issue.

The design of the GHEX in the heating mode is based on the space heating ground load, a combination of GSHP heating capacity, COP, and run fraction. However, if hot water is being generated using a dedicated GSHP or a GSHP with a desuperheater, the GHEX design in the heating mode must also account for the additional heating ground load and run fraction imposed by the GSHP-based domestic hot water generation system during the design heating month. The heating capacity of the GSHP will be the heating capacity of the water-to-water GSHP for the dedicated or combination unit, or 10 to 15% of the heating capacity of the GSHP when utilizing the desuperheater option. The design month hot water generation run fraction for the water-to-water GSHP is calculated using Equation 9.16 by first estimating the design month hot water energy (Equation 9.14) and the design month hot water run hours (Equation 9.15) as inputs.

$$DMHWE = \frac{\%HWG_{GSHP-DM} \bullet MHWU \bullet 62.3 \bullet 1}{7.48} \bullet (T_{tank} - T_{supply}) \bullet \left(1 + \frac{STL}{100}\right) \qquad \textbf{Equation 9.14}$$

Where:

$DMHWE$ = Design month hot water energy (Btu)
$\%HWG_{GSHP-DM}$ = Percent of design month hot water generated by GSHP (%)

$$DMHWRH = \frac{DMHWE}{HC_{HWG}} \qquad \textbf{Equation 9.15}$$

Where:

$DMHWRH$ = Design month hot water run hours (hrs)
HC_{HWG} = Heating capacity of GSHP (Btu/hr)

$$F_{HWG} = \frac{DMHWRH}{31 \cdot 24}$$ **Equation 9.16**

Where:

F_{HWG} = Design month run fraction of GSHP

The design of the GHEX in the heating mode will take the conservative approach and include the heating ground load and run fraction imposed by the GSHP-based domestic hot water generation system during the design heating month. The operation of a GSHP to generate hot water during the design heating month will increase the heating ground load and run fraction. In addition, the flow rate and annual hot water generation ground load effects on the design of the GHEX will be accounted for, which depend on the type of GSHP water heating being utilized.

1. **Dedicated domestic hot water generation system utilizing a water-to-water GSHP** The heating capacity of the water-to-water GSHP will be selected for recommended hot water recovery rates (Table 9.3), and will add to the total heating capacity that the GHEX must deliver at design heating conditions. The electric demand of the water-to-water GSHP will add to the total demand that the combined GSHP system will require, resulting in an average efficiency for all connected GSHP units at design conditions. The design month hot water generation run fraction for the water-to-water GSHP is calculated using Equation 9.16 by first estimating the design month hot water energy (Equation 9.14) and the design month hot water run hours (Equation 9.15) as inputs. Because the water-to-water GSHP will run independently of the space heating GSHP unit(s), and all may run at the same time, the GHEX must accommodate the combined water flow rates of all connected units.
2. **Dedicated domestic hot water generation system utilizing a combination GSHP** The heating capacity of the GSHP will be determined by either heating or cooling space load, and the unit will be controlled to heat water as a priority (making 100% of domestic hot water requirements) or as secondary to the space heating or cooling loads (making less than 100% of hot water requirements). Therefore, there will be no additional heating capacity, demand, or water flow rate that the GHEX will need to accommodate beyond those of the combination GSHP. The design month hot water generation run fraction for the combination GSHP is calculated using Equation 9.16 by first estimating the design month hot water energy (Equation 9.14) and the design month hot water run hours (Equation 9.15) as inputs.
3. **Supplemental (Desuperheater) domestic hot water generation** The heating capacity of the desuperheater will be determined from the manufacturer's literature or estimated as 10 to 15% of the heating capacity of the GSHP at design heating conditions. There will be no additional heating capacity, demand, or water flow rate that the GHEX will need to accommodate beyond those of the GSHP. The design month hot water generation run fraction for the combination GSHP is calculated using Equation 9.16 by first estimating the design month hot water energy (Equation 9.14) and the design month hot water run hours (Equation 9.15) as inputs.

The design of the GHEX in the cooling mode will take the conservative approach and ignore the heating ground load and run fraction imposed by the GSHP-based domestic hot water generation system during the design cooling month. Although the operation of a GSHP to generate hot water during the design cooling month will reduce the cooling ground load and run fraction, there is no guarantee that there will be a need for generation of domestic hot water during peak cooling periods. Only flow rate and annual hot water generation ground load effects on the design of the GHEX will be accounted for, which depend on the type of GSHP water heating being utilized.

1. **Dedicated domestic hot water generation system utilizing a water-to-water GSHP** The heating capacity, electric demand, and design month hot water generation run fraction of the water-to-water GSHP will be ignored relative to the GHEX design. Because the water-to-water GSHP will run independently of the space cooling GSHP unit(s), and all may run at the same time, the GHEX must accommodate the combined water flow rates of all connected units.
2. **Dedicated domestic hot water generation system utilizing a combination GSHP** The heating capacity, electric demand, and design month hot water generation run fraction of the GSHP will be ignored relative to the GHEX design. Also, there will be no additional water flow rate beyond that of the combination GSHP that the GHEX will need to accommodate.
3. **Supplemental (Desuperheater) domestic hot water generation** The heating capacity of the desuperheater, the electric demand, and the design month hot water generation run fraction will be ignored at design cooling conditions. There will be no additional heating capacity, demand, or water flow rate beyond those of the GSHP that the GHEX will need to accommodate.

The calculations to estimate the annual cost of hot water generation for each of the four common hot water heating systems, the annual ground heating energy load, and the design month data for input to the GHEX design have been summarized in Figure 9.16. The first set of calculations provides the cost of generating the annual hot water needs using each fuel source independently. These numbers can be compared to determine the opportunity for cost savings using a dedicated GSHP water heating system (100% of hot water generation) or a desuperheater GSHP water heating system (percentage of hot water generation can only be estimated, and ranges are provided based on the GSHP water heating configuration in a table in Figure 9.16). The annual ground heating energy load and the design month data for input to the GHEX design are summarized for direct input to the GHEX design worksheets.

Hot Water Generation Worksheet

Annual Cost of Operation Calculations

Monthly Hot Water Usage (MHWU) = ___1___ gal/month (Table 9.1)
Storage Tank Setpoint Temperature = ___2___ F (Table 9.2)
Cold Water Supply Temperature = ___3___ F (Table 5.11)
Storage Tank Standing Losses = ___4___ %
Annual Hot Water Usage (AHWU) (Equation 9.2) **(5)**

AHWU = MHWU • 12 = ________ • 12 = ________ gal/yr

Annual Hot Water Energy (AHWE) (Equation 9.3) **(6)**

$$AHWE = \frac{AHWU \bullet 62.3 \bullet 1}{7.48} \bullet (T_{tank} - T_{supply}) \bullet \left(1 + \frac{STL}{100}\right) = \frac{\_\_\_\_ \bullet 62.3 \bullet 1}{7.48} \bullet (\_\_\_\_ - \_\_\_\_) \bullet \left(1 + \frac{\_\_\_\_}{100}\right) = \_\_\_\_\_\_ \text{Btu/yr}$$

Annual Cost for 100% of Hot Water Generation by Individual Fuel Source

Fuel Source	Fuel Cost		Eff or COP_{Avg}	Annual Fuel Usage / Cost: Annual Usage		Annual Cost ($/yr)
Electric Resistance	**7**	($/kWh)	**8**	**9**	(kWh)	**10**
Propane	**11**	($/gal)	**12**	**13**	(Gal)	**14**
Natural Gas	**15**	($/Therm)	**16**	**17**	(Therm)	**18**
GSHP	**19**	($/kWh)	**20**	**21**	(kWh)	**22**

Annual Cost for Selected Percentage of Hot Water Generated by GSHP

Check One	System	HWG by GSHP (%) Guidelines: Annual	Guidelines: Design Month	Assumed: Annual	Assumed: Design Month
23	1) Dedicated GSHP w/ HWG Priority	100	100	**24**	**25**
	2) Dedicated GSHP w/ Space Htg/Clg Priority				
26	1 Tank Storage	40-60	50-70	**27**	**28**
29	2 Tank Storage	60-80	70-90	**30**	**31**
	3) GSHP w/desuperheater				
32	1 Tank Storage	20-40	30-50	**33**	**34**
35	2 Tank Storage	40-60	50-70	**36**	**37**

Other Fuel = ___38___ $Cost_{OtherFuel\_A}$ = ___39___ $/yr

$$Cost_{Combined} = \frac{\%HWG_{GSHP-A}}{100} \bullet Cost_{GSHP-A} + \left(1 - \frac{\%HWG_{GSHP-A}}{100}\right) \bullet Cost_{OtherFuel-A}$$ (Equation 9.12) **(40)**

$$Cost_{Combined} = \frac{\_\_\_\_}{100} \bullet \_\_\_\_ + \left(1 - \frac{\_\_\_\_}{100}\right) \bullet \_\_\_\_ = \_\_\_\_ \$/yr$$

$Savings_{GSHP} = Cost_{OtherFuel-A} - Cost_{Combined}$ = ________ − ________ = ________ $/yr **(41)**

Annual Ground Heating Energy Load for Input to GHEX Design (Equation 9.13) **(42)**

$$AGL_{HWG} = \%HWG_{GSHP-A} \bullet AHWE \bullet \left(\frac{COP_{Avg} - 1}{COP_{Avg}}\right) = \_\_\_\_ \bullet \left(\frac{\_\_\_\_ - 1}{\_\_\_\_}\right) = \_\_\_\_ \text{ Btu}$$ **(A)**

Design Month Data for Input to GHEX Design

HC_{HWG} = ___43___ Btu/hr
DMD_{HWG} = ___44___ kW
GPM_{HWG} = ___45___ gpm

DMHWE = ___46___ Btu (Equation 9.14)
DMHWRU = ___47___ hrs (Equation 9.15)
F_{HWG} = ___48___ (Equation 9.16) **(B)**

Figure 9.16. Hot Water Generation Worksheet

Inputs to Figure 9.16 are as described below using reference numbers to identify each input or calculation. A copy of Figure 9.16 without the reference numbers is provided in Appendix B.

1. Monthly hot water energy usage is estimated from Table 9.1 or determined from an engineering analysis of the facility.
2. The hot water storage tank setpoint temperature can be estimated from Table 9.2 for various hot water uses.
3. Cold water supply temperature may be estimated as the local well water temperature or, for buried water distribution systems, the average annual temperature in the top 10 feet of the earth (T_M from Table 5.11).
4. Standing losses for storage tank water heating systems may be estimated as 10 to 20% of the annual hot water use and less for a GSHP with single storage tank water heating system combined with an in-line water heating system.
5. Annual hot water usage is calculated using Equation 9.2.
6. Annual hot water energy is calculated using Equation 9.3.
7. The cost of electric energy provided to the electric resistance water heating system.
8. The efficiency of the electric resistance water heating system (100%).
9. The annual resistance heat electric energy to heat water (Equation 9.4).
10. The annual cost to heat water using an electric resistance water heating system (Equation 9.8).
11. The cost of propane provided to the propane water heating system.
12. The efficiency of the propane water heating system (80-90%).
13. The annual gallons of propane to heat water (Equation 9.5).
14. The annual cost to heat water using a propane water heating system (Equation 9.9).
15. The cost of natural gas provided to the natural gas water heating system.
16. The efficiency of the natural gas water heating system (85-95%).
17. The annual therms of natural gas to heat water (Equation 9.6).
18. The annual cost to heat water using a natural gas water heating system (Equation 9.10).
19. The cost of electric energy provided to the GSHP water heating system.
20. The average COP of the GSHP water heating system (3.0-4.0).
21. The annual GSHP electric energy to heat water (Equation 9.7).
22. The annual cost to heat water using a GSHP water heating system (Equation 9.11).
23. Check this box if using a dedicated water-to-water GSHP for hot water generation only or if using a water-to-water GSHP or a combination GSHP with hot water generation priority. Either case means that all hot water needs of the facility will be provided by the GSHP hot water generation system.
24. The annual percentage of hot water generated by the GSHP water heating system will be 100%.
25. The design month percentage of hot water generated by the GSHP water heating system will be 100%.
26. Check this box if using a water-to-water GSHP or a combination GSHP with single-tank storage and with space heating and cooling set as priority. Hot water generation will occur only when the space heating or cooling loads are satisfied and the percentage of hot water generated with the GSHP will be significantly less than 100%.
27. An estimate for the annual percentage of hot water generated by the GSHP water heating system. A suggested range is provided in the "Annual" sub-column of the "Guidelines" column.
28. An estimate for the design month percentage of hot water generated by the GSHP water heating system. A suggested range is provided in the "Design Month" sub-column of the "Guidelines" column.
29. Check this box if using a water-to-water GSHP or a combination GSHP with double-tank storage and with space heating and cooling set as priority. Hot water generation will occur only when the space heating or cooling loads are satisfied and the percentage of hot water generated with the GSHP will be more than the single-tank storage system, but still less than 100%.

30. An estimate for the annual percentage of hot water generated by the GSHP water heating system. A suggested range is provided in the "Annual" sub-column of the "Guidelines" column.
31. An estimate for the design month percentage of hot water generated by the GSHP water heating system. A suggested range is provided in the "Design Month" sub-column of the "Guidelines" column.
32. Check this box if using a water-to-water or water-to-air GSHP with a desuperheater and with single-tank storage. Hot water generation will occur during space heating or cooling operation of the GSHP. With the single-tank storage system, the percentage of hot water generated with the GSHP will be significantly less than 100%.
33. An estimate for the annual percentage of hot water generated by the GSHP water heating system. A suggested range is provided in the "Annual" sub-column of the "Guidelines" column.
34. An estimate for the design month percentage of hot water generated by the GSHP water heating system. A suggested range is provided in the"Design Month" sub-column of the "Guidelines" column.
35. Check this box if using a water-to-water or water-to-air GSHP with a desuperheater and with double-tank storage. Hot water generation will occur during space heating or cooling operation of the GSHP. With the double-tank storage system the percentage of hot water generated with the GSHP will be more than with the single-tank storage system, but still far less than 100%.
36. An estimate for the annual percentage of hot water generated by the GSHP water heating system. A suggested range is provided in the "Annual" sub-column of the "Guidelines" column.
37. An estimate for the design month percentage of hot water generated by the GSHP water heating system. A suggested range is provided in the "Design Month" sub-column of the "Guidelines" column.
38. The fuel source for the water heating system backing up the GSHP system.
39. The annual cost to generate hot water using the other fuel source water heating system.
40. The annual cost to generate hot water using a combined GSHP and other fuel source water heating system. Calculated using Equation 9.12.
41. The annual cost savings using a GSHP water heating system (stand-alone or combined with another fuel source) relative to the other fuel source water heating system alone.
42. The annual ground heating energy load to provide the portion of hot water generated using the GSHP water heating system. Calculated using Equation 9.13 and an input into the vertically and horizontally-bored GHEX heating mode design worksheets. This is the annual energy removed from the GHEX for heating water and will add to the space load annual ground heating energy on the GHEX heating mode design worksheet for calculating the effect of unbalanced ground load on vertically and horizontally-bored GHEX systems.
43. The heating capacity of the GSHP for generating hot water, which is used to calculate the run fraction for hot water generation and may be an input into the GHEX heating mode design worksheet. For a water-to-water GSHP, this is the heating capacity from the single-capacity, water-to-water GSHP selection worksheet at design heating conditions for the design hot water generation rate. For a combination GSHP, this is the heating capacity for the GSHP when operating in water-to-water mode at design heating conditions. For a water-to-air or water-to-water GSHP with a desuperheater, this is the published desuperheater capacity at design heating conditions or 10 to 15% of the heating capacity of the GSHP at design heating conditions.
44. The electric demand of the GSHP for generating hot water, which will be an input into the GHEX heating mode design worksheet for a water-to-water GSHP that is dedicated only to heating water.
45. The water flow rate through the GSHP for generating hot water, which will be an input into the GHEX heating mode design worksheet for a water-to-water GSHP that is dedicated only to heating water.
46. The design month hot water energy, which is calculated using Equation 9.14.
47. The design month hot water run hours, which is calculated using Equation 9.15.
48. The design month run fraction for hot water generation, which is calculated using Equation 9.16. This number is combined with the space heating load run fraction on the GHEX heating mode design worksheets to determine a weighted-average design month run fraction for heating.

9.3 Integration of Hot Water Generation GHEX Design Parameters into GHEX Design Worksheets

Each GHEX Design Worksheet has a section at the top-middle of the form titled "HWG GSHP Design Data" that contains pertinent GHEX design parameters (HC_{HWG}, DMD_{HWG}, GPM_{HWG}, AGL_{HWG}, and F_{HWG}) that should be included to properly account for the effect of hot water generation on the heating and cooling design lengths of the selected GHEX configuration. In addition, the type of GSHP hot water generation system (dedicated, combination, or desuperheater) and whether the GHEX design is based on heating or cooling will both have significant influence on the values for those design parameters. Because there are many combinations of GHEX configurations and GSHP hot water generation systems, tables have been developed to aid in determination of the design parameters and how they are integrated into the "Total Design Data" parameters that are used to calculate the GHEX design lengths.

Inputs to the heating mode GHEX design worksheets are summarized in Table 9.4. For each combination of GHEX Configuration and GSHP HWG System recommended values for HC_{HWG}, DMD_{HWG}, and GPM_{HWG} (based on GSHP HWG System type) and AGL_{HWG} and F_{HWG} (based on GHEX Configuration type) are provided. Proper calculation of each total design data parameter has been defined to reflect the operating characteristics of each GHEX Configuration and GSHP HWG System combination. Inputs to the cooling mode GHEX design

Table 9.4. Inputs to Heating Mode GHEX Design Worksheets

		Inputs to HWG GSHP Design Data[1]			Inputs from HWG Worksheet[5]		
GHEX Configuration	GSHP HWG System	HC_{HWG}	DMD_{HWG}	GPM_{HWG}	AGL_{HWG}	F_{HWG}	Total Design Data Parameters[6]
Vertically-Bored & Horizontally-Bored	Water-Water[2]	HC_C	DMD_C	GPM	A	B	$HC_D=HC_C+HC_{HWG}$ $DMD_D=DMD_C+DMD_{HWG}$ $GPM_D=GPM+GPM_{HWG}$ $AGL_{HD}=AGL_H+AGL_{HWG}$ $AGL_{CD}=AGL_C$
Vertically-Bored & Horizontally-Bored	Combination[3]	HC_C	0	0	A	B	$HC_D=HC_C$ $DMD_D=DMD_C$ $GPM_D=GPM$ $AGL_{HD}=AGL_H+AGL_{HWG}$ $AGL_{CD}=AGL_C$
Vertically-Bored & Horizontally-Bored	Desuperheater[4]	$HC_{Desuper}$	0	0	A	B	$HC_D=HC_C$ $DMD_D=DMD_C$ $GPM_D=GPM$ $AGL_{HD}=AGL_H+AGL_{HWG}$ $AGL_{CD}=AGL_C$
Horizontally-Trenched	Water-Water[2]	HC_C	DMD_C	GPM	---	B	$HC_D=HC_C+HC_{HWG}$ $DMD_D=DMD_C+DMD_{HWG}$ $GPM_D=GPM+GPM_{HWG}$
Horizontally-Trenched	Combination[3]	HC_C	0	0	---	B	$HC_D=HC_C$ $DMD_D=DMD_C$ $GPM_D=GPM$
Horizontally-Trenched	Desuperheater[4]	$HC_{Desuper}$	0	0	---	B	$HC_D=HC_C$ $DMD_D=DMD_C$ $GPM_D=GPM$

1. On the Vertically-Bored, Horizontally-Bored, and Horizontally-Trenched GHEX Design Worksheets - Heating Mode. A dashed line indicates that this parameter does not apply for the GHEX Configuration and GSHP HWG System combination.
2. HC_C, DMD_C, and GPM data from Water-to-Water GSHP Selection Worksheet at design water heating conditions.
3. HC_C is the water-to-water mode heating capacity at design heating conditions for the combination GSHP unit.
4. $HC_{Desuper}$ is the water heating capacity of the desuperheater at design heating conditions from the manufacturer's literature. $HC_{Desuper}$ can be estimated to be 10 to 15% of the heating capacity of the GSHP at design heating conditions.
5. AGL_{HWG} is obtained from Figure 9.16 (labeled A) and F_{HWG} is obtained from Figure 9.16 (labeled B).
6. All parameters in this column are on the GHEX Design Worksheet – Heating Mode.

worksheets are summarized in Table 9.5. For each combination of GHEX Configuration and GSHP HWG System, recommended values for GPM_{HWG} (based on GSHP HWG System type) and AGL_{HWG} (based on GHEX Configuration type) are provided. Proper calculation of each total design data parameter has been defined to reflect the operating characteristics of each GHEX Configuration and GSHP HWG System combination.

Table 9.5. Inputs to Cooling Mode GHEX Design Worksheets

		Inputs to HWG GSHP Design Data[1]			Inputs from HWG Worksheet[3]		
GHEX Configuration	GSHP HWG System	HC_{HWG}	DMD_{HWG}	GPM_{HWG}	AGL_{HWG}	F_{HWG}	Total Design Data Parameters[4]
Vertically-Bored & Horizontally-Bored	Water-Water[2]	---	---	GPM	A	---	$TC_D=TC_C$ $DMD_D=DMD_C$ $GPM_D=GPM+GPM_{HWG}$ $AGL_{HD}=AGL_H+AGL_{HWG}$ $AGL_{CD}=AGL_C$
Vertically-Bored & Horizontally-Bored	Combination	---	---	0	A	---	$TC_D=TC_C$ $DMD_D=DMD_C$ $GPM_D=GPM$ $AGL_{HD}=AGL_H+AGL_{HWG}$ $AGL_{CD}=AGL_C$
Vertically-Bored & Horizontally-Bored	Desuperheater	---	---	0	A	---	$TC_D=TC_C$ $DMD_D=DMD_C$ $GPM_D=GPM$ $AGL_{HD}=AGL_H+AGL_{HWG}$ $AGL_{CD}=AGL_C$
Horizontally-Trenched	Water-Water[2]	---	---	GPM	---	---	$TC_D=TC_C$ $DMD_D=DMD_C$ $GPM_D=GPM+GPM_{HWG}$
Horizontally-Trenched	Combination	---	---	0	---	---	$TC_D=TC_C$ $DMD_D=DMD_C$ $GPM_D=GPM$
Horizontally-Trenched	Desuperheater	---	---	0	---	---	$TC_D=TC_C$ $DMD_D=DMD_C$ $GPM_D=GPM$

1. On the Vertically-Bored, Horizontally-Bored, and Horizontally-Trenched GHEX Design Worksheets - Cooling Mode. A dashed line indicates that this parameter does not apply for the GHEX Configuration and GSHP HWG System combination.
2. GPM data at design heating conditions from Water-to-Water GSHP Selection Worksheet.
3. AGL_{HWG} is obtained from Figure 9.16 (labeled A). F_{HWG} is ignored for all cooling mode GHEX designs.
4. All parameters in this column are on the GHEX Design Worksheet – Cooling Mode.

Hot Water Generation Worksheet (Blank Copy) – See Appendix B

Figure 9.16 Hot Water Generation Worksheet

10 RESISTANCE HEAT

In This Section

10.1 Resistance Heat Operation

In the heating mode, a heat pump moves heat by extracting it from a low temperature source and "pumping" it to a high temperature sink. A ground source heat pump extracts heat from the earth to provide heating to the conditioned space. An air source heat pump extracts heat from the outside ambient air to provide heating to the conditioned space. Electricity is consumed during heat pump operation to operate the compressor, fan, and pumps to move the heat. For a GSHP system, approximately 3/4 of the heat delivered to the space will be extracted from the earth and the remainder will come from electric consumption of the compressor, fan, and pumps.

There are three key differences between air source heat pumps and ground source heat pumps:

- Moderate temperatures provided by ground-based heat exchange rather than outside air heat exchange
- Superior heat transfer capabilities of water over air
- Resistance heat operation during cold weather periods and during the defrost cycle on air source heat pumps

Air source heat pumps extract heat from the outside air in the heating mode as opposed to extracting heat from the earth (via the GHEX) with a GSHP. Extracting heat from the air when outdoor temperatures fall below 38-40 F becomes difficult with air source equipment. Both heating capacity and efficiency of such equipment diminishes to the point where supply air temperatures to the conditioned space become too low to provide satisfactory occupant comfort levels. During these periods, supplemental resistance heat operation will be necessary.

Additionally, an air source heat pump condensing coil typically operates at temperatures below the dew point temperature during operation in the heating mode (http://high-performance-hvac.com). Because of this, moisture removed from the air (38-40 F or colder) will cause ice to build up on the outdoor condensing coil. As heat is extracted from the outdoor air and absorbed by the refrigerant in the evaporator coil, the cold temperature of the refrigerant causes ice to form and build up on the outdoor coil's exterior surfaces. It is necessary to defrost the heat pump condenser coils periodically to prevent impeding air flow across the coil and degrading performance due to the ice and frost buildup. Initiation of the defrost cycle is typically controlled by a defrost timer in the control circuit. The intervals between defrost cycle operation can be adjusted to accommodate the climatic region and outdoor weather conditions. During operation in the defrost cycle, in spite of thermostat call for heating, the air source heat pump will switch from the heating mode to the cooling mode. The direction of heat flow is reversed, and heat is absorbed from the warmer indoor air and rejected to the outdoor coil to melt the ice buildup. To prevent discharging chilled air into the house, the air source heat pump will energize its electric resistance heating elements to reheat the air. During this cycle, the unit runs in cooling and heating at the same time. This type of air source heat pump operation occurs as needed throughout the annual heating season, increasing overall annual operating cost in the heating mode.

A GSHP sized for 100% of the peak heating load will not require resistance heat operation to provide heating to the space. As discussed in Chapter 3, GSHP systems can be designed to provide all required heating without the use of resistance heat. Use of resistance heat is purely a design decision for GSHP systems, not a requirement as with air source heat pump systems.

To heat with electric resistance heating elements, heat is generated by electricity consumption of the elements. All of the heat delivered to the space comes from electricity consumption. Electric resistance heat is the most expensive electrically-driven heating method available. In many cases, heating with electric resistance elements is the most expensive heating method of all the methods available, depending on energy prices. Excessive use of electric resistance heat can increase heating costs significantly. In a well-designed GSHP system, the heat pump handles as much of the heating as possible with the supplemental resistance heating elements operating relatively few hours during the course of the year.

10.1.1 SUPPLEMENTAL HEAT Resistance heat can be used to supplement the capacity of a GSHP in heating dominant applications to prevent over sizing in cooling capacity and during extreme weather conditions. When applied in this manner, resistance heat is a supplement to the GSHP's capacity. The small capacity of the supplemental resistance heater used in this application is not capable of conditioning the space to the desired temperature without the heating capacity of the GSHP unit.

Single-capacity GSHPs are typically sold as having two stages of heating capacity and one stage of cooling capacity. The compressor of a single-capacity GSHP unit can only provide one stage of heating capacity. The second stage of heating for a single-capacity GSHP is provided by supplemental resistance heat. Similarly, two-capacity GSHPs are typically sold as having three stages of heating capacity and two stages of cooling capacity, where the third stage of heating capacity is provided by supplemental resistance heat.

10.1.2 EMERGENCY HEAT Resistance heat can be used in an emergency capacity, to protect the structure in the event of a heat pump outage. In this emergency capacity, the resistance heat unit can be sized to either maintain the structure at a temperature suitable to prevent freeze damage or have sufficient capacity to maintain the dwelling at or near the thermostat set point temperature. However, this only protects in the event of a heat pump outage. In the event of an electric power outage, any fossil fuel, electric forced-air, or hydronic system will not have electricity available to operate circulating fans, pumps, control circuits, or other electrically-driven systems unless adequate standby generation capacity is available.

During electric heat operation (either supplemental or emergency), fan speed will typically adjust to the pre-set factory blower speed setting. Figure 10.1 displays the blower performance data for ClimateMaster's Tranquility 27™ Series GSHPs. For example, as shown in Figure 10.1, when the ClimateMaster Tranquility 27™ Model 049 switches into either auxiliary (supplementary) heating mode or emergency heating mode as dictated by the system thermostat, the ECM blower speed will adjust to produce 1660 CFM (for blower tap setting 3).

Airflow in CFM with wet coil and clean air filter

Model	Max ESP (in. wg)	Fan Motor (hp)	Tap Setting	Cooling Mode			Dehumid Mode			Heating Mode			Residential Units Only	
				Stg 1	Stg 2	Fan	Stg 1	Stg 2	Fan	Stg 1	Stg 2	Fan	AUX CFM	Aux/ Emerg Mode
026	0.50	1/2	4	810	950	475	630	740	475	920	1060	475	4	1060
	0.50	1/2	3	725	850	425	560	660	425	825	950	425	3	950
	0.50	1/2	2	620	730	370	490	570	370	710	820	370	2	820
	0.50	1/2	1	520	610	300				600	690	300	1	690
038	0.50	1/2	4	1120	1400	700	870	1090	700	1120	1400	700	4	1400
	0.50	1/2	3	1000	1250	630	780	980	630	1000	1250	630	3	1350
	0.50	1/2	2	860	1080	540	670	840	540	860	1080	540	2	1350
	0.50	1/2	1	730	900	450				730	900	450	1	1350
049	0.75	1	4	1460	1730	870	1140	1350	870	1560	1850	870	4	1850
	0.75	1	3	1300	1550	780	1020	1210	780	1400	1650	780	3	1660
	0.75	1	2	1120	1330	670	870	1040	670	1200	1430	670	2	1430
	0.75	1	1	940	1120	560				1010	1200	560	1	1350
064	0.75	1	4	1670	2050	1020	1300	1600	1020	1860	2280	1020	4	2280
	0.75	1	3	1500	1825	920	1160	1430	920	1650	2050	920	3	2040
	0.75	1	2	1280	1580	790	1000	1230	790	1430	1750	790	2	1750
	0.75	1	1	1080	1320	660				1200	1470	660	1	1470
072	0.75	1	4	1620	2190	1050	1270	1650	1050	1690	2230	1050	4	2230
	0.75	1	3	1500	1950	980	1170	1520	980	1600	2100	980	3	2100
	0.75	1	2	1400	1830	910	1100	1420	910	1400	1850	910	2	1870
	0.75	1	1	1320	1700	850				1240	1620	850	1	1670

SOURCE: ClimateMaster

Figure 10.1. ECM Blower Performance Data for ClimateMaster Tranquility 27™ Series GSHPs

10.2 Installing Resistance Heat

All GSHP manufacturers provide optional resistance heat packages that are either externally mounted to the discharge of the GSHP unit as shown in Figure 10.2 or internally mounted between the fan housing and discharge air opening within the GSHP cabinet as shown in Figure 10.3.

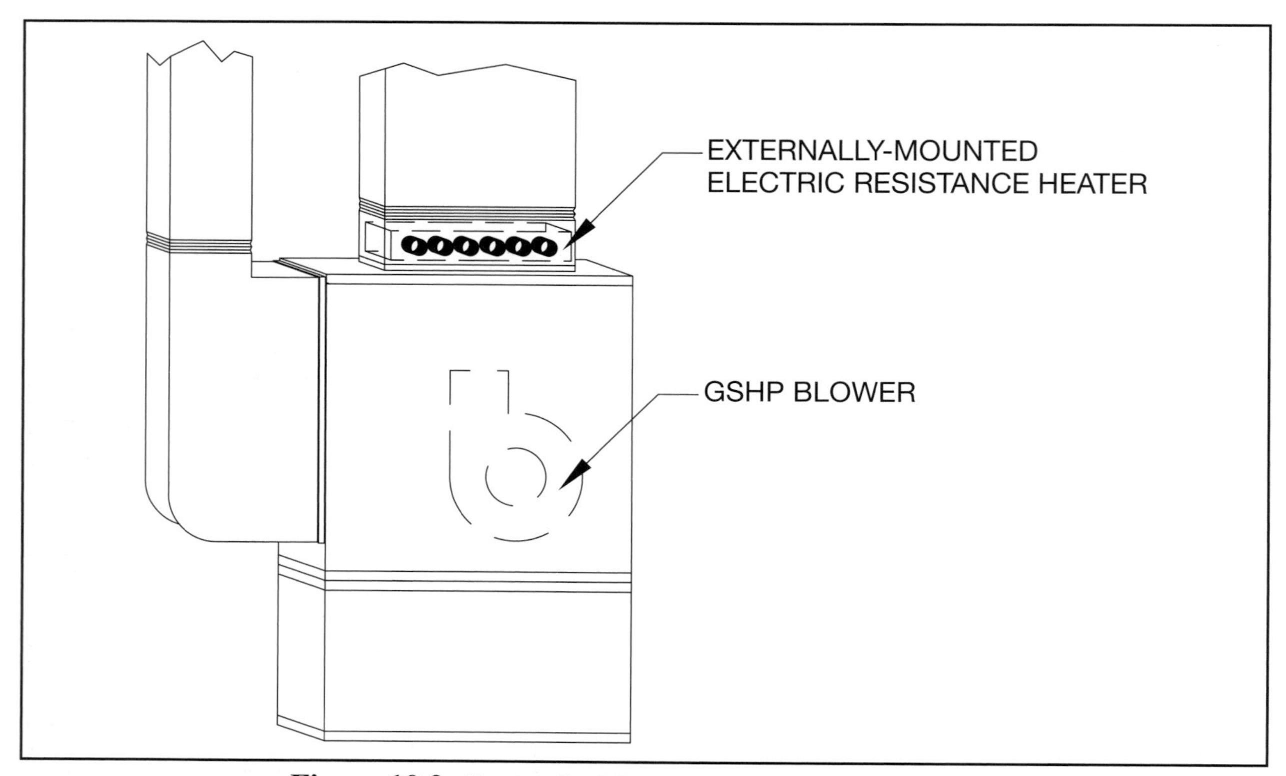

Figure 10.2. Externally-Mounted Electric Resistance Heat

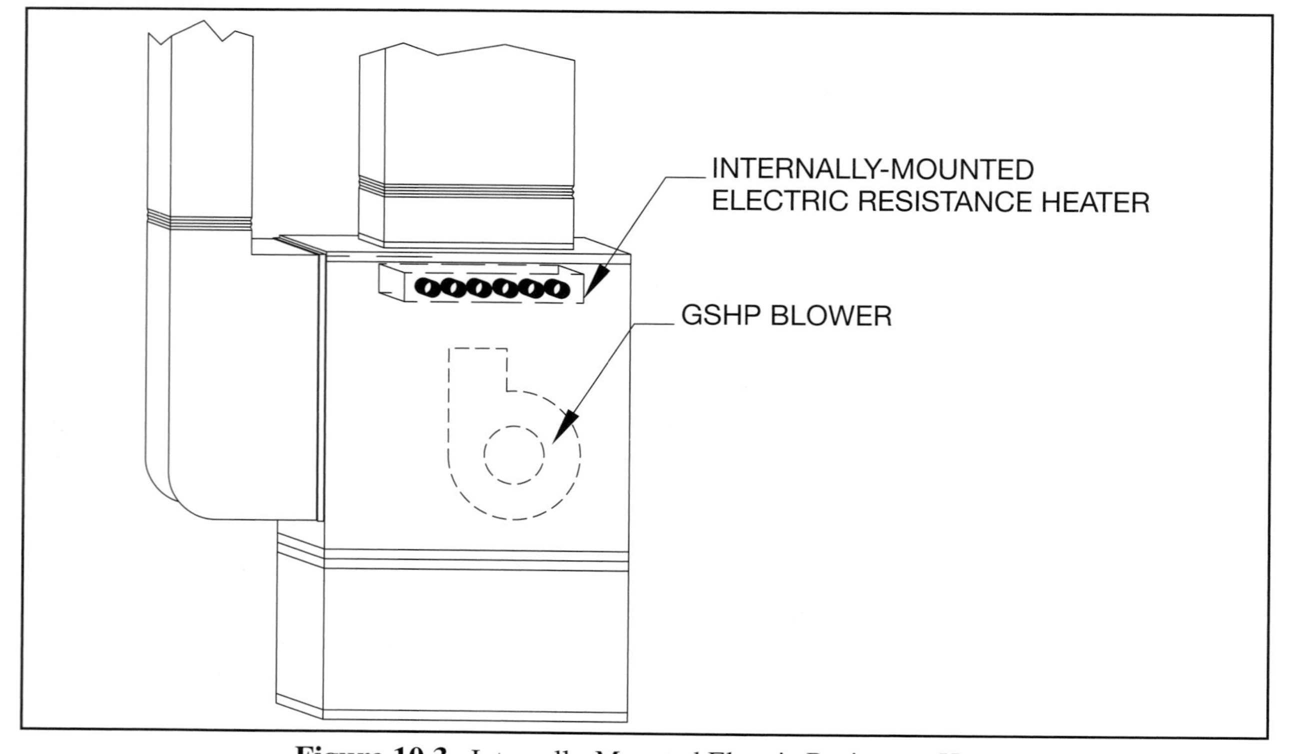

Figure 10.3. Internally-Mounted Electric Resistance Heat

These optional resistance heaters are offered in various incremental sizes and are designed to match the GSHP and its control circuit. Figure 10.4 gives the electric resistance heater ratings for the auxiliary electric heat models available through ClimateMaster. Note that the installed heating capacity of the heaters (kW rating shown in Figure 10.2) will depend on the power being supplied to the unit. For example, electric heat model AGM10A will provide 9.6 kW (32,700 Btu/hr) of heating capacity when 230V power is applied and only 7.2 kW (24,600 Btu/hr) of heating capacity when 208V power is applied. Notice also that the minimum CFM required to prevent overheating of the model AGM10A electric heater is 650 CFM (provided in the Figure 10.4).

Auxiliary Electric Heat Model	TT					kW Rating		Btuh Rating		Minimum CFM Required
	026	038	049	064	072	230V	208V	230V	208V	
AGM5A	■					4.8	3.6	16300	12300	500
AGM8A	■					7.6	5.7	25900	19400	650
AGM10A	■					9.6	7.2	32700	24600	650
AGM12A	■					11.4	8.6	38900	29200	750
AGL10A		■	■	■	■	9.6	7.2	32700	24600	1300
AGL15A		■	■	■	■	14.4	10.8	49100	36900	1350
AGL20A			■	■	■	19.2	14.4	65500	49200	1350

SOURCE: ClimateMaster

Figure 10.4. Electric Resistance Heat Ratings

Use Equation 10.1 to convert electric heating capacity from kW to Btu/hr.

$$CAP_{RH} = 3{,}412 \cdot CAP_{kW}$$ **Equation 10.1**

Where:

CAP_{RH} = Electric resistance heating capacity (Btu/hr)
CAP_{kW} = Nominal electric heating capacity (kW)

Typically, the nominal heating capacity of an electric resistance heater will be based on single-phase 230V power ratings. For example, the AGM10A electric heat model in Figure 10.2 is a 10-kW nominal electric heater, which is the approximate heating capacity of that electric heat model when 230V power is applied.

Because of the variation between manufacturers, the manufacturer's instructions and recommendations for sizing must be followed when installing and connecting any resistance heater to the system. The resistance heater control circuit is connected to the heat pump control circuit. Both are controlled by the system thermostat. Resistance heaters will require their own independent power supply with disconnect per the manufacturer's specifications as illustrated in Figure 10.5. Do not attempt to wire a resistance heater into the same power circuit as the GSHP unit.

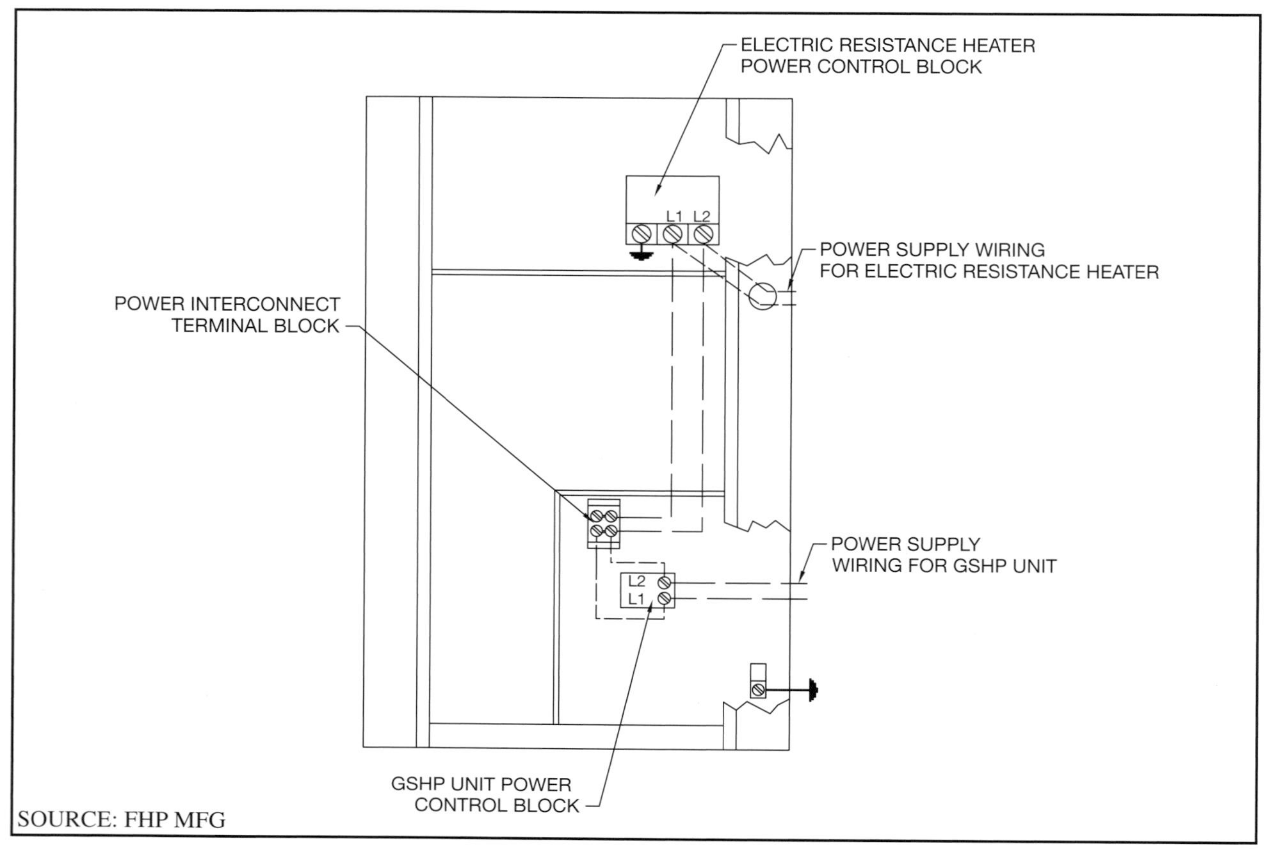

Figure 10.5. Power Supply for Resistance Heater & GSHP Unit

Auxiliary Electric Heat Model	Supply Circuit	Heater Amps		Minimum Circuit Amps		Maximum Fuse		Supply Wire	
		230V	208V	230V	208V	230V	208V	Min AWG	Max Ft
AGM5A	Single	20.0	17.3	25.0	21.6	25	25	10	70
AGM8A	Single	31.7	27.5	39.6	34.4	40	35	8	70
AGM10A	Single	40.0	34.7	50.0	43.4	50	45	6	90
AGM12A	Single	47.5	41.2	59.4	51.5	60	60	6	70
	Dual - L1/L2 Dual - L3/L4	31.7 15.8	27.5 13.7	39.6 19.8	34.4 17.1	40 20	35 20	8 12	70 50
AGL10A	Single	40.0	34.7	50.0	43.4	50	45	6	80
AGL15A	Single	60.0	52.0	75.0	65.0	80	70	6	50
	Dual - L1/L2 Dual - L3/L4	40.0 20.0	34.7 17.3	50.0 25.0	43.4 21.6	50 25	45 25	6 10	80 70
AGL20A	Single	80.0	69.3	100.0	86.6	100	90	2	100
	Dual - L1/L2 Dual - L3/L4	40.0 40.0	34.7 34.7	50.0 50.0	43.4 43.4	50 50	45 45	6 6	80 80

SOURCE: ClimateMaster

Figure 10.6. Electric Resistance Heat Data for Single-Phase Power

Figure 10.6 gives the electric resistance heat data for the electric heat models given in Figure 10.4. For electric heat model AGM10A, the minimum circuit amp rating is 50 amps. A 50-amp circuit breaker will need to be installed for the electric heater along with the power supply necessary for the GSHP unit itself. Additionally, 6-gage wire will need to be used for the electric heat power supply and the maximum length of the wire is limited to 90 feet.

10.2.1 STANDARD RESISTANCE HEAT CONTROL WIRING The stage of heating provided by the resistance heat is energized when the thermostat senses a specified temperature differential below the set point when the heat pump is either starting or running. Proper electric resistance heater operation will rely on proper control wiring from the terminal strip in the GSHP unit to the thermostat. Terminal strip and thermostat designations will vary by manufacturer. Refer to the manufacturer's installation specifications for proper control wiring from the GSHP to the thermostat. Table 10.1 displays the control wiring recommendations for ECONAR's GeoSource Vara 2 Plus dual-capacity GSHPs. This manufacturer typically provides internally-mounted, 10-kW nominal electric resistance heaters with all of its GSHP units.

Function	Terminal Strip	Thermostat
24 Volt Power	R	R
Common	X	C (X)
Blower	G	G
Reversing Valve	O	O
1st Stage Heat/Cool	Y	Y1
2nd Stage Heat/Cool	Y2	Y2
Lockout Signal	L	L
3rd Stage Heat	W2	W1
Emergency Heat	E	E

SOURCE: ECONAR GeoSource Vara 2 Plus Installation and Operating Instructions Manual

Table 10.1. Control Wiring Connections from GSHP Unit to Thermostat

For the control wiring recommendations given in Table 10.1, a W1 signal (3rd stage heat) from the thermostat with cause the GSHP unit to energize 5 kW (of the 10-kW nominal heater) of the electric heating elements to supplement its heating capacity. The electric resistance heating elements remain energized until Y2 (2nd stage heat) is satisfied on the thermostat. During electric resistance operation, the blower runs on high speed (ECONAR Energy Systems).

An E signal (emergency heat) from the thermostat will cause the GSHP unit to energize 10 kW of the electric heating elements. This signal from the thermostat will also cause the thermostat to lock out all compressor signals to the GSHP to prevent the compressor from operating. As with the 3rd stage of supplemental heat, the blower will run on high speed (ECONAR Energy Systems).

Figure 10.7 displays the electric wiring diagram for ECONAR's GeoSource Vara 2 Plus dual-capacity (3-heat/2-cool) GSHPs. The control wiring connections from the terminal strip in the GSHP to the thermostat are circled in the figure.

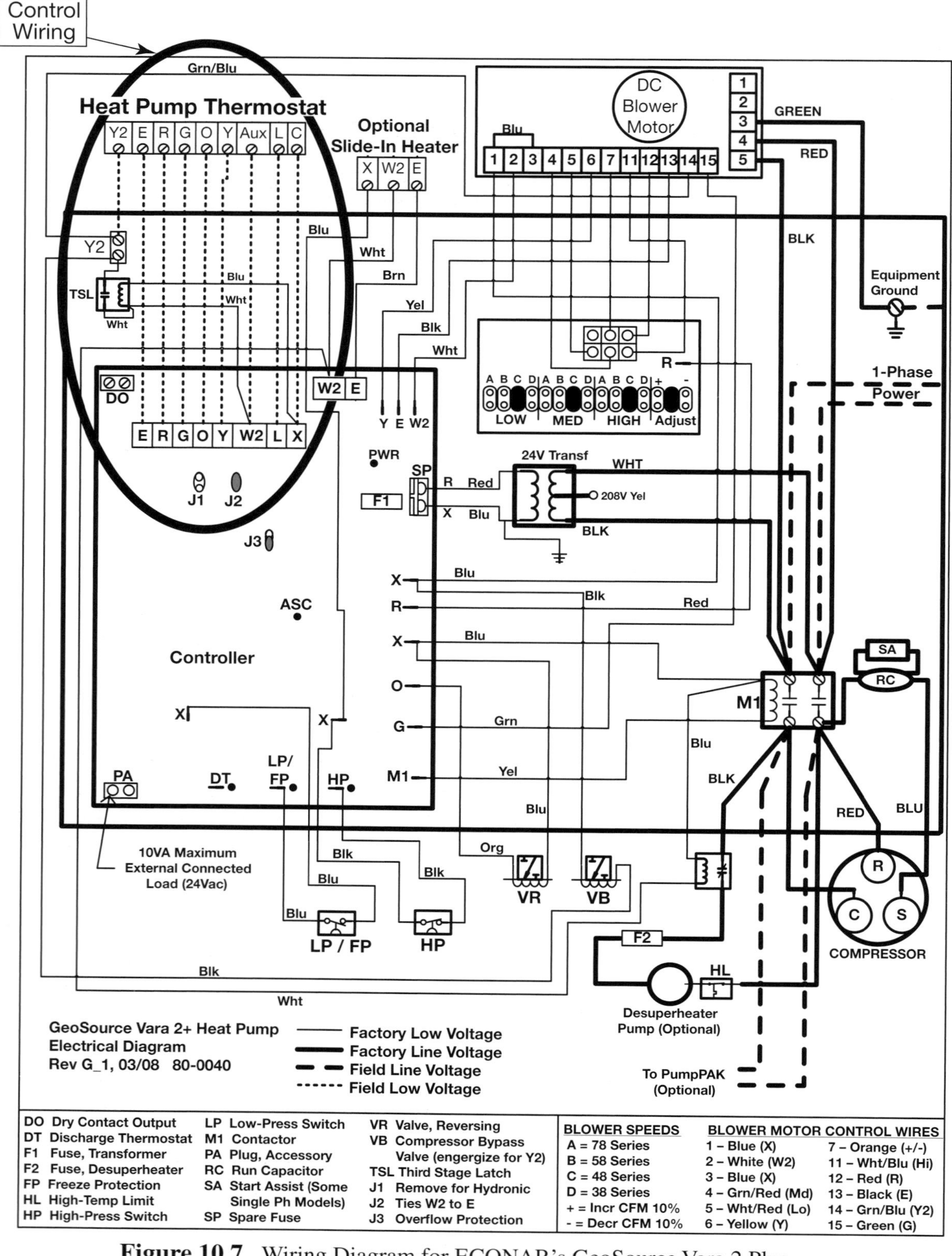

Figure 10.7. Wiring Diagram for ECONAR's GeoSource Vara 2 Plus

11 GSHP STARTUP, PERFORMANCE CHECKING, AND TROUBLESHOOTING

In This Section

11.1 GSHP System Startup

The purpose of this chapter is to help contractors prepare a system for initial startup, check for proper system operation after startup, and to help diagnose problems that may be encountered during the startup and checkout process. It is important that proper precautions be taken at every installation to help increase customer satisfaction and contractor success rate, but also to prevent possible injury to personnel and damage to equipment and the structure in which the equipment is installed.

System startup includes two different check procedures: a static system check and a dynamic system check. A static system check is performed before the system is turned on to identify obvious problems that must be resolved before proper operation can be expected. The dynamic system check is performed after the system has been started and consists of measuring various system performance parameters to ensure that the system is performing as promised.

11.1.1 STATIC SYSTEM CHECK PROCEDURE The recommended static system checkout procedure provided in this section was adapted from *LeClaire and Lafferty (1996)*. The recommended procedure is provided below, and Table 11.1 contains a checklist that should be filled out in conjunction with the static system check procedure.

1) Check that the shipping tie-downs and compressor mounting bolts are loosened and packing pads have been removed. Check to see if the blower wheel rotates freely and whether instruction manuals have been placed inside the unit.
2) Ensure that either power has been applied to the unit for 24 hours or the ambient air temperature has been 55 F or warmer for the same length of time. This will ensure that the refrigerant has been boiled out of the compressor crankcase. Applying power activates the crankcase heater (if used).
3) Verify that the GSHP unit is level (or properly tilted if horizontal).

4) Verify that the unit is properly wired and all lugs are tight.
   a) Check to see that backup resistance heater power wiring is complete and that fuses are installed, if necessary.
   b) Check to make sure that the unit is properly grounded.
   c) Make sure that the correct blower speed (PSC) or tap setting (ECM) has been selected.
5) While the thermostat is disconnected, measure the line (power) and low (control) voltages to verify they are within tolerances.
6) Check that ductwork and diffusers are free of blockages and/or debris. At this time, check to make sure that all dampers are open. Gaps in ductwork must be sealed so that air will not short circuit and give false readings. Return air headers are often found open at this time.
7) Ensure that all water flow valves are open to the unit. If it's an open-loop system, the well pump must be ready to deliver water to the unit.
8) Ensure that water in and water out lines are connected correctly on water and domestic hot water systems.
9) Check the entire system for leaks. Repair any leaks before startup.
10) If a desuperheater is used, disable it at this time.
11) If the system is closed loop, check the loop pump circuit for proper wiring. Wiring should be such that the pumps engage simultaneously with the compressor on the heat pump. Also check to be sure that the valves are open to the loop.
12) For closed-loop systems, check to be sure that the loopfield has been flushed and purged of air, leak tested, and pressurized (if a pressurized flow center has been installed).
13) Ensure that the filters are in their proper place, clean, and that the access panel is closed.
14) Ensure that the door panels are installed on the unit. Most manufacturers design their units such that the panels must be in place for air to be drawn across the air coil. In such models, dramatic equipment malfunction may occur if the panels are not in place.
15) Ensure that piping insulation is installed and sealed tightly. This will prevent unnecessary heat loss/gain and sweating.
16) Check the antifreeze solution (if used) for adequate levels of freeze protection. Remember that the evaporator will be the coldest component in the refrigeration system. Tag the system to indicate antifreeze type and temperature protection.
17) Set all DIP switches at this time, if used in solid-state boards.

11.1.2 DYNAMIC SYSTEM CHECK PROCEDURE Once the static check has been performed and completed, the unit is ready for startup and the dynamic system check can begin. The recommended dynamic system checkout procedure provided in this section was adapted from the *LeClaire and Lafferty (1996)*. The recommended procedure is provided below, and Table 11.2 is a checklist that should be filled out in conjunction with the dynamic system check procedure.

1) Switch the line voltage off to the unit.
2) Set the thermostat sub-base function switch to "Off" so that the compressor will not be called for. Check to make sure that the thermostat setpoint is well below (heating) or above (cooling) the room temperature.
3) Switch the line voltage on to the unit.

4) Set the blower to the "On" position at the thermostat and check to make sure that:
   a) Air is moving out of all diffusers.
   b) No improper noises are being generated.
   c) Blower motor amp draw is within specifications. Record this data.
5) Set the blower to the "Auto" position at the thermostat. The blower should stop.
6) Prepare to measure the compressor amperage at the common terminal of the compressor.
7) Disable the strip heater control circuit (if used) to simplify the startup procedure.
8) Set the thermostat sub-base function to "Heat" or "Cool" as appropriate. Next, move the thermostat set point to well above/below room temperature as appropriate. Check to make sure that:
   a) The compressor is running. Compressor will usually be held off by a time delay in the unit. Wait for the time delay relay (TDR) to reset before expecting the compressor to run.
   b) The blower is running.
   c) The source water is flowing. If the compressor is held off by the TDR, most systems hold the water from flowing during this time.
9) When the compressor starts, note the amperage. It should be within specifications. If not, refer to the troubleshooting section under the appropriate section (high or low amperage). Closely monitor the compressor amperage and note if it rises or falls. Amp draw of the compressor is a prime indicator of the condition of the refrigeration circuit. Record this data.

Table 11.1. Static System Check Procedure Checklist

Static System Check Procedure - Checklist	
1.__________	Unit correctly unpacked/uncrated, including the tie-down bolts, thermostat and blower wheel
2.__________	Unit warm - from crankcase heater or 24-hour ambient
3.__________	Unit level (or tilted correctly if horizontal)
4.__________	Electrical wiring complete, lugs tight, correctly grounded, correct blower speed selected
5.__________	Line voltage checked
6.__________	Ductwork complete and free of blockages
7.__________	Water flow valves open
8.__________	Correct water connections to unit (water in, water out)
9.__________	Check for leaks
10.__________	Temporarily disable the desuperheater at this time (if applicable)
11.__________	Closed-loop system wired properly - flows open if valves are used
12.__________	Closed-loop system filled, flushed, charged and leak tested
13.__________	Air filters in place, clean and access panel closed
14.__________	Door panels installed
15.__________	Piping insulation installed and sealed tightly
16.__________	Check antifreeze solution for adequate protection (if used) - tag the system to indicate antifreeze type and temperature protection
17.__________	Set all DIP switches at this time

LeClaire and Lafferty (1996)

Table 11.2. Dynamic System Check Procedure Checklist

Dynamic System Check Procedure - Checklist	
1.___________	Line voltage off to unit
2.___________	Thermostat sub-base to "off" and setpoint well below (heating) or above (cooling) room temperature
3.___________	Switch on line voltage to unit
4.___________	Call for "blower on" - check for air moving out all air diffusers and unusual noises
5.___________	Check blower amp draw
6.___________	Set blower switch to "auto" - blower should stop
7.___________	Prepare to measure compressor amperage
8.___________	Disable strip heater control circuit (if used)
9.___________	Set thermostat sub-base function to "Heat" or "Cool" as appropriate - move setpoint to well above/below room temperature as appropriate
10.___________	Is compressor running? Are there any unusual noises?
11.___________	Is blower running?
12.___________	Is water source flowing?
13.___________	If open loop, is pump motor cycling properly?
14.___________	Compressor amperage on common lead
15.___________	Activate hot water generator circuit
Notes___________	

LeClaire and Lafferty (1996)

11.2 GSHP System Operation and Performance Checking

Once the static and dynamic startup checkout procedures have been carried out, then performance checking the system can begin. If possible, a newly installed GSHP unit should be performance checked in both heating and cooling modes of operation to ensure proper equipment operation.

11.2.1 GSHP PERFORMANCE CHECKING TOOLS Performance checking a GSHP system requires special tools to make the necessary temperature, pressure, flow, and electrical measurements along with verifying antifreeze levels in the circulating fluid (if necessary) along with gaining access to the inside of the GSHP cabinet. Below is a list of basic tools that are required to adequately evaluate the performance of a GSHP system.

- Pressure gauge or flowmeter tool – Pressure gauge capable of reading up to 60 psi, flowmeter capable of reading up to 30 gpm (3 gpm per ton for a 10-ton unit)
- Thermocouple Needle/Digital Thermometer compatible for use with P/T ports – capable of reading to the nearest 0.1 F
- Multimeter – capable of reading voltages up to 480 volts (A/C)
- Clamp meter – capable of reading current up to 50 amps (A/C)
- Antifreeze testers
- Inclined Manometer – capable of measuring up to 1 inch of H_2O
- Tools to gain access to electrical panels

Figure 11.1 identifies important parameters that must be measured during performance checking of a water-to-air GSHP unit, and Figure 11.2 identifies important parameters that must be measured during performance checking of a water-to-water GSHP unit. Figure 11.3 provides a schematic of the power connection terminals on a GSHP unit where electric service is brought into the system from the fuse panel or service disconnect.

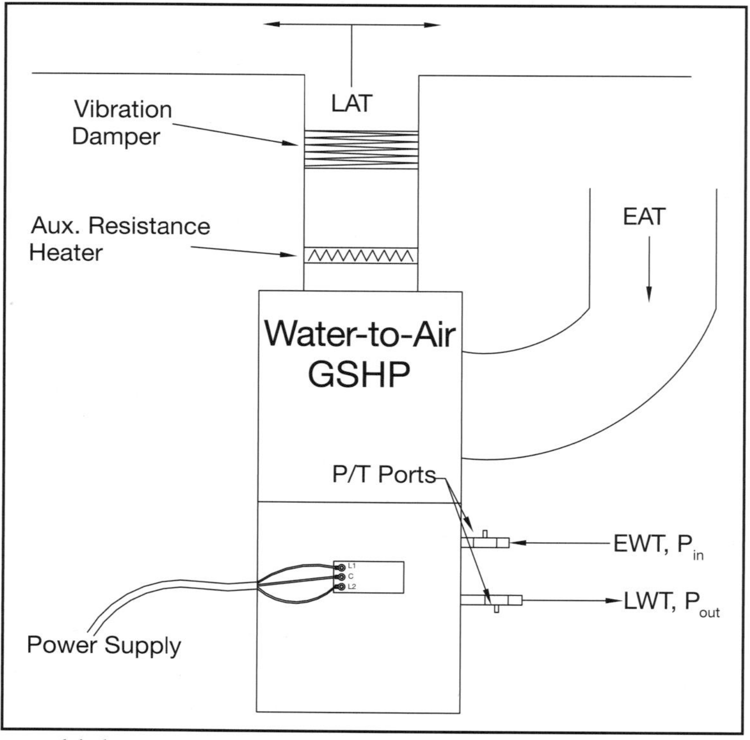

Figure 11.1. Water-to-Air GSHP Performance Checking Parameters

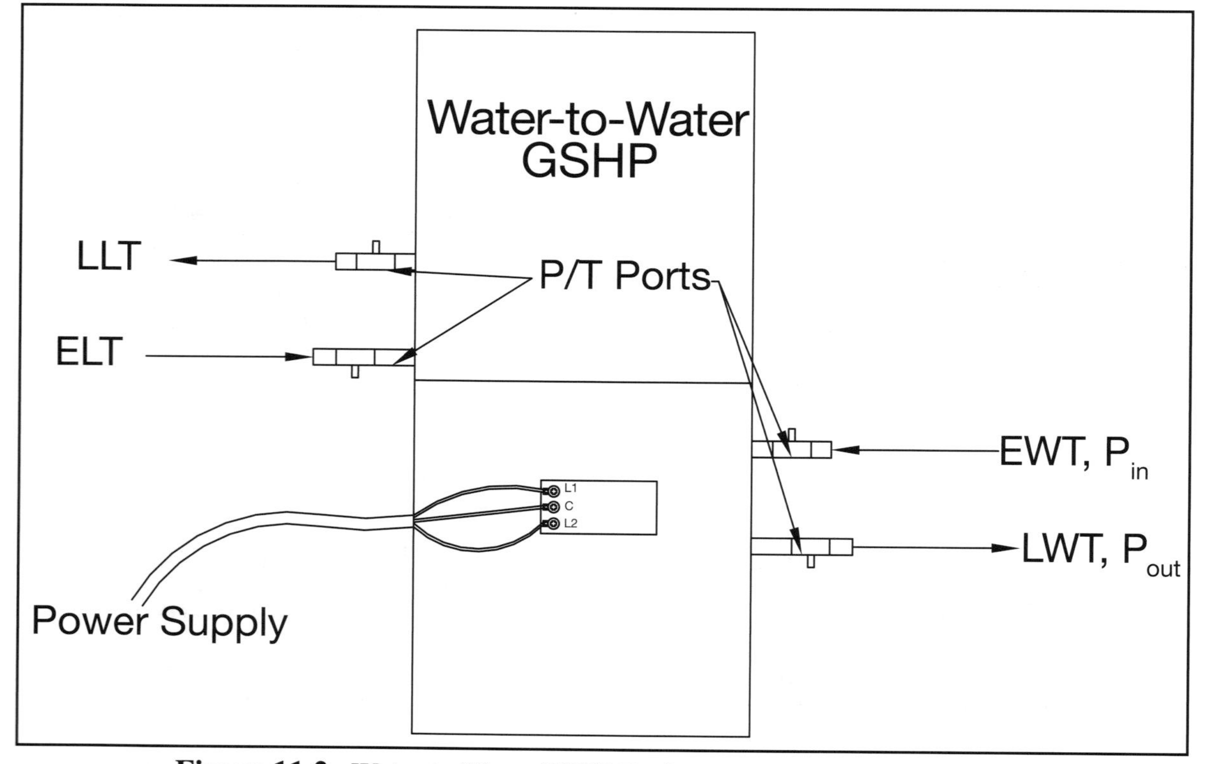

Figure 11.2. Water-to-Water GSHP Performance Checking Parameters

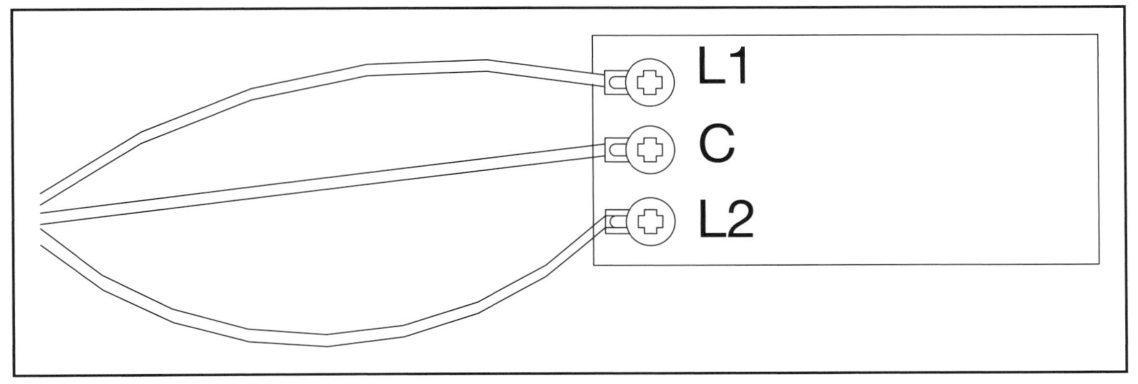

Figure 11.3. GSHP Electric Connection Details (From Fuse Panel to GSHP)

MULTIMETER USE A multimeter will be necessary to measure the voltage supplied to the GSHP unit during operation. Figure 11.4 displays an autoranging multimeter, which is the simplest type to use. Figure 11.5 displays a switched range multimeter, which is also common, but the user must select the proper range of voltage expected for proper measurement.

Measurement of electrical voltage supplied to the GSHP will require different measurement techniques for single-phase or three-phase service. Before attempting to take any voltage measurements, set the multimeter to take A/C voltage readings and ensure that the leads are plugged into the COM and Voltage inputs. For single-phase power, hold the leads of the multimeter tool across the two hot leads on the power input terminal of the GSHP (across L1 and L2 in Figure 11.3) and record the measurement. For three-phase power, hold the leads across all three possible electrical lead combinations (across L1 and L2, across L1 and C, and across L2 and C in Figure 11.3). Two of the three possible combinations will yield the same measurement, and that reading will be the phase voltage that will be recorded.

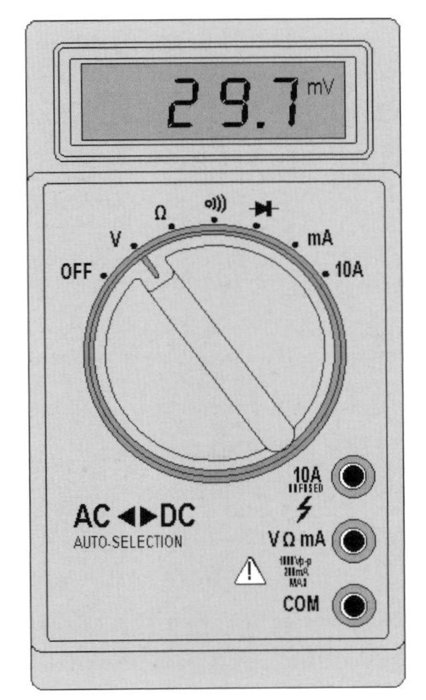

Figure 11.4. Autoranging Multimeter example

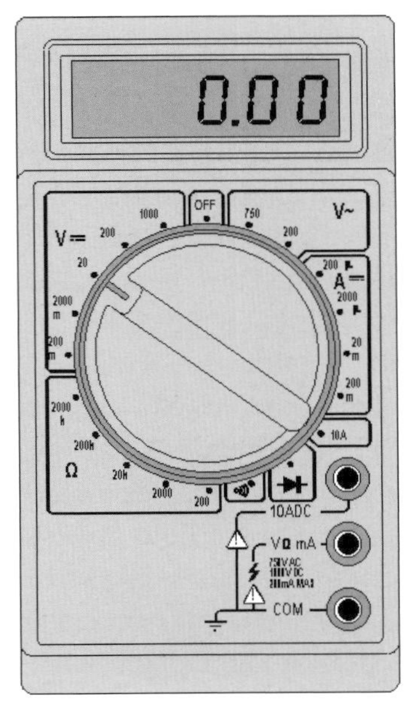

Figure 11.5. Switched Range Multimeter example

CLAMP METER USE A clamp meter can be used to measure the amp draw of the GSHP unit during operation. Figure 11.6 displays a common clamp meter. Ensure that the clamp meter is set to take A/C current readings before attempting any measurements. The method of measuring amp draw is the same regardless of whether single-phase or three-phase electrical power is supplied to the equipment, but the calculation of electrical power consumption will differ as will be discussed later. To properly measure GSHP unit amperage, two individual readings will be made by placing the tongs of the clamp meter around each of the wires connected to the hot leads (around the wire connected to L1 (Figure 11.3) to measure the current through L1 and around the wire connected to L2 (Figure 11.3) to measure the current through L2). Record the higher of the two amperage measurements for use in electrical power calculations.

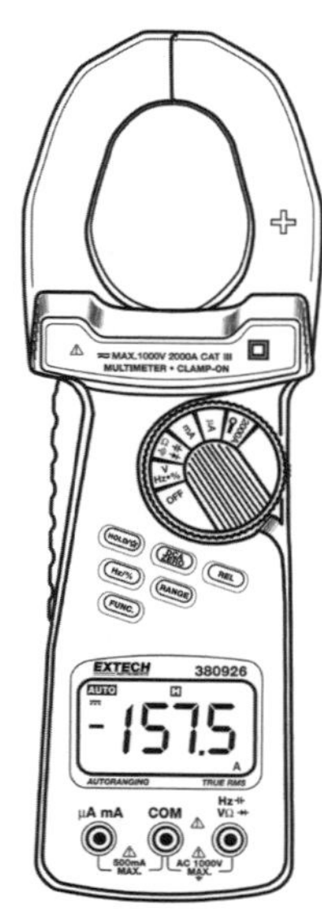

Figure 11.6. Clamp Meter Example

PRESSURE GAUGE/FLOWMETER USE A pressure gauge will be necessary to estimate the water flow rate through the GSHP unit. To use the pressure gauge, simply insert the needle into each of the P/T ports and record the pressure reading generated through each (P_{in} and P_{out} in Figures 11.1 and 11.2, use only one pressure gauge for both readings to prevent instrument calibration issues). The pressure drop through the coil will be the difference between the readings (Pressure Drop = $\Delta P = P_{in} - P_{out}$). Source-side flow rate through the GSHP can be estimated by comparing the measured pressure drop to the GSHP manufacturer's published data for pressure drop through the coil. For example, if a manufacturer publishes a 5-psi pressure drop through their GSHP unit at a flow rate of 10 gpm, and the measured pressure drop through the GSHP during operation is ~5-psi, then the operating flow rate can be taken to be ~10 gpm. Figure 6.6a depicts a typical analog, or needle-type, pressure gauge. Digital pressure gauges are also available in the marketplace.

Alternatively, flow rate may be directly measured via an inline-flowmeter or flowmeter tool. A flowmeter tool must be placed in the fluid flow path for measurements to be taken. Flowmeter tools are commonly used in conjunction with non-pressurized flow centers. Figure 11.7 shows a common flowmeter tool and how it is used with a non-pressurized flow center. An inline-flowmeter (Figure 11.8) may be permanently installed in GSHP piping, though such installation is generally not recommended due to unnecessary maintenance issues that may result because of leakage from the flowmeter.

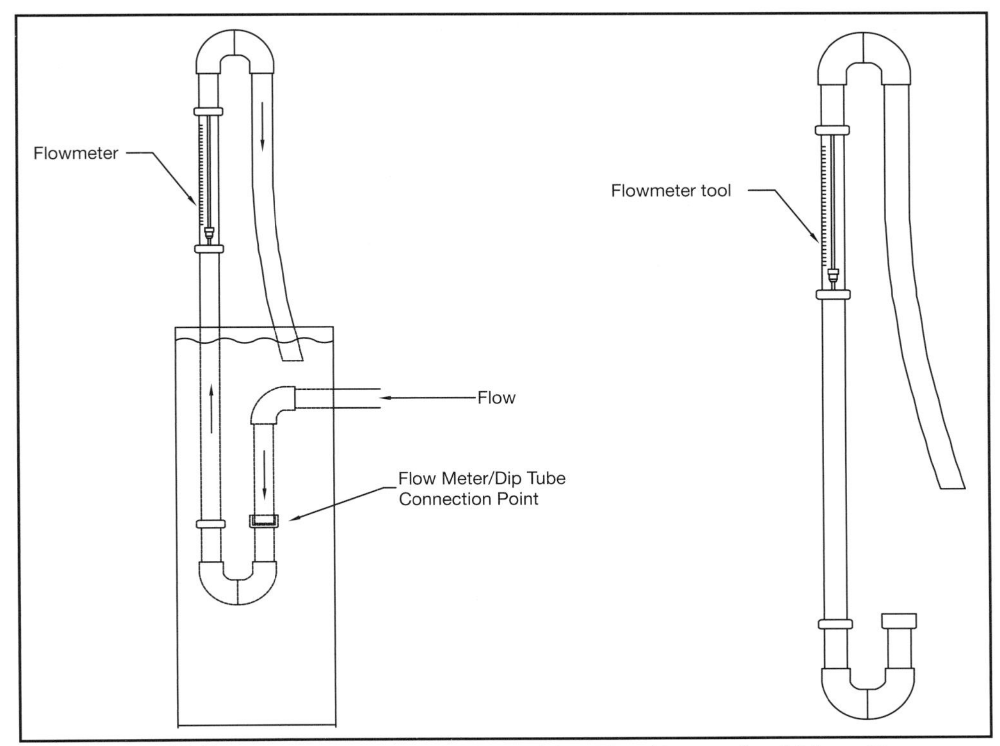

Figure 11.7. Flowmeter Tool & Use with a Non-Pressurized Flow Center

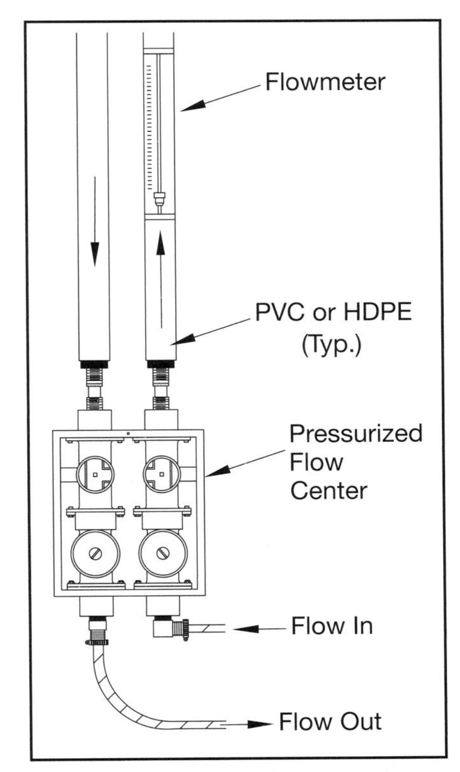

Figure 11.8. Inline Flowmeter

THERMOCOUPLE/DIGITAL THERMOMETER USE A thermocouple/digital thermometer (Figure 6.6a) will be necessary to measure entering and leaving water temperatures to/from the GSHP unit. The thermal probe is inserted into each of the P/T ports to record the temperature in each (EWT and LWT in Figure 11.1 and EWT, LWT, ELT, and LLT in Figure 11.2). When taking a temperature measurement, allow the thermal probe to reach thermal equilibrium with the circulating fluid. A proper temperature measurement can only be recorded once the thermometer/thermocouple reading stops fluctuating with time.

In addition to circulating water temperatures, a thermocouple/digital thermometer will be necessary to measure entering (EAT) and leaving (LAT) air temperatures to/from the GSHP unit. The EAT is measured on the return-air side of the heat pump, and the LAT is measured on the supply side of the heat pump, as shown in Figure 11.1. When taking the air temperature measurements, insert the temperature probe through the vibration break on the return-air side of the heat pump and after the first turn in the ductwork on the supply-air side of the heat pump. This is discussed in more detail in Section 11.2.6.

ANTIFREEZE TESTER USE Antifreeze testers will be necessary to determine the approximate antifreeze concentration and freeze protection level of the water-antifreeze mixture (if used) in the closed-loop GHEX piping system. The type of antifreeze tester will depend on antifreeze type. To test the concentration levels of a propylene glycol-water mixture, a refractometer will be required. For alcohol-water mixtures (methanol-water or ethanol-water), a hydrometer will be required. The contractor should contact their antifreeze supplier and ask for a "propylene glycol tester" or a "methanol tester."

INCLINED MANOMETER USE An inclined manometer will be required to perform ductwork static pressure drop measurements, if necessary (see Section 11.2.6). To measure the static pressure drop in the ductwork, simply insert the manometer tubes into the supply and return side of the ductwork and measure the height difference in the fluid caused by the pressure differential due to system air flow. Figure 11.9 shows a typical incline manometer. Digital manometers like that shown in Figure 11.10 are also available.

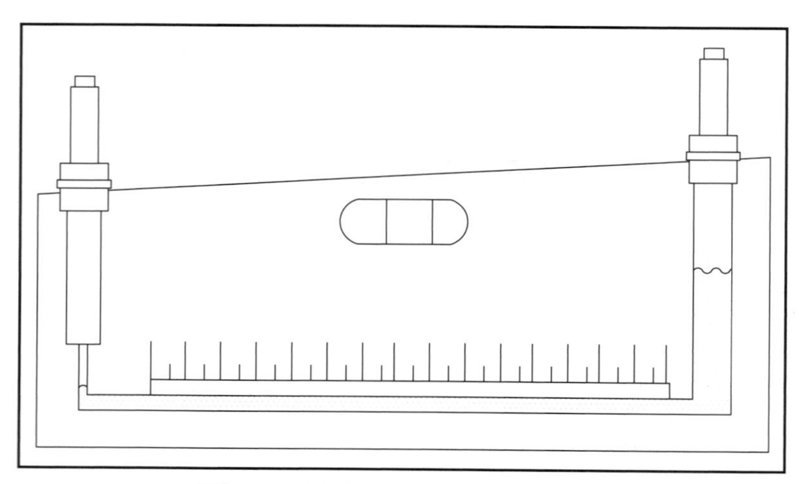

Figure 11.9. Incline Manometer

Figure 11.10. Digital Manometer

11.2.2 GSHP HEATING MODE OPERATION During the heating cycle, the performance of a GSHP is defined by its heating capacity (HC), electrical demand (DMD), and coefficient of performance (COP). Heating capacity can be checked by measuring the heat of extraction (HE) from the GHEX and adding that to the amount of heat added by the compressor, fan, and all other electrical inputs (DMD), as described by Equation 11.1. The performance of the GSHP in heating mode is estimated using water-side measurements because air-side measurements in the field are difficult to obtain with an acceptable degree of accuracy due to the difficulties in measuring air flow rates and air temperatures accurately.

$$HC = HE + (3.412 \cdot DMD)$$ **Equation 11.1**

Where:

HC = Heating capacity (Btu/hr)
HE = Heat of extraction (Btu/hr)
DMD = Electrical demand (W)

Heat of extraction is calculated using data obtained from field measurements on the water-side of the GSHP. The flow rate and circulating fluid temperature at the inlet and outlet of the GSHP unit are first measured, as discussed earlier, and then heat of extraction is calculated using Equation 11.2. The flow constant in Equation 11.2 depends on the GHEX circulating fluid type and temperature as provided in Table 11.3 for water and Table 11.4 for antifreeze mixtures.

$$HE = GPM \cdot (EWT - LWT) \cdot K$$ **Equation 11.2**

Where:

HE = Heat of extraction (Btu/hr)
GPM = System flow rate (gpm)
EWT = Entering water temperature to GSHP (F)
LWT = Leaving water temperature from GSHP (F)
K = Flow constant (Tables 11.3 and 11.4)

Table 11.3. Flow Constants for Water at Selected Temperatures

Flow Constant for Water	
Temp (˚F)	K
32	505
40	503
50	502
60	500
70	499
80	498
90	498
100	497

Table 11.4. Flow Constants for Various Antifreeze Solutions at Selected Temperatures

Flow Constants (K) for Propylene Glycol						
Concentration By Volume	Freeze Point (°F)	Temperature (°F)				
		30	35	40	45	50
10%	25.8	503	503	503	502	502
15%	22.5	501	501	500	500	500
20%	18.7	499	499	498	498	497
25%	14.0	499	499	498	498	497
30%	8.4	499	498	498	497	497
Flow Constants (K) for Methanol						
Concentration By Volume	Freeze Point (°F)	Temperature (°F)				
		30	35	40	45	50
5%	26.2	NA	512	511	511	511
10%	19.8	505	505	505	505	505
12.5%	16.2	501	501	501	501	501
15%	12.5	500	500	499	499	499
17.5%	8.8	496	496	495	495	495
20%	5.0	492	492	491	491	491
Flow Constants (K) for Ethanol						
Concentration By Volume	Freeze Point (°F)	Temperature (°F)				
		30	35	40	45	50
10%	23.8	516	516	516	517	517
15%	17.4	511	511	510	510	510
20%	10.9	510	510	509	509	509
25%	9.4	509	508	508	508	507
30%	7.8	506	505	505	504	504

Electric demand (DMD) is the rate at which the GSHP consumes electrical power and its calculation depends on whether the electrical service is single or three-phase, as described by Equation 11.3a (single-phase) and Equation 11.3b (three-phase) (ECONAR Energy Systems).

Single-Phase:

$$DMD = 0.85 \bullet V \bullet I \qquad \textbf{Equation 11.3a}$$

Three-Phase:

$$DMD = 0.85 \bullet 1.732 \bullet V \bullet I = 1.472 \bullet V \bullet I \qquad \textbf{Equation 11.3b}$$

Where:

- DMD = Electric demand (W)
- V = Voltage supplied to GSHP (V)
- I = Amperage draw through GSHP (A)
- 0.85 = Power factor for inductive loads, assumed average value
- 1.732 = Line voltage = 2 · (Phase voltage) · sin(120 F) = $\sqrt{3}$

Coefficient of Performance (COP) is a measure of the GSHP efficiency in the heating mode and is calculated using Equation 11.4.

$$COP = \frac{HC}{3.412 \bullet DMD}$$ **Equation 11.4**

Where:

COP = Coefficient of performance, dimensionless
HC = Heating capacity (Btu/hr)
DMD = Electric demand (W)

11.2.3 GSHP HEATING MODE PERFORMANCE CHECKING For a water-to-air GSHP, there are two levels of heating mode performance check. The first is to determine the heat of extraction from the GHEX and then compare to manufacturer's data for the measured operating conditions (EWT, GPM, and EAT). This test requires the least number of measurements and generally provides sufficient performance data to assure that the GSHP is operating properly. The second is to determine the heating capacity of the GSHP (which is the sum of the heat of extraction and the electrical demand) and then compare to manufacturer's data for the measured operating conditions (EWT, GPM, and EAT). This test requires additional measurements for electrical voltage and current, which requires more costly test equipment. In either case, if the measured heat of extraction does not agree with manufacturer's data, then the next step is to verify the CFM, which is not generally measured unless there is concern that a reduction in CFM is reducing the measured performance of the GSHP.

With the backup resistance heat disabled (if used) and the thermostat set well above the room temperature, start the GSHP in heating mode and allow approximately 10 minutes of operation for the EWT conditions to stabilize. Then, perform the following steps utilizing Table 11.5, a worksheet for measuring heating mode performance of water-to-air GSHPs, to document the performance check procedure and results.

FOR HEAT OF EXTRACTION PERFORMANCE CHECK:

1) Measure the temperature drop of the circulating fluid through the GSHP (EWT - LWT) on the ground-loop water side of the heat pump. Temperature measurements are taken directly through the P/T ports, which should be installed as close as possible to the ground-loop inlet and outlet connections of the GSHP. It is recommended that the same temperature probe be used for measuring both EWT and LWT, which reduces the possibility of introducing measurement error due to measurement differences associated with different probes.

2) Measure the water flow rate (GPM) through the unit. Flow rate can be measured directly if either a flow meter is installed in the circulation loop or if a flow meter tool is used in conjunction with a non-pressurized flow center. Because direct flow measurement is not always an option, flow can be estimated by measuring pressure drop through the water-to-refrigerant coil in the GSHP, and then, comparing to the manufacturer's published data. Equation 11.5 can be utilized to convert pressure drop measurements from pounds per square inch (psi) to feet of water, or vice versa.

$$wpd = 2.31 \bullet psi$$

-or-

$$psi = 0.433 \bullet wpd$$

Equation 11.5

Where:

wpd = Pressure drop in feet of water (ft)

psi = Pressure drop in pounds per square inch (psi)

3) For a water-to-air unit, measure the temperature of the air entering (EAT) the GSHP. This temperature measurement is needed because heat of extraction can be significantly impacted by the temperature of the entering air.

FOR HEATING CAPACITY PERFORMANCE CHECK, ADD THE FOLLOWING STEP:

4) Measure the electrical demand (DMD) of the GSHP by measuring the amperage through the unit and total voltage supplied following the procedures given in Section 11.2.1. Calculate the electrical demand of the unit according to Equation 11.3a for single-phase power and Equation 11.3b for three-phase power.

After determining the water temperatures entering (EWT) and leaving (LWT) the heat pump and the water flow rate (GPM) through the heat pump, the heat of extraction (HE) for the GHEX can be calculated. The measured heat of extraction should then be compared to the manufacturer's value for heat of extraction for verification that the GSHP is working properly. With the additional measurement of the electrical demand (DMD) to the GSHP, the heating capacity (HC) of the unit can be calculated. The measured heating capacity should then be compared to the manufacturer's value for heating capacity for verification that the GSHP is working properly. In either case, a measured value for heat of extraction that is significantly lower (10% or more) than the manufacturer's value may be due to restricted air flow through the GSHP, and the next step would be to verify the air flow through the GSHP as described in Section 11.2.6.

For a water-to-water GSHP, the procedures described above are used except that the EAT measurement in Step 3 is replaced with an entering water temperature (EWT) on the load side (ELT) of the GSHP (leaving load temperature (LLT) for some manufacturer's equipment if performance is based on that). If the measured heat of extraction does not agree with manufacturer's data, then the water flow rate on the load side of the GSHP must be verified using either direct measurement of water flow or pressure drop measurement across the load-side coil and head loss data from the manufacturer, following the same procedures described above for the measurement of water flow through the GHEX side of the GSHP. Table 11.6 provides a worksheet for measuring heating mode performance for water-to-water GSHPs.

Table 11.5. Heating Mode Water-to-Air GSHP Performance Check Worksheet

W-A GSHP Performance Check: Heating Mode[1]

Heat of Extraction

Fluid Flow / Temperature Information

1) Set T-Stat to Heating, Operate Continuously for 10-Min, Take Temp and Pressure Readings Across the Water-Refrigerant Coil

EWT ________ °F P_{in} ________ psi[3]

LWT ________ °F P_{out} ________ psi[3]

ΔT = (EWT - LWT) = ________ °F **$\Delta P = (P_{in} - P_{out})$ = ________ psi**

EAT ________ °F[2] **Est./Measured Flow Rate (GPM) = ________ gpm[3]**

Circulating Fluid Type Information

2) Measure/Record Fluid Type, Temperature, and Antifreeze Concentration (% By Volume, If Necessary)

Circulating Fluid Type ________ **Fluid Temp (°F)** ________ [4] **Concentration (By Vol)** ________ %

Estimated K-Value = ________ [5]

Heat of Extraction (HE) Calculation

3) Calculate the Heat of Extraction (HE) from Loopfield - Check for Agreement w/ Manufacturer's Specs

HE = GPM · (EWT - LWT) · K (Btu/hr)

HE = ________ · (______ - ______) · ________ = ________ · ________ · ________ = ________ Btu/hr

Electrical Demand

4) Measure Voltage Supplied to GSHP Unit and Measure Amp Draw

Voltage (V) = ________ Volts[6] **Amp Draw (I) = ________ Amps[7]**

5) Calculate Electrical Demand (DMD) of GSHP (Watts) Using One of the Following:

Single Phase Power: DMD = 0.85 · (V · I) -or- Three Phase Power: DMD = 1.732 · 0.85 · (V · I) (Watts)

1Ø: DMD = 0.85 · ________ · ________ = ________ Watts[8]

3Ø: DMD = 1.732 · 0.85 · ________ · ________ = ________ Watts[9]

GSHP Heating Capacity

6) Calculate Heating Capacity (Btu/hr) of GSHP Under Operating Conditions - Check for Agreement w/ Manufacturer's Specs

HC = HE + (3.412 · DMD) (Btu/hr)

HC = ________ + (3.412 · ________) = ________ + ________ = ________ Btu/hr

GSHP Efficiency

7) Calculate GSHP COP Under Operating Conditions - Check for Agreement w/ Manufacturer's Specs

COP = HC / (DMD · 3.412)

COP = ________ / (________ · 3.412) = ________ / ________ = ________

Overall Performance Check

Check Measured/Calculated Parameters Against Manufacturer's Specifications

Parameter	Calculated/Measured	Manufacturer's Data[10]	Agree (Y/N)
Heat of Extraction (HE - Btu/hr)			
Heating Capacity (HC - Btu/hr)			
Electrical Demand (DMD - kW)			
COP			

1) Refer to Section 11.2 in the manual for complete performance checkout procedure
2) Entering air temperature (EAT) is needed at this point for comparison of measured data against manufacturer's published data. Heating capacity depends on the temperature of the air entering the GSHP.
3) If flowmeter is available, directly measure system flow during operation. Estimate flow via pressure drop measurements and manufacturer's published specifications if no flowmeter is available
4) Take the operating fluid temp to be the average of the entering and leaving temp => Fluid Temp = (EWT + LWT) / 2
5) Refer to Tables 11.3 and 11.4 for K values for various types/concentrations of antifreeze at selected temps. Use K = 500 if no info is available. Interpolation of K-values is permissible, extrapolation is not.
6) Measure voltage supplied (using multimeter) to GSHP unit across hot terminals according to Section 11.2.1.
7) Measure amp draw (using clampmeter) through each positive power terminal of GSHP (do not measure on the common leg). Amp draw to use for calculation is the higher of the two measurements.
8) Use this equation if single-phase power is connected to the GSHP.
9) Use this equation if three-phase power is connected to the GSHP.
10) Ensure that the manufacturer's data is properly selected for comparison at the operating EAT, EWT, and flow rate (GPM).

Table 11.6. Heating Mode Water-to-Water GSHP Performance Check Worksheet

W-W GSHP Performance Check: Heating Mode[1]

Heat of Extraction

Fluid Flow / Temperature Information

1) Set T-Stat to Heating, Operate Continuously for 10-Min, Take Temp and Pressure Readings Across the Water-Refrigerant Coils

EWT ____________ °F

LWT ____________ °F

ΔT = (EWT - LWT) = ____________ °F

P_{in} ____________ psi[3]

P_{out} ____________ psi[3]

$\Delta P = (P_{in} - P_{out})$ = ____________ psi

Est./Measured Source-Side Flow Rate (GPM) = ____________ gpm[3]

ELT ____________ °F[2]

LLT ____________ °F[2]

$P_{in\text{-}load}$ ____________ psi[2,3]

$P_{out\text{-}load}$ ____________ psi[2,3]

$\Delta P_{load} = (P_{in\text{-}load} - P_{out\text{-}load})$ = ____________ psi

Est./Measured Load-Side Flow Rate (GPM_{load}) = ____________ gpm[2,3]

Circulating Fluid Type Information

2) Measure/Record Fluid Type, Temperature, and Antifreeze Concentration (% By Volume, If Necessary)

Circulating Fluid Type ____________ **Fluid Temp (°F)** ____________ [4] **Concentration (By Vol)** ____________ **%**

Estimated K-Value = ____________ [5]

Heat of Extraction (HE) Calculation

3) Calculate the Heat of Extraction (HE) from Loopfield - Check for Agreement w/ Manufacturer's Specs

$$HE = GPM \cdot (EWT - LWT) \cdot K \quad \text{(Btu/hr)}$$

HE = ________ · (______ - ______) · ________ = ________ · ________ · ________ = ________ Btu/hr

Electrical Demand

4) Measure Voltage Supplied to GSHP Unit and Measure Amp Draw

Voltage (V) = ____________ **Volts[6]**

Amp Draw (I) = ____________ **Amps[7]**

5) Calculate Electrical Demand (D_{heat}) of GSHP (Watts) Using One of the Following:

Single Phase Power: $D_{heat} = 0.85 \cdot (V \cdot I)$ -or- Three Phase Power: $D_{heat} = 1.732 \cdot 0.85 \cdot (V \cdot I)$ (Watts)

1Ø: D_{heat} = 0.85 · ________ · ________ = ________ Watts[8]

3Ø: D_{heat} = 1.732 · 0.85 · ________ · ________ = ________ Watts[9]

GSHP Heating Capacity

6) Calculate Heating Capacity (Btu/hr) of GSHP Under Operating Conditions - Check for Agreement w/ Manufacturer's Specs

$$HC = HE + (3.412 \cdot D_{heat}) \quad \text{(Btu/hr)}$$

HC = ________ + (3.412 · ________) = ________ + ________ = ________ Btu/hr

GSHP Efficiency

7) Calculate GSHP COP Under Operating Conditions - Check for Agreement w/ Manufacturer's Specs

$$COP = HC / (D_{heat} \cdot 3.412)$$

COP = ________ / (________ · 3.412) = ________ / ________ = ________

Overall Performance Check

Check Measured/Calculated Parameters Against Manufacturer's Specifications

Parameter	Calculated/Measured	Manufacturer's Data[10]	Agree (Y/N)
Heat of Extraction (HE - Btu/hr)			
Heating Capacity (HC - Btu/hr)			
Electrical Demand (D_{heat} - kW)			
COP			

1) Refer to Section 11.2 in the manual for complete performance checkout procedure
2) Entering load temperature (ELT) and leaving load temperature (LLT) are needed at this point for comparison of measured data against manufacturer's published data. Heating capacity depends on the temperature of the load side water temperatures of the GSHP. Load-side flowrate is also needed for comparison of measured to published data.
3) If flowmeter is available, directly measure flow during operation. Estimate flow via pressure drop measurements and manufacturer's published specifications if no flowmeter is available
4) Take the operating fluid temp to be the average of the entering and leaving temp => Fluid Temp = (EWT + LWT) / 2
5) Refer to Tables 11.3 and 11.4 for K values for various types/concentrations of antifreeze at selected temps. Use K = 500 if no info is available. Interpolation of K-values is permissible, extrapolation is not.
6) Measure voltage supplied (using multimeter) to GSHP unit across hot terminals according to Section 11.2.1.
7) Measure amp draw (using clampmeter) through each positive power terminal of GSHP (do not measure on the common leg). Amp draw to use for calculation is the higher of the two measurements.
8) Use this equation if single-phase power is connected to the GSHP.
9) Use this equation if three-phase power is connected to the GSHP.
10) Ensure that the manufacturer's data is properly selected for comparison at the operating ELT or LLT, EWT, source-side flowrate (GPM) and load-side flow rate (GPM_{load}).

11.2.4 GSHP COOLING MODE OPERATION During the cooling cycle, the performance of a GSHP is defined by its total cooling capacity (TC), electrical demand (DMD), and energy efficiency ratio (EER). Total cooling capacity can be checked by measuring the amount of heat rejected (HR) to the GHEX, and then, subtracting the amount of heat added by the compressor, fan, and all other electrical inputs DMD from HR, as described by Equation 11.6. The performance of the GSHP in cooling mode is estimated using water-side measurements because air-side measurements in the field are difficult to obtain with an acceptable degree of accuracy due to the difficulties in measuring air flow rates and air temperature accurately. In addition, determining the amount of total cooling capacity that is due to condensation of water from the air stream (latent cooling) requires specialized equipment.

$$TC = HR - (3.412 \bullet DMD)$$ **Equation 11.6**

Where:

TC = Total cooling capacity (Btu/hr)
HR = Heat of rejection (Btu/hr)
DMD = Electrical demand (W)

Heat of rejection is calculated using data obtained from field measurements on the water side of the GSHP. The flow rate and circulating fluid temperature at the inlet and outlet of the GSHP unit are first measured, as discussed earlier, and then heat of rejection is calculated using Equation 11.7. The flow constant in Equation 11.7 depends on the GHEX circulating fluid type and temperature as provided in Table 11.3 for water and Table 11.4 for antifreeze mixtures.

$$HR = GPM \bullet (LWT - EWT) \bullet K$$ **Equation 11.7**

Where:

HR = Heat of rejection (Btu/hr)
GPM = System flow rate (gpm)
LWT = Leaving water temperature from GSHP (F)
EWT = Entering water temperature to GSHP (F)
K = Flow constant (Tables 11.3 and 11.4)

Electric demand (DMD) is the rate at which the GSHP consumes electrical power, and its calculation depends on whether the electrical service is single or three-phase, as described by Equation 11.3a (single-phase) and Equation 11.3b (three-phase) (ECONAR Energy Systems).

Energy Efficiency Ratio (EER) is a measure of the GSHP efficiency in the cooling mode and is calculated using Equation 11.8.

$$EER = \frac{TC}{DMD}$$ **Equation 11.8**

Where:

EER = Energy efficiency ratio, dimensionless
TC = Total cooling capacity (Btu/hr)
DMD = Electric demand (W)

11.2.5 GSHP COOLING MODE PERFORMANCE CHECKING For a water-to-air GSHP, there are two levels of cooling mode performance check. The first is to determine the heat of rejection to the GHEX, and then, compare to manufacturer's data for the measured operating conditions (EWT, GPM, and EAT). This test requires the least number of measurements and generally provides sufficient performance data to assure that the GSHP is operating properly. The second is to determine the total cooling capacity of the GSHP (which is the reat of rejection minus the electrical demand), and then, compare to manufacturer's data for the measured operating conditions (EWT, GPM, and EAT). This test requires additional measurements for electrical voltage and current, which requires more costly test equipment. In either case, if the measured heat of rejection does not agree with manufacturer's data, then the next step is to verify the CFM, which is not generally measured unless there is concern that a reduction in CFM is reducing the measured performance of the GSHP.

With the thermostat set well below the room temperature, start the GSHP in cooling mode and allow approximately 10 minutes of operation for the EWT conditions to stabilize. Then, perform the following steps utilizing Table 11.7, a worksheet for measuring cooling mode performance of water-to-air GSHPs, to document the performance check procedure and results.

FOR HEAT OF REJECTION PERFORMANCE CHECK:

1) Measure the temperature rise of the circulating fluid through the GSHP (LWT - EWT) on the ground loop water side of the heat pump. Temperature measurements are taken directly through the P/T ports, which should be installed as close as possible to the ground-loop inlet and outlet connections of the GSHP. It is recommended that the same temperature probe be used for measuring both LWT and EWT, which reduces the possibility of introducing measurement error due to measurement differences associated with different probes.
2) Measure the water flow rate (GPM) through the unit. Flow rate can be measured directly if either a flow meter is installed in the circulation loop or if a flow meter tool is used in conjunction with a non-pressurized flow center. Because direct flow measurement is not always an option, flow can be estimated by measuring pressure drop through the water-to-refrigerant coil in the GSHP, and then, comparing to the manufacturer's published data. Equation 11.5 can be utilized to convert pressure drop measurements from pounds per square inch (psi) to feet of water, or vice versa.
3) For a water-to-air unit, measure the temperature of the air entering (EAT) the GSHP. This temperature measurement is needed because heat of rejection can be significantly impacted by the temperature of the entering air.

FOR TOTAL COOLING CAPACITY PERFORMANCE CHECK, ADD THE FOLLOWING STEP:

4) Measure the electrical demand (DMD) of the GSHP by measuring the amperage through the unit and total voltage supplied following the procedures given in Section 11.2.1. Calculate the electrical demand of the unit according to Equation 11.3a for single-phase power and Equation 11.3b for three-phase power.

After determining the water temperatures leaving (LWT) and entering (EWT) the heat pump and the water flow rate (GPM) through the heat pump, the heat of rejection (HR) for the GHEX can be calculated.

The measured heat of rejection should then be compared to the manufacturer's value for heat of rejection for verification that the GSHP is working properly. With the additional measurement of the electrical demand (DMD) to the GSHP, the total cooling capacity (TC) of the unit can be calculated. The measured total cooling capacity should then be compared to the manufacturer's value for total cooling capacity for verification that the GSHP is working properly. In either case, a measured value for heat of rejection that is significantly lower (10% or more) than the manufacturer's value may be due to restricted air flow through the GSHP, and the next step would be to verify the air flow through the GSHP as described in Section 11.2.6.

Table 11.7. Cooling Mode Water-to-Air GSHP Performance Check Worksheet

W-A GSHP Performance Check: Cooling Mode[1]

Heat of Rejection

Fluid Flow / Temperature Information

1) Set T-Stat to Cooling, Operate Continuously for 10-Min, Take Temp and Pressure Readings Across the Water-Refrigerant Coil

EWT ____________ ˚F P_{in} ____________ psi[3]

LWT ____________ ˚F P_{out} ____________ psi[3]

ΔT = (LWT - EWT) = ____________ **˚F** **ΔP = (P_{in} - P_{out}) =** ____________ **psi**

EAT ____________ ˚F[2] **Est./Measured Flow Rate (GPM) =** ____________ **gpm[3]**

Circulating Fluid Type Information

2) Measure/Record Fluid Type, Temperature, and Antifreeze Concentration (% By Volume, If Necessary)

Circulating Fluid Type ____________ **Fluid Temp (˚F)** ____________ [4] **Concentration (By Vol)** ____________ %

Estimated K-Value = ____________ [5]

Heat of Rejection (HR) Calculation

3) Calculate the Heat of Rejection (HR) to Loopfield - Check for Agreement w/ Manufacturer's Specs

HR = GPM · (LWT - EWT) · K (Btu/hr)

HR = ________ **· (** ______ **-** ______ **) ·** ________ **=** ________ **·** ________ **·** ________ **=** __________ **Btu/hr**

Electrical Demand

4) Measure Voltage Supplied to GSHP Unit and Measure Amp Draw

Voltage (V) = ____________ **Volts[6]** **Amp Draw (I) =** ____________ **Amps[7]**

5) Calculate Electrical Demand (DMD) of GSHP (Watts) Using One of the Following:

Single Phase Power: DMD = 0.85 · (V · I) -or- Three Phase Power: D_{cool} = 1.732 · 0.85 · (V · I) (Watts)

1Ø: DMD = 0.85 · __________ **·** __________ **=** __________ **Watts[8]**

3Ø: DMD = 1.732 · 0.85 · __________ **·** __________ **=** __________ **Watts[9]**

GSHP Total Cooling Capacity

6) Calculate Total Cooling Capacity (Btu/hr) of GSHP Under Operating Conditions - Check for Agreement w/ Manufacturer's Specs

TC = HR - (3.412 · DMD) (Btu/hr)

TC = __________ **- (3.412 ·** __________ **) =** __________ **-** __________ **=** __________ **Btu/hr**

GSHP Efficiency

7) Calculate GSHP EER Under Operating Conditions - Check for Agreement w/ Manufacturer's Specs

EER = TC / (DMD)

EER = __________ **/** __________ **=** __________

Overall Performance Check

Check Measured/Calculated Parameters Against Manufacturer's Specifications

Parameter	Calculated/Measured	Manufacturer's Data[10]	Agree (Y/N)
Heat of Rejection (HR - Btu/hr)			
Total Cooling Capacity (TC - Btu/hr)			
Electrical Demand (DMD - kW)			
EER			

1) Refer to Section 11.2 in the manual for complete performance checkout procedure
2) Entering air temperature (EAT) is needed at this point for comparison of measured data against manufacturer's published data. Cooling capacity depends on the temperature of the air entering the GSHP.
3) If flowmeter is available, directly measure system flow during operation. Estimate flow via pressure drop measurements and manufacturer's published specifications if no flowmeter is available
4) Take the operating fluid temp to be the average of the entering and leaving temp => Fluid Temp = (EWT + LWT) / 2
5) Refer to Tables 11.3 and 11.4 for K values for various types/concentrations of antifreeze at selected temps. Use K = 500 if no info is available. Interpolation of K-values is permissible, extrapolation is not.
6) Measure voltage supplied (using multimeter) to GSHP unit across hot terminals according to Section 11.2.1.
7) Measure amp draw (using clampmeter) through each positive power terminal of GSHP (do not measure on the common leg). Amp draw to use for calculation is the higher of the two measurements.
8) Use this equation if single-phase power is connected to the GSHP.
9) Use this equation if three-phase power is connected to the GSHP.
10) Ensure that the manufacturer's data is properly selected for comparison at the operating EAT, EWT, and flow rate (GPM).

Table 11.8. Cooling Mode Water-to-Water GSHP Performance Check Worksheet

W-W GSHP Performance Check: Cooling Mode[1]

Heat of Rejection

Fluid Flow / Temperature Information

1) Set T-Stat to Cooling, Operate Continuously for 10-Min, Take Temp and Pressure Readings Across the Water-Refrigerant Coils

EWT ______________ ˚F

LWT ______________ ˚F

ΔT = (LWT - EWT) = ______________ ˚F

P_{in} ______________ psi[3]

P_{out} ______________ psi[3]

$\Delta P = (P_{in} - P_{out})$ = ______________ psi

Est./Measured Source-Side Flow Rate (GPM) = ______________ gpm[3]

ELT ______________ ˚F[2]

LLT ______________ ˚F[2]

$P_{in\text{-}load}$ ______________ psi[2,3]

$P_{out\text{-}load}$ ______________ psi[2,3]

$\Delta P_{load} = (P_{in\text{-}load} - P_{out\text{-}load})$ = ______________ psi

Est./Measured Load-Side Flow Rate (GPM_{load}) = ______________ gpm[2,3]

Circulating Fluid Type Information

2) Measure/Record Fluid Type, Temperature, and Antifreeze Concentration (% By Volume, If Necessary)

Circulating Fluid Type ______________ **Fluid Temp (˚F)** ______________[4] **Concentration (By Vol)** ______________ %

Estimated K-Value = ______________[5]

Heat of Rejection (HR) Calculation

3) Calculate the Heat of Rejection (HR) to Loopfield - Check for Agreement w/ Manufacturer's Specs

HR = GPM · (LWT - EWT) · K (Btu/hr)

HR = _________ **· (** _______ **-** _______ **) ·** _________ **=** _________ **·** _________ **·** _________ **=** ___________ **Btu/hr**

Electrical Demand

4) Measure Voltage Supplied to GSHP Unit and Measure Amp Draw

Voltage (V) = ______________ **Volts**[6] **Amp Draw (I) =** ______________ **Amps**[7]

5) Calculate Electrical Demand (DMD) of GSHP (Watts) Using One of the Following:

Single Phase Power: DMD= 0.85 · (V · I) -or- Three Phase Power: D_{cool} = 1.732 · 0.85 · (V · I) (Watts)

1Ø: DMD = 0.85 · ___________ **·** ___________ **=** ___________ **Watts**[8]

3Ø: DMD = 1.732 · 0.85 · ___________ **·** ___________ **=** ___________ **Watts**[9]

GSHP Total Cooling Capacity

6) Calculate Total Cooling Capacity (Btu/hr) of GSHP Under Operating Conditions - Check for Agreement w/ Manufacturer's Specs

TC = HR - (3.412 · DMD) (Btu/hr)

TC = ___________ **- (3.412 ·** ___________ **) =** ___________ **-** ___________ **=** ___________ **Btu/hr**

GSHP Efficiency

7) Calculate GSHP EER Under Operating Conditions - Check for Agreement w/ Manufacturer's Specs

EER = TC / (DMD)

EER = ___________ **/** ___________ **=** ___________

Overall Performance Check

Check Measured/Calculated Parameters Against Manufacturer's Specifications

Parameter	Calculated/Measured	Manufacturer's Data[10]	Agree (Y/N)
Heat of Rejection (HR - Btu/hr)			
Total Cooling Capacity (TC - Btu/hr)			
Electrical Demand (DMD - kW)			
EER			

1) Refer to Section 11.2 in the manual for complete performance checkout procedure
2) Entering load temperature (ELT) and leaving load temperature (LLT) are needed at this point for comparison of measured data against manufacturer's published data. Cooling capacity depends on the temperature of the load side water temperatures of the GSHP. Load-side flowrate is also needed for comparison of measured to published data.
3) If flowmeter is available, directly measure flow during operation. Estimate flow via pressure drop measurements and manufacturer's published specifications if no flowmeter is available
4) Take the operating fluid temp to be the average of the entering and leaving temp => Fluid Temp = (EWT + LWT) / 2
5) Refer to Tables 11.3 and 11.4 for K values for various types/concentrations of antifreeze at selected temps. Use K = 500 if no info is available. Interpolation of K-values is permissible, extrapolation is not.
6) Measure voltage supplied (using multimeter) to GSHP unit across hot terminals according to Section 11.2.1.
7) Measure amp draw (using clampmeter) through each positive power terminal of GSHP (do not measure on the common leg). Amp draw to use for calculation is the higher of the two measurements.
8) Use this equation if single-phase power is connected to the GSHP.
9) Use this equation if three-phase power is connected to the GSHP.
10) Ensure that the manufacturer's data is properly selected for comparison at the operating ELT or LLT, EWT, source-side flowrate (GPM) and load-side flow rate (GPM_{load}).

For a water-to-water GSHP, the procedures described above are used except that the EAT measurement in Step 3 is replaced with an entering water temperature on the load side (ELT) of the GSHP (leaving load temperature (LLT) for some manufacturer's equipment if performance is based on that). If the measured heat of rejection does not agree with manufacturer's data, then the water flow rate on the load side of the GSHP must be verified using either direct measurement of water flow or pressure drop measurement across the load-side coil and head loss data from the manufacturer, following the same procedures described above for the measurement of water flow through the GHEX side of the GSHP. Table 11.8 provides a worksheet for measuring cooling mode performance for water-to-water GSHPs.

11.2.6 GSHP BLOWER PERFORMANCE CHECKING Generally, air-side blower performance checking measures will not be required. However, these measurements are necessary if the measured performance of a GSHP unit differs significantly from manufacturers' published data (see the Overall Performance Check section on Tables 11.5 through 11.8). The first step in analyzing differences between measured and published data is to check for proper air flow, which requires the use of the backup electric resistance plenum heater (if used). First, set the thermostat to "emergency heating" mode and raise the set point to activate the blower and strip resistance plenum heater while the compressor and circulating pumps are deactivated. Allow 2-3 minutes of operation for air flow conditions to stabilize, and then, perform the following steps.

1) Measure the temperature differential (ΔT_{air}) of the air across the coil by measuring the entering air temperature (EAT) on the return side of the heat pump and the leaving air temperature (LAT) after it flows across the strip heater. Measure EAT at the vibration break on the return side of the heat pump and LAT after the first turn in the ductwork on the supply side of the heat pump. The LAT on the supply side should be made after the first turn in the ductwork to ensure uniform air stream temperatures by allowing sufficient distance for the air to mix after crossing the strip resistance heater. Also, additional distance between the thermometer and the resistance heater, and elimination of direct line-of-sight between the two, minimizes radiant effects from the heater on the temperature probe. Calculate the temperature differential of the air across the coil according to Equation 11.9.

$$\Delta T_{air} = EAT - LAT$$ **Equation 11.9**

Where:

ΔT_{air} = Air temperature differential across resistance heating coil (F)
EAT = Entering air temperature (F)
LAT = Leaving air temperature (F)

2) Estimate the air flow produced by the system blower (CFM) according to Equation 11.10. Check manufacturer's specifications to determine the nominal capacity of the electric strip heater, which can range from 5 to 20 kW. Also, verify the capacity of the electric strip heater in Emergency Heating Mode.

$$CFM = \frac{3{,}412 \bullet kW_{res}}{1.08 \bullet \Delta T_{air}}$$ **Equation 11.10**

Where:

CFM = Estimated air flow rate produced by blower (CFM)
kW_{res} = Installed capacity of backup plenum resistance heater (kW)

3) Utilization of Equation 11.10 is an approximation of system air flow, which ignores heat added to the air by the blower itself and also neglects any variability of the power supplied to the resistance heating coil. A more accurate method for estimating system air flow (while still ignoring fan energy) is to actually measure the voltage and amp draw of the strip resistance heater by using Equation 11.11a for single-phase power and 11.11b for three-phase power:

$$CFM = \frac{3{,}412 \bullet V_{res} \bullet I_{res}}{1{,}080 \bullet \Delta T_{air}}$$ **Equation 11.11a**

$$CFM = \frac{3{,}412 \bullet 1.732 \bullet V_{res} \bullet I_{res}}{1{,}080 \bullet \Delta T_{air}}$$ **Equation 11.11b**

Where:

V_{res} = Voltage supplied to backup plenum resistance heater (Volts)
I_{res} = Amp draw of the backup plenum resistance heater (Amps)

A correction factor will need to be applied to manufacturers' published data if measured air flow differs from the rated air flow. Table 11.9 provides a worksheet for measuring and documenting the air-side performance check of a GSHP unit.

Table 11.9. Air-Side Performance Check Worksheet

W-A GSHP Blower Performance Check[1]

System Air Flow Check[2]

Air-Side Measurements

1) Set T-Stat to Emergency Heating, Allow for 2-3 Minutes Continous Resistance Heating Operation, Measure Air Temperatures

EAT ______________ °F[3] V_{res} ______________ volts[4]
LAT ______________ °F[3] I_{res} ______________ amps[4]
ΔT_{air} = (LAT - EAT) = ______________ **°F**   **Static Pressure Drop (IWG)** ______________ **in. H_2O**[5]

2) Estimate System Air Flow During Emergency Heating Operation According to One of the Following Equations:

$$CFM = 3{,}412 \cdot kW_{res} / [\,1.08 \cdot (LAT - EAT)\,]$$

CFM = 3,412 · __________ **/ [1.08 · (**__________ **-** __________**)] =** __________ **/** __________ **=** ______________ **cfm**[6]

$$1\emptyset:\ CFM = 3{,}412 \cdot V_{res} \cdot I_{res} / [\,1{,}080 \cdot (LAT - EAT)\,]\ (cfm)$$

$$3\emptyset:\ CFM = 3{,}412 \cdot 1.732 \cdot V_{res} \cdot I_{res} / [\,1{,}080 \cdot (LAT - EAT)\,]\ (cfm)$$

1Ø: CFM = 3,412 · _________ **·** _________ **/ [1,080 · (**_________ **-** _________**)] =** _________ **/** _________ **=** ____________ **cfm**[7a]
3Ø: CFM = 3,412 · 1.732 · ________ **·** ________ **/ [1,080 · (**________ **-** ________**)] =** ________ **/** ________ **=** ____________ **cfm**[7b]

Airflow Comparison

3) Compare Calculated Airflow to Rated Airflow

Actual Airflow ______________ **cfm**   **Rated Airflow** ______________ **cfm**
Measured Static ______________ **in. H_2O**[5]   **Rated Static** ______________ **in. H_2O**
% of Rated Capacity ______________ **%**

Performance Correction Factors

Htg Cap (HC) CF ______________   **Total Clg Cap (TC) CF** ______________
Htg DMD CF ______________   **Clg DMD CF** ______________

Corrected Heating Performance Data

4) Calculate Heating Capacity, Electrical Demand, Heat of Extraction, and COP Based on Airflow Correction Factors

Corrected HC ______________ **Btu/hr**   **Corrected DMD** ______________ **Watts**

5) Calculate Corrected Heat of Extraction and COP:

$$COP_{corr} = \text{Corrected HC} / (\text{Corrected DMD} \cdot 3.412)$$

COP_{corr} = ____________ **/ (**____________ **· 3.412) =** ____________ **/** ____________ **=** ____________

$$HE_{corr} = (\text{Corrected HC}) \cdot (COP_{corr} - 1) / (COP_{corr})$$

HE_{corr} = ____________ **· (**____________ **- 1) / (**____________**) =** ____________ **/** ____________ **=** ____________

Overall Heating Performance Check

Check Measured/Calculated Parameters Against Manufacturer's Specifications

Parameter	Calculated/Measured	Corrected Manufacturer's Data	Agree (Y/N)
Heat of Extraction (HE - Btu/hr)			
Htg Electrical Demand (DMD - kW)			
COP			

Corrected Cooling Performance Data

4) Calculate Total Cooling Capacity, Electrical Demand, Heat of Rejection, and EER Based on Airflow Correction Factors

Corrected TC ______________ **Btu/hr**   **Corrected DMD** ______________ **Watts**

5) Calculate Corrected Heat of Rejection and EER:

$$EER_{corr} = \text{Corrected TC} / \text{Corrected DMD}$$

EER_{corr} = ____________ **/** ____________ **=** ____________

$$HR_{corr} = (\text{Corrected TC}) \cdot (EER_{corr} + 3.412) / (EER_{corr})$$

HR_{corr} = ____________ **· (**____________ **+ 3.412) / (**____________**) =** ____________ **/** ____________ **=** ____________

Overall Cooling Performance Check

Check Measured/Calculated Parameters Against Manufacturer's Specifications

Parameter	Calculated/Measured	Corrected Manufacturer's Data	Agree (Y/N)
Heat of Rejection (HR - Btu/hr)			
Total Cooling Capacity (TC - Btu/hr)			
Clg Electrical Demand (DMD - kW)			
EER			

1) Blower performance checking needs to be performed only when discrepancies between measured and published data are significant.
2) Perform air-side check independent of GSHP compressor operation to check for adequate air flow. To do so necessitates the use of backup electric resistance plemun heater.
3) Measure entering air temp (EAT) directly on the return side of the GSHP. Measure leaving air temp (LAT) after the first turn in the ductwork on the supply side of the GSHP.
4) Only measure voltage supply (V_{res}) and amp draw (I_{res}) of resistance heater if Equation 11.11a-11.11b (Section 11.2.6) is to be used (as opposed to Equation 11.10).
5) Measure the static pressure drop of the ductwork due to airflow with a manometer.
6) To use this equation, estimate the heating capacity of the electric resistance heater to be equal to its nominal capacity (e.g. - 10-kW).
7a) To use this equation, directly measure the heating capacity of the electric resistance by measuring the voltage supplied and amp draw during operation (1-phase).
7b) To use this equation, directly measure the heating capacity of the electric resistance by measuring the voltage supplied and amp draw during operation (3-phase).

11.3 Troubleshooting

After performance checking the GSHP system according to Sections 11.2.2 and 11.2.4, it will need to be examined to determine the root of any detected problems. When troubleshooting a GSHP system, always remember that every part of the heat pump system is interrelated (air side, water side, refrigeration circuit, and electrical side). If a problem occurs on the air side of the heat pump (such as extremely restricted air flow), it will affect temperature difference across the coil on the water side, refrigeration temperatures and pressures, and electrical demand of the system as a whole. Each system needs to properly operate in order for the entire system to deliver its expected performance.

11.3.1 AIR SIDE TROUBLESHOOTING If the blower does not deliver the desired amount of air flow to condition the space, the system will perform unacceptably. Generally, poor system performance will lead to comfort issues and unsatisfied customers. In most cases, poor system performance is the result of either poor design or poor installation practice. The following list of air flow problems was adapted from *LeClaire and Lafferty (1996):*

INSUFFICIENT AIR FLOW

1) **Blower wheel not turning fast enough.** This can be caused by simply selecting the wrong tap speed on the motor. An undersized blower motor will also give the same symptoms.
2) **Restricted ductwork.** Plugged or dirty filters are common examples. Check for foreign objects in the ductwork, such as instruction books, tool boxes, etc. In commercial installations, fire dampers might be closed. Visually check for these items to eliminate their possibility and check to ensure that all supply diffusers and return grilles are open and free from obstruction.
3) **Ductwork too small.** This problem will typically cause high return-to-supply air temperature differences. The main trunkline could be too small, which will cause the unit to become stressed and will also bring about whole house complaints (temperatures that vary from room to room). Additionally, one or more branchlines could be too small, which will cause uneven room-to-room temperatures, high-pitched wind noises, and static pressures above recommendations. When one or two branchlines are too small, unit problems are unlikely.
4) **Incorrect ECM motor tap.** ECM motors select CFM, not motor speed. This results in low air velocities much of the time. ECM motors select CFM based on the amp draw of the motor. With a given blower wheel the motor relates to CFM and holds the speed to plus or minus 5%. Check to ensure proper blower tap speed settings.
5) **Cold air temperatures from the supply ducts in cooling.** This happens when the ECM motor delivers low air flow rates in the cooling mode. Large temperature differences in the air side will result.

EXCESSIVE AIR FLOW

Though it is unusual, problems could possibly be created by having too much air flow. Fortunately, most of these problems can easily be alleviated.

1) **Ductwork too large.** If ductwork is grossly oversized in trunklines and branchlines, the centrifugal blower could be overloaded and cause motor problems. Diagnose this condition by measuring the amperage of the blower motor. Do not measure the motor amperage with the blower door off. This will

cause a false high amperage reading and distress the unit. Selecting a slower blower speed will generally cure the problem. Broken, open ductwork, or doors or panels left off will produce the same problems.

2) **Ductwork leaky or not completed.** Occasionally, the return air system is temporarily left unfinished during installation. If the unit is installed in an unheated/unfinished area of the house, the customer will complain of high electric bills and discomfort if this has not been accounted for and proper provisions made.

The size of the ductwork is of the utmost importance with respect to air-side GSHP performance. Problems on the air side of the GSHP system will either result in excessive noise, uneven room temperatures, inadequate levels of heating or cooling, or low water temperature differentials. Excessive noise is caused either by equipment vibration that is transferred throughout the entire house via the ductwork or by an undersized duct system. Insulating the ductwork for sound, as shown in Section 3.5, will greatly reduce system noise. Uneven room temperatures are generally caused by undersized or leaky ductwork. Setting the blower to "on" to constantly circulate the air through the space will help alleviate this problem. Inadequate levels of heating or cooling and low water temperature differentials are generally caused by low system air flow.

11.3.2 WATER SIDE TROUBLESHOOTING Similar to the air side on the GSHP, if the circulating pump(s) do not deliver the desired amount of water flow to satisfy the equipment, the end result will be unacceptable system performance. The following list of water flow problems was adapted from *LeClaire and Lafferty (1996):*

1) **Low flow (GPM).** Low flow is generally caused by an inoperative or improperly sized pump, a restricted piping system, or a low supply pressure. Always perform piping head loss calculations when designing a closed-loop system, and always remember to apply viscosity multipliers to the calculations for the specific level/type of freeze protection used in the system (if applicable). Small pipe and fittings add considerable friction to the system. Compare the results of the head loss calculations to the manufacturers' pump curves to ensure that the circulating pumps have been properly selected.
2) **No flow (GPM).** This problem is caused by blown fuses or inadequate power supplied to the pumps, pump failure due to shaft seizure or failure, air lock, or lack of positive suction side pressure (causing pump cavitation, only an issue with pressurized flow centers). Pump cavitation due to insufficient suction side pressure will most likely occur in the spring or summer when pipe expansion (due to higher fluid temperatures) causes the loopfield pressure to drastically decrease.
3) **Low incoming fluid temperature in heating or high incoming temperature in cooling in closed-loop systems (with adequate levels of fluid flow).** This problem is the result of an undersized ground-loop, poor contact between the ground-loop piping and the earth due to substandard grouting or backfilling practices, or pipe velocities that are too slow to cause turbulent flow in the loops. Always refer to the GHEX design documentation and compare the calculated building loads to the installed ground loop. Ensure that the ground loop was installed per specifications and that the specifications were within reason.
4) **Restricted/fouled heat exchanger.** This problem is most common in open-loop, water well systems. Minerals or sand from the water can be deposited on the water surface of the brazed-plate heat exchanger. This will degrade the rate of heat exchange in the water-to-refrigerant heat exchanger. This condition will exhibit a low temperature differential between the inlet and exit of the water side of the heat pump.

11.3.3 ELECTRICAL SIDE TROUBLESHOOTING Problems with the electrical side of a GSHP system will be the toughest to diagnose because the circuitry can be very complex. Training and certification is available through most GSHP manufacturers and such certification is recommended for all technicians who will be regularly working with a specific manufacturer. It is through the certification process that a technician will become most familiar with the products they work with. Additionally, they will become more efficient when diagnosing electrical problems whenever they arise.

The first step to take when troubleshooting the electrical side of a GSHP system is to check for broken, burned, or loose wires, blown fuses, tripped circuit breakers, or incorrectly connected wiring. Once that simple check has been completed, commence checking the lockout relays, solenoid valves, transformers, etc., that may be causing problems according to manufacturer specifications and initial system diagnosis.

11.3.4 REFRIGERATION CIRCUIT TROUBLESHOOTING Almost every problem that can occur in a GSHP installation will affect the refrigeration side of the system. It is recommended that all other components of the heat pump be checked before checking the refrigeration circuit to prevent any unnecessary trouble caused by doing so. Installation of refrigeration gauges should be the last step taken during the troubleshooting process. In most cases, diagnosis of the air side, water side, and electrical side will fix the problems affecting the refrigeration circuit without using refrigerant gauges. If every other component on the heat pump has been determined to be free from problems and the performance of the system is still erratic, check the refrigeration circuit. Always refer to manufacturer's specifications to determine proper refrigerant pressures both on the suction and discharge sides of the compressor. Table 11.10 summarizes how to diagnose common refrigeration system malfunctions and the probable cause of each.

11.3.5 GENERAL SYSTEM TROUBLESHOOTING There are some other common problems with GSHP systems not previously discussed, which include water leaking off of the heat pump unit itself, equipment short cycling, and periods when the GSHP unit runs constantly without reaching the thermostat setpoint.

1) **Water leaking off of unit.** If there is water leaking from the GSHP unit and there are no leaks on the water side of the heat pump, the root of the problem will be the condensation line. Make sure that the condensate line is not plugged, that the unit is properly pitched to allow for drainage (if horizontal), that the condensate line has adequate slope to allow for drainage (typically ¼-inch drop per foot of length), and check for leaks in the condensate line. Additionally, condensate lines should always have a trap. A trap will prevent condensation from being drawn back into the unit during operation. Some manufacturers internally incorporate a trap on their equipment. In such cases, an external trap will not be necessary. Condensate drains must not be double-trapped or have drain lines that are improperly routed/supported in a manner that can create a second trap. Double traps will prevent proper condensate drainage and will cause a safety trip on the GSHP unit or condensate overflow. Always refer to manufacturer's specifications to determine whether a condensate trap is necessary.
2) **Equipment short cycling.** Equipment short cycling (when subjected to off-peak conditions) is a sign of equipment over sizing. In northern climates, GSHP equipment will tend to be sized for heating while being significantly over sized in cooling. When equipment is significantly over sized in cooling, it will not run for extended periods of time causing issues such as inadequate dehumidification. Conversely, in southern climates, GSHP equipment will tend to be sized for cooling while being over sized in heating. Equipment short cycling reduces overall system

efficiency and life expectancy. In such cases, two-capacity equipment is recommended.

3) **Constant GSHP equipment operation without satisfying thermostat setpoint.** This problem is caused by undersized equipment, an undersized ground loop, or a combination of both (assuming that there are no issues on the air, water, electrical, or refrigerant sides of the heat pump). This condition will also cause higher than normal electric utility bills. If this problem exists, refer to all documentation and validate peak heating/cooling load calculations, equipment selection, and GHEX design. Also, thoroughly inspect all aspects of the equipment installation to ensure that everything was installed according to specifications.

Table 11.11 gives a general guide to help in the overall system troubleshooting process:

Table 11.10. Refrigeration System Troubleshooting Guide

Refrigeration System Malfunctions and Probable Causes							
	Head Pressure	Suction Pressure	Compressor Amp Draw	Superheat	Subcool	Air Temp. Difference	Water Temp. Difference
Undercharged System (Leak)	↓	↓	↓	↑	↓	↓	↓
Overcharged System	↑	↑	↑	↓	↑	↑	↑
Low Air Flow (Htg)	↑	↑	↑	↑	↓	↑	↓
Low Air Flow (Clg)	↓	↓	↓	↓	↑	↑	↓
Low Water Flow (Htg)	↓	↓	↓	↓	↑	↓	↑
Low Water Flow (Clg)	↑	↑	↑	↑	↓	↓	↑
High Air Flow (Htg)	↓	↑	↑	↑	↓	↑	↓
High Air Flow (Clg)	↓	↑	↑	↑	↓	↓	↑
High Water Flow (Htg)	↑	↑	↑	↑	↓	↑	↓
High Water Flow (Clg)	↓	↓	↓	↓	↑	↑	↓
Low Indoor Air Temperature (Htg)	↓	↓	↓	↓	↑	↑	↓
Low Indoor Air Temperature (Clg)	↓	↓	↓	↓	↑	↓	↓
High Indoor Air Temperature (Htg)	↑	↑	↑	↑	↓	↓	↓
High Indoor Air Temperature/Humidity (Clg)	↑	↑	↑	↑	↓	↓	↓
Restricted Filter/Drier	↓	↓	↓	↑	↑	↓	↓
Inefficient Compressor	↓	↑	↓	↑	↑	↓	↓
TXV Superheat Spring - No Bulb Charge	↓	↓	↓	↑	↑	↓	↓
Plugged Cap Tube	↓	↓	↓	↑	↑	↓	↓
Scaled Heat Exchanger (Htg)	↓	↓	↓	↓	↑	↓	↓
Scaled Heat Exchanger (Clg)	↑	↑	↑	↑	↓	↓	↓

LeClaire and Lafferty (1996)

Table 11.11. Troubleshooting Guide

Problem	Possible Cause	Checks & Corrections
Excessive noise	Equipment vibration	Check mountings & tighten as needed Check for proper installation of vibration breaks and sound insulation in ductwork
	Air in ground loop piping	Flush loop to remove air
	Defective pump	Replace pump
Water leaking out of or off of unit	Condensate drain	Check that the unit is pitched correctly to allow for proper drainage, that condensate piping is not restricted, that proper condensate line slope exists, and that a trap has been installed [if necessary]
	Ground loop lines/connections & condensate lines/connections	Check to ensure that the connections are tight and that the piping is insulated
Unit trips out on low pressure	Low flow/no flow	See low flow/no flow section below
Loopfield pressure loss	Leak	Find and repair leak as needed
	Temperature change	Pressure should vary 10-30 psi as seasons change (pressurized systems only) - normal operation
	Pipe expansion	Plastic pipe thermally expanding (again, pressurized systems only) - normal operation
Low flow/no flow in loopfield	No power	Check for loose or broken wires and connections and check power supply
	Blown fuse	Replace fuse and check for cause
	Pump shaft siezed up	Remove indicator plug from center of circulator pump or use magnetic indicator to verify shaft rotation
	Air lock	Flush loop to remove air
	Pump sized too small	Increase pump capacity [add pump(s)]
	Defective pump	Replace pump
	No positive pressure (pressurized systems only)	Check for leaks, add fluid, and flush loop
	High solution viscosity	Change antifreeze type/concentration to accommodate the operating loop temperature
	Loop fluid freezing/frozen	Switch thermostat to "cooling" mode to see if flow improves at warmer temperatures; add antifreeze and verify freeze protection levels
	Restriction in loop	Replace/repair kinked section of pipe
Low imcoming loop temp (htg)	Undersized ground loop	Add additional bores or horizontal trenches to loopfield
	Poor soil/grout contact with pipe	Fill voids by rebackfilling if possible
	Flow velocity through pipe is too low	Add pump(s) to increase fluid velocity
Restricted/fouled heat exchanger (most common in open loop water-well systems)	Minerals/sand deposited on the heat exchanger	Flush/clean heat exchanger or replace
Equipment short cycles	Oversized GSHP unit	Recalculate building load to verify GHSP size needed
Inadequate heating/cooling	Low air flow	Check fan for problems, duct work for obstructions, and ensure that the filter is clean
	Low water flow	Check pump to ensure it is sized and working properly, and check heat exchanger for fouling
Low water temperature differential	Open loop system	Check for heat exchanger fouling or excessive water flow
	Closed loop system	Check for inadequate antifreeze in loop or ice forming in the heat exchanger
	Low air flow	Check for fan problems, duct work for obstructions, and ensure that the filter is clean

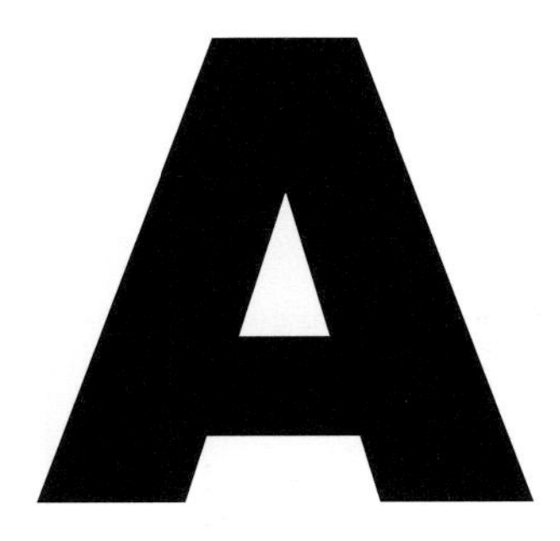

APPENDIX A: DESIGN EXAMPLES

Example A.1.1 Heating Load Single-Capacity, Water-to-Air GSHP Selection – Grand Forks, ND

For illustration purposes, a 3,500 ft^2 single-story home with a finished basement was selected for analysis (unconditioned, attached two-stall garage). Grand Forks, ND was the geographic location chosen for this example because it has an extreme heating climate. The Wrightsoft software package Right-Suite Residential was used for all peak load analysis and duct design for this example. The software adheres to Manual J principles. Other software packages that utilize Manual J will produce similar results. The blueprint for the home is displayed in Figures A.1 and A.2 with the orientation set as the bottom being the north side of the home:

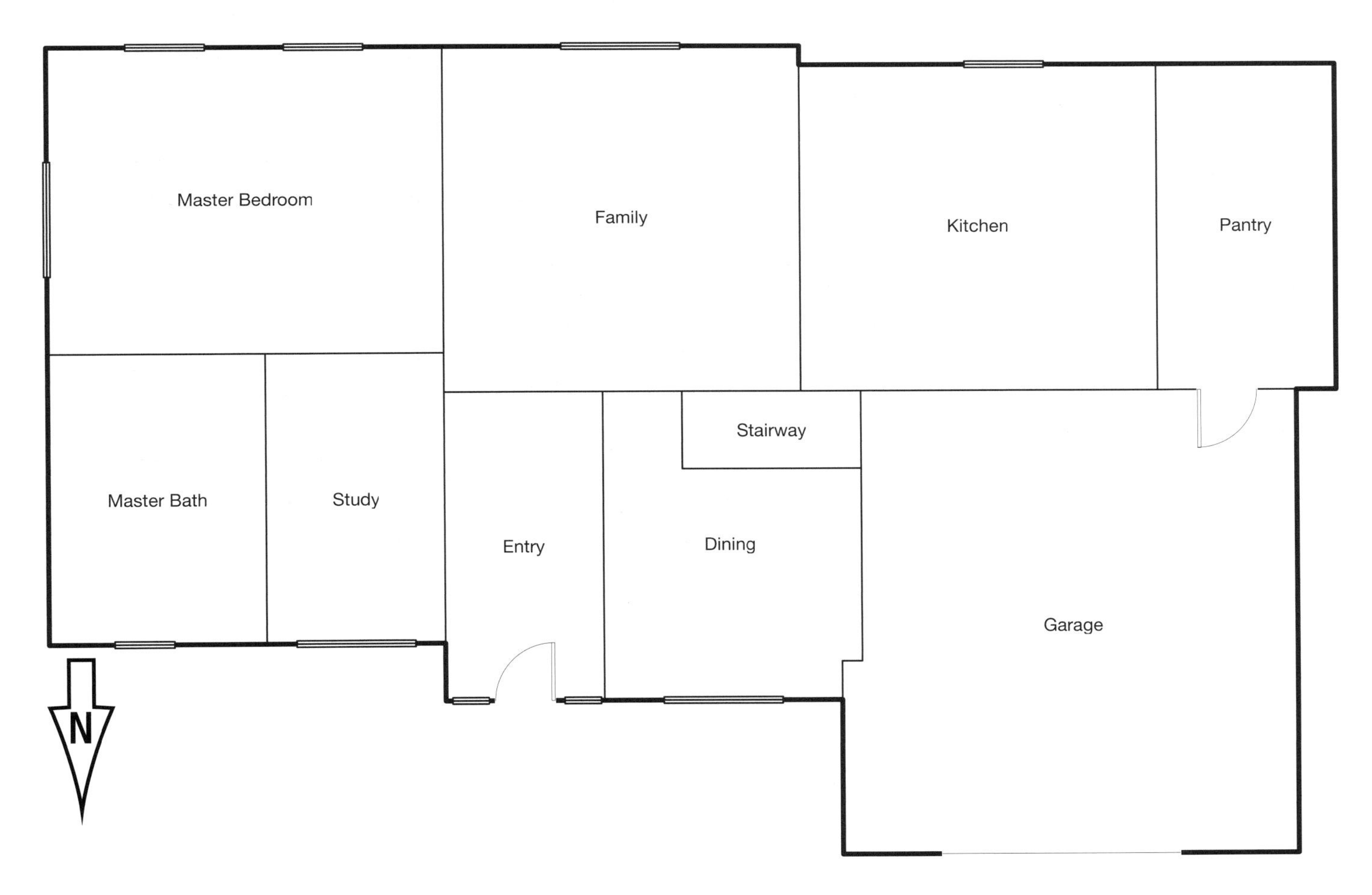

Figure A.1. Load Calculation Example Main Level Blueprint

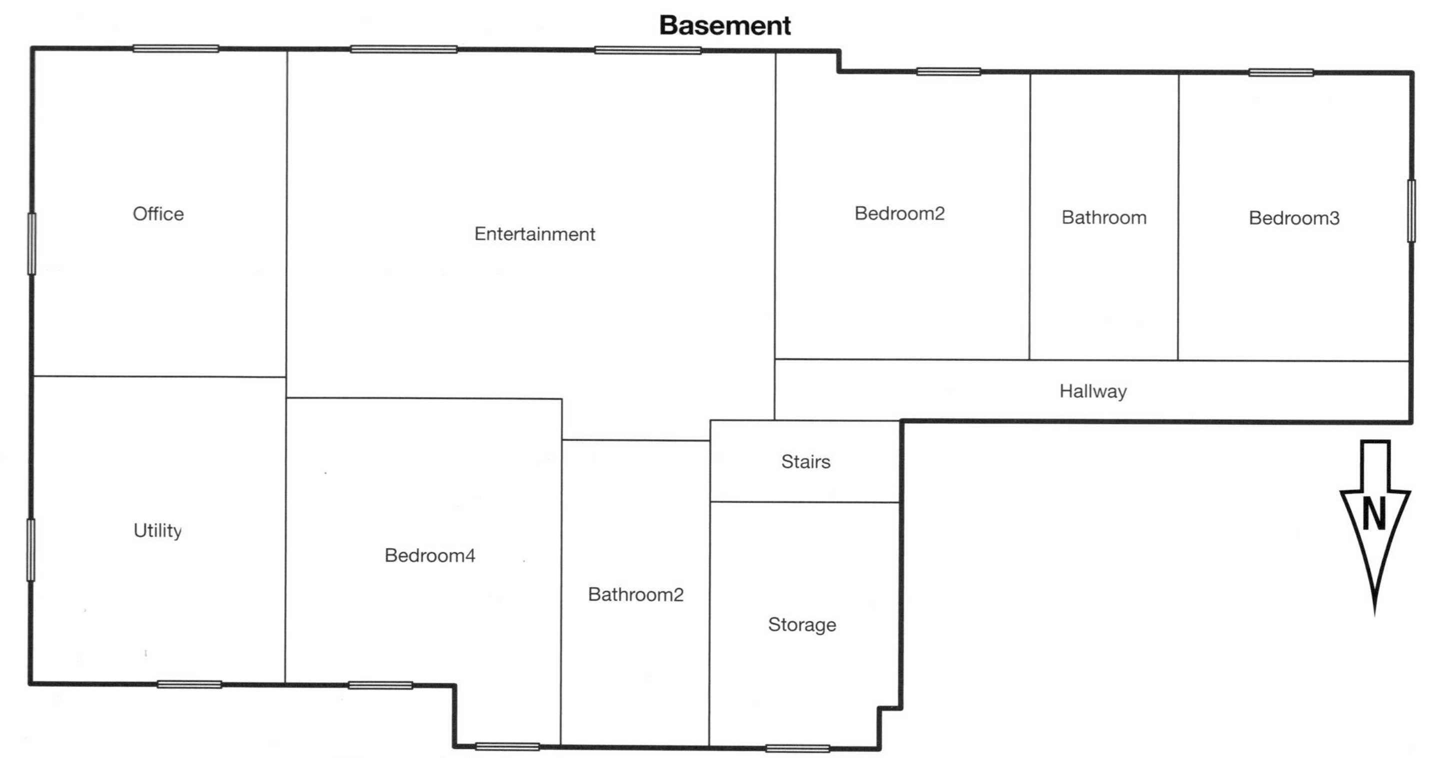

Figure A.2. Load Calculation Example Basement Blueprint

The first step in calculating peak heating and cooling loads is to determine outdoor design conditions according to geographic location as given by Chapter 28 in the *ASHRAE Handbook of Fundamentals, 2005*. The outdoor design conditions for Grand Forks, ND are given by Figure A.3:

Location:		
Grand Forks, AFB, ND, US		
Elevation: 912 ft		
Latitude: 48 °N		
Outdoor:	**Heating**	**Cooling**
Dry bulb (°F)	-16	91
Daily range (°F)	-	23 (M)
Wet bulb (°F)	-	71
Wind speed (mph)	15.0	7.5

Figure A.3. Outdoor Design Conditions for Grand Forks, ND

The outdoor design temperatures listed in Figure A.3 are the 99% annual percentile temperature for heating and 0.4% annual percentile temperature for cooling. Utilizing the 99% annual percentile temperature for heating means that 1% (87 hours based on 8760 hours in a year) of the annual hours are, on average, colder than the heating design temperature. Additionally, utilizing the 0.4% annual percentile temperature for cooling means that 0.4% (35 hours) of the annual hours are, on average, warmer than the design cooling temperature.

The next step in calculating peak loads is to collect information about the construction quality of the structure in question and make assumptions about the level of internal heat gains for appliances and occupants. This particular home was assumed to have been built according to standard construction techniques. The assumptions were as follows:

- Ceilings - Under vented attic space, no radiant barrier, dark asphalt shingles, R-30 blown-in insulation, U=0.032 Btu/hr-ft^2-F
- Basement Floor - Below grade, carpeted, no insulation, width of shortest side 32 feet, U=0.020 Btu/hr-ft^2-F
- Basement Walls - 8-inch concrete, wood stud framing, R-13 cavity insulation, no board insulation, U=0.053 Btu/hr-ft^2-F

- Upstairs Walls - Vinyl siding exterior, wood stud framing, no board insulation, R-19 cavity insulation (6-inch walls), U=0.068 Btu/hr-ft^2-F
- Windows – Wood frame, operable, two pane, clear, U=0.57 Btu/hr-ft^2-F
- Doors - Metal, polystyrene core, no storm, U=0.35 Btu/hr-ft^2-F
- 8-foot high ceilings both floors
- Ductwork located in conditioned space, 0% duct losses
- Cooling loads only – four occupants in Family Room, two occupants and two appliances in Kitchen, four occupants in Entertainment Room, one appliance in the Utility Room, and one appliance in the Office (sensible gain=230 Btuh/person, latent gain=200 Btuh/person, sensible gain=1200 Btuh/appliance)
- Unconditioned Garage
- All conditioned spaces assumed to have a thermostat set point of 70 F in heating and 75 F in cooling with 50% relative humidity
- Average construction quality, simplified infiltration method assuming ACH=0.28 heating, ACH=0.15 cooling (ACH = air changes per hour, the volume of air inside the home that is exchanged with outside air via infiltration)

Once the assumptions for construction quality, thermostat set point, and internal gains have been made, the load calculation process can begin. Typical to most smaller residential installations, this home will be served by a single piece of equipment. The home will then be defined as a single zone, resulting in the entire building peak heating and cooling loads being the block loads served by the equipment. Additionally, the home in this example will not utilize a ventilation system of any type, but will rely on infiltration to satisfy fresh air requirements. The assumption can then be made that the equipment loads for this example will be the peak block loads. No additional loads are imposed on the equipment by ventilation, duct losses, etc.

Following the procedures set forth by *ACCA Manual J* will result in a Manual J worksheet being filled out similar to Figure A.4:

Line							Entire House				Master Bedroom			
1	Room name						Entire House				Master Bedroom			
2	Exposed wall						357.0 ft				36.0 ft			
3	Ceiling height						8.0 ft			d	8.0 ft		heat/cool	
4	Room dimensions										20.0 x 16.0 ft			
5	Room area						3510.0 ft²				320.0 ft²			
	Ty	Construction number	U-value (Btuh/ft²-°F)	Or	HTM (Btuh/ft²)		Area (ft²) or perimeter (ft)		Load (Btuh)		Area (ft²) or perimeter (ft)		Load (Btuh)	
					Heat	Cool	Gross	N/P/S	Heat	Cool	Gross	N/P/S	Heat	Cool
6	W	15B13-0wc-6	0.093	nn	4.56	0.94	520	484	2529	89	0	0	0	0
.	G	1D-c2ow	0.570	nn	49.02	19.42	36	0	1765	699	0	0	0	0
.	W	12E-0sw	0.068	nn	5.85	1.15	336	246	1441	284	0	0	0	0
.	G	1D-c2ow	0.570	nn	49.02	19.42	69	0	3366	1334	0	0	0	0
11	D	11N0	0.350	nn	30.10	9.12	21	21	632	191	0	0	0	0
	W	12E-0sw	0.068	ee	5.85	1.15	272	248	1450	286	128	104	608	120
	G	1D-c2ow	0.570	ee	49.02	60.82	24	0	1176	1460	24	0	1176	1460
	W	15B13-0wc-6	0.093	ee	4.56	0.94	272	254	1330	47	0	0	0	0
	G	1D-c2ow	0.570	ee	49.02	60.82	18	0	882	1095	0	0	0	0
	W	12E-0sw	0.068	ss	5.85	1.15	520	460	2690	530	160	136	795	157
	G	1D-c2ow	0.570	ss	49.02	36.86	60	0	2941	2212	24	0	1176	885
	W	15B13-0wc-6	0.093	ss	4.56	0.94	520	460	2337	66	0	0	0	0
	G	1D-c2ow	0.570	ss	49.02	36.86	60	0	2941	2212	0	0	0	0
	W	15B13-0wc-6	0.093	ww	4.56	0.94	272	263	1402	56	0	0	0	0
	G	1D-c2ow	0.570	ww	49.02	60.82	9	0	441	547	0	0	0	0
	WW	12E-0sw	0.068	ww	5.85	1.15	144	144	842	166	0	0	0	0
	P	12C-0sw	0.091	- -	7.83	1.01	344	344	2692	349	0	0	0	0
	CC	16B-30ad	0.032	- -	2.75	1.59	1773	1773	4879	2811	320	320	881	507
	F F	21A-32c	0.020	- -	1.72	0.00	1737	198	2988	0	0	0	0	0
6	c) AED excursion									79				-115
	Envelope loss/gain								38726	14513			4637	3013
12	a) Infiltration								7542	752			1302	130
	b) Room ventilation								0	0			0	0
13	Internal gains:		Occupants @		230		10			2300	0			0
			Appliances @		1200		4			4800	0			0
	Subtotal (lines 6 to 13)								46268	22364			5939	3143
	Less external load								0	0			0	0
	Less transfer								0	0			0	0
	Redistribution								0	0			0	0
14	Subtotal								46268	22364			5939	3143
15	Duct loads						0%	0%	0	0	0%	0%	0	0
	Total room load								46268	22364			5939	3143
	Air required (cfm)								1650	1550			212	218

Figure A.4. Form J Worksheet for Grand Forks, ND Example

This particular worksheet displays the calculations performed for the entire house in addition to individual calculations for the master bedroom. A worksheet will need to be filled out for each room to calculate individual room loads to obtain air flow requirements for duct sizing. Once the load calculations are executed and checked for accuracy, equipment sizing can be performed. For the sake of example, we will assume the homeowner wanted to purchase a ClimateMaster Tranquility 20™ Model 048 (single capacity) ground source heat pump system with ECM blower to serve the home based on the total heating load shown on the Form J Worksheet. The air flow volume rates given in all of the worksheets have been adjusted to match the rated capacity of the specified heat pump (ECM blower tap setting 3 of 4 in performance data). The room-by-room load breakdown in addition to individual air flow requirements are given by Figure A.5:

ROOM NAME	Area (ft^2)	Htg load (Btuh)	Clg load (Btuh)	Htg AVF (cfm)	Clg AVF (cfm)
Master Bedroom	320	5939	3143	212	218
Family	324	3504	2680	125	186
Kitchen	306	3793	4098	135	284
Dining	206	3404	914	121	63
Master Bath	165	3129	785	112	54
Bedroom2	171	1344	435	48	30
Bedroom3	154	2313	1032	82	72
Bathroom	98	535	19	19	1
Entertainment	408	3137	2283	112	158
Office	192	2659	2168	95	150
Bedroom4	197	1915	364	68	25
Utility	180	2462	1910	88	132
Pantry	153	3182	635	113	44
Entry	128	2407	752	86	52
Study	135	2154	741	77	51
Bathroom2	105	547	19	20	1
Storage	106	1652	217	59	15
Stairs	36	271	11	10	1
Stairway	36	350	86	12	6
Hallway	90	1569	73	56	5
Entire House d	3510	46268	22364	1650	***1550***
Other equip loads		0	0		
Equip. @ 1.00 RSM			22364		
Latent cooling			2551		
TOTALS	3510	46268	24916	1650	***1550***

Figure A.5. Room-By-Room Peak Load Breakdown for Grand Forks, ND Example

The equipment sizing loads are also shown in Figure A.6, where they are broken down to show the contribution of each building component:

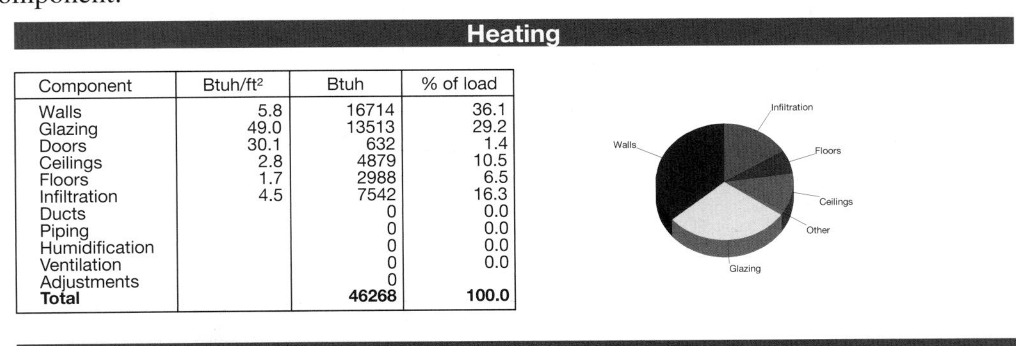

Heating

Component	Btuh/ft²	Btuh	% of load
Walls	5.8	16714	36.1
Glazing	49.0	13513	29.2
Doors	30.1	632	1.4
Ceilings	2.8	4879	10.5
Floors	1.7	2988	6.5
Infiltration	4.5	7542	16.3
Ducts		0	0.0
Piping		0	0.0
Humidification		0	0.0
Ventilation		0	0.0
Adjustments		0	
Total		**46268**	**100.0**

Cooling

Component	Btuh/ft²	Btuh	% of load
Walls	0.6	1873	8.4
Glazing	35.0	9637	43.1
Doors	9.1	191	0.9
Ceilings	1.6	2811	12.6
Floors	0.0	0	0.0
Infiltration	0.5	752	3.4
Ducts		0	0.0
Ventilation		0	0.0
Internal gains		7100	31.7
Blower		0	0.0
Adjustments		0	
Total		**22364**	**100.0**

Walls
Internal Gains
Glazing
Infiltration
Ceilings
Other

Figure A.6. Peak Equipment Loads for Grand Forks, ND Example

The peak block heating load is 46,268 Btu/hr for this example. The peak block sensible cooling load is 22,364 Btu/hr and the space SHF (sensible heat factor) is 0.90 [Peak Sensible Cooling Load = 22,364 Btu/hr, Peak Total Cooling Load = 24,916 Btu/hr, SHF = (Sensible Cooling Load / Total Cooling Load)]. At first glance, it can be seen that this particular system will have to be sized according to the peak heating load. However, the heat pump capacity will be driven by the type of equipment selected. If single-capacity equipment is utilized, it would probably be prudent to select a heat pump with capacity equaling 75% of the heating load because the cooling load is half as much (air moisture removal issues as discussed in Chapter 3). If a two-capacity heat pump was selected, sizing for 100% of the heating load is recommended. A heat pump selection worksheet would need to be filled out to ensure the heat pump unit will properly serve the home. Figure A.7 displays the performance data, and Figure A.8 displays the selection worksheet for the selected single-capacity ClimateMaster heat pump filled out according to the results of the calculations for this example (based on air flow of 1650 CFM in heating, 1550 CFM in cooling, ECM tap setting=3):

1,550 CFM Nominal (Rated) Air flow Cooling, 1,650 CFM Nominal (Rated) Airflow Heating

Performance capacities shown in thousands of Btuh

EWT °F	GPM	WPD		Cooling - EAT 80/67°F							Heating - EAT 70°F						
		PSI	FT	Airflow CFM	TC	SC	kW	HR	EER	HW	Airflow CFM	HC	kW	HE	LAT	COP	HW
20	12.0	4.8	11.1	Operation not recommended							1200	32.9	3.41	21.9	95	2.82	4.2
	12.0	4.8	11.1								1650	33.9	3.15	23.2	89	3.15	3.6
30	6.0	1.3	3.0	1120	53.5	31.9	1.91	59.9	28.0	-	1200	35.4	3.47	24.2	97	2.99	4.9
	6.0	1.3	3.0	1550	55.6	37.1	2.02	62.5	27.5	-	1650	36.5	3.21	25.7	91	3.34	4.2
	9.0	2.6	6.0	1120	54.1	31.9	1.81	60.1	29.9	-	1200	36.8	3.51	25.4	98	3.08	4.9
	9.0	2.6	6.0	1550	56.2	37.1	1.91	62.7	29.4	-	1650	38.0	3.24	27.0	91	3.44	4.2
	12.0	4.5	10.4	1120	54.8	32.1	1.76	60.6	31.1	-	1200	37.6	3.52	26.2	99	3.13	4.9
	12.0	4.5	10.4	1550	56.9	[illegible]	1.86	63.2	30.5		1650	38.8	3.26	27.8	92	3.49	4.2
40	6.0	1.2	2.8	1120	54.3	32.8	2.09	61.3	26.0	-	1200	40.0	3.58	28.6	101	[illegible]	5.6
	6.0	1.2	2.8	1550	56.5	38.2	2.21	64.0	25.6	-	1650	41.6	3.31	30.4	93	3.68	4.8
	9.0	2.6	6.0	1120	54.8	32.8	1.97	61.4	27.9	-	1200	42.2	3.62	30.3	103	3.42	5.5
	9.0	2.6	6.0	1550	57.0	38.2	2.08	64.1	27.4	-	1650	43.6	3.34	32.2	94	3.82	4.8
	12.0	4.4	10.2	1120	55.1	32.8	1.91	61.4	28.8	-	1200	43.3	3.64	31.3	103	3.49	5.5
	12.0	4.4	10.2	1550	57.3	38.2	2.02	64.1	28.3	-	1650	44.7	3.36	33.3	95	3.89	4.8
50	6.0	1.1	2.5	1120	53.5	32.9	2.31	61.3	23.2	2.7	1200	45.8	3.69	33.5	105	3.64	6.2
	6.0	1.1	2.5	1550	55.6	38.3	2.44	63.9	22.8	2.9	1650	47.2	3.41	35.6	96	4.06	5.3
	9.0	2.5	5.8	1120	54.4	33.1	2.16	61.6	25.2	2.6	1200	48.2	3.74	35.7	107	3.78	6.2
	9.0	2.5	5.8	1550	56.5	38.5	2.28	64.3	24.8	2.7	1650	49.7	3.45	37.9	98	4.22	5.3
	12.0	4.2	9.7	1120	54.7	33.1	2.09	61.7	26.2	2.4	1200	49.5	3.76	36.9	108	3.86	6.2
	12.0	4.2	9.7	1550	56.8	38.5	2.21	64.3	25.7	2.5	1650	51.1	3.48	39.3	99	4.30	5.3
60	6.0	1.0	2.3	1120	51.7	32.4	2.55	60.3	20.2	3.3	1200	51.5	3.80	38.7	110	3.97	6.8
	6.0	1.0	2.3	1550	53.7	37.7	2.70	62.9	19.9	3.5	1650	53.1	3.52	41.1	100	4.43	5.9
	9.0	2.4	5.5	1120	53.0	32.8	2.38	61.0	22.3	3.1	1200	54.4	3.86	41.3	112	4.13	6.8
	9.0	2.4	5.5	1550	55.1	38.2	2.52	63.7	21.9	3.2	1650	56.1	3.57	43.9	101	4.60	5.9
	12.0	4.0	9.2	1120	53.6	32.9	2.30	61.3	23.3	2.9	1200	56.0	3.89	42.8	113	4.21	6.8
	12.0	4.0	9.2	1550	55.7	38.3	2.43	64.0	22.9	3.0	1650	57.7	3.60	45.4	102	4.70	5.8
70	6.0	1.0	2.3	1120	49.3	31.6	2.83	58.9	17.4	4.1	1200	57.3	3.92	44.0	114	4.28	7.4
	6.0	1.0	2.3	1550	51.3	36.7	3.00	61.5	17.1	4.2	1650	59.1	3.63	46.7	103	4.78	6.4
	9.0	2.3	5.3	1120	51.0	32.2	2.64	59.9	19.3	3.8	1200	60.6	3.99	46.9	117	4.45	7.4
	9.0	2.3	5.3	1550	53.0	37.4	2.79	62.5	19.0	3.9	1650	62.4	3.69	49.8	105	4.96	6.4
	12.0	3.8	8.8	1120	51.8	32.4	2.54	60.4	20.3	3.5	1200	62.3	4.03	48.5	118	4.53	7.4
	12.0	3.8	8.8	1550	53.8	37.7	2.69	63.0	20.0	3.6	1650	64.2	3.73	51.5	106	5.05	6.4
80	6.0	0.9	2.1	1120	46.6	30.6	3.15	57.3	14.8	5.0	1200	63.0	4.05	49.1	119	4.56	8.0
	6.0	0.9	2.1	1550	48.4	35.6	3.33	59.8	14.5	5.2	1650	65.0	3.74	52.2	106	5.09	6.9
	9.0	2.3	5.3	1120	48.4	31.3	2.93	58.4	16.5	4.6	1200	66.4	4.13	52.2	121	4.71	8.0
	9.0	2.3	5.3	1550	50.4	36.4	3.10	60.9	16.2	4.8	1650	68.5	3.82	55.4	108	5.26	6.9
	12.0	3.6	8.3	1120	49.3	31.6	2.83	58.9	17.4	4.3	1200	68.2	4.18	53.7	123	4.79	8.0
	12.0	3.6	8.3	1550	51.3	36.8	2.99	61.5	17.1	4.4	1650	70.3	3.86	57.1	109	5.34	6.9
85	6.0	0.9	2.1	1120	45.1	30.1	3.33	56.5	13.5	5.6	1200	65.7	4.12	51.5	121	4.68	8.3
	6.0	0.9	2.1	1550	46.9	35.0	3.52	59.0	13.3	5.8	1650	67.8	3.80	54.7	108	5.22	7.1
	9.0	2.3	5.2	1120	47.0	30.7	3.10	57.6	15.2	5.1	1200	69.1	4.20	54.5	123	4.81	8.3
	9.0	2.3	5.2	1550	48.9	35.8	3.28	60.1	14.9	5.4	1650	71.2	3.89	57.9	110	5.37	7.1
	12.0	3.6	8.2	1120	48.0	31.1	2.99	58.1	16.0	4.7	1200	70.8	4.25	56.0	125	4.88	8.3
	12.0	3.6	8.2	1550	49.8	36.2	3.17	60.6	15.7	4.9	1650	73.0	3.93	59.5	111	5.44	7.1
90	6.0	0.9	2.1	1120	43.7	29.5	3.51	55.7	12.4	6.1	1200	68.4	4.18	53.9	123	4.79	8.6
	6.0	0.9	2.1	1550	45.4	34.4	3.72	58.1	12.2	6.4	1650	70.5	3.86	57.3	110	5.35	7.4
	9.0	2.2	5.1	1120	45.6	30.2	3.27	56.8	13.9	5.7	1200	71.7	4.28	56.8	125	4.91	8.6
	9.0	2.2	5.1	1550	47.4	35.2	3.46	59.2	13.7	5.9	1650	73.9	3.95	60.4	111	5.48	7.4
	12.0	3.5	8.1	1120	46.6	30.6	3.16	57.3	14.8	5.2	1200	73.3	4.33	58.2	127	4.96	8.5
	12.0	3.5	8.1	1550	48.4	35.6	3.34	59.8	14.5	5.4	1650	75.6	4.00	61.9	112	5.54	7.3
100	6.0	0.8	1.8	1120	40.8	28.5	3.92	54.3	10.4	7.4	Operation not recommended						
	6.0	0.8	1.8	1550	42.4	33.2	4.15	56.6	10.2	7.7							
	9.0	2.1	4.9	1120	42.6	29.2	3.66	55.2	11.7	6.8							
	9.0	2.1	4.9	1550	44.3	33.9	3.87	57.5	11.5	7.1							
	12.0	3.3	7.6	1120	43.6	29.5	3.53	55.7	12.3	6.3							
	12.0	3.3	7.6	1550	45.3	34.3	3.73	58.1	12.1	6.5							
110	6.0	0.8	1.8	1120	38.0	27.7	4.39	53.1	8.6	8.8							
	6.0	0.8	1.8	1550	39.5	32.2	4.65	55.4	8.5	9.2							
	9.0	2.0	4.6	1120	39.7	28.2	4.10	53.8	9.7	8.2							
	9.0	2.0	4.6	1550	41.3	32.8	4.33	56.1	9.5	8.5							
	12.0	3.2	7.4	1120	40.6	28.5	3.95	54.2	10.3	7.5							
	12.0	3.2	7.4	1550	42.2	33.1	4.18	56.5	10.1	7.8							
120	6.0	0.7	1.6	1120	35.5	27.1	4.93	52.5	7.2	10.5							
	6.0	0.7	1.6	1550	36.9	31.6	5.21	54.8	7.1	10.9							
	9.0	1.9	4.4	1120	37.0	27.4	4.59	52.8	8.0	9.7							
	9.0	1.9	4.4	1550	38.4	31.9	4.86	55.1	7.9	10.1							
	12.0	3.0	6.9	1120	37.8	27.6	4.44	53.0	8.5	8.9							
	12.0	3.0	6.9	1550	39.3	32.1	4.69	55.3	8.4	9.3							

Figure A.7. ClimateMaster Tranquility 20^{TM} Model 048 w/ ECM Blower Performance Data

Single Capacity Water-Air GSHP Selection Worksheet

Block Load Description = Grand Forks, ND Example Zone # ____

Heating			Cooling		
Total Heat Loss =	46,268	Btu/hr @ -16 F OAT	Total Heat Gain =	24,915	Btu/hr @ 91 F OAT
			Sen Heat Gain =	22,364	Btu/hr
			Lat Heat Gain =	2,551	Btu/hr
T-Stat Set Point =	70 / 75 / 50	(Htg dB / Clg dB / %-RH)	SHF_{space}=	Sen Heat Gain / Total Heat Gain	
			SHF_{space}=	0.90	

Brand ClimateMaster

Model Tranquility 20™ w/ ECM Model 048

Heating Mode Performance Data

Performance Parameters			Correction Factors[A] HC	Correction Factors[A] DMD
EWT_{min}=	30	F		
GPM=	12	gpm		
EAT=	70	F	CF_{EAT}=1.00	CF_{EAT}=1.00
CFM=	1650	cfm	CF_{CFM}=1.00	CF_{CFM}=1.00
Antifreeze / %[B]		PG / 20%	CF_{AF}=0.989	CF_{AF}=0.996
$CF_{EAT} \times CF_{CFM} \times CF_{AF}$==>			CF_{HC}=0.989	CF_{DMD}=0.996

Cooling Mode Performance Data

Performance Parameters			Correction Factors[A] TC	Correction Factors[A] SC	Correction Factors[A] DMD
EWT_{max}=	90	F			
GPM=	12	gpm			
EAT=	75 / 63	F/F (dB/wB)	CF_{EAT}=0.941	CF_{EAT}=0.938	CF_{EAT}=0.995
CFM=	1550	cfm	CF_{CFM}= 1.00	CF_{CFM}= 1.00	CF_{CFM}= 1.00
Antifreeze / %[B]		PG / 20%	CF_{AF}= 0.992	CF_{AF}= 0.992	CF_{AF}= 1.004
$CF_{EAT} \times CF_{CFM} \times CF_{AF}$==>			CF_{TC}=0.933	CF_{SC}=0.930	CF_{DMD}=0.999

Tabled Heating Mode Performance Data

HC =	38,800	Btu/hr
DMD =	3.26	kW
COP =	3.49	
HE =	27,800	Btu/hr (= HC x (COP-1)/COP)[C]

Tabled Cooling Mode Performance Data

TC =	48,400	Btu/hr
SC =	35,600	Btu/hr (= TC x 0.75 approximately)[C]
DMD =	3.34	kW
EER =	14.5	
HR =	59,800	Btu/hr (= TC x (EER+3.412)/EER)[C]

Corrected Heating Mode Performance Data

HC_c=	$HC \times CF_{HC}$=	38,373	Btu/hr
DMD_c=	$DMD \times CF_{DMD}$=	3.25	kW
COP_c=	$HC_c/(DMD_c \times 3412)$ =	3.46	
HE_c =	$HC_c \times (COP_c-1)/COP_c$ =	27,283	Btu/hr
LWT_{min}=	$EWT_{min} - HE_c/(500xGPM)$ =	25.45	F

Corrected Cooling Mode Performance Data

TC_c=	$TC \times CF_{TC}$=	45,157	Btu/hr
SC_c=	$SC \times CF_{SC}$=	33,108	Btu/hr
DMD_c=	$DMD \times CF_{DMD}$=	3.34	kW
EER_c=	$TC_c/(DMD_c \times 1000)$ =	13.52	
HR_c =	$TC_c \times (EER_c+3.412)/EER_c$ =	56,553	Btu/hr
LWT_{max}=	$EWT_{max} + HR_c/(500xGPM)$ =	99.43	F

% Sizing in Heating Mode

% Sizing = $\frac{HC_c}{\text{Total Heat Loss}}$ x 100%

= 82.94 %

% Sizing in Cooling Mode

% Oversizing = $\frac{SC_c - \text{Sen Heat Gain}}{\text{Sen Heat Gain}}$ x 100%

= 48.04 % (> 0 & ≤ 25%)____OK

SHF_{unit}= $\frac{SC_c}{TC_c}$

= 0.73 (≤ SHF_{space}) x OK

Latent Clg Cap = $TC_c - SC_c$

= 12,049 Btu/hr (> Lat Ht Gain) x OK

Blower Performance Summary

Type	Speed	CFM	Static (in)
PSC	L M H		
ECM	On	775	0.75 max
	w/ Comp	1650 (htg)	0.75 max
	w/ Emer	1650	0.75 max

Water Coil Performance Summary

EWT_{min}	GPM	WPD (psi)	WPD (ft)	Circulating Fluid
30	12	4.5	10.4	15% Methanol Solution

A. Apply correction factors for CFM and EAT as needed depending on format of Manufacturer's Literature. (No correction - CF=1)

B. $CF_{AF} = CF_{Used}/CF_{Tested}$. CF_{Used} for antifreeze being used. CF_{Tested} for circulating fluid performance was determined with.

C. Use this equation if performance parameter not provided by Manufacturer in the Engineering Specifications.

NOTES: 1) CF_{Used}-->20% Propylene Glycol by Volume (See Manufacturer Provided Tables)

2) CF_{Tested}-->15% Methanol by Volume (See Manufacturer Provided Tables)

Figure A.8. Single-Capacity, Water-to-Air GSHP Selection Worksheet for Grand Forks, ND Example

Typically, the design parameters EWT_{min} and EWT_{max} are determined according to whether the residence is located in a heating or cooling dominated climate. For heating dominant climates, values for EWT_{min} between 25 and 40 F are typical. For cooling dominant climates, values for EWT_{max} between 85 and 105 F are typical. Further discussion on determination of these parameters and their effects on heat pump performance, required ground-loop capacity, etc., can be found in Chapter 2. Refer to Figure 2.13 for Antifreeze Correction Factors as provided by ClimateMaster. The heating capacity antifreeze correction factor listed in Figure 3.16 was found as a result of the following calculation:

$$CF_{Used} / CF_{Tested} = 0.9575 / 0.968 = 0.989$$

The value for CF_{Used} was interpolated from the correction factor values given for 15% and 25% propylene glycol in Figure 2.13 and the value for CF_{Tested} was read directly from the same figure for 15% methanol by volume. The remaining correction factors were calculated in similar fashion.

Example A.1.2 Heating Load Two-Capacity, Water-to-Air GSHP Selection – Grand Forks, ND

As shown by the heat pump selection worksheet in Example A.1.1, the heat pump is oversized in sensible cooling capacity by 48.04% and is sized to handle only 82.94% of the peak heating load under worst-case conditions (EWT_{min} = 30 F). Because of the amount of over sizing in sensible cooling capacity, the heat pump unit will not run for extended periods of time and air dehumidification may become an issue. A two-capacity unit would probably be better suited to serve this residence. In addition, a 5-ton (nominal) capacity heat pump would be better suited to serve the residence in heating under peak conditions. Increasing nominal GSHP system capacity from 4 to 5 tons will not increase the initial system cost by a large amount because essentially the same amount of ground loop will be required for both systems (refer to examples A.1.7 and A.1.8). The majority of the incremental cost will then be in the higher cost of the larger GSHP unit.

The manufacturer's part-load and full-load performance data tables and heat pump selection worksheet for the ClimateMaster Tranquility 27™ Model 064 (two-capacity heat pump) are shown by Figures A.9, A.10, and A.11, respectively. The air flow volume rates are given in the figures (ECM blower tap setting 3 of 4 in performance data).

1500 CFM Nominal (Rated) Airflow Cooling, 1650 CFM Nominal (Rated) Airflow Heating

Performance capacities shown in thousands of Btuh

EWT °F	GPM	WPD		Cooling - EAT 80/67°F							Heating - EAT 70°F						
		PSI	FT	Airflow CFM	TC	SC	kW	HR	EER	HWC	Airflow CFM	HC	kW	HE	LAT	COP	HWC
20	14.0	4.1	9.4	Operation not recommended							1430	28.7	2.85	19.5	88.6	2.95	3.4
	14.0	4.1	9.4								1650	29.0	2.77	19.9	86.3	3.07	3.0
30	7.0	0.5	1.1	1280	49.1	33.2	1.54	54.2	31.8	-	1430	31.7	2.87	22.5	90.5	3.24	3.5
	7.0	0.5	1.1	1500	49.7	35.0	1.56	55.0	31.8	-	1650	32.0	2.78	22.9	88.0	3.37	3.1
	10.5	1.9	4.4	1280	50.1	33.6	1.50	55.1	33.4	-	1430	32.7	2.87	23.4	91.2	3.34	3.5
	10.5	1.9	4.4	1500	50.8	35.3	1.52	55.9	33.4	-	1650	33.0	2.79	23.9	88.5	3.47	3.0
	14.0	3.9	9.0	1280	51.6	34.4	1.48	56.6	34.7	-	1430	33.2	2.87	24.0	91.5	3.39	3.4
	14.0	3.9	9.0	1500	52.3	36.2	1.51	57.4	34.7	-	1650	33.6	2.79	24.4	88.8	3.53	3.0
40	7.0	0.4	0.9	1280	51.9	35.8	1.68	57.6	30.9	-	1430	36.1	2.89	26.8	93.4	3.67	3.6
	7.0	0.4	0.9	1500	52.7	37.6	1.71	58.4	30.9	-	1650	36.5	2.80	27.4	90.5	3.82	3.1
	10.5	1.8	4.3	1280	52.3	35.8	1.60	57.7	32.7	-	1430	37.4	2.89	28.1	94.2	3.79	3.6
	10.5	1.8	4.3	1500	53.1	37.7	1.63	58.5	32.7	-	1650	37.8	2.81	28.6	91.2	3.95	3.1
	14.0	3.7	8.6	1280	52.7	35.9	1.57	57.9	33.6	-	1430	38.1	2.90	28.8	94.7	3.86	3.5
	14.0	3.7	8.6	1500	53.4	37.7	1.59	58.8	33.6	-	1650	38.5	2.81	29.3	91.6	4.02	3.1
50	7.0	0.3	0.7	1280	52.0	36.5	1.88	58.4	27.7	1.0	1430	40.9	2.91	31.5	96.5	4.11	3.7
	7.0	0.3	0.7	1500	52.8	38.4	1.91	59.2	27.7	1.0	1650	41.3	2.83	32.1	93.2	4.28	3.2
	10.5	1.8	4.1	1280	52.6	36.5	1.76	58.5	29.8	0.9	1430	42.4	2.92	33.0	97.5	4.26	3.7
	10.5	1.8	4.1	1500	53.3	38.4	1.79	59.3	29.8	0.9	1650	42.9	2.83	33.6	94.1	4.43	3.2
	14.0	3.6	8.2	1280	52.7	36.5	1.71	58.5	30.8	0.8	1430	43.2	2.92	33.8	98.0	4.33	3.6
	14.0	3.6	8.2	1500	53.5	38.4	1.74	59.4	30.8	0.8	1650	43.7	2.84	34.4	94.5	4.51	3.2
60	7.0	0.3	0.6	1280	50.2	35.8	2.12	57.3	23.7	1.7	1430	45.7	2.94	36.2	99.6	4.55	3.9
	7.0	0.3	0.6	1500	50.9	37.6	2.15	58.2	23.7	1.7	1650	46.2	2.86	36.9	95.9	4.74	3.4
	10.5	1.7	4.0	1280	51.3	36.2	1.98	58.0	25.9	1.5	1430	47.5	2.96	37.9	100.8	4.71	3.8
	10.5	1.7	4.0	1500	52.0	38.1	2.01	58.8	25.9	1.6	1650	48.0	2.87	38.7	97.0	4.91	3.3
	14.0	3.4	7.8	1280	51.8	36.4	1.91	58.3	27.1	1.4	1430	48.5	2.96	38.9	101.4	4.79	3.8
	14.0	3.4	7.8	1500	52.5	38.3	1.94	59.1	27.1	1.4	1650	49.0	2.88	39.6	97.5	4.99	3.3
70	7.0	0.2	0.5	1280	47.5	34.5	2.41	55.7	19.8	2.3	1430	50.6	2.98	40.9	102.8	4.97	4.1
	7.0	0.2	0.5	1500	48.2	36.3	2.44	56.5	19.8	2.4	1650	51.2	2.90	41.7	98.7	5.18	3.6
	10.5	1.7	3.9	1280	49.1	35.2	2.24	56.7	21.9	2.1	1430	52.6	3.01	42.9	104.1	5.13	4.1
	10.5	1.7	3.9	1500	49.7	37.1	2.28	57.5	21.9	2.2	1650	53.2	2.92	43.7	99.9	5.35	3.5
	14.0	3.3	7.5	1280	49.8	35.6	2.17	57.1	23.0	1.9	1430	53.7	3.02	43.9	104.8	5.22	4.0
	14.0	3.3	7.5	1500	50.5	37.4	2.20	57.9	23.0	2.0	1650	54.3	2.93	44.8	100.5	5.43	3.5
80	7.0	0.2	0.4	1280	44.5	33.0	2.74	53.8	16.2	3.0	1430	55.4	3.04	45.6	105.9	5.35	4.4
	7.0	0.2	0.4	1500	45.1	34.7	2.78	54.5	16.2	3.0	1650	56.1	2.95	46.5	101.5	5.57	3.8
	10.5	1.6	3.8	1280	46.1	33.8	2.56	54.8	18.0	2.7	1430	57.6	3.07	47.7	107.3	5.50	4.3
	10.5	1.6	3.8	1500	46.8	35.6	2.59	55.6	18.0	2.7	1650	58.3	2.98	48.6	102.7	5.73	3.7
	14.0	3.1	7.2	1280	46.9	34.2	2.47	55.3	19.0	2.4	1430	58.8	3.09	48.8	108.1	5.58	4.3
	14.0	3.1	7.2	1500	47.6	36.0	2.51	56.1	19.0	2.5	1650	59.5	3.00	49.7	103.4	5.81	3.7
85	7.0	0.2	0.4	1280	42.8	32.3	2.93	52.8	14.6	3.3	1430	57.8	3.08	47.8	107.4	5.51	4.6
	7.0	0.2	0.4	1500	43.5	34.0	2.97	53.6	14.6	3.3	1650	58.4	2.99	48.7	102.8	5.73	4.0
	10.5	1.6	3.7	1280	44.5	33.1	2.74	53.8	16.2	3.0	1430	60.0	3.11	49.9	108.9	5.65	4.5
	10.5	1.6	3.7	1500	45.1	34.8	2.78	54.6	16.2	3.0	1650	60.7	3.02	50.9	104.1	5.88	3.9
	14.0	3.0	7.0	1280	45.3	33.5	2.65	54.3	17.1	2.7	1430	61.2	3.14	51.0	109.6	5.72	4.5
	14.0	3.0	7.0	1500	46.0	35.2	2.68	55.1	17.1	2.7	1650	61.9	3.04	52.0	104.7	5.96	3.9
90	7.0	0.1	0.3	1280	41.2	31.6	3.12	51.8	13.2	3.5	1430	60.1	3.11	50.0	108.9	5.66	4.7
	7.0	0.1	0.3	1500	41.8	33.2	3.16	52.6	13.2	3.6	1650	60.8	3.02	51.0	104.1	5.90	4.1
	10.5	1.5	3.6	1280	42.9	32.3	2.92	52.8	14.7	3.2	1430	62.4	3.16	52.1	110.4	5.80	4.6
	10.5	1.5	3.6	1500	43.5	34.0	2.96	53.6	14.7	3.3	1650	63.1	3.06	53.2	105.4	6.04	4.0
	14.0	3.0	6.8	1280	43.7	32.7	2.82	53.3	15.5	2.9	1430	63.6	3.18	53.2	111.2	5.86	4.6
	14.0	3.0	6.8	1500	44.3	34.4	2.86	54.1	15.5	2.9	1650	64.3	3.09	54.3	106.1	6.10	4.0
100	7.0	0.1	0.2	1280	38.1	30.3	3.54	50.2	10.8	4.1	Operation not recommended						
	7.0	0.1	0.2	1500	38.7	31.9	3.59	50.9	10.8	4.2							
	10.5	1.5	3.5	1280	39.6	30.9	3.33	51.0	11.9	3.7							
	10.5	1.5	3.5	1500	40.2	32.5	3.38	51.7	11.9	3.8							
	14.0	2.8	6.5	1280	40.4	31.2	3.22	51.4	12.5	3.3							
	14.0	2.8	6.5	1500	41.0	32.8	3.27	52.1	12.5	3.4							
110	7.0	0.1	0.2	1280	35.5	29.6	4.02	49.3	8.8	4.6							
	7.0	0.1	0.2	1500	36.0	31.1	4.08	49.9	8.8	4.7							
	10.5	1.5	3.3	1280	36.7	29.8	3.79	49.6	9.7	4.1							
	10.5	1.5	3.3	1500	37.2	31.4	3.84	50.3	9.7	4.2							
	14.0	2.7	6.2	1280	37.3	30.0	3.67	49.9	10.2	3.7							
	14.0	2.7	6.2	1500	37.9	31.6	3.73	50.6	10.2	3.8							
120	7.0	0.1	0.1	1280	33.7	29.2	4.57	49.3	7.4	5.0							
	7.0	0.1	0.1	1500	34.2	30.7	4.63	50.0	7.4	5.1							
	10.5	1.4	3.2	1280	34.4	29.3	4.30	49.1	8.0	4.5							
	10.5	1.4	3.2	1500	34.9	30.8	4.36	49.8	8.0	4.6							
	14.0	2.6	6.0	1280	34.9	29.5	4.18	49.1	8.3	4.1							
	14.0	2.6	6.0	1500	35.4	31.0	4.24	49.8	8.3	4.1							

Interpolation is permissible; extrapolation is not.
All entering air conditions are 80°F DB and 67°F WB in cooling, and 70°F DB in heating.
ARI/ISO certified conditions are 80.6°F DB and 66.2°F WB in cooling and 68°F DB in heating.
Table does not reflect fan or pump power corrections for ARI/ISO conditions.
All performance is based upon the lower voltage of dual voltage rated units.
Operation below 40°F EWT is based upon a 15% antifreeze solution.
Operation below 60°F EWT requires optional insulated water/refrigerant circuit.
See performance correction tables for operating conditions other than those listed above.
For operation in the shaded areas, please see the Performance Data Selection Notes on Page TT-22

Figure A.9. ClimateMaster Tranquility 27™ Model 064 Part-Load Performance Data

1825 CFM Nominal (Rated) Airflow Cooling, 2050 CFM Nominal (Rated) Airflow Heating

Performance capacities shown in thousands of Btuh

EWT °F	GPM	WPD		Cooling - EAT 80/67°F							Heating - EAT 70°F						
		PSI	FT	Airflow CFM	TC	SC	kW	HR	EER	HWC	Airflow CFM	HC	kW	HE	LAT	COP	HWC
20	15.0	5.0	11.6	Operation not recommended							1750	41.0	3.87	28.3	91.7	3.10	4.0
	15.0	5.0	11.6								2050	41.8	3.71	29.2	88.9	3.30	3.5
30	7.5	0.6	1.5	1580	65.8	41.6	2.78	75.1	23.7	-	1750	44.6	3.96	31.5	93.6	3.29	4.1
	7.5	0.6	1.5	1825	67.3	45.6	2.90	77.2	23.2	-	2050	45.4	3.8	32.6	90.5	3.50	3.6
	11.3	2.3	5.3	1580	66.7	42.1	2.65	75.7	25.2	-	1750	46.4	4.01	33.1	94.6	3.39	4.1
	11.3	2.3	5.3	1825	68.3	46.2	2.77	77.8	24.7	-	2050	47.3	3.85	34.3	91.4	3.60	3.5
	15.0	4.8	11	1580	68.1	42.9	2.60	76.8	26.2	-	1750	47.4	4.04	34.6	95.1	3.44	4.0
	15.0	4.8	11	1825	69.7	47.1	2.71	78.0	25.7		2050	48.3	3.88	35.2	91.8	3.65	3.5
40	7.5	0.5	1.2	1580	67.5	43.1	3.00	77.6	22.5	-	1750	50.0	4.13	36.9	96.8	3.59	4.3
	7.5	0.5	1.2	1825	69.1	47.3	3.13	79.8	22.0	-	2050	51.5	3.96	38.1	93.3	3.82	3.7
	11.3	2.2	5.1	1580	68.4	43.4	2.85	78.0	24.0	-	1750	52.8	4.19	38.8	97.9	3.69	4.2
	11.3	2.2	5.1	1825	70.0	47.6	2.98	80.2	23.5	-	2050	53.8	4.02	40.2	94.3	3.92	3.6
	15.0	4.5	10.4	1580	68.7	43.5	2.78	78.0	24.7	-	1750	53.9	4.22	39.9	98.5	3.74	4.1
	15.0	4.5	10.4	1825	70.3	47.7	2.90	80.2	24.2	-	2050	55.0	4.05	41.3	94.8	3.98	3.6
50	7.5	0.4	1.0	1580	67.7	43.9	3.27	78.7	20.7	2.0	1750	56.7	4.30	42.4	100.0	3.86	4.5
	7.5	0.4	1.0	1825	69.3	48.1	3.41	80.9	20.3	2.1	2050	57.8	4.12	43.8	96.1	4.11	3.9
	11.3	2.1	4.9	1580	68.4	43.9	3.08	78.8	22.2	1.8	1750	59.3	4.37	44.7	101.4	3.97	4.4
	11.3	2.1	4.9	1825	70.1	48.1	3.21	81.0	21.8	1.8	2050	60.4	4.19	46.2	97.3	4.22	3.8
	15.0	4.3	9.9	1580	68.8	43.9	2.99	78.9	23.0	1.5	1750	60.7	4.41	45.9	102.1	4.03	4.3
	15.0	4.3	9.9	1825	70.4	48.2	3.13	81.1	22.5	1.5	2050	61.8	4.23	47.5	97.9	4.28	3.8
60	7.5	0.4	0.8	1580	65.8	43.4	3.56	77.9	18.5	2.9	1750	63.0	4.48	48.0	103.3	4.12	4.8
	7.5	0.4	0.8	1825	67.3	47.6	3.72	80.0	18.1	3.0	2050	64.2	4.30	49.6	99.0	4.38	4.2
	11.3	2.1	4.8	1580	67.2	43.8	3.35	78.6	20.1	2.5	1750	66.0	4.57	50.6	104.9	4.23	4.7
	11.3	2.1	4.8	1825	68.8	48.0	3.49	80.7	19.7	2.5	2050	67.3	4.38	52.4	100.4	4.50	4.1
	15.0	4.1	9.4	1580	67.8	43.9	3.25	78.8	20.9	2.1	1750	67.6	4.62	52.1	105.8	4.29	4.7
	15.0	4.1	9.4	1825	69.4	48.1	3.39	80.9	20.5	2.1	2050	68.9	4.43	53.9	101.1	4.57	4.0
70	7.5	0.3	0.7	1580	63.1	42.5	3.91	76.4	16.1	4.0	1750	69.4	4.67	53.7	106.7	4.36	5.3
	7.5	0.3	0.7	1825	64.6	46.7	4.08	78.5	15.8	4.0	2050	70.8	4.48	55.6	102.0	4.63	4.6
	11.3	2.0	4.6	1580	65.0	43.2	3.66	77.5	17.8	3.4	1750	72.9	4.78	56.8	108.6	4.48	5.2
	11.3	2.0	4.6	1825	66.6	47.3	3.82	79.6	17.4	3.4	2050	74.4	4.58	58.8	103.6	4.76	4.5
	15.0	3.9	8.9	1580	65.9	43.4	3.54	77.9	18.6	2.8	1750	74.9	4.83	58.6	109.6	4.54	5.1
	15.0	3.9	8.9	1825	67.5	47.6	3.70	80.1	18.2	2.9	2050	76.3	4.63	60.6	104.5	4.83	4.4
80	7.5	0.2	0.5	1580	59.8	41.4	4.31	74.5	13.9	5.2	1750	76.1	4.87	59.7	110.3	4.58	5.8
	7.5	0.2	0.5	1825	61.2	45.4	4.50	76.6	13.6	5.3	2050	77.6	4.67	61.7	105.1	4.87	5.1
	11.3	2.0	4.5	1580	62.1	42.2	4.03	75.8	15.4	4.4	1750	80.2	5.00	63.3	112.4	4.70	5.7
	11.3	2.0	4.5	1825	63.6	46.3	4.21	77.9	15.1	4.5	2050	81.8	4.79	65.4	106.9	5.00	5.0
	15.0	3.7	8.4	1580	63.2	42.6	3.90	76.4	16.2	3.6	1750	82.5	5.07	65.3	113.6	4.77	5.6
	15.0	3.7	8.4	1825	64.7	46.7	4.07	78.6	15.9	3.7	2050	84.1	4.86	67.5	108.0	5.07	4.9
85	7.5	0.2	0.5	1580	58.0	40.7	4.54	73.6	12.8	5.9	1750	79.6	4.98	62.8	112.1	4.69	6.2
	7.5	0.2	0.5	1825	59.4	44.7	4.74	75.6	12.5	6.0	2050	81.2	4.78	64.9	106.7	4.98	5.4
	11.3	1.9	4.4	1580	60.4	41.6	4.24	74.9	14.2	5.0	1750	84.0	5.12	66.6	114.5	4.81	6.1
	11.3	1.9	4.4	1825	61.8	45.6	4.43	76.9	14.0	5.1	2050	85.6	4.91	68.9	108.7	5.11	5.3
	15.0	3.6	8.2	1580	61.5	42.0	4.10	75.5	15.0	4.1	1750	86.5	5.20	68.8	115.8	4.88	6.0
	15.0	3.6	8.2	1825	63.0	46.1	4.28	77.6	14.7	4.2	2050	88.1	4.98	71.2	109.8	5.19	5.2
90	7.5	0.2	0.4	1580	56.2	40.1	4.78	72.6	11.8	6.5	1750	83.1	5.09	65.9	114.0	4.79	6.5
	7.5	0.2	0.4	1825	57.5	43.9	4.99	74.6	11.5	6.6	2050	84.7	4.88	68.1	108.3	5.09	5.6
	11.3	1.9	4.3	1580	58.7	41.0	4.46	73.9	13.2	5.5	1750	87.8	5.24	70.0	116.5	4.91	6.4
	11.3	1.9	4.3	1825	60.0	44.9	4.65	75.9	12.9	5.6	2050	89.5	5.02	72.4	110.4	5.22	5.5
	15.0	3.5	8.0	1580	59.9	41.4	4.31	74.6	13.9	4.6	1750	90.5	5.32	72.3	117.9	4.98	6.3
	15.0	3.5	8.0	1825	61.3	45.4	4.49	76.6	13.6	4.6	2050	92.2	5.10	74.8	111.7	5.30	5.4
100	7.5	0.1	0.3	1580	52.4	38.6	5.32	70.7	9.8	8.0	Operation not recommended						
	7.5	0.1	0.3	1825	53.6	42.3	5.55	72.6	9.7	8.2							
	11.3	1.8	4.2	1580	54.9	39.6	4.96	71.9	11.1	6.8							
	11.3	1.8	4.2	1825	56.2	43.4	5.18	73.9	10.9	6.9							
	15.0	3.3	7.6	1580	56.1	40.1	4.79	72.5	11.7	5.6							
	15.0	3.3	7.6	1825	57.5	43.9	5.00	74.6	11.5	5.7							
110	7.5	0.1	0.2	1580	48.6	37.1	5.95	69.0	8.2	9.7							
	7.5	0.1	0.2	1825	49.7	40.7	6.21	71.0	8.0	9.9							
	11.3	1.8	4.0	1580	51.0	38.1	5.54	70.0	9.2	8.2							
	11.3	1.8	4.0	1825	52.2	41.7	5.78	72.0	9.0	8.4							
	15.0	3.1	7.2	1580	52.2	38.6	5.34	70.6	9.8	6.7							
	15.0	3.1	7.2	1825	53.5	42.3	5.58	72.5	9.6	6.9							
120	7.5	0.1	0.1	1580	44.9	35.7	6.67	67.9	6.7	11.5							
	7.5	0.1	0.1	1825	46.0	39.2	6.97	69.8	6.6	11.7							
	11.3	1.7	3.9	1580	47.1	36.6	6.21	68.5	7.6	9.8							
	11.3	1.7	3.9	1825	48.3	40.1	6.48	70.5	7.4	9.9							
	15.0	2.9	6.8	1580	48.3	37.0	5.99	69.0	8.1	8.0							
	15.0	2.9	6.8	1825	49.5	40.6	6.25	70.9	7.9	8.1							

Interpolation is permissible; extrapolation is not.
All entering air conditions are 80°F DB and 67°F WB in cooling, and 70°F DB in heating.
ARI/ISO certified conditions are 80.6°F DB and 66.2°F WB in cooling and 68°F DB in heating.
Table does not reflect fan or pump power corrections for ARI/ISO conditions.
All performance is based upon the lower voltage of dual voltage rated units.
Operation below 40°F EWT is based upon a 15% antifreeze solution.
Operation below 60°F EWT requires optional insulated water/refrigerant circuit.
See performance correction tables for operating conditions other than those listed above.
For operation in the shaded areas, please see the Performance Data Selection Notes on Page TT-22

Figure A.10. ClimateMaster Tranquility 27™ Model 064 Full-Load Performance Data

Two Capacity Water-Air GSHP Selection Worksheet

Block Load Description = Grand Forks, ND Example | Zone #

Total Heat Loss = 46,268 Btu/hr @ -16 F OAT	Total Heat Gain = 24,915 Btu/hr @ 91 F OAT
	Sen Heat Gain = 22,364 Btu/hr
	Lat Heat Gain = 2,551 Btu/hr
T-Stat Set Point = 70 / 75 / 50 (Htg dB / Clg dB / %-RH)	SHF_{space} = Sen Heat Gain / Total Heat Gain
	SHF_{space} = 0.90

Brand ClimateMaster

Model Tranquility 27™ Model 064

Sized for High Capacity in (Heating) / Cooling

((High) / Low) Capacity Heating Mode Performance Data

Performance Parameters			Correction Factors[A] HC	Correction Factors[A] DMD
EWT_{min}=	30	F		
GPM=	15	gpm		
EAT=	70	F	CF_{EAT}=1.00	CF_{EAT}=1.00
CFM=	2050	cfm	CF_{CFM}=1.00	CF_{CFM}=1.00
Antifreeze / %[B]	PG / 20%		CF_{AF}=0.989	CF_{AF}=0.996
CF_{EAT} x CF_{CFM} x CF_{AF}==>			CF_{HC}=0.989	CF_{DMD}=0.996

(High / (Low)) Capacity Cooling Mode Performance Data

Performance Parameters			Correction Factors[A] TC	Correction Factors[A] SC	Correction Factors[A] DMD
EWT_{max}=	90	F			
GPM=	14	gpm			
EAT=	75/63	F/F (dB/wB)	CF_{EAT}=0.954	CF_{EAT}=0.963	CF_{EAT}=0.987
CFM=	1500	cfm	CF_{CFM}= 1.00	CF_{CFM}= 1.00	CF_{CFM}= 1.00
Antifreeze / %[B]	PG / 20%		CF_{AF}= 0.992	CF_{AF}= 0.992	CF_{AF}= 1.004
CF_{EAT} x CF_{CFM} x CF_{AF}==>			CF_{TC}=0.946	CF_{SC}=0.955	CF_{DMD}=0.991

Tabled Heating Mode Performance Data

HC =	48,300	Btu/hr
DMD =	3.88	kW
COP =	3.65	
HE =	35,200	Btu/hr (= HC x (COP-1)/COP)[C]

Tabled Cooling Mode Performance Data

TC =	44,300	Btu/hr
SC =	34,400	Btu/hr (= TC x 0.75 approximately)[C]
DMD =	2.86	kW
EER =	15.50	
HR =	54,100	Btu/hr (= TC x (EER+3.412)/EER)[C]

Corrected Heating Mode Performance Data

HC_c=	HC x CF_{HC}=	47,769	Btu/hr
DMD_c=	DMD x CF_{DMD}=	3.86	kW
COP_c=	HC_c/(DMD_c x 3412) =	3.62	
HE_c =	HC_c x (COP_c-1)/COP_c =	34,573	Btu/hr
LWT_{min}=	EWT_{min} - HE_c/(500xGPM) =	25.39	F

Corrected Cooling Mode Performance Data

TC_c=	TC x CF_{TC}=	41,908	Btu/hr
SC_c=	SC x CF_{SC}=	32,852	Btu/hr
DMD_c=	DMD x CF_{DMD}=	2.83	kW
EER_c=	TC_c/(DMD_c x 1000) =	14.81	
HR_c =	TC_c x (EER_c+3.412)/EER_c =	51,563	Btu/hr
LWT_{max}=	EWT_{max} + HR_c/(500xGPM) =	97.34	F

% Sizing in Heating Mode

% Sizing = HC_c / Total Heat Loss x 100%

= 103.24 %

% Sizing in Cooling Mode

% Oversizing = (SC_c - Sen Heat Gain) / Sen Heat Gain x 100%

= 46.90 % (> 0 & ≤ 25%) __ OK

SHF_{unit}= SC_c / TC_c

= 0.784 (≤ SHF_{space}) x OK

Latent Clg Cap = TC_c - SC_c

= 9,056 Btu/hr (> Lat Ht Gain) x OK

Blower Performance Summary

Type	Speed	CFM	Static (in)
ECM	On	920	0.75 Max
	w/ Low	1650	0.75 Max
	w/ High	2050	0.75 Max
	w/ Emer	2040	0.75 Max

Water Coil Performance Summary

EWT_{min}	GPM	WPD (psi)	WPD (ft)	Circulating Fluid
30	15	4.8	11	15% Methanol Solution

A. Apply correction factors for CFM and EAT as needed depending on format of Manufacturer's Literature. (No correction - CF=1)

B. CF_{AF} = CF_{Used}/CF_{Tested}. CF_{Used} for antifreeze being used. CF_{Tested} for circulating fluid performance was determined with.

C. Use this equation if performance parameter not provided by Manufacturer in the Engineering Specifications.

NOTES: 1) CF_{Used}-->20% Propylene Glycol by Volume (See Manufacturer Provided Tables)

2) CF_{Tested}-->15% Methanol by Volume (See Manufacturer Provided Tables)

Figure A.11. Two-Capacity, Water-to-Air GSHP Selection Worksheet for Grand Forks, ND Example

Notice that this model is well suited for this application in heating capacity under worst-case conditions (EWT_{min} = 30 F - 103.24% sizing in heating). However, this model is still significantly oversized in sensible cooling capacity (low speed - 46.90% over sizing). Refer to Figure 2.13 for Antifreeze Correction Factors as provided by ClimateMaster. One possible solution to remedy the significant over sizing in cooling capacity would be to slow the fan down. To do so, the fan on the ClimateMaster Tranquility 27™ Model 064 would have to be wired to tap setting 2 rather than tap setting 3. This would slow the fan down from 1500 CFM to 1280 CFM (85% of rated air flow) when in low-speed cooling mode and from 2050 CFM to 1750 CFM (85% of rated air flow) in high-speed heating mode. Heating capacity of the unit would be minimally affected by the reduction in air flow and the sensible cooling capacity would be reduced to a more acceptable level.

Airflow in CFM with wet coil and clean airfilter

Model	Max ESP (in. wg)	Fan Motor (hp)	Tap Setting	Cooling Mode			Dehumid Mode			Heating Mode			Residential Units Only	
				Stg 1	Stg 2	Fan	Stg 1	Stg 2	Fan	Stg 1	Stg 2	Fan	AUX CFM	Aux/ Emerg Mode
026	0.50	1/2	4	810	950	475	630	740	475	920	1060	475	4	1060
	0.50	1/2	3	725	850	425	560	660	425	825	950	425	3	950
	0.50	1/2	2	620	730	370	490	570	370	710	820	370	2	820
	0.50	1/2	1	520	610	300				600	690	300	1	690
038	0.50	1/2	4	1120	1400	700	870	1090	700	1120	1400	700	4	1400
	0.50	1/2	3	1000	1250	630	780	980	630	1000	1250	630	3	1350
	0.50	1/2	2	860	1080	540	670	840	540	860	1080	540	2	1350
	0.50	1/2	1	730	900	450				730	900	450	1	1350
049	0.75	1	4	1460	1730	870	1140	1350	870	1560	1850	870	4	1850
	0.75	1	3	1300	1550	780	1020	1210	780	1400	1650	780	3	1660
	0.75	1	2	1120	1330	670	870	1040	670	1200	1430	670	2	1430
	0.75	1	1	940	1120	560				1010	1200	560	1	1350
064	0.75	1	4	1670	2050	1020	1300	1600	1020	1860	2280	1020	4	2280
	0.75	1	3	1500	1825	920	1160	1430	920	1650	2050	920	3	2040
	0.75	1	2	1280	1580	790	1000	1230	790	1430	1750	790	2	1750
	0.75	1	1	1080	1320	660				1200	1470	660	1	1470
072	0.75	1	4	1620	2190	1050	1270	1650	1050	1690	2230	1050	4	2230
	0.75	1	3	1500	1950	980	1170	1520	980	1600	2100	980	3	2100
	0.75	1	2	1400	1830	910	1100	1420	910	1400	1850	910	2	1870
	0.75	1	1	1320	1700	850				1240	1620	850	1	1670

Low-Speed Clg.

High-Speed Htg.

Figure A.12. ClimateMaster Tranquility 27™ Model 064 Fan Performance Data

Airflow	Cooling				Heating		
% of Rated	Total Capacity	Sensible Capacity	Power	Heat of Rejection	Heating Capacity	Power	Heat of Extraction
60%	0.920	0.781	0.959	0.927	0.946	1.241	0.881
69%	0.942	0.832	0.964	0.946	0.960	1.163	0.915
75%	0.956	0.867	0.696	0.959	0.969	1.115	0.937
81%	0.969	0.901	0.975	0.970	0.978	1.076	0.956
88%	0.981	0.934	0.982	0.981	0.986	1.043	0.973
94%	0.991	0.967	0.990	0.991	0.993	1.018	0.988
100%	1.000	1.000	1.000	1.000	1.000	1.000	1.000
106%	1.007	1.033	1.011	1.008	1.006	0.990	1.010
113%	1.013	1.065	1.023	1.015	1.012	0.986	1.017
119%	1.018	1.098	1.036	1.021	1.017	0.983	1.024
125%	1.021	1.131	1.051	1.026	1.021	0.981	1.030
130%	1.023	1.159	1.063	1.030	1.024	0.979	1.034

Figure A.13. ClimateMaster Air Flow Part-Load Correction Factors

Airflow	Cooling				Heating		
% of Rated	Total Capacity	Sensible Capacity	Power	Heat of Rejection	Heating Capacity	Power	Heat of Extraction
60%	0.925	0.788	0.913	0.922	0.946	1.153	0.896
69%	0.946	0.829	0.926	0.942	0.959	1.107	0.924
75%	0.960	0.861	0.937	0.955	0.969	1.078	0.942
81%	0.972	0.895	0.950	0.968	0.977	1.053	0.959
88%	0.983	0.930	0.965	0.979	0.985	1.032	0.974
94%	0.992	0.965	0.982	0.990	0.993	1.014	0.988
100%	1.000	1.000	1.000	1.000	1.000	1.000	1.000
106%	1.007	1.033	1.020	1.009	1.006	0.989	1.011
113%	1.012	1.064	1.042	1.018	1.012	0.982	1.019
119%	1.016	1.092	1.066	1.025	1.018	0.979	1.027
125%	1.018	1.116	1.091	1.032	1.022	0.977	1.033
130%	1.019	1.132	1.112	1.037	1.026	0.975	1.038

Figure A.14. ClimateMaster Air Flow Full-Load Correction Factors

As shown in Figures A.13 and A.14, reducing the air flow rate to 85% of the rated values will reduce high heating capacity and low sensible cooling capacity by 1.8% and 8.0%, respectively (interpolated high heating CF_{cfm}=0.982, high HC_c=46,909 Btu/hr, interpolated low cooling CF_{cfm}=0.920, low SC_c=30,224 Btu/hr). Changing the fan motor from tap setting 3 to tap setting 2 would affect the GSHP unit's capacity such that it would still be sized to handle the building's peak heating load (101.38% sizing in heating), but it would only be oversized by 35.14% in cooling capacity (low-speed). This would be a much better scenario in terms of occupant comfort levels, equipment run time, and dehumidification.

Because the GSHP unit has been sized for 100% of the peak heating load, a backup heating system such as strip resistance heat installed in the plenum will only be necessary during extreme weather conditions, generally occurring very few hours in a given year, when the outdoor temperatures are lower than the temperature for which the peak heating load was calculated (Grand Forks, ND - t_{dh}= -16 F). As previously stated (Example A.1.1), the loopfield size for this home will be driven by the heat pump capacity as it relates to the equipment run time to satisfy the actual building load. The amount of bore required for the project will be similar for the 4-ton, single-capacity unit specified in Example A.1.1 as it would be for the 5-ton heat pump selected for this example because equipment run time would be less for the larger unit. Therefore, the incremental cost of stepping up to a larger system would primarily be driven by the difference in heat pump cost as loopfield cost will essentially be the same.

Example A.1.3 Heating Load Single-Capacity, Water-to-Water GSHP Selection – Grand Forks, ND

Now, for the sake of example, assume that the customer wanted to explore the possibility of conditioning the home with water-to-water GSHP equipment utilizing in-floor radiant heating panels coupled with fan coils for forced air heating and cooling. For this example, the building was assumed to have the same peak equipment heating and cooling loads. However, in a real situation, the peak loads would need to be recalculated because of the additional equipment loads imposed on the system by hydronic supply-return piping losses to/from the fan coils, in addition to piping losses from the in-floor radiant heating panels. Assuming that the customer wanted to utilize the same brand of heat pump, a single-capacity, water-to-water ClimateMaster Genesis GSW heat pump will need to be selected for analysis. The manufacturer performance data tables and heat pump selection worksheet for the ClimateMaster Genesis GSW Model 060 (single-capacity heat pump) are shown by Figures A.15, A.16, and A.17, respectively.

SOURCE				LOAD																					
EWT	Flow			EWT	Flow 7.5 gpm							Flow 11.3 gpm							Flow 15.0 gpm						
°F	GPM	WPD PSI	WPD FT	°F	TC MBtuh	Power KW	HR MBtuh	LWT °F	EER	WPD PSI	WPD FT	TC MBtuh	Power KW	HR MBtuh	LWT °F	EER	WPD PSI	WPD FT	TC MBtuh	Power KW	HR MBtuh	LWT °F	EER	WPD PSI	WPD FT
				50	46.3	2.43	54.5	37.7	19.1	1.8	4.2	48.6	2.45	56.9	41.4	19.9	3.4	8.0	49.7	2.45	58.1	43.4	20.3	5.6	12.9
				60	51.4	2.47	59.8	46.3	20.8	1.6	3.7	53.7	2.48	62.1	50.5	21.6	3.1	7.1	54.8	2.49	63.3	52.7	22.0	5.0	11.6
	7.5	2.2	5.2	70	56.0	2.50	64.5	55.1	22.4	1.4	3.3	58.2	2.52	66.8	59.7	23.1	2.8	6.5	59.3	2.52	67.9	62.1	23.5	4.6	10.6
				80	60.1	2.53	68.7	64.0	23.8	1.3	3.0	62.2	2.54	70.8	69.0	24.5	2.6	5.9	63.1	2.55	71.8	71.6	24.8	4.2	9.7
				90	63.6	2.55	72.3	73.0	25.0	1.1	2.6	65.5	2.56	74.2	78.4	25.6	2.2	5.2	66.4	2.56	75.1	81.2	25.9	3.8	8.8
				50	46.7	2.32	54.6	37.5	20.1	1.8	4.2	49.0	2.34	57.0	41.3	21.0	3.4	8.0	50.2	2.35	58.2	43.3	21.4	5.6	12.9
				60	51.9	2.36	59.9	46.2	22.0	1.6	3.7	54.2	2.38	62.3	50.4	22.8	3.1	7.1	55.4	2.38	63.5	52.6	23.2	5.0	11.6
50	11.3	4.0	9.2	70	56.5	2.39	64.7	54.9	23.6	1.4	3.3	58.8	2.41	67.0	59.6	24.4	2.8	6.5	59.9	2.41	68.1	62.0	24.8	4.6	10.6
				80	60.7	2.42	68.9	63.8	25.1	1.3	3.0	62.8	2.43	71.1	68.9	25.8	2.6	5.9	63.8	2.44	72.1	71.5	26.2	4.2	9.7
				90	64.3	2.44	72.6	72.9	26.4	1.1	2.6	66.2	2.45	74.5	78.3	27.1	2.2	5.2	67.0	2.45	75.4	81.1	27.4	3.8	8.8
				50	47.4	2.27	55.2	37.4	20.9	1.8	4.2	49.8	2.29	57.6	41.2	21.8	3.4	8.0	51.0	2.29	58.8	43.2	22.2	5.6	12.9
				60	52.6	2.31	60.5	46.0	22.8	1.6	3.7	55.0	2.32	62.9	50.3	23.7	3.1	7.1	56.2	2.33	64.1	52.5	24.1	5.0	11.6
	15.0	6.1	14.1	70	57.4	2.34	65.4	54.7	24.6	1.4	3.3	59.7	2.35	67.7	59.4	25.4	2.8	6.5	60.8	2.36	68.8	61.9	25.8	4.6	10.6
				80	61.6	2.36	69.6	63.6	26.1	1.3	3.0	63.7	2.37	71.8	68.7	26.8	2.6	5.9	64.7	2.38	72.9	71.4	27.2	4.2	9.7
				90	65.3	2.38	73.4	72.6	27.4	1.1	2.6	67.2	2.39	75.3	78.1	28.1	2.2	5.2	68.1	2.39	76.2	80.9	28.4	3.8	8.8
				50	43.1	3.04	53.5	38.5	14.2	1.8	4.2	45.4	3.06	55.8	42.0	14.8	3.4	8.0	46.5	3.07	57.0	43.8	15.2	5.6	12.9
				60	48.5	3.09	59.0	47.1	15.7	1.6	3.7	50.8	3.11	61.4	51.0	16.3	3.1	7.1	52.0	3.12	62.6	53.1	16.6	5.0	11.6
	7.5	1.9	4.5	70	53.4	3.14	64.1	55.8	17.0	1.4	3.3	55.8	3.16	66.5	60.1	17.7	2.8	6.5	56.9	3.17	67.7	62.4	18.0	4.6	10.6
				80	57.9	3.18	68.8	64.5	18.2	1.3	3.0	60.2	3.20	71.1	69.3	18.8	2.6	5.9	61.3	3.21	72.3	71.8	19.1	4.2	9.7
				90	62.0	3.22	73.0	73.5	19.3	1.1	2.6	64.2	3.24	75.2	78.6	19.8	2.2	5.2	65.2	3.25	76.3	81.3	20.1	3.8	8.8
				50	43.6	2.91	53.5	38.4	15.0	1.8	4.2	45.8	2.93	55.8	41.9	15.7	3.4	8.0	47.0	2.94	57.0	43.7	16.0	5.6	12.9
				60	49.0	2.96	59.1	46.9	16.6	1.6	3.7	51.3	2.98	61.5	50.9	17.2	3.1	7.1	52.5	2.99	62.7	53.0	17.6	5.0	11.6
70	11.3	3.6	8.2	70	54.0	3.00	64.2	55.6	18.0	1.4	3.3	56.3	3.02	66.6	60.0	18.6	2.8	6.5	57.5	3.03	67.8	62.3	19.0	4.6	10.6
				80	58.5	3.04	68.9	64.4	19.2	1.3	3.0	60.8	3.06	71.3	69.2	19.9	2.6	5.9	62.0	3.07	72.4	71.7	20.2	4.2	9.7
				90	62.6	3.08	73.1	73.3	20.4	1.1	2.6	64.8	3.10	75.4	78.5	20.9	2.2	5.2	65.9	3.11	76.5	81.2	21.2	3.8	8.8
				50	44.2	2.84	53.9	38.2	15.6	1.8	4.2	46.5	2.86	56.3	41.8	16.3	3.4	8.0	47.7	2.87	57.5	43.6	16.6	5.6	12.9
				60	49.7	2.89	59.6	46.7	17.2	1.6	3.7	52.1	2.91	62.0	50.8	17.9	3.1	7.1	53.3	2.92	63.3	52.9	18.3	5.0	11.6
	15.0	5.5	12.7	70	54.8	2.93	64.8	55.4	18.7	1.4	3.3	57.2	2.95	67.2	59.9	19.4	2.8	6.5	58.4	2.96	68.5	62.2	19.7	4.6	10.6
				80	59.4	2.97	69.5	64.2	20.0	1.3	3.0	61.7	2.99	71.9	69.1	20.6	2.6	5.9	62.9	3.00	73.1	71.6	21.0	4.2	9.7
				90	63.6	3.01	73.8	73.0	21.1	1.1	2.6	65.8	3.03	76.1	78.4	21.7	2.2	5.2	66.8	3.03	77.2	81.1	22.0	3.8	8.8
				50	39.9	3.87	53.1	39.4	10.3	1.8	4.2	42.0	3.90	55.3	42.6	10.8	3.4	8.0	43.0	3.91	56.4	44.3	11.0	5.6	12.9
				60	45.2	3.94	58.6	47.9	11.5	1.6	3.7	47.3	3.96	60.8	51.6	11.9	3.1	7.1	48.4	3.97	62.0	53.5	12.2	5.0	11.6
	7.5	1.5	3.5	70	50.0	3.99	63.6	56.7	12.5	1.4	3.3	52.1	4.02	65.8	60.8	13.0	2.8	6.5	53.2	4.03	66.9	62.9	13.2	4.6	10.6
				80	54.4	4.04	68.1	65.5	13.5	1.3	3.0	56.4	4.05	70.2	70.0	13.9	2.6	5.9	57.4	4.06	71.2	72.4	14.1	4.2	9.7
				90	58.2	4.07	72.1	74.5	14.3	1.1	2.6	60.1	4.08	74.0	79.4	14.7	2.2	5.2	60.9	4.08	74.9	81.9	14.9	3.8	8.8
				50	40.3	3.70	53.0	39.2	10.9	1.8	4.2	42.4	3.73	55.1	42.5	11.4	3.4	8.0	43.5	3.74	56.2	44.2	11.6	5.6	12.9
				60	45.7	3.77	58.5	47.8	12.1	1.6	3.7	47.8	3.79	60.7	51.5	12.6	3.1	7.1	48.9	3.80	61.9	53.5	12.9	5.0	11.6
90	11.3	3.1	7.1	70	50.5	3.82	63.6	56.5	13.2	1.4	3.3	52.7	3.84	65.8	60.7	13.7	2.8	6.5	53.7	3.85	66.9	62.8	14.0	4.6	10.6
				80	54.9	3.86	68.1	65.4	14.2	1.3	3.0	56.9	3.88	70.2	69.9	14.7	2.6	5.9	57.9	3.88	71.2	72.3	14.9	4.2	9.7
				90	58.8	3.89	72.0	74.3	15.1	1.1	2.6	60.7	3.90	74.0	79.3	15.6	2.2	5.2	61.6	[illegible]	[illegible]	[illegible]	[illegible]	3.8	8.8
				50	[illegible]	[illegible]	[illegible]	[illegible]	[illegible]	[illegible]	[illegible]	[illegible]	[illegible]	[illegible]	[illegible]	[illegible]	[illegible]	[illegible]	44.1	3.65	56.6	44.1	12.1	5.6	12.9
				60	46.4	3.68	58.9	47.6	12.6	1.6	3.7	48.5	3.70	61.2	51.4	13.1	3.1	7.1	[illegible]	[illegible]	[illegible]	[illegible]	[illegible]	[illegible]	11.6
	15.0	5.0	11.6	70	51.3	3.73	64.0	56.3	13.7	1.4	3.3	53.4	3.75	66.3	60.5	14.2	2.8	6.5	54.5	3.76	67.4	62.7	14.5	4.6	10.6
				80	55.7	3.77	68.6	65.1	14.8	1.3	3.0	57.8	3.79	70.7	69.8	15.3	2.6	5.9	58.8	3.80	71.8	72.2	15.5	4.2	9.7
				90	59.7	3.80	72.6	74.1	15.7	1.1	2.6	61.6	3.81	74.6	79.1	16.2	2.2	5.2	62.5	3.81	75.5	81.7	16.4	3.8	8.8
				50	35.1	4.87	51.8	40.6	7.2	1.8	4.2	37.0	4.90	53.7	43.5	7.5	3.4	8.0	38.0	4.92	54.7	44.9	7.7	5.6	12.9
				60	40.6	4.96	57.5	49.2	8.2	1.6	3.7	42.5	4.99	59.6	52.5	8.5	3.1	7.1	43.5	5.01	60.6	54.2	8.7	5.0	11.6
	7.5	1.3	3.0	70	45.6	5.04	62.8	57.8	9.0	1.4	3.3	47.6	5.07	64.9	61.6	9.4	2.8	6.5	48.7	5.08	66.0	63.5	9.6	4.6	10.6
				80	50.2	5.10	67.6	66.6	9.8	1.3	3.0	52.2	5.13	69.7	70.8	10.2	2.6	5.9	53.3	5.14	70.8	72.9	10.4	4.2	9.7
				90	54.4	5.16	72.0	75.5	10.5	1.1	2.6	56.4	5.18	74.1	80.0	10.9	2.2	5.2	57.3	5.19	75.1	82.4	11.0	3.8	8.8
				50	35.5	4.66	51.4	40.5	7.6	1.8	4.2	37.4	4.69	53.4	43.4	8.0	3.4	8.0	38.3	4.71	54.4	44.9	8.1	5.6	12.9
				60	41.0	4.75	57.2	49.1	8.6	1.6	3.7	43.0	4.77	59.3	52.4	9.0	3.1	7.1	44.0	4.79	60.3	54.1	9.2	5.0	11.6
110	11.3	2.8	6.4	70	46.0	4.82	62.5	57.7	9.6	1.4	3.3	48.1	4.85	64.6	61.5	9.9	2.8	6.5	49.1	4.86	65.7	63.4	10.1	4.6	10.6
				80	50.7	4.88	67.4	66.5	10.4	1.3	3.0	52.8	4.91	69.5	70.7	10.8	2.6	5.9	53.8	4.92	70.6	72.8	10.9	4.2	9.7
				90	55.0	4.93	71.8	75.3	11.1	1.1	2.6	56.9	4.96	73.9	79.9	11.5	2.2	5.2	57.9	4.97	74.9	82.3	11.7	3.8	8.8
				50	36.0	4.55	51.6	40.4	7.9	1.8	4.2	37.9	4.58	53.6	43.3	8.3	3.4	8.0	38.9	4.60	54.6	44.8	8.5	5.6	12.9
				60	41.6	4.64	57.4	48.9	9.0	1.6	3.7	43.6	4.67	59.5	52.3	9.3	3.1	7.1	44.6	4.68	60.6	54.0	9.5	5.0	11.6
	15.0	4.6	10.5	70	46.7	4.71	62.8	57.5	9.9	1.4	3.3	48.8	4.74	65.0	61.4	10.3	2.8	6.5	49.9	4.75	66.1	63.3	10.5	4.6	10.6
				80	51.5	4.77	67.7	66.3	10.8	1.3	3.0	53.6	4.80	69.9	70.5	11.2	2.6	5.9	54.6	4.81	71.0	72.7	11.4	4.2	9.7
				90	55.8	4.82	72.2	75.1	11.6	1.1	2.6	57.8	4.84	74.3	79.8	11.9	2.2	5.2	58.8	4.85	75.3	82.2	12.1	3.8	8.8

Figure A.15. ClimateMaster Genesis GSW Model 060 Cooling Performance Data

Notice that the single-capacity, water-to-water GSHP was selected for its capacity at an entering load temperature of 100 F in heating and 50 F in cooling. In heating, the radiant piping layout in addition to fan coil sizing should be based on a 105 F entering water temperature from the heat pump (LLT_H on the Single-Capacity, Water-to-Water Heat Pump Selection Worksheet). Additionally, sizing of the fan coil by way of its sensible and latent cooling capacity should be based on an entering water temperature of 44 F from the heat pump (LLT_C on the Single-Capacity, Water-to-Water Heat Pump Selection Worksheet).

As shown by the heat pump selection worksheet, the heat pump is undersized in heating capacity by 16.21% under worst-case conditions (EWT_{min} = 30 F, ELT_{max} = 100 F). Typically, under sizing of water-to-water equipment is not recommended. As stated in Chapter 3, equipment over sizing of 25% is recommended for water-to-water GSHP systems (heating or cooling-based designs) to allow for quicker storage tank temperature recovery and "ramp up" operation subsequent to thermostat setback periods. For this example, stepping up from a 5-ton (nominal) to a 6-ton (nominal) unit would be suggested. However, for the brand and model of heat pump chosen, only 3-ton, 5-ton, and 10-ton units are available. A different brand/model of GSHP would have to be utilized for this application.

SOURCE				LOAD																					
EWT	Flow			EWT	Flow 7.5 gpm							Flow 11.3 gpm							Flow 15.0 gpm						
	GPM	WPD			HC	Power	HE	LWT	COP	WPD		HC	Power	HE	LWT	COP	WPD		HC	Power	HE	LWT	COP	WPD	
°F		PSI	FT	°F	MBtuh	KW	MBtuh	°F		PSI	FT	MBtuh	KW	MBtuh	°F		PSI	FT	MBtuh	KW	MBtuh	°F		PSI	FT
20	15.0	7.9	18.2	60	36.8	2.53	28.2	69.8	4.26	1.6	3.7	37.2	2.44	28.9	66.6	4.48	3.1	7.1	37.4	2.39	29.3	65.0	4.58	5.0	11.6
				80	35.0	3.17	24.2	89.3	3.24	1.3	3.0	35.2	3.04	24.9	86.2	3.39	2.6	5.9	35.4	2.98	25.2	84.7	3.48	4.2	9.7
				100	33.9	4.01	20.2	109.0	2.48	0.9	2.1	34.0	3.86	20.9	106.0	2.59	1.9	4.5	34.1	3.78	21.2	104.5	2.64	3.5	8.1
				120	33.6	5.08	16.3	129.0	1.94	0.6	1.5	33.6	4.89	17.0	126.0	2.02	1.6	3.6	33.6	4.79	17.3	124.5	2.06	2.9	6.7
30	7.5	3.1	7.2	60	38.2	2.55	29.5	70.2	4.39	1.6	3.7	38.7	2.46	30.3	66.8	4.62	3.1	7.1	38.9	2.41	30.7	65.2	4.73	5.0	11.6
				80	36.3	3.19	25.4	89.7	3.34	1.3	3.0	36.6	3.06	26.1	86.5	3.50	2.6	5.9	36.7	3.00	26.5	84.9	3.58	4.2	9.7
				100	35.2	4.04	21.4	109.4	2.55	0.9	2.1	35.4	3.88	22.1	106.3	2.67	1.9	4.5	35.4	3.81	22.4	104.7	2.73	3.5	8.1
				120	35.0	5.12	17.5	129.3	2.00	0.6	1.5	34.9	4.92	18.1	126.2	2.08	1.6	3.6	34.9	4.83	18.5	124.7	2.12	2.9	6.7
	11.3	5.0	11.5	60	40.5	2.57	31.7	70.8	4.61	1.6	3.7	41.0	2.48	32.5	67.2	4.84	3.1	7.1	41.2	2.43	32.9	65.5	4.96	5.0	11.6
				80	38.5	3.22	27.5	90.3	3.50	1.3	3.0	38.8	3.09	28.2	86.9	3.68	2.6	5.9	38.9	3.03	28.6	85.2	3.76	4.2	9.7
				100	37.3	4.08	23.4	109.9	2.68	0.9	2.1	37.4	3.92	24.1	106.6	2.80	1.9	4.5	37.5	3.84	24.4	105.0	2.86	3.5	8.1
				120	37.0	5.16	19.4	129.9	2.10	0.6	1.5	37.0	4.97	20.1	126.5	2.18	1.6	3.6	37.0	4.87	20.4	124.9	2.23	2.9	6.7
	15.0	7.4	17.0	60	42.3	2.58	33.5	71.3	4.80	1.6	3.7	42.8	2.49	34.3	67.6	5.04	3.1	7.1	43.0	2.44	34.7	65.7	5.16	5.0	11.6
				80	40.2	3.23	29.2	90.7	3.65	1.3	3.0	40.5	3.10	29.9	87.2	3.82	2.6	5.9	40.7	3.04	30.8	85.4	3.92	4.2	9.7
				100	39.6	4.09	25.6	110.4	2.79	0.9	2.1	39.1	3.93	25.7	106.9	2.91	1.9	4.5	39.2	3.86	26.0	105.2	2.98	3.5	8.1
				120	38.7	5.18	21.0	130.3	2.19	0.6	1.5	38.7	4.99	21.6	126.8	2.27	1.6	3.6	38.7	4.89	22.0	125.2	2.32	2.9	6.7
40	7.5	2.7	6.2	60	44.1	2.62	35.1	71.8	4.93	1.6	3.7	44.3	2.48	35.8	67.8	5.23	3.1	7.1	44.3	2.42	36.1	65.9	5.37	5.0	11.6
				80	43.1	3.32	31.8	91.5	3.80	1.3	3.0	43.4	3.14	32.6	87.7	4.05	2.6	5.9	43.5	3.06	33.1	85.8	4.17	4.2	9.7
				100	41.8	4.21	27.5	111.2	2.91	0.9	2.1	42.1	3.99	28.5	107.5	3.10	1.9	4.5	42.3	3.89	29.0	105.6	3.19	3.5	8.1
				120	40.3	5.28	22.2	130.7	2.23	0.6	1.5	40.6	5.03	23.5	127.2	2.37	1.6	3.6	40.8	4.91	24.1	125.4	2.44	2.9	6.7
	11.3	4.4	10.2	60	46.7	2.64	37.7	72.4	5.17	1.6	3.7	46.9	2.50	38.3	68.3	5.49	3.1	7.1	47.0	2.44	38.6	66.3	5.64	5.0	11.6
				80	45.6	3.35	34.2	92.2	3.99	1.3	3.0	45.9	3.17	35.1	88.1	4.25	2.6	5.9	46.0	3.08	35.5	86.1	4.38	4.2	9.7
				100	44.3	4.25	29.8	111.8	3.06	0.9	2.1	44.6	4.03	30.9	107.9	3.25	1.9	4.5	44.8	3.92	31.4	106.0	3.35	3.5	8.1
				120	42.6	5.33	24.5	131.4	2.35	0.6	1.5	43.0	5.07	25.7	127.6	2.49	1.6	3.6	43.2	4.95	26.3	125.8	2.56	2.9	6.7
	15.0	6.6	15.2	60	48.8	2.66	39.7	73.0	5.38	1.6	3.7	49.0	2.51	40.4	68.7	5.71	3.1	7.1	49.1	2.45	40.7	66.5	5.87	5.0	11.6
				80	47.7	3.36	36.2	92.7	4.16	1.3	3.0	48.0	3.18	37.1	88.5	4.42	2.6	5.9	48.1	3.10	37.5	86.4	4.56	4.2	9.7
				100	46.3	4.26	31.7	112.3	3.18	0.9	2.1	46.6	4.04	32.8	108.3	3.38	1.9	4.5	46.8	3.94	33.4	106.2	3.48	3.5	8.1
				120	44.5	5.35	26.3	131.9	2.44	0.6	1.5	45.0	5.09	27.6	128.0	2.59	1.6	3.6	45.1	4.97	28.2	126.0	2.66	2.9	6.7
50	7.5	2.2	5.2	60	50.0	2.67	40.9	73.3	5.50	1.6	3.7	50.3	2.53	41.6	68.9	5.83	3.1	7.1	50.3	2.46	41.9	66.7	5.99	5.0	11.6
				80	48.9	3.38	37.4	93.0	4.24	1.3	3.0	49.2	3.20	38.3	88.7	4.51	2.6	5.9	49.4	3.11	38.7	86.6	4.65	4.2	9.7
				100	47.5	4.28	32.9	112.7	3.25	0.9	2.1	47.8	4.06	34.0	108.5	3.45	1.9	4.5	48.0	3.95	34.5	106.4	3.56	3.5	8.1
				120	45.7	5.37	27.4	132.2	2.49	0.6	1.5	46.1	5.12	28.7	128.2	2.64	1.6	3.6	46.3	5.00	29.3	126.2	2.72	2.9	6.7
	11.3	4.0	9.2	60	53.0	2.69	43.8	74.1	5.77	1.6	3.7	53.2	2.55	44.5	69.4	6.12	3.1	7.1	53.3	2.48	44.8	67.1	6.29	5.0	11.6
				80	51.8	3.41	40.2	93.8	4.45	1.3	3.0	52.1	3.22	41.1	89.2	4.74	2.6	5.9	52.3	3.14	41.6	87.0	4.88	4.2	9.7
				100	50.3	4.32	35.6	113.4	3.41	0.9	2.1	50.7	4.10	36.7	109.0	3.63	1.9	4.5	50.9	3.99	37.2	106.8	3.74	3.5	8.1
				120	48.4	5.42	29.9	132.9	2.62	0.6	1.5	48.9	5.16	31.2	128.6	2.77	1.6	3.6	49.1	5.04	31.9	126.5	2.85	2.9	6.7
	15.0	6.1	14.1	60	55.4	2.70	46.1	74.8	6.00	1.6	3.7	55.6	2.56	46.9	69.8	6.37	3.1	7.1	55.7	2.49	47.2	67.4	6.55	5.0	11.6
				80	54.1	3.42	42.5	94.4	4.64	1.3	3.0	54.5	3.24	43.4	89.6	4.93	2.6	5.9	54.6	3.15	43.9	87.3	5.08	4.2	9.7
				100	52.5	4.34	37.7	114.0	3.55	0.9	2.1	52.9	4.11	38.9	109.4	3.77	1.9	4.5	53.1	4.01	39.5	107.1	3.89	3.5	8.1
				120	50.6	5.44	32.0	133.5	2.72	0.6	1.5	51.0	5.18	33.3	129.0	2.88	1.6	3.6	51.3	5.06	34.0	126.8	2.97	2.9	6.7
60	7.5	2.1	4.8	60	55.5	2.71	46.3	74.8	6.00	1.6	3.7	55.8	2.57	47.0	69.9	6.37	3.1	7.1	55.9	2.50	47.3	67.4	6.55	5.0	11.6
				80	54.3	3.43	42.6	94.5	4.64	1.3	3.0	54.6	3.25	43.5	89.7	4.93	2.6	5.9	54.8	3.16	44.0	87.3	5.08	4.2	9.7
				100	52.7	4.35	37.9	114.1	3.55	0.9	2.1	53.1	4.13	39.0	109.4	3.77	1.9	4.5	53.3	4.02	39.6	107.1	3.89	3.5	8.1
				120	50.7	5.46	32.1	133.5	2.72	0.6	1.5	51.2	5.20	33.4	129.1	2.88	1.6	3.6	51.4	5.08	34.1	126.9	2.97	2.9	6.7
	11.3	3.8	8.7	60	58.8	2.73	49.5	75.7	6.31	1.6	3.7	59.1	2.59	50.2	70.5	6.69	3.1	7.1	59.2	2.52	50.6	67.9	6.88	5.0	11.6
				80	57.5	3.46	45.7	95.3	4.87	1.3	3.0	57.9	3.27	46.7	90.2	5.18	2.6	5.9	58.0	3.19	47.1	87.7	5.34	4.2	9.7
				100	55.8	4.39	40.8	114.9	3.73	0.9	2.1	56.2	4.16	42.0	110.0	3.96	1.9	4.5	56.4	4.05	42.6	107.5	4.08	3.5	8.1
				120	53.7	5.51	34.9	134.3	2.86	0.6	1.5	54.2	5.24	36.3	129.6	3.03	1.6	3.6	54.4	5.12	37.0	127.3	3.12	2.9	6.7
	15.0	5.7	13.3	60	61.4	2.74	52.1	76.4	6.56	1.6	3.7	61.7	2.60	52.8	70.9	6.96	3.1	7.1	61.8	2.53	53.2	68.2	7.15	5.0	11.6
				80	60.1	3.48	48.2	96.0	5.06	1.3	3.0	60.4	3.29	49.2	90.7	5.39	2.6	5.9	60.6	3.20	49.7	88.1	5.55	4.2	9.7
				100	58.3	4.41	43.3	115.5	3.88	0.9	2.1	58.8	4.18	44.5	110.4	4.12	1.9	4.5	59.0	4.07	45.1	107.9	4.25	3.5	8.1
				120	56.1	5.53	37.3	135.0	2.98	0.6	1.5	56.6	5.27	38.7	130.0	3.15	1.6	3.6	56.9	5.14	39.3	127.6	3.24	2.9	6.7
70	7.5	1.9	4.5	60	59.7	2.79	50.2	75.9	6.27	1.6	3.7	60.0	2.61	51.1	70.6	6.73	3.1	7.1	60.2	2.53	51.5	68.0	6.96	5.0	11.6
				80	58.7	3.53	46.7	95.7	4.87	1.3	3.0	59.0	3.31	47.7	90.4	5.23	2.6	5.9	59.2	3.20	48.2	87.9	5.42	4.2	9.7
				100	57.7	4.46	42.5	115.4	3.80	0.9	2.1	58.0	4.18	43.7	110.3	4.06	1.9	4.5	58.1	4.05	44.3	107.8	4.20	3.5	8.1
				120	OPERATION NOT RECOMMENDED							57.0	5.24	39.1	130.1	3.19	1.6	3.6	57.1	5.09	39.7	127.6	3.29	2.9	6.7
	11.3	3.6	8.2	60	63.3	2.82	53.7	76.9	6.59	1.6	3.7	63.6	2.64	54.6	71.3	7.07	3.1	7.1	63.7	2.55	55.0	68.5	7.31	5.0	11.6
				80	62.2	3.56	50.0	96.6	5.12	1.3	3.0	62.5	3.33	51.1	91.1	5.49	2.6	5.9	62.6	3.23	51.6	88.4	5.69	4.2	9.7
				100	61.1	4.49	45.8	116.3	3.99	0.9	2.1	61.4	4.22	47.0	110.9	4.27	1.9	4.5	61.6	4.09	47.6	108.2	4.41	3.5	8.1
				120	OPERATION NOT RECOMMENDED							60.4	5.29	42.3	130.7	3.35	1.6	3.6	60.5	5.14	43.0	128.1	3.45	2.9	6.7
	15.0	5.5	12.7	60	66.1	2.83	56.5	77.6	6.85	1.6	3.7	66.4	2.65	57.4	71.8	7.35	3.1	7.1	66.6	2.56	57.8	68.9	7.61	5.0	11.6
				80	65.0	3.58	52.8	97.3	5.32	1.3	3.0	65.3	3.35	53.9	91.6	5.72	2.6	5.9	65.4	3.24	54.4	88.7	5.92	4.2	9.7
				100	63.9	4.51	48.5	117.0	4.15	0.9	2.1	64.2	4.24	49.7	111.4	4.44	1.9	4.5	64.3	4.11	50.3	108.6	4.59	3.5	8.1
				120	OPERATION NOT RECOMMENDED							63.1	5.31	44.9	131.2	3.48	1.6	3.6	63.2	5.16	45.6	128.4	3.59	2.9	6.7

Figure A.16. ClimateMaster Genesis GSW Model 060 Heating Performance Data

Single Capacity Water-Water GSHP Selection Worksheet

Block Load Description = Grand Forks, ND Example   Zone #

Heating		Cooling	
Total Htg Load =	46,268 Btu/hr @ -16 F OAT	Total Clg Load =	24,915 Btu/hr @ 91 F OAT
Htg Load Design Water Temp =	100 F (HLDWT)	Clg Load Design Water Temp =	50 F (CLDWT)
Htg Load Water Flow Rate =	15 gpm	Clg Load Water Flow Rate =	15 gpm
T-Stat Set Point =	70/75/50 (Htg DB / Clg DB / %-RH)		

Brand: ClimateMaster

Model: Genesis GSW 060

Heating Mode Performance Data

Performance Parameters			Correction Factors[A] HC	Correction Factors[A] DMD
EWT_{min}=	30	F		
GPM=	15	gpm		
GPM_{load}=	15	gpm		
ELT_{max}=	100	F		
Antifreeze / %[B]	PG / 20%		CF_{AF}=0.989	CF_{AF}=0.996

Cooling Mode Performance Data

Performance Parameters			Correction Factors[A] TC	Correction Factors[A] DMD
EWT_{max}=	90	F		
GPM=	15	gpm		
GPM_{load}=	15	gpm		
ELT_{min}=	50	F		
Antifreeze / %[B]	PG / 20%		CF_{AF}= 0.992	CF_{AF}= 1.004

Tabled Heating Mode Performance Data

HC =	39,200	Btu/hr
DMD =	3.86	kW
COP =	2.98	
HE =	26,000	Btu/hr (= HC x (COP-1)/COP)[C]
LLT_H =	105.22	F (= ELT_{max} + HC / (GPM_{load} x 500)) (≥ HLDWT)

Tabled Cooling Mode Performance Data

TC =	44,100	Btu/hr
DMD =	3.65	kW
EER =	12.1	
HR =	56,600	Btu/hr (= TC x (EER+3.412)/EER)[C]
LLT_C =	44.12	F (= ELT_{min} - TC / (GPM_{load} x 500)) (≤ CLDWT)

Corrected Heating Mode Performance Data

HC_c=	HC x CF_{AF} =	38,769	Btu/hr
DMD_c=	DMD x C_{AF} =	3.84	kW
COP_c=	HC_c/(DMD_c x 3412) =	2.96	
HE_c =	HC_c x (COP_c-1)/COP_c =	25,671	Btu/hr
LWT_{min}=	EWT_{min} - HE_c/(500xGPM) =	26.58	F

Corrected Cooling Mode Performance Data

TC_c=	TC x CF_{AF} =	43,747	Btu/hr
DMD_c=	DMD x CF_{AF} =	3.66	kW
EER_c=	HC_c/(DMD_c x 1000) =	11.95	
HR_c =	HC_c x (EER_c+3.412)/EER_c =	56,238	Btu/hr
LWT_{max}=	EWT_{max} + HR_c/(500xGPM) =	97.50	F

% Sizing in Heating Mode

$$\% \text{ Sizing} = \frac{HC_c}{\text{Total Heat Loss}} \times 100\%$$

% Sizing = 83.79 %

% Sizing in Cooling Mode

$$\% \text{ Oversizing} = \frac{TC_c - \text{Total Clg Ld}}{\text{Total Clg Ld}} \times 100\%$$

= 75.58 %

Tabled Water Coil Performance Summary

Side	Mode	EWT_{min}	GPM	WPD (psi)	WPD (ft)	Circulating Fluid
Loop	Htg	30	15	7.4	17.0	15% Methanol Solution
Load	Clg	50	15	5.6	12.9	Water

A. Apply correction factors for Antifreeze (AF) as needed depending on Manufacturer's Literature. (No correction - CF=1)

B. CF_{AF} = CF_{Used}/CF_{Tested}. CF_{Used} for antifreeze being used. CF_{Tested} for circulating fluid performance was determined with.

C. Use this equation if performance parameter not provided by Manufacturer in the Engineering Specifications.

NOTES: 1) CF_{Used}-->20% Propylene Glycol by Volume (See Manufacturer Provided Tables)

2) CF_{Tested}-->15% Methanol by Volume (See Manufacturer Provided Tables)

Figure A.17. Single-Capacity, Water-to-Water GSHP Selection Worksheet for Grand Forks, ND Example

Example A.1.4 Heating Load Single-Capacity, Water-to-Air GSHP Duct Sizing – Grand Forks, ND

Once a decision has been made as to the exact model of heat pump to be used, ductwork sizing is the next step in the design process. Because the air flow requirements for each room have already been determined for the 4-ton, single-capacity heat pump (shown in the Figure A.5), those are the values to be used for ductwork sizing in this example. As stated in Example A.1.1, the heat pump has been sized according to its ratings at an air flow rate of 1650 CFM in heating. According to the manufacturer's fan performance data, this particular model can produce 1650 CFM with a maximum of 0.75 inches wg of available external static pressure. According to guidelines, ductwork should be sized to cause pressure drop at the design air flow rate somewhere in the range of 0.10-0.35 IWC. For this example, the ductwork will be sized such that the longest circulation path causes a pressure drop of approximately 0.30 IWC at the given air flow rate. Remember to ensure that maximum air velocity guidelines are followed to prevent excessive system noise during operation. The calculated air flow requirements for both heating and cooling are given in Figure A.5. Ductwork should be sized according to pressure drop and maximum flow velocity guidelines for the larger of the two air flow requirements given for each room. For example, because of internal gains in the kitchen, the air flow volume requirement for cooling is larger than the requirement for heating (284 CFM versus 135 CFM, respectively). Conversely, the air flow volume requirement for heating in the dining room is the larger than the value for cooling (121 CFM versus 63 CFM, respectively). The duct run to the kitchen needs to be sized to handle 284 CFM in cooling whereas the duct run to the dining room needs to be sized to handle 121 CFM in heating.

Before the sizing process can begin, details of the construction of the duct system will need to be determined. For this example, standard sheet metal ductwork was used. Specifically, standard rectangular ductwork sizes were used for the supply and return trunks and standard round ductwork sizes were used for the branches. Typical to this type of installation, supply registers are assumed to be located in the floors on the main level and in the ceilings in the basement. Return grilles are assumed to be located in the floors on both levels. On the main level, because the supply registers and return grilles are all located at floor level, they should be placed on opposite sides of the room to avoid short-circuiting. Ductwork design practices and guidelines are given in complete detail in ACCA Manual D.

Using the Wrightsoft software package, Right-Suite Residential, perform the duct design according to the individual room air flow requirements and the external static pressure available with the selected piece of equipment yields ductwork sizing and layout as given in Figures A.18 and A.19. Right-Suite Residential utilizes ACCA Manual D to perform duct sizing calculations. Programs utilizing the same guidelines will produce similar results.

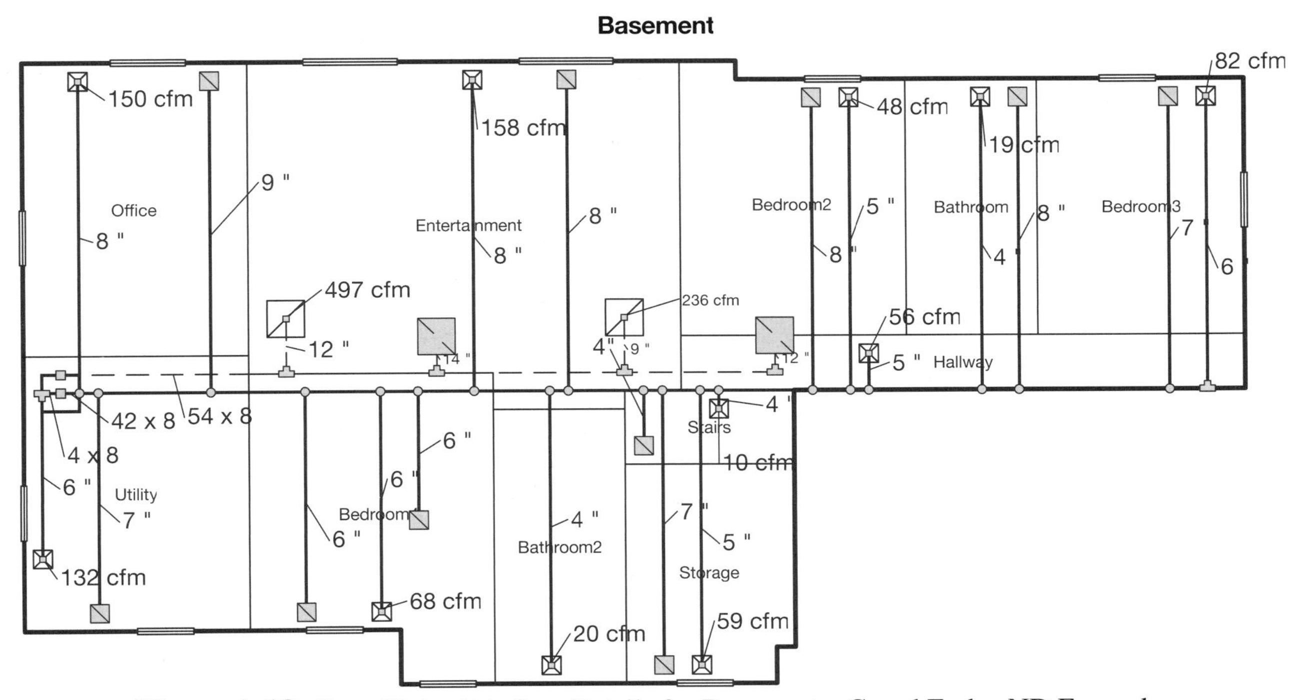

Figure A.18. Duct Sizing/Air flow Details for Basement – Grand Forks, ND Example

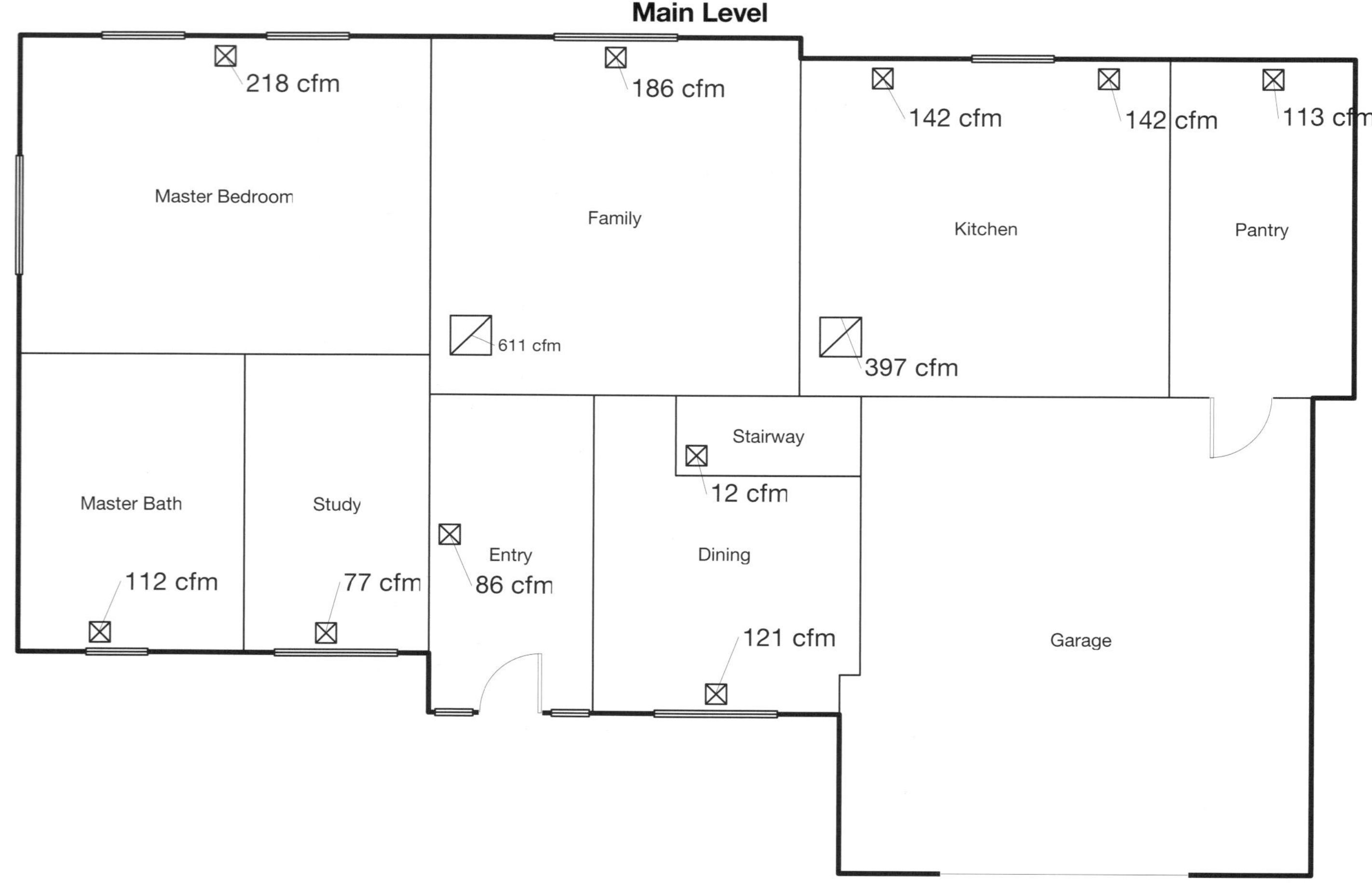

Figure A.19. Duct Sizing/Air flow Details for Main Level – Grand Forks, ND Example

DUCTWORK LAYOUT LEGEND:

- Supply Register Serving Main Level (Main Level View)
- Return Grille Serving Main Level (Main Level View)
- Supply Register Serving Main Level (Basement View)
- Return Grille Serving Main Level (Basement View)
- Supply Register Serving Basement (Basement View)
- Return Grille Serving Basement (Basement View)

Example A.1.5 Heating Load Single-Capacity, Water-to-Air GSHP Energy Analysis via Simplified Bin Method – Grand Forks, ND

After the heat pump sizing and selection process is complete, a customer may request information as to what the approximate payback period will be for the heat pump system when compared to another type of heating/cooling system. The simplified bin method as previously described (Chapter 3, Section 3.6) will yield reasonable approximations for annual operating cost for any type of system for the geographic location of the residence.

For this example, the simplified bin method will be performed for the residence utilizing the ClimateMaster Tranquility 20™ Model 048 (single-capacity) ground source heat pump system. Additionally, the analysis will be performed to calculate annual system operating cost for a typical year in Grand Forks. Only a single, simplified load profile is needed for analysis because internal gains do not fluctuate much when compared to a commercial application. The weather bin data of interest is then the annual number of hours spent in each temperature bin, not broken down according to time of day. The assumptions utilized to perform the bin analysis are summarized by Tables A.1-A.2

Table A.1. Bin Method Assumptions for Grand Forks, ND Example

Heating			Cooling		
Q_{PH}	46,268	Btu/hr	Q_{PC}	24,916	Btu/hr
T-Stat Set Pt.	70	˚F	T-Stat Set Pt.	75	˚F
t_{dh}	-16	˚F	t_{dc}	91	˚F
t_{bh}	62	˚F	t_{bc}	67	˚F
r_e	0.049	$/kWh	r_e	0.049	$/kWh
EWT_{min}	30	˚F	EWT_{max}	90	˚F
EWT_{mean}	46	˚F	EWT_{mean}	46	˚F
t_{mean}	40.2	˚F	t_{mean}	40.2	˚F
D_c	0.15		D_c	0.15	
Space SHF	NA		Space SHF	0.90	

Table A.2. Heat Pump Performance Assumptions for Bin Method (Grand Forks Example)

	EWT (˚F)	Total CAP (Btu/hr)	Sensible CAP (Btu/hr)	Demand (kW)
Htg.	30	38,373	NA	3.25
	50	50,538	NA	3.47
Clg.	70	50,195	35,061	2.69
	90	45,157	33,108	3.34

The performance data for the ClimateMaster Tranquility 20™ Model 048 has been corrected according to the correction factors listed in Table 3.15. This is consistent with the definitions given in Section 3.4, EWT_{lth}=30 F, EWT_{hth}=50 F, EWT_{ltc}=70 F, EWT_{htc}=90 F, etc. The results of the calculations performed via the bin method are summarized by Table A.3:

Table A.3. Bin Method Calculation Summary for Grand Forks, ND Example

Grand Forks, ND																			
Temp Bin (˚F)	t_{air} (˚F)	h_b (hrs)	Q_B (Btu/hr)	EWT (˚F)	HC/SC (Btu/hr)	TC (Cooling Only) Btu/hr	RF	PLF	RF_A	DMD_h (kW)	DMD_c (kW)	DMD_{res} (kW)	DMD_{htotal} (kW)	RT_A (hrs)	E (kWh)	C_e ($)	gl_h (Btu)	gl_c (Btu)
105/109	107	0	37,374	103.9	31,755	41,666	1.00	1.00	1.00	0.00	0.00	0.00	0.00	0.0	0	$0.00	0	0
100/104	102	1	32,702	99.5	32,178	42,757	1.00	1.00	1.00	0.00	3.65	0.00	0.00	1.0	4	$0.18	0	-55,210
95/99	97	9	28,031	95.2	32,601	43,848	0.86	0.98	0.88	0.00	3.51	0.00	0.00	7.9	28	$1.36	0	-431,957
90/94	92	43	23,359	90.9	33,023	44,939	0.71	0.96	0.74	0.00	3.37	0.00	0.00	31.8	107	$5.25	0	-1,716,379
85/89	87	105	18,687	86.5	33,446	46,030	0.56	0.93	0.60	0.00	3.23	0.00	0.00	62.8	203	$9.94	0	-3,346,358
80/84	82	223	14,015	82.2	33,869	47,121	0.41	0.91	0.45	0.00	3.09	0.00	0.00	101.2	312	$15.30	0	-5,320,069
75/79	77	327	9,343	77.9	34,292	48,212	0.27	0.89	0.31	0.00	2.95	0.00	0.00	100.0	295	$14.44	0	-5,191,057
70/74	72	457	4,672	73.5	34,715	49,302	0.13	0.87	0.15	0.00	2.81	0.00	0.00	70.7	198	$9.71	0	-3,620,758
65/69	67	568	0	69.2	35,138	50,393	0.00	0.85	0.00	0.00	2.66	0.00	0.00	0.0	0	$0.00	0	0
60/64	62	646	0	52.2	51,880	0	0.00	0.85	0.00	3.49	0.00	0.00	3.49	0.0	0	$0.00	0	0
55/59	57	627	2,966	50.8	51,014	0	0.06	0.86	0.07	3.48	0.00	0.00	3.48	42.5	148	$7.24	1,426,957	0
50/54	52	570	5,932	49.4	50,148	0	0.12	0.87	0.14	3.46	0.00	0.00	3.46	77.7	269	$13.18	2,584,487	0
45/49	47	513	8,898	47.9	49,283	0	0.18	0.88	0.21	3.45	0.00	0.00	3.45	105.6	364	$17.84	3,475,112	0
40/44	42	501	11,864	46.5	48,417	0	0.25	0.89	0.28	3.43	0.00	0.00	3.43	138.4	475	$23.28	4,506,287	0
35/39	37	575	14,829	45.1	47,551	0	0.31	0.90	0.35	3.42	0.00	0.00	3.42	200.0	683	$33.47	6,436,890	0
30/34	32	632	17,795	43.7	46,685	0	0.38	0.91	0.42	3.40	0.00	0.00	3.40	265.6	903	$44.25	8,451,723	0
25/29	27	544	20,761	42.2	45,819	0	0.45	0.92	0.49	3.38	0.00	0.00	3.38	268.5	909	$44.53	8,447,514	0
20/24	22	457	23,727	40.8	44,953	0	0.53	0.93	0.57	3.37	0.00	0.00	3.37	259.6	875	$42.85	8,070,571	0
15/19	17	390	26,693	39.4	44,088	0	0.61	0.94	0.64	3.35	0.00	0.00	3.35	251.0	842	$41.24	7,708,614	0
10/14	12	330	29,659	38.0	43,222	0	0.69	0.95	0.72	3.34	0.00	0.00	3.34	237.6	793	$38.86	7,208,630	0
5/9	7	307	32,625	36.5	42,356	0	0.77	0.97	0.80	3.32	0.00	0.00	3.32	244.9	814	$39.87	7,335,518	0
0/4	2	274	35,591	35.1	41,490	0	0.86	0.98	0.88	3.31	0.00	0.00	3.31	240.2	794	$38.91	7,100,290	0
-5/-1	-3	187	38,557	33.7	40,624	0	0.95	0.99	0.96	3.29	0.00	0.00	3.29	178.8	589	$28.84	5,217,335	0
-10/-6	-8	185	41,523	32.3	39,758	0	1.00	1.00	1.00	3.28	0.00	0.52	3.79	185.0	702	$34.38	5,288,013	0
-15/-11	-13	137	44,488	30.9	38,893	0	1.00	1.00	1.00	3.26	0.00	1.64	4.90	137.0	671	$32.89	3,804,688	0
-20/-16	-18	87	47,454	29.4	38,027	0	1.00	1.00	1.00	3.24	0.00	2.76	6.01	87.0	523	$25.61	2,345,436	0
-25/-21	-23	42	50,420	28.0	37,161	0	1.00	1.00	1.00	3.23	0.00	3.89	7.11	42.0	299	$14.64	1,098,158	0
-30/-26	-28	17	53,386	26.6	36,295	0	1.00	1.00	1.00	3.21	0.00	5.01	8.22	17.0	140	$6.85	430,682	0
-35/-31	-33	0	56,352	25.2	35,429	0	1.00	1.00	1.00	0.00	0.00	0.00	0.00	0.0	0	$0.00	0	0
-40/-36	-38	0	59,318	23.7	34,563	0	1.00	1.00	1.00	0.00	0.00	0.00	0.00	0.0	0	$0.00	0	0
												Totals	Heating	2,978	10,790	$528.72	90,936,906	
													Cooling	375	1,146	$56.18		-19,681,789

The following conclusions can be drawn from Table A.3:

- The estimated amount of heat pump full-load run hours in heating mode for this example is 2,978 hours, found by summing the number of run hours (RT_A) when the outdoor air temperature is lower than the balance point temperature for heating.
- The estimated amount of heat pump full-load run hours in cooling mode for this example is 375 hours, found by summing the number of run hours (RT_A) when the outdoor air temperature is higher than the balance point temperature for cooling.
- The estimated amount of supplemental resistance heat run hours for this example is 468 hours, found by summing the number of run hours (RT_A) when the heat pump capacity is less than the building load (driven by outdoor air temperature).
- The total estimated annual operating cost for the ClimateMaster Tranquility 20™ Model 048 ground source heat pump in heating (heat pump operation plus supplemental resistance heat operation) is $528.72 per year for this example (at the assumed electric rate).
- The total estimated annual operating cost for this system in cooling is $56.18 per year for this example.
- The estimated amount of heat removed from the ground (heating mode) via the heat pump system is 90,936,906 Btu.
- The estimated amount of heat rejected to the ground (cooling mode) via the heat pump system is 19,681,789 Btu.
- Because of the severely unbalanced ground loads, the potential exists to draw enough energy from the soil to decrease the deep earth temperature in the center of the loopfield over time, which will detrimentally affect system performance (more of a factor for large systems). It is because of this that an accurate ground-loop design is critical.

Example A.1.6 Heating Load Single-Capacity (4-ton), Water-to-Air GSHP Vertical Loopfield Sizing – Grand Forks, ND

After the peak building loads have been calculated and GSHP equipment selected for the load, the loopfield design can be performed. As discussed in Chapter 5, the first step in designing the loopfield for any installation is to perform a comprehensive site survey. For the sake of example, this Grand Forks, ND residence was assumed to be located on a typical city lot where the best option is to utilize a vertically-bored GHEX for the ground source loopfield. As a starting point for the calculations, the following important design assumptions were made:

- GHEX consists of four vertically-bored U-bends in a 1 x 4 straight-line configuration, U-bend diameter=3/4 inch
- Minimum entering water temperature from the loopfield, EWT_{min}=30 F
- Deep earth temperature, T_G=46 F (provided by the results of a formation thermal conductivity test run in the area)
- Formation thermal conductivity, k_G=1.00 Btu/hr-ft-F (Determined according to known soil properties for the geographic location and from Table 5.6 – assumed heavy clay formation – 15% water)
- C-C bore spacing=15 feet
- Bore diameter, D_{bore}=5 inch
- k_{grout}=0.40 Btu/hr-ft-F (20% solids mixture – slurry consisting of 20% bentonite & 80% water by mass)
- System flow rate, GPM=12

Figure A.20 displays the vertically-bored GHEX design worksheet (heating based design) for this residence. All of the data in the "Heat Pump Design Data" column were taken directly from the Single-Capacity, Water-to-Air Heat Pump Selection Worksheet (Figure A.8) for the ClimateMaster Tranquility 20™ Model 048 GSHP. The "Borefield Layout" column was filled out according to the design assumptions that were made. The "Bin Analysis Data" column was filled out with data taken directly from the results of the bin analysis performed in Example A.1.5, Table A.3. The run fraction in heating was found on Figure 3.10 (heating degree days for Grand Forks, H_{DD}=9,500, Percent-Sizing=83% ➔F_H=0.80). The "Vertical Bore Design Data" column was filled out according to the design assumptions made. The borehole resistance ($R_B + R_G \cdot F_H$) was found on Table 5.8 according to the design assumptions made.

Vertically-Bored GHEX Design Worksheet – Heating Mode

Space Heating GSHP Design Data

HC_C[1]	=	**38,373**	Btu/hr
DMD_C[1]	=	**3.25**	kW
GPM[1]	=	**12**	gpm
AGL_H	=	**90,936,906**	Btu
AGL_C	=	**19,681,789**	Btu
Total Heat Loss[1]	=	**46,268**	Btu/hr
Percent Sizing	=	**82.94**	%
F_{Jan}	=	**.80**	

HWG GSHP Design Data

HC_{HWG}	=	**N/A**	Btu/hr
DMD_{HWG}	=	**N/A**	kW
GPM_{HWG}	=	**N/A**	gpm
AGL_{HWG}	=	**N/A**	Btu
F_{HWG}	=	**N/A**	

Total Design Data

HC_D	=	**38,373**	Btu/hr
DMD_D	=	**3.25**	kW
GPM_D	=	**12**	gpm
AGL_{HD}	=	**90,936,906**	Btu
AGL_{CD}	=	**19,681,789**	Btu

GHEX Design Data

COP_D[2]	$(= HC_D / (DMD_D \times 3{,}412))$	=	**3.46**	
HE_D[2]	$(= HC_D \times (COP_D - 1)/COP_D)$	=	**27,283**	Btu/hr
F_H	$(= (F_{Jan} \times HC_C + F_{HWG} \times HC_{HWG}) / HC_D)$	=	**.80**	
EWT_{min}[3]		=	**30**	F
LWT_{min}[2]	$(= EWT_{min} - HE_D/(500 \times GPM_D))$	=	**25.45**	F

1. For single heat pump installation use data directly from the appropriate GSHP Selection Worksheet. For multiple heat pumps in the installation sum all heat pump Total Heat Losses, HC_C's, DMD_C's and GPM's that will be connected to a single GHEX using the table provided below.
2. For single heat pump installation use data directly from appropriate GSHP Selection Worksheet. For multiple heat pumps use the equation provided.
3. EWT_{min} is obtained directly from the appropriate GSHP Selection Worksheet and must be the same for the selection of all heat pumps for a multiple heat pump installation connected to a single GHEX.

	Zone 1	Zone 2	Zone 3	Zone 4	Zone 5	Zone 6	Zone 7	Zone 8	Zone 9	Zone 10	Total
Total Heat Loss											
HC_C											
DMD_C											
GPM											

Vertical Borefield Layout

U-bend D_P	=	**3/4**	in
GPM_{FP}	=	**3**	gpm/flowpath
N_{FP}	=	**4**	flowpaths
N_{BIS}	=	**1**	bores in series
N_B	=	**4**	bores
Layout	=	**1** x **4**	($N_{Rows} \times N_{Bores/Row}$)
Bore Spacing	=	**15**	ft

Vertical Borefield Design Data

T_G	=	**46**	F
k_G	=	**1.00**	Btu/hr ft F (Table 5.6)
D_{bore}	=	**5.0**	in
k_{Grout}	=	**0.40**	Btu/hr ft F
$(R_B + R_G \cdot F_H)$	=	**0.791**	hr ft F/Btu (Table 5.8, 5.9 or 5.10)

Borehole Design Lengths (Equations 5.1 and 5.2)

$$L_{H,T} = \frac{HC_D \cdot \left(\frac{COP_D - 1}{COP_D}\right) \cdot \left(R_B + R_G \cdot F_H\right)}{T_G - \left(\frac{EWT_{\min} + LWT_{\min}}{2}\right)} = \frac{38{,}373 \cdot \left(\frac{3.46 - 1}{3.46}\right) \cdot (.791)}{46 - \left(\frac{30 + 25.45}{2}\right)} = \frac{38{,}373 \cdot (0.711) \cdot (0.791)}{46 - (27.73)} = 1181 \; \textit{ft of bore}$$

$$L_{H,B} = \frac{L_{H,T}}{N_B} = \frac{1181}{4} = 295 \; \textit{ft per bore}$$

Unbalanced Ground Load Borehole Design Lengths (Equations 5.9a, 5.10, and 5.11)

$$NNAGL = \frac{AGL_{HD} - AGL_{CD}}{L_{H,T} \cdot \left(T_G - \left(\frac{EWT_{\min} + LWT_{\min}}{2}\right)\right)} = \frac{90{,}936{,}906 - 19{,}681{,}789}{1181 \cdot \left(46 - \left(\frac{30 + 25.45}{2}\right)\right)} = 3{,}302 \; Btu/ftF$$

B_M = 1.33 (Figures 5.13 through 5.18)

$$L_{H,T,UGL} = B_M \cdot L_{H,T} = 1.33 \cdot 1181 = 1571 \; \textit{ft of bore}$$

$$L_{H,B,UGL} = \frac{L_{H,T,UGL}}{N_B} = \frac{1571}{4} = 393 \; \textit{ft per bore}$$

Heating Design Length Calculations Summary Table

Layout for Flow and Number of Bores						Design Length Calculations				Unbalanced Ground Load Design Lengths						
D_P	GPM_{FP}	N_{FP}	N_{BIS}	N_B	Layout	k_{Grout}	$R_B+R_G \cdot F_H$	$L_{H,T}$	$L_{H,B}$	Layout	Spacing	NNAGL	B_M	$L_{H,T,UGL}$	$L_{H,B,UGL}$	Annotation
3/4	3	4	1	4	1 x 4	0.40	0.791	1181	295	1 x 4	15	3302	1.33	1571	393	
3/4	3	4	1	4	1 x 4	0.88	0.647	966	242	1 x 4	15	4037	1.45	1401	350	k_{Grout} = 0.88
3/4	3	4	1	4	1 x 4	0.88	0.647	966	242	1 x 4	20	4037	1.27	1227	307	Spacing = 20ft.
3/4	2.8	5	1	5	1 x 5	0.88	0.647	966	193	1 x 5	20	4037	1.27	1227	245	N_B = 5
3/4	3	4	1	4	1 x 4	0.88	0.670	726	182	1 x 4	20	4135	1.27	922	231	EWT_{min} = 25°F

Figure A.20. Vertically-Bored GHEX Design Worksheet for Grand Forks, ND Example (4-ton)

Upon performing the calculations summarized in the "Heating Design Length Calculations Summary Table" on the Vertically-Bored GHEX Design Worksheet (Figure A.20), the following conclusions can be made:

1) The total amount of bore required to satisfy the system constraints imposed by the original design assumptions was found to be $L_{H,T,UGL}$=1,571 ft (393 ft/bore – 4 bores).
2) Using a thermally-enhanced grouting material (thermal conductivity, k_{Grout}=0.88 Btu/hr-ft-F) as the backfill material instead of the 20% solids backfill (k_{Grout}=0.40 Btu/hr-ft-F) reduces the amount of bore required by 10.8% ($L_{H,T,UGL}$=1,401 ft, 350 ft/bore – 4 bores).
3) Increasing the center-center (C-C) bore spacing from 15 feet to 20 feet further reduces the amount of required bore length by 11.8% ($L_{H,T,UGL}$=1,227 ft, 307 ft/bore – 4 bores).
4) Utilizing ¾-inch diameter U-bends for borehole depths greater than 200-250 feet (depending on flow through each U-bend) will tend to consume excessive pumping power. Dividing the total bore length ($L_{H,T,UGL}$=1,227 ft) into five bores rather than four reduces individual bore lengths from 307 ft/bore to 245 ft/bore. Keep in mind that dividing the system flow equally into five parallel paths (rather than four) will reduce the system head loss at the same flow. If the same pump were to be used for the 5-loop system as would have been used for the 4-loop system, the end result would be an increase in total system flow. For this example, it was assumed that total system flow would increase from 12 gpm to 14 gpm. Heat pump performance effects on the increased system flow were neglected.
5) In some cases (extreme climates coupled with poor soil), maintaining a minimum entering water temperature, EWT_{min} of 30 F will not prove to be economically feasible because of the large amount of bore length necessary to do so. It is in such instances that a designer must weigh costs and benefits (reduced system performance, increased operating cost, reduced bore length requirements) of designing a loopfield that allows the entering water temperature to reach lower temperatures under design conditions. The last calculation listed on the "Heating Design Length Calculations Summary Table" was performed for EWT_{min}=25 F. The effects of heat pump performance were accounted for in this calculation (though they are not shown) and the results of the calculations show a 24.9% reduction in required bore length versus designing around EWT_{min}=30 F (k_{Grout}=0.88 Btu/hr-ft-F, bore spacing=20ft, 1x4 straight-line configuration). For EWT_{min}=25 F, the selected equipment will only be sized such that it will handle 78% of the peak heating load under design conditions (HC_C=35,950 Btu/hr), and it will operate with lower efficiency (DMD_C=3.19 kW, COP_{Hc}=3.30) to produce the required heating. In reality, the recommended approach to solving this problem would be to utilize a 5-ton (nominal) capacity heat pump (rather than the 4-ton unit selected) coupled with a vertically-bored U-bend GHEX configuration (five bores – refer to Examples A.1.2 and A.1.7). *Remember, utilizing EWT_{min}=25 F is only recommended in extreme situations.* Utilizing EWT_{min}=25 F also necessitates the use of higher levels of freeze protection in a system and in some cases, larger circulating pumps. For the example, the need for higher levels of freeze protection was neglected. Utilizing EWT_{min}=25 F and propylene glycol as the circulating fluid will cause issues in attaining turbulent flow in the U-bends. If possible (check state regulations), the use of a methanol-water solution as the circulating fluid is recommended because less fluid flow is necessary to attain turbulent flow as compared to propylene glycol-water solutions that offer similar freeze protection levels. The calculations show that 231 feet of active bore length (x4 bores – total active length = 924 feet) is necessary to satisfy the heating load for the home. Actual bore depth will have to be at least 5-6 feet more than the minimum required active bore length required because active bore length is measured from the bottom of the header trench to the bottom of the vertical bore. For head loss calculation purposes, the GHEX would be assumed to consist of four vertically-bored U-bends each 240 feet in depth (235 feet active depth) in an outside close-header, parallel configuration.

Example A.1.7 Heating Load Two-Capacity (5-ton), Water-to-Air GSHP Vertical Loopfield Sizing – Grand Forks, ND

To illustrate the argument that bore length requirements will not be significantly different for a 5-ton GSHP unit (versus the originally selected 4-ton unit) for the same residence, the calculations performed in Example A.1.6 will be performed in this example with the assumed use of ClimateMaster Tranquility 27™ Model 064 GSHP system. Again, this Grand Forks, ND residence was assumed to be located on a typical city lot where the best option is to utilize a vertically-bored GHEX for the ground source loopfield. As a starting point for the calculations, the following important design assumptions were made:

- GHEX consists of five vertically-bored U-bends in a 1 x 5 straight-line configuration, U-bend diameter=3/4 inch
- Minimum entering water temperature from the loopfield, EWT_{min}=30 F
- Deep earth temperature, T_G=46 F (provided by the results of a formation thermal conductivity test run in the area)
- Formation thermal conductivity, k_G=1.00 Btu/hr-ft-F (Determined according to known soil properties for the geographic location and from Table 5.6 – assumed clay formation)
- C-C bore spacing=15 feet
- Bore diameter, D_{bore}=5 inch
- k_{grout}=0.40 Btu/hr-ft-F (20% solids mixture – slurry consisting of 20% bentonite & 80% water by mass)
- System flow rate, GPM=15

Figure A.21 displays the vertically-bored GHEX design worksheet (heating based design) for this residence. All of the data in the "Heat Pump Design Data" column were taken directly from the Two-Capacity, Water-to-Air Heat Pump Selection Worksheet (Figure A.11) for the ClimateMaster Tranquility 27™ Model 064 GSHP. The "Borefield Layout" column was filled out according to the design assumptions that were made. The data in the "Bin Analysis Data" column was taken directly from the results of a bin analysis performed in similar fashion as Example A.1.5, Table A.3, although the calculations are not shown. The run fraction in heating was found on Figure 3.10 (heating degree days for Grand Forks, H_{DD}=9,500, Percent-Sizing=100% → F_H=0.64). The "Vertical Bore Design Data" column was filled out according to the design assumptions made. The borehole resistance ($R_B+R_G \cdot F_H$) was found on Table 5.8 according to the design assumptions made.

Vertically-Bored GHEX Design Worksheet – Heating Mode

Space Heating GSHP Design Data

HC_C[1]	= 47,769	Btu/hr
DMD_C[1]	= 3.86	kW
GPM[1]	= 15	gpm
AGL_H	= 93,791,906	Btu
AGL_C	= 19,977,376	Btu
Total Heat Loss[1]	= 46,268	Btu/hr
Percent Sizing	= 103.24	%
F_{Jan}	= .64	

HWG GSHP Design Data

HC_{HWG}	= N/A	Btu/hr
DMD_{HWG}	= N/A	kW
GPM_{HWG}	= N/A	gpm
AGL_{HWG}	= N/A	Btu
F_{HWG}	= N/A	

Total Design Data

HC_D	= 46,268	Btu/hr
DMD_D	= 3.86	kW
GPM_D	= 15	gpm
AGL_{HD}	= 93,791,906	Btu
AGL_{CD}	= 19,977,376	Btu

GHEX Design Data

COP_D[2]	(= HC_D / (DMD_D x 3,412)	= 3.63	
HE_D[2]	(= HC_D x (COP_D-1)/COP_D)	= 34,609	Btu/hr
F_H	(= (F_{Jan} x HC_C + F_{HWG} x HC_{HWG}) / HC_D)	= .64	
EWT_{min}[3]		= 30	F
LWT_{min}[2]	(= EWT_{min} – HE_D/(500xGPM_D))	= 25.39	F

1. For single heat pump installation use data directly from the appropriate GSHP Selection Worksheet. For multiple heat pumps in the installation sum all heat pump Total Heat Losses, HC_C's, DMD_C's and GPM's that will be connected to a single GHEX using the table provided below.
2. For single heat pump installation use data directly from appropriate GSHP Selection Worksheet. For multiple heat pumps use the equation provided.
3. EWT_{min} is obtained directly from the appropriate GSHP Selection Worksheet and must be the same for the selection of all heat pumps for a multiple heat pump installation connected to a single GHEX.

	Zone 1	Zone 2	Zone 3	Zone 4	Zone 5	Zone 6	Zone 7	Zone 8	Zone 9	Zone 10	Total
Total Heat Loss											
HC_C											
DMD_C											
GPM											

Vertical Borefield Layout

U-bend D_P	= 3/4	in
GPM_{FP}	= 3	gpm/flowpath
N_{FP}	= 5	flowpaths
N_{BIS}	= 1	bores in series
N_B	= 5	bores
Layout	= 1 x 5	(N_{Rows} x $N_{Bores/Row}$)
Bore Spacing	= 15	ft

Vertical Borefield Design Data

T_G	= 46	F
k_G	= 1.00	Btu/hr ft F (Table 5.6)
D_{bore}	= 5.0	in
k_{Grout}	= 0.40	Btu/hr ft F
(R_B + $R_G \cdot F_H$)	= 0.700	hr ft F/Btu (Table 5.8, 5.9 or 5.10)

Borehole Design Lengths (Equations 5.1 and 5.2)

$$L_{H,T} = \frac{HC_D \cdot \left(\frac{COP_D - 1}{COP_D}\right) \cdot (R_B + R_G \cdot F_H)}{T_G - \left(\frac{EWT_{min} + LWT_{min}}{2}\right)} = \frac{47{,}769 \cdot \left(\frac{3.62 - 1}{3.62}\right) \cdot (.700)}{46 - \left(\frac{30 + 25.39}{2}\right)} = \frac{47{,}769 \cdot (0.724) \cdot (0.700)}{46 - (27.70)} = 1323 \textit{ ft of bore}$$

$$L_{H,B} = \frac{L_{H,T}}{N_B} = \frac{1323}{5} = 265 \textit{ ft per bore}$$

Unbalanced Ground Load Borehole Design Lengths (Equations 5.9a, 5.10, and 5.11)

$$NNAGL = \frac{AGL_{HD} - AGL_{CD}}{L_{H,T} \cdot \left(T_G - \left(\frac{EWT_{min} + LWT_{min}}{2}\right)\right)} = \frac{93{,}791{,}906 - 19{,}977{,}376}{1323 \cdot \left(46 - \left(\frac{30 + 25.39}{2}\right)\right)} = 3{,}048 \; Btu/ftF$$

B_M = 1.32 (Figures 5.13 through 5.18)

$$L_{H,T,UGL} = B_M \cdot L_{H,T} = 1.32 \cdot 1323 = 1746 \textit{ ft of bore} \qquad L_{H,B,UGL} = \frac{L_{H,T,UGL}}{N_B} = \frac{1746}{5} = 349 \textit{ ft per bore}$$

Heating Design Length Calculations Summary Table

Layout for Flow and Number of Bores						Design Length Calculations				Unbalanced Ground Load Design Lengths					
D_P	GPM_{FP}	N_{FP}	N_{BIS}	N_B	Layout	k_{Grout}	$R_B + R_G \cdot F_H$	$L_{H,T}$	$L_{H,B}$	Layout	Spacing	NNAGL	B_M	$L_{H,T,UGL}$	$L_{H,B,UGL}$
3/4	3	5	1	5	1 x 5	0.40	0.700	1322	265	1 x 5	15	3048	1.32	1746	349
3/4	3	5	1	5	1 x 5	0.88	0.556	1050	210	1 x 5	15	3841	1.42	1492	298
3/4	3	5	1	5	1 x 5	0.88	0.556	1050	210	1 x 5	20	3841	1.26	1324	265
3/4	3	5	1	5	1 x 5	0.88	0.584	798	160	1 x 5	20	3944	1.27	1013	203

k_{Grout} = 0.88 →

EWT_{min} = 25°F →

← Spacing = 20ft.

Figure A.21. Vertically-Bored GHEX Design Worksheet for Grand Forks, ND Example (5-ton)

Upon performing the calculations summarized in the "Heating Design Length Calculations Summary Table" on the Vertically-Bored GHEX Design Worksheet (Figure A.21), the following conclusions can be made:

1) The total amount of bore required to satisfy the system constraints imposed by the original design assumptions was found to be $L_{H,T,UGL}$=1,746 ft (349 ft/bore – five bores).
2) Using a thermally-enhanced grouting material (thermal conductivity, k_{Grout}=0.88 Btu/hr-ft-F) as the backfill material instead of the 20% solids backfill (k_{Grout}=0.40 Btu/hr-ft-F) reduces the amount of bore required by 14.5% ($L_{H,T,UGL}$=1,492 ft, 298 ft/bore – five bores).
3) Increasing the C-C bore spacing from 15 feet to 20 feet further reduces the amount of required bore length by 11.3% ($L_{H,T,UGL}$=1,324 ft, 265 ft/bore – five bores).
4) In some cases (extreme climates coupled with poor soil), maintaining a minimum entering water temperature (EWT_{min}) of 30 F will not prove to be economically feasible because of the large amount of bore length necessary to do so. It is in such instances that a designer must weigh costs and benefits (reduced system performance, increased operating cost, reduced bore length requirements) of designing a loopfield that allows the entering water temperature to reach lower temperatures under design conditions. The last calculation listed on the "Heating Design Length Calculations Summary Table" was performed for EWT_{min}=25 F. The effects of heat pump performance were accounted for in this calculation (though they are not shown) and the results of the calculations show a 23.5% reduction in required bore length versus designing around EWT_{min}=30 F (k_{Grout}=0.88 Btu/hr-ft-F, bore spacing=20 ft, 1x5 straight-line configuration). For EWT_{min}=25 F, the selected equipment will only be sized such that it will handle 96% of the peak heating load under design conditions (HC_C=44,431 Btu/hr), and it will operate with lower efficiency (DMD_C=3.78 kW, COP_{Hc}=3.45) to produce the required heating. Remember, utilizing EWT_{min}=25 F is only recommended in extreme situations. Utilizing EWT_{min}=25 F also necessitates the use of higher levels of freeze protection in a system and in some cases, larger circulating pumps. For the example, the need for higher levels of freeze protection was neglected. Utilizing EWT_{min}=25 F and propylene glycol as the circulating fluid will cause issues in attaining turbulent flow in the U-bends. If possible (check state regulations), use of a methanol-water solution as the circulating fluid is recommended because less fluid flow is necessary to attain turbulent flow as compared to propylene glycol-water solutions that offer similar freeze protection levels. The calculations show that 203 feet of active bore length (x5 bores – total active length=1,013 feet) is necessary to satisfy the heating load for the home. Actual bore depth will have to be at least 5-6 feet more than the minimum required active bore length required because active bore length is measured from the bottom of the header trench to the bottom of the vertical bore.
5) The difference in active bore length between this 5-ton system and the 4-ton system in Example A.1.6 is merely 91 feet (1,013 feet versus 922 feet active bore length, +9.9.%). Under sizing the system is a better situation in terms of cooling capacity.

For head loss calculation purposes, the GHEX would be assumed to consist of five vertically-bored U-bends each 200 feet in depth (195 feet active depth) in the step-down, step-up reverse-return header configuration shown in Figure 5.8(d).

Example A.1.8 Heating Load Two-Capacity, Water-to-Air GSHP System Head Loss Calculations – Grand Forks, ND

Once the loopfield design has been completed, interior piping layout and system head loss calculations need to be performed. The assumptions utilized to perform system head loss calculations are as follows:

- Pressurized flow center utilized
- Outside piping brought to the interior via wall penetration shown by Figure 7.2a, heat pump assumed to be located in close proximity to where wall penetrations are made
- GHEX piping assumed to be connected in the step-down, step-up reverse-return header configuration shown in Figure 5.8(d)
- Supply-return pipe run distance from loopfield to the flow center inside the building assumed to be 50 feet
- Supply-return piping assumed to be 1-1/4-inch DR-11 HDPE
- Reinforced rubber hose from flow center-heat pump assumed to be 1-inch diameter
- Heat pump/flow center/GHEX connection made according to Figure 7.3a
- Circulating fluid assumed to be water at 40 F for head loss calculations, antifreeze multipliers (Table 4.21) then applied to head loss calculations for 25% propylene glycol at 25 F

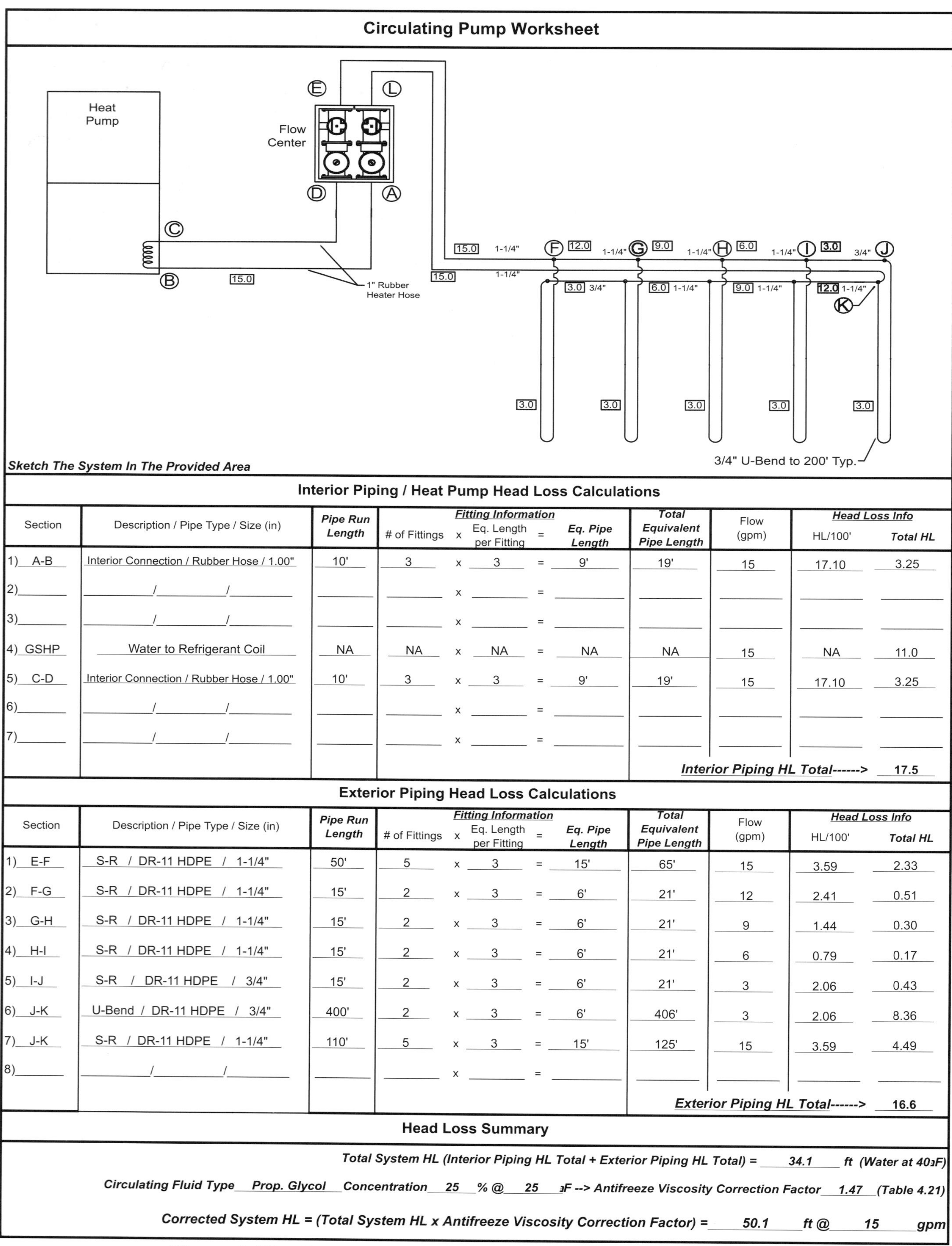

Circulating Pump Worksheet

Sketch The System In The Provided Area

Interior Piping / Heat Pump Head Loss Calculations

Section	Description / Pipe Type / Size (in)	Pipe Run Length	# of Fittings	x	Fitting Information: Eq. Length per Fitting	=	Eq. Pipe Length	Total Equivalent Pipe Length	Flow (gpm)	Head Loss Info: HL/100'	Total HL
1) A-B	Interior Connection / Rubber Hose / 1.00"	10'	3	x	3	=	9'	19'	15	17.10	3.25
2)	/ /			x		=					
3)	/ /			x		=					
4) GSHP	Water to Refrigerant Coil	NA	NA	x	NA	=	NA	NA	15	NA	11.0
5) C-D	Interior Connection / Rubber Hose / 1.00"	10'	3	x	3	=	9'	19'	15	17.10	3.25
6)	/ /			x		=					
7)	/ /			x		=					

Interior Piping HL Total------> 17.5

Exterior Piping Head Loss Calculations

Section	Description / Pipe Type / Size (in)	Pipe Run Length	# of Fittings	x	Fitting Information: Eq. Length per Fitting	=	Eq. Pipe Length	Total Equivalent Pipe Length	Flow (gpm)	Head Loss Info: HL/100'	Total HL
1) E-F	S-R / DR-11 HDPE / 1-1/4"	50'	5	x	3	=	15'	65'	15	3.59	2.33
2) F-G	S-R / DR-11 HDPE / 1-1/4"	15'	2	x	3	=	6'	21'	12	2.41	0.51
3) G-H	S-R / DR-11 HDPE / 1-1/4"	15'	2	x	3	=	6'	21'	9	1.44	0.30
4) H-I	S-R / DR-11 HDPE / 1-1/4"	15'	2	x	3	=	6'	21'	6	0.79	0.17
5) I-J	S-R / DR-11 HDPE / 3/4"	15'	2	x	3	=	6'	21'	3	2.06	0.43
6) J-K	U-Bend / DR-11 HDPE / 3/4"	400'	2	x	3	=	6'	406'	3	2.06	8.36
7) J-K	S-R / DR-11 HDPE / 1-1/4"	110'	5	x	3	=	15'	125'	15	3.59	4.49
8)	/ /			x		=					

Exterior Piping HL Total------> 16.6

Head Loss Summary

Total System HL (Interior Piping HL Total + Exterior Piping HL Total) = 34.1 *ft (Water at 40₃F)*

Circulating Fluid Type Prop. Glycol *Concentration* 25 *% @* 25 *₃F --> Antifreeze Viscosity Correction Factor* 1.47 *(Table 4.21)*

Corrected System HL = (Total System HL x Antifreeze Viscosity Correction Factor) = 50.1 *ft @* 15 *gpm*

Figure A.22. Circulating Pump Worksheet for Grand Forks, ND Example

The Circulating Pump Worksheet shown by Figure A.22 summarizes the system head loss calculations for this example.

As is shown by the head loss calculations in Figure A.22, the circulating pump(s) will have to be large enough to provide 50.1 feet of head at a system flow rate of 15 gpm. Because standing column flow centers are typically equipped with circulating pumps manufactured by Grundfos, two Grundfos Series-UP 26-116F circulating pumps were selected to provide system flow for this example. A single Grundfos Series-UP 26-116F circulator is capable of providing approximately 26 feet of head at 15 gpm. Therefore, to use this circulating pump model, two of the 26-116F circulators (connected in series) will have to be used. Two UP 26-116F pumps in series will be able to provide 52 feet of head at a system flow of 15 gpm. Because the circulating pumps will be slightly oversized for a system flow of 15 gpm, the actual operating flow rate will be somewhat higher. The flow in the system will be greatly restricted by the refrigerant-to-water (source side) coil. Head loss is directly proportional to the square of system flow. In other words, doubling flow will increase head loss in the system by four times. Tripling system flow will increase head loss by nine times. Because of the proportional relationship of head loss to flow rate, slightly over sizing pumps will cause slight increases in system flow.

A second circulating pump worksheet (shown by Figure A.23) was filled out at a decreased flow rate to generate the system curve displayed in Figure A.24. The system curve can be used to determine the actual system operating flow rate given that 2 x Grundfos UP 26-116F circulating pumps (series connected) were selected to provide system flow.

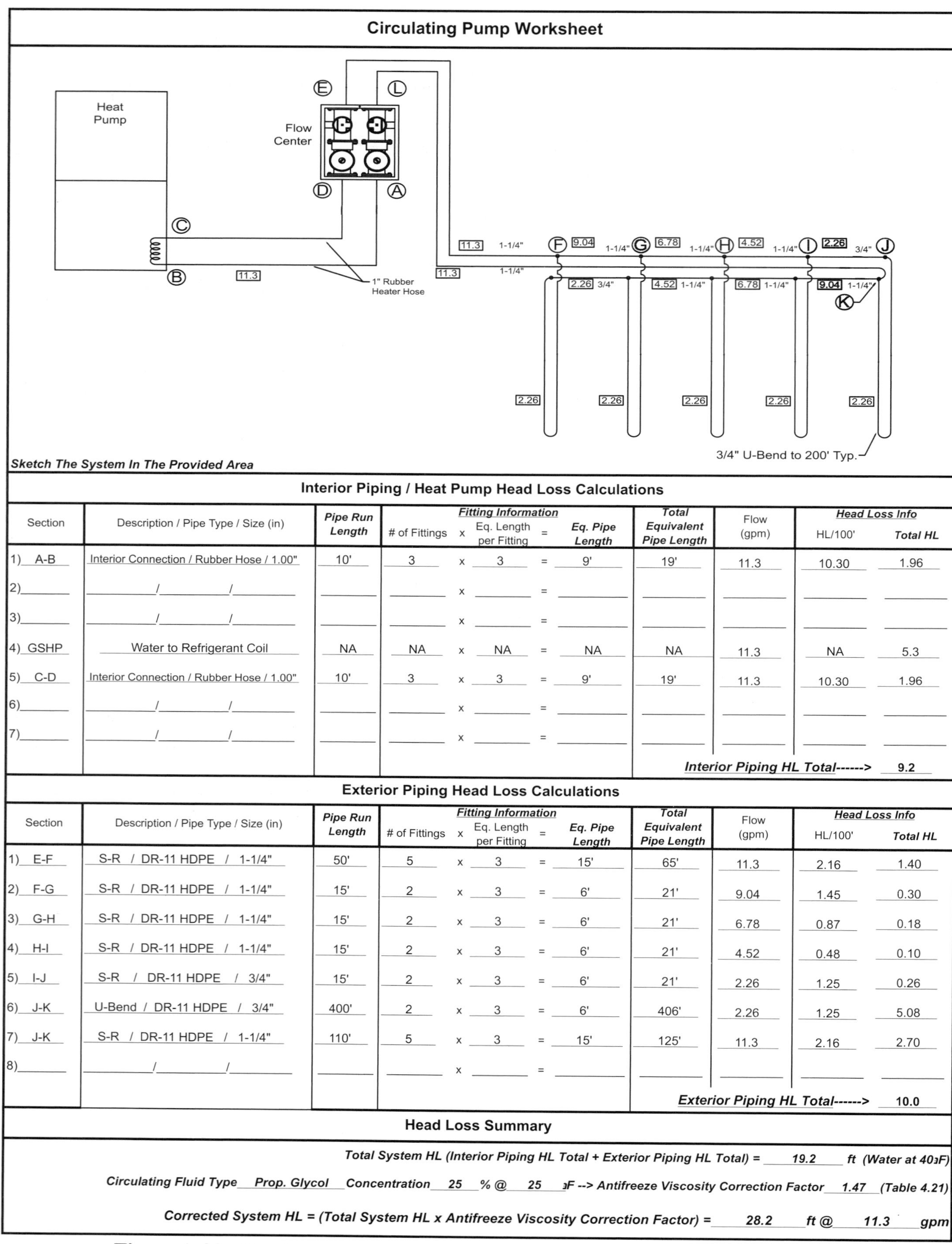

Circulating Pump Worksheet

Heat Pump
Flow Center
E L D A C B K F G H I J
11.3 1-1/4" 9.04 1-1/4" 6.78 1-1/4" 4.52 1-1/4" 2.26 3/4"
11.3 1-1/4" 2.26 3/4" 4.52 1-1/4" 6.78 1-1/4" 9.04 1-1/4"
11.3
1" Rubber Heater Hose
2.26 2.26 2.26 2.26 2.26
3/4" U-Bend to 200' Typ.

Sketch The System In The Provided Area

Interior Piping / Heat Pump Head Loss Calculations

Section	Description / Pipe Type / Size (in)	Pipe Run Length	# of Fittings	x	Eq. Length per Fitting	=	Eq. Pipe Length	Total Equivalent Pipe Length	Flow (gpm)	HL/100'	Total HL
			Fitting Information							*Head Loss Info*	
1) A-B	Interior Connection / Rubber Hose / 1.00"	10'	3	x	3	=	9'	19'	11.3	10.30	1.96
2)	/ /			x		=					
3)	/ /			x		=					
4) GSHP	Water to Refrigerant Coil	NA	NA	x	NA	=	NA	NA	11.3	NA	5.3
5) C-D	Interior Connection / Rubber Hose / 1.00"	10'	3	x	3	=	9'	19'	11.3	10.30	1.96
6)	/ /			x		=					
7)	/ /			x		=					

Interior Piping HL Total------> 9.2

Exterior Piping Head Loss Calculations

Section	Description / Pipe Type / Size (in)	Pipe Run Length	# of Fittings	x	Eq. Length per Fitting	=	Eq. Pipe Length	Total Equivalent Pipe Length	Flow (gpm)	HL/100'	Total HL
			Fitting Information							*Head Loss Info*	
1) E-F	S-R / DR-11 HDPE / 1-1/4"	50'	5	x	3	=	15'	65'	11.3	2.16	1.40
2) F-G	S-R / DR-11 HDPE / 1-1/4"	15'	2	x	3	=	6'	21'	9.04	1.45	0.30
3) G-H	S-R / DR-11 HDPE / 1-1/4"	15'	2	x	3	=	6'	21'	6.78	0.87	0.18
4) H-I	S-R / DR-11 HDPE / 1-1/4"	15'	2	x	3	=	6'	21'	4.52	0.48	0.10
5) I-J	S-R / DR-11 HDPE / 3/4"	15'	2	x	3	=	6'	21'	2.26	1.25	0.26
6) J-K	U-Bend / DR-11 HDPE / 3/4"	400'	2	x	3	=	6'	406'	2.26	1.25	5.08
7) J-K	S-R / DR-11 HDPE / 1-1/4"	110'	5	x	3	=	15'	125'	11.3	2.16	2.70
8)	/ /			x		=					

Exterior Piping HL Total------> 10.0

Head Loss Summary

Total System HL (Interior Piping HL Total + Exterior Piping HL Total) = 19.2 *ft (Water at 40ãF)*

Circulating Fluid Type Prop. Glycol *Concentration* 25 *% @* 25 *ãF --> Antifreeze Viscosity Correction Factor* 1.47 *(Table 4.21)*

Corrected System HL = (Total System HL x Antifreeze Viscosity Correction Factor) = 28.2 *ft @* 11.3 *gpm*

Figure A.23. Circulating Pump Worksheet for Grand Forks, ND Example (Lower Flow)

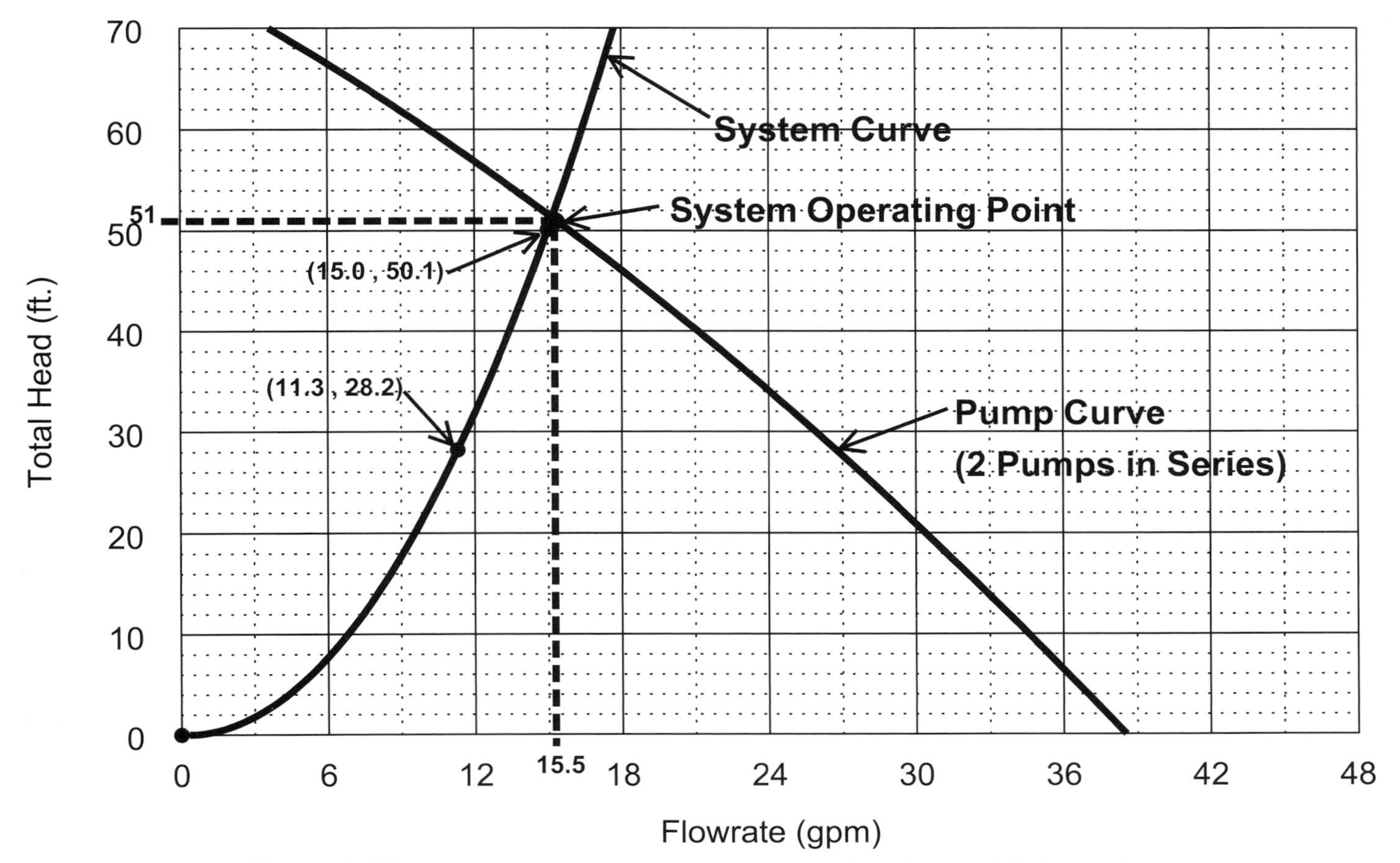

Figure A.24. System Curve & Operating Point for Grand Forks, ND Example

The pump curve shown in Figure A.24 is for two Grundfos UP 26-116F connected in series. The system curve shown in the figure was generated by calculating the head loss for system flow rates of 0 gpm (0 feet of head), 11.3 gpm (28.2 feet of calculated head loss), and 15 gpm (50.1 feet of calculated head loss). To generate the system curve, a second order polynomial curve fit through the three points given [(0 gpm, 0 feet), (11.3 gpm, 28.2 feet) and (15 gpm, 50.1 feet)] was generated via use of software (SlideWrite, Microsoft Excel, etc.).

As is shown in Figure A.24, the system operating point will be somewhere around a system flow of 15.5 gpm at 51 feet of head loss, which is slightly above the optimum flow of 3 gpm per ton of installed capacity.

Example A.1.9 Heating Load Two-Capacity, Water-to-Air GSHP System Piping Volume Calculations – Grand Forks, ND

Once the loopfield and interior piping designs have been completed, total system pipe volume calculations should be performed to determine the amount of antifreeze needed for adequate freeze protection. The assumptions utilized to perform system piping volume calculations are as follows:

- Non-pressurized flow center utilized
- Outside piping brought to the interior via wall penetration shown by Figure 7.2a, heat pump assumed to be located in close proximity to where wall penetrations are made
- GHEX piping assumed to be connected in the step-down, step-up reverse-return header configuration shown in Figure 5.8(d)
- Supply-return pipe run distance from loopfield to the flow center inside the building assumed to be 50 feet
- Supply-return piping assumed to be 1-1/4-inch DR-11 HDPE
- Reinforced rubber hose from flow center-heat pump assumed to be 1-inch diameter, distance assumed to be 10 feet
- Heat pump/flow center/GHEX connection made according to Figure 7.3a
- Circulating fluid necessary for adequate freeze protection is 25% propylene glycol by volume

The Antifreeze Volume Worksheet shown by Figure A.25 summarizes the system piping volume calculations for this example.

Antifreeze Volume Worksheet

Job/Client Name Grand Forks, ND Example **GSHP Technician** John Doe

Loopfield Description 5 vertically installed 200' bores, 3/4" u-bends, 1-1/4" S-R piping

Design EWT_{min} 25˚F **Antifreeze Type** Propylene Glycol **Concentration (By Vol.)** 25%

Piping System Information

Pipe Section Description	Pipe Type & Size	V_i (gal/100')	L_i (ft)
1) Loop Volume	3/4" DR-11 HDPE	3.01	5 x 410 = 2050
2) Exterior GHEX S-R Piping Volume	1-1/4" DR-11 HDPE	7.53	100
3) Interior S-R Piping Volume	N/A	0	0
4) Header Piping Volume	1-1/4" DR-11 HDPE	7.53	15 x 8 = 120
5) Piping Volume from Flow Center to GSHP	1" Rubber Heater Hose (200 psi)	4.08	20

Total Piping System Volume Calculations

Pipe Section Description	(V_i)	x	(L_i / 100)	=	Pipe Volume
1) Loop Volume	3.01	x	20.5	=	61.5
2) Exterior GHEX S-R Piping Volume	7.53	x	1.00	=	7.5
3) Interior S-R Piping Volume		x		=	
4) Header Piping Volume	7.53	x	1.20	=	9.0
5) Piping Volume from Flow Center to GSHP	4.08	x	0.20	=	0.8
			Total Pipe Volume	=	78.8
(Total Pipe Volume) · (Concentration / 100) = Volume of Antifreeze Needed				=	19.7

Figure A.25. Antifreeze Volume Worksheet – Grand Forks, ND Example

Example A.1.10 Heating Load Two-Capacity (5-ton), Water-to-Air GSHP Horizontally-Trenched Loopfield Sizing – Grand Forks, ND

Now, assume that the Grand Forks, ND example residence is located where available land area is not a constraint on the type of GHEX selected. More than likely, a horizontally-trenched system will prove to be more economically feasible in such cases. For the sake of example, assume that the homeowner wanted to utilize a horizontal slinky system. As a starting point for the calculations, the following important design assumptions were made:

- GHEX consists of five horizontally-trenched slinky loops, pipe diameter=3/4-inch, slinky pitch=36-inch, slinky diameter=36-inch, C-C spacing=11 feet, trench depth=8 feet, slinky laying horizontally in bottom of trench (5-pipe laying configuration)
- Minimum entering water temperature from the loopfield, EWT_{min}=30 F
- Mean soil temperature, T_M=42 F (provided in Table 5.11)
- Annual temperature swing, A_s=33 F
- Offset, T_O=35 Days
- Soil thermal conductivity assumed to be k_{soil}=0.75 Btu/hr-ft-F, soil diffusivity assumed to be α_{soil}=0.50 ft^2/day
- System flow rate, GPM=15gpm
- Circulating fluid = 15% methanol by volume

Figure A.26 displays the horizontally-trenched GHEX design worksheet (heating-based design) for this residence. All of the data in the "Heat Pump Design Data" column were taken directly from the Two-Capacity, Water-to-Air Heat Pump Selection Worksheet (Figure A.11). The "Loopfield Layout" column was filled out according to the design assumptions that were made. The run fraction in heating in the "Bin Analysis Data" column was found on Figure 3.10 (heating degree days for Grand Forks, H_{DD}=9,500, Percent-Sizing=100% ➔ F_H=0.64). The "Horizontal Trench Design Data" column was filled out according to the design assumptions made. The normalized minimum soil temperature $T'(d_{avg},T_O)$ was found on Table 5.12 according to the design assumptions made.

Methanol (15% by volume) was chosen as the circulating fluid because it has lower viscosity (lower pumping pressure requirements and lower flow requirements for turbulence). Before selecting a circulating fluid for a GSHP system, ensure that state and local regulations allow for use of that fluid in an underground piping system.

The first two sets of calculations displayed in the "Heating Design Length Summary Table" were performed assuming use of five and six loops, respectively. Similar to Example A.1.6, it was assumed that total system flow would increase from 15 gpm to 17 gpm because the fluid flow would be divided equally into six horizontal loops rather than five. Heat pump performance increase due to the increased system flow was neglected.

The last set of calculations were performed using EWT_{min}=25 F. The effects of heat pump performance were accounted for in this calculation (though they are not shown) and the results of the calculations show a 36.6% reduction in required trench length versus designing around EWT_{min}=30 F. For EWT_{min}=25 F, the selected equipment will only be sized such that it will handle 96.3% of the peak heating load under design conditions, and it will operate with lower efficiency (COP_{Hc}=3.45) to produce the required heating. Remember, utilizing EWT_{min}=25 F is only recommended in extreme situations. Utilizing EWT_{min}=25 F also necessitates the use of higher levels of freeze protection in a system and in some cases, larger circulating pumps. For the example, the need for use of higher levels of freeze protection was neglected. To satisfy the heating load for this residence, six horizontally-trenched slinky loops will need to be installed [Trench Dimensions: 3 feet Wide x 8 feet Deep x 180 feet Long x 11 feet C-C trench spacing, Slinky Details: 36-inch Diameter x 36-inch Pitch x 900 feet coil, laying horizontally in bottom of trench, ¾-inch (nominal) DR-11 HDPE pipe].

Horizontally-Trenched GHEX Design Worksheet – Heating Mode

Space Heating GSHP Design Data

HC_C[1]	= 47,769	Btu/hr
DMD_C[1]	= 3.86	kW
GPM[1]	= 15	gpm
Total Heat Loss[1]	= 46,268	Btu/hr
Percent Sizing	= 103.24	%
F_{Jan}	= .64	

HWG GSHP Design Data

HC_{HWG}	= N/A	Btu/hr
DMD_{HWG}	= N/A	kW
GPM_{HWG}	= N/A	gpm
F_{HWG}	= N/A	

Total Design Data

HC_D	= 46,268	Btu/hr
DMD_D	= 3.86	kW
GPM_D	= 15	gpm

Horizontal GHEX Design Data

COP_D[2] (= HC_D / (DMD_D x 3,412)) = **3.63**

HE_D[2] (= HC_D x (COP_D-1)/COP_D) = **34,609** Btu/hr

F_H (= (F_{Jan} x HC_C + F_{HWG} x HC_{HWG}) / HC_D) = **.64**

EWT_{min}[3] = **30** F

LWT_{min}[2] (= EWT_{min} – HE_D/(500xGPM_D)) = **25.39** F

1. For single heat pump installation use data directly from the appropriate GSHP Selection Worksheet. For multiple heat pumps in the installation sum all heat pump Total Heat Losses, HC_C's, DMD_C's and GPM's that will be connected to a single GHEX using the table provided below.
2. For single heat pump installation use data directly from appropriate GSHP Selection Worksheet. For multiple heat pumps use the equation provided.
3. EWT_{min} is obtained directly from the appropriate GSHP Selection Worksheet and must be the same for the selection of all heat pumps for a multiple heat pump installation connected to a single GHEX.

	Zone 1	Zone 2	Zone 3	Zone 4	Zone 5	Zone 6	Zone 7	Zone 8	Zone 9	Zone 10	Total
Total Heat Loss											
HC_C											
DMD_C											
GPM											

Horizontal Trench Design Data

Location Grand Forks , ND (City, State) **Soil Type** ________ (Section 5.3.2.2)

T_M = **42** F (Table 5.11) k_{soil} = 0.75 Btu/hr ft F

A_s = **33** F (Table 5.11) α_{soil} = 0.50 ft²/day

T_O = **35** Days (Table 5.11)

Trench and Pipe Configuration ==> Table **5.24** (Select from Tables 5.17 – 5.29)

N_P = **5** ft_{pipe}/ft_{trench} S_H = NA ft R_s = 2.74 hr ft F/Btu

Conf. = Laying (Stand / Laying / Rect) S_{cc} = 11 ft S_M = 1.10

d_{bp} = **8.0** ft d_{avg} = 8.0 ft

D_P = 3/4 in (Nom.)

GPM_{FP} = **3** gpm/flowpath

N_{FP} = **3** flowpaths

$N_{FP/T}$ = **1** flowpaths/trench

N_T = **5** trenches

Design Soil Temperature for Heating

$T'(d_{avg},T_O)$ = **-0.1743** F (Table 5.12)

$$T_{S,L} = T'(d_{avg}, T_O) \bullet A_S + T_M = -0.1743 \cdot 33 + 42 = 36.25 \text{ F} \quad \text{(Eq. 5.25)}$$

Pipe Resistance and Pipe Multiplier

R_p = 0.141 hr ft F/Btu (Table 4.6)

P_M = 1.00 (Figure 5.24)

Horizontal Trench Design Length (Equations 5.14, 5.15, and 5.16)

$$L_{H,P} = \frac{HC_D \cdot \left(\frac{COP_D - 1}{COP_D}\right) \cdot \left(R_P + R_S \cdot P_M \cdot S_M \cdot F_H\right)}{T_{S,L} - \left(\frac{EWT_{min} + LWT_{min}}{2}\right)} = \frac{47{,}769 \cdot \left(\frac{3.63-1}{3.63}\right) \cdot (0.141 + 2.74 \cdot 1.00 \cdot 1.10 \cdot 0.64)}{36.25 - \left(\frac{30 + 25.39}{2}\right)} = \frac{47{,}769 \cdot (0.725) \cdot (2.070)}{36.25 - (27.70)} = 8385 \text{ ft of pipe}$$

$$L_{H,P/T} = \frac{L_{H,P}}{N_T} = \frac{8385}{5} = 1677\, ft_{pipe/trench} \qquad L_{H,T} = \frac{L_{H,P/T}}{N_P} = \frac{1677}{5} = 335\, ft_{trench}$$

Heating Design Length Calculations Summary Table

Layout for Flow and Number of Trenches						Pipe and Trench Design Length Calculations								
D_P	GPM_{FP}	N_{FP}	Trench/N_P	$N_{FP/T}$	N_T	d_{avg}	$T'(d_{avg},T_O)$	$T_{S.L}$	R_s	P_M	S_{cc}/S_M	$L_{H.P}$	$L_{H.P/T}$	$L_{H.T}$
3/4	3.0	5	Laying/5	1	5	8	-0.1743	36.25	2.74	1.0	11/1.1	8385	1677	335
3/4	2.8	6	Laying/5	1	6	8	-0.1743	36.25	2.74	1.0	11/1.1	8385	1398	280
3/4	2.8	6	Laying/5	1	6	8	-0.1743	36.25	2.74	1.0	11/1.1	5318	886	177
			/								/			
			/								/			

$N_T = 5$ → (row 1); $EWT_{min} = 25°F$ → (row 3); $N_T = 6$ → (row 2)

Figure A.26. Horizontally-Trenched GHEX Design Worksheet for Grand Forks, ND Example

Example A.1.11 Heating Load Two-Capacity (5-ton), Water-to-Air GSHP Horizontally-Bored Loopfield Sizing – Grand Forks, ND

As in Example A.1.10, assume that the Grand Forks, ND example residence is located on acreage where available land area is not a constraint on the type of GHEX selected. However, the homeowner wants to explore the option of a horizontally-bored system because of the large amount of excavation necessary to install the horizontally-trenched system. As a starting point for the calculations, the following important design assumptions were made:

- GHEX consists of five horizontally-bored U-bends in a 1 x 5 configuration, U-bend diameter=3/4-inch, burial depth = 15 feet
- Minimum entering water temperature from the loopfield, EWT_{min}=30 F
- Mean soil temperature, T_M=42 F (provided in Table 5.11, used because burial depth < 20 feet)
- Annual temperature swing, A_s=33 F (used because burial depth < 20 feet)
- Offset (in the top 10-15 feet), T_O=35 Days (used because burial depth < 20 feet)
- Deep earth temperature, T_G=46 F (provided by the results of a formation thermal conductivity test run in the area – used when average pipe burial depth > 20-feet)
- Soil thermal conductivity assumed to be k_{soil}=0.85 Btu/hr-ft-F, soil diffusivity assumed to be α_{soil}=0.60 ft^2/day
- C-C bore spacing=15 feet
- Bore diameter, D_{bore}=5-inch
- k_{grout}=0.40 Btu/hr-ft-F (20% solids mixture – slurry consisting of 20% bentonite & 80% water by mass)
- System flow rate, GPM=15gpm
- Circulating fluid = 15% methanol by volume

Figure A.27 displays the horizontally-bored GHEX design worksheet (heating based design) for this residence. All of the data in the "Heat Pump Design Data" column were taken directly from the Two-Capacity, Water-to-Air Heat Pump Selection Worksheet (Figure A.11). The "Loopfield Layout" column was filled out according to the design assumptions that were made. The run fraction in heating in the "Bin Analysis Data" column was found on Figure 3.10 (heating degree days for Grand Forks, H_{DD}=9,500, Percent-Sizing=100% ➔F_H=0.64). The "HorizontallyBored Design Data" column was filled out according to the design assumptions made. The normalized minimum soil temperature $T'(d_{avg},T_O)$ was found on Table 5.12 according to the design assumptions made.

Methanol (15% by volume) was chosen as the circulating fluid because it has lower viscosity (lower pumping pressure requirements and lower flow requirements for turbulence). Before selecting a circulating fluid for a GSHP system, ensure that state and local regulations allow for use of that fluid in an underground piping system.

Horizontally-Bored GHEX Design Worksheet – Heating Mode

Space Heating GSHP Design Data

HC_C[1] = 47,769 Btu/hr
DMD_C[1] = 3.86 kW
GPM[1] = 15 gpm
AGL_H = 93,791,906 Btu
AGL_C = 19,977,376 Btu
Total Heat Loss[1] = 46,268 Btu/hr
Percent Sizing = 103.24 %
F_{Jan} = .64

HWG GSHP Design Data

HC_{HWG} = N/A Btu/hr
DMD_{HWG} = N/A kW
GPM_{HWG} = N/A gpm
AGL_{HWG} = N/A Btu
F_{HWG} = N/A

Total Design Data

HC_D = 47,769 Btu/hr
DMD_D = 3.86 kW
GPM_D = 15 gpm
AGL_{HD} = 93,791,906 Btu
AGL_{CD} = 19,977,376 Btu

GHEX Design Data

COP_D[2] $(= HC_D / (DMD_D \times 3{,}412))$ = 3.63
HE_D[2] $(= HC_D \times (COP_D-1)/COP_D)$ = 34,609 Btu/hr
F_H $(= (F_{Jan} \times HC_C + F_{HWG} \times HC_{HWG}) / HC_D)$ = .64
EWT_{min}[3] = 30 F
LWT_{min}[2] $(= EWT_{min} - HE_D/(500 \times GPM_D))$ = 25.39 F

1. For single heat pump installation use data directly from the appropriate GSHP Selection Worksheet. For multiple heat pumps in the installation sum all heat pump Total Heat Losses, HC_C's, DMD_C's and GPM's that will be connected to a single GHEX using the table provided below.
2. For single heat pump installation use data directly from appropriate GSHP Selection Worksheet. For multiple heat pumps use the equation provided.
3. EWT_{min} is obtained directly from the appropriate GSHP Selection Worksheet and must be the same for the selection of all heat pumps for a multiple heat pump installation connected to a single GHEX.

	Zone 1	Zone 2	Zone 3	Zone 4	Zone 5	Zone 6	Zone 7	Zone 8	Zone 9	Zone 10	Total
Total Heat Loss											
HC_C											
DMD_C											
GPM											

Soil Design Data

Location Grand Forks, ND (City, State)
T_M = 42 F (Table 5.11)
A_s = 33 F (Table 5.11)
T_O = 35 Days (Table 5.11)
Soil Type ______ (Section 5.3.2.2)
k_{soil} = 0.85 Btu/hr ft F
α_{soil} = 0.60 ft²/day

Horizontally-Bored GHEX Layout

U-bend D_P = 3/4 in
GPM_{FP} = 3 gpm/flowpath
N_{FP} = 5 flowpaths
N_{BIS} = 1 bores in series
N_B = 5 bores
Layout = 1 x 5 ($N_{Layers} \times N_{Bores/Layer}$)
Bore Spacing = 15 ft

Horizontally-Bored GHEX Design Data

D_{bore} = 5 in
k_{Grout} = 0.40 Btu/hr ft F
$(R_B + R_G \cdot F_H)$ = 0.765 hr ft F/Btu
d_{avg} = 15 ft
$T'(d_{avg},T_O)$ = 0.0371 F (Table 5.12)
$T_{S,L} = T'(d_{avg},T_O) \cdot A_s + T_M =$ 0.0371 • 33 + 42 = 43.22F

Borehole Design Lengths (Equations 5.28 and 5.29)

$$L_{H,T} = \frac{HC_D \cdot \left(\frac{COP_D - 1}{COP_D}\right) \cdot (R_B + R_G \cdot F_H)}{T_{S,L} - \left(\frac{EWT_{min} + LWT_{min}}{2}\right)} = \frac{47{,}769 \cdot \left(\frac{3.63-1}{3.63}\right) \cdot (0.765)}{43.22 - \left(\frac{30+25.39}{2}\right)} = \frac{47{,}769 \cdot (0.725) \cdot (0.765)}{43.22-(27.7)} = 1707 \textit{ ft of bore}$$

$$L_{H,B} = \frac{L_{H,T}}{N_B} = \frac{1707}{5} = 341 \textit{ ft per bore}$$

Unbalanced Ground Load Borehole Design Lengths (Equations 5.9a, 5.10, and 5.11)

$$NNAGL = \frac{AGL_{HD} - AGL_{CD}}{L_{H,T} \cdot \left(T_{S,L} - \left(\frac{EWT_{min} + LWT_{min}}{2}\right)\right)} = \frac{93{,}791{,}906 - 19{,}977{,}376}{1707 \cdot \left(43.22 - \left(\frac{30+27.7}{2}\right)\right)} = 2786 \ Btu/ftF$$

B_M = 1.35 (Figures 5.13 through 5.18)

$$L_{H,T,UGL} = B_M \cdot L_{H,T} = 1.35 \cdot 1707 = 2304 \textit{ ft of bore} \qquad L_{H,B,UGL} = \frac{L_{H,T,UGL}}{N_B} = \frac{2304}{5} = 461 \textit{ ft per bore}$$

Heating Design Length Calculations Summary Table

Layout for Flow and Number of Bores						Design Length Calculations				Unbalanced Ground Load Design Lengths					
D_P	GPM_{FP}	N_{FP}	N_{BIS}	N_B	Layout	k_{Grout}	$R_B+R_G \cdot F_H$	$L_{H,T}$	$L_{H,B}$	Layout	Spacing	NNAGL	B_M	$L_{H,T,UGL}$	$L_{H,B,UGL}$
3/4	3	5	1	5	1x5	0.40	0.765	1707	341	1x5	15	2786	1.35	2304	461
3/4	3	5	1	5	1x5	0.88	0.620	1383	277	1x5	15	3439	1.45	2005	401
3/4	3	5	1	5	1x5	0.88	0.620	1383	277	1x5	20	3439	1.27	1756	351
3/4	3	5	1	5	1x5	0.88	0.620	1173	235	1x5	20	3439	1.27	1490	298
3/4	3	5	1	5	1x5	0.88	0.648	885	177	1x5	20	3556	1.28	1133	227

Side annotations: k_{Grout} = 0.88 (row 2); d_{avg} = 25ft, T_{soil} = 46°F (row 4); Spacing = 20ft. (row 3); EWT_{min} = 25°F (row 5)

Figure A.27. Horizontally-Bored GHEX Design Worksheet for Grand Forks, ND Example

Upon performing the calculations summarized in the "Heating Design Length Calculations Summary Table" on the Horizontally-Bored GHEX Design Worksheet (Figure A.27), the following conclusions can be made:

1) The total amount of bore required to satisfy the system constraints imposed by the original design assumptions was found to be $L_{H,T,UGL}$=2,304 ft (461 ft/bore – five bores).
2) Using a thermally-enhanced grouting material (thermal conductivity, k_{Grout}=0.88 Btu/hr-ft-F) as the backfill material instead of the 20% solids backfill (k_{Grout}=0.40 Btu/hr-ft-F) reduces the amount of bore required by 13.0% ($L_{H,T,UGL}$=2,005 ft, 401 ft/bore – five bores).
3) Increasing the C-C bore spacing from 15 feet to 20 feet further reduces the amount of required bore length by 12.4% ($L_{H,T,UGL}$=1,756 ft, 351 ft/bore – five bores).
4) Increasing the average pipe burial depth from 15 feet to 25 feet (and using a soil design temperature of 46 F rather than $T_{S,L}$=43.22 F) further reduces the amount of required bore length by 15.1% ($L_{H,T,UGL}$=1,490 ft, 298 ft/bore – five bores).
5) In some cases (extreme climates coupled with poor soil), maintaining a minimum entering water temperature, (EWT_{min}) of 30 F will not prove to be economically feasible because of the large amount of bore length necessary to do so. It is in such instances that a designer must weigh costs and benefits (reduced system performance, increased operating cost, reduced bore length requirements) of designing a loopfield that allows the entering water temperature to reach lower temperatures under design conditions. The last calculation listed on the "Heating Design Length Calculations Summary Table" was performed for EWT_{min}=25 F. The effects of heat pump performance were accounted for in this calculation (though they are not shown) and the results of the calculations show a 24.0% reduction in required bore length versus designing around EWT_{min}=30 F (k_{Grout}=0.88 Btu/hr-ft-F, bore spacing=20 ft, 1x5 configuration, average pipe burial depth=25 ft). For EWT_{min}=25 F, the selected equipment will only be sized such that it will handle 96% of the peak heating load under design conditions (HC_C=44,431 Btu/hr), and it will operate with lower efficiency (DMD_C=3.78 kW, COP_{Hc}=3.45) to produce the required heating. Remember, utilizing EWT_{min}=25 F is only recommended in extreme situations. Utilizing EWT_{min}=25 F also necessitates the use of higher levels of freeze protection in a system and in some cases, larger circulating pumps. For the example, the need for higher levels of freeze protection was neglected. Utilizing EWT_{min}=25 F and propylene glycol as the circulating fluid will cause issues attaining turbulent flow in the U-bends. If possible (check state regulations), use of a methanol-water solution as the circulating fluid is recommended because less fluid flow is necessary to attain turbulent flow as compared to propylene glycol-water solutions that offer similar freeze protection levels. The calculations show that 227 feet of active bore length (x5 bores – total active length=1,133 feet) is necessary to satisfy the heating load for the home.

Example A.2.1 Cooling Load Single-Capacity, Water-to-Air GSHP Selection – Phoenix, AZ

For illustration purposes, a 3,500 ft^2 single-story home with a finished basement was selected for analysis (unconditioned, attached two-stall garage). Phoenix, AZ was the geographic location chosen for this example because it has an extreme cooling climate. The Wrightsoft software package Right-Suite Residential was used for all peak load analysis for this example. The software adheres to Manual J principles. Other software packages that utilize Manual J will produce similar results. The blueprint for the home is displayed in Figures A.28 and A.29 with the orientation set as the bottom being the north side of the home:

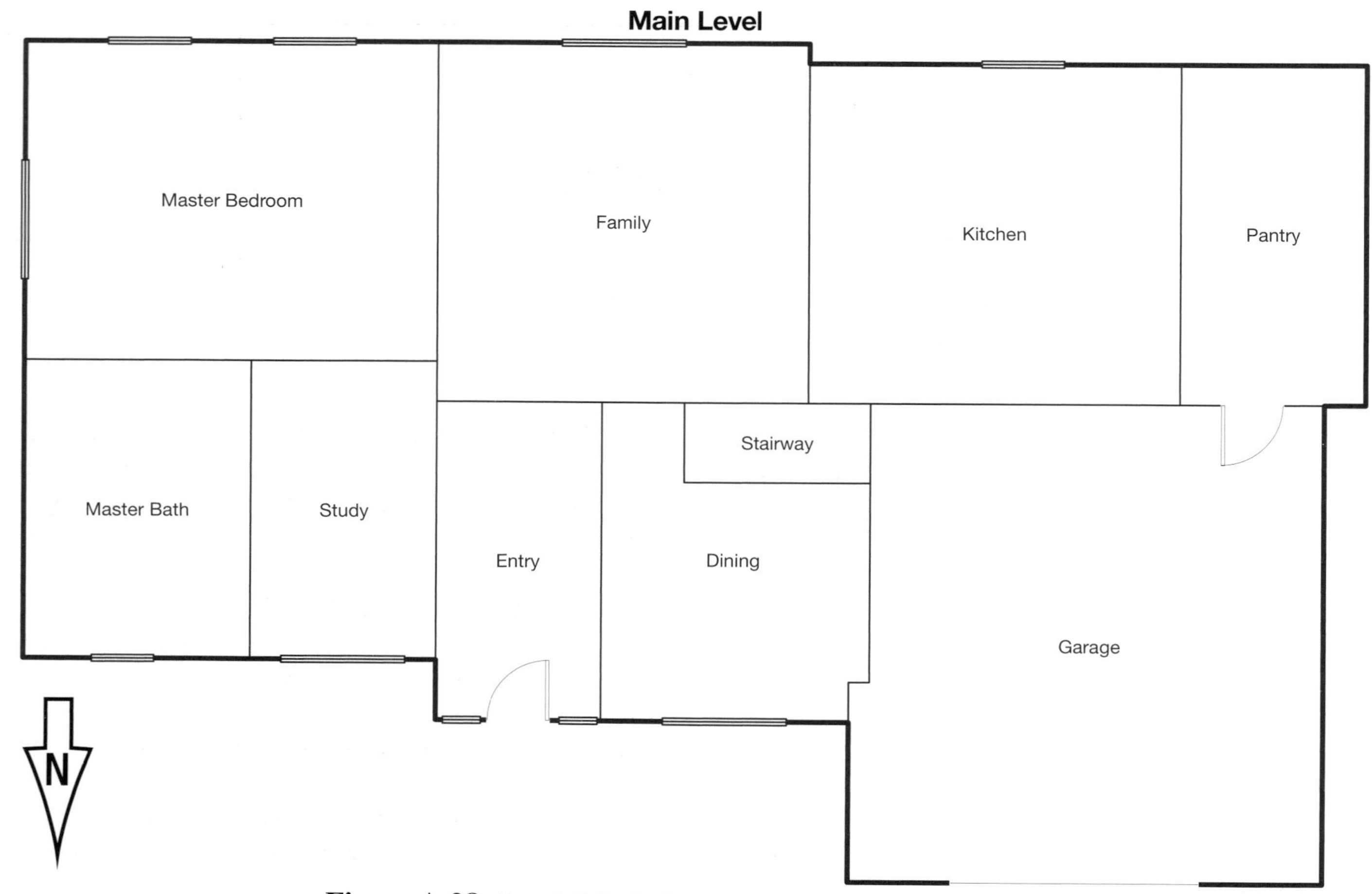

Figure A.28. Load Calculation Example Main Level Blueprint

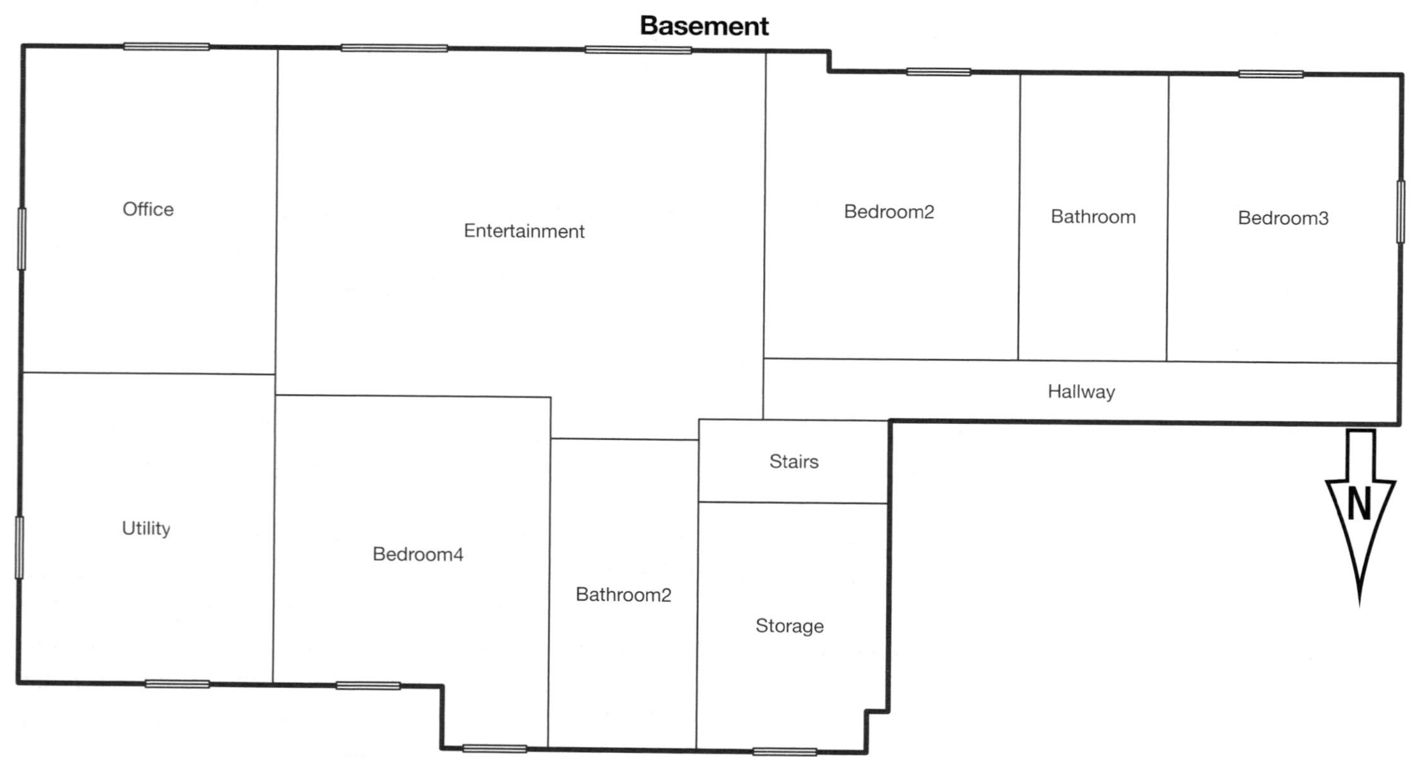

Figure A.29. Load Calculation Example Basement Blueprint

The first step in calculating peak heating and cooling loads is to determine outdoor design conditions according to geographic location as given by Chapter 28 in the *ASHRAE Handbook of Fundamentals, 2005*. The outdoor design conditions for Phoenix, AZ are given by Figure A.30:

Location:		
Phoenix, Luke AFB, AZ, US		
Elevation: 1089 ft		
Latitude: 34 °N		
Outdoor:	**Heating**	**Cooling**
Dry bulb (°F)	38	110
Daily range (°F)	-	25 (M)
Wet bulb (°F)	-	71
Wind speed (mph)	15.0	7.5

Figure A.30. Outdoor Design Conditions for Phoenix, AZ

The outdoor design temperatures listed in Figure A.30 are the 99% annual percentile temperature for heating and 0.4% annual percentile temperature for cooling. Utilizing the 99% annual percentile temperature for heating means that 1% (87 hours based on 8760 hours in a year) of the annual hours are, on average, colder than the heating design temperature. Additionally, utilizing the 0.4% annual percentile temperature for cooling means that 0.4% (35 hours) of the annual hours are, on average, warmer than the design cooling temperature.

The next step in calculating peak loads is to collect information about the construction quality of the structure in question and make assumptions about the level of internal heat gains for appliances and occupants. This particular home was assumed to have been built according to standard construction techniques. The assumptions were as follows:

- Ceilings - Under vented attic space, no radiant barrier, dark asphalt shingles, R-30 blown-in insulation, U=0.032 Btu/hr-ft^2-F
- Basement Floor - Below grade, carpeted, no insulation, width of shortest side 32 feet, U=0.020 Btu/hr-ft^2-F
- Basement Walls – 8-inch concrete, wood stud framing, R-13 cavity insulation, no board insulation, U=0.053 Btu/hr-ft^2-F
- Upstairs Walls - Vinyl siding exterior, wood stud framing, no board insulation, R-19 cavity insulation (6-inch walls), U=0.068 Btu/hr-ft^2-F
- Windows – Wood frame, operable, 2 pane, clear, U=0.57 Btu/hr-ft^2-F
- Doors - Metal, polystyrene core, no storm, U=0.35 Btu/hr-ft^2-F
- 8-foot high ceilings both floors
- Ductwork located in conditioned space, 0% duct losses
- Cooling loads only – four occupants in Family Room, two occupants and two appliances in Kitchen, four occupants in Entertainment Room, one appliance in the Utility Room, and one appliance in the Office (sensible gain=230 Btuh/person, latent gain=200 Btuh/person, sensible gain=1200 Btuh/appliance)
- Unconditioned Garage
- All conditioned spaces assumed to have a thermostat set point of 70 F in heating and 75 F in cooling with 50% relative humidity
- Average construction quality, simplified infiltration method assuming ACH=0.28 heating, ACH=0.15 cooling (ACH = air changes per hour, the volume of air inside the home that is exchanged with outside air via infiltration)

Once the assumptions for construction quality, thermostat set point, and internal gains have been made, the load calculation process can begin. Typical to most smaller residential installations, this home will be served by a single piece of equipment. The home will then be defined as a single zone, resulting in the entire building peak heating and cooling loads being the block loads served by the equipment. In addition, the home in this example will not utilize a ventilation system of any type, but will rely on infiltration to satisfy fresh air requirements. The assumption can then be made that the equipment loads for this example will be the peak block loads. No additional loads are imposed on the equipment by ventilation, duct losses, etc.

Following the procedures set forth by *ACCA Manual J* will result in a Manual J worksheet being filled out similar to Figure A.31:

#	Ty	Construction number	U-value (Btuh/ft²-°F)	Or	HTM (Btuh/ft²) Heat	HTM (Btuh/ft²) Cool	Entire House Area (ft²) or perimeter (ft) Gross	Entire House Area (ft²) or perimeter (ft) N/P/S	Entire House Load (Btuh) Heat	Entire House Load (Btuh) Cool	Master Bedroom Area (ft²) or perimeter (ft) Gross	Master Bedroom Area (ft²) or perimeter (ft) N/P/S	Master Bedroom Load (Btuh) Heat	Master Bedroom Load (Btuh) Cool
1	Room name						Entire House				Master Bedroom			
2	Exposed wall						357.0 ft				36.0 ft			
3	Ceiling height						8.0 ft				8.0 ft			heat/cool
4	Room dimensions								d		20.0 x 16.0 ft			
5	Room area						3510.0 ft²				320.0 ft²			
6	W	15B13-0wc-6	0.093	nn	1.70	2.60	520	484	941	245	0	0	0	0
.	G	1D-c2ow	0.570	nn	18.24	31.41	36	0	657	1131	0	0	0	0
.	W	12E-0sw	0.068	nn	2.18	2.37	336	246	536	583	0	0	0	0
.	G	1D-c2ow	0.570	nn	18.24	31.41	69	0	1252	2157	0	0	0	0
11	D	11N0	0.350	nn	11.20	15.37	21	21	235	323	0	0	0	0
	W	12E-0sw	0.068	ee	2.18	2.37	272	248	540	587	128	104	226	246
	G	1D-c2ow	0.570	ee	18.24	73.54	24	0	438	1765	24	0	438	1765
	W	15B13-0wc-6	0.093	ee	1.70	2.60	272	254	495	130	0	0	0	0
	G	1D-c2ow	0.570	ee	18.24	73.54	18	0	328	1324	0	0	0	0
	W	12E-0sw	0.068	ss	2.18	2.37	520	460	1001	1089	160	136	296	322
	G	1D-c2ow	0.570	ss	18.24	38.28	60	0	1094	2297	24	0	438	919
	W	15B13-0wc-6	0.093	ss	1.70	2.60	520	460	870	182	0	0	0	0
	G	1D-c2ow	0.570	ss	18.24	38.28	60	0	1094	2297	0	0	0	0
	W	15B13-0wc-6	0.093	ww	1.70	2.60	272	263	522	154	0	0	0	0
	G	1D-c2ow	0.570	ww	18.24	73.54	9	0	164	662	0	0	0	0
	WW	12E-0sw	0.068	ww	2.18	2.37	144	144	313	341	0	0	0	0
	P	12C-0sw	0.091	- -	2.91	2.64	344	344	1002	908	0	0	0	0
	CC	16B-30ad	0.032	- -	1.02	2.16	1773	1773	1816	3824	320	320	328	690
	F F	21A-32c	0.020	- -	0.64	0.00	1737	198	1112	0	0	0	0	0
6	c) AED excursion									0				-64
	Envelope loss/gain								14410	19995			1725	3878
12	a) Infiltration								2788	1634			481	282
	b) Room ventilation								0	0			0	0
13	Internal gains:		Occupants @		230		10			2300	0			0
			Appliances @		1200		4			4800	0			0
	Subtotal (lines 6 to 13)								17198	28729			2207	4160
	Less external load								0	0			0	0
	Less transfer								0	0			0	0
	Redistribution								0	0			0	0
14	Subtotal								17198	28729			2207	4160
15	Duct loads						0%	0%	0	0	0%	0%	0	0
	Total room load								17198	28729			2207	4160
	Air required (cfm)								1400	1400			180	203

Figure A.31. Form J Worksheet for Phoenix, AZ Example

This particular worksheet displays the calculations performed for the entire house in addition to individual calculations for the master bedroom. A worksheet will need to be filled out for each room to calculate individual room loads to obtain air flow requirements for duct sizing. Once the load calculations are executed and checked for accuracy, equipment sizing can be performed. For the sake of example, we will assume the homeowner wanted to purchase a Florida Heat Pump Envirosaver ES042 (single capacity) ground source heat pump system to serve the home based on the total cooling load shown on the Form J Worksheet. The air flow volume rates given in all of the worksheets have been adjusted to match the rated capacity of the specified heat pump (ECM blower). The room-by-room load breakdown, in addition to individual air flow requirements, is given by Figure A.32:

ROOM NAME	Area (ft²)	Htg load (Btuh)	Clg load (Btuh)	Htg AVF (cfm)	Clg AVF (cfm)
Master Bedroom	320	2207	4160	180	203
Family	324	1302	3038	106	148
Kitchen	306	1410	4725	115	230
Dining	206	1266	1622	103	79
Master Bath	165	1162	1379	95	67
Bedroom2	171	500	443	41	22
Bedroom3	154	860	1262	70	61
Bathroom	98	199	49	16	2
Entertainment	408	1167	2240	95	109
Office	192	989	2430	80	118
Bedroom4	197	712	624	58	30
Utility	180	916	2256	75	110
Pantry	153	1182	1209	96	59
Entry	128	895	1250	73	61
Study	135	801	1210	65	59
Bathroom2	105	203	49	17	2
Storage	106	614	404	50	20
Stairs	36	101	28	8	1
Stairway	36	130	160	11	8
Hallway	90	583	191	47	9
Entire House d	3510	17198	28729	*1400*	*1400*
Other equip loads		0	0		
Equip. @ 1.00 RSM			28729		
Latent cooling			1678		
TOTALS	3510	17198	30407	*1400*	*1400*

Figure A.32. Room-By-Room Peak Load Breakdown for Phoenix, AZ Example

The equipment sizing loads are also shown in Figure A.33, where they are broken down to show the contribution of each building component:

Heating

Component	Btuh/ft²	Btuh	% of load
Walls	2.1	6219	36.2
Glazing	18.2	5028	29.2
Doors	11.2	235	1.4
Ceilings	1.0	1816	10.6
Floors	0.6	1112	6.5
Infiltration	1.7	2788	16.2
Ducts		0	0.0
Piping		0	0.0
Humidification		0	0.0
Ventilation		0	0.0
Adjustments		0	
Total		**17198**	**100.0**

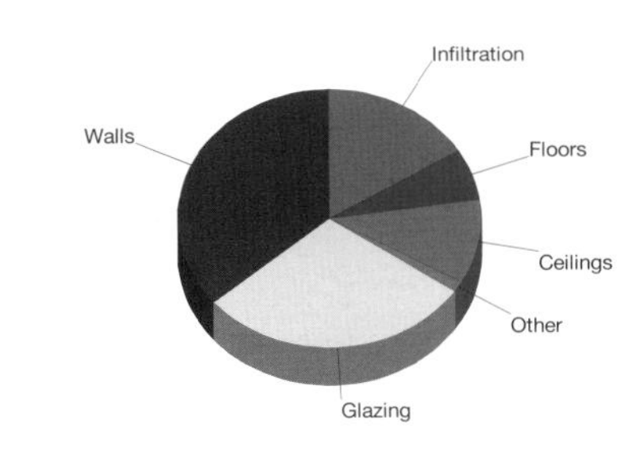

Cooling

Component	Btuh/ft²	Btuh	% of load
Walls	1.5	4218	14.7
Glazing	42.2	11631	40.5
Doors	15.4	323	1.1
Ceilings	2.2	3824	13.3
Floors	0.0	0	0.0
Infiltration	1.0	1634	5.7
Ducts		0	0.0
Ventilation		0	0.0
Internal gains		7100	24.7
Blower		0	0.0
Adjustments		0	
Total		**28729**	**100.0**

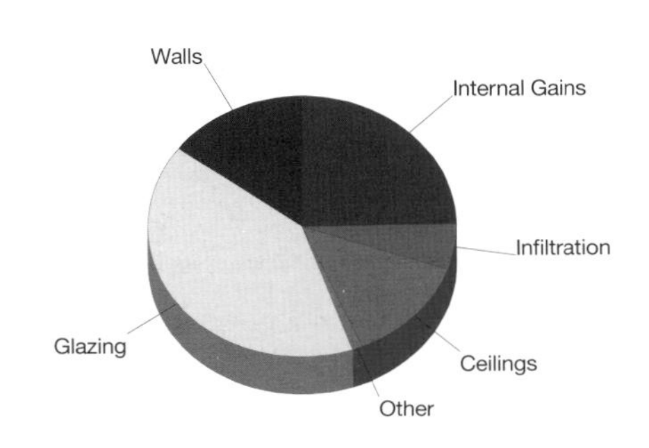

Figure A.33. Peak Equipment Loads for Phoenix, AZ Example

The peak block heating load is 17,198 Btu/hr for this example. The peak block sensible cooling load is 28,729 Btu/hr and the space SHF (sensible heat factor) is 0.94 [Peak Sensible Cooling Load = 28,729 Btu/hr, Peak Total Cooling Load = 30,407 Btu/hr, SHF = (Sensible Cooling Load / Total Cooling Load)]. It can be seen that this particular system will have to be sized according to the peak cooling load. A heat pump selection worksheet would need to be filled out to ensure the heat pump unit will properly serve the home. Figure A.34 displays the performance data, and Figure A.35 displays the selection worksheet for the selected single-capacity Florida Heat Pump Envirosaver ES042 filled out according to the results of the calculations for this example (based on air flow of 1400 CFM in heating and cooling, ECM blower, tap setting=A):

CAPACITY DATA All performance at 1400 CFM and 9.0 GPM

COOLING EFT Range (Standard) 50°F to 100°F

Entering Fluid Temp. (°F)	Entering Air Temp. (°F)	Total Capacity (MBtuH)	Sensible Capacity (MBtuH)	Sensible to Total Ratio	Power Input (kW)	Heat of Reject (MBtuH)	EER
50°	70°db 61°wb	41.43	27.17	0.66	1.90	47.91	21.8
60°		40.23	26.61	0.66	2.20	47.74	18.3
70°		39.03	26.14	0.67	2.50	47.57	15.6
85°		37.23	25.59	0.69	2.95	47.31	12.6
100°		35.43	25.19	0.71	3.40	47.05	10.4
50°	75°db 63°wb	44.37	32.41	0.73	1.91	50.89	23.2
60°		43.08	31.74	0.74	2.21	50.63	19.5
70°		41.79	31.17	0.75	2.52	50.38	16.6
85°		39.87	30.52	0.77	2.97	50.00	13.4
100°		37.94	30.05	0.79	3.42	49.63	11.1
50°	80°db 67°wb	48.66	35.74	0.73	1.93	55.23	25.3
60°		47.25	35.00	0.74	2.23	54.86	21.2
70°		45.84	34.38	0.75	2.54	54.50	18.1
85°		43.73	33.66	0.77	2.99	53.95	14.6
100°		41.61	33.14	0.80	3.45	53.39	12.1
50°	85°db 71°wb	52.95	39.11	0.74	1.94	59.57	27.3
60°		51.41	38.29	0.74	2.25	59.09	22.9
70°		49.88	37.62	0.75	2.56	58.61	19.5
85°		47.58	36.83	0.77	3.02	57.89	15.8
100°		45.28	36.26	0.80	3.48	57.16	13.0

Units are complete packages containing compressor, reversing valve, expansion valve metering device, ECM fan motor and heat exchangers. Also included are safety controls: Overload protection for motors, high and low refrigerant pressure switches and solid state lock-out circuit. Optional UL approved internal electric heater, factory installed with primary thermal overload protection and magnetic contactors (208/230-1-60 only) optional UL approved internal Heat Recovery Package and/or Ground Loop Pump with purge connections available

Performance based on ARI/ISO rated air flow, fluid flow and voltage. For conditions other than rated, consult the FHP EAD selection software. Due to variations in installation actual performance may vary marginally from tabulated values.

As a result of continuing research and development, specifications are subject to change without notice.
ES042IP60 Rev: 10-06

HEATING EFT Range (Standard) 25°F to 80°F

Entering Fluid Temp. (°F)	Entering Air Temp. (°F)	Total Capacity (MBtuH)	Power Input (kW)	Heat of Abs. (MBtuH)	COP
50°	60°	41.02	2.80	31.48	4.3
60°		45.64	2.86	35.88	4.7
70°		50.25	2.92	40.29	5.0
80°		54.87	2.98	44.70	5.4
50°	70°	38.77	2.85	29.04	4.0
60°		43.13	2.91	33.19	4.3
70°		47.49	2.97	37.34	4.7
80°		51.85	3.04	41.49	5.0
50°	80°	36.12	2.92	26.17	3.6
60°		40.18	2.98	30.02	4.0
70°		44.24	3.04	33.86	4.3
80°		48.30	3.11	37.70	4.6

LOW TEMP HEATING Antifreeze Required

Entering Fluid Temp. (°F)	Entering Air Temp. (°F)	Total Capacity (MBtuH)	Power Input (kW)	Heat of Abs. (MBtuH)	COP
25°	60°	28.91	2.65	19.88	3.2
30°		31.17	2.68	22.04	3.4
40°		35.70	2.74	26.35	3.8
25°	70°	27.33	2.70	18.12	3.0
30°		29.46	2.73	20.16	3.2
40°		33.74	2.79	24.22	3.5
25°	80°	25.46	2.76	16.05	2.7
30°		27.45	2.79	17.93	2.9
40°		31.43	2.85	21.70	3.2

FHP MANUFACTURING COMPANY
601 N.W. 65th Court - Fort Lauderdale, FL 33309
Phone: (954) 776-5471 - Fax: (800) 776-5529
http://www.fhp-mfg.com

Figure A.34. Florida Heat Pump Envirosaver ES042 Manufacturer Performance Data

Typically, the design parameters EWT_{min} and EWT_{max} are determined according to whether the residence is located in a heating or cooling dominated climate. For heating dominant climates, values for EWT_{min} between 25 and 40 F are typical. For cooling dominant climates, values for EWT_{max} between 85 and 105 F are typical. Further discussion on determination of these parameters and their effects on heat pump performance, required ground-loop capacity, etc., can be found in Chapter 2.

Single Capacity Water-Air GSHP Selection Worksheet

Block Load Description = Phoenix, AZ Example Zone #

Total Heat Loss =	17,198	Btu/hr @ 38 F OAT
T-Stat Set Point =	70 / 75 / 50	(Htg DB / Clg DB / %-RH)

Total Heat Gain =	30,407	Btu/hr @ 110 F OAT
Sen Heat Gain =	28,729	Btu/hr
Lat Heat Gain =	1,678	Btu/hr
SHF_{space} =	Sen Heat Gain / Total Heat Gain	
SHF_{space} =	0.94	

Brand Florida Heat Pump

Model Envirosaver ES042

Heating Mode Performance Data

Performance Parameters			Correction Factors[A] HC	Correction Factors[A] DMD
EWT_{min}=	50	F	CF_{EAT}=1.00	CF_{EAT}=1.00
GPM=	9	gpm	CF_{CFM}=1.00	CF_{CFM}=1.00
EAT=	70	F	CF_{AF}=1.00	CF_{AF}=1.00
CFM=	1400	cfm		
Antifreeze / %[B]	None			
CF_{EAT} x CF_{CFM} x CF_{AF}==>			CF_{HC}=1.00	CF_{DMD}=1.00

Cooling Mode Performance Data

Performance Parameters			Correction Factors[A] TC	Correction Factors[A] SC	Correction Factors[A] DMD
EWT_{max}=	100	F	CF_{EAT}=1.00	CF_{EAT}=1.00	CF_{EAT}=1.00
GPM=	9	gpm	CF_{CFM}= 1.00	CF_{CFM}= 1.00	CF_{CFM}= 1.00
EAT=	75 / 63	F/F (dB/wB)	CF_{AF}= 1.00	CF_{AF}= 1.00	CF_{AF}= 1.00
CFM=	1400	cfm			
Antifreeze / %[B]	None				
CF_{EAT} x CF_{CFM} x CF_{AF}==>			CF_{TC}=1.00	CF_{SC}=1.00	CF_{DMD}=1.00

Tabled Heating Mode Performance Data

HC =	38,770	Btu/hr
DMD =	2.85	kW
COP =	4.00	
HE =	29,040	Btu/hr (= HC x (COP-1)/COP)[C]

Tabled Cooling Mode Performance Data

TC =	37,940	Btu/hr
SC =	30,050	Btu/hr (= TC x 0.75 approximately)[C]
DMD =	3.42	kW
EER =	11.1	
HR =	49,630	Btu/hr (= TC x (EER+3.412)/EER)[C]

Corrected Heating Mode Performance Data

HC_c=	HC x CF_{HC}=	38,770	Btu/hr
DMD_c=	DMD x CF_{DMD}=	2.85	kW
COP_c=	HC_c/(DMD_c x 3412) =	4.00	
HE_c =	HC_c x (COP_c-1)/COP_c =	29,040	Btu/hr
LWT_{min}=	EWT_{min} - HE_c/(500xGPM) =	23.55	F

Corrected Cooling Mode Performance Data

TC_c=	TC x CF_{TC}=	37,940	Btu/hr
SC_c=	SC x CF_{SC}=	30,050	Btu/hr
DMD_c=	DMD x CF_{DMD}=	3.42	kW
EER_c=	TC_c/(DMD_c x 1000) =	11.1	
HR_c =	TC_c x (EER_c+3.412)/EER_c =	49,630	Btu/hr
LWT_{max}=	EWT_{max} + HR_c/(500xGPM) =	111.03	F

% Sizing in Heating Mode

% Sizing = $\frac{HC_c}{\text{Total Heat Loss}}$ x 100%

= 225.43 %

% Sizing in Cooling Mode

% Oversizing = $\frac{SC_c - \text{Sen Heat Gain}}{\text{Sen Heat Gain}}$ x 100%

= 4.60 % (> 0 & ≤ 25%) x OK

SHF_{unit}= $\frac{SC_c}{TC_c}$

= 0.79 (≤ SHF_{space}) x OK

Latent Clg Cap = TC_c - SC_c

= 7,890 Btu/hr (> Lat Ht Gain) x OK

Blower Performance Summary

Type	Speed	CFM	Static (in)
PSC	L M H		
ECM	On		
	w/ Comp	1400	0.60 Max
	w/ Emer		

Water Coil Performance Summary

EWT_{min}	GPM	WPD (psi)	WPD (ft)	Circulating Fluid
45	9	2.47	5.7	Water

A. Apply correction factors for CFM and EAT as needed depending on format of Manufacturer's Literature. (No correction - CF=1)

B. CF_{AF} = CF_{Used}/CF_{Tested}. CF_{Used} for antifreeze being used. CF_{Tested} for circulating fluid performance was determined with.

C. Use this equation if performance parameter not provided by Manufacturer in the Engineering Specifications.

NOTES: 1) CF_{Used}-->1.00 (No Antifreeze Used Because EWT_{min}>45°F)

2) CF_{Tested}-->Not Supplied By Manufacturer, Assumed use of Water as Circulating Fluid

Figure A.35. Single-Capacity, Water-to-Air GSHP Selection Worksheet for Phoenix, AZ Example

Example A.2.2 Cooling Load Two-Capacity, Water-to-Air GSHP Selection – Phoenix, AZ

As shown by the heat pump selection worksheet, the heat pump is oversized in heating capacity by 125.43%. Because of the amount of over sizing in heating capacity, the heat pump unit will not run for extended periods of time and equipment cycling may reduce system efficiency and life expectancy. A two-capacity unit would be better suited to serve this residence. The manufacturer performance data tables and heat pump selection worksheet for the Florida Heat Pump Envirosaver ES049 (two-capacity heat pump) are shown by Figures A.36, A.37, and A.38, respectively:

CAPACITY DATA - PART LOAD

COOLING All performance at 1000 CFM and 12.0 GPM EFT Range (Standard) 50°F to 100°F

Entering Fluid Temp. (°F)	Entering Air Temp. (°F)	Total Capacity (MBtuH)	Sensible Capacity (MBtuH)	Sensible to Total Ratio	Power Input (kW)	Heat of Reject (MBtuH)	EER
50°	70°db 61°wb	33.66	22.07	0.66	1.32	38.17	25.5
60°		32.44	21.44	0.66	1.49	37.54	21.7
70°		31.21	20.89	0.67	1.67	36.90	18.7
85°		29.37	20.17	0.69	1.92	35.94	15.3
100°		27.53	19.56	0.71	2.18	34.98	12.6
50°	75°db 63°wb	36.05	26.33	0.73	1.33	40.59	27.1
60°		34.74	25.58	0.74	1.50	39.86	23.1
70°		33.42	24.92	0.75	1.68	39.14	19.9
85°		31.45	24.07	0.77	1.94	38.06	16.2
100°		29.48	23.34	0.79	2.20	36.98	13.4
50°	80°db 67°wb	39.54	29.04	0.73	1.34	44.11	29.5
60°		38.10	28.21	0.74	1.51	43.27	25.2
70°		36.66	27.49	0.75	1.69	42.42	21.7
85°		34.50	26.55	0.77	1.95	41.16	17.7
100°		32.34	25.75	0.80	2.21	39.89	14.6
50°	85°db 71°wb	43.03	31.77	0.74	1.35	47.64	31.9
60°		41.46	30.87	0.74	1.53	46.67	27.2
70°		39.90	30.08	0.75	1.70	45.71	23.4
85°		37.55	29.05	0.77	1.97	44.26	19.1
100°		35.20	28.18	0.80	2.23	42.81	15.8

Units are complete packages containing compressor, reversing valve, expansion valve metering device, ECM fan motor and heat exchangers. Also included are safety controls: Overload protection for motors, high and low refrigerant pressure switches and solid state lock-out circuit. Optional UL approved internal electric heater, factory installed with primary thermal overload protection and magnetic contactors (208/230-1-60 only) optional UL approved internal Heat Recovery Package and/or Ground Loop Pump with purge connections available

Performance based on ARI/ISO rated air flow, fluid flow and voltage. For conditions other than rated, consult the FHP EAD selection software. Due to variations in installation actual performance may vary marginally from tabulated values.

As a result of continuing research and development, specifications are subject to change without notice.
ES049.1IP60 Rev: 10-06

HEATING EFT Range (Standard) 25°F to 80°F

Entering Fluid Temp. (°F)	Entering Air Temp. (°F)	Total Capacity (MBtuH)	Power Input (kW)	Heat of Abs. (MBtuH)	COP
50°	60°	34.88	2.02	27.98	5.1
60°		37.93	2.04	30.97	5.4
70°		40.98	2.06	33.95	5.8
80°		44.02	2.08	36.94	6.2
50°	70°	32.96	2.06	25.94	4.7
60°		35.84	2.08	28.75	5.1
70°		38.72	2.10	31.57	5.4
80°		41.60	2.12	34.38	5.8
50°	80°	30.71	2.10	23.53	4.3
60°		33.40	2.12	26.15	4.6
70°		36.08	2.14	28.76	4.9
80°		38.76	2.16	31.38	5.2

LOW TEMP HEATING Antifreeze Required

Entering Fluid Temp. (°F)	Entering Air Temp. (°F)	Total Capacity (MBtuH)	Power Input (kW)	Heat of Abs. (MBtuH)	COP
25°	60°	26.72	1.97	19.99	4.0
30°		28.22	1.98	21.45	4.2
40°		31.20	2.00	24.37	4.6
25°	70°	25.26	2.01	18.40	3.7
30°		26.67	2.02	19.78	3.9
40°		29.49	2.04	22.53	4.2
25°	80°	23.53	2.06	16.52	3.4
30°		24.85	2.07	17.80	3.5
40°		27.48	2.08	20.36	3.9

FHP MANUFACTURING COMPANY
601 N.W. 65th Court - Fort Lauderdale, FL 33309
Phone: (954) 776-5471 - Fax: (800) 776-5529
http://www.fhp-mfg.com

Figure A.36. Florida Heat Pump Envirosaver ES049 Part-Load Performance Data

CAPACITY DATA - FULL LOAD

COOLING All performance at 1600 CFM and 12.0 GPM

EFT Range (Standard) 50°F to 100°F

Entering Fluid Temp. (°F)	Entering Air Temp. (°F)	Total Capacity (MBtuH)	Sensible Capacity (MBtuH)	Sensible to Total Ratio	Power Input (kW)	Heat of Reject (MBtuH)	EER
50°	70°db 61°wb	47.65	31.22	0.66	2.56	56.38	18.6
60°		45.68	30.18	0.66	2.80	55.24	16.3
70°		43.70	29.23	0.67	3.05	54.09	14.3
85°		40.73	27.96	0.69	3.41	52.37	11.9
100°		37.77	26.83	0.71	3.78	50.66	10.0
50°	75°db 63°wb	51.03	37.26	0.73	2.57	59.81	19.8
60°		48.92	36.01	0.74	2.82	58.53	17.4
70°		46.80	34.88	0.75	3.06	57.25	15.3
85°		43.62	33.37	0.77	3.43	55.33	12.7
100°		40.45	32.02	0.79	3.80	53.41	10.6
50°	80°db 67°wb	55.97	41.10	0.73	2.59	64.83	21.6
60°		53.65	39.72	0.74	2.84	63.35	18.9
70°		51.33	38.48	0.75	3.09	61.87	16.6
85°		47.86	36.82	0.77	3.46	59.66	13.8
100°		44.38	35.32	0.80	3.83	57.44	11.6
50°	85°db 71°wb	60.92	44.97	0.74	2.61	69.84	23.3
60°		58.39	43.47	0.74	2.86	68.17	20.4
70°		55.87	42.11	0.75	3.11	66.50	17.9
85°		52.09	40.29	0.77	3.49	63.99	14.9
100°		48.30	38.66	0.80	3.86	61.48	12.5

Units are complete packages containing compressor, reversing valve, expansion valve metering device, ECM fan motor and heat exchangers. Also included are safety controls: Overload protection for motors, high and low refrigerant pressure switches and solid state lock-out circuit. Optional UL approved internal electric heater, factory installed with primary thermal overload protection and magnetic contactors (208/230-1-60 only) optional UL approved internal Heat Recovery Package and/or Ground Loop Pump with purge connections available

Performance based on ARI/ISO rated air flow, fluid flow and voltage. For conditions other than rated, consult the FHP EAD selection software. Due to variations in installation actual performance may vary marginally from tabulated values.

As a result of continuing research and development, specifications are subject to change without notice.
ES049.2IP60 Rev: 10-06

HEATING

EFT Range (Standard) 25°F to 80°F

Entering Fluid Temp. (°F)	Entering Air Temp. (°F)	Total Capacity (MBtuH)	Power Input (kW)	Heat of Abs. (MBtuH)	COP
50°	60°	50.51	3.28	39.32	4.5
60°		56.32	3.42	44.65	4.8
70°		62.12	3.56	49.99	5.1
80°		67.93	3.70	55.32	5.4
50°	70°	47.74	3.34	36.34	4.2
60°		53.22	3.48	41.34	4.5
70°		58.71	3.62	46.35	4.7
80°		64.20	3.76	51.35	5.0
50°	80°	44.48	3.41	32.82	3.8
60°		49.59	3.56	37.44	4.1
70°		54.70	3.70	42.05	4.3
80°		59.81	3.85	46.67	4.6

LOW TEMP HEATING

Antifreeze Required

Entering Fluid Temp. (°F)	Entering Air Temp. (°F)	Total Capacity (MBtuH)	Power Input (kW)	Heat of Abs. (MBtuH)	COP
25°	60°	35.28	2.93	25.28	3.5
30°		38.13	3.00	27.89	3.7
40°		43.82	3.14	33.11	4.1
25°	70°	33.35	2.99	23.16	3.3
30°		36.04	3.06	25.61	3.5
40°		41.42	3.20	30.51	3.8
25°	80°	31.08	3.05	20.67	3.0
30°		33.59	3.12	22.92	3.1
40°		38.60	3.27	27.44	3.5

FHP MANUFACTURING COMPANY
601 N.W. 65th Court - Fort Lauderdale, FL 33309
Phone: (954) 776-5471 - Fax: (800) 776-5529
http://www.fhp-mfg.com

Figure A.37. Florida Heat Pump Envirosaver ES049 Full-Load Performance Data

Notice that this model is still significantly oversized in heating capacity under worst-case conditions (EWT_{min} = 50 F – 191.65% sizing in heating). A 3.5-ton, two-capacity unit would have been better suited to serve this home. However, a 3.5-ton capacity unit is only available as a single-capacity unit for this specific heat pump line, explaining why a 4-ton, dual-capacity unit was selected for this example.

Two Capacity Water-Air GSHP Selection Worksheet

Block Load Description = Phoenix, AZ Example   Zone #

Heating		Cooling	
Total Heat Loss =	17,198 Btu/hr @ 38 F OAT	Total Heat Gain =	30,407 Btu/hr @ 110 F OAT
		Sen Heat Gain =	28,729 Btu/hr
		Lat Heat Gain =	1,678 Btu/hr
T-Stat Set Point =	70 / 75 / 50 (Htg DB / Clg DB / %-RH)	SHF_{space} =	Sen Heat Gain / Total Heat Gain
		SHF_{space} =	0.94

Brand: Florida Heat Pump

Model: Envirosaver ES049

Sized for High Capacity in Heating / (Cooling) [Cooling circled]

(High / (Low)) Capacity Heating Mode Performance Data [Low circled]

Performance Parameters			Correction Factors[A] HC	Correction Factors[A] DMD
EWT_{min}=	50	F	CF_{EAT}= 1.00	CF_{EAT}= 1.00
GPM=	12	gpm	CF_{CFM}= 1.00	CF_{CFM}= 1.00
EAT=	70	F	CF_{AF}= 1.00	CF_{AF}= 1.00
CFM=	1000	cfm		
Antifreeze / %[B]	None			
CF_{EAT} x CF_{CFM} x CF_{AF}==>			CF_{HC}= 1.00	CF_{DMD}= 1.00

Tabled Heating Mode Performance Data

HC =	32,960	Btu/hr
DMD =	2.06	kW
COP =	4.7	
HE =	25,940	Btu/hr (= HC x (COP-1)/COP)[C]

Corrected Heating Mode Performance Data

HC_c=	HC x CF_{HC}=	32,960	Btu/hr
DMD_c=	DMD x CF_{DMD}=	2.06	kW
COP_c=	HC_c/(DMD_c x 3412) =	4.7	
HE_c =	HC_c x (COP_c-1)/COP_c =	25,940	Btu/hr
LWT_{min}=	EWT_{min} - HE_c/(500xGPM) =	45.68	F

% Sizing in Heating Mode

% Sizing = HC_c / Total Heat Loss x 100%

= 191.65 %

((High) / Low) Capacity Cooling Mode Performance Data [High circled]

Performance Parameters			Correction Factors[A] TC	Correction Factors[A] SC	Correction Factors[A] DMD
EWT_{max}=	100	F	CF_{EAT}= 1.00	CF_{EAT}= 1.00	CF_{EAT}= 1.00
GPM=	12	gpm	CF_{CFM}= 1.00	CF_{CFM}= 1.00	CF_{CFM}= 1.00
EAT=	75 / 63	F/F (dB/wB)	CF_{AF}= 1.00	CF_{AF}= 1.00	CF_{AF}= 1.00
CFM=	1600	cfm			
Antifreeze / %[B]	None				
CF_{EAT} x CF_{CFM} x CF_{AF}==>			CF_{TC}= 1.00	CF_{SC}= 1.00	CF_{DMD}= 1.00

Tabled Cooling Mode Performance Data

TC =	40,450	Btu/hr
SC =	32,020	Btu/hr (= TC x 0.75 approximately)[C]
DMD =	3.80	kW
EER =	10.60	
HR =	53,410	Btu/hr (= TC x (EER+3.412)/EER)[C]

Corrected Cooling Mode Performance Data

TC_c=	TC x CF_{TC}=	40,450	Btu/hr
SC_c=	SC x CF_{SC}=	32,020	Btu/hr
DMD_c=	DMD x CF_{DMD}=	3.80	kW
EER_c=	TC_c/(DMD_c x 1000) =	10.60	
HR_c =	TC_c x (EER_c+3.412)/EER_c =	53,410	Btu/hr
LWT_{max}=	EWT_{max} + HR_c/(500xGPM) =	108.9	F

% Sizing in Cooling Mode

% Oversizing = (SC_c - Sen Heat Gain) / Sen Heat Gain x 100%

= 11.46 % (> 0 & ≤ 25%) x OK

SHF_{unit}= SC_c / TC_c

= 0.79 (≤ SHF_{space}) x OK

Latent Clg Cap = TC_c - SC_c

= 8,430 Btu/hr (> Lat Ht Gain) x OK

Blower Performance Summary

Type	Speed	CFM	Static (in)
ECM	On		
	w/ Low		
	w/ High	1600	0.60 Max
	w/ Emer		

Water Coil Performance Summary

EWT_{min}	GPM	WPD (psi)	WPD (ft)	Circulating Fluid
45	12	4.42	10.2	Water

A. Apply correction factors for CFM and EAT as needed depending on format of Manufacturer's Literature. (No correction - CF=1)

B. CF_{AF} = CF_{Used}/CF_{Tested}. CF_{Used} for antifreeze being used. CF_{Tested} for circulating fluid performance was determined with.

C. Use this equation if performance parameter not provided by Manufacturer in the Engineering Specifications.

NOTES: 1) CF_{Used}-->1.00 (No Antifreeze Used Because EWT_{min}>45°F)

2) CF_{Tested}-->Not Supplied By Manufacturer, Assumed use of Water as Circulating Fluid

Figure A.38. Two-Capacity, Water-to-Air GSHP Selection Worksheet for Phoenix, AZ Example

Example A.2.3 Cooling Load Single-Capacity, Water-to-Water GSHP Selection – Phoenix, AZ

Now, for the sake of example, assume that the customer wanted to explore the possibility of conditioning the home with water-to-water GSHP equipment utilizing in-floor radiant heating panels coupled with fan coils for forced air heating and cooling. For this example, the building was assumed to have the same peak equipment heating and cooling loads. However, in a real situation, the peak loads would need to be recalculated because of the additional equipment loads imposed on the system by hydronic supply-return piping losses to/from the fan coils in addition to piping losses from the in-floor radiant heating panels. Assuming that the customer wanted to utilize the same brand of heat pump, a single-capacity, water-to-water Florida Heat Pump Aquarius unit will need to be selected for analysis. The manufacturer performance data tables and heat pump selection worksheet for the Florida Heat Pump Aquarius WW048 (single-capacity heat pump) are shown by Figures A.39 and A.40, respectively:

UNIT WEIGHT

Unit Weight (lbs) 300
Shipping Weight (lbs) 320

CHILLER PERFORMANCE

Based on 8.1 GPM load and 10.1GPM source fluid flow

Leaving Load Fluid (F)	Entering Source Fluid (F)	Total Capacity (Tons)	Total Capacity (MBtuH)	Power Input (kW)	EER	Heat Rejection (MBtuH)
40°	75°	3.31	39.71	2.65	14.99	48.75
	80°	3.23	38.72	2.81	13.77	48.32
	85°	3.14	37.69	2.98	12.64	47.86
	90°	3.05	36.61	3.16	11.58	47.39
	95°	2.96	35.48	3.35	10.58	46.92
42°	75°	3.42	41.07	2.65	15.51	50.11
	80°	3.34	40.05	2.81	14.24	49.64
	85°	3.25	38.98	2.98	13.07	49.15
	90°	3.16	37.87	3.16	11.97	48.66
	95°	3.06	36.71	3.35	10.94	48.16
44°	75°	3.48	41.76	2.65	15.77	50.79
	80°	3.39	40.72	2.81	14.48	50.31
	85°	3.30	39.64	2.98	13.29	49.81
	90°	3.21	38.51	3.16	12.17	49.30
	95°	3.11	37.33	3.36	11.13	48.78
45°	75°	3.54	42.46	2.65	16.04	51.49
	80°	3.45	41.41	2.81	14.73	51.00
	85°	3.36	40.30	2.98	13.51	50.48
	90°	3.26	39.16	3.16	12.38	49.95
	95°	3.16	37.97	3.36	11.31	49.42
46°	75°	3.66	43.89	2.64	16.60	52.91
	80°	3.57	42.80	2.81	15.23	52.39
	85°	3.47	41.66	2.98	13.97	51.84
	90°	3.37	40.48	3.16	12.79	51.28
	95°	3.27	39.26	3.36	11.69	50.71
48°	75°	3.72	44.62	2.64	16.88	53.63
	80°	3.63	43.51	2.81	15.49	53.10
	85°	3.53	42.36	2.98	14.20	52.53
	90°	3.43	41.16	3.16	13.01	51.96
	95°	3.33	39.91	3.36	11.88	51.37
50°	75°	3.84	46.10	2.64	17.46	55.10
	80°	3.75	44.96	2.81	16.01	54.54
	85°	3.65	43.77	2.98	14.68	53.94
	90°	3.54	42.53	3.16	13.44	53.33
	95°	3.44	41.25	3.36	12.28	52.71

HEATING PERFORMANCE

Based on 8.1 GPM load and 10.1 GPM source fluid flow

Leaving Load Fluid (F)	Entering Source Fluid (F)	Heating Capacity (MBtuH)	Power Input (kW)	COP	Heat of Absorb. (MBtuH)
100°	35°	40.12	3.19	3.69	29.24
	40°	42.57	3.17	3.93	31.74
	50°	47.91	3.14	4.48	37.20
	60°	53.84	3.09	5.10	43.29
	70°	60.37	3.03	5.84	50.03
110°	35°	39.76	3.60	3.24	27.48
	40°	42.11	3.58	3.45	29.89
	50°	47.22	3.54	3.91	35.14
	60°	52.91	3.49	4.44	40.99
	70°	59.21	3.44	5.05	47.48
120°	35°	39.51	4.08	2.84	25.58
	40°	41.72	4.06	3.01	27.88
	50°	46.57	4.01	3.41	32.90
	60°	52.00	3.96	3.85	38.50
	70°	58.03	3.90	4.36	44.73
125°	35°	39.42	4.66	2.48	23.52
	40°	41.49	4.62	2.63	25.71
	50°	46.05	4.56	2.96	30.49
	60°	51.18	4.50	3.33	35.83
	70°	56.90	4.43	3.76	41.78

Units are complete packages featuring 1 stage operation and containing refrigeration compressor, reversing valve, expansion valve metering device and water to refrigerant heat exchangers. Also included are safety controls: Overload protection for compressor, high and low refrigerant pressure switches and a lock-out control circuit.

FHP MANUFACTURING
601 N.W. 65th Court
Fort Lauderdale, FL 33309
Phone: (954) 776-5471 Fax: (800) 776-5529
http://www.fhp-mfg.com

As a result of continuing research and development, specifications are subject to change without notice. WW048IP6.P65 Rev:10-04

Figure A.39. Florida Heat Pump Aquarius WW048 Performance Data

Single Capacity Water-Water GSHP Selection Worksheet

Block Load Description = Phoenix, AZ Example | Zone #

Heating			Cooling		
Total Htg Load =	17,198	Btu/hr @ 38 F OAT	Total Clg Load =	30,407	Btu/hr @ 110 F OAT
Htg Load Design Water Temp =	90	F (HLDWT)	Clg Load Design Water Temp =	50	F (CLDWT)
Htg Load Water Flow Rate =	8.1	gpm	Clg Load Water Flow Rate =	8.1	gpm

T-Stat Set Point = 70/75/50 (Htg DB / Clg DB / %-RH)

Brand: Florida Heat Pump

Model: Aquarius WW048

Heating Mode Performance Data

Performance Parameters			Correction Factors[A] HC	Correction Factors[A] DMD
EWT_{min}=	50	F		
GPM=	10.1	gpm		
GPM_{load}=	8.1	gpm		
ELT_{max}=	90	F		
Antifreeze / %[B]	None		CF_{AF}=1.00	CF_{AF}=1.00

Cooling Mode Performance Data

Performance Parameters			Correction Factors[A] TC	Correction Factors[A] DMD
EWT_{max}=	95	F		
GPM=	10.1	gpm		
GPM_{load}=	8.1	gpm		
ELT_{min}=	50	F		
Antifreeze / %[B]	None		CF_{AF}=1.00	CF_{AF}=1.00

Tabled Heating Mode Performance Data

HC =	47,910	Btu/hr
DMD =	3.14	kW
COP =	4.48	
HE =	37,200	Btu/hr (= HC x (COP-1)/COP)[C]
LLT_H =	101.83	F (= ELT_{max} + HC / (GPM_{load} x 500)) (≥ HLDWT)

Tabled Cooling Mode Performance Data

TC =	35,480	Btu/hr
DMD =	3.35	kW
EER =	10.58	
HR =	46,920	Btu/hr (= TC x (EER+3.412)/EER)[C]
LLT_C =	41.24	F (= ELT_{min} - TC / (GPM_{load} x 500)) (≤ CLDWT)

Corrected Heating Mode Performance Data

HC_c=	HC x CF_{AF} =	47,910	Btu/hr
DMD_c=	DMD x C_{AF} =	3.14	kW
COP_c=	HC_c/(DMD_c x 3412) =	4.48	
HE_c =	HC_c x (COP_c-1)/COP_c =	37,200	Btu/hr
LWT_{min}=	EWT_{min} - HE_c/(500xGPM) =	42.63	F

Corrected Cooling Mode Performance Data

TC_c=	TC x CF_{AF} =	35,480	Btu/hr
DMD_c=	DMD x CF_{AF} =	3.35	kW
EER_c=	HC_c/(DMD_c x 1000) =	10.58	
HR_c =	HC_c x (EER_c+3.412)/EER_c =	46,920	Btu/hr
LWT_{max}=	EWT_{max} + HR_c/(500xGPM) =	104.29	F

% Sizing in Heating Mode

$$\% \text{ Sizing} = \frac{HC_c}{\text{Total Heat Loss}} \times 100\%$$

% Sizing = 278.58 %

% Sizing in Cooling Mode

$$\% \text{ Oversizing} = \frac{TC_c - \text{Total Clg Ld}}{\text{Total Clg Ld}} \times 100\%$$

= 16.68 %

Tabled Water Coil Performance Summary

Side	Mode	EWT_{min}	GPM	WPD (psi)	WPD (ft)	Circulating Fluid
Loop	Htg	55	10	3.4	7.9	Water
Load	Clg	55	8	2.3	5.3	Water

A. Apply correction factors for Antifreeze (AF) as needed depending on Manufacturer's Literature. (No correction - CF=1)

B. CF_{AF} = CF_{Used}/CF_{Tested}. CF_{Used} for antifreeze being used. CF_{Tested} for circulating fluid performance was determined with.

C. Use this equation if performance parameter not provided by Manufacturer in the Engineering Specifications.

NOTES: 1) CF_{Used}-->1.00 (No Antifreeze Used Because EWT_{min}>45˚F)

2) CF_{Tested}-->Not Supplied By Manufacturer, Assumed use of Water as Circulating Fluid

Figure A.40. Single-Capacity, Water-to-Water GSHP Selection Worksheet for Phoenix, AZ Example

Notice that the single-capacity, water-to-water GSHP was selected for its capacity at an entering load temperature of 90 F in heating and 50 F in cooling. In heating, the radiant piping layout in addition to fan coil sizing should be based on a 100 F entering water temperature from the heat pump (LLT_H on the Single-Capacity, Water-to-Water Heat Pump Selection Worksheet). Additionally, sizing of the fan coil by way of its sensible and latent cooling capacity should be based on an entering water temperature of 40 F from the heat pump (LLT_C on the Single-Capacity, Water-to-Water Heat Pump Selection Worksheet).

As shown by the heat pump selection worksheet, this heat pump would be well suited to serve the home in cooling. However, because of the significant over sizing in heating capacity (278.58% sizing under worst case conditions), reduced system efficiency may be a result of equipment short cycling. Use of either a two-capacity, water-to-water heat pump or use of a hydronic storage tank would be recommended to alleviate this problem. Typically, hydronic storage tanks are sized to provide 6-10 gallons of storage capacity per nominal ton of installed heat pump capacity.

Example A.2.4 Cooling Load Single-Capacity, Water-to-Air GSHP Duct Sizing – Phoenix, AZ

Once a decision has been made as to the exact model of heat pump to be used, ductwork sizing is the next step in the design process. Because the air flow requirements for each room have already been determined for the 4-ton, single-capacity heat pump (shown in the Figure A.32), those are the values to be used for ductwork sizing in this example. As stated in Example A.2.1, the heat pump has been sized according to its ratings at an air flow rate of 1400 CFM in heating and cooling. According to the manufacturer's fan performance data, this particular model can produce 1400 CFM with a maximum of 0.60 IWC of available external static pressure. According to guidelines, ductwork should be sized to cause pressure drop at the design air flow rate somewhere in the range of 0.10-0.35 IWC. For this example, the ductwork will be sized such that the longest circulation path causes a pressure drop of approximately 0.30 IWC at the given air flow rate. Remember to ensure that maximum air velocity guidelines are followed to prevent excessive system noise during operation. The calculated air flow requirements for both heating and cooling are given in Figure A.32. Ductwork should be sized according to pressure drop and maximum flow velocity guidelines for the larger of the two given air flow requirements for each room. For example, because of high heat loss in the dining room, the air flow volume requirement for heating is larger than the requirement for cooling (103 CFM versus 79 CFM, respectively). Conversely, the air flow volume requirement for cooling in the master bedroom is the larger than the value for heating (203 CFM versus 180 CFM, respectively). The duct run to the dining room needs to be sized to handle 103 CFM in heating whereas the duct run to the master bedroom needs to be sized to handle 203 CFM in cooling.

Before the sizing process can begin, details of the construction of the duct system will need to be determined. For this example, standard sheet metal ductwork was used. Specifically, standard rectangular ductwork sizes were used for the supply and return trunks and standard round ductwork sizes were used for the branches. Typical to this type of installation, supply registers are assumed to be located in the floors on the main level and in the ceilings in the basement. Return grilles are assumed to be located in the floors on both levels. On the main level, because supply registers and return grilles are all located at floor level, they should be placed on opposite sides of the room to avoid short-circuiting. Ductwork design practices and guidelines are given in complete detail in *ACCA Manual D*.

Using the Wrightsoft software package, Right-Suite Residential, perform the duct design according to the individual room air flow requirements and the external static pressure available with the selected piece of equipment yields ductwork sizing and layout as given in Figures A.41 and A.42. Right-Suite Residential utilizes *ACCA Manual D* to perform duct sizing calculations. Programs utilizing the same guidelines will produce similar results.

DUCTWORK LAYOUT LEGEND:

- Supply Register Serving Main Level (Main Level View)
- Return Grille Serving Main Level (Main Level View)
- Supply Register Serving Main Level (Basement View)
- Return Grille Serving Main Level (Basement View)
- Supply Register Serving Basement (Basement View)
- Return Grille Serving Basement (Basement View)

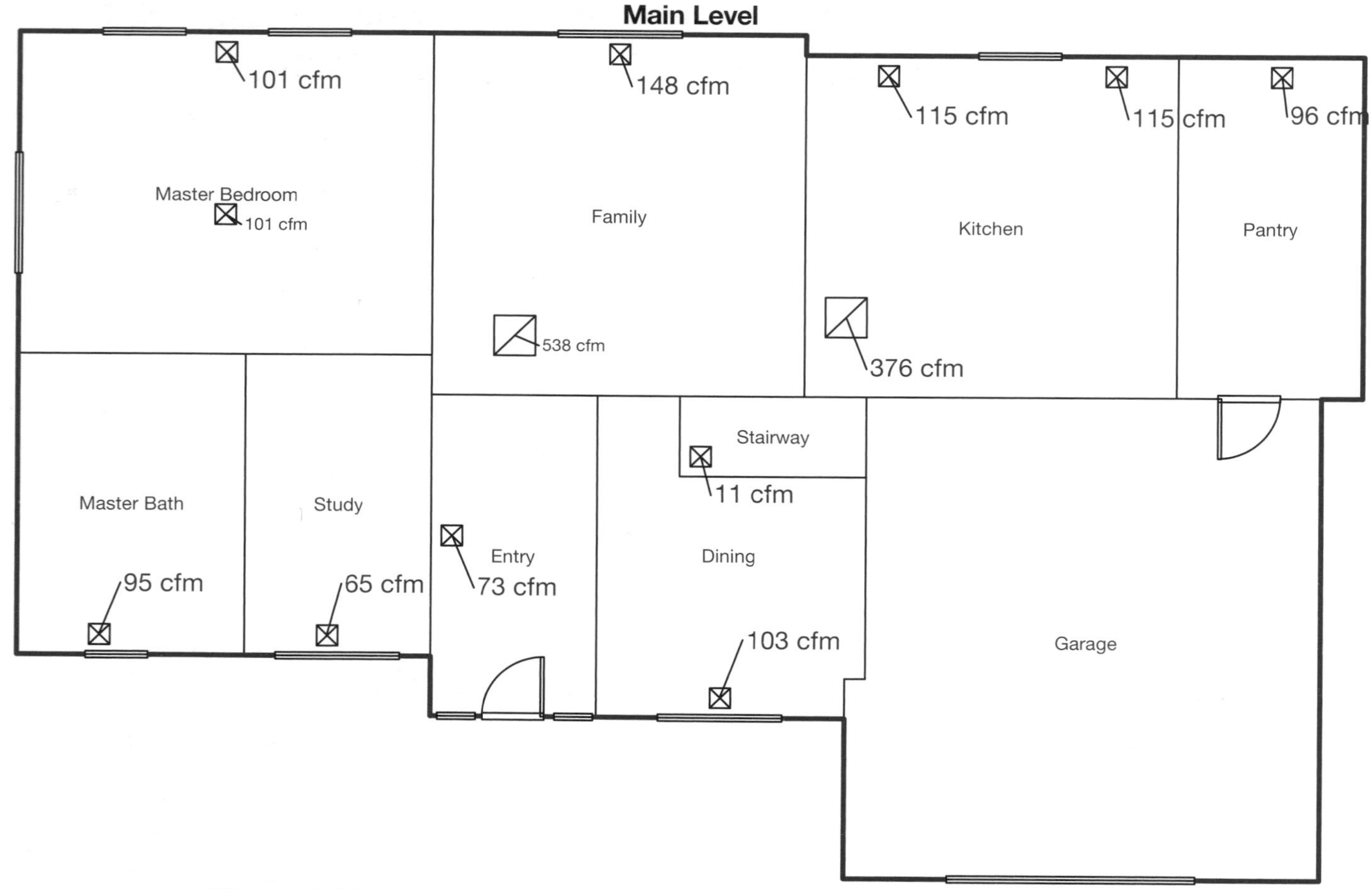

Figure A.41. Duct Sizing/Air flow Details for Main Level – Phoenix, AZ Example

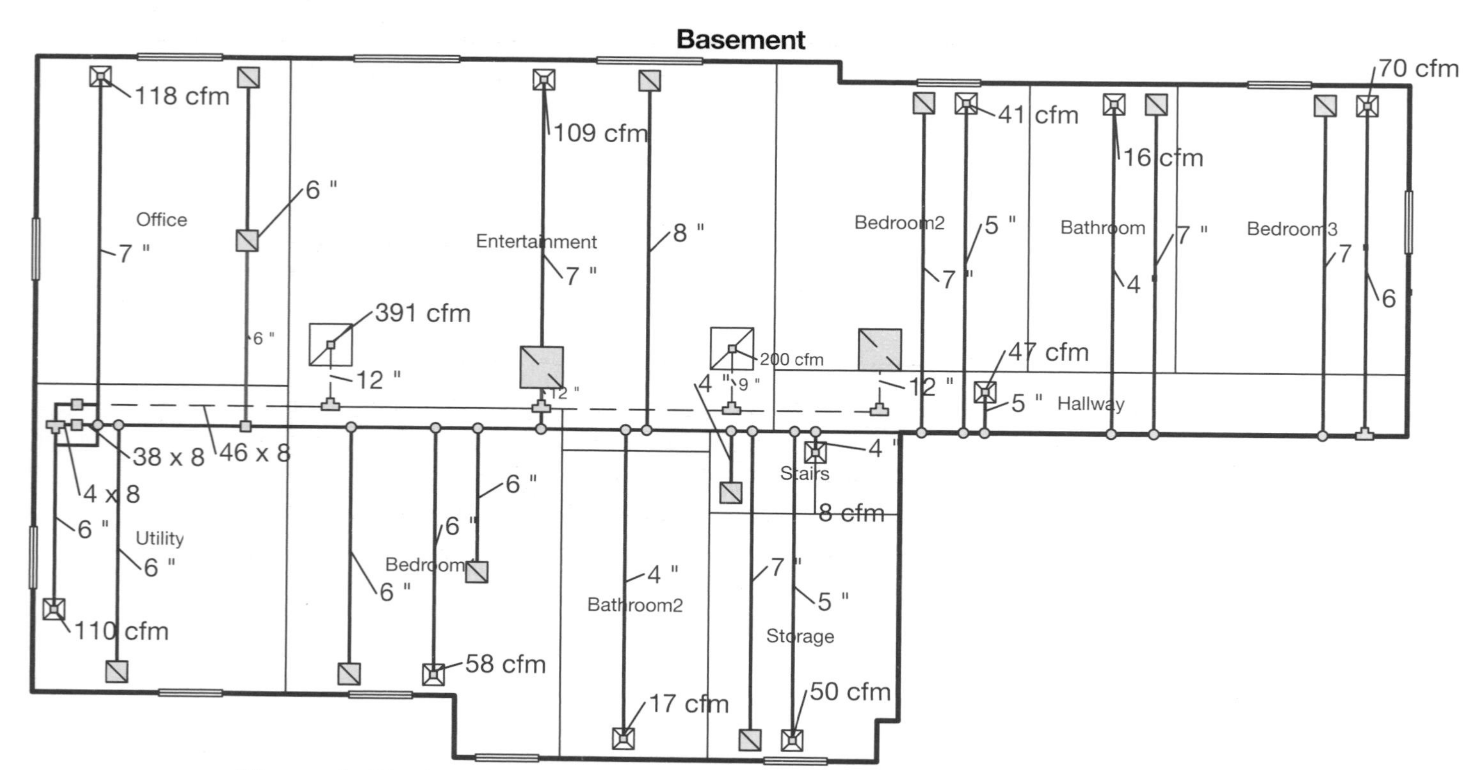

Figure A.42. Duct Sizing/Air flow Details for Basement – Phoenix, AZ Example

Example A.2.5 Cooling Load Single-Capacity, Water-to-Air GSHP Energy Analysis via Simplified Bin Method – Phoenix, AZ

After the heat pump sizing and selection process is complete, a customer may request information as to what the approximate payback period will be for the heat pump system when compared to another type of heating/cooling system. The simplified bin method as previously described will yield reasonable approximations for annual operating cost for any type of system for the geographic location of the residence.

For this example, the simplified bin method will be performed for the residence utilizing the Florida Heat Pump Envirosaver ES042 (single capacity) ground source heat pump system. Additionally, the analysis will be performed to calculate annual system operating cost for a typical year in Phoenix. Only a single, simplified load profile is needed for analysis because internal gains do not fluctuate much when compared to a commercial application. The weather bin data of interest is then the annual number of hours spent in each temperature bin, not broken down according to time of day. The assumptions utilized to perform the bin analysis are summarized by Tables A.4-A.5:

Heating			Cooling		
Q_{PH}	17,198	Btu/hr	Q_{PC}	30,407	Btu/hr
T-Stat Set Pt.	70	°F	T-Stat Set Pt.	75	°F
t_{dh}	38	°F	t_{dc}	110	°F
t_{bh}	62	°F	t_{bc}	67	°F
r_e	0.049	$/kWh	r_e	0.049	$/kWh
EWT_{min}	50	°F	EWT_{max}	100	°F
EWT_{mean}	76	°F	EWT_{mean}	76	°F
t_{mean}	71.8	°F	t_{mean}	71.8	°F
D_c	0.15		D_c	0.15	
Space SHF	NA		Space SHF	0.94	

Table A.4. Bin Method Assumptions for Phoenix, AZ Example

	EWT (°F)	Total CAP (Btu/hr)	Sensible CAP (Btu/hr)	Demand (kW)
Htg.	30	29,460	NA	2.73
	50	38,770	NA	2.85
Clg.	85	39,870	30,520	2.97
	100	37,940	30,050	3.42

Table A.5. Heat Pump Performance Assumptions for Bin Method (Phoenix Example)

In being consistent with the definitions given in Section 3.4, EWT_{lth}=30 F, EWT_{hth}=50 F, EWT_{ltc}=85 F, EWT_{htc}=100 F, etc. The results of the calculations performed via the bin method are summarized by Table A.6:

Table A.6. Bin Method Calculation Summary for Phoenix, AZ Example

Phoenix, AZ																			
Temp Bin (˚F)	t_{air} (˚F)	h_b (hrs)	Q_B (Btu/hr)	EWT (˚F)	HC/SC (Btu/hr)	TC (Cooling Only) Btu/hr	RF	PLF	RF_A	D_{heat} (kW)	D_{cool} (kW)	D_{res} (kW)	D_{htotal} (kW)	RT_A (hrs)	E (kWh)	C_e ($)	gl_h (Btu)	gl_c (Btu)
115/119	117	4	33,236	104.4	29,912	37,374	1.00	1.00	1.00	0.00	3.55	0.00	0.00	4.0	14	$0.70	0	-197,973
110/114	112	46	29,912	101.3	30,011	37,778	1.00	1.00	1.00	0.00	3.46	0.00	0.00	45.9	159	$7.77	0	-2,273,001
105/109	107	183	26,588	98.1	30,109	38,183	0.88	0.98	0.90	0.00	3.36	0.00	0.00	164.5	553	$27.11	0	-8,024,936
100/104	102	370	23,265	95.0	30,207	38,587	0.77	0.97	0.80	0.00	3.27	0.00	0.00	295.1	965	$47.28	0	-14,174,408
95/99	97	504	19,941	91.8	30,306	38,991	0.66	0.95	0.69	0.00	3.17	0.00	0.00	349.6	1110	$54.38	0	-16,523,235
90/94	92	618	16,618	88.7	30,404	39,395	0.55	0.93	0.59	0.00	3.08	0.00	0.00	362.4	1117	$54.71	0	-16,857,115
85/89	87	714	13,294	85.5	30,503	39,799	0.44	0.92	0.48	0.00	2.99	0.00	0.00	340.0	1015	$49.75	0	-15,555,996
80/84	82	751	9,971	82.4	30,601	40,203	0.33	0.90	0.36	0.00	2.89	0.00	0.00	272.2	787	$38.58	0	-12,252,339
75/79	77	722	6,647	79.3	30,700	40,608	0.22	0.88	0.25	0.00	2.80	0.00	0.00	177.1	496	$24.29	0	-7,840,549
70/74	72	730	3,324	76.1	30,798	41,012	0.11	0.87	0.12	0.00	2.70	0.00	0.00	90.9	246	$12.05	0	-3,957,555
65/69	67	748	0	73.0	30,896	41,416	0.00	0.85	0.00	0.00	2.61	0.00	0.00	0.0	0	$0.00	0	0
60/64	62	779	0	68.5	47,364	47,364	0.00	0.85	0.00	2.96	0.00	0.00	2.96	0.0	0	$0.00	0	0
55/59	57	813	3,583	64.6	45,573	45,573	0.08	0.86	0.09	2.94	0.00	0.00	2.94	74.2	218	$10.68	2,272,247	0
50/54	52	750	7,166	60.8	43,783	43,783	0.16	0.87	0.19	2.91	0.00	0.00	2.91	140.4	409	$20.05	4,153,667	0
45/49	47	563	10,749	56.9	41,993	41,993	0.26	0.89	0.29	2.89	0.00	0.00	2.89	162.2	469	$22.98	4,629,772	0
40/44	42	318	14,332	53.1	40,202	40,202	0.36	0.90	0.39	2.87	0.00	0.00	2.87	125.5	360	$17.64	3,447,961	0
35/39	37	123	17,915	49.2	38,412	38,412	0.47	0.92	0.51	2.85	0.00	0.00	2.85	62.4	177	$8.69	1,646,570	0
30/34	32	22	21,498	45.4	36,622	36,622	0.59	0.94	0.63	2.82	0.00	0.00	2.82	13.8	39	$1.90	348,583	0
25/29	27	3	25,080	41.5	34,831	34,831	0.72	0.96	0.75	2.80	0.00	0.00	2.80	2.3	6	$0.31	54,610	0
20/24	22	0	28,663	37.7	33,041	33,041	0.87	0.98	0.89	0.00	0.00	0.00	0.00	0.0	0	$0.00	0	0
15/19	17	0	32,246	33.8	31,250	31,250	1.00	1.00	1.00	0.00	0.00	0.00	0.00	0.0	0	$0.00	0	0
												Totals	Heating	581	1,679	$82.25	16,553,409	
													Cooling	2,102	6,461	$316.61		-97,657,108

The following conclusions can be drawn from Table A.6:

- The estimated amount of heat pump full-load run hours in heating mode for this example is 581 hours, found by summing the number of run hours (RT_A) when the outdoor air temperature is lower than the balance point temperature for heating.
- The estimated amount of heat pump full-load run hours in cooling mode for this example is 2,102 hours, found by summing the number of run hours (RT_A) when the outdoor air temperature is higher than the balance point temperature for cooling.
- The estimated amount of supplemental resistance heat run hours for this example is 0 hours, found by summing the number of run hours (RT_A) when the heat pump capacity is less than the building load (driven by outdoor air temperature).
- The total estimated annual operating cost for the Florida Heat Pump Envirosaver ES042 ground source heat pump in heating is $82.25 per year for this example (at the assumed electric rate).
- The total estimated annual operating cost for this system in cooling is $316.61 per year for this example.
- The estimated amount of heat removed from the ground (heating mode) via the heat pump system is 16,553,409 Btu.
- The estimated amount of heat rejected to the ground (cooling mode) via the heat pump system is 97,657,108 Btu.
- Because of the severely unbalanced ground loads, potential exists to reject enough energy to the soil to raise the deep earth temperature in the center of the loopfield over time, which will detrimentally affect system performance (more of a factor for large systems). It is because of this that an accurate ground-loop design is critical.

Example A.2.6 Cooling Load Single-Capacity, Water-to-Air GSHP Vertical Loopfield Sizing – Phoenix, AZ

After the peak building loads have been calculated and GSHP equipment selected for the load, the loopfield design can be performed. As discussed in Chapter 5, the first step in designing the loopfield for any installation is to perform a comprehensive site survey. For the sake of example, this Phoenix, AZ residence was assumed to be located on a typical city lot where the best option is to utilize a vertically-bored GHEX for the ground source loopfield. As a starting point for the calculations, the following important design assumptions were made:

- GHEX consists of three vertically-bored U-bends in a 1 x 3 straight-line configuration, U-bend diameter=3/4-inch
- Maximum entering water temperature from the loopfield, EWT_{max}=100 F
- Deep earth temperature, T_G=76 F (provided by the results of a formation thermal conductivity test run in the area)
- Formation thermal conductivity, k_G=1.15 Btu/hr-ft-F (Determined according to known soil properties for the geographic location and from Table 5.6 – assumed light sand formation – 15% water)
- C-C bore spacing=15 feet
- Bore diameter, D_{bore}=5-inch
- k_{grout}=0.40 Btu/hr-ft-F (20% solids mixture – slurry consisting of 20% bentonite & 80% water by mass)
- System flow rate, GPM=9gpm

Figure A.43 displays the vertically-bored GHEX design worksheet (cooling based design) for this residence. All of the data in the "Heat Pump Design Data" column were taken directly from the Single-Capacity, Water-to-Air Heat Pump Selection Worksheet (Figure A.35) for the Florida Heat Pump Envirosaver ES042 GSHP. The "Borefield Layout" column was filled out according to the design assumptions that were made. The "Bin Analysis Data" column was filled out with data taken directly from the results of the bin analysis performed in Example A.2.5, Table A.6. The run fraction in cooling was found on Figure 3.11 (cooling degree days for Phoenix, C_{DD}=4,200, Percent-Sizing=104.6% (based on sensible capacity) ➔ F_C=0.58). The "Vertical Bore Design Data" column was filled out according to the design assumptions made. The borehole resistance ($R_B+R_G \cdot F_C$) was found on Table 5.8 according to the design assumptions made.

Vertically-Bored GHEX Design Worksheet – Cooling Mode

Space Cooling GSHP Design Data

TC_C[1] = 37,940 Btu/hr
DMD_C[1] = 3.42 kW
GPM[1] = 9.0 gpm
AGL_H = 16,553,409 Btu
AGL_C = 97,657,108 Btu
Total Heat Gain[1] = 30,407 Btu/hr
Percent Sizing = 104.60 %
F_{Jul} = 0.58

HWG GSHP Design Data

GPM_{HWG} = N/A gpm
AGL_{HWG} = N/A Btu

Total Design Data

TC_D = 37,940 Btu/hr
DMD_D = 3.42 kW
GPM_D = 9.0 gpm
AGL_{HD} = 16,553,409 Btu
AGL_{CD} = 97,657,108 Btu

Vertical GHEX Design Data

EER_D[2] (= TC_D / (DMD_D x 1000)) = 11.09
HR_D[2] (= TC_D x (EER_D+3.412)/EER_D) = 49,613 Btu/hr
F_C (= F_{Jul}) = 0.58
EWT_{max}[3] = 100.0 F
LWT_{max}[2] (= EWT_{max} + HR_D/(500xGPM_D)) = 111.0 F

1. For single heat pump installation use data directly from the appropriate GSHP Selection Worksheet. For multiple heat pumps in the installation sum all heat pump Total Heat Losses, HC_C's, DMD_C's and GPM's that will be connected to a single GHEX using the table provided below.
2. For single heat pump installation use data directly from appropriate GSHP Selection Worksheet. For multiple heat pumps use the equation provided.
3. EWT_{max} is obtained directly from the appropriate GSHP Selection Worksheet and must be the same for the selection of all heat pumps for a multiple heat pump installation connected to a single GHEX.

	Zone 1	Zone 2	Zone 3	Zone 4	Zone 5	Zone 6	Zone 7	Zone 8	Zone 9	Zone 10	Total
Total Heat Gain											
TC_C											
DMD_C											
GPM											

Borefield Layout

U-bend D_P = 3/4 in
GPM_{FP} = 3 gpm/flowpath
N_{FP} = 3 flowpaths
N_{BIS} = 1 bores in series
N_B = 3 bores
Layout = 1 x 3 (N_{Rows} x $N_{Bores/Row}$)
Bore Spacing = 15 ft

Vertical Borehole Design Data

T_G = 76 F
k_G = 1.15 Btu/hr ft F (Table 5.6)
D_{bore} = 5.0 in
k_{Grout} = 0.40 Btu/hr ft F
($R_B + R_G \cdot F_C$) = 0.623 hr ft F/Btu (Table 5.8, 5.9 or 5.10)

Total Borehole Design Length (Equations 5.3 and 5.4)

$$L_{C,T} = \frac{TC_D \cdot \left(\frac{EER_D + 3.412}{EER_D}\right) \cdot (R_B + R_G \cdot F_C)}{\left(\frac{EWT_{max} + LWT_{max}}{2}\right) - T_G} = \frac{37{,}940 \cdot \left(\frac{11.09 + 3.412}{11.09}\right) \cdot (0.623)}{\left(\frac{100 + 111}{2}\right) - 76} = \frac{37{,}940 \cdot (1.308) \cdot (0.623)}{(105.5) - 76} = 1048 \text{ ft of bore}$$

$$L_{C,B} = \frac{L_{C,T}}{N_B} = \frac{1048}{3} = 349 \text{ ft per bore}$$

Unbalanced Ground Load Borehole Design Lengths (Equations 5.9b, 5.12, and 5.13)

$$NNAGL = \frac{AGL_{CD} - AGL_{HD}}{L_{C,T} \cdot \left(\left(\frac{EWT_{max} + LWT_{max}}{2}\right) - T_G\right)} = \frac{97{,}657{,}409 - 16{,}553{,}409}{1048 \cdot \left(\left(\frac{100 + 111}{2}\right) - 76\right)} = 2623 \text{ Btu/ftF}$$

B_M = 1.23 (Figures 5.13 through 5.18)

$$L_{C,T,UGL} = B_M \cdot L_{C,T} = 1.23 \cdot 1048 = 1289 \text{ ft of bore}$$

$$L_{C,B,UGL} = \frac{L_{C,T,UGL}}{N_B} = \frac{1289}{3} = 430 \text{ ft per bore}$$

Cooling Design Length Calculations Summary Table

Layout for Flow and Number of Bores						Design Length Calculations				Unbalanced Ground Load Design Lengths					
D_P	GPM_{FP}	N_{FP}	N_{BIS}	N_B	Layout	k_{Grout}	R_B+R_G·F_C	$L_{C,T}$	$L_{C,B}$	Layout	Spacing	NNAGL	B_M	$L_{C,T,UGL}$	$L_{C,B,UGL}$
3/4	3	3	1	3	1x3	0.40	0.623	1048	349	1x3	15	2623	1.23	1289	430
3/4	3	3	1	3	1x3	0.88	0.478	804	268	1x3	15	3419	1.32	1061	354
3/4	3	3	1	3	1x3	0.88	0.478	804	268	1x3	20	3419	1.19	956	319
3/4	2.75	4	1	4	1x4	0.88	0.478	804	201	1x4	20	3419	1.19	956	239

k_{Grout} = 0.88
N_B = 4
Spacing = 20ft.

Figure A.43. Vertically-Bored GHEX Design Worksheet for Phoenix, AZ Example

Upon performing the calculations summarized in the “Cooling Design Length Calculations Summary Table” on the Vertically-Bored GHEX Design Worksheet (Figure A.43), the following conclusions can be made:

1) The total amount of bore required to satisfy the system constraints imposed by the original design assumptions was found to be $L_{C,T,UGL}$=1,289 feet (430 ft/bore – three bores).
2) Using a thermally-enhanced grouting material (thermal conductivity, k_{Grout}=0.88 Btu/hr-ft-F) as the backfill material instead of the 20% solids backfill (k_{Grout}=0.40 Btu/hr-ft-F) reduces the amount of bore required by 17.7% ($L_{C,T,UGL}$=1,061 ft, 354 ft/bore – three bores).
3) Increasing the C-C bore spacing from 15 feet to 20 feet further reduces the amount of required bore length by 9.9% ($L_{C,T,UGL}$=956 ft, 319 ft/bore – three bores).
4) Utilizing ¾-inch diameter U-bends for borehole depths greater than 200-250 feet (depending on flow through each U-bend) will tend to consume excessive pumping power. Dividing the total bore length ($L_{C,T,UGL}$=956 ft) into four bores rather than three reduces individual bore lengths from 319 ft/bore to 239 ft/bore. Keep in mind that dividing the system flow equally into four parallel paths (rather than three) will reduce the system head loss at the same flow. If the same pump were to be used for the 4-loop system as would have been used for the 3-loop system, the end result would be an increase in total system flow. For this example, it was assumed that total system flow would increase from 9 gpm to 11 gpm. Heat pump performance effects on the increased system flow were neglected. The calculations show that 239 feet of active bore length (x4 bores – total active length=956 feet) is necessary to satisfy the cooling load for the home. Actual bore depth will have to be at least 5-6 feet more than the minimum required active bore length required because active bore length is measured from the bottom of the header trench to the bottom of the vertical bore.

For head loss calculation purposes, the GHEX would be assumed to consist of four vertically-bored U-bends each 245 feet in depth (240 feet active depth) in an outside close-header, parallel configuration.

Example A.2.7 Cooling Load Single-Capacity, Water-to-Air GSHP System Head Loss Calculations – Phoenix, AZ

Once the loopfield design has been completed, interior piping layout and system head loss calculations need to be performed. The assumptions utilized to perform system head loss calculations are as follows:

- Pressurized flow center utilized
- Outside piping brought to the interior underneath the floor slab as shown by Figure 7.2b, heat pump assumed to be located in close proximity to where floor slab penetrations are made
- GHEX piping assumed to be connected in an outside, close-header configuration
- Supply-return pipe run distance from loopfield to the pressurized flow center inside the building assumed to be 30 feet
- Supply-return piping assumed to be 1-1/4-inch DR-11 HDPE
- Reinforced rubber hose from flow center-heat pump assumed to be 1-inch diameter
- Heat pump/flow center connections made according to Figure 7.3a
- Circulating fluid assumed to be water at 40 F for head loss calculations. Antifreeze will not be necessary for this cooling-based design because EWT_{min}>45 F

The Circulating Pump Worksheet shown by Figure A.44 summarizes the system head loss calculations for this example.

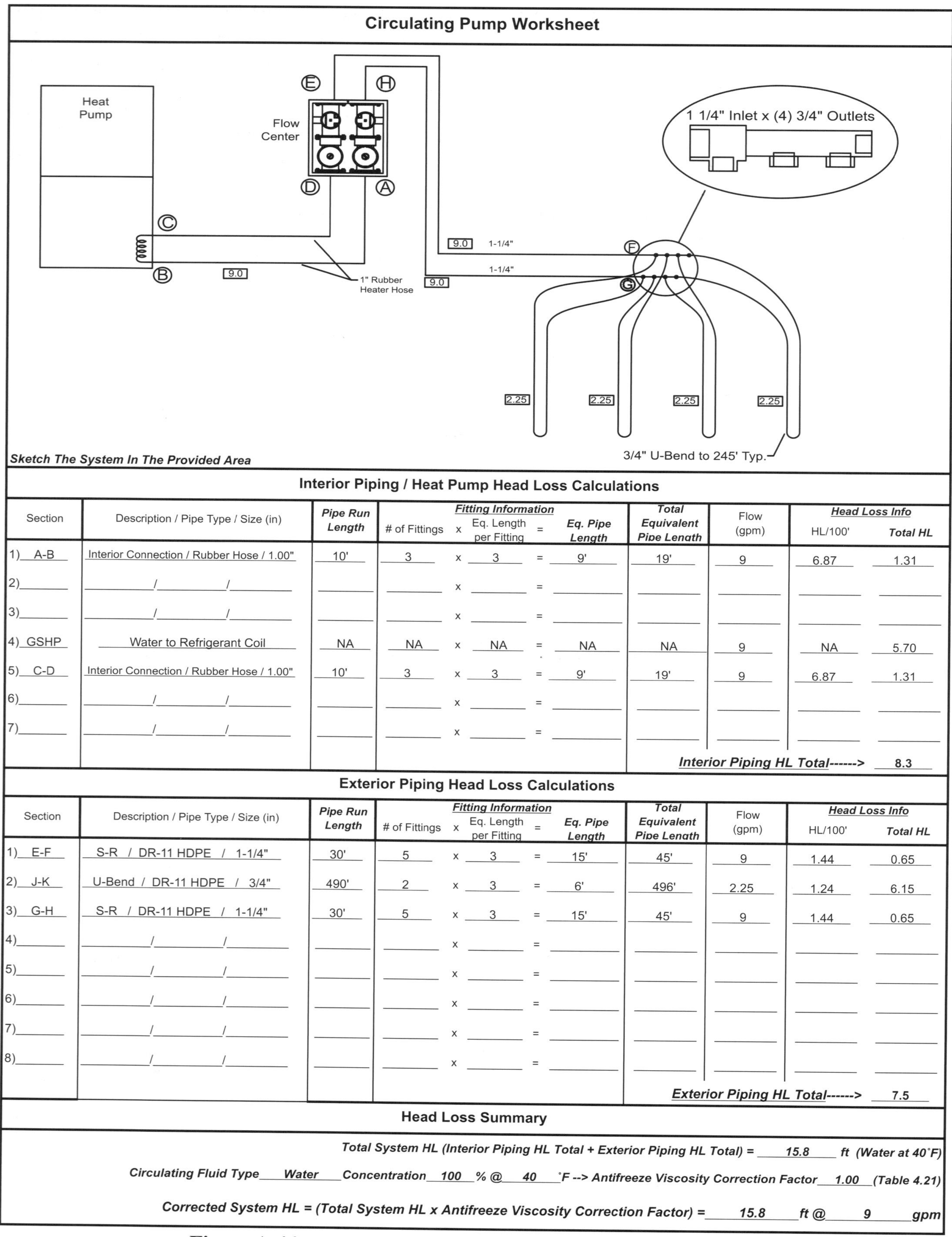

Circulating Pump Worksheet

Sketch The System In The Provided Area

Interior Piping / Heat Pump Head Loss Calculations

Section	Description / Pipe Type / Size (in)	Pipe Run Length	# of Fittings	x	Eq. Length per Fitting	=	Eq. Pipe Length	Total Equivalent Pipe Length	Flow (gpm)	HL/100'	Total HL
1) A-B	Interior Connection / Rubber Hose / 1.00"	10'	3	x	3	=	9'	19'	9	6.87	1.31
2)	/ /			x		=					
3)	/ /			x		=					
4) GSHP	Water to Refrigerant Coil	NA	NA	x	NA	=	NA	NA	9	NA	5.70
5) C-D	Interior Connection / Rubber Hose / 1.00"	10'	3	x	3	=	9'	19'	9	6.87	1.31
6)	/ /			x		=					
7)	/ /			x		=					

Interior Piping HL Total------> 8.3

Exterior Piping Head Loss Calculations

Section	Description / Pipe Type / Size (in)	Pipe Run Length	# of Fittings	x	Eq. Length per Fitting	=	Eq. Pipe Length	Total Equivalent Pipe Length	Flow (gpm)	HL/100'	Total HL
1) E-F	S-R / DR-11 HDPE / 1-1/4"	30'	5	x	3	=	15'	45'	9	1.44	0.65
2) J-K	U-Bend / DR-11 HDPE / 3/4"	490'	2	x	3	=	6'	496'	2.25	1.24	6.15
3) G-H	S-R / DR-11 HDPE / 1-1/4"	30'	5	x	3	=	15'	45'	9	1.44	0.65
4)	/ /			x		=					
5)	/ /			x		=					
6)	/ /			x		=					
7)	/ /			x		=					
8)	/ /			x		=					

Exterior Piping HL Total------> 7.5

Head Loss Summary

Total System HL (Interior Piping HL Total + Exterior Piping HL Total) = 15.8 *ft (Water at 40°F)*

Circulating Fluid Type Water *Concentration* 100 *% @* 40 *°F --> Antifreeze Viscosity Correction Factor* 1.00 *(Table 4.21)*

Corrected System HL = (Total System HL x Antifreeze Viscosity Correction Factor) = 15.8 *ft @* 9 *gpm*

Figure A.44. Circulating Pump Worksheet for Phoenix, AZ Example

As is shown by the head loss calculations in Figure A.44, the circulating pump(s) will have to be large enough to provide 15.8 feet of head at a system flow rate of 9 gpm. A single Armstrong Arm*flo* E-8 circulating pump was selected to provide system flow for this example. The selected circulator is capable of providing approximately 30 feet of head at 9 gpm. Because the circulating pumps will be slightly oversized for a system flow of 9 gpm, the actual operating flow rate will be somewhat higher. The flow in the system will be greatly restricted by the refrigerant-to-water (source side) coil. Head loss is directly proportional to the square of system flow. In other words, doubling flow will increase head loss in the system by four times. Tripling system flow will increase head loss by nine times. Because of the proportional relationship of head loss to flow rate, slightly over sizing pumps will cause slight increases in system flow.

A second circulating pump worksheet (shown by Figure A.45) was filled out at a higher flow rate to generate the system curve displayed in Figure A.46. The system curve can be used to determine the actual system operating flow rate given that the Armstrong Arm*flo* E-8 circulating pump was selected to provide system flow.

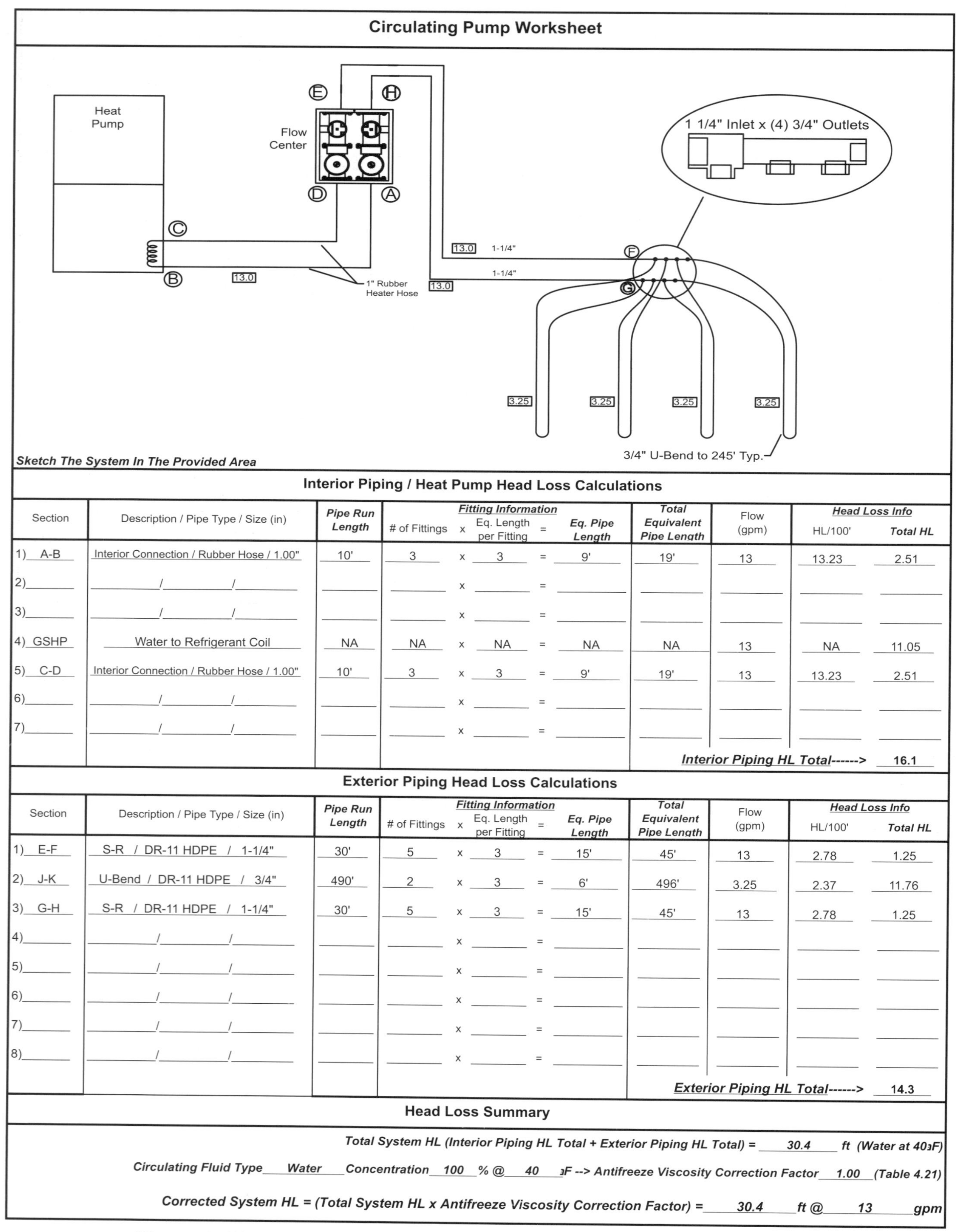

Circulating Pump Worksheet

Interior Piping / Heat Pump Head Loss Calculations

Section	Description / Pipe Type / Size (in)	Pipe Run Length	Fitting Information: # of Fittings	x Eq. Length per Fitting	= Eq. Pipe Length	Total Equivalent Pipe Length	Flow (gpm)	Head Loss Info: HL/100'	Total HL
1) A-B	Interior Connection / Rubber Hose / 1.00"	10'	3	3	9'	19'	13	13.23	2.51
2)	/ /								
3)	/ /								
4) GSHP	Water to Refrigerant Coil	NA	NA	NA	NA	NA	13	NA	11.05
5) C-D	Interior Connection / Rubber Hose / 1.00"	10'	3	3	9'	19'	13	13.23	2.51
6)	/ /								
7)	/ /								

Interior Piping HL Total------> 16.1

Exterior Piping Head Loss Calculations

Section	Description / Pipe Type / Size (in)	Pipe Run Length	Fitting Information: # of Fittings	x Eq. Length per Fitting	= Eq. Pipe Length	Total Equivalent Pipe Length	Flow (gpm)	Head Loss Info: HL/100'	Total HL
1) E-F	S-R / DR-11 HDPE / 1-1/4"	30'	5	3	15'	45'	13	2.78	1.25
2) J-K	U-Bend / DR-11 HDPE / 3/4"	490'	2	3	6'	496'	3.25	2.37	11.76
3) G-H	S-R / DR-11 HDPE / 1-1/4"	30'	5	3	15'	45'	13	2.78	1.25
4)	/ /								
5)	/ /								
6)	/ /								
7)	/ /								
8)	/ /								

Exterior Piping HL Total------> 14.3

Head Loss Summary

Total System HL (Interior Piping HL Total + Exterior Piping HL Total) = 30.4 *ft (Water at 40°F)*

Circulating Fluid Type Water *Concentration* 100 *% @* 40 *°F --> Antifreeze Viscosity Correction Factor* 1.00 *(Table 4.21)*

Corrected System HL = (Total System HL x Antifreeze Viscosity Correction Factor) = 30.4 *ft @* 13 *gpm*

Figure A.45. Circulating Pump Worksheet for Phoenix, AZ Example (Higher Flow)

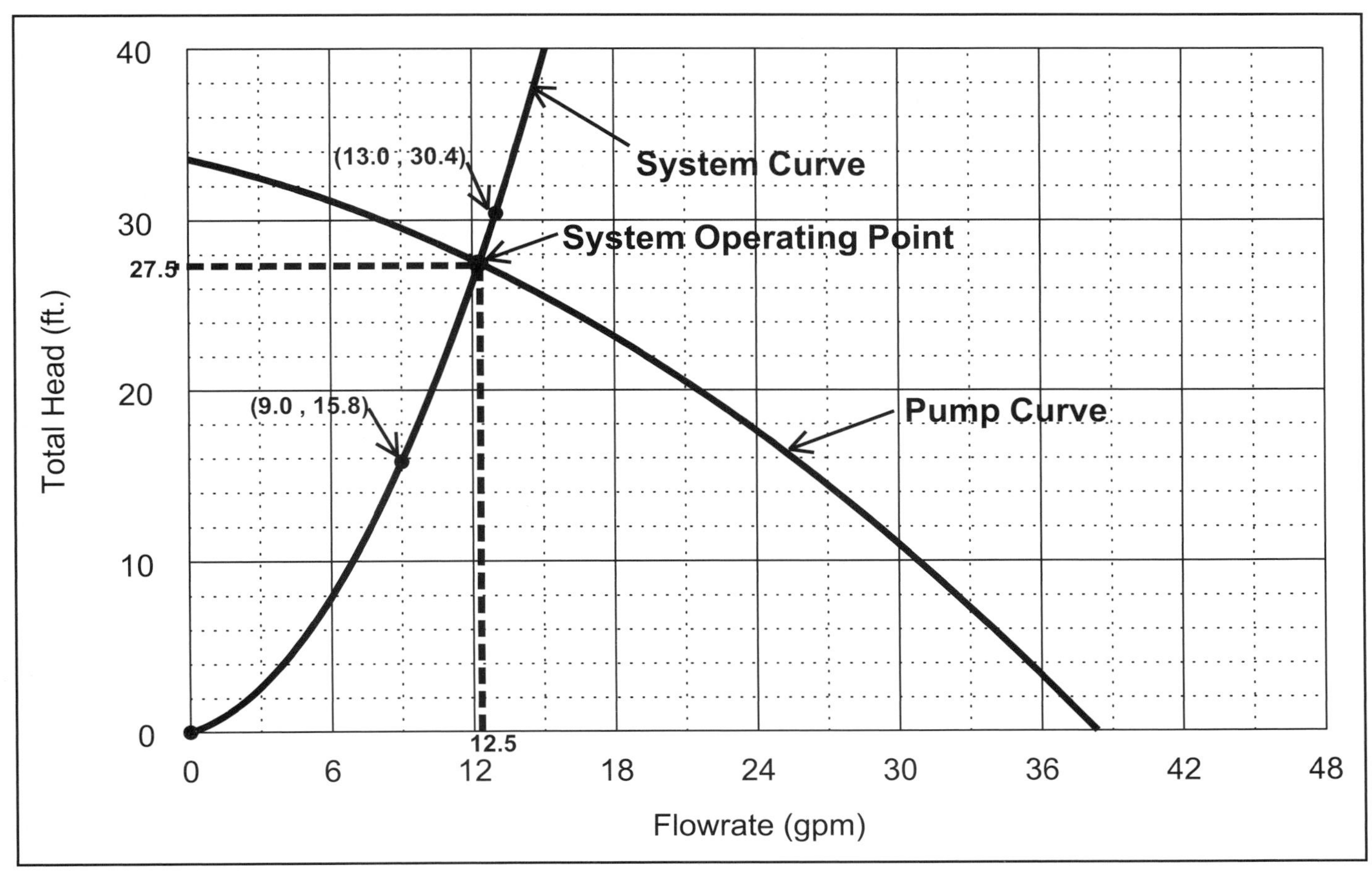

Figure A.46. System Curve & Operating Point for Phoenix, AZ Example

The pump curve shown in Figure A.46 is for a single Armstrong Arm*flo* E-8 circulator. The system curve shown in the figure was generated by calculating the head loss for system flow rates of 0 gpm (0 feet of head), 9 gpm (15.8 feet of calculated head loss), and 13 gpm (30.4 feet of calculated head loss). To generate the system curve, a second order polynomial curve fit through the three points given [(0 gpm, 0 feet), (9 gpm, 15.8 feet) and (13 gpm, 30.4 feet)] was generated via use of software (SlideWrite, Microsoft Excel, etc.).

As is shown in Figure A.46, the system operating point will be somewhere around a system flow of 12.5 gpm at 27.5 feet of head loss, which is slightly above the optimum flow of 3 gpm per ton of installed capacity.

Example A.2.8 Cooling Load Single-Capacity, Water-to-Air GSHP Horizontally-Trenched Loopfield Sizing – Phoenix, AZ

Now, assume that the Phoenix, AZ example residence is located where available land area is not a constraint on the type of GHEX selected. More than likely, a horizontally-trenched system will prove to be more economically feasible in such cases. For the sake of example, assume that the homeowner wanted to utilize a horizontal slinky system. As a starting point for the calculations, the following important design assumptions were made:

- GHEX consists of four horizontally-trenched slinky loops, pipe diameter=3/4-inch, slinky pitch=36-inch, slinky diameter=36-inch, C-C spacing=10 feet, trench depth=8 feet, slinky laying horizontally in bottom of trench
- Maximum entering water temperature from the loopfield, EWT_{max}=100 F
- Mean soil temperature, T_M=73 F (provided in Table 5.11)
- Annual temperature swing, A_s=23 F
- Offset, T_O=33 Days
- Soil thermal conductivity, k_{soil}=0.75 Btu/hr-ft-F, diffusivity, α_{soil}=0.50 ft^2/day
- System flow rate, GPM=9gpm
- Circulating fluid = Water at 40 F

Figure A.47 displays the horizontally-trenched GHEX design worksheet (cooling based design) for this residence. All of the data in the "Heat Pump Design Data" column were taken directly from the Single-Capacity, Water-to-Air Heat Pump Selection Worksheet (Figure A.35). The "Loopfield Layout" column was filled out according to the design assumptions that were made. The run fraction in cooling in the "Bin Analysis Data" column was found on Figure 3.11 (cooling degree days for Phoenix, C_{DD}=4,200, Percent-Sizing=100% → F_C=0.58). The "Horizontal Trench Design Data" column was filled out according to the design assumptions made. The normalized maximum soil temperature $T'(d_{avg},T_O)$ was found on Table 5.13 according to the design assumptions made.

The first set of calculations displayed in the "Cooling Design Length Summary Table" was performed assuming use of four loops with 9-foot C-C trench spacing. Increasing the spacing from 9 feet C-C to 11 feet C-C decreased the amount of pipe necessary by 5.6% (3,417 feet versus 3,226 feet respectively).

The last two sets of calculations displayed in the "Cooling Design Length Summary Table" were performed assuming use of four and five loops, respectively. To satisfy the cooling load for this residence, five horizontally-trenched slinky loops will need to be installed [Trench Dimensions: 3 feet Wide x 8 feet Deep x 160 feet Long x 11 feet C-C trench spacing, Slinky Details: 36-inch Diameter x 36-inch Pitch x 800 foot coil, laying horizontally in bottom of trench, ¾-inch (nominal) DR-11 HDPE pipe].

It is highly recommended that a design such as the one illustrated in this example be evaluated to determine its economic feasibility. In warmer, southern climates, it may be prudent to bury the horizontally-trenched loopfield at an average pipe burial depth of 5-6 feet or less. The design procedure given in Chapter 5 gives the designer the capability to design for a large range of pipe burial depths. Burying the horizontally-trenched loopfield at shallower depths will expose the system to greater annual soil temperature swings, less moisture stability through the annual cycle, and greater risk of being damaged during other activities (utility installation or repair, fence installation, etc). The designer must weigh the costs of digging shallower trenches and installing longer GHEX lengths versus digging deeper trenches and installing shorter GHEX lengths.

Horizontally-Trenched GHEX Design Worksheet – Cooling Mode

Space Cooling GSHP Design Data

TC_C[1] = 37,940 Btu/hr
DMD_C[1] = 3.42 kW
GPM[1] = 9 gpm
Total Heat Gain[1] = 30,407 Btu/hr
Percent Sizing = 104.60 %
F_{Jul} = 0.58

HWG GSHP Design Data

GPM_{HWG} = N/A gpm

Total Design Data

TC_D = 37,940 Btu/hr
DMD_D = 3.42 kW
GPM_D = 9 gpm

GHEX Design Data

EER_D[2] ($= TC_D / (DMD_D \times 1000)$) = 11.10
HR_D[2] ($= TC_D \times (EER_D+3.412)/EER_D$) = 49,630 Btu/hr
F_C ($= F_{Jul}$) = 0.58

EWT_{max}[3] = 100 F
LWT_{max}[2] ($= EWT_{max} + HR_D/(500 \times GPM_D)$) = 111.03 F

1. For single heat pump installation use data directly from the appropriate GSHP Selection Worksheet. For multiple heat pumps in the installation sum all heat pump Total Heat Gains, TC_C's, DMD_C's and GPM's that will be connected to a single GHEX using the table provided below.
2. For single heat pump installation use data directly from appropriate GSHP Selection Worksheet. For multiple heat pumps use the equation provided.
3. EWT_{max} is obtained directly from the appropriate GSHP Selection Worksheet and must be the same for the selection of all heat pumps for a multiple heat pump installation connected to a single GHEX.

	Zone 1	Zone 2	Zone 3	Zone 4	Zone 5	Zone 6	Zone 7	Zone 8	Zone 9	Zone 10	Total
Total Heat Gain											
HC_C											
DMD_C											
GPM											

Horizontal Trench Design Data

Location Phoenix, AZ (City, State)

T_M = 73 F (Table 5.11)
A_s = 23 F (Table 5.11)
T_O = 33 Days (Table 5.11)

Soil Type Sandy Clay (Section 5.3.2.2)

k_{soil} = 0.75 Btu/hr ft F
α_{soil} = 0.50 ft^2/day

Trench and Pipe Configuration ==> Table ________ (Select from Tables 5.17 – 5.29)

N_P = 5 ft_{pipe}/ft_{trench}
Conf. = Laying (Stand / Laying / Rect)

S_H = 3 ft
S_{cc} = 9 ft
d_{bp} = 8 ft

R_s = 2.74 hr ft F/Btu
S_M = 1.16
d_{avg} = 8 ft

D_P = 3/4 in (Nom.)
GPM_{FP} = 2.25 gpm/flowpath
N_{FP} = 4 flowpaths
$N_{FP/T}$ = 1 flowpaths/trench
N_T = 4 trenches

Design Soil Temperature for Cooling

$T'(d_{avg}, T_O+180)$ = 0.1611 F (Table 5.13)

$T_{S,H} = T'(d_{avg}, T_O+180) \cdot A_s + T_M = 0.1611 \cdot 23 + 73 = 76.71$ F (Eq. 5.26)

Pipe Resistance and Pipe Multiplier

R_p = 0.141 hr ft F/Btu (Table 4.6)
P_M = 1.0 (Figure 5.24)

Horizontal Trench Design Lengths (Equations 5.17, 5.18, and 5.19)

$$L_{C,P} = \frac{TC_D \cdot \left(\frac{EER_D + 3.412}{EER_D}\right) \cdot \left(R_P + R_S \cdot P_M \cdot S_M \cdot F_C\right)}{\left(\frac{EWT_{max} + LWT_{max}}{2}\right) - T_{S,H}} = \frac{37{,}940 \cdot \left(\frac{11.10 + 3.412}{11.10}\right) \cdot (0.141 + 2.74 \cdot 1.00 \cdot 1.16 \cdot 0.58)}{\left(\frac{100 + 111.0}{2}\right) - 76.71} = \frac{37{,}940 \cdot (1.307) \cdot (1.984)}{(105.5) - 76.71} = 3417 \text{ ft of pipe}$$

$$L_{C,P/T} = \frac{L_{C,P}}{N_T} = \frac{3417}{4} = 854\ ft_{pipe}/trench \qquad L_{C,T} = \frac{L_{C,P/T}}{N_P} = \frac{807}{5} = 171\ ft_{trench}$$

Cooling Design Length Calculations Summary Table

Layout for Flow and Number of Trenches						Pipe and Trench Design Length Calculations								
D_P	GPM_{FP}	N_{FP}	Trench/N_P	$N_{FP/T}$	N_T	d_{avg}	$T'(d_{avg}, T_O+180)$	$T_{S,H}$	R_s	P_M	S_{cc}/S_M	$L_{C,P}$	$L_{C,P/T}$	$L_{C,T}$
¾	2.25	4	Laying/5	1	4	8	0.1611	76.71	2.74	1	9/1.16	3,417	854	171
¾	2.25	4	Laying/5	1	4	8	0.1611	76.71	2.74	1	11/1.09	3,226	807	161
¾	1.80	5	Laying/5	1	5	8	0.1611	76.71	2.74	1	11/1.09	3,226	645	129
			/								/			
			/								/			

S_{cc} = 11' (second row); N_T = 5 (third row)

Figure A.47. Horizontally-Trenched GHEX Design Worksheet for Phoenix, AZ Example

Example A.2.9 Cooling Load Single-Capacity, Water-to-Air GSHP Horizontally-Bored Loopfield Sizing – Phoenix, AZ

As in Example A.2.8, assume that the Phoenix, AZ example residence is located where available land area is not a constraint on the type of GHEX selected. However, the homeowner wants to explore the option of a horizontally-bored system because of the large amount of excavation necessary to install the horizontally-trenched system. As a starting point for the calculations, the following important design assumptions were made:

- GHEX consists of four horizontally-bored U-bends in a 1 x 4 configuration, U-bend diameter=3/4-inch, burial depth = 10 feet
- Maximum entering water temperature from the loopfield, EWT_{max}=100 F
- Mean soil temperature, T_M=72 F (provided in Table 5.11, used because burial depth < 20 feet)
- Annual temperature swing, A_s=23 F (used because burial depth < 20 feet)
- Offset (in the top 10-15 ft), T_O=33 Days (used because burial depth < 20 feet)
- Soil thermal conductivity assumed to be k_{soil}=0.85 Btu/hr-ft-F, soil diffusivity assumed to be α_{soil}=0.60 ft^2/day
- C-C bore spacing=15 feet
- Bore diameter, D_{bore}=5-inch
- k_{grout}=0.40 Btu/hr-ft-F (20% solids mixture – slurry consisting of 20% bentonite & 80% water by mass)
- System flow rate, GPM=9gpm
- Circulating Fluid = Water at 40 F

Figure A.48 displays the horizontally-bored GHEX design worksheet (cooling based design) for this residence. All of the data in the “Heat Pump Design Data” column were taken directly from the Single-Capacity, Water-to-Air Heat Pump Selection Worksheet (Figure A.35). The “Loopfield Layout” column was filled out according to the design assumptions that were made. The run fraction in cooling in the “Bin Analysis Data” column was found on Figure 3.11 (cooling degree days for Phoenix, C_{DD}=4,200, Percent-Sizing=100% ➔ F_C=0.58). The “Horizontal Trench Design Data” column was filled out according to the design assumptions made. The normalized maximum soil temperature $T'(d_{avg},T_O)$ was found on Table 5.13 according to the design assumptions made.

It was assumed that total system flow would increase from 9 gpm to 11 gpm because the total system flow was divided into four bores (rather than three). Heat pump performance effects on the increased system flow were neglected.

Horizontally-Bored GHEX Design Worksheet – Cooling Mode

Space Cooling GSHP Design Data

TC_C[1] = **37,940** Btu/hr
DMD_C[1] = **3.42** kW
GPM[1] = **11.0** gpm
AGL_H = **16,553,409** Btu
AGL_C = **97,657,108** Btu
Total Heat Gain[1] = **30,407** Btu/hr
Percent Sizing = **104.60** %
F_{Jul} = **0.58**

HWG GSHP Design Data

GPM_{HWG} = **N/A** gpm
AGL_{HWG} = **N/A** Btu

Total Design Data

TC_D = **37,940** Btu/hr
DMD_D = **3.42** kW
GPM_D = **11.0** gpm
AGL_{HD} = **16,553,409** Btu
AGL_{CD} = **97,657,108** Btu

GHEX Design Data

EER_D[2] (= TC_D / (DMD_D x 1000)) = **11.09**
HR_D[2] (= TC_D x (EER_D+3.412)/EER_D) = **49,613** Btu/hr
F_C (= F_{Jul}) = **0.58**
EWT_{max}[3] = **100.0** F
LWT_{max}[2] (= EWT_{max} + HR_D/(500xGPM_D)) = **111.0** F

1. For single heat pump installation use data directly from the appropriate GSHP Selection Worksheet. For multiple heat pumps in the installation sum all heat pump Total Heat Gains, TC_C's, DMD_C's and GPM's that will be connected to a single GHEX using the table provided below.
2. For single heat pump installation use data directly from appropriate GSHP Selection Worksheet. For multiple heat pumps use the equation provided.
3. EWT_{max} is obtained directly from the appropriate GSHP Selection Worksheet and must be the same for the selection of all heat pumps for a multiple heat pump installation connected to a single GHEX.

	Zone 1	Zone 2	Zone 3	Zone 4	Zone 5	Zone 6	Zone 7	Zone 8	Zone 9	Zone 10	Total
Total Heat Gain											
TC_C											
DMD_C											
GPM											

Soil Design Data

Location **Phoenix** , **AZ** (City, State)
T_M = **73** F (Table 5.11)
A_s = **23** F (Table 5.11)
T_O = **33** Days (Table 5.11)
Soil Type ______________ (Section 5.3.2.2)
k_{soil} = **0.85** Btu/hr ft F
α_{soil} = **0.60** ft²/day

Horizontally-Bored GHEX Layout

U-bend D_P = **3/4** in
GPM_{FP} = **3** gpm/flowpath
N_{FP} = **4** flowpaths
N_{BIS} = **1** bores in series
N_B = **4** bores
Layout = **1** x **4** (N_{Layers} x $N_{Bores/Layer}$)
Bore Spacing = **15** ft

Horizontally-Bored GHEX Design Data

D_{bore} = **5** in
k_{Grout} = **0.40** Btu/hr ft F
$(R_B + R_G \cdot F_C)$ = **0.724** hr ft F/Btu
d_{avg} = **10** ft
$T'(d_{avg}, T_O+180)$ = **0.0978** F (Table 5.13)
$T_{S,H} = T'(d_{avg},T_O+180) \cdot A_s + T_M$ = **0.0978 • 23 + 73 = 75.25 F**

Borehole Design Lengths (Equations 5.30 and 5.31)

$$L_{C,T} = \frac{TC_D \cdot \left(\frac{EER_D + 3.412}{EER_D}\right) \cdot (R_B + R_G \cdot F_C)}{\left(\frac{EWT_{max} + LWT_{max}}{2}\right) - T_{S,H}} = \frac{37{,}940 \cdot \left(\frac{11.09 + 3.412}{11.09}\right) \cdot (0.724)}{\left(\frac{100 + 111}{2}\right) - 75.25} = \frac{37{,}940 \cdot (1.308) \cdot (0.724)}{(105.5) - 75.25} = 1188 \textit{ ft of bore}$$

$$L_{C,B} = \frac{L_{C,T}}{N_B} = \frac{1188}{4} = 297 \textit{ ft per bore}$$

Unbalanced Ground Load Borehole Design Lengths (Equations 5.9b, 5.12, and 5.13)

$$NNAGL = \frac{AGL_{CD} - AGL_{HD}}{L_{C,T} \cdot \left(\left(\frac{EWT_{max} + LWT_{max}}{2}\right) - T_{S,H}\right)} = \frac{97{,}657{,}108 - 16{,}553{,}409}{1188 \cdot \left(\left(\frac{100 + 111}{2}\right) - 75.25\right)} = 2257 \; Btu/ftF$$

B_M = **1.28** (Figures 5.13 through 5.18)

$$L_{C,T,UGL} = B_M \cdot L_{C,T} = 1.28 \cdot 1188 = 1521 \textit{ ft of bore} \qquad L_{C,B,UGL} = \frac{L_{C,T,UGL}}{N_B} = \frac{1521}{4} = 380 \textit{ ft per bore}$$

Cooling Design Length Calculations Summary Table

Layout for Flow and Number of Bores						Design Length Calculations				Unbalanced Ground Load Design Lengths					
D_P	GPM_{FP}	N_{FP}	N_{BIS}	N_B	Layout	k_{Grout}	$R_B+R_G \cdot F_C$	$L_{C,T}$	$L_{C,B}$	Layout	Spacing	NNAGL	B_M	$L_{C,T,UGL}$	$L_{C,B,UGL}$
3/4	2.75	4	1	4	1x4	0.40	0.724	1188	297	1x4	15	2257	1.28	1521	380
3/4	2.75	4	1	4	1x4	0.88	0.580	951	238	1x4	15	2819	1.35	1284	321
3/4	2.75	4	1	4	1x4	0.88	0.580	951	238	1x4	20	2819	1.22	1160	290

k_{Grout} = 0.88 → (second row)
Spacing = 20ft. → (third row)

Figure A.48. Horizontally-Bored GHEX Design Worksheet for Phoenix, AZ Example

Upon performing the calculations summarized in the "Cooling Design Length Calculations Summary Table" on the Horizontally-Bored GHEX Design Worksheet (Figure A.48), the following conclusions can be made:

1) The total amount of bore required to satisfy the system constraints imposed by the original design assumptions was found to be $L_{C,T,UGL}$=1,521 feet (380 ft/bore – four bores).
2) Using a thermally-enhanced grouting material (thermal conductivity, k_{Grout}=0.88 Btu/hr-ft-F) as the backfill material instead of the 20% solids backfill (k_{Grout}=0.40 Btu/hr-ft-F) reduces the amount of bore required by 15.6% ($L_{C,T,UGL}$=1,284 feet, 321 ft/bore – four bores).
3) Increasing the C-C bore spacing from 15 feet to 20 feet further reduces the amount of required bore length by 9.7% ($L_{C,T,UGL}$=1,160 feet, 290 ft/bore – four bores).

Example A.3.1 Peak Load Breakdown & GSHP Selection for Multiple Heat Pump System – Louisville, KY

For illustration purposes, a large home in Louisville, KY was selected for analysis. Because of the size of the home, it was split into four zones (two zones on the main level, one on the upper level, and one in the basement). The GSHP selection and performance data breakdown is given in Table A.7. The peak heating load and equipment sizing data is given in Table A.8, and the peak cooling load and equipment sizing data is given in Table A.9. For the sake of example, assume the homeowner wanted to purchase GeoComfort Serenity GS (single-capacity) water-to-air GSHP equipment for the home. The heating performance data in Table A.7 is based on the water and air flow rates given in the table, 30 F EWT_{min} from the loopfield and 70 F entering air temperatures from the space. The cooling performance data in Table A.7 is based on the water and air flow rates given in the table, 90 F EWT_{max} from the loopfield and 80/67 F (db/wb) entering air temperatures from the space. All GSHP units were selected based on the principles given in Chapter 2 and utilizing the Single-Capacity, Water-to-Air GSHP Selection Worksheet in given in Chapter 3 (Figure 3.2).

Table A.7. GSHP Selection & Performance Data

Zone	Heat Pump Model	Heating Performance Data			Cooling Performance Data				Air Flow (cfm)	Water Flow (gpm)
		HC_D (Btu/hr)	DMD_D (kW)	COP_D	TC_D (Btu/hr)	SC_D (Btu/hr)	DMD_D (kW)	EER_D		
Main Level-1	GeoComfort Serenity GS042	35,800	3.07	3.42	43,600	34,000	3.25	13.42	1,400	11.0
Main Level-2	GeoComfort Serenity GS048	43,500	3.83	3.33	53,000	41,300	3.99	13.28	1,700	13.0
Upper Level	GeoComfort Serenity GS030	26,400	2.20	3.52	32,200	25,100	2.33	13.82	1,000	7.0
Basement	GeoComfort Serenity GS024	20,500	1.71	3.51	24,900	19,400	1.86	13.39	800	6.0

Table A.8. Peak Heating Load & Equipment Sizing Data

Zone	Heating				
	Heat Loss (Btu/hr)	HC_D (Btu/hr)	DMD_D (kW)	COP_D	% Sizing
Main Level-1	35,063	35,800	3.07	3.42	102%
Main Level-2	42,642	43,500	3.83	3.33	102%
Upper Level	23,102	26,400	2.20	3.52	114%
Basement	18,206	20,500	1.71	3.51	113%
Total	119,013	126,200	10.81	3.42	106%

Table A.9. Peak Cooling Load & Equipment Sizing Data

Zone	Cooling						
	Total Heat Gain (Btu/hr)	Sensible Heat Gain (Btu/hr)	TC_D (Btu/hr)	SC_D (Btu/hr)	DMD_D (kW)	EER_D	% Sizing
Main Level-1	28,702	24,397	43,600	34,000	3.25	13.42	139%
Main Level-2	21,408	18,197	53,000	41,300	3.99	13.28	227%
Upper Level	30,172	25,646	32,200	25,100	2.33	13.82	98%
Basement	8,708	7,402	24,900	19,400	1.86	13.39	262%
Total	88,990	75,642	153,700	119,800	11.43	13.45	158%

As shown in Table A.8, the total installed heating capacity is 126,200 Btu/hr (106% sizing at EWT_{min}=30 F) and the system installed COP will be 3.42 (DMD_D=10.81 kW).

Example A.3.2 Energy Analysis for Multiple Heat Pump System – Louisville, KY

After the heat pump sizing and selection process is complete, an energy analysis will need to be performed to estimate annual operating cost savings of the GSHP system over a conventional HVAC system and to estimate the annual ground loads for the vertically-bored GHEX design process. The simplified bin method as described in Chapter 3 will be used for this example. The assumptions utilized to perform the bin analysis are summarized by Tables A.10-A.11:

Table A.10. Bin Method Assumptions for Louisville, KY Example

Heating			Cooling		
Q_{PH}	119,013	Btu/hr	Q_{PC}	88,990	Btu/hr
T-Stat Set Pt.	70	°F	T-Stat Set Pt.	75	°F
t_{dh}	12	°F	t_{dc}	93	°F
t_{bh}	62	°F	t_{bc}	67	°F
r_e	0.085	$/kWh	r_e	0.085	$/kWh
EWT_{min}	30	°F	EWT_{max}	90	°F
EWT_{mean}	58	°F	EWT_{mean}	58	°F
t_{mean}	56.7	°F	t_{mean}	56.7	°F
D_c	0.15		D_c	0.15	
Space SHF	NA		Space SHF	0.85	

Table A.11. Heat Pump Performance Assumptions for Bin Method (Louisville, KY Example)

	EWT (°F)	Total CAP (Btu/hr)	Sensible CAP (Btu/hr)	Demand (kW)
Htg.	30	126,200	NA	10.81
	50	165,400	NA	11.48
Clg.	70	168,000	131,100	9.31
	90	153,700	119,800	11.43

To be consistent with the definitions given in Section 3.4, EWT_{lth}=30 F, EWT_{hth}=50 F, EWT_{ltc}=70 F, EWT_{htc}=90 F, etc. The results of the calculations performed using the bin method are summarized by Table A.12:

Table A.12. Bin Method Calculation Summary for Louisville, KY Example

Louisville, KY																			
Temp Bin (°F)	t_{air} (°F)	h_b (hrs)	Q_B (Btu/hr)	EWT (°F)	HC/SC (Btu/hr)	TC (Cooling Only) Btu/hr	RF	PLF	RF_A	DMD_h (kW)	DMD_c (kW)	DMD_{res} (kW)	DMD_{htotal} (kW)	RT_A (hrs)	E (kWh)	C_e ($)	gl_h (Btu)	gl_c (Btu)
105/109	107	0	116,372	102.3	112,827	144,876	1.00	1.00	1.00	0.00	0.00	0.00	0.00	0.0	0	$0.00	0	0
100/104	102	1	101,825	97.9	115,317	148,027	0.88	0.98	0.90	0.00	12.27	0.00	0.00	0.9	11	$0.94	0	-167,678
95/99	97	16	87,279	93.5	117,808	151,179	0.74	0.96	0.77	0.00	11.80	0.00	0.00	12.3	146	$12.37	0	-2,269,431
90/94	92	119	72,732	89.1	120,298	154,330	0.60	0.94	0.64	0.00	11.34	0.00	0.00	76.5	867	$73.70	0	-13,886,613
85/89	87	314	58,186	84.7	122,788	157,482	0.47	0.92	0.51	0.00	10.87	0.00	0.00	161.5	1756	$149.25	0	-28,950,789
80/84	82	547	43,639	80.3	125,279	160,633	0.35	0.90	0.39	0.00	10.40	0.00	0.00	211.2	2197	$186.72	0	-37,369,872
75/79	77	747	29,093	75.9	127,769	163,785	0.23	0.88	0.26	0.00	9.93	0.00	0.00	192.4	1911	$162.46	0	-33,624,059
70/74	72	959	14,546	71.5	130,260	166,936	0.11	0.87	0.13	0.00	9.47	0.00	0.00	123.6	1170	$99.43	0	-21,337,461
65/69	67	810	0	67.1	132,750	170,088	0.00	0.85	0.00	0.00	9.00	0.00	0.00	0.0	0	$0.00	0	0
60/64	62	749	0	61.3	187,587	0	0.00	0.85	0.00	11.86	0.00	0.00	11.86	0.0	0	$0.00	0	0
55/59	57	667	11,901	58.2	181,448	0	0.07	0.86	0.08	11.75	0.00	0.00	11.75	50.9	598	$50.84	6,183,588	0
50/54	52	664	23,803	55.1	175,310	0	0.14	0.87	0.16	11.65	0.00	0.00	11.65	103.6	1207	$102.57	12,221,501	0
45/49	47	621	35,704	51.9	169,171	0	0.21	0.88	0.24	11.54	0.00	0.00	11.54	148.7	1716	$145.87	17,009,576	0
40/44	42	624	47,605	48.8	163,032	0	0.29	0.89	0.33	11.44	0.00	0.00	11.44	203.9	2332	$198.22	22,593,792	0
35/39	37	645	59,507	45.7	156,894	0	0.38	0.91	0.42	11.33	0.00	0.00	11.33	269.8	3058	$259.89	28,920,744	0
30/34	32	527	71,408	42.5	150,755	0	0.47	0.92	0.51	11.23	0.00	0.00	11.23	271.0	3043	$258.70	28,067,422	0
25/29	27	330	83,309	39.4	144,616	0	0.58	0.94	0.62	11.12	0.00	0.00	11.12	203.0	2258	$191.97	20,276,118	0
20/24	22	179	95,210	36.3	138,477	0	0.69	0.95	0.72	11.02	0.00	0.00	11.02	129.1	1423	$120.95	12,415,199	0
15/19	17	115	107,112	33.1	132,339	0	0.81	0.97	0.83	10.91	0.00	0.00	10.91	95.8	1046	$88.90	8,851,455	0
10/14	12	67	119,013	30.0	126,200	0	0.94	0.99	0.95	10.81	0.00	0.00	10.81	63.7	689	$58.56	5,643,395	0
5/9	7	32	130,914	26.9	120,061	0	1.00	1.00	1.00	10.71	0.00	3.18	13.89	32.0	444	$37.77	2,673,138	0
0/4	2	14	142,816	23.7	113,923	0	1.00	1.00	1.00	10.60	0.00	8.47	19.07	14.0	267	$22.69	1,088,568	0
-5/-1	-3	6	154,717	20.6	107,784	0	1.00	1.00	1.00	10.50	0.00	13.76	24.25	6.0	146	$12.37	431,845	0
-10/-6	-8	4	166,618	17.5	101,645	0	1.00	1.00	1.00	10.39	0.00	19.04	29.43	4.0	118	$10.01	264,774	0
-15/-11	-13	1	178,520	14.3	95,506	0	1.00	1.00	1.00	10.29	0.00	24.33	34.62	1.0	35	$2.94	60,413	0
-20/-16	-18	0	190,421	11.2	89,368	0	1.00	1.00	1.00	0.00	0.00	0.00	0.00	0.0	0	$0.00	0	0
-25/-21	-23	0	202,322	8.1	83,229	0	1.00	1.00	1.00	0.00	0.00	0.00	0.00	0.0	0	$0.00	0	0
-30/-26	-28	0	214,223	4.9	77,090	0	1.00	1.00	1.00	0.00	0.00	0.00	0.00	0.0	0	$0.00	0	0
-35/-31	-33	0	226,125	1.8	70,952	0	1.00	1.00	1.00	0.00	0.00	0.00	0.00	0.0	0	$0.00	0	0
-40/-36	-38	0	238,026	-1.3	64,813	0	1.00	1.00	1.00	0.00	0.00	0.00	0.00	0.0	0	$0.00	0	0
												Totals	Heating	1,596	18,379	$1,562.23	166,701,528	
													Cooling	778	8,057	$684.88		-137,445,100

The following conclusions can be drawn from Table A.12:

- The estimated amount of heat pump run hours in heating mode for this example is 1,596 hours, found by summing the number of run hours (RT_A) when the outdoor air temperature is lower than the balance point temperature for heating.
- The estimated amount of heat pump run hours in cooling mode for this example is 778 hours, found by summing the number of run hours (RT_A) when the outdoor air temperature is higher than the balance point temperature for cooling.
- The estimated amount of supplemental resistance heat run hours for this example is 57 hours, found by summing the number of run hours (RT_A) when the heat pump capacity is less than the building load (driven by outdoor air temperature).
- The total estimated annual operating cost for the selected equipment in heating (heat pump operation plus supplemental resistance heat operation) is $1,562.23 per year for this example (at the assumed electric rate).
- The total estimated annual operating cost for this system in cooling is $684.88 per year for this example.
- The estimated amount of heat removed from the ground (heating mode) via the heat pump system is 166,701,528 Btu.
- The estimated amount of heat rejected to the ground (cooling mode) via the heat pump system is 137,445,100 Btu.

Because the bin analysis for this system was performed for the block load for the building (the sum of the zone loads), the calculated equipment run hours in heating and cooling as well as the resistance heat run hours represent the AVERAGE number of hours that each GSHP unit will run throughout the annual cycle.

Example A.3.3 Multiple Heat Pump System Vertical Loopfield Sizing – Louisville, KY

After the peak building loads have been calculated and GSHP equipment selected for the load, the loopfield design can be performed. As discussed in Chapter 5, the first step in designing the loopfield for any installation is to perform a comprehensive site survey. For the sake of example, this Louisville, KY residence was assumed to be located where the best option is to utilize a vertically-bored GHEX for the ground source loopfield. As a starting point for the calculations, the following important design assumptions were made:

- GHEX consists of eight vertically-bored U-bends (1 bore per 1.5 tons of installed nominal GSHP capacity) in a 2 x 4 rectangular configuration, U-bend diameter=1-inch
- Minimum entering water temperature from the loopfield, EWT_{min}=30 F
- Maximum entering water temperature from the loopfield, EWT_{max}=90 F
- Deep earth temperature, T_G=58 F (determined from Figure 5.11)
- Formation thermal conductivity, k_G=1.30 Btu/hr-ft-F (assumed value)
- C-C bore spacing=15 feet
- Bore diameter, D_{bore}=5-inch
- k_{grout}=0.40 Btu/hr-ft-F (20% solids mixture – slurry consisting of 20% bentonite & 80% water by mass)
- System flow rate, GPM=37 gpm (Table A.7)

Although the peak heating load is greater than the peak total cooling load (119,013 Btu/hr vs. 88,990 Btu/hr, respectively), the annual ground load will be fairly balanced (as shown in Table A.12). Because the ground load for this example is fairly balanced and because the average annual air temperature is between 50-60 F, both the heating and cooling-based GHEX design calculations will need to be performed before a final GHEX design decision can be made.

Figure A.49 displays the vertically-bored GHEX design worksheet (heating based design) for this residence. All of the data in the "Heat Pump Design Data" column were taken from Table A.8. The "Borefield Layout" column was filled out according to the design assumptions that were made. The "Bin Analysis Data" column was filled out with data taken directly from the results of the bin analysis performed in Example A.3.2, Table A.12. The run fraction in heating was found on Figure 3.10 (heating degree days for Louisville, H_{DD}=4,352, Percent-Sizing=106% ➔ F_H=0.50). The "Vertical Bore Design Data" column was filled out according to the design assumptions made. The borehole resistance ($R_B+R_G \cdot F_H$) was found on Table 5.9 according to the design assumptions made.

Figure A.50 displays the vertically-bored GHEX design worksheet (cooling-based design) for this residence. All of the data in the "Heat Pump Design Data" column were taken from Table A.8. The "Borefield Layout" column was filled out according to the design assumptions that were made. The "Bin Analysis Data" column was filled out with data taken directly from the results of the bin analysis performed in Example A.3.2, Table A.12. The run fraction in cooling was found on Figure 3.11 (cooling degree days for Louisville, C_{DD}=1,443, Percent-Sizing=158% ➔ F_C=0.30). The "Vertical Bore Design Data" column was filled out according to the design assumptions made. The borehole resistance ($R_B+R_G \cdot F_C$) was found on Table 5.9 according to the design assumptions made.

Vertically-Bored GHEX Design Worksheet – Heating Mode

Space Heating GSHP Design Data

HC_C[1]	=	**126,200**	Btu/hr
DMD_C[1]	=	**10.81**	kW
GPM[1]	=	**37**	gpm
AGL_H	=	**166,701,528**	Btu
AGL_C	=	**137,445,100**	Btu
Total Heat Loss[1]	=	**119,013**	Btu/hr
Percent Sizing	=	**106.03**	%
F_{Jan}	=	**.50**	

HWG GSHP Design Data

HC_{HWG}	=	**N/A**	Btu/hr
DMD_{HWG}	=	**N/A**	kW
GPM_{HWG}	=	**N/A**	gpm
AGL_{HWG}	=	**N/A**	Btu
F_{HWG}	=	**N/A**	

Total Design Data

HC_D	=	**126,200**	Btu/hr
DMD_D	=	**10.81**	kW
GPM_D	=	**37**	gpm
AGL_{HD}	=	**166,701,528**	Btu
AGL_{CD}	=	**137,445,100**	Btu

GHEX Design Data

COP_D[2]	(= HC_D / (DMD_D x 3,412)	=	**3.42**	
HE_D[2]	(= HC_D x (COP_D-1)/COP_D)	=	**89,299**	Btu/hr
F_H	(= (F_{Jan} x HC_C + F_{HWG} x HC_{HWG}) / HC_D)	=	**.50**	
EWT_{min}[3]		=	**30**	F
LWT_{min}[2]	(= EWT_{min} – HE_D/(500xGPM_D))	=	**25.17**	F

1. For single heat pump installation use data directly from the appropriate GSHP Selection Worksheet. For multiple heat pumps in the installation sum all heat pump Total Heat Losses, HC_C's, DMD_C's and GPM's that will be connected to a single GHEX using the table provided below.
2. For single heat pump installation use data directly from appropriate GSHP Selection Worksheet. For multiple heat pumps use the equation provided.
3. EWT_{min} is obtained directly from the appropriate GSHP Selection Worksheet and must be the same for the selection of all heat pumps for a multiple heat pump installation connected to a single GHEX.

	Zone 1	Zone 2	Zone 3	Zone 4	Zone 5	Zone 6	Zone 7	Zone 8	Zone 9	Zone 10	Total
Total Heat Loss	35,063	42,642	23,102	18,206							119,013
HC_C	35,800	43,500	26,400	20,500							126,200
DMD_C	3.07	3.83	2.20	1.71							10.81
GPM	11.0	13.0	7.0	6.0							37.0

Vertical Borefield Layout

U-bend D_P	=	**1**	in
GPM_{FP}	=	**4.6**	gpm/flowpath
N_{FP}	=	**8**	flowpaths
N_{BIS}	=	**1**	bores in series
N_B	=	**8**	bores
Layout	=	**2** x **4**	(N_{Rows} x $N_{Bores/Row}$)
Bore Spacing	=	**15**	ft

Vertical Borefield Design Data

T_G	=	**58**	F
k_G	=	**1.30**	Btu/hr ft F (Table 5.6)
D_{bore}	=	**5.0**	in
k_{Grout}	=	**0.40**	Btu/hr ft F
($R_B + R_G \cdot F_H$)	=	**0.528**	hr ft F/Btu (Table 5.8, 5.9 or 5.10)

Borehole Design Lengths (Equations 5.1 and 5.2)

$$L_{H,T} = \frac{HC_D \cdot \left(\frac{COP_D - 1}{COP_D}\right) \cdot \left(R_B + R_G \cdot F_H\right)}{T_G - \left(\frac{EWT_{min} + LWT_{min}}{2}\right)} = \frac{47{,}769 \cdot \left(\frac{3.62 - 1}{3.62}\right) \cdot (0.700)}{58 - \left(\frac{30 + 25.39}{2}\right)} = \frac{126{,}200 \cdot (0.708) \cdot (0.528)}{58 - (27.59)} = 1551 \textit{ ft of bore}$$

$$L_{H,B} = \frac{L_{H,T}}{N_B} = \frac{1551}{8} = 194 \textit{ ft per bore}$$

Unbalanced Ground Load Borehole Design Lengths (Equations 5.9a, 5.10, and 5.11)

$$NNAGL = \frac{AGL_{HD} - AGL_{CD}}{L_{H,T} \cdot \left(T_G - \left(\frac{EWT_{min} + LWT_{min}}{2}\right)\right)} = \frac{166{,}701{,}528 - 137{,}445{,}100}{1551 \cdot \left(58 - \left(\frac{30 + 25.17}{2}\right)\right)} = 620 \; Btu/ftF$$

B_M = **1.05** (Figures 5.13 through 5.18)

$$L_{H,T,UGL} = B_M \cdot L_{H,T} = 1.05 \cdot 1551 = 1629 \textit{ ft of bore} \qquad L_{H,B,UGL} = \frac{L_{H,T,UGL}}{N_B} = \frac{1629}{8} = 204 \textit{ ft per bore}$$

Heating Design Length Calculations Summary Table

Layout for Flow and Number of Bores						Design Length Calculations				Unbalanced Ground Load Design Lengths					
D_P	GPM_{FP}	N_{FP}	N_{BIS}	N_B	Layout	k_{Grout}	$R_B + R_G \cdot F_H$	$L_{H,T}$	$L_{H,B}$	Layout	Spacing	NNAGL	B_M	$L_{H,T,UGL}$	$L_{H,B,UGL}$
1	4.6	8	1	8	2x4	0.40	0.528	1551	194	2x4	15	620	1.05	1629	204
1	4.6	8	1	8	2x4	0.88	0.398	1169	146	2x4	15	822	1.07	1251	156
1	4.6	8	1	8	2x4	0.88	0.390	1960	245	2x4	15	782	1.07	2097	262
							F_H=0.48								

k_{Grout} = 0.88 → (row 2)

EWT_{min} = 40° F → (row 3)

Figure A.49. Vertically-Bored GHEX Heating Design Worksheet for Louisville, KY Example

Vertically-Bored GHEX Design Worksheet – Cooling Mode

Space Cooling GSHP Design Data

TC_C[1] = 153,700 Btu/hr
DMD_C[1] = 11.43 kW
GPM[1] = 37.0 gpm
AGL_H = 166,701,528 Btu
AGL_C = 137,445,100 Btu
Total Heat Gain[1] = 88,990 Btu/hr
Percent Sizing = 158.38 %
F_{Jul} = 0.30

HWG GSHP Design Data

GPM_{HWG} = N/A gpm
AGL_{HWG} = N/A Btu

Total Design Data

TC_D = 153,700 Btu/hr
DMD_D = 11.43 kW
GPM_D = 37.0 gpm
AGL_{HD} = 166,701,528 Btu
AGL_{CD} = 137,445,100 Btu

Vertical GHEX Design Data

EER_D[2] (= TC_D / (DMD_D x 1000)) = 13.45
HR_D[2] (= TC_D x (EER_D+3.412)/EER_D) = 192,690 Btu/hr
F_C (= F_{Jul}) = 0.30

EWT_{max}[3] = 90.0 F
LWT_{max}[2] (= EWT_{max} + HR_D/(500xGPM_D)) = 100.4 F

1. For single heat pump installation use data directly from the appropriate GSHP Selection Worksheet. For multiple heat pumps in the installation sum all heat pump Total Heat Gains, TC_C's, DMD_C's and GPM's that will be connected to a single GHEX using the table provided below.
2. For single heat pump installation use data directly from appropriate GSHP Selection Worksheet. For multiple heat pumps use the equation provided.
3. EWT_{max} is obtained directly from the appropriate GSHP Selection Worksheet and must be the same for the selection of all heat pumps for a multiple heat pump installation connected to a single GHEX.

	Zone 1	Zone 2	Zone 3	Zone 4	Zone 5	Zone 6	Zone 7	Zone 8	Zone 9	Zone 10	Total
Total Heat Gain	28,702	21,408	30,172	8,708							88,990
TC_C	43,600	53,000	32,200	24,900							153,700
DMD_C	3.25	3.99	2.33	1.86							11.43
GPM	11.0	13.0	7.0	6.0							37.0

Vertical Borefield Layout

U-bend D_P = 1 in
GPM_{FP} = 4.6 gpm/flowpath
N_{FP} = 8 flowpaths
N_{BIS} = 1 bores in series
N_B = 8 bores
Layout = 2 x 4 (N_{Rows} x $N_{Bores/Row}$)
Bore Spacing = 15 ft

Vertical Borefield Design Data

T_G = 58 F
k_G = 1.30 Btu/hr ft F (Table 5.6)
D_{bore} = 5.0 in
k_{Grout} = 0.40 Btu/hr ft F
($R_B + R_G \cdot F_C$) = 0.441 hr ft F/Btu (Table 5.8, 5.9 or 5.10)

Total Borehole Design Length (Equations 5.3 and 5.4)

$$L_{C,T} = \frac{TC_D \cdot \left(\frac{EER_D + 3.412}{EER_D}\right) \cdot (R_B + R_G \cdot F_C)}{\left(\frac{EWT_{max} + LWT_{max}}{2}\right) - T_G} = \frac{153{,}700 \cdot \left(\frac{13.45 + 3.412}{13.45}\right) \cdot (0.441)}{\left(\frac{90 + 100.4}{2}\right) - 58} = \frac{153{,}700 \cdot (1.254) \cdot (0.441)}{(95.2) - 58} = 2285 \textit{ ft of bore}$$

$$L_{C,B} = \frac{L_{C,T}}{N_B} = \frac{2285}{8} = 285 \textit{ ft per bore}$$

Unbalanced Ground Load Borehole Design Lengths (Equations 5.9b, 5.12, and 5.13)

$$NNAGL = \frac{AGL_{CD} - AGL_{HD}}{L_{C,T} \cdot \left(\left(\frac{EWT_{max} + LWT_{max}}{2}\right) - T_G\right)} = \frac{137{,}445{,}100 - 166{,}701{,}528}{2285 \cdot \left(\left(\frac{90 + 100.4}{2}\right) - 58\right)} = -344 \; Btu/ftF$$

Negative → B_M = 1.00

B_M = 1.00 (Figures 5.13 through 5.18)

$$L_{C,T,UGL} = B_M \cdot L_{C,T} = 1.00 \cdot 2285 = 2285 \textit{ ft of bore}$$

$$L_{C,B,UGL} = \frac{L_{C,T,UGL}}{N_B} = \frac{2285}{8} = 285 \textit{ ft per bore}$$

Cooling Design Length Calculations Summary Table

Layout for Flow and Number of Bores						Design Length Calculations				Unbalanced Ground Load Design Lengths					
D_P	GPM_{FP}	N_{FP}	N_{BIS}	N_B	Layout	k_{Grout}	$R_B+R_G \cdot F_C$	$L_{C,T}$	$L_{C,B}$	Layout	Spacing	NNAGL	B_M	$L_{C,T,UGL}$	$L_{C,B,UGL}$
1	4.6	8	1	8	2x4	0.40	0.441	2285	285	2x4	15	N/A	1.00	2285	285
1	4.6	8	1	8	2x4	0.88	0.311	1611	201	2x4	15	N/A	1.00	1611	201
1	4.6	8	1	8	2x4	0.88	0.306	1838	230	2x4	15	N/A	1.00	1838	230
							F_C=0.29								

k_{Grout} = 0.88 → (row 2)
EWT_{max} = 85°F ← (row 3)

Figure A.50. Vertically-Bored GHEX Cooling Design Worksheet for Louisville, KY Example

Upon performing the calculations summarized in the "Heating Design Length Calculations Summary Table" on the Vertically-Bored GHEX Design Worksheet (Figure A.49), the following conclusions can be made:

1) The total amount of bore required to satisfy the system constraints imposed by the original design assumptions was found to be $L_{H,T,UGL}$=1,629 feet (204 ft/bore – eight bores).
2) Using a thermally-enhanced grouting material (thermal conductivity, k_{Grout}=0.88 Btu/hr-ft-F) as the backfill material instead of the 20% solids backfill (k_{Grout}=0.40 Btu/hr-ft-F) reduces the amount of bore required by 23.2% ($L_{H,T,UGL}$=1,251 feet, 156 ft/bore – eight bores).
3) The last calculation listed on the "Heating Design Length Calculations Summary Table" was performed for EWT_{min}=40 F. The effects of heat pump performance were accounted for in this calculation (though they are not shown) and the results of the calculations show a 67.6% increase in required bore length versus designing around EWT_{min}=30 F (k_{Grout}=0.88 Btu/hr-ft-F, bore spacing=15 feet, 2x4 rectangular configuration). For EWT_{min}=40 F, the selected equipment will be sized such that it will handle the peak heating load under design conditions with higher efficiency (HC_D=145,800 Btu/hr, DMD_D=11.15 kW, COP_D=3.83). The calculations were performed for EWT_{min}=40 F because, as shown in Figure A.50, the cooling design lengths for EWT_{max}=90 F were longer than the heating design lengths for EWT_{min}=30 F. The bore length multiplier, B_M, determined by calculating the NNAGL, was applied to the heating design lengths because the results of the energy analysis (Example A.3.2) displayed an annual heat of extraction from the loopfield (AGL_{HD}>AGL_{CD}).

Upon performing the calculations summarized in the "Cooling Design Length Calculations Summary Table" on the Vertically-Bored GHEX Design Worksheet (Figure A.50), the following conclusions can be made:

1) The total amount of bore required to satisfy the system constraints imposed by the original design assumptions was found to be $L_{C,T,UGL}$=2,285 feet (285 ft/bore – eight bores).
2) Using a thermally-enhanced grouting material (thermal conductivity, k_{Grout}=0.88 Btu/hr-ft-F) as the backfill material instead of the 20% solids backfill (k_{Grout}=0.40 Btu/hr-ft-F) reduces the amount of bore required by 29.5% ($L_{C,T,UGL}$=1,611 feet, 201 ft/bore – eight bores).
3) The last calculation listed on the "Cooling Design Length Calculations Summary Table" was performed for EWT_{max}=85 F. The effects of heat pump performance were accounted for in this calculation (though they are not shown) and the results of the calculations show a 14.1% increase in required bore length versus designing around EWT_{max}=90 F (k_{Grout}=0.88 Btu/hr-ft-F, bore spacing=15 feet, 2x4 rectangular configuration). For EWT_{max}=85 F, the selected equipment will be sized such that it will handle the peak cooling load under design conditions with higher efficiency (TC_D=157,275 Btu/hr, DMD_D=10.90 kW, EER_D=14.43). The calculations were performed for EWT_{max}=85 F because, as shown in Figure A.49, the heating design lengths for EWT_{min}=40 F were longer than the cooling design lengths for EWT_{max}=90 F. The bore length multiplier, B_M, determined by calculating the NNAGL, was applied to the heating design lengths because the results of the energy analysis (Example A.3.2) displayed an annual heat of extraction from the loopfield (AGL_{HD}>AGL_{CD}).

Because the design heating and cooling lengths are close to one another (depending on the EWT_{min}/EWT_{max} combination used for design), further analysis should be done to determine the optimum design length and EWT_{min}/EWT_{max} combination. Table A.13 shows the calculations for various EWT_{min}/EWT_{max} combinations and the effects on GHEX design lengths.

Table A.13. Design Length Comparisons for Various Min/Max EWT Combinations

Heating							Cooling				
EWT_{min} (·F)	F_H	$(R_B+R_G \cdot F_H)$ (hr ft °F/Btu)	AGL_{HD} (Btu)	NNAGL	B_M	$L_{H,T}$ (ft)	EWT_{max} (·F)	F_C	$(R_B+R_G \cdot F_C)$ (hr ft °F/Btu)	AGL_{CD} (Btu)	$L_{C,T}$ (ft)
30	0.50	0.398	166,701,528	882	1.08	1,264	80	0.28	0.302	135,326,462	2,169
35	0.50	0.398	168,680,267	850	1.08	1,648	80	0.28	0.302	135,326,462	2,169
40	0.48	0.390	170,295,153	833	1.08	2,160	80	0.28	0.302	135,326,462	2,169
45	0.46	0.381	171,731,178	817	1.07	2,962	80	0.28	0.302	135,326,462	2,169
30	0.50	0.398	166,701,528	853	1.08	1,261	85	0.29	0.306	136,369,401	1,838
35	0.50	0.398	168,680,267	823	1.08	1,644	85	0.29	0.306	136,369,401	1,838
40	0.48	0.390	170,295,153	808	1.07	2,156	85	0.29	0.306	136,369,401	1,838
45	0.46	0.381	171,731,178	794	1.07	2,956	85	0.29	0.310	136,369,401	1,838
30	0.50	0.398	166,701,528	822	1.07	1,251	90	0.30	0.311	137,445,100	1,611
35	0.50	0.398	168,680,267	796	1.07	1,641	90	0.30	0.311	137,445,100	1,611
40	0.48	0.390	170,295,153	782	1.07	2,097	90	0.30	0.311	137,445,100	1,611
45	0.46	0.381	171,731,178	770	1.07	2,950	90	0.30	0.311	137,445,100	1,611
30	0.50	0.398	166,701,528	791	1.07	1,254	95	0.31	0.315	138,555,230	1,406
35	0.50	0.398	168,680,267	767	1.07	1,637	95	0.31	0.315	138,555,230	1,406
40	0.48	0.390	170,295,153	756	1.07	2,146	95	0.31	0.315	138,555,230	1,406
45	0.46	0.381	171,731,178	745	1.07	2,944	95	0.31	0.315	138,555,230	1,406

All of the calculations given in Table A.13 are for a GHEX with 1-inch diameter DR-11 HDPE U-bend piping, 15 feet C-C bore spacing (2x4 rectangular configuration), 5-inch borehole diameter, and an 0.88 Btu/hr-ft-F grout thermal conductivity value. The purpose of Table A.13 is to display the effects of varying EWT_{min} (from 30 F to 45 F) and EWT_{max} (from 80 F to 95 F) on GHEX design lengths.

The final GHEX design lengths chosen will depend on the designer's preferred EWT_{min}/EWT_{max} combination. In most cases, it is not recommended that the GHEX be designed such that EWT_{max}>90 F unless the deep earth temperature is higher than 65-70 F (refer to Section 2.7 for EWT_{min}/EWT_{max} selection guidelines). For this example, to satisfy the constraint of EWT_{max}=90 F, the total active bore length will need to be at least 1,611 feet. Therefore, utilizing an EWT_{min} value of 30 F in heating will not be an option for this example because the GHEX design length to do so is 1,264 feet, which is less that the required active bore length for EWT_{max}=90 F.

To avoid using antifreeze, the GHEX will need to be designed for EWT_{min}=45 F. As shown in Table A.13, the required total active bore length for this parameter is 2,950 feet, which is 83.1% greater than the active bore length required for EWT_{max}=90 F. Installing 2,950 total feet of active bore to avoid the use of antifreeze will not be economical, as the installation cost savings from avoiding the use of antifreeze will not offset the additional cost to install 83.1% more GHEX piping.

The most economical solution for this example would be to install 1,611 total active bore length to satisfy the constraint of EWT_{max}=90 F. If the 2x4 grid (eight bores) is used, the active bore length will be 201 ft/bore. However, because 1-inch U-bends are typically used for bore depths between 225 feet and 300 feet, a 1x7 grid could also be used. If this configuration were used, the active length per bore would increase from 210 ft/bore to 230 ft/bore and the flow through each bore would increase from 4.63 gpm to 5.29 gpm per bore (37 gpm total system flow rate).

For head loss calculation purposes, the GHEX will be assumed to consist of seven vertically-bored, 1-inch diameter U-bends each 235 feet in depth (230 feet active depth) in the step-down, step-up reverse-return configuration given in Figure 5.9. Because EWT_{min}~30 F, loop freeze protection will be required down to 17 F (10 F below the average loop temperature under design conditions). A water-antifreeze solution of 20% propylene glycol by volume will provide adequate freeze protection for this example.

Example A.3.4 Interior Piping Layout for Multiple Heat Pump System – Louisville, KY

After the heat pump sizing and selection process is complete, the placement of each GSHP unit in the residence will need to be determined so that the interior piping and air distribution systems can be designed. For the interior piping system, a common return line from the heat pumps back to the GHEX was used and is shown in Figure A.52. The pipe sizes given in the figure were based on 1-3 feet of head loss per 100 feet of pipe length. When a common return line is used for a distributed pumping system, check valves must be incorporated in the piping system to prevent "short circuiting" flow through an inactive GSHP in the system and to force flow through the active GSHP units and out to the GHEX. As shown in Figure A.51, the non-pressurized, distributed pumping system used for this example incorporates the use of check valves in each set of pumps. No field-installed check valves will be required for this system.

The supply lines from the flow center to the heat pump units are not displayed in Figure A.52 for clarity. The non-pressurized distributed pumping system in Figure 6.11 was used for this example. As shown in Figure 6.11 and again in Figure A.51, individual supply lines were run from each set of pumps to the respective heat pump served by those pumps. Running the individual supply lines (as opposed to running a single supply line with branches to each GSHP unit) eliminates the need for solenoid zone valves, pressure transducers, etc. The drawback, however, will be an increase in total installed piping material to serve the system.

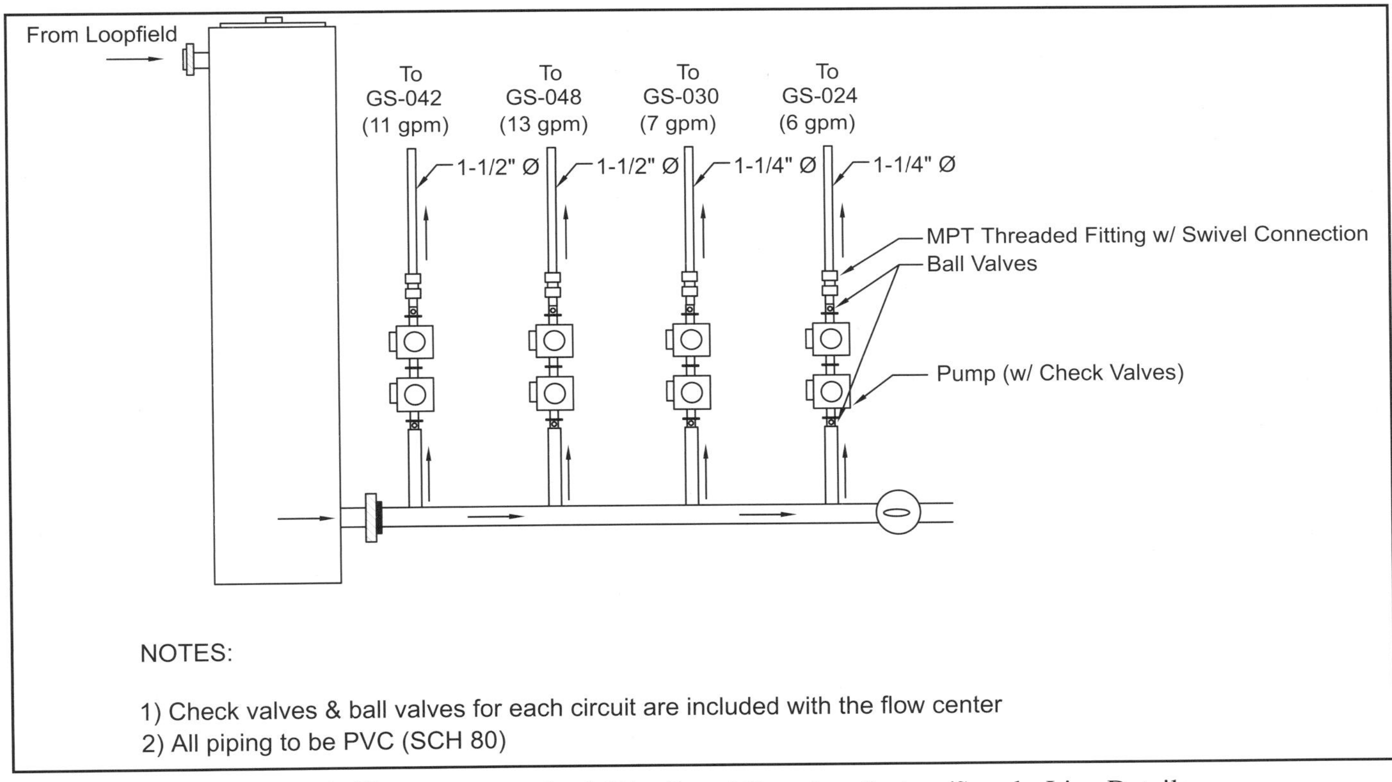

Figure A.51. Non-Pressurized, Distributed Pumping System/Supply-Line Details

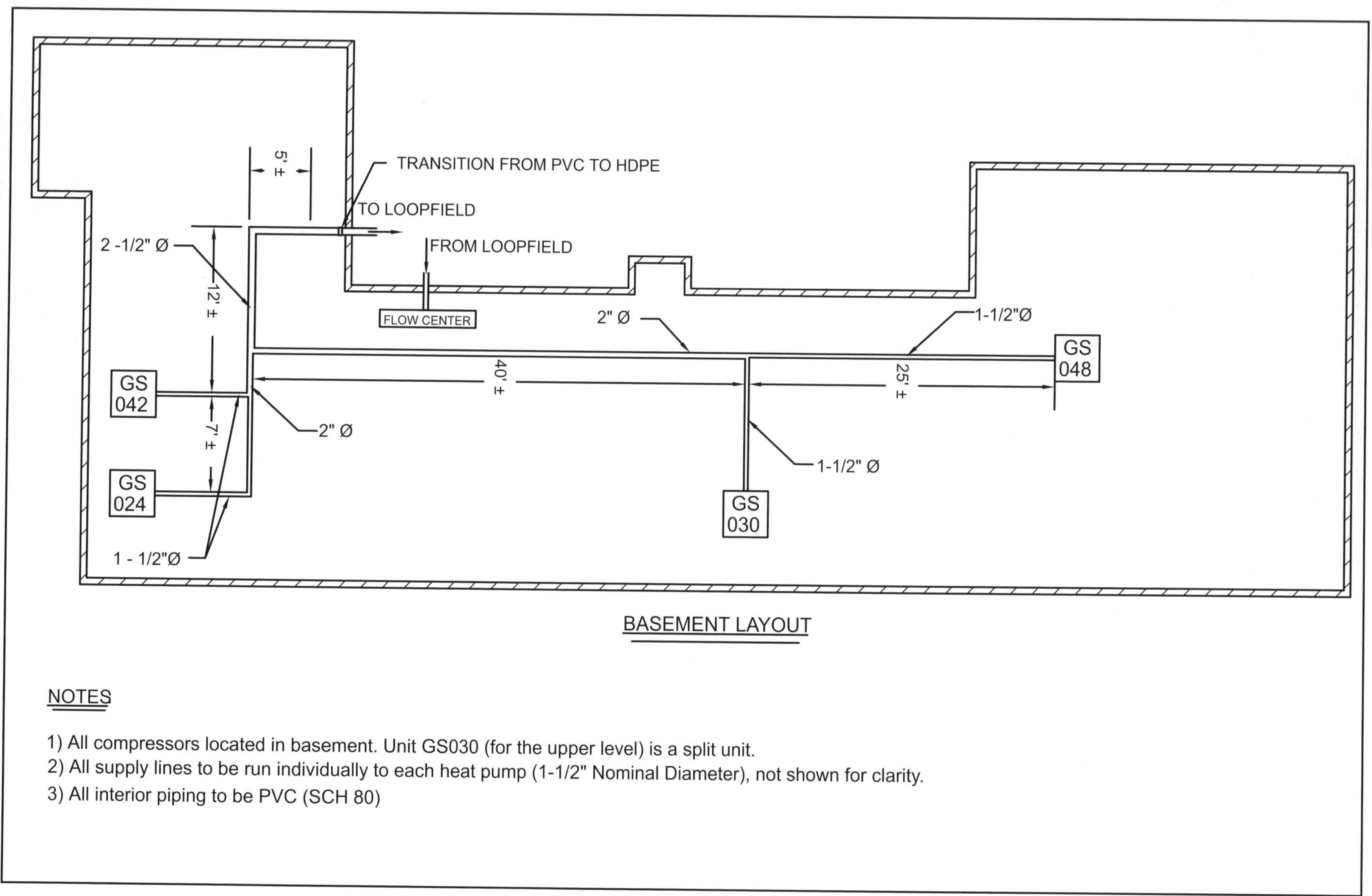

Figure A.52. Multiple Heat Pump Example GSHP Unit/Interior Piping Layout

Example A.3.5 Exterior Piping Layout for Multiple Heat Pump System – Louisville, KY

As discussed in Example A.3.3, the GHEX will consist of seven vertically-bored, 1-inch diameter U-bends each 235 feet in depth (230 feet active depth) in the step-down, step-up reverse-return configuration given in Figure 5.9 (operating flow rates shown). Figure A.53 displays the step-down, step-up reverse-return piping system for this example. Figure A.54 displays a preliminary site plan for this system with the proposed vertical-bore locations.

For head loss calculations, the supply-return piping run from the building to the first bore in the loopfield will be assumed to be 70 feet. As displayed in Figure A.53, the system operating flow rate will be 37 gpm and the supply-return pipe sizing will be 2-inch (DR-11 HDPE).

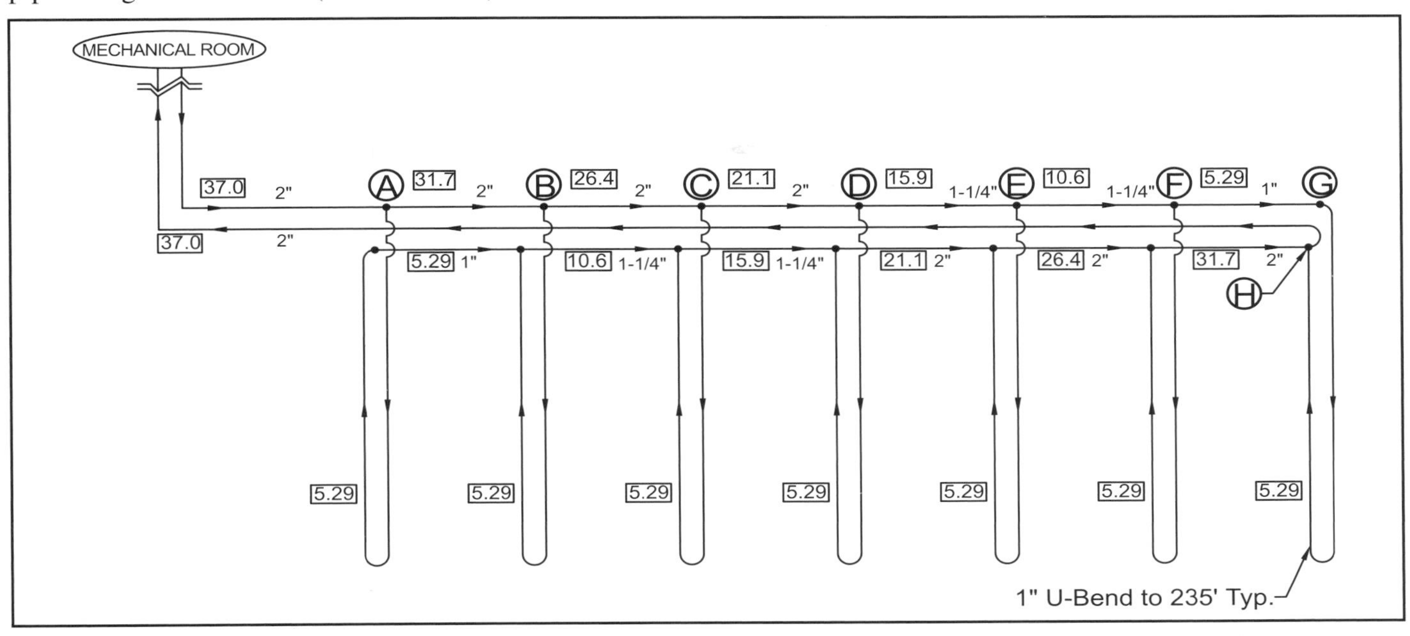

Figure A.53. Step-Down, Step-Up Reverse-Return Layout for Multiple Heat Pump Example

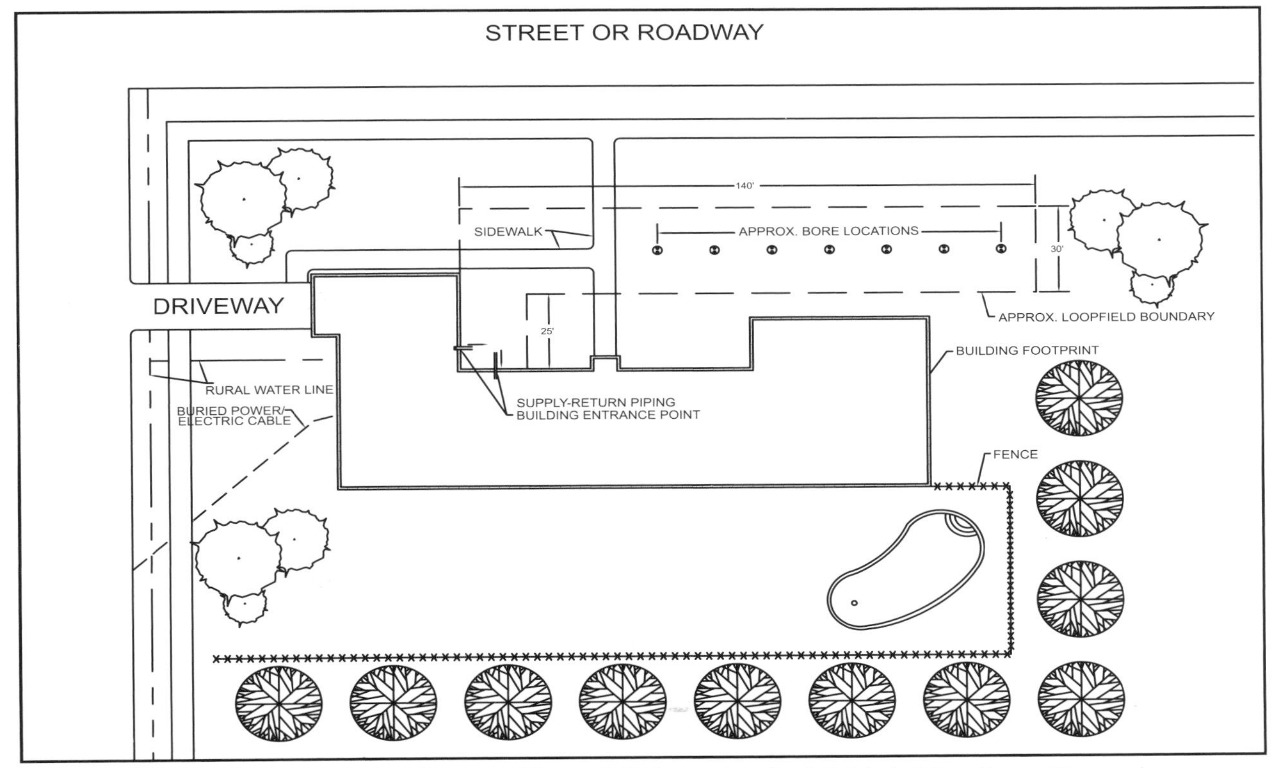

Figure A.54. Site Plan/Proposed Loopfield Location for Multiple Heat Pump Example

Example A.3.6 System Head Loss Calculations for Multiple Heat Pump System – Louisville, KY

As discussed in Example A.3.5, the GHEX will consist of seven vertically-bored, 1-inch diameter U-bends each 235 feet in depth (230 feet active depth) in the step-down, step-up reverse-return configuration given in Figure A.53. Additionally, the 2-inch DR-11 HDPE supply-return piping run from the building to the first bore in the loopfield was assumed to be 70 feet. The system operating flow rate is 37 gpm and the individual flow rate through each U-bend is 5.3 gpm.

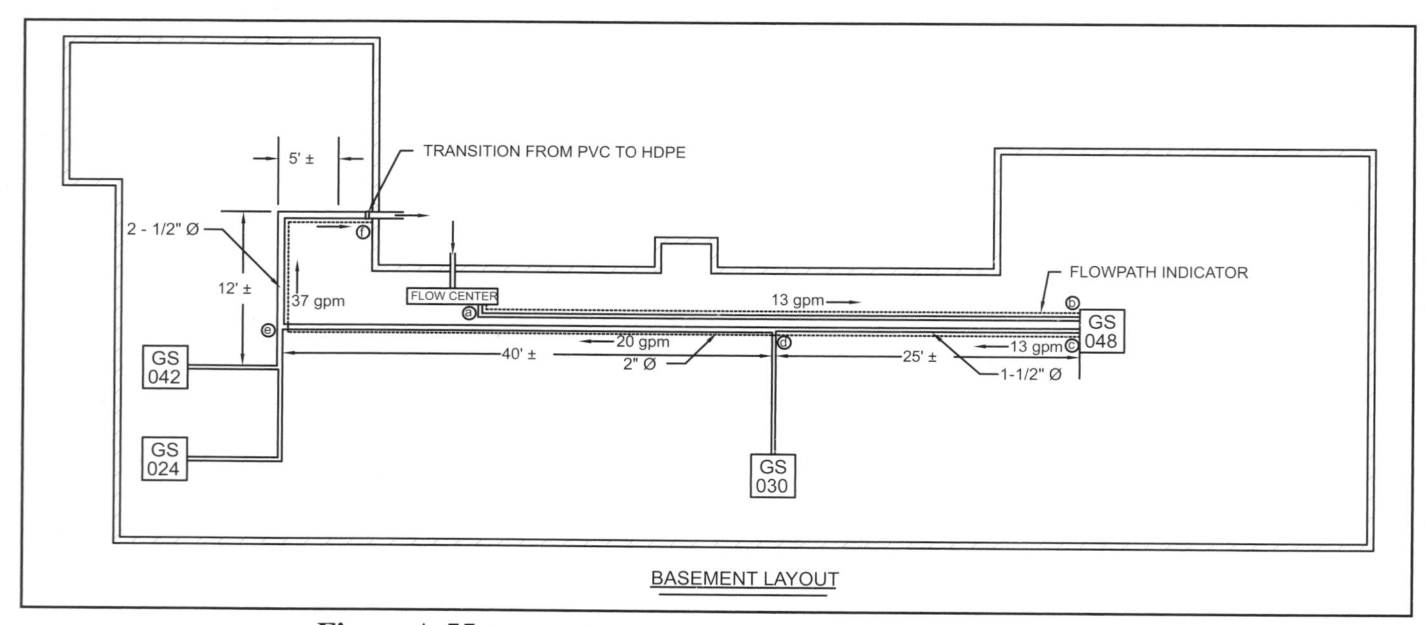

Figure A.55. Interior Piping Flow Path for Head Loss Calculations

In addition to identifying a single parallel flow path for the interior piping system, a single parallel flow path in the exterior GHEX piping system must also be identified for head loss calculation purposes. The single parallel flow path through the GHEX identified for head loss calculations is displayed in the Circulating Pump Worksheet in Figure A.56.

As shown in Figure A.56, the circulating system will need to be capable of providing 37 gpm at a total system head loss of 49.3 feet. Because the distributed pumping configuration was utilized for this example, each set of pumps will need to be capable of providing the flow that is required by the GSHP unit they serve at the total system head loss. The flow capability requirements for each set of pumps are as follows:

- Flow Requirements for the GS-042: 11 gpm @ 49.3 feet of head
- Flow Requirements for the GS-048: 13 gpm @ 49.3 feet of head
- Flow Requirements for the GS-030: 7 gpm @ 49.3 feet of head
- Flow Requirements for the GS-024: 6 gpm @ 49.3 feet of head

For simplicity of design and because no circulating pump will be able to exactly match the flow capacity and head requirements of a given system, select the circulating pumps that will match the flow capacity and head requirements for the worst case (in this example, flow capacity and head requirements for the GS-048) and utilize the same make/model circulators for each heat pump. Then, to balance the system and provide the required flow to each GSHP unit, the ball valves (incorporated in the non-pressurized distributed flow center displayed in Figure A.51) can be used as balancing valves to throttle the flow to the smaller heat pumps. Throttle flow through the ball valves until the pressure drop through the water-to-refrigerant coil matches the pressure drop in the manufacturer's performance data tables.

Circulating Pump Worksheet

NOTE: See Figure A.55 for Interior Piping Flow Path Used for Head Loss Calculations

MECHANICAL ROOM

A B C D E F G H

37.0 2" | 31.7 2" | 26.4 2" | 21.1 2" | 15.9 1-1/4" | 10.6 1-1/4" | 5.29 1"

37.0 2" | 5.29 1" | 10.6 1-1/4" | 15.9 1-1/4" | 21.1 2" | 26.4 2" | 31.7 2"

5.29 5.29 5.29 5.29 5.29 5.29 5.29

1" U-Bend to 235' Typ.

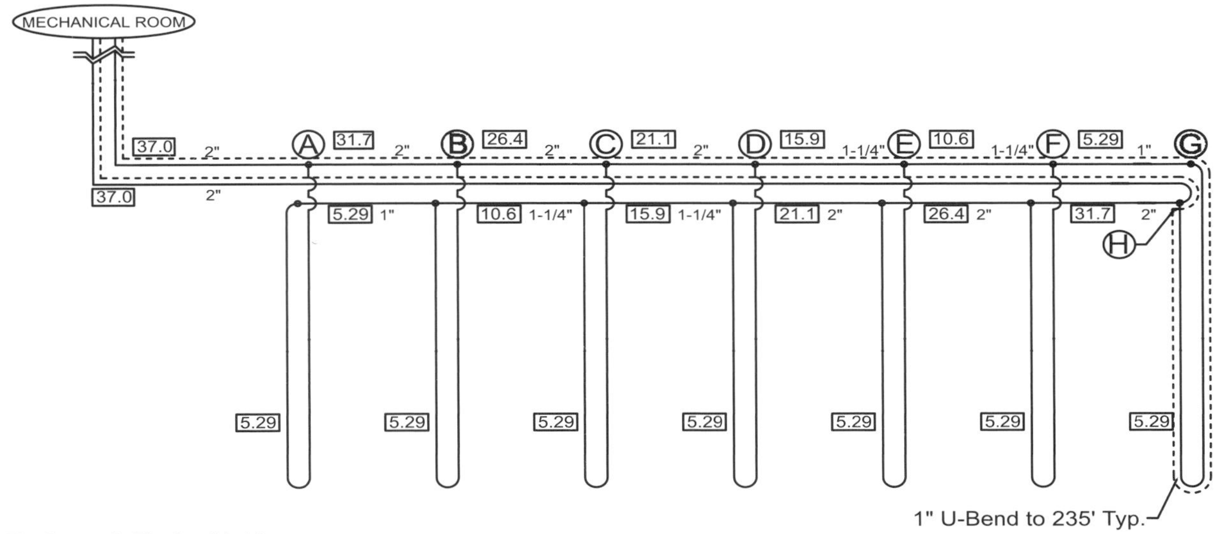

Sketch The System In The Provided Area

Interior Piping / Heat Pump Head Loss Calculations

Section	Description / Pipe Type / Size (in)	Pipe Run Length	Fitting Information: # of Fittings	x	Eq. Length per Fitting	=	Eq. Pipe Length	Total Equivalent Pipe Length	Flow (gpm)	Head Loss Info: HL/100'	Total HL
1) a-b	Flow Center - GS048/PVC (SCH80)/1.50"	45'	3	x	3	=	9'	54'	13	1.73	0.93
2)	/ /			x		=					
3)	/ /			x		=					
4) GSHP	Water to Refrigerant Coil	NA	NA	x	NA	=	NA	NA	13	NA	15.0
5) c-d	GS-048 - GS-030 / PVC (SCH80) / 1.50"	25'	2	x	3	=	6'	31'	13	1.73	0.54
6) d-e	GS-030 - GS-042 / PVC (SCH80) / 2.00"	40'	2	x	3	=	6'	46'	20	1.09	0.50
7) e-f	GS-042 - Exterior / PVC (SCH80) / 2.50"	12'	3	x	3	=	9'	21'	37	1.39	0.29
										Interior Piping HL Total------>	17.3

Exterior Piping Head Loss Calculations

Section	Description / Pipe Type / Size (in)	Pipe Run Length	Fitting Information: # of Fittings	x	Eq. Length per Fitting	=	Eq. Pipe Length	Total Equivalent Pipe Length	Flow (gpm)	Head Loss Info: HL/100'	Total HL
1) Mech-A	S-R / DR-11 HDPE / 2"	70'	5	x	3	=	15'	85'	37.0	3.26	2.77
2) A-B	S-R / DR-11 HDPE / 2"	15'	2	x	3	=	6'	21'	31.7	2.47	0.52
3) B-C	S-R / DR-11 HDPE / 2"	15'	2	x	3	=	6'	21'	26.4	1.78	0.37
4) C-D	S-R / DR-11 HDPE / 2"	15'	2	x	3	=	6'	21'	21.1	1.19	0.25
5) D-E	S-R / DR-11 HDPE / 1-1/4"	15'	2	x	3	=	6'	21'	15.9	3.97	0.83
6) E-F	S-R / DR-11 HDPE / 1-1/4"	15'	2	x	3	=	6'	21'	10.6	1.92	0.40
7) F-G-H	S-R & U-Bend / DR-11 HDPE / 1"	485'	2	x	3	=	6'	491'	5.29	1.69	8.30
8) H-Mech	S-R / DR-11 HDPE / 2"	160'	5	x	3	=	15'	175'	37.0	3.26	5.71
										Exterior Piping HL Total------>	19.2

Head Loss Summary

Total System HL (Interior Piping HL Total + Exterior Piping HL Total) = 36.5 *ft (Water at 40°F)*

Circulating Fluid Type Prop. Glycol *Concentration* 20 *% @* 25 *°F --> Antifreeze Viscosity Correction Factor* 1.35 *(Table 4.21)*

Corrected System HL = (Total System HL x Antifreeze Viscosity Correction Factor) = 49.3 *ft @* 37 *gpm*

Figure A.56. Circulating Pump Worksheet for Multiple Heat Pump Example

The pump curve for two Grundfos UP 26-99F circulators in series (Figure A.57) shows that they can produce approximately 46 feet of head at a flow rate of 13 gpm. Using two series-connected UP 26-99F circulators for each heat pump (a total of eight circulators will be used) will satisfy the system flow capacity and head requirements. Using two UP 26-99F circulators in series will result in a circulating system that will exceed the flow capacity and head requirements for the two smaller heat pumps (GS-030 and GS-024). As previously stated, ball valves will be necessary to throttle flow through the smaller heat pumps until the pressure drop through the water-to-refrigerant coil matches the pressure drop in the manufacturer's performance data tables in order to provide adequate flow to each GSHP unit in the system.

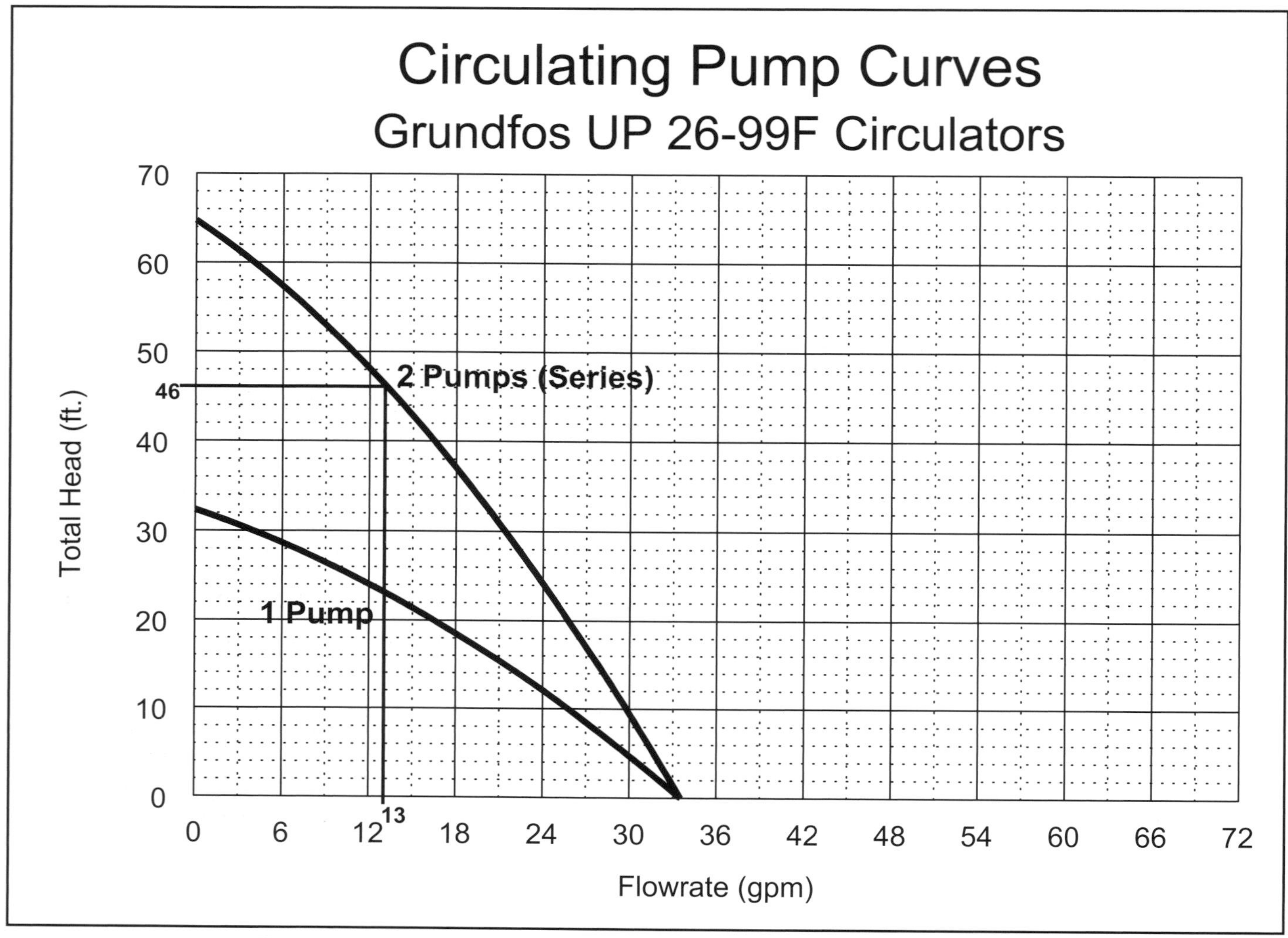

Figure A.57. Circulating Pump Curves for Multiple Heat Pump Example

Example A.3.7 System Piping Volume Calculations for Multiple Heat Pump System – Louisville, KY

The last step in designing this system would be to perform total piping volume calculations to determine the amount of antifreeze necessary for adequate freeze protection. Figures A.51-A.54 and A.58 were utilized to determine total run lengths of each type/size of pipe in order to perform the system volume calculations. Because the Antifreeze Volume Worksheet in Chapter 8 (Table 8.8) is to be used for single heat pump systems primarily, a spreadsheet program such as Microsoft Excel would be better suited to perform the antifreeze volume calculations for a large system such as this. When using a spreadsheet program to perform the calculations, they will need to be done in similar fashion as Example 8.3 using the pipe volumes given in Tables 8.3-8.6.

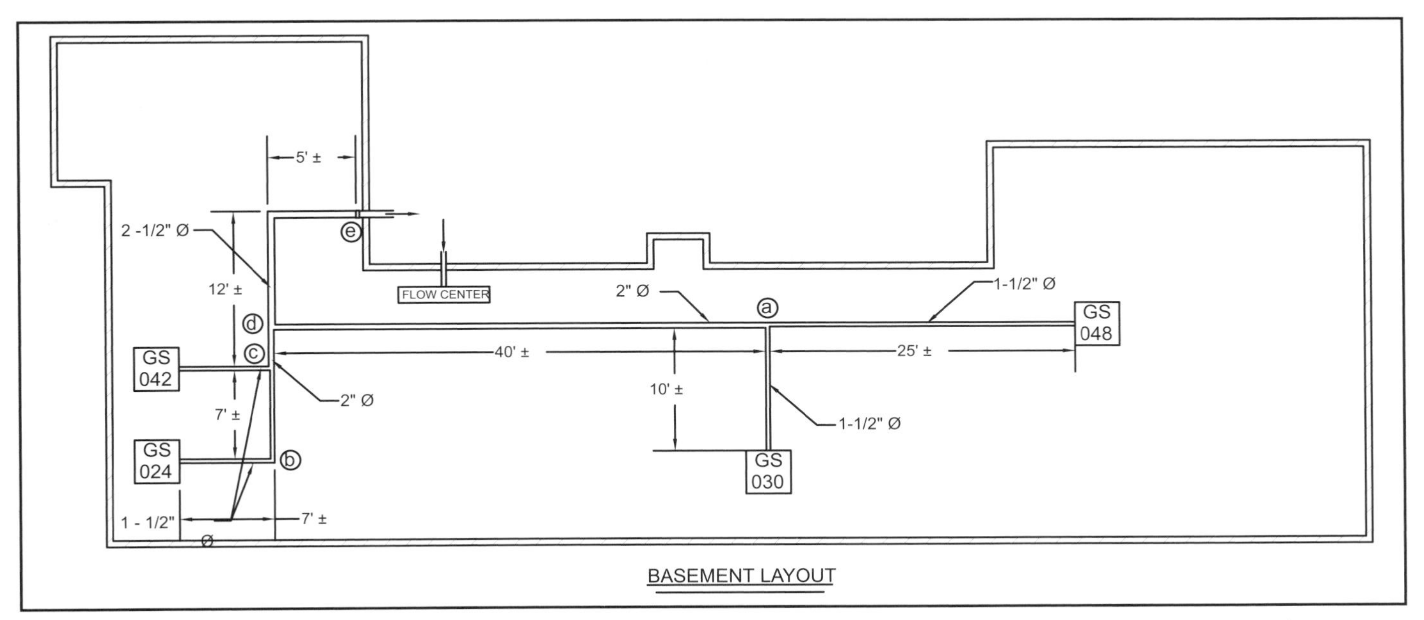

Figure A.58. Basement Layout for Antifreeze Volume Calculations - Multiple Heat Pump Example

Table A.14 was filled out according to Figures A.51 and A.58:

Table A.14. Interior Piping Volume Calculations for Multiple Heat Pump Example

Interior Piping System						
Run Type	Run Description	Pipe Type	Pipe Size (in)	Volume per Unit Length (gal/100')	Run Length (ft)	Pipe Volume (gal)
Supply	Flow Center - GS-042	SCH 80 PVC	1 1/2	9.18	20'	1.84
	Flow Center - GS-048	SCH 80 PVC	1 1/2	9.18	55'	5.05
	Flow Center - GS-030	SCH 80 PVC	1 1/4	6.66	40'	2.66
	Flow Center - GS-024	SCH 80 PVC	1 1/4	6.66	25'	1.67
Return	GS-048 - (a)	SCH 80 PVC	1 1/2	9.18	25'	2.30
	GS-030 - (a)	SCH 80 PVC	1 1/2	9.18	10'	0.92
	(a) - (d) - (c)	SCH 80 PVC	2	15.34	43'	6.60
	GS-042 - (c)	SCH 80 PVC	1 1/2	9.18	5'	0.46
	GS-024 - (b)	SCH 80 PVC	1 1/2	9.18	5'	0.46
	(b) - (c)	SCH 80 PVC	1 1/2	9.18	7'	0.64
	(d) - (e)	SCH 80 PVC	2 1/2	22.02	17'	3.74

Antifreeze Type/Concentration = 20% Propylene Glycol by Volume

Total Piping Volume = 26.3 Gallons
Total Antifreeze Volume = 5.3 Gallons

Table A.15 was filled out according to Figures A.53-A.54

Table A.15. Ground-Loop Piping Volume Calculations for Multiple Heat Pump Example

Ground-Loop Piping System						
Run Type	Run Description	Pipe Type	Pipe Size (in)	Volume per Unit Length (gal/100')	Run Length (ft)	Pipe Volume (gal)
S-R	SDSURR*	DR-11 HDPE	2	15.41	320'	49.31
	SDSURR*	DR-11 HDPE	1 1/4	7.53	60'	4.52
	SDSURR*	DR-11 HDPE	1	4.72	30'	1.42
GHEX	U-Bend	DR-11 HDPE	1	4.72	3290' **	155.29

Antifreeze Type/Concentration = 20% Propylene Glycol by Volume
**-SDSURR = Step-Down, Step-Up, Reverse-Return*
***-(7 u-bends) x (470' per u-bend) = 3290'*

Total Piping Volume = 210.5 Gallons
Total Antifreeze Volume = 42.1 Gallons

As displayed in Tables A.14-A.15, 47.4 gallons (5.3 gal + 42.1 gal = 47.4 gal) of propylene glycol will need to be installed in the piping system for a 20% propylene glycol concentration level (by volume) to freeze protect down to 17 F.

APPENDIX B: WORKSHEETS

In This Section

Single Capacity Water-Air GSHP Selection Worksheet

Block Load Description = ____________ Zone # ______

Total Heat Loss = ______ Btu/hr @ ____F OAT	Total Heat Gain = ______ Btu/hr @ ____F OAT
	Sen Heat Gain = ______ Btu/hr
	Lat Heat Gain = ______ Btu/hr
T-Stat Set Point = / / (Htg dB / Clg dB / %-RH)	SHF_{space} = Sen Heat Gain / Total Heat Gain
	SHF_{space} = ______

Brand ______

Model ______

Heating Mode Performance Data

Performance Parameters			Correction Factors[A] HC	Correction Factors[A] DMD
EWT_{min}=	______	F	CF_{EAT}=____	CF_{EAT}=____
GPM=	______	gpm	CF_{CFM}=____	CF_{CFM}=____
EAT=	______	F	CF_{AF}=____	CF_{AF}=____
CFM=	______	cfm		
Antifreeze / %[B]	/			
$CF_{EAT} \times CF_{CFM} \times CF_{AF}$==>			CF_{HC}=____	CF_{DMD}=____

Cooling Mode Performance Data

Performance Parameters			Correction Factors[A] TC	Correction Factors[A] SC	Correction Factors[A] DMD
EWT_{max}=	______	F	CF_{EAT}=____	CF_{EAT}=____	CF_{EAT}=____
GPM=	______	gpm	CF_{CFM}=____	CF_{CFM}=____	CF_{CFM}=____
EAT=	/	F/F (dB/wB)	CF_{AF}=____	CF_{AF}=____	CF_{AF}=____
CFM=	______	cfm			
Antifreeze / %[B]	/				
$CF_{EAT} \times CF_{CFM} \times CF_{AF}$==>			CF_{TC}=____	CF_{SC}=____	CF_{DMD}=____

Tabled Heating Mode Performance Data

HC = ______ Btu/hr

DMD = ______ kW

COP = ______

HE = ______ Btu/hr (= HC x (COP-1)/COP)[C]

Tabled Cooling Mode Performance Data

TC = ______ Btu/hr

SC = ______ Btu/hr (= TC x 0.75 approximately)[C]

DMD = ______ kW

EER = ______

HR = ______ Btu/hr (= TC x (EER+3.412)/EER)[C]

Corrected Heating Mode Performance Data

HC_c=	$HC \times CF_{HC}$=	______	Btu/hr
DMD_c=	$DMD \times CF_{DMD}$=	______	kW
COP_c=	$HC_c/(DMD_c \times 3412)$ =	______	
HE_c =	$HC_c \times (COP_c-1)/COP_c$ =	______	Btu/hr
LWT_{min}=	$EWT_{min} - HE_c/(500xGPM)$ =	______	F

Corrected Cooling Mode Performance Data

TC_c=	$TC \times CF_{TC}$=	______	Btu/hr
SC_c=	$SC \times CF_{SC}$=	______	Btu/hr
DMD_c=	$DMD \times CF_{DMD}$=	______	kW
EER_c=	$TC_c/(DMD_c \times 1000)$ =	______	
HR_c =	$TC_c \times (EER_c+3.412)/EER_c$ =	______	Btu/hr
LWT_{max}=	$EWT_{max} + HR_c/(500xGPM)$ =	______	F

% Sizing in Heating Mode

% Sizing = $\frac{HC_c}{\text{Total Heat Loss}}$ x 100%

= ______ %

% Sizing in Cooling Mode

% Oversizing = $\frac{SC_c - \text{Sen Heat Gain}}{\text{Sen Heat Gain}}$ x 100%

= ______ % (> 0 & ≤ 25%) ____ OK

SHF_{unit}= $\frac{SC_c}{TC_c}$

= ______ (≤ SHF_{space}) ____ OK

Latent Clg Cap = $TC_c - SC_c$

= ______ Btu/hr (> Lat Ht Gain) ____ OK

Blower Performance Summary

Type	Speed	CFM	Static (in)
PSC	L M H		
ECM	On		
	w/ Comp		
	w/ Emer		

Tabled Water Coil Performance Summary

EWT_{min}	GPM	WPD (psi)	WPD (ft)	Test Circulating Fluid

A. Apply correction factors for CFM and EAT as needed depending on format of Manufacturer's Literature. (No correction - CF=1)

B. $CF_{AF} = CF_{Used}/CF_{Tested}$. CF_{Used} for antifreeze being used. CF_{Tested} for circulating fluid performance was determined with.

C. Use this equation if performance parameter not provided by Manufacturer in the Engineering Specifications.

NOTES:

Figure 3.2. Single-Capacity, Water-to-Air Heat Pump Selection Worksheet

Two Capacity Water-Air GSHP Selection Worksheet

Block Load Description = ________ Zone # ____

Total Heat Loss = ________ Btu/hr @ _____F OAT

T-Stat Set Point = / / (Htg dB / Clg dB / %-RH)

Total Heat Gain = ________ Btu/hr @ _____F OAT
Sen Heat Gain = ________ Btu/hr
Lat Heat Gain = ________ Btu/hr

SHF_{space}= Sen Heat Gain / Total Heat Gain

SHF_{space}= ________

Brand ________
Model ________

Sized for High Capacity in Heating / Cooling

(High / Low) Capacity Heating Mode Performance Data

Performance Parameters		Correction Factors[A] HC	Correction Factors[A] DMD
EWT_{min}=	F	CF_{EAT}=____	CF_{EAT}=____
GPM=	gpm	CF_{CFM}=____	CF_{CFM}=____
EAT=	F	CF_{AF}=____	CF_{AF}=____
CFM=	cfm		
Antifreeze / %[B]	/		
CF_{EAT} x CF_{CFM} x CF_{AF}==>		CF_{HC}=____	CF_{DMD}=____

(High / Low) Capacity Cooling Mode Performance Data

Performance Parameters		Correction Factors[A] TC	Correction Factors[A] SC	Correction Factors[A] DMD
EWT_{max}=	F	CF_{EAT}=____	CF_{EAT}=____	CF_{EAT}=____
GPM=	gpm	CF_{CFM}=____	CF_{CFM}=____	CF_{CFM}=____
EAT= /	F/F (dB/wB)	CF_{AF}=____	CF_{AF}=____	CF_{AF}=____
CFM=	cfm			
Antifreeze / %[B]	/			
CF_{EAT} x CF_{CFM} x CF_{AF}==>		CF_{TC}=____	CF_{SC}=____	CF_{DMD}=____

Tabled Heating Mode Performance Data

HC = ________ Btu/hr
DMD = ________ kW
COP = ________
HE = ________ Btu/hr (= HC x (COP-1)/COP)[C]

Tabled Cooling Mode Performance Data

TC = ________ Btu/hr
SC = ________ Btu/hr (= TC x 0.75 approximately)[C]
DMD = ________ kW
EER = ________
HR = ________ Btu/hr (= TC x (EER+3.412)/EER)[C]

Corrected Heating Mode Performance Data

HC_c= HC x CF_{HC}= ________ Btu/hr
DMD_c= DMD x CF_{DMD}= ________ kW
COP_c= $HC_c/(DMD_c \times 3412)$ = ________
HE_c = $HC_c \times (COP_c-1)/COP_c$ = ________ Btu/hr
LWT_{min}= $EWT_{min} - HE_c/(500xGPM)$ = ________ F

Corrected Cooling Mode Performance Data

TC_c= TC x CF_{TC}= ________ Btu/hr
SC_c= SC x CF_{SC}= ________ Btu/hr
DMD_c= DMD x CF_{DMD}= ________ kW
EER_c= $TC_c/(DMD_c \times 1000)$ = ________
HR_c = $TC_c \times (EER_c+3.412)/EER_c$ = ________ Btu/hr
LWT_{max}= $EWT_{max} + HR_c/(500xGPM)$ = ________ F

% Sizing in Heating Mode

% Sizing = HC_c / Total Heat Loss x 100%

= ________ %

% Sizing in Cooling Mode

% Oversizing = (SC_c - Sen Heat Gain) / Sen Heat Gain x 100%

= ________ % (> 0 & ≤ 25%) ____ OK

SHF_{unit}= SC_c / TC_c

= ________ (≤ SHF_{space}) ____ OK

Latent Clg Cap = TC_c - SC_c

= ________ Btu/hr (> Lat Ht Gain) ____ OK

Blower Performance Summary

Type	Speed	CFM	Static (in)
ECM	On		
	w/ Low		
	w/ High		
	w/ Emer		

Tabled Water Coil Performance Summary

EWT_{min}	GPM	WPD (psi)	WPD (ft)	Test Circulating Fluid

A. Apply correction factors for CFM and EAT as needed depending on format of Manufacturer's Literature. (No correction - CF=1)

B. $CF_{AF} = CF_{Used}/CF_{Tested}$. CF_{Used} for antifreeze being used. CF_{Tested} for circulating fluid performance was determined with.

C. Use this equation if performance parameter not provided by Manufacturer in the Engineering Specifications.

NOTES:

Figure 3.3. Two-Capacity, Water-to-Air Heat Pump Selection Worksheet

Single Capacity Water-Water GSHP Selection Worksheet

Block Load Description = ______________________ Zone # ______

Total Htg Load = ______ Btu/hr @ _____F OAT
Htg Load Design Water Temp = ______ F (HLDWT)
Htg Load Water Flow Rate = ______ gpm

T-Stat Set Point = / / (Htg dB / Clg dB / %-RH)

Total Clg Load = ______ Btu/hr @ _____F OAT
Clg Load Design Water Temp = ______ F (CLDWT)
Clg Load Water Flow Rate = ______ gpm

Brand ______________________
Model ______________________

Heating Mode Performance Data

Performance Parameters

EWT_{min}= ______ F
GPM= ______ gpm
GPM_{load}= ______ gpm
ELT_{max}= ______ F
Antifreeze / %[B] ______ /

Correction Factors[A]

HC	DMD
CF_{AF}=______	CF_{AF}=______

Tabled Heating Mode Performance Data

HC = ______ Btu/hr
DMD = ______ kW
COP = ______
HE = ______ Btu/hr (= HC x (COP-1)/COP)[C]
LLT_H = ______ F (= ELT_{max} + HC / (GPM_{load} x 500)) (≥ HLDWT)

Corrected Heating Mode Performance Data

HC_c= HC x CF_{AF} = ______ Btu/hr
DMD_c= DMD x C_{AF} = ______ kW
COP_c= HC_c/(DMD_c x 3412) = ______
HE_c = HC_c x (COP_c-1)/COP_c = ______ Btu/hr
LWT_{min}= EWT_{min} - HE_c/(500xGPM) = ______ F

% Sizing in Heating Mode

$$\% \text{ Sizing} = \frac{HC_c}{\text{Total Heat Loss}} \times 100\%$$

% Sizing = ______ %

Cooling Mode Performance Data

Performance Parameters

EWT_{max}= ______ F
GPM= ______ gpm
GPM_{load}= ______ gpm
ELT_{min}= ______ F
Antifreeze / %[B] ______ /

Correction Factors[A]

TC	DMD
CF_{AF}=______	CF_{AF}=______

Tabled Cooling Mode Performance Data

TC = ______ Btu/hr
DMD = ______ kW
EER = ______
HR = ______ Btu/hr (= TC x (EER+3.412)/EER)[C]
LLT_C = ______ F (= ELT_{min} - TC / (GPM_{load} x 500)) (≤ CLDWT)

Corrected Cooling Mode Performance Data

TC_c= TC x CF_{AF} = ______ Btu/hr
DMD_c= DMD x CF_{AF} = ______ kW
EER_c= HC_c/(DMD_c x 1000) = ______
HR_c = HC_c x (EER_c+3.412)/EER_c = ______ Btu/hr
LWT_{max}= EWT_{max} + HR_c/(500xGPM) = ______ F

% Sizing in Cooling Mode

$$\% \text{ Oversizing} = \frac{TC_c - \text{Total Clg Ld}}{\text{Total Clg Ld}} \times 100\%$$

= ______ %

Tabled Water Coil Performance Summary

Side	Mode	EWT_{min}	GPM	WPD (psi)	WPD (ft)	Test Circulating Fluid
Loop						
Load						

A. Apply correction factors for Antifreeze (AF) as needed depending on Manufacturer's Literature. (No correction - CF=1)
B. $CF_{AF} = CF_{Used}/CF_{Tested}$. CF_{Used} for antifreeze being used. CF_{Tested} for circulating fluid performance was determined with.
C. Use this equation if performance parameter not provided by Manufacturer in the Engineering Specifications.

NOTES:

Figure 3.4. Single-Capacity, Water-to-Water Heat Pump Selection Worksheet

Vertically-Bored GHEX Design Worksheet – Heating Mode

Space Heating GSHP Design Data

HC_C [1] = ________ Btu/hr
DMD_C [1] = ________ kW
GPM [1] = ________ gpm
AGL_H = ________ Btu
AGL_C = ________ Btu
Total Heat Loss [1] = ________ Btu/hr
Percent Sizing = ________ %
F_{Jan} = ________

HWG GSHP Design Data

HC_{HWG} = ________ Btu/hr
DMD_{HWG} = ________ kW
GPM_{HWG} = ________ gpm
AGL_{HWG} = ________ Btu
F_{HWG} = ________

Total Design Data

HC_D = ________ Btu/hr
DMD_D = ________ kW
GPM_D = ________ gpm
AGL_{HD} = ________ Btu
AGL_{CD} = ________ Btu

GHEX Design Data

COP_D [2] (= HC_D / (DMD_D x 3,412)) = ________
HE_D [2] (= HC_D x (COP_D-1)/COP_D) = ________ Btu/hr
F_H (= (F_{Jan} x HC_C + F_{HWG} x HC_{HWG}) / HC_D) = ________
EWT_{min} [3] = ________ F
LWT_{min} [2] (= EWT_{min} – HE_D/(500xGPM_D) = ________ F

1. For single heat pump installation use data directly from the appropriate GSHP Selection Worksheet. For multiple heat pumps in the installation sum all heat pump Total Heat Losses, HC_C's, DMD_C's and GPM's that will be connected to a single GHEX using the table provided below.
2. For single heat pump installation use data directly from appropriate GSHP Selection Worksheet. For multiple heat pumps use the equation provided.
3. EWT_{min} is obtained directly from the appropriate GSHP Selection Worksheet and must be the same for the selection of all heat pumps for a multiple heat pump installation connected to a single GHEX.

	Zone 1	Zone 2	Zone 3	Zone 4	Zone 5	Zone 6	Zone 7	Zone 8	Zone 9	Zone 10	Total
Total Heat Loss											
HC_C											
DMD_C											
GPM											

Vertical Borefield Layout

U-bend D_P = ________ in
GPM_{FP} = ________ gpm/flowpath
N_{FP} = ________ flowpaths
N_{BIS} = ________ bores in series
N_B = ________ bores
Layout = ____ x ____ (N_{Rows} x $N_{Bores/Row}$)
Bore Spacing = ________ ft

Vertical Borefield Design Data

T_G = ________ F
k_G = ________ Btu/hr ft F (Table 5.6)
D_{bore} = ________ in
k_{Grout} = ________ Btu/hr ft F
$(R_B + R_G \cdot F_H)$ = ________ hr ft F/Btu (Table 5.8, 5.9 or 5.10)

Borehole Design Lengths (Equations 5.1 and 5.2)

$$L_{H,T} = \frac{HC_D \cdot \left(\frac{COP_D - 1}{COP_D}\right) \cdot (R_B + R_G \cdot F_H)}{T_G - \left(\frac{EWT_{min} + LWT_{min}}{2}\right)} = \frac{\_\_\_ \cdot \left(\frac{\_\_\_ - 1}{\_\_\_}\right) \cdot (\_\_\_)}{\_\_\_ - \left(\frac{\_\_\_ + \_\_\_}{2}\right)} = \frac{\_\_\_ \cdot (\_\_\_) \cdot (\_\_\_)}{\_\_\_ - (\_\_\_)} = \_\_\_$$

ft of bore (Below Header Trench)

$$L_{H,B} = \frac{L_{H,T}}{N_B} = \frac{\_\_\_}{\_\_\_} = \_\_\_$$

ft per bore (Below Header Trench)

Unbalanced Ground Load Borehole Design Lengths (Equations 5.9a, 5.10, and 5.11)

$$NNAGL = \frac{AGL_{HD} - AGL_{CD}}{L_{H,T} \cdot \left(T_G - \left(\frac{EWT_{min} + LWT_{min}}{2}\right)\right)} = \frac{\_\_\_ - \_\_\_}{\_\_\_ \cdot \left(\_\_\_ - \left(\frac{\_\_\_ + \_\_\_}{2}\right)\right)} = \_\_\_ \text{ Btu/ftF}$$

B_M = ________ (Figures 5.13 through 5.18)

$$L_{H,T,UGL} = B_M \cdot L_{H,T} = \_\_\_ \cdot \_\_\_ = \_\_\_ \text{ ft of bore}$$

$$L_{H,B,UGL} = \frac{L_{H,T,UGL}}{N_B} = \frac{\_\_\_}{\_\_\_} = \_\_\_ \text{ ft per bore}$$

Heating Design Length Calculations Summary Table

Layout for Flow and Number of Bores						Design Length Calculations				Unbalanced Ground Load Design Lengths					
D_P	GPM_{FP}	N_{FP}	N_{BIS}	N_B	Layout	k_{Grout}	$R_B+R_G \cdot F_H$	$L_{H,T}$	$L_{H,B}$	Layout	Spacing	NNAGL	B_M	$L_{H,T,UGL}$	$L_{H,B,UGL}$
					/										
					/										
					/										
					/										
					/										

Figure 5.19. Vertically-Bored GHEX Design Worksheet – Heating Mode

Vertically-Bored GHEX Design Worksheet – Cooling Mode

Space Cooling GSHP Design Data

TC_C[1] =________ Btu/hr
DMD_C[1] =________ kW
GPM[1] =________ gpm
AGL_H =________ Btu
AGL_C =________ Btu
Total Heat Gain[1] =________ Btu/hr
Percent Sizing =________ %
F_{Jul} =________

HWG GSHP Design Data

GPM_{HWG} =________ gpm
AGL_{HWG} =________ Btu

Total Design Data

TC_D =________ Btu/hr
DMD_D =________ kW
GPM_D =________ gpm
AGL_{HD} =________ Btu
AGL_{CD} =________ Btu

GHEX Design Data

EER_D[2] (= TC_D / (DMD_D x 1000) =________
HR_D[2] (= TC_D x (EER_D+3.412)/EER_D) =________ Btu/hr
F_C (= F_{Jul}) =________
EWT_{max}[3] =________ F
LWT_{max}[2] (= EWT_{max} + HR_D/(500xGPM_D) =________ F

1. For single heat pump installation, use data directly from the appropriate GSHP Selection Worksheet. For multiple heat pumps in the installation, sum all heat pump Total Heat Gains, TC_C's, DMD_C's, and GPM's that will be connected to a single GHEX using the table provided below.
2. For single heat pump installation, use data directly from appropriate GSHP Selection Worksheet. For multiple heat pumps, use the equation provided.
3. EWT_{max} is obtained directly from the appropriate GSHP Selection Worksheet and must be the same for the selection of all heat pumps for a multiple heat pump installation connected to a single GHEX.

	Zone 1	Zone 2	Zone 3	Zone 4	Zone 5	Zone 6	Zone 7	Zone 8	Zone 9	Zone 10	Total
Total Heat Gain											
TC_C											
DMD_C											
GPM											

Vertical Borefield Layout

U-bend D_P =________ in
GPM_{FP} =________ gpm/flowpath
N_{FP} =________ flowpaths
N_{BIS} =________ bores in series
N_B =________ bores
Layout =____ x ____ (N_{Rows} x $N_{Bores/Row}$)
Bore Spacing =________ ft

Vertical Borefield Design Data

T_G =________ F
k_G =________ Btu/hr ft F (Table 5.6)
D_{bore} =________ in
k_{Grout} =________ Btu/hr ft F
$(R_B + R_G \cdot F_C)$ =________ hr ft F/Btu (Table 5.8, 5.9 or 5.10)

Borehole Design Lengths (Equations 5.3 and 5.4)

$$L_{C,T} = \frac{TC_D \cdot \left(\frac{EER_D + 3.412}{EER_D}\right) \cdot (R_B + R_G \cdot F_C)}{\left(\frac{EWT_{max} + LWT_{max}}{2}\right) - T_G} = \frac{\_\_\_ \cdot \left(\frac{\_\_\_ + 3.412}{\_\_\_}\right) \cdot (\_\_\_)}{\left(\frac{\_\_\_ + \_\_\_}{2}\right) - \_\_\_} = \frac{\_\_\_ \cdot (\_\_\_) \cdot (\_\_\_)}{(\_\_\_) - \_\_\_} = \_\_\_$$

ft of bore (Below H[eader] T[rench])

$$L_{C,B} = \frac{L_{C,T}}{N_B} = \frac{\_\_\_}{\_\_\_} = \_\_\_$$ ft per bore (Below Header Trench)

Unbalanced Ground Load Borehole Design Lengths (Equations 5.9b, 5.12, and 5.13)

$$NNAGL = \frac{AGL_{CD} - AGL_{HD}}{L_{C,T} \cdot \left(\left(\frac{EWT_{max} + LWT_{max}}{2}\right) - T_G\right)} = \frac{\_\_\_ - \_\_\_}{\_\_\_ \cdot \left(\left(\frac{\_\_\_ + \_\_\_}{2}\right) - \_\_\_\right)} = \_\_\_ \text{ Btu / ftF}$$

B_M =________ (Figures 5.13 through 5.18)

$L_{C,T,UGL} = B_M \cdot L_{C,T} = \_\_\_ \cdot \_\_\_ = \_\_\_$ ft of bore

$L_{C,B,UGL} = \frac{L_{C,T,UGL}}{N_B} = \frac{\_\_\_}{\_\_\_} = \_\_\_$ ft per bore

Cooling Design Length Calculations Summary Table

Layout for Flow and Number of Bores						Design Length Calculations				Unbalanced Ground Load Design Lengths					
D_P	GPM_{FP}	N_{FP}	N_{BIS}	N_B	Layout	k_{Grout}	$R_B+R_G \cdot F_C$	$L_{C,T}$	$L_{C,B}$	Layout	Spacing	NNAGL	B_M	$L_{C,T,UGL}$	$L_{C,B,UGL}$
					/										
					/										
					/										
					/										
					/										

Figure 5.20. Vertically-Bored GHEX Design Worksheet – Cooling Mode

Horizontally-Trenched GHEX Design Worksheet – Heating Mode

Space Heating GSHP Design Data

HC_C[1] =__________ Btu/hr
DMD_C[1] =__________ kW
GPM[1] =__________ gpm
Total Heat Loss[1] =__________ Btu/hr
Percent Sizing =__________ %
F_{Jan} =__________

HWG GSHP Design Data

HC_{HWG} =__________ Btu/hr
DMD_{HWG} =__________ kW
GPM_{HWG} =__________ gpm
F_{HWG} =__________

Total Design Data

HC_D =__________ Btu/hr
DMD_D =__________ kW
GPM_D =__________ gpm

GHEX Design Data

COP_D[2] (= HC_D / (DMD_D x 3,412) =__________
HE_D[2] (= HC_D x (COP_D-1)/COP_D) =__________ Btu/hr
F_H (= (F_{Jan} x HC_C + F_{HWG} x HC_{HWG}) / HC_D) =__________
EWT_{min}[3] =__________ F
LWT_{min}[2] (= EWT_{min} – HE_D/(500xGPM_D)) =__________ F

1. For single heat pump installation, use data directly from the appropriate GSHP Selection Worksheet. For multiple heat pumps in the installation, sum all heat pump Total Heat Losses, HC_C's, DMD_C's, and GPMs that will be connected to a single GHEX using the table provided below.
2. For single heat pump installation, use data directly from appropriate GSHP Selection Worksheet. For multiple heat pumps, use the equation provided.
3. EWT_{min} is obtained directly from the appropriate GSHP Selection Worksheet and must be the same for the selection of all heat pumps for a multiple heat pump installation connected to a single GHEX.

	Zone 1	Zone 2	Zone 3	Zone 4	Zone 5	Zone 6	Zone 7	Zone 8	Zone 9	Zone 10	Total
Total Heat Loss											
HC_C											
DMD_C											
GPM											

Horizontal Trench Design Data

Location ____________, ______ (City, State)

T_M =__________ F (Table 5.11)
A_s =__________ F (Table 5.11)
T_O =__________ Days (Table 5.11)

Soil Type ____________ (Section 5.3.2.2)

k_{soil} =__________ Btu/hr ft F
α_{soil} =__________ ft²/day

Trench and Pipe Configuration ==> Table __________ (Select from Tables 5.17 – 5.29)

N_P =__________ ft_{pipe}/ft_{trench}
Conf. =__________ (Stand / Laying / Rect)
S_H =__________ ft
S_{cc} =__________ ft
d_{bp} =__________ ft
R_s =__________ hr ft F/Btu
S_M =__________
d_{avg} =__________ ft

D_P =__________ in (Nom.)
GPM_{FP} =__________ gpm/flowpath
N_{FP} =__________ flowpaths
$N_{FP/T}$ =__________ flowpaths/trench
N_T =__________ trenches

Design Soil Temperature for Heating

$T'(d_{avg},T_O)$ =__________ F (Table 5.12)
$T_{S,L} = T'(d_{avg},T_O) \cdot A_s + T_M$ = ______•______+______=______F (Eq. 5.25)

Pipe Resistance and Pipe Multiplier

R_p =__________ hr ft F/Btu (Table 4.6)
P_M =__________ (Figure 5.24)

Horizontal Trench Design Lengths (Equations 5.14, 5.15, and 5.16)

$$L_{H,P} = \frac{HC_D \cdot \left(\frac{COP_D - 1}{COP_D}\right)(R_P + R_S \cdot P_M \cdot S_M \cdot F_H)}{T_{S,L} - \left(\frac{EWT_{min} + LWT_{min}}{2}\right)} = \frac{\_\_\_ \cdot \left(\frac{\_\_\_ - 1}{\_\_\_}\right)(\_\_\_ + \_\_\_ \cdot \_\_\_ \cdot \_\_\_ \cdot \_\_\_)}{\_\_\_ - \left(\frac{\_\_\_ + \_\_\_}{2}\right)} = \frac{\_\_\_ \cdot (\_\_\_)(\_\_\_)}{\_\_\_ - (\_\_\_)} = \_\_\_ \text{ ft of pipe}$$

$$L_{H,P/T} = \frac{L_{H,P}}{N_T} = \frac{\_\_\_}{\_\_\_} = \_\_\_ \; ft_{pipe} / \text{trench} \qquad L_{H,T} = \frac{L_{H,P/T}}{N_P} = \frac{\_\_\_}{\_\_\_} = \_\_\_ \; ft_{trench}$$

Heating Design Length Calculations Summary Table

Layout for Flow and Number of Trenches						Pipe and Trench Design Length Calculations								
D_P	GPM_{FP}	N_{FP}	Trench/N_P	$N_{FP/T}$	N_T	d_{avg}	$T'(d_{avg},T_o)$	$T_{S,L}$	R_s	P_M	S_{cc}/S_M	$L_{H,P}$	$L_{H,P/T}$	$L_{H,T}$
			/								/			
			/								/			
			/								/			
			/								/			
			/								/			

Figure 5.25. Horizontally-Trenched GHEX Design Worksheet – Heating Mode

Horizontally-Trenched GHEX Design Worksheet – Cooling Mode

Space Cooling GSHP Design Data

TC_C[1] =__________ Btu/hr
DMD_C[1] =__________ kW
GPM[1] =__________ gpm
Total Heat Gain[1] =__________ Btu/hr
Percent Sizing =__________ %
F_{Jul} =__________

HWG GSHP Design Data

GPM_{HWG} =__________ gpm

Total Design Data

TC_D =__________ Btu/hr
DMD_D =__________ kW
GPM_D =__________ gpm

GHEX Design Data

EER_D[2] (= TC_D / (DMD_D x 1000) =__________
HR_D[2] (= TC_D x (EER_D+3.412)/EER_D) =__________ Btu/hr
F_C (= F_{Jul}) =__________

EWT_{max}[3] =__________ F
LWT_{max}[2] (= EWT_{max} + HR_D/(500xGPM_D) =__________ F

1. For single heat pump installation, use data directly from the appropriate GSHP Selection Worksheet. For multiple heat pumps in the installation, sum all heat pump Total Heat Gains, TC_C's, DMD_C's, and GPM's that will be connected to a single GHEX using the table provided below.
2. For single heat pump installation, use data directly from appropriate GSHP Selection Worksheet. For multiple heat pumps, use the equation provided.
3. EWT_{max} is obtained directly from the appropriate GSHP Selection Worksheet and must be the same for the selection of all heat pumps for a multiple heat pump installation connected to a single GHEX.

	Zone 1	Zone 2	Zone 3	Zone 4	Zone 5	Zone 6	Zone 7	Zone 8	Zone 9	Zone 10	Total
Total Heat Gain											
HC_C											
DMD_C											
GPM											

Horizontal Trench Design Data

Location ________________, ______ (City, State)

T_M =__________ F (Table 5.11)
A_s =__________ F (Table 5.11)
T_O =__________ Days (Table 5.11)

Soil Type____________________ (Section 5.3.2.2)

k_{soil} =__________ Btu/hr ft F
α_{soil} =__________ ft²/day

Trench and Pipe Configuration ==> Table ____________ (Select from Tables 5.17 – 5.29)

N_P =__________ ft_{pipe}/ft_{trench}
Conf. =__________ (Stand / Laying / Rect)

S_H =__________ ft
S_{cc} =__________ ft
d_{bp} =__________ ft

R_s =__________ hr ft F/Btu
S_M =__________
d_{avg} =__________ ft

D_P =__________ in (Nom.)
GPM_{FP} =__________ gpm/flowpath
N_{FP} =__________ flowpaths
$N_{FP/T}$ =__________ flowpaths/trench
N_T =__________ trenches

Design Soil Temperature for Cooling

$T'(d_{avg},T_O+180)$ =____________ F (Table 5.13)
$T_{S,H} = T'(d_{avg},T_O+180)\bullet A_s+T_M$=______•______+______=______F (Eq. 5.26)

Pipe Resistance and Pipe Multiplier

R_p =__________ hr ft F/Btu (Table 4.6)
P_M =__________ (Figure 5.24)

Horizontal Trench Design Lengths (Equations 5.17, 5.18, and 5.19)

$$L_{C,P} = \frac{TC_D \cdot \left(\frac{EER_D + 3.412}{EER_D}\right)(R_P + R_S \cdot P_M \cdot S_M \cdot F_C)}{\left(\frac{EWT_{max} + LWT_{max}}{2}\right) - T_{S,H}} = \frac{\_\_\_ \cdot \left(\frac{\_\_\_ + 3.412}{\_\_\_}\right)(\_\_\_ + \_\_\_ \cdot \_\_\_ \cdot \_\_\_ \cdot \_\_\_)}{\left(\frac{\_\_\_ + \_\_\_}{2}\right) - \_\_\_} = \frac{\_\_\_ \cdot (\_\_\_)(\_\_\_)}{(\_\_\_) - \_\_\_} = \_\_\_ \text{ ft of pipe}$$

$$L_{C,P/T} = \frac{L_{C,P}}{N_T} = \frac{\_\_\_}{\_\_\_} = \_\_\_ \; ft_{pipe} / \text{trench}$$

$$L_{C,T} = \frac{L_{C,P/T}}{N_P} = \frac{\_\_\_}{\_\_\_} = \_\_\_ \; ft_{trench}$$

Cooling Design Length Calculations Summary Table

Layout for Flow and Number of Trenches						Pipe and Trench Design Length Calculations								
D_P	GPM_{FP}	N_{FP}	Trench/N_P	$N_{FP/T}$	N_T	d_{avg}	$T'(d_{avg},T_o+180)$	$T_{S,H}$	R_s	P_M	S_{cc}/S_M	$L_{C,P}$	$L_{C,P/T}$	$L_{C,T}$
			/								/			
			/								/			
			/								/			
			/								/			
			/								/			

Figure 5.26. Horizontally-Trenched GHEX Design Worksheet – Cooling Mode

Horizontally-Bored GHEX Design Worksheet – Heating Mode

Space Heating GSHP Design Data

HC_C[1] = ______ Btu/hr
DMD_C[1] = ______ kW
GPM[1] = ______ gpm
AGL_H = ______ Btu
AGL_C = ______ Btu
Total Heat Loss[1] = ______ Btu/hr
Percent Sizing = ______ %
F_{Jan} = ______

HWG GSHP Design Data

HC_{HWG} = ______ Btu/hr
DMD_{HWG} = ______ kW
GPM_{HWG} = ______ gpm
AGL_{HWG} = ______ Btu
F_{HWG} = ______

Total Design Data

HC_D = ______ Btu/hr
DMD_D = ______ kW
GPM_D = ______ gpm
AGL_{HD} = ______ Btu
AGL_{CD} = ______ Btu

GHEX Design Data

COP_D[2] (= HC_D / (DMD_D x 3,412)) = ______
HE_D[2] (= HC_D x (COP_D-1)/COP_D) = ______ Btu/hr
F_H (= (F_{Jan} x HC_C + F_{HWG} x HC_{HWG}) / HC_D) = ______
EWT_{min}[3] = ______ F
LWT_{min}[2] (= EWT_{min} – HE_D/(500xGPM_D)) = ______ F

1. For single heat pump installation, use data directly from the appropriate GSHP Selection Worksheet. For multiple heat pumps in the installation, sum all heat pump Total Heat Losses, HC_C's, DMD_C's, and GPM's that will be connected to a single GHEX using the table provided below.
2. For single heat pump installation, use data directly from appropriate GSHP Selection Worksheet. For multiple heat pumps, use the equation provided.
3. EWT_{min} is obtained directly from the appropriate GSHP Selection Worksheet and must be the same for the selection of all heat pumps for a multiple heat pump installation connected to a single GHEX.

	Zone 1	Zone 2	Zone 3	Zone 4	Zone 5	Zone 6	Zone 7	Zone 8	Zone 9	Zone 10	Total
Total Heat Loss											
HC_C											
DMD_C											
GPM											

Soil Design Data

Location ______ , ______ (City, State)
T_M = ______ F (Table 5.11)
A_s = ______ F (Table 5.11)
T_O = ______ Days (Table 5.11)
Soil Type ______ (Section 5.3.2.2)
k_{soil} = ______ Btu/hr ft F
α_{soil} = ______ ft²/day

Horizontally-Bored GHEX Layout

U-bend D_P = ______ in
GPM_{FP} = ______ gpm/flowpath
N_{FP} = ______ flowpaths
N_{BIS} = ______ bores in series
N_B = ______ bores
Layout = ____ x ____ (N_{Layers} x $N_{Bores/Layer}$)
Bore Spacing = ______ ft

Horizontally-Bored GHEX Design Data

D_{bore} = ______ in
k_{Grout} = ______ Btu/hr ft F
$(R_B + R_G \cdot F_H)$ = ______ hr ft F/Btu
d_{avg} = ______ ft
$T'(d_{avg},T_O)$ = ______ F (Table 5.12)
$T_{S,L} = T'(d_{avg},T_O) \bullet A_s + T_M$ = ______ • ______ + ______ = ______ F

Borehole Design Lengths (Equations 5.28 and 5.29)

$$L_{H,T} = \frac{HC_D \cdot \left(\frac{COP_D - 1}{COP_D}\right) \cdot (R_B + R_G \cdot F_H)}{T_{S,L} - \left(\frac{EWT_{min} + LWT_{min}}{2}\right)} = \frac{\_\_\_ \cdot \left(\frac{\_\_\_ - 1}{\_\_\_}\right) \cdot (\_\_\_)}{\_\_\_ - \left(\frac{\_\_\_ + \_\_\_}{2}\right)} = \frac{\_\_\_ \cdot (\_\_\_) \cdot (\_\_\_)}{\_\_\_ - (\_\_\_)} = \_\_\_ \text{ ft of bore}$$

$$L_{H,B} = \frac{L_{H,T}}{N_B} = \frac{\_\_\_}{\_\_\_} = \_\_\_ \text{ ft per bore}$$

Unbalanced Ground Load Borehole Design Lengths (Equations 5.9a, 5.10, and 5.11)

$$NNAGL = \frac{AGL_{HD} - AGL_{CD}}{L_{H,T} \cdot \left(T_{S,L} - \left(\frac{EWT_{min} + LWT_{min}}{2}\right)\right)} = \frac{\_\_\_ - \_\_\_}{\_\_\_ \cdot \left(\_\_\_ - \left(\frac{\_\_\_ + \_\_\_}{2}\right)\right)} = \_\_\_ \text{ Btu / ftF}$$

B_M = ______ (Figures 5.13 through 5.18)

$$L_{H,T,UGL} = B_M \cdot L_{H,T} = \_\_\_ \cdot \_\_\_ = \_\_\_ \text{ ft of bore}$$

$$L_{H,B,UGL} = \frac{L_{H,T,UGL}}{N_B} = \frac{\_\_\_}{\_\_\_} = \_\_\_ \text{ ft per bore}$$

Heating Design Length Calculations Summary Table

Layout for Flow and Number of Bores						Design Length Calculations				Unbalanced Ground Load Design Lengths					
D_P	GPM_{FP}	N_{FP}	N_{BIS}	N_B	Layout	k_{Grout}	$R_B+R_G \cdot F_H$	$L_{H,T}$	$L_{H,B}$	Layout	Spacing	NNAGL	B_M	$L_{H,T,UGL}$	$L_{H,B,UGL}$
					/										
					/										
					/										
					/										

Figure 5.27. Horizontally-Bored GHEX Design Worksheet – Heating Mode

Horizontally-Bored GHEX Design Worksheet – Cooling Mode

Space Cooling GSHP Design Data

TC_C[1] = ______ Btu/hr
DMD_C[1] = ______ kW
GPM[1] = ______ gpm
AGL_H = ______ Btu
AGL_C = ______ Btu
Total Heat Gain[1] = ______ Btu/hr
Percent Sizing = ______ %
F_{Jul} = ______

HWG GSHP Design Data

GPM_{HWG} = ______ gpm
AGL_{HWG} = ______ Btu

Total Design Data

TC_D = ______ Btu/hr
DMD_D = ______ kW
GPM_D = ______ gpm
AGL_{HD} = ______ Btu
AGL_{CD} = ______ Btu

GHEX Design Data

EER_D[2] (= TC_D / (DMD_D x 1000)) = ______
HR_D[2] (= TC_D x (EER_D+3.412)/EER_D) = ______ Btu/hr
F_C (= F_{Jul}) = ______
EWT_{max}[3] = ______ F
LWT_{max}[2] (= EWT_{max} + HR_D/(500xGPM_D)) = ______ F

1. For single heat pump installation, use data directly from the appropriate GSHP Selection Worksheet. For multiple heat pumps in the installation, sum all heat pump Total Heat Gains, TC_C's, DMD_C's, and GPM's that will be connected to a single GHEX using the table provided below.
2. For single heat pump installation, use data directly from appropriate GSHP Selection Worksheet. For multiple heat pumps, use the equation provided.
3. EWT_{max} is obtained directly from the appropriate GSHP Selection Worksheet and must be the same for the selection of all heat pumps for a multiple heat pump installation connected to a single GHEX.

	Zone 1	Zone 2	Zone 3	Zone 4	Zone 5	Zone 6	Zone 7	Zone 8	Zone 9	Zone 10	Total
Total Heat Gain											
TC_C											
DMD_C											
GPM											

Soil Design Data

Location ______ , ______ (City, State)
T_M = ______ F (Table 5.11)
A_s = ______ F (Table 5.11)
T_O = ______ Days (Table 5.11)
Soil Type ______ (Section 5.3.2.2)
k_{soil} = ______ Btu/hr ft F
α_{soil} = ______ ft^2/day

Horizontally-Bored GHEX Layout

U-bend D_P = ______ in
GPM_{FP} = ______ gpm/flowpath
N_{FP} = ______ flowpaths
N_{BIS} = ______ bores in series
N_B = ______ bores
Layout = ______ x ______ (N_{Layers} x $N_{Bores/Layer}$)
Bore Spacing = ______ ft

Horizontally-Bored GHEX Design Data

D_{bore} = ______ in
k_{Grout} = ______ Btu/hr ft F
(R_B + R_G·F_H) = ______ hr ft F/Btu
d_{avg} = ______ ft
T'(d_{avg},T_O+180) = ______ F (Table 5.13)
$T_{S,H}$ = T'(d_{avg},T_O+180)•A_s+T_M= ______•______+______=______F

Borehole Design Lengths (Equations 5.30 and 5.31)

$$L_{C,T} = \frac{TC_D \cdot \left(\frac{EER_D + 3.412}{EER_D}\right) \cdot (R_B + R_G \cdot F_C)}{\left(\frac{EWT_{max} + LWT_{max}}{2}\right) - T_{S,H}} = \frac{\_\_\_ \cdot \left(\frac{\_\_\_ + 3.412}{\_\_\_}\right) \cdot (\_\_\_)}{\left(\frac{\_\_\_ + \_\_\_}{2}\right) - \_\_\_} = \frac{\_\_\_ \cdot (\_\_\_) \cdot (\_\_\_)}{(\_\_\_) - \_\_\_} = \_\_\_ \text{ ft of bore}$$

$$L_{C,B} = \frac{L_{C,T}}{N_B} = \frac{\_\_\_}{\_\_\_} = \_\_\_ \text{ ft per bore}$$

Unbalanced Ground Load Borehole Design Lengths (Equations 5.9b, 5.12, and 5.13)

$$NNAGL = \frac{AGL_{CD} - AGL_{HD}}{L_{C,T} \cdot \left(\left(\frac{EWT_{max} + LWT_{max}}{2}\right) - T_{S,H}\right)} = \frac{\_\_\_ - \_\_\_}{\_\_\_ \cdot \left(\left(\frac{\_\_\_ + \_\_\_}{2}\right) - \_\_\_\right)} = \_\_\_ \text{ Btu/ftF}$$

B_M = ______ (Figures 5.13 through 5.18)

$$L_{C,T,UGL} = B_M \cdot L_{C,T} = \_\_\_ \cdot \_\_\_ = \_\_\_ \text{ ft of bore}$$

$$L_{C,B,UGL} = \frac{L_{C,T,UGL}}{N_B} = \frac{\_\_\_}{\_\_\_} = \_\_\_ \text{ ft per bore}$$

Cooling Design Length Calculations Summary Table

Layout for Flow and Number of Bores						Design Length Calculations				Unbalanced Ground Load Design Lengths					
D_P	GPM_{FP}	N_{FP}	N_{BIS}	N_B	Layout	k_{Grout}	R_B+R_G·F_C	$L_{C,T}$	$L_{C,B}$	Layout	Spacing	NNAGL	B_M	$L_{C,T,UGL}$	$L_{C,B,UGL}$
					/										
					/										
					/										
					/										

Figure 5.28. Horizontally-Bored GHEX Design Worksheet – Cooling Mode.

Circulating Pump Worksheet

Heat Pump

Flow Center

Sketch The System In The Provided Area

Interior Piping / Heat Pump Head Loss Calculations

Section	Description / Pipe Type / Size (in)	*Pipe Run Length*	*Fitting Information*					*Total Equivalent Pipe Length*	Flow (gpm)	*Head Loss Info*	
			# of Fittings	x	Eq. Length per Fitting	=	*Eq. Pipe Length*			HL/100'	*Total HL*
1)_____	_____/_____/_____			x		=					
2)_____	_____/_____/_____			x		=					
3)_____	_____/_____/_____			x		=					
4) GSHP	Water to Refrigerant Coil	NA	NA	x	NA	=	NA	NA		NA	
5)_____	_____/_____/_____			x		=					
6)_____	_____/_____/_____			x		=					
7)_____	_____/_____/_____			x		=					
								Interior Piping HL Total------>			_____

Exterior Piping Head Loss Calculations

Section	Description / Pipe Type / Size (in)	*Pipe Run Length*	*Fitting Information*					*Total Equivalent Pipe Length*	Flow (gpm)	*Head Loss Info*	
			# of Fittings	x	Eq. Length per Fitting	=	*Eq. Pipe Length*			HL/100'	*Total HL*
1)_____	_____/_____/_____			x		=					
2)_____	_____/_____/_____			x		=					
3)_____	_____/_____/_____			x		=					
4)_____	_____/_____/_____			x		=					
5)_____	_____/_____/_____			x		=					
6)_____	_____/_____/_____			x		=					
7)_____	_____/_____/_____			x		=					
8)_____	_____/_____/_____			x		=					
								Exterior Piping HL Total------>			_____

Head Loss Summary

Total System HL (Interior Piping HL Total + Exterior Piping HL Total) = __________ ft (Water at 40°F)

Circulating Fluid Type__________ Concentration_____% @ _____ °F --> Antifreeze Viscosity Correction Factor_______ (Table 4.21)

Corrected System HL = (Total System HL x Antifreeze Viscosity Correction Factor) = __________ ft @ ________ gpm

Figure 6.17. Circulating Pump Worksheet for Head Loss Calculations

Hot Water Generation Worksheet

Annual Cost of Operation Calculations

Monthly Hot Water Usage (MHWU) = ________ gal/month (Table 9.1)
Storage Tank Setpoint Temperature = ________ F (Table 9.2)
Cold Water Supply Temperature = ________ F (Table 5.11)
Storage Tank Standing Losses = ________ %
Annual Hot Water Usage (AHWU) (Equation 9.2)

AHWU = MHWU • 12 = ________ • 12 = ________ gal/yr

Annual Hot Water Energy (AHWE) (Equation 9.3)

$$AHWE = \frac{AHWU \bullet 62.3 \bullet 1}{7.48} \bullet (T_{tank} - T_{supply}) \bullet \left(1 + \frac{STL}{100}\right) = \frac{\_\_\_\_ \bullet 62.3 \bullet 1}{7.48} \bullet (\_\_\_\_ - \_\_\_\_) \bullet \left(1 + \frac{\_\_\_\_}{100}\right) = \_\_\_\_\_\_ \text{ Btu/yr}$$

Annual Cost for 100% of Hot Water Generation by Individual Fuel Source

Fuel Source	Fuel Cost		Eff or COP_{Avg}	Annual Fuel Usage / Cost		
				Annual Usage		Annual Cost ($/yr)
Electric Resistance		($/kWh)			(kWh)	
Propane		($/gal)			(Gal)	
Natural Gas		($/Therm)			(Therm)	
GSHP		($/kWh)			(kWh)	

Annual Cost for Selected Percentage of Hot Water Generated by GSHP

Check One	System	HWG by GSHP (%)			
		Guidelines		Assumed	
		Annual	Design Month	Annual	Design Month
	1) Dedicated GSHP w/ HWG Priority	100	100		
	2) Dedicated GSHP w/ Space Htg/Clg Priority				
	1 Tank Storage	40-60	50-70		
	2 Tank Storage	60-80	70-90		
	3) GSHP w/desuperheater				
	1 Tank Storage	20-40	30-50		
	2 Tank Storage	40-60	50-70		

Other Fuel = ________ $Cost_{OtherFuel\_A}$ = ________ $/yr

$$Cost_{Combined} = \frac{\%HWG_{GSHP-A}}{100} \bullet Cost_{GSHP-A} + \left(1 - \frac{\%HWG_{GSHP-A}}{100}\right) \bullet Cost_{OtherFuel-A}$$ (Equation 9.12)

$$Cost_{Combined} = \frac{\_\_\_\_}{100} \bullet \_\_\_\_ + \left(1 - \frac{\_\_\_\_}{100}\right) \bullet \_\_\_\_ = \_\_\_\_ \text{ \$/yr}$$

$$Savings_{GSHP} = Cost_{OtherFuel-A} - Cost_{Combined} = \_\_\_\_ - \_\_\_\_ = \_\_\_\_ \text{ \$/yr}$$

Annual Ground Heating Energy Load for Input to GHEX Design (Equation 9.13)

$$AGL_{HWG} = \%HWG_{GSHP-A} \bullet AHWE \bullet \left(\frac{COP_{Avg} - 1}{COP_{Avg}}\right) = \_\_\_\_ \bullet \left(\frac{\_\_\_\_ - 1}{\_\_\_\_}\right) = \_\_\_\_ \text{ Btu}$$ **(A)**

Design Month Data for Input to GHEX Design

HC_{HWG} = ________ Btu/hr
DMD_{HWG} = ________ kW
GPM_{HWG} = ________ gpm

DMHWE = ________ Btu (Equation 9.14)
DMHWRU = ________ hrs (Equation 9.15)
F_{HWG} = ________ (Equation 9.16) **(B)**

Figure 9.16. Hot Water Generation Worksheet

REFERENCES

1-1	EPA. 1993. Space Conditioning: The Next Frontier. The Potential of Advanced Residential Space Conditioning Technologies for Reducing Pollution and Saving Money. Environmental Protection Agency. EPA 430-R-93-004. April 1993.
1-3	EIA. 2007a. Residential Electricity Prices: A Consumer's Guide. Energy Information Agency. eia.doe.gov. EIA Brochures.
1-3	EIA. 2007b. Residential Natural Gas Prices: Information for Consumers. Energy Information Agency. eia.doe.gov. EIA Brochures.
1-4	EIA. 2007c. Propane Prices: What Consumers Should Know. Energy Information Agency. eia.doe.gov.
1-4	EIA. 2007d. Residential Heating Oil Prices: What Consumers Should Know. Energy Information Agency. eia.doe.gov. EIA Brochures.
1-23	Klaassen. 2007. Heat Pump Technologies in Perspective. Iowa Energy Center - Energy Resource Station. Iowa State University.
2-37,60	ISO. 1998a. ISO 13256-1. Water Source Heat Pumps – Testing and Rating for Performance – Part 1: Water-to-air and brine-to-air heat pumps. International Standards Organization. ISO 13256-1:1998(E).
2-68	ISO. 1998b. ISO 13256-2. Water Source Heat Pumps – Testing and Rating for Performance – Part 2: Water-to-water and brine-to-water heat pumps. International Standards Organization. ISO 13256-2:1998(E).
3-2,3,5,8	ACCA. 2006. Manual J – HVAC Residential Load Calculation, 8th Edition. Air Conditioning Contractors of America. Washington, DC.
3-5,8,28,29 5-60	ASHRAE. 2009. Handbook of Fundamentals. American Society of Heating, Refrigerating and Air-Conditioning Engineers, Inc. Atlanta, GA.
3-26,27,28	ACCA. 2009. Manual D – Residential Duct Systems. Air Conditioning Contractors of America. Washington, DC.
3-29	NOAA. 2000. Enginering Weather Data Supplement. National Oceanic and Atmospheric Administration. National Climatic Data Center. Asheville, NC.
3-29	ASHRAE. 1997. WYEC2 - Weather Year for Energy Calculations 2. American Society of Heating, Refrigerating and Air-Conditioning Engineers, Inc. Atlanta, GA.

3-29	USAF. 1978. Engineering Weather Data. Department of the Air Force Manual AFM 88-29. U.S. Government Printing Office, Washington, DC.
4-1	ASHRAE. 2008. Handbook of Systems and Equipment. American Society of Heating, Refrigerating and Air-Conditioning Engineers, Inc. Atlanta, GA.
4-1	ASTM. 1996. Standard Specification for Seamless Copper Water Tube. Standard B88. American Society for Testing and Materials, West Conshohocken, PA.
4-1	ASTM. 1997. Standard Specification for Seamless Copper Tube for Air Conditioning and Refrigeration Field Service. Standard B280. American Society for Testing and Materials, West Conshohocken, PA.
4-4	ASTM. 2010. Standard Specification for Polyethylene Plastics Pipe and Fittings Materials. Standard D-3350. American Society for Testing and Materials, West Conshohocken, PA.
4-4, 5-28	IGSHPA, 2009. Closed-Loop/Geothermal Heat Pump Systems - Design and Installation Standards. 2009 Edition. International Ground Source Heat Pump Association. Stillwater, OK.
5-24	Southern Methodist University Geothermal Lab. 2009. http://smu.edu/geothermal/ Southern Methodist University. Dallas, TX.
5-25	NWWA. 1985. Water Source Heat Pump Handbook. National Water Well Association. Dublin, OH.
5-26	NOAA. 2009. NOAA Satellite and Information Service. http://www.ncdc.noaa.gov/ National Environmental Satellite, Data and Information Service. U.S. Department of Commerce. Washington, DC.
5-27,28	ASHRAE. 2011. Handbook of HVAC Applications. American Society of Heating, Refrigerating and Air-Conditioning Engineers, Inc. Atlanta, GA.
5-27	Kavanaugh, S.P. and K. Rafferty. 1997. Ground-Source Heat Pumps - Design of Geothermal Systems for Commercial and Institutional Buildings. ASHRAE.
5-28	Kavanaugh, S.P. 2000. Field Tests for Ground Thermal Properties – Methods and Impact on GSHP System Design. ASHRAE Transactions 106(1):DA00-13-4.
5-28	Kavanaugh, S.P. 2001. Investigations of Methods for Determining Soil Formation Thermal Characteristics form Short Term Field Tests. ASHRAE RP1118, Final Report.
5-30, 7-16,18,21,27	IGSHPA. 2000. Grouting for Vertical Geothermal Heat Pump Systems – Engineering Design and Field Procedures Manual. International Ground Source Heat Pump Association. Stillwater, OK.
5-52	Kusuda, T. and .R. Achenbach. 1965. Earth Temperature and thermal Diffusivity at Selected Stations in the United States. ASHRAE. Chicago.
5-60	Salomone, L.A. and J.I. Marlow. 1989. Soil and rock Classification According to Thermal Conductivity: Design of Ground-Coupled Heat Pump Systems. EPRI CU-6482. Electric Power Research Institute, Palo Alto, Ca.
5-60,61	Remund, C.P. 1998a. Thermal Performance of Soils and Backfills in Horizontal Ground Coupled Heat Pump System Applications. Electric Power Research Institute. TR-110480. Palo Alto, CA.
5-60,61	Remund, C.P. 1998b. Soil Thermal Conductivity Experimental Results Report. Electric Power Research Institute. Final Report WO3024-34. Palo Alto, CA.
5-62	IGSHPA. 1994. Closed-Loop Geothermal Systems – Slinky Installation Guide. International Ground Source Heat Pump Association. Stillwater, OK.
6-14	ASHRAE. 2008. Handbook of HVAC Systems and Equipment. American Society of Heating, Refrigerating and Air-Conditioning Engineers, Inc. Atlanta, GA.
6-14	Kavanaugh, S.P. and K. Rafferty. 1997. Ground-Source Heat Pumps: Design of Geothermal Systems for Commercial and Institutional Buildings. American Society of Heating, Refrigerating and Air-Conditioning Engineers, Inc. Atlanta, GA.
6-23, 11-1,2,3,4,24,25,27	LeClaire, M.R. and M.J. Lafferty. 1996. ACCA, FSU and MGEA Geothermal Heat Pump Training Certification Program. Air Conditioning Contractors of America. Washington, DC.

GLOSSARY

Absolute Pressure: A pressure reading relative to absolute zero pressure, expressed as the sum of atmospheric pressure and gauge pressure (psia).

Active Borehole Length: The length of the U-bend in the borehole below the header trench (usually 4 to 6 feet less than the total borehole length from the surface).

Air Source Heat Pump (ASHP): A heat pump that uses an air-to-refrigerant heat exchanger to extract and reject heat to the outside air.

Ambient Air: The surrounding air (usually outdoor air or the air in a specific location).

Annual Fuel Utilization Efficiency (AFUE): The average efficiency for fuel-burning equipment for an entire heating season, expressed as seasonal heating energy delivered (Btu) divided by the seasonal fuel energy consumed (Btu).

Annual Ground Load: Defined to be the difference between the annual amount of heat rejected to the GHEX in the cooling mode and the annual amount of heat extracted from the GHEX in the heating mode.

As-Built Drawing: A detailed drawing that shows everything included on the site plan in addition to the exact location, dimensions, and other pertinent details for a given GHEX installation after the installation is complete.

Balance Point Temperature: The outdoor air temperatures where internal heat gains from people, appliances, etc. offset the envelope heat loss to the atmosphere. It is at the balance point temperatures where no indoor heating or cooling will be required to maintain the temperature of the home at the thermostat set point.

Bend: A fitting either molded separately or formed from pipe for the purpose of accommodating a directional change.

Bin: In the bin method, a temperature increment, usually 5 F, into which the range of temperatures for an area are divided. Bins are used to produce a frequency distribution of hourly, monthly, or annual outdoor temperature occurrences for a specified location.

Block Load: Defined to be the sum of the zone loads. A block load calculation is necessary for a building with multiple zones served by a centralized heating/cooling system.

Blowers: Fans used to force air across the heat exchanger. With a ground source heat pump, the only blower used is to force air through the central heating system.

British Thermal Unit (Btu): The quantity of heat required to raise the temperature of one pound of water one degree Fahrenheit at a specified temperature.

Cavitation: The formation of bubbles due to partial vacuums in a flowing liquid as a result of separation of fluid particles.

Centralized Pumping System: Flow centers located centrally in a ground source system that produce flow to all heat pump units in that system.

Circulating Pump(s): The pump(s) that circulates the fluid in the closed-loop system during normal operation.

Closed-Loop System: The heat exchange loop in a GSHP system that is comprised of the ground heat exchanger, the circulating pump, and the water-source heat pump in which the heat transfer fluid is not exposed to the atmosphere.

Coefficient of Performance (COP): A measure of heating efficiency for heat pump equipment, expressed as the heating energy provided to the space (Btu) divided by the electric energy consumed to provide that heating (Btu).

Coil: A heat exchanger used to transfer energy from one source to another. In ground source heat pumps, water-to-refrigerant and refrigerant-to-air coils are used.

Combination GSHP Unit: A GSHP that has the ability to heat or cool air at full capacity or to heat or cool water at full capacity, but not both at the same time.

Compressor: The central component of a heat pump system. The compressor increases the pressure of a refrigerant fluid, and simultaneously reduces its volume, while causing the fluid to move through the system.

Compressor, Reciprocating: A positive displacement compressor in which the change in internal volume of the compression chamber(s) is accomplished by the reciprocating motion of one or more pistons.

Compressor, Rotary: A positive displacement compressor in which the change in internal volume of the compression chamber(s) is accomplished by the rotation of a rolling piston or rotating sliding vanes.

Concentric Fitting: A fitting used to connect the GSHP desuperheater piping to the domestic hot water tank. The concentric fitting is one for which the suction and discharge lines coming into and out of the hot water tank are through the same opening.

Condenser: A heat exchanger in which hot, pressurized (gaseous) refrigerant is condensed by transferring heat to cooler surrounding air, water, or earth.

Cycling Efficiency: Ratio of actual efficiency to steady-state efficiency. The actual efficiency of a heating or cooling system is somewhat lower due to start-up and shut-down losses.

Damper: A device used to vary the air flow rate through an air outlet, inlet, or duct.

Defrost Cycle: The control-activated process of removing accumulated frost and ice from the outdoor heat exchanger of air source heat pumps.

Degradation Coefficient: A dimensionless number used to help quantify the amount of efficiency loss due to startup, shutdown, and other part-load operation. For practical purposes in the case of a ground source heat pump system, a value within the range of 0.10 - 0.15 is a reasonable approximation for this coefficient.

Degree Day: A measure of the severity and duration of an outdoor temperature deviation above or below a fixed temperature (65 F), used in estimating the heating or cooling requirement and fuel consumption of a building for either summer or winter conditions.

Demand (DMD): The electrical input required to operate a GSHP unit for space conditioning.

Design Loads: The peak heating or cooling load used to select the equipment for a system (such as a heat pump) and to design the air distribution system (supply air diffusers, return air grilles, and the duct system). Design loads are based on standard or accepted conditions for a given locality (a design day).

Design Temperature, Summer: A specific temperature used in calculating the cooling load of a building. The summer design temperature is typically the outdoor air temperature that is exceeded 0.4% or 1.0% of the time.

Design Temperature, Winter: A specific temperature used in calculating the heating load of a building. The winter design temperature is typically the outdoor temperature that is exceeded 99.0% or 99.6% of the time.

Desuperheater: A device for recovering superheat from the compressor discharge gas of a heat pump or central air conditioner for use in heating or preheating water. Also known as a heat recovery water heater.

Diffuser (Air): A supply air outlet composed of deflecting elements discharging air in various directions and patterns to accomplish mixing of supply and room air.

Dimension Ratio (DR): A specific ratio of the average specified outside diameter to the minimum specified wall thickness (OD/t) for outside-diameter controlled plastic pipe.

Direct Expansion (DX) Earth-Coupled Heat Pump: A heat pump system in which the refrigerant is circulated in pipes buried underground.

Distributed Pumping System: A system made up of smaller, individual pumping stations (one flow center for each heat pump) each controlled individually by the operation of the specific heat pump unit that they serve.

Dual Circuit GSHP Unit: A GSHP which utilizes two compressors (generally of different capacities) connected to two refrigeration circuits to allow multiple modes of operation. This unit may use one compressor in heating or cooling of ducted air only, one compressor to heat water only, one compressor heating or cooling air while the other is heating water, or both compressors either heating or cooling air.

Dynamic System Check: A GSHP system startup check procedure performed after the system has been started and consists of measuring various system performance parameters to ensure that the system is performing as promised.

Effective Length: The design parameter for calculating the friction loss for a run. Effective length, expressed in feet, is the sum of the actual length and the equivalent fitting lengths.

Efficiency: A measure of the useful output of a system divided by the input required to drive the system.

Energy Efficiency Ratio (EER): A measure of cooling efficiency for heat pump equipment, expressed as the cooling energy removed from the space (Btu) divided by the electric energy consumed to provide that cooling (W).

Energy Loads: Used in predicting the energy necessary to operate the system for some prescribed time such as a month, year, or season. The calculation methodology may be the same as for the design load; however, the actual operating and weather data are used instead of design conditions.

Equipment Loads: Loads served by the heating/cooling system that are not included in peak heating/cooling block load calculations. These loads include duct and hydronic piping losses/gains as well as ventilation loads.

Equivalent Length: Used in duct system calculation procedures to express the same pressure loss for pipes of various diameters.

Envelope Load: The total heating or cooling load through any building component surrounding a conditioned space, including fenestration and infiltration loads.

Environmental Stress Cracking: The development of cracks in a material that is subjected to stress or strain in the presence of specific chemicals.

Evaporator: A heat exchanger in which cold, low-pressure (liquid) refrigerant is vaporized to absorb heat from the warmer surrounding air, earth, or water.

EWT_{max}: The maximum temperature that the circulating fluid will reach in a closed-loop ground heat exchanger during the entire cooling season over the life of the equipment. This parameter is specified by the designer during the closed-loop GHEX design process.

EWT_{min}: The minimum temperature that the circulating fluid will reach in a closed-loop ground heat exchanger during the entire heating season over the life of the equipment. This parameter is specified by the designer during the closed-loop GHEX design process.

Expansion Valve: A device that reduces the pressure of liquid refrigerant entering the evaporator and meters and regulates the flow of refrigerant so that it can properly absorb heat.

Fitting: A piping component used to join or terminate sections of pipe or to provide changes of direction or branching in a pipe system.

Flow Center: A packaged set of circulating pumps mounted in a cabinet, which often includes valves and ports for flushing/purging, antifreeze charging (if used), and loop pressurization (if a pressurized flow center is used).

Flow Regime: A term used in fluid mechanics to define the nature of fluid flow in any situation. Flow regime can either be regarded as laminar, transitional, or turbulent.

Flush Cart: A system which integrates the purge pump with the valving, hose connections, electrical connections, filtration, and reservoir tank on a hand cart for maximum portability and ease of use during operation. Flush carts fabricated for residential or light commercial use will typically utilize high-head, high-volume purge pumps from 1-1/2 hp to 3 hp in size.

Fuse: To make a plastic pipe joint by heat and pressure.

Gauge Pressure: Pressure reading directly taken from a pressure sensor or gauge (psi).

Grille: A louvered or perforated covering of an inlet or outlet opening through which air flows.

Ground Loads: Associated with ground source systems and related to the design of the GHEX. In principle, these calculations are similar to the energy loads except the ground load is heat rejected to the ground (cooling mode) or removed from it (heating mode).

Ground Source Heat Pump (GSHP): A heat pump that uses the earth itself as a heat source and heat sink. It is coupled to the ground by means of a closed-loop heat exchanger (ground coil) installed horizontally or vertically underground.

Grout: A material used during the grouting process specifically designed to form a hydraulic barrier in the borehole and to promote transfer between the GHEX piping and the earth. Most grouting products are bentonite-based with fewer being cement-based.

Grouting: The practice of making a conscious effort to form a hydraulic barrier in a borehole to protect the integrity of the deep earth environment. Proper grouting implies that an approved grouting material is used and that it is placed in the hole starting through a tremie line, filling it from bottom to top.

Head Loss: The pressure drop due to the flow of a specific fluid at a given temperature and flow rate through a specific type and size of pipe.

Headering: The process of connecting individual GHEX loops to the supply-return piping to be taken into a mechanical room or vault.

Heat Exchanger: A device, often a coil, specifically designed to transfer heat between two physically separated fluids of different temperatures.

Heat Fusion: Making a joint by heating the mating surfaces of the pipe components to be joined and pressing them together so that they fuse and become essentially one piece.

Heat of Extraction: The portion of a GSHP's heating capacity that is extracted from the earth in the heating mode. Heat of extraction is always smaller than the heating capacity of the heat pump because the electrical power consumption of the compressor, fan, and pumps add to the heating capacity of the GSHP.

Heat of Rejection: The amount of heat that must be rejected to the earth in the cooling mode to provide cooling to the space. The heat of rejection is always larger than the cooling capacity of the heat pump because the electrical power consumption of the compressor, fan, and pumps must also be rejected to the heat sink (ground connection).

Heat Pump: A mechanical device used for heating and cooling which operates by pumping heat from a cooler to a warmer location. Heat pumps can draw heat from a number of sources, e.g., air, water, or earth, and are most often either air source or water source.

Heat Sink: The medium—air, water, earth, etc.—which receives heat from a heat pump.

Heat Source: The medium—air, water, earth, etc.—from which heat is extracted by a heat pump.

Heating Seasonal Performance Factor (HSPF): A measure of heating efficiency for air source heat pump equipment on an annual basis, expressed as the heating energy provided to the space (Btu) divided by the electric energy consumed (Watt-hour) over the entire heating season.

Heat Transfer Resistance: A system's resistance to heat flow resulting from the specific thermal properties and dimensions of the system.

Hoop Stress: The tensile stress in the wall of the pipe in the circumferential orientation due to internal hydrostatic pressure.

Hose Kit: A packaged set of hose, clamps, and fittings used to connect the GHEX piping to the GSHP unit as well as the flow center.

Hydronic: A heating or cooling distribution system using liquid piped throughout the house to radiators or convectors.

Inhibitor: A fluid additive specifically designed to decrease the rate of oxidation in metal (rust) and promotion of microbial life (bacteria) in the closed-loop circulating fluid.

Isolation Hangers: Insulated, tubular holders for refrigerant piping used to prevent transmission of vibration from pipes to the structure.

Joint: The location at which two pieces of pipe or a pipe and a fitting are connected.

Joint, Butt-Fused: A joint in which the prepared ends of the joint components are heated and then placed in contact to form the joint.

Joint, Clamped Insert-Fitting: A mechanical joint using external metal clamps or other mechanical devices to form a pressure seal between an insert fitting and the outside surface of the pipe.

Joint, Heat-Fused: A joint made using heat and pressure only. There are three basic types of heat-fused joints: butt-fused, socket- or insert-fused, and saddle-fused.

Joint, Saddle-Fused: A joint in which the curved base of the saddle fitting and a corresponding area of the pipe surface are heated and then placed together to form the joint.

Joint, Socket-Fused: A joint in which the two pieces to be heat fused are connected using a third fitting or coupling with a female end.

Laminar Flow Regime: A term used to describe the flow condition where fluid flow is streamlined and smooth.

Latent Cooling Load: The amount of moisture that must be removed to maintain the space at the desired humidity level.

Life-Cycle Cost: A method of analyzing the cost of HVAC systems that considers all the significant costs of ownership, including the time value of money, initial capital investment, energy costs, and maintenance costs over the service life of each system under consideration.

Non-Pressurized Flow Center: A flow center that maintains positive suction pressure on the pumps via a standing column of water. These systems do not require that the loopfield piping be pressurized to ensure proper operation.

Open-Loop System: In a ground-coupled heat pump, a heat exchanger system in which the transfer fluid is exposed to the atmosphere.

P/T Ports: Pressure/temperature ports used to monitor the performance of a heat pump system. A small-diameter opening that allows a probe to be inserted during system operation.

Packaged Heat Pump: A self-contained heat pump unit available as either a free-delivery, ductless system for single rooms or as a larger, central ducted unit that can heat or cool an entire home.

Packaged Terminal Heat Pump: A window or through-the-wall-mounted, air-to-air heat pump unit designed to heat or cool a single room or zone.

Parallel System: A flow condition where two or more fluid paths are possible in the closed-loop circuit.

Performance Factor: The ratio of useful output capacity of a system to the input required to obtain it. Units of capacity and input need not be consistent.

Positive Displacement Pump: A pump that moves a set volume of fluid through the system for each revolution of the driving shaft. Positive displacement pumps are commonly used in conjunction with high-solids grouting materials.

Power Flushing: A higher than normal water flow and pressure in a ground heat exchanger used to flush air and debris from the closed-loop piping system.

Pressure: When expressed with reference to pipe, the force per unit area exerted by the medium in the pipe.

Pressure Drop: The decrease in pressure down the length of a pipe resulting from fluid flow.

Pressure Rating: The estimated maximum pressure that the medium in the pipe can exert continuously with a high degree of certainty that failure of the pipe will not occur.

Pressurized Flow Center: A flow center that typically consists of circulating pumps mounted in a cabinet. Positive pressure must be maintained on the system at all times (via pressurization of the lines) to ensure positive suction-side pressure on the pumps in order to produce flow.

Pump Curve: A curve used to display the amount of back pressure (head loss, feet) that a given circulating pump would be able to overcome at a given flow rate, typically provided by the pump manufacturer.

Purge Pump: A high-pressure and high-flow-rate pump used to flush air and debris from the closed-loop circuit of a closed-loop/ground-source (cl/gs) heat pump system.

Refrigerant: A fluid of extremely low boiling point used to transfer heat between the heat source and heat sink. It absorbs heat at low temperature and low pressure and rejects heat at a higher temperature and higher pressure, usually involving changes of state in the fluid (i.e., from liquid to vapor and back).

Register: A combination grille and damper device covering an air inlet or outlet opening.

Return (Air): Air returned to the space conditioning unit from the conditioned space.

Reynolds Number: A dimensionless parameter calculated to describe internal pipe flow condition. Defined to be the ratio of inertia forces to viscous forces. A Reynolds Number less than 2,500 indicates that the internal pipe flow condition is laminar. A Reynolds Number greater than 2,500 indicates that the flow condition is turbulent.

Riprap: The heavy cobbles used at the waterfront of oceans/lakes/rivers, storm water discharge outflows, drainage ditches, etc. to prevent erosion. Particle size can range from 4-6-inch gravel to natural or man-made boulders several feet in diameter, depending on the requirement.

Run Fraction: The fraction of time that a GSHP system operates to condition a space for a given period of time, expressed as a decimal.

Run Time: The number of hours that a GSHP system operates to condition a space for a given period of time.

Saturated Liquid: The temperature and pressure where refrigerant is all liquid, but will immediately begin to evaporate with addition of heat.

Saturated Vapor: The temperature and pressure where refrigerant is all vapor, but will immediately begin to condense with removal of heat.

Saturation Temperature: The temperature at which refrigerant will either immediately condense with the removal of heat (if in the vapor phase) or evaporate with the addition of heat (if in the liquid phase) at a given pressure.

Scaling: The build up of water impurities on the inside surface of the water-to-refrigerant heat exchanger in a GSHP, primarily caused by hardness and alkalinity of the water. This problem occurs primarily in open-loop systems and can cause fouling in the heat exchanger, diminishing the overall effectiveness and efficiency of the system.

Schedule: A pipe size and wall thickness classification system (outside diameters and wall thicknesses) originated by the iron pipe industry.

Seasonal Energy Efficiency Ratio (SEER): A measure of cooling efficiency for air source heat pump equipment on an annual basis, expressed as the cooling energy removed from the space divided by the electric energy consumed over the entire cooling season.

Sensible Cooling Load: The amount of sensible heat that must be removed to maintain the space at the thermostat set point temperature.

Sensible Heat Factor (SHF): The percentage of the total cooling load that can be attributed to the sensible load. Defined to be sensible cooling load divided by the total load, expressed as a decimal.

Series System: A system in which the circulating fluid from the heat pump(s) has a single flow path through the ground heat exchanger.

Simple Payback Method: A method for analyzing the cost of HVAC systems which considers only the time it takes for annual energy and maintenance cost savings to offset an initial difference in cost between two systems.

Site Plan: A detailed drawing that shows where buildings, buried utilities, landscaping, permanent fencing, etc. are located on a property and also where a potential GHEX could be installed.

Slinky: A horizontally-trenched or pond-loop configuration where loop piping is coiled into a slinky shape to reduce the amount of surface area necessary for a given GHEX installation.

Soil/Field Resistance: The resistance to heat flow resulting from soil thermal properties and underground pipe placement.

Standing Column Flow Center: See Non-Pressurized Flow Center.

Static System Check: A GSHP system startup check procedure performed before the system is turned on to identify obvious problems that must be resolved before proper operation can be expected.

Suction Line: The tube of pipe that carries the refrigerant vapor from the evaporator to the compressor inlet.

Supplemental Heating: A heating system component used when a heat pump cannot satisfy the space heating requirements by itself, during the defrost cycle (for air source equipment only), or as an emergency backup when the main system is inoperable. Usually electric resistance heat, but natural gas, LPG, or oil heating systems are also used.

Therm: A quantity of heat equivalent to 100,000 Btu.

Thermostat: An instrument that responds to changes in temperature and is used to directly or indirectly control indoor temperature by operating a space conditioning system.

Throw: The distance an airstream travels after leaving a supply outlet before the velocity is reduced to a specific terminal velocity (usually 50 ft/min).

Ton of Refrigeration: A measure of the amount of heat absorption required to melt 1 ton of ice in 24 hours. A ton of refrigeration is a measure of the amount of cooling delivered by a heat pump (or other air conditioning system). One ton of refrigeration is equivalent to a cooling rate of 12,000 Btu per hour.

Total Cooling Load: The total amount of heat energy that must be removed from a space to keep it at the thermostat set point temperature as well as at the desired humidity level, defined to be the sum of the sensible cooling load and the latent cooling load.

Tremie Line: The pipe used to pump an appropriate grouting material into a borehole from the bottom of the hole to the top. A tremie line will commonly be made of 1-inch or 1-¼-inch diameter HDPE pipe.

Turbulent Flow Regime: A term used to describe the flow condition where fluid flow becomes chaotic and disordered. The mixing effect caused by turbulent flow maximizes heat transfer between the fluid and pipe walls in the closed-loop GHEX while also increasing the system pumping pressure.

U-Bend: A prefabricated closed-return pipe assembly used in vertical heat exchangers to connect the two pipes at the bottom of the bore hole.

Unitary Heat Pump: A complete factory-assembled heat pump.

Valve, Expansion: A device for regulating the flow of liquid refrigerant to the evaporator. Two types of valves are commonly used: an electronic valve that responds to variation in electric resistance reflecting changes in refrigerant temperature, and a thermostatic valve that uses a refrigerant-filled bulb to sense changes in refrigerant temperature.

Valve, Reversing: An electrically operated valve that allows the heat pump to switch from heating to cooling, or vice versa, by changing the refrigerant's direction of flow.

Water Source Heat Pump: A heat pump that uses a water-to-refrigerant heat exchanger to extract heat from the heat source.

Water Source Heat Pump, Closed Loop: Closed-loop systems circulate a heat transfer fluid (such as water or a water-antifreeze mixture) continuously to extract or reject heat from a ground or water heat source or sink.

Water Source Heat Pump, Open Loop: Open-loop systems pump groundwater or surface water from a well, river, or lake through a water-to-refrigerant heat exchanger and return the water to its source, a drainage basin, pond, or storm sewer.

Zone Load: The amount of heating or cooling that the delivery system must provide to satisfy the peak loads for a specific zone, and a single thermostat is used to control the delivery system for that zone.

List of Common Acronyms & Abbreviations:

ACCA: Air Conditioning Contractors of America
AEE: The Association of Energy Engineers
AGL: Annual Ground Load
α (Alpha): Soil Thermal Diffusivity
ARI: Air-Conditioning and Refrigeration Institute
A_s: Annual Soil Temperature Swing Above and Below the Mean (In the top 10 feet of earth)
ASHP: Air Source Heat Pump
ASHRAE: American Society of Heating, Refrigerating, and Air-Conditioning Engineers
ASTM: American Society for Testing and Materials
BM: Bore Length Multiplier
C-C: Center to Center Spacing
CF: Correction Factor
CFM: Cubic Feet per Minute
d_{avg}: Average Pipe Burial Depth
DMD: Electrical Demand
DOE: U.S. Department of Energy
d_{bp}: Depth to the Bottom of the Pipe
DR: Dimension Ratio
DSIRE: Database of State Incentives for Renewables & Efficiency
EAT: Entering Air Temperature
ECM: Electrically Commutated Motor
EER: Energy Efficiency Ratio
EPA: Environmental Protection Agency
ELT: Entering Load Temperature
EPRI: Electric Power Research Institute
ESP: External Static Pressure
EST: Entering Source Temperature
EWT: Entering Water Temperature
F_C: Run-Fraction during the cooling design month
FFR: Flushing Flow Rate
F_H: Run-Fraction during the heating design month
FLRH: Full-Load Run Hours
FPS: Feet per Second
GHEX: Ground Heat Exchanger
GHPC: Geothermal Heat Pump Consortium
GPM: Gallons per Minute
GSHP: Ground Source Heat Hump
HC: Heating Capacity
HE: Heat of Extraction
HL: Head Loss
HR: Heat of Rejection
HSPF: Heating Seasonal Performance Factor
HVAC: Heating, Ventilating, and Air Conditioning
HWG: Hot Water Generation
IGSHPA: International Ground Source Heat Pump Association
IN. WG: Inches of Water Gauge

ISO: International Organization for Standardization
k_G: Formation Thermal Conductivity
k_{Grout}: Grout Thermal Conductivity
LAT: Leaving Air Temperature
LLT: Leaving Load Temperature
LST: Leaving Source Temperature
LWT: Leaving Water Temperature
NNAGL: Net Normalized Annual Ground Load
OAT: Outdoor Air Temperature
P/T Port: Pressure/Temperature Measurement Port
PM: Pipe Multiplier
PSC: Permanent Split Capacitor
PSI: Pounds per Square Inch
Re: Reynolds Number
ROI: Return on Investment
R_p: Pipe Thermal Resistance
R_S: Soil Resistance
SC: Sensible Cooling Capacity
SCC: Center to Center Trench Spacing
SDSURR: Step-Down, Step-Up Reverse-Return Header
SEER: Seasonal Energy Efficiency Ratio
SHF: Sensible Heat Factor
S_M: Spacing Multiplier
TC: Total Cooling Capacity
T_M: Mean Earth Temperature (In the top 10 feet)
T_O: Number of Days After January 1 to Reach the Minimum Earth Surface Temperature
WPD: Water Pressure Drop
WSHP: Water Source Heat Pump

Index

NOTES

NOTES

NOTES

NOTES